9th International Conference on Monitoring, Modeling & Management of Emergent Economy (M3E2 2021)

SHS Web of Conferences Volume 107 (2021)

Odessa, Ukraine
26 - 28 May 2021

Editors:

Serhiy Semerikov
Vladimir Soloviev
Liubov Kibalnyk
Oleksandr Chernyak
Hanna Danylchuk

ISBN: 978-1-7138-3008-5

Printed from e-media with permission by:

Curran Associates, Inc.
57 Morehouse Lane
Red Hook, NY 12571

Some format issues inherent in the e-media version may also appear in this print version.

For additional information, please contact EDP Sciences – Web of Conferences
at the address below.

EDP Sciences – Web of Conferences
17, Avenue du Hoggar
Parc d'Activité de Courtabœuf
BP 112
F-91944 Les Ulis Cedex A
France

Phone: +33 (0) 1 69 18 75 75
Fax: +33 (0) 1 69 28 84 91

contact-edps@webofconferences.org

Additional copies of this publication are available from:

Curran Associates, Inc.
57 Morehouse Lane
Red Hook, NY 12571 USA
Phone: 845-758-0400
Fax: 845-758-2633
Email: curran@proceedings.com
Web: www.proceedings.com

TABLE OF CONTENTS

PREFACE: M3E2 2021 – ECONOMIC AFTERSHOCK OF COVID-19 1
Semerikov Serhiy, Soloviev Vladimir, Kibalnyk Liubov, Chernyak Oleksandr, Danylchuk Hanna

RECURRENCE ANALYSIS OF INNOVATION BEHAVIOR OF BITCOIN MARKET AGENTS IN CONDITIONS OF COVID-19 15
Kucherova Hanna, Los Vita, Ocheretin Dmytro, Bilska Olha, Makazan Evgenia

ECONOMIC PRODUCTIVITY AND COMPETITIVENESS OF UKRAINIAN EXPORTS DUE TO THE GLOBAL CHALLENGES 24
Petkova Lesia, Berezina Olena, Honcharenko Iryna, Osadchenko Ihor

COMPARATIVE ANALYSIS OF THE STOCK QUOTES DYNAMICS FOR IT-SECTOR AND THE ENTERTAINMENT INDUSTRY COMPANIES BASED ON THE CHARACTERISTICS OF MEMORY DEPTH 31
Maksyshko Nataliia, Vasylieva Oksana

APPLICATION OF BLOCKCHAIN TECHNOLOGY IN ACCOUNTING AND AUDIT: INTERNATIONAL AND DOMESTIC EXPERIENCE 40
Nezhyva Mariia, Zaremba Olha, Nehodenko Viktoria

MANAGEMENT OF FINANCIAL AND ECONOMIC SECURITY OF CRITICAL INFRASTRUCTURE OBJECTS IN THE CONDITIONS OF RISKS OF QUARANTINE RESTRICTIONS: STRATEGIC AND PERSONNEL ASPECTS 47
Zachosova Nataliia, Kutsenko Dmytro, Koval Oleksii, Kovalenko Andrii

ECONOPHYSICS OF CRYPTOCURRENCY CRASHES: AN OVERVIEW 55
Bielinskyi Andrii, Serdyuk Oleksandr, Semerikov Serhiy, Soloviev Vladimir

THE LACK OF REVERSIBILITY DURING FINANCIAL CRISIS AND ITS IDENTIFICATION 114
Bielinskyi Andrii, Hushko Serhii, Matviychuk Andriy, Serdyuk Oleksandr, Semerikov Serhiy, Soloviev Vladimir

UKRAINE'S FOREIGN TRADE: RESPONSES TO GLOBAL CHALLENGES 124
Radzivivska Svitlana, Us Ivan

ASSESSMENT OF FACTORS INFLUENCING THE STABILITY OF UKRAINIAN EXPORT BASED ON THE FRACTAL ANALYSIS 133
Kainara Dmytro, Kaliuzhna Iuliia

DECISION SUPPORT INFORMATION AND ANALYTICAL TECHNOLOGY IN DISCHARGE MILITARY PERSONNEL EMPLOYMENT 139
Medvid Mykhailo, Ivashchenko Peter, Britchenko Igor, Trubavina Iryna, Liutyi Volodymyr

MARKETING FORECASTING BASED ON BIG DATA INFORMATION 146
Ivanov Sergey, Ivanov Mykola

FUZZY ARITHMETIC IN ECONOMIC RESEARCH 152
Gorbatiuk Kateryna, Zavhorodnia Tetiana, Proskurovych Oksana, Mantalyuk Olha

THE DEVELOPMENT OF ENERGY CONSUMPTION FORECASTING MODEL FOR A METALLURGICAL ENTERPRISE 160
Bakurova Anna, Yus'kiv Olesia, Shyrokorad Dima, Tereschenko Elina, Riabenko Anton

A FLEXIBLE EVOLUTIONARY MODEL OF MACHINE LEARNING OF THE MOST SUCCESSFUL STRATEGIES OF HUMAN CAPITAL DEVELOPMENT 167
Porokhnya Vasyl, Penev Vladyslav

OPTIMIZATION MODELS FOR THE USE OF AGRICULTURAL AVIATION WHEN PERFORMING CHEMICAL WORKS 175
Saparbayev Abdizhapar, Makulova Aiymzhan, Kaziyeva Assel, Klyshbayeva Zaure, Saparbayeva Elmira

IMPROVING THE INNOVATIVE DEVELOPMENT MANAGEMENT OF ZAPORIZHZHIA REGION'S INDUSTRIAL COMPLEX 182
Pulina Tetiana, Shmygol Nadiia, Narivs'kyi Oleksii, uczka W adys awa, Shmygol Yeugeniia

MODELING GRAIN TRANSPORTATION IN THE SYSTEM OF GRAIN PROCESSING INDUSTRIES 189
Saparbayev Abdizhapar, Makulova Aiymzhan, Bayboltaeva Nesipkul, Sarsenbieva Nagima, Imatayeva Aliya

USING A COMPREHENSIVE INDEX TECHNOLOGY TO ANALYZE STRUCTURAL CHANGES IN THE REGIONS' ECONOMIC DEVELOPMENT IN A COVID-19 PANDEMIC: THE CASE OF UKRAINE 195
Hryhoruk Pavlo, Khrushch Nila, Grygoruk Svitlana

MULTICRITERIA OPTIMIZATION OF OIL AND GAS ENTERPRISES FINANCIAL STABILITY USING THE GENETIC ALGORITHM METHOD 205
Shkvaryliuk Marta, Horal Liliana, Khvostina Inesa, Yashcheritsyna Natalia, Shiyko Vira
IMPROVEMENT OF THE METHODS OF ASSESSING THE INFLUENCE OF EXTERNAL FACTORS ON THE STRENGTHS AND WEAKNESSES OF THE ENTERPRISE 211
Lozovik Yriij, Tishkov Bogdan, Pomazun Oksana, Derbentsev Vasily, Hostryk Alexey
ENVIRONMENTAL INVESTMENTS AS THE BASIS OF "GREEN" ECONOMY: EMPIRICAL EVIDENCE FROM UKRAINIAN COMPANIES 225
Vasylchuk Iryna, Suprun Anatolij, Slyusarenko Kateryna, Sadovenko Maryna
SCIENTIFIC APPROACH TO QUANTITATIVE MEASUREMENT AND ECONOMIC PROCESSES RESEARCH IN CORPORATE MANAGEMENT 234
Orlov Oliver, Ryasnykh Yevheniya, Dumanska Kathrin, Savchenko Olena
THE IMPACT OF INFLATION ON THE UNEMPLOYMENT RATE IN EGYPT: A VAR APPROACH 241
Attia Mohamed Omran Emad, Bilan Yuriy
MODELING OF STRATEGIES FOR THE DEVELOPMENT OF THE RESORT-RECREATION SPHERE OF UKRAINE 246
Zakharchenko Pavel, Kostenko Ganna, Kungurtseva-Mashchenko Tatyana, Zhvanenko Svitlana, Mukhin Viktor
MODELING THE DEVELOPMENT OF THE TOURISM INDUSTRY IN THE SMART AGE OF GLOBALIZATION THROUGH TRANSNATIONAL COOPERATION AND CAPACITY BUILDING 253
Shmygol Nadiia, Cherniavska Olena, Pawliszczy Dariusz, Shmygol Yeugeniia, Cherniavska Alexandra
HUMAN POTENTIAL AND THE SYSTEM OF ITS COMMERCIALIZATION MANAGEMENT 260
Shkoda Tetiana, Gernego Iuliia, Tepliuk Mariia, Stefanita Anastasia
GLOBAL MIGRATION PROCESSES ANALYSIS AND MODELLING OF MIGRATION ATTRACTIVENESS OF COUNTRIES THROUGH FUZZY LOGIC 270
Danylchuk Hanna, Kibalnyk Liubov, Kovtun Oksana, Pursky Oleg, Stachowiak Zenon
FOREIGN DIRECT INVESTMENT TRENDS: AN ANALYSIS OF THE STRUCTURE AND DYNAMICS IN THE CONTEXT OF GLOBALIZATION 278
Kibalnyk Liubov, Leshchenko Maryna, Feshchenko Olena, Wang Haiyu
ANALYSIS OF MONETARY REGULATION OF THE INVESTMENT SPHERE IN THE REPUBLIC OF BELARUS 287
Zelenkevich Marina, Bandarenka Natallia
ECONOMETRIC MODELLING OF BANK ACTIVITIES: VALUE-BASED APPROACH TO THE PROBLEM LOANS TERMS' RESCHEDULING 297
Onikiienko Serhii, Dyba Mykhailo, Gernego Iuliia
COMPUTATIONAL METHOD FOR RISK ASSESSMENT OF REGIONAL SOCIO-ECONOMIC DEVELOPMENT 306
Pursky Oleg, Dubovyk Tatiana, Buchatska Iryna, Lutsenko Iryna, Danylchuk Hanna
A FORECASTING THE CONSUMER PRICE INDEX USING TIME SERIES MODEL 312
Shinkarenko Volodymyr, Hostryk Alexey, Shynkarenko Larysa, Dolinskyi Leonid
MONITORING THE QUALITY OF E-LEARNING IMPLEMENTATION IN EDUCATIONAL INSTITUTIONS 318
Habrusiev Valerii, Tereshchuk Hryhoriy, Tsidylo Ivan, Martyniuk Serhii, Kulyanda Olena
MATHEMATICAL MODEL FOR IMITATION OF MANAGEMENT OF THE ENTERPRISE'S LOGISTICAL SYSTEM 325
Ivanov Roman, Sherstennikov Yuriy, Porokhnya Vasyl, Grynko Tetyana
ANALYSIS, FORECASTING, AND MANAGEMENT OF INDICATORS OF THE FOREST INDUSTRY OF THE REGION 334
Ovchynikova Olena, Dupliak Olha
THE CONCEPT OF ASSESSING THE CITY COUNCIL MANAGEMENT CONTROL LEVEL FOR EFFECTIVE DECISION MAKING 343
Melikhova Tetiana, Mikhailutsa Olena
THE PROBLEM OF ESTIMATING THE SUSTAINABLE DEVELOPMENT OF THE INTEGRATED TECHNOGENIC INDUSTRIAL SYSTEM 350
Ramazanov Sultan, Honcharenko Oleksandr
MACHINE LEARNING METHODS APPLICATION FOR CONSUMER BANKING 356
Kaminskyi Andrii, Nehrey Maryna, Zomchak Larysa
ECONOMIC-MATHEMATICAL MODEL FOR COMPLEX RISK ASSESSMENT OF THE ENTERPRISE INVESTMENT PROJECT USING FUZZY LOGIC 364
Chaikovska Inna, Hryhoruk Pavlo, Chaikovskyi Maksym
Author Index

Statement of Peer review

In submitting conference proceedings to *Web of Conferences*, the editors of the proceedings certify to the Publisher that

1. They adhere to its **Policy on Publishing Integrity** in order to safeguard good scientific practice in publishing.
2. All articles have been subjected to peer review administered by the proceedings editors.
3. Reviews have been conducted by expert referees, who have been requested to provide unbiased and constructive comments aimed, whenever possible, at improving the work.
4. Proceedings editors have taken all reasonable steps to ensure the quality of the materials they publish and their decision to accept or reject a paper for publication has been based only on the merits of the work and the relevance to the journal.

Title, date and place of the conference

9th International Conference on Monitoring, Modeling & Management of Emergent Economy (M3E2 2021)

May 26-28, 2021 | Odessa, Ukraine

Proceedings editor(s):

Serhiy Semerikov

Vladimir Soloviev

Liubov Kibalnyk

Oleksandr Chernyak

Hanna Danylchuk

Date and editor's signature

May 13, 2021

M3E2 2021 – Economic aftershock of COVID-19

Serhiy Semerikov[1,2,3,4,*], *Vladimir* Soloviev[1,**], *Liubov* Kibalnyk[5,***], *Oleksandr* Chernyak[6,****], and *Hanna* Danylchuk[5,†]

[1]Kryvyi Rih State Pedagogical University, 54 Gagarin Ave., Kryvyi Rih, 50086, Ukraine
[2]Kryvyi Rih National University, 11 Vitalii Matusevych Str., Kryvyi Rih, 50027, Ukraine
[3]Institute of Information Technologies and Learning Tools of NAES of Ukraine, 9 M. Berlynskoho Str., Kyiv, 04060, Ukraine
[4]University of Educational Management, 52-A Sichovykh Striltsiv Str., Kyiv, 04053, Ukraine
[5]The Bohdan Khmelnytsky National University of Cherkasy, 81 Shevchenko Blvd., Cherkasy, 18031, Ukraine
[6]Taras Shevchenko National University of Kyiv, 60 Volodymyrska Str., Kyiv, 01033, Ukraine

Abstract. This is an introductory text to a collection of papers from the M3E2 2021: 9th International Conference on Monitoring, Modeling & Management of Emergent Economy, which held in Odessa National University of Economics, Odessa, Ukraine, on the May 26-28, 2021. It consists of short introduction, conference review and some observations about the event and its future.

1 Introduction

The development of human society, its various spheres of activity and functioning in recent years is characterized by the emergence of new challenges and threats. Humanity is facing many global problems of development of technology and the increasing scientific and technological progress. In addition to the usual environmental and man-made disasters, financial crises, the new ones appeared. The COVID-19 pandemic, for example. As a result, technologies and online communication instruments began to develop at an increasing pace. The need to work and study remotely has forced people to master new methods and tools of digital technology.

Due to the limited movement and transportation, as well as due to the reduction of production and business closure, enterprises, regions, national economies and transnational capital found themselves in a difficult situation. They experienced a decline in production and trade, bankruptcy, reduced profitability, slower growth etc. This is just a small list of the effects of the pandemic on the world economy.

The causes, mechanisms and consequences of such processes are of particular concern to the scientific community, which is the first to try to understand the depth and importance of these changes. This year's theme of the international conference International Conference on Monitoring, Modeling & Management of Emergent Economy has especially relevant issues that experts of various fields are trying to raise and solve in their works.

The authors' attempt to find out the causes of the crisis and the possibility of using modern ICT to solve existing problems or prevent economic, political and environmental threats, need to be especially considered. The countries with emerging economies are particularly vulnerable to new challenges, so participation in constructive scientific discourse is very important. Modern challenges contribute to the search for new approaches to solving these problems.

In their research scientists focus on economic, financial security and sustainability of enterprises and regions; digitalization of all spheres of human life; modern methods of management and marketing activities; development and analysis of the financial market and cryptocurrency market; modeling and forecasting of international economic activity of various business entities; especially relevant methods of machine learning and fuzzy logic; solving the problems of various sectors of the economy of Ukraine and other countries, especially countries with emerging economies.

The subject of the works included in the proceedings it necessary to search for a new scientific paradigm in a constantly changing environment. After all, new challenges and threats are a certain stimulus for the development of scientific thought. We expect that the research of the participants of this conference will be useful for scientists, teachers, students and representatives of the business community.

2 M3E2 2021: At a glance

The **Monitoring, Modeling & Management of Emergent Economy** (M3E2, https://m3e2.ccjournals.eu/2021/) is a peer-reviewed international conference focusing on research advances and applications of nonlinear dynamics

*e-mail: semerikov@gmail.com
**e-mail: vnsoloviev2016@gmail.com
***e-mail: liubovkibalnyk@gmail.com
****e-mail: chernyak@univ.kiev.ua
†e-mail: abdanilchuk@gmail.com

methods, econophysics and complex systems methodology of emergent economy.

The M3E2 Conference occupies contributions in all aspects of Computational Finance, Economics, Risk Management, Statistical Finance, Trading and Market Microstructure, (Deep) Machine Learning technologies and tools, paradigms and models, relevant to modern financial engineering and technological decisions in the modern age. There is urgent general need for principled changes in postclassic economy elicited by current models, tools, services, networks and IT communication.

M3E2 2021 topics of interest since 2019 [1–3]:

- Complex cyberphysical systems, synergy, econophysics, economy of agents (papers [4, 5])
- Dynamics of emergent markets in crisis and post-crisis period (papers [6–9])
- Economic security (papers [10, 11])
- Global challenges for economic theory and practice in Europe (papers [8, 12, 13])
- Information systems and technologies in economics (papers [10, 14–18])
- Innovation models of economic development (papers [8, 12, 19–28])
- Machine learning for prediction of emergent economy dynamics (papers [18, 29])
- Management of the state's economic safety and economic safety of economic agents (papers [10, 11])
- Methods and models of artificial intelligence in economic systems (papers [18, 29])
- Modeling of hospitality sphere development (papers [30, 31])
- Models of global transformations (papers [12, 32, 33])
- Monitoring, modeling and forecasting in the banking sector (papers [19, 34])
- Monitoring, modeling, forecasting and preemption of crisis in socio-economic systems (papers [4, 5, 10, 17, 19, 23, 25, 32, 35–39])
- Optimal management of socio-economic processes (papers [8, 10, 12, 17, 20, 22, 25, 38, 40])
- Risk management models in emergent economy (papers [19, 35, 41, 42])

This volume contains the papers presented at M3E2 2021: 9th International Conference on Monitoring, Modeling & Management of Emergent Economy held on May 26-28, 2021 in Odessa, Ukraine.

There were 49 submissions. Each submission was reviewed by at least 3, and on the average 3.1, program committee members. The committee decided to accept 39 papers.

3 M3E2 2021: Program committee

Dr. **George Abuselidze**, Professor of Economics and Business, Department of Finance, Banking and Insurance, Batumi Shota Rustaveli State University, Batumi, Georgia

George Abuselidze, from the Batumi Shota Rustaveli State University (in 2001) and Ivane Javakhishvili Tbilisi State University, Georgia, in 2005, and a Doctor of Economics Sciences degree (Dr. habil.) from the National Academy of Sciences of Georgia, in 2005. Since 2002, he has been working in the Finance and Banking at the Ivane Javakhishvili Tbilisi State University, since 2006 - Batumi Shota Rustaveli State University where he is currently Head department of Finance, Banking and Insurance. His research interests include Economics, Econometrics, Finance and Social Sciences (miscellaneous). He has published a number of papers in international journals and volumes in book series, is a member of editorial or/and review boards of Oeconomia Copernicana, Journal of Financial Economic Policy, International Journal of Economics and Finance, journal of Science and studies of accounting and finance: problems and perspectives, Management Studies and etc. He also played instrumental role in different prestigious internal collaborative research project with USA, Canada, Lithuania, Poland, Ukraine, Turkey and etc.

WWW: https://orcid.org/0000-0002-5834-1233
e-mail: george.abuselidze@bsu.edu.ge

Prof. Dr. **Vitalina Babenko** is Professor (Full) of International E-commerce and Hotel&Restaurant Business Department, V. N. Karazin Kharkiv National University (Ukraine), Dr. Sci. (habil.) in Economics, PhD in Technical Sciences

Her sphere of scientific interests includes the management of innovation processes, economic-mathematical modelling, information systems and technologies. She has published around 300 papers in peer-reviewed journals and in the proceedings of conferences. She has taken part in more than 100 conferences and workshops as a Chair, Keynote Speaker, member of the Scientific Program Committee, an Organizer and Session Chair. She is Editor-in-Chief of Journal of International Relations, Economics, Country Studies, Tourism; International Journal of Economics and Management Systems and member of editorial board of some scientific journals. She is the Member of

Doctoral Academic Councils in Economics, Expert of the Ministry of Education of Ukraine and Expert of Ukrainian Institute Scientific Technical Information. Council many national and international informatics associations.

WWW: http://international-relations-tourism.karazin.ua/Babenko

e-mail: vitalinababenko@karazin.ua

Dr. **Oleksandr Chernyak** was the Professor, Head of Economic Cybernetics Department, Taras Shevchenko National University of Kyiv

WWW: http://science.univ.kiev.ua/en/researchgroups/research.php?ELEMENT_ID=2594

Dr. **Hanna Danylchuk**, Associate Professor of the Department of Economics and Business Modeling, The Bohdan Khmelnytsky National University of Cherkasy, Cherkasy, Ukraine

Hanna Danylchuk, born in 1969, received her PhD in Economics (East European University of Economics and Management) in 2015. Since 2006 she has been working in the field of economic systems modeling at The Bohdan Khmelnytsky National University of Cherkasy. Research interests: modeling of crisis phenomena in financial markets. She has published a number of papers in international journals.

e-mail: abdanilchuk@gmail.com

Dr. **Irina Georgescu**, Lecturer of Computational Intelligence, Department of Informatics and Economic Cybernetics, Bucharest University of Economics, Bucharest, Romania

Irina GEORGESCU holds a PhD in Economics from Turku Centre for Computer Science, Turku, Finland. Currently she is a lecturer at the Department of Economic Informatics and Cybernetics, Bucharest Academy of Economic Studies. Her research interests lie in the areas of fuzzy economics, computational intelligence and econo-

 https://doi.org/10.1051/shsconf/202110700001

metrics. She is the author of about 40 journal papers and 2 books published in Springer Verlag.

e-mail: irina.georgescu@csie.ase.ro

Dr. **Lidiya Guryanova**, PhD, Doctor of Economics (Habilitation in Economics), Associate Professor, professor of Economic Cybernetics Department, Simon Kuznets Kharkiv National University of Economics

Coordinator of the Master's program "Assessment, analysis and forecasting of socio-economic processes" in the specialty 8.18010024 "Applied Economics", a member of the Ukrainian Association of Economic Cybernetics, a reviewer of the journals "Economics of Development" (Ukraine), "Mathematical Problems in Engineering" (USA), Head of Section, a member of the program, organizational and technical committees of international scientific-practical conference "Modern problems of modelling the socio-economic systems" (Ukraine, Slovenia, Poland, the USA, Bulgaria), "Innovation and information technology in the development of business and education" (Russian Federation, Ukraine, Germany, Bulgaria, France), scientific director of training of highly qualified personnel (specialty 08.00.11 – mathematical methods, models and information technology in the economy).

WWW: https://ek.hneu.edu.ua/en/professors/guryanova-l-s/

e-mail: guryanovalidiya@gmail.com

Dr. **Alexey Hostryk**, Odessa National Economic University, Ukraine

WWW: https://orcid.org/0000-0002-5834-1233

e-mail: AlexeyGostrik@gmail.com

Dr. **Pavlo Hryhoruk**, Professor of Department of Automated Systems and Modeling in Economics of Khmelnytskyi National University, Khmelnytskyi, Ukraine

Pavlo Hryhoruk, received a Doctor of Economic Sciences degree (Dr. habil.) from the Khmelnytskyi National University, in 2013. Since 1997, he has been working in the field of economic and mathematical modeling at the Khmelnytskyi National University, where he is currently Head of Department of Automated Systems and Modeling in Economics. Directions of studies are related to decision-making, multidimensional modeling of socio-economic systems, comprehensive assessment of economic phenomena latent characteristics, financial security, sustainable development, information technologies in education. He has published a number of papers in domestic and international journals, monographs and volumes in book series. Since December 2020 he is Editor-in-Chief of *Herald of Khmelnytskyi National University. Economic Sciences*.

e-mail: violete@ukr.net,hryhoruk@khnu.km.ua

Dr. **Serhii Hushko**, State University of Economics and Technology, Ukraine
WWW: https://www.duet.edu.ua/en/persons/12
e-mail: gushko77@gmail.com

Dr. **Nila Khrushch**, Professor, Head of Department of Finance, Banking and Insurance of Khmelnytskyi National University, Khmelnytskyi, Ukraine

Nila Khrushch, received a Doctor of Economic Sciences degree (Dr. habil.) from the Institute of Market Problems and Economic & Ecological Researched of National Academy of Sciences of Ukraine, Odesa in 2007. Since 1993, she has been working in the field of innovation-investment and strategic management at the Khmelnytskyi National University, where she is currently Head of Department of Finance, Banking and Insurance. Area of studies includes financial and economic research and modeling of business entities activity, financial security, sustainable development, strategic and financial management. She has published a number of papers in domestic and international journals, monographs and volumes in book series, is a member of editorial board of *Herald of Khmelnytskyi National University. Economic Sciences*.

e-mail: nila.ukr@gmail.com,khrushch@khnu.km.ua

Dr. **Liubov Kibalnyk**, Head and Professor of Department of Economics and Business Modelling, Cherkasy Bohdan Khmelnytsky National University, Cherkasy, Ukraine

Liubov Kibalnyk, born in 1969, received a Candidate of Economic Sciences degree (Dr. phil.) from the Taras Shevchenko National University of Kyiv, Ukraine, in 2002, and a Doctor of Economic Sciences degree (Dr. habil.) from the Institute of International Relations of Taras Shevchenko National University of Kyiv, in 2015. Since 1994, she has been working in the field of economics, international economic relations and modeling at the Cherkasy Bohdan Khmelnytsky National University, where he is currently Head of Department.

She specialized in the field of modeling economic processes in the global environment. She is the author of more than 150 scientific and methodological works published in international and national editions.

WWW: http://econom-law.cdu.edu.ua/?page_id=804
e-mail: liubovkibalnyk@gmail.com

Dr. **Arnold Kiv**, Ben-Gurion University of the Negev, Israel

Arnold Kiv received the D. Sc. (Dr. Hab.) degree in solid state physics from Tartu Institute of Physics, Tartu, Estonia, in 1978. From 1964 to 1982, he was a Senior Researcher and a Head of the Laboratory of Radiation Effects, Institute of Nuclear Physics, Academy

of Sciences, Tashkent, Uzbekistan. From 1983 to 1998, he was a Head of the Department of Theoretical Physics, South-Ukrainian National Pedagogical University, Odessa, Ukraine. In 1997, he was an Invited Professor, Western Ontario University, Canada. From 1999 to the present, he is a Professor-Researcher in the Department of Materials Engineering, Ben-Gurion University of the Negev, Israel. In 1996 and 2011 he was co-Director of NATO Advanced research Workshops and an Editor of two NATO Series books. He has about 200 publications, three monographs and three Invention Certificates in the field of radiation effects in solid state electronics. His research interests include mechanisms of formation of radiation defects in solids, interaction of fast particles with materials, radiation methods in microelectronics, including computer simulation, analytical calculations and experimental studies.

PhD of Public Administration, **Oksana Kovtun**, associate professor, Department of Public Administration and Project Management, State Higher Educational Institution "University of Educational Management", Kyiv, Ukraine

Oksana Kovtun, born in 1971, received a Candidate of Public Administration degree (Dr. phil.) from the Council for the Study of Productive Forces of Ukraine at the National Academy of Sciences of Ukraine, Ukraine, in 2010. Since 2000, she has been working in the field of public finance research, household financial behavior and mechanisms for their activation in Classic Private University (Zaporizhzhya). I am currently an associate professor at the Department of Public Administration and Project Management, State Higher Educational Institution «University of Educational Management». Shies research interests: modeling of financial markets, study of structural changes in the economy, innovations in the financial sector. She has published a number of papers in international journals and workshops.

WWW: http://umo.edu.ua/institutes/imp/struktura-institutu/kaf-upravl-proekt/sklad/kovtun-oksana-anatolijivna

e-mail: kovtun.oa71@gmail.com

Dr. **Hanna Kucherova**, Professor of economy, Department of Economy, Classic Private University, Zaporizhzhia, Ukraine

Hanna Kucherova, born in 1983, received a Candidate of Economical Sciences degree from the Classic Private University, in 2011, and a Doctor of Economical Sciences degree (Dr. habil.) from the Classic Private University, in 2017. Since 2008, she has been working in the field of pricing, economy, tax, economy behavior at the Classic Private University of Ukraine, where he is currently professor Chair of Economics. Her research interests include modeling the country's business climate, behavior of socio-economic agents, tax consciousness and information transparency.

WWW: https://scholar.google.ru/citations?user=FhBOVExn1foC

e-mail: kucherovahanna@gmail.com

Dr. **Serhii Lehenchuk**, head of the department of management information systems and accounting, Zhytomyr Polytechnic State University, Zhytomyr, Ukraine.

Serhii Lehenchuk, born in 1981, received a Candidate of Economic Sciences degree (Dr. phil.) from the National Agrarian University, Kyiv, Ukraine, in 2006, and a Doctor of Economic Sciences degree (Dr. habil.) from the Zhytomyr State Technological University, in 2011. Since 2004, he has been working in the field of accounting and information systems at the Zhytomyr Polytechnic State University. His research interests include general accounting theory, positive accounting theory and accounting choice, intangible assets accounting, corporate and integrated reporting. He has published a number of papers in international and Ukrainian journals, is a member of editorial boards of "Economics, Management and Administration", "Problems of Theory and Methodology of Accounting, Control and Analysis", "Socio-Economic Research Bulletin", "Public Policy and Accounting".

WWW: https://ztu.edu.ua/ua/structure/faculties/fbso/bok.php

e-mail: legenchyk2014@gmail.com

Dr. **Nataliia Maksyshko**, Doctor of Economic Sciences, Professor, Head of Department of Economic Cybernetics, Professor, Zaporizhzhia National University, Zaporizhzhia, Ukraine

Nataliia Maksyshko, received a Candidate in Physics and Mathematics Sciences degree (Dr. phil.) from the V.M. Glushkov Institute of Cybernetics of National Academy of Sciences of Ukraine, in 1990 (Mathematical modeling and computational methods), and a Doctor of Economics Sciences degree (Dr. habil.) (Economic and mathematical modeling) from the Kyiv National Economic University named after Vadym Hetman, in 2010. Since 1983, she has been working in the field of economic and mathematical modeling at the Zaporizhzhia National University, where she is currently Head of Department of Economic Cybernetics. Her research interests include economic and mathematical modeling of complex economic systems, economic dynamics, discrete optimization, information technology in economics and education. She has published a number of papers in international journals and volumes in book series, is a member of editorial board of Bulletin of Zaporizhzhia National University. Economic sciences.

e-mail: maxishko@ukr.net

Dr. **Andriy Matviychuk**, Professor of Economic and Mathematical Modeling, Kyiv National Economic University named after Vadym Hetman, Kyiv, Ukraine

Andriy Matviychuk, born in 1978, became a Candidate of Science in Economics (PhD) by specialty Economic and Mathematical Modelling at the Technological University of Podillia (Khmelnitskiy) in 2003, and a Doctor of Economics (DSc) with a degree in Mathematical Methods, Models and Information Technologies in Economics at the Kyiv National Economic University named after Vadym Hetman in 2008.

Since 1990 he has been working in the field of AI & ML at the Kyiv National Economic University named after Vadym Hetman, where he is currently professor of Economic and Mathematical Modeling Department, director

of Institute of Modeling and Informational Technologies in Economics and CEO at KNEU Science Park. His research interests cover the field of mathematical modeling of complex systems, primarily using tools of neural networks and fuzzy logic. He is a Chief Editor of scientific and analytical journal Neuro-Fuzzy Modeling Techniques in Economics and a member of editorial boards of a number of scientific journals.

WWW: https://kneu.edu.ua/ua/depts9/k_ekon_matematychn_modeljuvannja/vykladachi_kmm/Matvijchuk.A.V/

e-mail: editor@nfmte.com

Dr. **Inese Mavlutova**, Professor (Full) at Department of Economics and Finance, BA School of Business and Finance, Latvia

e-mail: inese.mavlutova@ba.lv

Dr. **Iveta Mietule**, Rector of Rezekne Academy of Technologies, Latvia, Professor of Economics

Iveta Mietule was born in 1972 in Latvia. Received her PhD in Economics from University of Latvia in 2009. Member of the Association of University Professors of

Latvia, professor of Rezekne Academy of Technologies, member of the Senate of RTA, expert of the Latvian Council of Science in the field of Economic and Management Sciences. Author of more than 50 publications, 3 monographs. Participant and contributor of more than 15 projects and programmes, included FRONTEX, ERDF, ESF, INTERREG, ERASMUS and the Latvian National Research Programme. International experience of organizing and participating in more than 50 events (conferences, lectures, seminars etc.).

WWW: https://orcid.org/0000-0001-7662-9866

e-mail: iveta.mietule@rta.lv,mietule@inbox.lv

Dr. **Dariusz Pawliszczy**, Mayor of Gromadka, Poland

e-mail: pawliszczy@interia.pl

Dr. **Oleg Pursky**, Professor of Computer Science and Information Systems, Head of Department of Computer Science and Information Systems, Kyiv National University of Trade and Economics, Kyiv, Ukraine

Oleg Pursky, born in 1967, received a Candidate of Sciences in Physics and Mathematics degree (Dr. phil.) from the Institute for Low Temperature Physics and Engineering of the National Academy of Sciences of Ukraine, in 2001, and a Doctor of Sciences in Physics and Mathematics degree (Dr. habil.) from the Taras Shevchenko National University of Kyiv, Ukraine, in 2010. His research interests include informational systems development, computer simulation and modeling of socio-economic systems. He has published a number of papers in international journals, monographs and volumes in book series, is a member of editorial board of International Journal of Economic Theory and Application, reviewer of scientific journals International Journal of Modern Physics (B) and Heat Transfer and certified Data Science&Machine Learning specialist. He is a member of Scientific Council section of Ukrainian Ministry of Education and Science on the specialty "Informatics and Cybernetics". Currently, he is working as a Head of Department of Computer Science and Information Systems, Kyiv National University of Trade and Economics.

WWW: https://knute.edu.ua/blog/read/?pid=12695&uk

e-mail: Pursky_O@ukr.net

Dr. **Sultan Ramazanov**, Kyiv National Economic University named after Vadym Hetman, Ukraine

Born 12/13/1949. Mathematician, cybernetic, environmental economist; design specialist and development of integrated management information systems complex ecological-economic objects with elements of artificial intelligence. Professor, Doctor of Engineering, Doctor of Economics Science, Honored Worker of Science and Technology of Ukraine. Honored Worker of Education of Ukraine. Excellence in education of Ukraine. Professor of the Department "Information systems in the economy" of Kyiv National Economic University named after Vadym Hetman. Professor Emeritus of East Ukrainian National University named after V. Dahl and Poltava University of Economics and Trade.

Prepared 1 DSc and 14 PhDs (7 scientific fields). He is the author of 495 scientific and methodological works, including 41 monographs and 14 textbooks. Expert of the Ministry of Education and Science of Ukraine ("Informatics and Cybernetics"). Academician of Academies: International Academy of Informatics, The International Academy of Environmental Sciences and Life Safety, The Academy of Technological Sciences of Ukraine, the Academy of Economic Sciences of Ukraine, Transport Academy of Ukraine. Total work experience in Ukrainian universities - 47 years (since 1973). 28 years – head of the department and at the same time 13 years – dean of the faculty (EUNU named after V. Dahl, Luhansk and Severodonetsk) ([]from 1973 to 2014). Since 2014 - Professor of Departments at PUET (Poltava) and KNEU (Kyiv).

e-mail: sramazanov@i.ua,sultan.ramazanov@kneu.edu.ua

Dr. **Serhiy Semerikov**, Professor of Computer Science and Educational technology, Kryvyi Rih State Pedagogical University, Ukraine

Serhiy Semerikov is professor of Department of Computer Science and Applied Mathematics at Kryvyi Rih State Pedagogical University. He got both PhD and DSc in education (informatics) from the National Pedagogical Dragomanov University in 2001 and 2009, respectively. The main directions of Dr. Semerikov' research is methods of learning and educational technology.

https://doi.org/10.1051/shsconf/202110700001

WWW: https://kdpu.edu.ua/semerikov/
e-mail: semerikov@gmail.com

Dr. **Nadiia Shmygol**, Professor of Management Department, National University "Zaporizhzhia Polytechnic", Zaporizhzhia, Ukraine

Nadiia Shmygol, born in Pavlodar city in Kazakhstan 1972. In 2005 defended a thesis and obtained a diploma of Candidate of Sciences (comparable to the academic degree of Doctor of Philosophy, Ph.D.) Economics in Economic and Mathematical Modeling. In 2013 defended a thesis and obtained a diploma of Doctor of Sciences in Economy and Management of enterprises. Certificate of professor of Accounting and Audit Department (2014). Author of over 200 scientific and practical publications in the field of economics and management.

Sphere of scientific interests: management accounting, CSR, sustainable development, circular economy, sustainable marketing, resource-efficient production, decision-making methods, environmental management.

WWW: https://zp.edu.ua/nadiya-mikolayivna-shmigol
e-mail: nadezdashm@gmail.com

Dr. **Kateryna Shymanska**, Head of the Department, Department of Digital Economy and International Economic Relations, Zhytomyr Polytechnic State University, Zhytomyr, Ukraine

Kateryna Shymanska, born in 1985, received a Candidate of Economic Sciences degree (Dr. phil.) from the Zhytomyr State Technological University, Ukraine, in 2010, and a Doctor of Economic Sciences degree (Dr. habil.) from the Vasyl' Stus Donetsk National University, in 2019. Since 2014, she has been working in the field of international economic relations at the Zhytomyr Polytechnic State University, where she is currently Head of the Department of Digital Economy and International Economic Relations. Her research interests include international migration consequences and regulation, but also international trade and the challenges of the digital economy. She has published a number of papers on the specified problems.

WWW: https://ztu.edu.ua/ua/structure/faculties/fbso/kme.php

e-mail: kv.shymanska@gmail.com

Vladimir N. Soloviev received the D. Sc. (Dr. Hab.) degree in solid state physics from Institute of Physics of the National Academy of Sciences of Ukraine, in 1993. From 1992 to 2000 and from 2016 to the present head of the Department of Informatics and Applied Mathematics of Kryvyi Rih State Pedagogical University. In the period from 2000 to 2016, he carry out research on critical and crisis phenomena in the financial markets at various universities in Kyiv, Cherkassy and Kryvyi Rih. He has about 300 publications in the field of solid state physics, complex systems and quantitative methods of constructing precursors of crisis phenomena in systems of different nature.

Dr. **Victoria Solovieva**, Associate Professor, Head of Department, State University of Economics and Technology, Ukraine

She graduated from the Faculty of Physics and Mathematics of the Kryvyi Rih Pedagogical Institute. She has a Ph.D., specializing in Economic and Mathematical Modeling. He has about 130 publications in the field of complex systems and quantitative methods for constructing precursors of crisis phenomena in systems of various nature.

e-mail: vikasolovieva2027@gmail.com

Dr. **Galyna Velykoivanenko**, Head of the Economic and Mathematical Modeling Department, Kyiv National Economic University named after Vadym Hetman, Ukraine

e-mail: ivanenko@kneu.edu.ua

Dr. **Eugene Yakub**, Professor of Economic Information and Computer Technology, Department of Economic Cybernetics and Information Technology, Odessa National Economic University, Odessa, Ukraine

Eugene Yakub, born in 1946, received a Candidate of Technical Sciences degree (Dr.Phil.) from the Odessa Technological Institute of Refrigeration, USSR, in 1973, and a Doctor of Physical and Mathematical Sciences degree (Dr. habil.) from the Institute of High Temperatures of the Academy of Sciences of USSR, in 1990. Since 1977, he has been working in the field of mathematical and computer modelling at the Odessa National Economic University, where he is currently head of Department of Economic Cybernetics and Information Technology. His research interests include modelling and computer simulation of various complex natural and economic systems. He has published a number of papers in international journals and monographs, is a member of several Academic Councils.

WWW: http://oneu.edu.ua/ru/yakub-yevgen-solomonovich
e-mail: yakub@oneu.edu.ua

Oleh Yatsiuk, Department of Economics and Management Theory, Ivano-Frankivsk National Technical University of Oil and Gas, Ivano-Frankivsk, Ukraine

Oleh Yatsiuk, born in 1978, received a Master degree in Economics from the Ivano-Frankivsk National Technical University of Oil and Gas, in 2000. Since 2001, he has been working in the field of enterprise' financial sanation at the Ivano-Frankivsk National Technical University of Oil and Gas. His research interests include finances of the enterprise, anti-crisis management and HR management. He has more than 200 published scientific and educational works, including monographs, textbooks, papers in domestic and international journals.

WWW: https://nung.edu.ua/person/yacyuk-oleg-stepanovich-0
e-mail: olegstya@gmail.com

Dr. **Nataliia Zachosova**, Professor of Management and Economic Security Department, Bohdan Khmelnytsky National University of Cherkasy, Cherkasy, Ukraine

Nataliia Zachosova, born in 1986, received a Candidate of Economic Sciences degree (Dr. phil.) in 2011, and a Doctor of Economic Sciences degree (Dr. habil.) in 2017. In 2018 she received the Scholarship of the Cabinet of Ministers of Ukraine for Young Scientists (and in 2020 she got it once again) and in 2019 – the Nominal Scholarship of the Verkhovna Rada of Ukraine for the Most Talented Young Scientists. Since 2014, she has been working in the field of financial and economic security management at the Bohdan Khmelnytsky National University of Cherkasy, where she is currently Professor and the Chairman of the Council of Young Scientists. Her research interests include security oriented management, strategic and personnel management. She has published a number of papers in international journals and volumes in book series, is a member of editorial boards of Bulletin of the Cherkasy Bohdan Khmelnytsky National University, Series "Economic Sciences".

WWW: https://orcid.org/0000-0001-8469-3681
e-mail: natazachosova@gmail.com

Dr. **Pavel Zakharchenko**, Professor of Economic Cybernetics, Head of Department of Economic Cybernetics and Finances, Berdyansk State Pedagogical University, Berdyansk, Ukraine

Pavel Zakharchenko, born in 1957, received a Candidate of Technical Sciences (Dr. phil.) from the Kharkov Institute of Radio Electronics, USSR, in 1982, and a Doctor of Economical Sciences degree (Dr. habil.) from the

Kiev National Economic University of Ukraine, in 2010. Since 2000, he has been working in the field of finance and economic cybernetics at the Berdyansk State Pedagogical University, where he is currently Head of Department of Economic Cybernetics and Finances. His research interests include health economics models, chaos and catastrophe theory. He has published a number of articles in international journals and volumes in book series, and is a member of the editorial board of scientific journals.

WWW: https://bdpu.org.ua/faculties/gef/structure-gef/kaf-fin/

e-mail: pvzz1957@gmail.com

4 M3E2 2021: Conclusion and outlook

The vision of the M3E2 2021 is provides a premier interdisciplinary platform for researchers, practitioners and educators to present and discuss the most recent innovations, trends, and concerns as well as practical challenges encountered and solutions adopted in the fields of emergent economy.

The conference has successfully performing forum to transferring and discussing research result among the researcher, students, government, private sector or industries. Participants and presenters from several countries such as Belarus, China, Czechia, Kazakhstan, Moldova, Poland and Ukraine have attended the conference in-person and online to share their significant contribution in research related to Monitoring, Modeling & Management of Emergent Economy.

We are thankful to all the authors who submitted papers and the delegates for their participation and their interest in M3E2 as a platform to share their ideas and innovation. Also, we are also thankful to all the program committee members for providing continuous guidance and efforts taken by peer reviewers contributed to improve the quality of papers provided constructive critical comments, improvements and corrections to the authors are gratefully appreciated for their contribution to the success of the conference. Moreover, we would like to thank the developers and other professional staff of *Not So Easy Science Education* platform (https://notso.easyscience.education), who made it possible for us to use the resources of this excellent and comprehensive conference management system, from the call of papers and inviting reviewers, to handling paper submissions, communicating with the authors etc.

We are looking forward to excellent presentations and fruitful discussions, which will broaden our professional horizons. We hope all participants enjoy this conference and meet again in more friendly, hilarious, and happiness of further M3E2 2022.

References

[1] S. Semerikov, V. Soloviev, L. Kibalnyk, O. Chernyak, H. Danylchuk, SHS Web of Conferences **65**, 00001 (2019)

[2] A. Kiv, V. Soloviev, S. Semerikov, H. Danylchuk, L. Kibalnyk, A. Matviychuk, CEUR Workshop Proceedings **2422**, 1 (2019)

[3] A. Kiv, P. Hryhoruk, I. Khvostina, V. Solovieva, V. Soloviev, S. Semerikov, CEUR Workshop Proceedings **2713**, 1 (2020)

[4] E. of cryptocurrency crashes: an overview, SHS Web of Conferences (2021, in press)

[5] A. Bielinskyi, S. Hushko, A. Matviychuk, O. Serdyuk, S. Semerikov, V. Soloviev, SHS Web of Conferences (2021, in press)

[6] H. Kucherova, V. Los, D. Ocheretin, O. Bilska, E. Makazan, SHS Web of Conferences (2021, in press)

[7] L. Petkova, O. Berezina, I. Honcharenko, I. Osadchenko, SHS Web of Conferences (2021, in press)

[8] D. Kainara, I. Kaliuzhna, SHS Web of Conferences (2021, in press)

[9] N. Maksyshko, O. Vasylieva, SHS Web of Conferences (2021, in press)

[10] M. Nezhyva, O. Zaremba, V. Nehodenko, SHS Web of Conferences (2021, in press)

[11] N. Zachosova, D. Kutsenko, O. Koval, A. Kovalenko, SHS Web of Conferences (2021, in press)

[12] T. Shkoda, I. Gernego, M. Tepliuk, A. Stefanita, SHS Web of Conferences (2021, in press)

[13] S. Radzivivska, I. Us, SHS Web of Conferences (2021, in press)

[14] M. Medvid, P. Ivashchenko, I. Britchenko, I. Trubavina, V. Liutyi, SHS Web of Conferences (2021, in press)

[15] S. Ivanov, M. Ivanov, SHS Web of Conferences (2021, in press)

[16] K. Gorbatiuk, T. Zavhorodnia, O. Proskurovych, O. Mantalyuk, SHS Web of Conferences (2021, in press)

[17] S. Ramazanov, O. Honcharenko, SHS Web of Conferences (2021, in press)

[18] V. Porokhnya, V. Penev, SHS Web of Conferences (2021, in press)

[19] S. Onikiienko, M. Dyba, I. Gernego, SHS Web of Conferences (2021, in press)

[20] A. Saparbayev, A. Makulova, A. Kaziyeva, Z. Klyshbayeva, E. Saparbayeva, SHS Web of Conferences (2021, in press)

[21] T. Pulina, N. Shmygol, O. Narivs'kyi, W. Łuczka, Y. Shmygol, SHS Web of Conferences (2021, in press)

[22] A. Saparbayev, A. Makulova, N. Bayboltaeva, N. Sarsenbieva, A. Imatayeva, SHS Web of Conferences (2021, in press)

[23] P. Hryhoruk, N. Khrushch, S. Grygoruk, SHS Web of Conferences (2021, in press)

[24] M. Shkvaryliuk, L. Horal, I. Khvostina, N. Yashcheritsyna, V. Shiyko, SHS Web of Conferences (2021, in press)

[25] Y. Lozovik, B. Tishkov, O. Pomazun, V. Derbentsev, A. Hostryk, SHS Web of Conferences (2021, in press)

[26] I. Vasylchuk, A. Suprun, K. Slyusarenko, M. Sadovenko, SHS Web of Conferences (2021, in press)

[27] O. Orlov, Y. Ryasnykh, K. Dumanska, O. Savchenko, SHS Web of Conferences (2021, in press)

[28] E.A.M. Omran, Y. Bilan, SHS Web of Conferences (2021, in press)

[29] A. Bakurova, O. Yus'kiv, D. Shyrokorad, E. Tereschenko, A. Riabenko, SHS Web of Conferences (2021, in press)

[30] P. Zakharchenko, G. Kostenko, T. Kungurtseva-Mashchenko, S. Zhvanenko, V. Mukhin, SHS Web of Conferences (2021, in press)

[31] N. Shmygol, O. Cherniavska, D. Pawliszczy, Y. Shmygol, A. Cherniavska, SHS Web of Conferences (2021, in press)

[32] H. Danylchuk, L. Kibalnyk, O. Kovtun, O. Pursky, Z. Stachowiak, SHS Web of Conferences (2021, in press)

[33] L. Kibalnyk, M. Leshchenko, O. Feshchenko, H. Wang, SHS Web of Conferences (2021, in press)

[34] M. Zelenkevich, N. Bandarenka, SHS Web of Conferences (2021, in press)

[35] O. Pursky, T. Dubovyk, I.B.I. Lutsenko, H. Danylchuk, SHS Web of Conferences (2021, in press)

[36] V. Shinkarenko, A. Hostryk, L. Shynkarenko, L. Dolinskyi, SHS Web of Conferences (2021, in press)

[37] V. Habrusiev, H. Tereshchuk, I. Tsidylo, S. Martyniuk, O. Kulyanda, SHS Web of Conferences (2021, in press)

[38] R. Ivanov, Y. Sherstennikov, V. Porokhnya, T. Grynko, SHS Web of Conferences (2021, in press)

[39] O. Ovchynikova, O. Dupliak, SHS Web of Conferences (2021, in press)

[40] T. Melikhova, O. Mikhailutsa, SHS Web of Conferences (2021, in press)

[41] A. Kaminskyi, M. Nehrey, L. Zomchak, SHS Web of Conferences (2021, in press)

[42] I. Chaikovska, P. Hryhoruk, M. Chaikovskyi, SHS Web of Conferences (2021, in press)

Recurrence analysis of innovation behavior of bitcoin market agents in conditions of COVID-19

Hanna Kucherova[1,*], *Vita* Los[2,**], *Dmytro* Ocheretin[2,***], *Olha* Bilska[1,****], and *Evgenia* Makazan[2,†]

[1]Classic Private University, 70B Zhukovsky Str., Zaporizhzhia, 69002, Ukraine
[2]Zaporizhzhia National University, 66 Zhukovsky Str., Zaporizhzhia, 69600, Ukraine

Abstract. The relevance of the research subject is explained by a fundamental change in the conditions of existence and development of agents of the digital economy, limited knowledge about their behavior under conditions of quarantine restrictions. The aim of the research is to study the series of the dynamics of the price of bitcoin and the frequency of online requests for bitcoin as an indicator of the behavior of agents of the digital economy using the methods of qualitative recurrent analysis. The types of constructed time series plots of the price of bitcoin and the frequency of requests for bitcoin are defined as drift with a superimposed linearly gradually increasing sequence, which indicates the unpredictability of the behavior of digital economy agents with a gradual stabilization in new quality trend. The scientific novelty of the research results lies in the proven connection between the series of bitcoin price dynamics and the frequency of online requests for bitcoin, tracking changes in the behavior of digital economy agents before and after the introduction of quarantine restrictions. The procedure for conducting a qualitative recurrence analysis of the series of dynamics is generalized, which takes into account the specifics of the formation of the frequency of online requests for bitcoin, the price and the behavioral aspect of its formation. The practical value lies in defining the characterization of the behavior model of digital economy agents under conditions of quarantine restrictions. The behavior of digital economy agents in the context of COVID-2019 requires further research, in particular, using cross-recurrent analysis methods.

1 Introduction

It is difficult to find a clear generalization of the processes that have begun and are currently taking place in the countries of the world in the COVID-pandemic [1, 2]. Unpreparedness for new risks of this type was demonstrated by both individuals and legal entities, authorities and society as a whole. And this is expected, since the collapse of development and interaction of individual sectors of the socio-economic environment has been forming for decades. The unsystematic nature of the government's programmatic actions and the devaluation of the priority issues of ensuring the basic conditions for the survival of society and business have led to the fact that we have become hostages of our own impotence in providing the necessary resistance to the expansion of the disease and the negative socio-economic consequences of the pandemic.

New living conditions draw attention directly to the model of our behavior and in the digital market, determine that the issues of adhering to the security of interactions and the possibility of development within the new established boundaries are priority basic goals for each of us. The established models of the existence of forms and ways of life in the countries of the world have confirmed the need to reduce them, reorient the volumes of consumption, production, change business models, ways of interaction, and so on. New challenges and updated values have actualized the demand for specific goods and services that provide a solution to complex issues of socio-economic security for each in pandemic, have led to fundamental changes in the behavior of entities in all online and offline business markets. Today, timely tracking of changes in behavior in the markets contributes to the formation of a new quality of management, quick adaptation of business, changes in the basic principles of interaction and functioning of subjects of all spheres, clarification of current trends and prediction of the formation of new trends in key indicators.

An immediate rethinking of the socio-economic behavior of everyone is fundamental; the adoption of its new restrictions should take place both at the level of the subjects' consciousness and taking into account the limiting security measures. A business that has been developing offline as a priority, being partially present in the online space, while ignoring the rapid development of Internet technologies, is now unable to overcome the financial and economic losses that were incurred as a result of quarantine restrictions. Therefore, it is logical to actualize the movement of most of the business to the online space, more active use of modern digital technologies and tools.

*e-mail: kucherovahanna@gmail.com
**e-mail: vitalos.2704@gmail.com
***e-mail: odvisua@gmail.com
****e-mail: lexusol1987@gmail.com
†e-mail: e.v.makazan@gmail.com

The parameters of consumer and business requests in the context of the pandemic have undergone fundamental changes both in structure and in quality, in the way of use. Unfortunately, modern socio-economic instruments are not able to describe the general picture of changes in demand and the behavior of economy agents, to demonstrate and reliably predict the trends of current changes in key instruments, the dynamics of which forms the profitability of the markets. Because the external environment during the pandemic has become too unpredictable. Thus, tools such as Google Trend have become useful, where the frequency of queries for keywords makes it possible to track trends in the interests of subjects of the digital market, and not only digital market in real time.

The frequency of online queries as an integrated indicator of information retrieval and the activity of economy agents determines the sphere of consumer interest, which is described by a tuple of parameters $\langle C_S, M, R_F, t\rangle$, where C_S – semantic search characteristics (semantic core), M – meaning of the semantic search, R_F – business activity of agents (frequency of requests), t – time period.

The behavioral model of economy agents is formed in such a way that with an increase in the subject's interest in the target search area, the frequency of online queries for keywords increases. If the agents' interest is not cognitive, but corresponds to the implementation of their strategy in the market, then an increase in the frequency of requests affects the price parameters of target instruments in the direction of their values increase. Thus, an upward trend in the dynamics of time series of key indicators is formed, followed by market monitoring, the results of which arouse interest in relevant modern tools.

This direct cyclical pattern of interdependence of interests in tools and the frequency of corresponding online requests requires system monitoring of their dynamics and an explanation of their changes in different periods of time, which will make it possible to determine or establish a model of agent behavior, factors influencing it and predict its development in new conditions of socio-economic system in offline and online.

Thus, the purpose of the research is the behavior of agents of the digital economy, which is described using the rate of requests for the bitcoin rate in the context of COVID-19. The object of the research is the time series: requests "bitcoin" in English and Russian of Ukrainian users of the digital market, and the price of bitcoin. The subject of research is the methods of nonlinear dynamics.

2 Related works

A significant number of scientific and practical works are devoted to the research of time series of cryptocurrencies, carried out by methods of machine learning and forecasting [3, 4], nonlinear autoregression [5], binomial logistic regression [6], recurrent neural network, ARIMA model [7], Bayesian neural networks [8], theory of complex systems [9–13], fractal analysis [9, 14–19] and others. However, the identification of the behavior of digital economy agents requires a search for specific parameters that would uniquely determine it. The search for such parameters is difficult, since their characterized by weak structuring and significant subjectivity. So, to date, the results of the influence of social networks on the dynamics of the indicators of the digital market [7] have been obtained, including a quantitative relationship between social networks and the bitcoin price [14–16], a connection between the parameters of bitcoin and online searches in Google has been confirmed [18], Google queries and exchange rates [18]. Based on the results of the research, the authors confirm the thesis that the indicators of online search and social networks really determine the dynamics of the rate of cryptocurrencies, therefore, this direction requires further research. Researchers [3, 7, 8, 20, 21] have proved that the complexity of the processes, the peculiarities of the formation time series of bitcoin and the frequency of requests for bitcoin require the use of the methodology of nonlinear dynamics, the methods of which are actively used in the study of processes during the manifestation of crisis phenomena [5, 9], which in fully complies with the conditions of today. However, a fairly powerful toolkit of nonlinear dynamics only partially used in the study of the behavior of economy agents in the online information space, which proves the timeliness of this study.

3 Matherials

According to Google Trends data [22], the search query "bitcoin" is characterized by a simple semantic core, which serves as the basis for the formation of a more complex, refined online information search. In particular, online information requests such as "bitcoin halving", "price bitcoin", "what is bitcoin", "bitcoin course", "bitcoin dollar", "bitcoin price usd", "freebitcoin" have shown the maximum popularity in the online environment over the past year. Interest in bitcoin is not evenly distributed in Ukraine, most often in Kyiv, Kyiv region, Kharkiv, Odesa and Chernivtsi regions are interested in its price and characteristics. Google Trends the values of the frequency of requests in the Ukrainian language as insufficient in importance. Google Trends noted that the frequency of Bitcoin requests was very popular in Russian and English. At the same time, the dynamics of the frequency of requests in the Ukrainian language showed a similarity of the dynamics of the frequency of requests in Russian. The behavior of market agents in terms of the frequency of Bitcoin requests in Ukrainian the same as the frequency of Bitcoin requests in Russian and English. The dynamics of the studied time series is shown in figure 1.

Figure 1 shows that the dynamics of indicators is characterized by a high value of volatility. The dynamic series of the frequency of requests "bitcoin" is described by a 70.58% coefficient of variation, "bitcoin" in Russian – 90.48%, and the price of bitcoin – 83.43%. In addition, the trend in the frequency of requests for bitcoin in English and Russian are similar, but differ in the values of the number of requests, which is quite logical, taking into account the dominant languages in the country.

The maximum value of the bitcoin price for the study period was observed on December 17, 2017 and amounted to \$19140.8 , the minimum – on August 23,

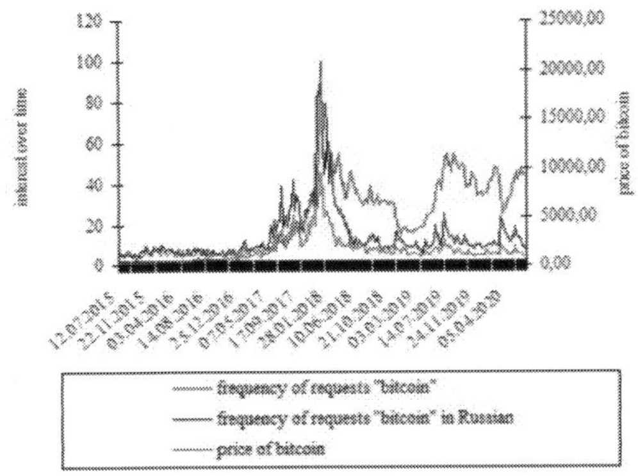

Figure 1. Frequency of requests "bitcoin", "bitcoin" in Russian on Google in Ukraine

2015 ($228.17). However, the maximum value of the frequency of requests "bitcoin" (46 points) and "bitcoin" in Russian (100 points) was also reached on December 17, 2017. The minimum value of the series of dynamics of the frequency of requests for "bitcoin" of Russian at the level of 3 points was observed on September 27, 2015 and August 07, 2016, and the series of dynamics of the frequency of requests for "bitcoin" on December 11, 2016.

The maximum increase in the price of bitcoin [23] for the study period was recorded at the level of 41.5% (July 23, 2017), that is, the price increased by 800.6 dollars relative to the previous period (July 16, 2017). The maximum decrease in the price of bitcoin was observed precisely at the beginning of quarantine measures in March 2020 (March 15, 2020). Bitcoin price fell by 33.5% or $2,715.8 relative to the previous period (March 08, 2020). In general, on average, over the period under study, the price of bitcoin increased by 1.9% weekly. So, in June 2020, the price of bitcoin increased 29.4 times compared to the same period in 2015.

Studying individual periods of the time series of dynamics, it was found that the maximum values of the price of bitcoin and the frequency of online requests for bitcoin either coincide in the period, or are very close in time. Also, sharp trends in the price of bitcoin are accompanied by a surge in interest in it, even if the price trend is decreasing. Whereas the stable fluctuations of the bitcoin price within the boundaries of a certain corridor, to a lesser extent, but still of interest to economy agents.

It is proposed to eliminate the unpredictability of the studied time series of dynamics using recurrent analysis, the implementation of which will make it possible to characterize the behavior of digital economy agents in new conditions of existence and development.

4 Methodology

To study the time series, the authors proposed a methodology based on the method of recurrence analysis for analyzing the behavior of digital economy agents in the context of COVID-19.

The input time series for the analysis were generated using data from Google Trends [22] and InvestFunds [23] for the period from July 12, 2015 to June 28, 2020. Authors investigate three time series, namely: the frequency of online requests for bitcoin in Ukraine on English and Russian, and the price of bitcoin. After the formation of the database, the behavior of the dynamics of the studied time series should be analyzed in order to establish the type of behavior and the presence of a trend or random behavior of the series.

To apply the proposed methodology, the data values of the initial series were normalized in the interval [0, 1] by the formula

$$x_{norm} = \frac{x_i - x_{\min}}{x_{\max} - x_{\min}}, \quad (1)$$

where x_{norm} – normalized time series value;

$x_{\max}$, $x_{\min}$ – maximum and minimum value of the time series, respectively.

After the formation of the set of initial data, the next step is the choice of the reasonable delay of the time series (G). Choosing G should take into account the following [18]:

– G must be large enough so that the value $x(t)$ different from the value $x(t + G)$;

– but if G is too large, then at time $(t + G)$ the system will lose information about what happened at time t.

Taking into account the above, the authors define the function of mutual information (AMI) S for the time series that analyzed, taking into account nonlinear correlations [24]. The time series delay time G corresponds to the first local minimum of the mutual information function (AMI). The time series reasonable delay time was calculated in the R environment using the tseriesChaos library.

To determine the dimension of the time series, the authors used the false nearest neighbor method given in the work [25]. This method is based on the assumption that in a correctly constructed attractor the neighboring points of the phase trajectory remain very close at the following iterations. If the nearest points diverge, then they are called false nearest neighbors. The problem is to choose a dimension of the phase space r at which the fraction of points with false neighbors is minimized. The calculation of false neighbors was calculated in the R environment using the fractal library.

The next step is to build a recurrence plot based on the calculated space dimension and time delay parameters. Recurrence plot is the projection of m-dimensional pseudophase space onto a surface. Let the point x_i correspond to the point of the phase trajectory $x(t)$ that describes the dynamical system in the m-dimensional space at the time $t = i$ ($i = 1, ..., N$), then the recurrence plot is an array of points in which a nonzero element with coordinates (i, j) corresponds to the case when the distance between x_i and x_j is less than γ:

$$RP_{i,j} = \theta\left(\gamma - \left\|x_i - x_j\right\|\right) x_i,\ x_j \in R^m\ (i, j = 1, ..., N), \quad (2)$$

where γ – the size of the neighborhood of the point x_i,

$\|x_i - x_j\|$ – distance between points,
$\theta(\cdot)$ – Heaviside function.

The next stage of the proposed methodology is analysis of the measures of recurrence plot. This analysis makes it possible to determine the measures of complexity of the structures of recurrence plot, such as: recurrence rate, percent recurrence, percent determinism, average diagonal line length, maximum diagonal line length [16, 26, 27]. Let's consider these measures in more detail. Recurrence rate (RR) shows the density of recurrent points, that is, it characterizes the probability of repetition of system states:

$$\mathrm{RR} = \frac{1}{N^2} \sum_{i,\, j=1}^{N} R_{i,j}, \tag{3}$$

where N – number of points on the phase space trajectory.

Percent recurrence ($\%REC$) displays a decrease in the regularity of the system's behavior and is used to analyze the dynamic structure of a time series:

$$\%REC = \frac{\sum_{l=1}^{N} lP(l)}{N} \cdot 100. \tag{4}$$

If the percent recurrence value is less than 1%, then it is said that there is a regular, clearly defined dynamics of the time series behavior. For the regularity of the system's behaviors percent recurrence value from 1% to 5%. If the value of the percent recurrence lies in the range from 5% to 20% or more, then this indicates that the dynamics of the time series behavior is not regular and noisy interevent-type data. Depending on the parameters of the delay and the dimension of the phase space, the following average values of the percent recurrence can be obtained: a low ($\%REC \approx 1\%$ to 3%), moderate ($\%REC \approx 5\%$ to 10%), and high ($\%REC \approx 15\%$ to 20%) [28].

Percent determinism ($\%DET$) considers the diagonal lines of the recurrence plot. The frequency distribution of the lengths of the diagonal lines in the recurrence plot can be written as follows: $P^{\gamma}(l) = \{l_i, 1, ..., N_l\}$, where l_i – i-th diagonal line length, N_l – the number of diagonal lines (each line is counted only once). If the time series is stochastic, then the diagonal lines of the recurrence diagram are very short or completely absent. And deterministic time series are characterized by long diagonals and a small number of separate recurrent points.Percent determinism ($\%DET$) is defined using this formula:

$$\%DET = \frac{\sum_{l=l_{\min}}^{N} lP(l)}{\sum_{l=l_1}^{N} lP(l)} \cdot 100. \tag{5}$$

Percent determinism characterizes the level of predictability of the time series.

Diagonal structures show the time during which a section of a trajectory comes close enough to another section of a trajectory. Thus, these lines allow conclusions to be drawn regarding the divergence of the trajectory elements. These indicators are average diagonal line length (ADL) and maximum diagonal line length (MDL).

Average diagonal line length (ADL) characterizes the average time during which two sections of the trajectory pass close to each other, and can be considered as the average predictability time of the series. This measure is calculated using the formula:

$$ADL = \frac{\sum_{l=l_{\min}}^{N} lP(l)}{\sum_{l=l_{\min}}^{N} P(l)}. \tag{6}$$

Respectively, maximum diagonal line length (MDL) characterizes the length of the trend and is defined as:

$$MDL = \max(\{l_i;\ i = 1, ..., N_l\}). \tag{7}$$

Based on the analysis of the statistical characteristics of the recurrence plot, it is possible to determine the presence of homogeneous processes with independent random values, processes with slowly varying parameters, periodic or oscillating processes that correspond to nonlinear systems. Thus, the analysis of the recurrence plot allows one to evaluate the characteristics of a nonlinear object on relatively short time series, which makes it possible to make decisions regarding the object's control. The measures of the recurrence plot was calculated in the R environment using the nonlinearTseries library.

5 Results

Based on the results of the calculations, recurrence plots were obtained, the topological analysis of which makes it possible to determine the structure, type, changes in the behavior of the research object, the boundaries of phase transitions, and to establish the sensitivity of quantitative measures.

Recurrence quantification analysis (RQA) can be used not only to quantify the dynamics of a whole time-series, but also to study changes in the dynamics of a time series. Windowed RQA is potentially a very powerful tool for detecting changes in subsets of the whole time-series. To conduct a better study of the temporal structure, the dynamic series of the bitcoin request frequency (in English) and time-series of price of bitcoin were divided into four periods, the length of each of which was 65 observations: A) July 12, 2015 – October 02, 2016, B) October 09, 2016 – December 31, 2017, C) January 07, 2018 – March 31, 2019, D) April 07,2019 – June 28, 2020. Recurrence plot of the behavior of digital economy agents were constructed and RQA was calculated for each subset of the time series (adjacent non-overlapping windows of 65 data points) (figure 2).

Gradual changes in the parameters of the behavior of digital economy agents are clearly visible, in particular during period B (October 09, 2016 – December 31, 2017), where the drift of the attractor was revealed (white lower and upper corners of the plot, a diagonal line) and the emergence of a new structure during the period of influence of the consequences COVID-2019 in period D (April 07,2019 – June 28, 2020).

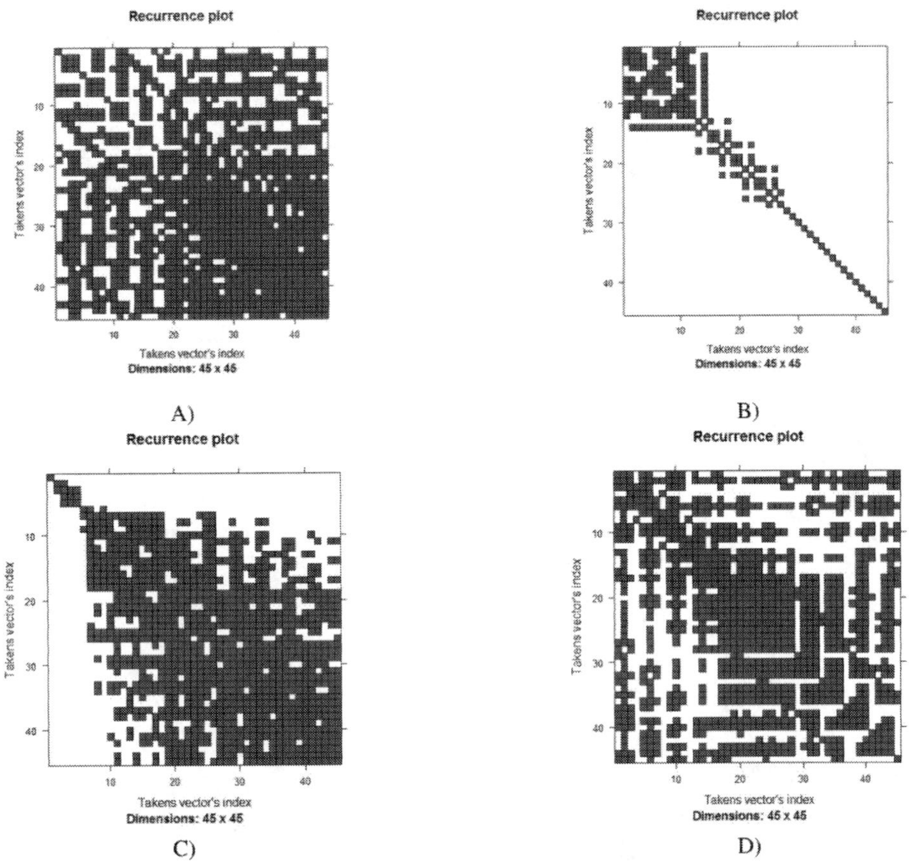

Figure 2. Recurrence plot of the behavior of digital economy agents (frequency of online requests "bitcoin" according to Google Trends in Ukraine): A) July 12, 2015 – October 02, 2016, B) October 09, 2016 – December 31, 2017, C) January 07, 2018 – March 31, 2019, D) April 07,2019 – June 28, 2020

Analyzed period on the recurrence plot is displayed by contrasting areas and stripes, which is explained by sharp changes in values, randomness, rarity, in general, stochastic behavior of digital market entities, unpredictability of changes in interest in it.

However, separately placed random recurrent points are not the result of randomness, their presence proves that interest in the price of bitcoin is simply not stable over time and can be caused by significant fluctuations. The measures of the recurrence plot are investigated such as: percent recurrence (*%REC*), percent determinism (*%DET*), average diagonal line length (*ADL*), maximum diagonal line length (*MDL*) (table 1).

The change in the *%REC* value from 71.46% to 57.93% in the periods under study proves that the behavior of digital market entities is unstable, noisy interevent-type data. The lowest value of *%REC* (9.83%) was recorded in period B, when the trajectory of behavior reaches a new level of development of processes. Starting from period C, the *%REC* value increases significantly and fluctuates within the range of 54-57%. The obtained values indicate an increase in irregularity and the presence of noisy interevent-type data in this time series. Percent determinism (*%DET*) ranges from 73-93%. In addition, in periods A and C the measure value is almost the same, while in period D it slightly decreased to 89.60%.

Table 1. Recurrence Quantification Analysis of the frequency of requests "bitcoin"

Measures of the recurrence plot	1 plot (A)	2 plot (B)	3 plot (C)	4 plot (D)
%REC	71.46	9.83	54.67	57.93
%DET	93.78	73.87	93.86	89.60
ADL	4.80	3.97	5.74	3.12
MDL	35.00	6.00	39.00	27.00

Average trend predictability time (*ADL*) and maximum diagonal line length (*MDL*) are lowest in periods B and D. Want in period D the *MDL* value reaches 27 points, which is 4.5 times more than the same value in period B. Thus, the trajectory of the behavior of digital economy agents demonstrates a phase transition to another level of development in period B and the formation of a new, approximate model to the existing one, but with new qualitative parameters in period D.

At the same time, the recurrence plots for the corresponding periods of the bitcoin price are also characterized by the attractor drift. However, the dynamics of the indicator adheres to the original trajectory from September 10, 2016 and takes into account the changes in processes during the D period as a result of the impact of COVID-2019 (figure 3). Despite the powerful volatility and extreme growth in the price of bitcoin (in particular in 2018), the symmetry of plots proves the constancy of the formation of bitcoin price over time, the tendency of points to a given trajectory, but with the inherent randomness of qualitative values. The measures of the recurrence plot for price of bitcoin are in table 2.

Table 2. Recurrence Quantification Analysis of the price of bitcoin

Measures of the recurrence plot	1 plot (A)	2 plot (B)	3 plot (C)	4 plot (D)
%REC	1	37.5	24.5	17.5
%DET	99.5	82.67	77.55	82.86
ADL	19	12	10	5
MDL	10.76	6.53	4.47	3.87

The $\%REC$ value of the series of bitcoin price dynamics is lower than the value of the similar measure of the frequency of requests for bitcoin, which proves the more constant behavior of digital economy agents. However, in period A, a time series of bitcoin price dynamics were characterized by regularity and clear definition of the dynamics of changes in the time series ($\%REC = 1\%$).

The indicators of predictability of the time series ($\%DET$) are similar with the dynamics of changes in requests for bitcoin in periods B and D, but differ significantly in values. In particular, the lowest values of the measure were recorded in period C.

However, the time series of bitcoin price dynamics are characterized by a high level of unpredictability, the average predictability time significantly decreased from 19 points in period A to 5 points in period D. The length of the trend decreased from 10.76 points in period A to 3.87 points in period D. Another result was obtained from the data of recurrence plot of agents of the digital economy for the indicator of the frequency of requests "bitcoin" in Russian (figure 4, figure 5). And this is logical, since Ukraine is a bilingual country (Ukrainian and Russian languages).

In figure 5, period A clearly shows the drift of the attractor, the formation of a strong trend in the price of bitcoin, which is displayed in the form of an arrow. The formation of a trend is accompanied by an increase in interest in price changes, the frequency of requests is not stable in time for the frequency and value (figure 4, A).

Black stripes characterize nonstationarity of behavior, which means the formation of a transitional period. Periodic patterns characterize the cyclical nature of certain changes in interest in the price of bitcoin, the distance between which determines the period. Black dots, which are repeated in isolation, characterize a random interest in the price of bitcoin, its rapid changes (figure 4, A). The formation of the price trend, which is presented on the recurrence plot for period B (figure 5, B), is characterized by a certain stationarity of fluctuations within the accepted range of values. The fading of bitcoin price fluctuations is accompanied by the systematization of interest in it, streamlining the behavior of digital economy agents, reducing it to the usual monitoring of prices (figure 4, B). The measures of the recurrence plot for price of bitcoin and frequency of online requests "bitcoin" in Russian are in table 3. The constancy and regularity of the bitcoin price dynamics significantly increases in period B compared to period A, the $\%REC$ value changed from 54.71% to 5.41%. However, the percent determinism ($\%DET$) is reduced from 97.97% to 82.61% in the corresponding periods. Over the selected periods, the average predictability time decreased from 19.08 points to 5.7 points, and the trend length decreased by half - from 77 points to 35 points. There is a change in the phase space, trends in general. Other trends show metrics for the time series of bitcoin requests. The percent determinism ($\%DET$) is decreasing, from 97.97% to 82.61% in the corresponding periods. The average predictability level slightly decreased from 4 points to 3.9 points, and the trend length increased from 13 points to 19 points, which confirms the formation of a clear trend and a strong trend.

6 Conclusions

As a result of the research, the relationship between the time series of the price of bitcoin and the frequency of online requests for bitcoin as an indicator that characterizes the behavior of agents of the digital economy. The behavior differs were confirmed in the periods before and after the introduction of the COVID-19 quarantine restrictions. An increase in the activity and internal disturbances in the behavior of agents of the digital economy during periods of significant volatility in the price of bitcoin and, on the contrary, its decrease during a period of stabilization and insignificant fluctuations in a certain price band was confirmed. This pattern is explained by the decentralization of the digital market tools, the complexity of the processes of internal self-organization of agents, the strategies they have chosen in the digital market, their role, emotional-volitional, behavioral and cognitive features that are uncommonly manifested during quarantine restrictions. The approach and procedure for the implementation of the Recurrence Quantification Analysis (RQA) of the time series of the price of bitcoin and the frequency of online requests about it are determined. The investigated recurrence plots of the bitcoin price and the frequency of online requests for bitcoin indicate the unpredictability of the behavior of digital economy agents with a pronounced linear trend. As a result of the research of windowed RQA, the authors came to the conclusion that the behavior of the agents of the digital economy (the frequency of requests for "bitcoin" in English) showed a change in trend precisely in the period October 09,2016 – March 31, 2019, where the drift of the attractor and the emergence of a new structure were revealed. During the period of COVID-2019 quarantine re-

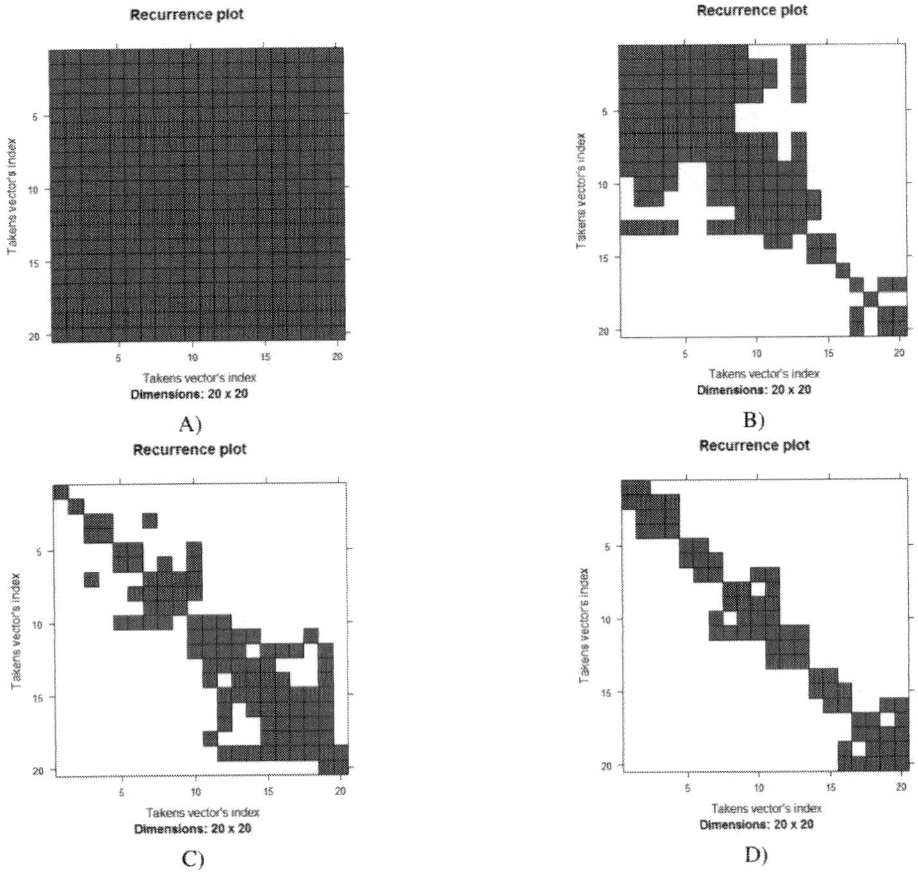

Figure 3. Recurrence plot of the price of bitcoin: A) July 12, 2015 – October 02, 2016, B) October 09, 2016 – December 31, 2017, C) January 07, 2018 – March 31, 2019, D) April 07,2019 – June 28, 2020

Table 3. Recurrence Quantification Analysis of the price of bitcoin and frequency of online requests "bitcoin" in Russian

Measures of the recurrence plot	Price of bitcoin		Frequency of online requests "bitcoin" in Russian	
	1 plot (A)	2 plot (B)	1 plot (A)	2 plot (B)
%REC	54.7	5.41	20.07	39.45
%DET	98	82.6	71.19	74.71
ADL	19.1	5.7	4	3,9
MDL	77	35	13	19

strictions, the behavior of economy agents is characterized by an increased level of stochastic, although interest in the price of bitcoin is unstable over time, but can be caused by significant fluctuations. A qualitative analysis of recurrence plots of the behavior of digital economy agents confirmed the formation of a new phase transition and destabilization during the period of quarantine restrictions, which forms a behavior model similar to the previous one, but with new qualitative parameters. The bitcoin price is characterized by the drift of the attractor, the constancy of price formation over time, adherence to the initial trajectory and the tendency of points towards it with a characteristic randomness of qualitative values, which is explained by changes in socio-economic processes, in particular as a result of the impact of COVID-2019. However, the time series of the bitcoin price and the frequency of requests "bitcoin" in Russian are qualitatively different from the similar series of the frequency of requests "bitcoin" in English for Ukrainian online-user. The investigated series of bitcoin price dynamics over two periods also demonstrates the drift of the attractor and the formation of a strong trend, which is accompanied by an increase and stabilization of interest in price fluctuations. The frequency of online requests is unstable over time only during the period July 12, 2015 – December 31, 2017. The non-stationarity of the behavior of economy agents was confirmed, the formation of a transitional period with a characteristic alternation of cyclical and random changes in interest in the

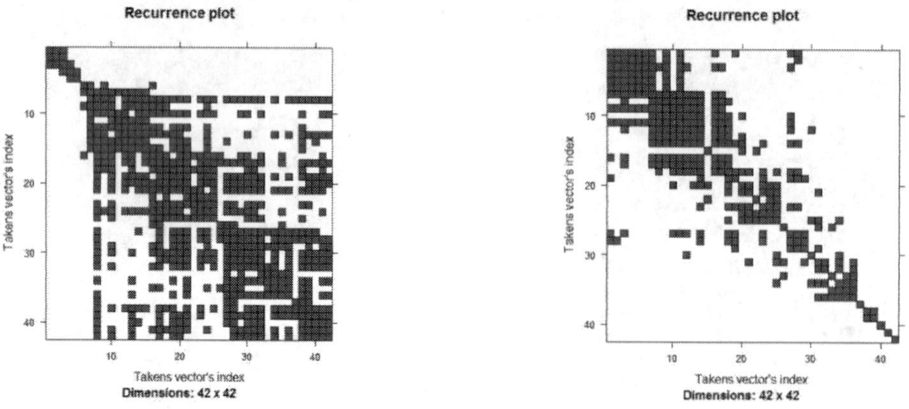

A) B)

Figure 4. Recurrence plot of the behavior of digital economy agents (frequency of online requests "bitcoin" in Russian according to Google Trends in Ukraine): A) July 12, 2015 – December 31, 2017, B) January 07, 2018 – June 28, 2020

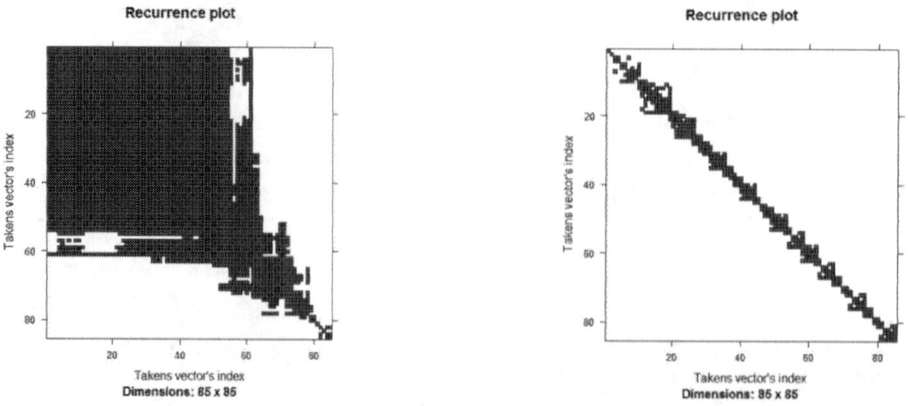

A) B)

Figure 5. Recurrence plot of the price of bitcoin: A) July 12, 2015 – December 31, 2017, B) January 07, 2018 – June 28, 2020

bitcoin price in the period July 12, 2015 – December 31, 2017 was recorded. Whereas the period January 07, 2018 – June 28, 2020 is characterized by a certain stationarity of fluctuations, which is accompanied by the systematization of interest in the price of bitcoin, streamlining the behavior of agents of the digital economy, reducing it to the usual monitoring of prices. Thus, the behavior of digital economy agents in the context of COVID-2019 has generated a new trend with updated qualitative indicators, which requires further research, including using cross-recurrent analysis methods.

References

[1] S. Semerikov, S. Chukharev, S. Sakhno, A. Striuk, V. Osadchyi, V. Solovieva, T. Vakaliuk, P. Nechypurenko, O. Bondarenko, H. Danylchuk, E3S Web of Conferences **166** (2020)

[2] S. Semerikov, H. Kucherova, V. Los, D. Ocheretin, CEUR Workshop Proceedings **2845**, 22 (2021)

[3] P.M. Krafft, N. Della Penna, A.S. Pentland, p. 1–13 (2018)

[4] M. Saad, J. Choi, D. Nyang, J. Kim, A. Mohaisen, IEEE Systems Journal **14**, 321 (2020)

[5] N. Indera, I. Yassin, A. Zabidi, Z. Rizman, Journal of Fundamental and Applied Sciences **9**, 791 (2017)

[6] I. Madan, S. Saluja, A. Zhao, *Automated bitcoin trading via machine learning algorithms* (2014), https://tinyurl.com/jb5r93w

[7] S. McNally, Master's thesis, Dublin, National College of Ireland (2016), http://norma.ncirl.ie/id/eprint/2496

[8] H. Jang, J. Lee, IEEE Access **6**, 5427 (2017)

[9] V. Soloviev, A. Belinskiy, Communications in Computer and Information Science **1007**, 276 (2019)

[10] A. Bielinskyi, V. Soloviev, CEUR Workshop Proceedings **2292**, 37 (2018)

[11] V. Soloviev, A. Belinskij, CEUR Workshop Proceedings **2104**, 116 (2018)

[12] V. Soloviev, S. Yevtushenko, V. Batareyev, CEUR Workshop Proceedings **2546**, 87 (2019)

[13] V. Soloviev, V. Solovieva, A. Tuliakova, A. Hostryk, L. Pichl, CEUR Workshop Proceedings **2713**, 53 (2020)

[14] L. Kristoufek, Scientific Reports **3**, 3415 (2013)

[15] D. Garcia, C.J. Tessone, P. Mavrodiev, N. Perony, Journal of The Royal Society Interface **11**, 20140623 (2014)

[16] L. Kirichenko, T. Radivilova, V. Bullakh, V. Chakryan, International Journal Information Technologies and Knowledge **12.1**, 43 (2018)

[17] H. Kucherova, Actual Problems of Economics **8**, 122 (2019)

[18] H. Kucherova, D. Ocheretin, V. Los, O. Bilska, *Recurrence Analysis of the Economic Behavior of Agents of the Frequency of Online Exchange Rate in the Information Environment*, in *Proceedings of the III International Scientific Congress Society of Ambient Intelligence 2020 (ISC-SAI 2020)* (Atlantis Press, 2020), pp. 279–288, ISBN 978-94-6252-933-5, ISSN 2352-5428, `https://doi.org/10.2991/aebmr.k.200318.035`

[19] H. Danylchuk, O. Kovtun, L. Kibalnyk, O. Sysoiev, E3S Web of Conferences **166** (2020)

[20] Z.H. Munim, M.H. Shakil, I. Alon, J. Risk Financial Manag. **12**, 103 (2019)

[21] G.C. Dumitrescu, Global Economic Observer **5**, 63 (2017)

[22] *Google Trends Home page* (2021), `https://trends.google.com.ua/trends/?geo=UA`

[23] *Bitcoin. InvestFunds. Home page* (2021), `https://investfunds.ru/indexes/9021`

[24] R. Hegger, H. Kantz, T. Schreiber, Chaos: An Interdisciplinary Journal of Nonlinear Science **9**, 413 (1999)

[25] M.B. Kennel, R. Brown, H.D.I. Abarbanel, Phys Rev A **45**, 3403–3411 (1992)

[26] A. Fabretti, M. Ausloos, Int J Mod Phys C **16**, 671 (2005)

[27] F. Birleanu, C. Ioana, C. Gervaise, J. Chanussot, A. Serbanescu, G. Serban, *On the recurrence plot analysis method behaviour under scaling transform*, in *2011 IEEE Statistical Signal Processing Workshop (SSP)* (IEEE, Nice, France, 2011), pp. 789–792, `https://ieeexplore.ieee.org/document/5967823`

[28] S. Wallot, Discourse Process **54**, 382 (2017)

Economic productivity and competitiveness of Ukrainian exports due to the global challenges

Lesia Petkova[1,*], *Olena* Berezina[1,**], *Iryna* Honcharenko[1,***], and *Ihor* Osadchenko[1,****]

[1]Cherkasy State Technological University, 460 Schevchenko Blvd., Cherkasy, 18006, Ukraine

Abstract. Productivity and economic growth are key factors to maintain and improve the competitiveness of nations in the global market. The paper analyzes the prospects for the competitiveness of Ukrainian exports in the terms of pandemic circumstances and post-pandemic recovery of the global economy. The prospects for strengthening the competitiveness of Ukraine's economy evaluating based on the modified approach for assessing the revealed comparative advantage. The dynamics and structure of major industries exports were estimating. The research result proved that the growth of innovative products in the iron and steel industry increases its competitiveness in the world market. The established reduction of the identified comparative advantages index for the main exports positions reflects the presence of structural and technological lags in the modern structure of the national economy and requires economic policy measures aimed at long-term action. Respectively, the main goals of contemporary national economic policy aimed at promoting the export competitiveness of Ukrainian products (goods & services) have to be the stimulating of R&D, infrastructure modernization and capital deepening.

1 Introduction

2020 marked the beginning of the global recession in global economic dynamics. This recession was quite expected due to the long period of previous growth, the accumulation of global imbalances, especially in the financial markets. Nowadays the forecasts of the future global situation in the context of the stability of existing national economic models, the capabilities of regional integration associations, the capacity and capacity of international organizations, their ability to effectively address external and internal challenges intensified. The cyclical nature of economic development objectively determines the change in the tools of state regulation for ensuring its stability. The challenges of the COVID-19 pandemic and the resulting large-scale economic constraints have significantly changed the target priorities of national and international economic policy [1, 2]. The dilemma of the priority of state support faced the countries of the European Union, is also the characteristic for the individual states' economic policy. The issue of choice in the EU was the focus of economic policy on reducing interstate socio-economic differentiation and the commitment of individual countries (for example, Poland and Hungary), using the temporary concentration of power at the national government level, to move away from EU fundamental values, including protection freedom, equality, rule of law. The approach of opponents of a common European anti-crisis policy leads, among other consequences, to non-transparent and economically unjustified support for near the government businesses through financial incentives.

Activation of competitive advantages of the national economy is determined by the available human, physical, natural potential, level and complementarity of economic development due to the basic theory of M. Porter [3]. H. Davies and P. Ellis [4] determined productivity of industries as the more important factor for country's sustainable economic growth. The effective means of strengthening the national economy in the era of domination of multinational companies is the ability to adjust the production of national firms and industries within global value chains, according to AJ Smith (2010) [5], S. Chopra and P. Meindl [6], J. A. Robinson and D. Acemoglu [7] noted that the basic issue of ensuring the preconditions for sustainable economic growth, strengthening international competitiveness and improving the quality of life in growing countries is the creation of inclusive economic and political institutions. The functioning of these institutions determines the possibility of progress in education and technology – factors that increase the productivity of the economy, industries and enterprises.

The most significant factors in increasing production productivity and competitiveness at the current stage of post-industrial development are the conditions for improving key economic factors – human capital and technology. The relationship between the quality of human capital and technology, the ability of employees to use new technologies and to improve them were studied by J. R. Baldwin, B. Diverty and D. Sabourin [8]. They grounded, that productivity of human capital is significantly higher in indus-

*e-mail: l_petkova@ukr.net
**e-mail: o.berezina@chdtu.edu.ua
***e-mail: irynahoncharenko@gmail.com
****e-mail: meb@chdtu.edu.ua

tries where modern information and communication technologies are used. Widespread use of these technologies leads to cost reductions, changes in business organizational structures. Wages growth faster in industries that actively use modern technology. Moreover, firms that support the training of their employees and carry out research and development are more technological and competitive.

Infrastructure factors are effective for the constant increase in capital and labour productivity. Analysis of public investment in infrastructure projects reflects the importance of their effective management, especially in emerging markets, to mitigate the threat of budget imbalances. A. Isaksson noted [9] that the basic preconditions for increasing productivity and competitiveness are investments in human capital (education and health care), improvement of infrastructure, financial security, and development of basic institutions (openness and competition). Good governance is the key to the success of quality, productivity and competitiveness policies.

The case analysis of some countries with growing markets made by B. F. Filip [10] proves the priority of information technology, increasing the share of R&D in total costs for increasing multifactor productivity and growth of competitiveness.

The interaction and role of social, human and technological capital in increasing TFP were studied by B. Dettori, E. Marrocu and R. Paci [11]. The results of their study confirm the high effectiveness of these factors supporting policy for providing productivity growth, economic cohesion in European regions. Simultaneous support and interaction of technological, human and social capital form the spatial poles of accelerated development.

The role of international trade in national competitiveness has been studied by G. Muratoğlu and Y. Muratoğlu [12]. They determined the determinants of export competitiveness and grounded that transnational FDI has not always made a positive impact on national productivity. Instead, the growth of the share of high-tech export has a clear positive effect on the growth of the country's comparative advantages.

The challenges of increasing national economic productivity are quite acute for Ukraine. Decades of transformational development have not formed a consistent trend of improving the quality of economic growth factors and long-term prerequisites for productivity gains. One of the main challenges for Ukraine is the creation of new competitive advantages associated with investments in the latest technologies, innovations, research, human capital, efficient allocation of resources and redistribution, which is accompanied by changes in the behaviour of economic entities [13].

Respectively, the main goals of contemporary national economic policy aimed at promoting the export competitiveness of Ukrainian products (goods & services) have to be the stimulating of R&D, infrastructure modernization and capital deepening.

In this issue the method for assessing the revealed comparative advantage used for estimating the prospects for strengthening the competitiveness of Ukraine's economy through the dynamics and structure of exports of major industries. The modified approach to the assessment of comparative advantages is based on the method of determining the coefficient of the revealed comparative advantage (Revealed Comparative Advantage, RCA) [14].

$$RCA = \frac{1000}{X_j^t + M_j^t} \times \left[X_{ij}^t + M_{ij}^t - \left(X_j^t - M_j^t \right) \times \frac{X_{ij}^t + M_{ij}^t}{X_j^t + M_j^t} \right] \quad (1)$$

where X_j^t – the total exports of the country (s) relatively to the group of other analyzed countries;

M_j^t – the total imports of the country (i) relatively to the group of other analyzed countries;

X_{ij}^t – the export of the industry or sector (j) of the country (s) relatively to the group of other analyzed countries;

M_{ij}^t – imports of goods of the sector (j) of the country (i) relatively to the group of other analyzed countries.

This work aims to analyze the prospects for the competitiveness of Ukrainian exports in the terms of pandemic circumstances and post-pandemic recovery of the global economy.

2 Results

The GDP of most countries reduced in 2020. According to IMF[15] estimates, the economic decline is expected at the global level (world) – 4.4%, in the group of developed countries (advanced economies, AE) – 5.8%, in growing and developing countries (emerging markets and developing economies, EMDE) – 3.3%. The expected decline in Ukraine's GDP in 2020 is 7.2% [15]. Overcoming the effects of the global economic crisis is difficult for advanced economies, but low-income countries (LIC) are particularly affected and burdened with significant external debts. About 1.5 billion people in the least developed countries feel the weakness of national health care systems and institutional failure. The total amount of state-to-business support for employees was about 20% of GDP in AE during the quarantine period. However, in LIC this support was only about 2% of GDP. It has to be noted that national governments implement active measures to minimize the effects of the crisis and pandemic. As in previous situations of global economic downturn, monetary easing measures are being taken, and businesses are receiving significant fiscal and credit support to minimize losses from the recession.

The EU's response to the latest threats is significant. It reflects a clear awareness of partnership and solidarity advantages in overcoming challenges, as well as strengths of collective and coordinated action in contrast to individual decisions. The European Commission is implementing the Coronavirus Response Investment Initiative (CRII) to mobilize cohesion policy and for speed response to threats[16]. The main objectives of European policy against the pandemic and economic downturn are the support for health systems, labour markets, small and medium-sized businesses and help the most affected regions. The list of national business support measures introduced by European countries can be divided into the following areas: tax incentives (deferral and reduction of tax payments), financial guarantees for companies that need the additional resources during the crisis, simplification of credit support, income support, subsidies for wages in forced part-time employment, etc. [17]. The example of AE and integration associations makes it possible to adjust the economic policy of other countries in a pandemic and economic recession.

Ukraine's economy is declining in key macroeconomic indicators. The decline in some of them was recorded before the global pandemic and significantly deepened by its effects. For example, the index of industrial production began the ongoing decline in October 2019, the turnover of transport companies began the ongoing decline in January 2020. The set of preconditions determines the complexity of this situation, both internal and external, and economic recovery requires an active public policy using the most effective and proven measures.

The dynamics of the Ukrainian economy competitiveness through the RCA index for the largest export goods and services categories during last seven years given in table 1.

Iron and steel were the largest export category of Ukraine until recently. Ukraine sold more grain crops than iron and steel at the total cost only in 2019. The main importers of Ukrainian steel are Italy, Egypt, Russia, Turkey and Poland. Products of this category are mainly raw materials. The index of Ukrainian iron and steel exports tends to decline steadily, which indicates a deterioration in the competitiveness of domestic metallurgy. For example, in 2013 the value of the indicator was 97.66, but in 2019 it was only 76.25, and further deterioration and decline expected. Iron or steel products are other types of metallurgical products that have a significant decrease in the index also. In 2013 it was 12.5, but in 2019 – only 1.35. Insufficient level of technological renewal of Ukrainian metallurgy against the background of increasing competition in the industry and aggressive policy of competitors to expand markets create a negative outlook for domestic exports.

Instead, exports and the competitiveness of the Ukrainian agricultural sector are growing. Indices of the categories "cereals", "seeds and fruits of oilseeds", "animal and vegetable fats", "residues and wastes of the food industry" are growing. RCA indicators doubled for each category. However, due to the lower level of the agricultural sector adaptability, its role in the structure of export specialization hasn't to be prevailing.

Ukraine has negative RCA indices for electric machines, nuclear reactors, boilers and machines – categories where imports are more than exports. At the same time, the dynamics indicate that the competitiveness of electric machines has halved in the last six years, and of nuclear reactors, boilers and machines – three times. The index of wood and wood products has almost doubled. The main product is lumber – the raw materials for the manufacture of furniture et al. Thus, the manufacturability of Ukrainian commodity exports and competitiveness of more technological categories of exports and industries has decreased over the past seven years.

The indicators of the RCA index for the largest categories of Ukrainian exports of services shown in table 2.

The situation with the competitiveness of Ukrainian services is much more optimistic than with goods. The trade balance of transportation, namely: sea, air, rail, road and pipeline services is positive. Ukraine is using its potential as a transit state. The pipeline transport services amount half of all transport services exports. The index of transport services competitiveness is volatile and hasn't an obvious tendency to increase or decrease.

The competitiveness of computer and information services tripled and the competitiveness of telecommunications services increased more than doubling. The main trade partners in computer, information and telecommunications services are the United States, the United Kingdom, Israel, Malta and Germany. Half of the services in this area are exported to the United States. IT sector is one of the most promising areas of further development and specialization with significant potential for value-added. Moreover, during this period, the competitiveness of construction, financial, insurance and other types of services has increased. Ukraine changed its status from an importer to an exporter in construction services, the index has increased 8 times. For financial services, this indicator doubled.

Ukraine's position in 2019 improved by 5 points in the Doing Business Ranking (71st of 190 countries). The country showed the largest growth in such categories as taxation (growth by 110 positions, 54th place), international trade (+70 positions, 78th place) and investor protection (+56 positions, 72nd place). In the Doing Business 2020 (DB 2020), Ukraine hold 64th place, improving its position compared to Doing Business-2019, by seven points [20]. The greatest progress was in the indicators "protection of minority shareholders" (the position improved by 27 points), and "obtaining the permissions for building and construction" by 10 points.

Meanwhile, the rating agency Moody's Investors Service upgraded Ukraine's sovereign rating from Caa2 to Caa1, which means a change in the prognosis from "positive" to "stable".

Ukraine ranked 83rd out of 140 in the Global Competitiveness Index (GCI 2020 Ukraine ranks 110th in terms of state institutions, 57th in terms of infrastructure development, 77th in terms of adaptation of modern technologies, as well as macroeconomic stability – 131st, healthcare – 94th, education – 46th place, goods market – 73rd, labour market – 66th, financial system – 117th, market size – 47th, business dynamics – 86th, ability to innovate – 58th) [21].

In the Global Innovation Index 2020 [22], Ukraine rose 7 places to 43rd, ahead of Thailand and behind Croatia and Greece. At the same time, Poland took 39th place, Russia – 46th, Moldova – 48th, Kazakhstan – 74th, Belarus – 86th. The best indicator in the group of the former USSR obtained by Estonia which took 24th place. The leaders of the ranking are the following ten most innovative countries: Switzerland, the Netherlands, Sweden, Great Britain, Singapore, the United States, Finland, Denmark, Germany and Ireland.

Table 1. Index of Ukrainian goods exports competitiveness (according to the data [18, 19])

Categories	Years						
	2013	2014	2015	2016	2017	2018	2019
Iron and steel	97.66	107.78	96.78	89.35	88.30	91.58	76.25
Cereal crops	47.87	57.33	77.37	81.51	73.01	73.77	93.85
Animal and vegetable fats	24.84	32.68	40.84	51.29	50.30	44.83	44.76
Ore, slag, ash	25.83	26.08	21.31	21.14	25.90	27.10	31.23
Electric machines	-11.27	-10.11	-9.96	-12.23	-12.07	-17.27	-26.76
Seeds and fruits of oil plants	13.51	12.79	16.48	17.01	20.10	17.51	22.10
Reactors, boilers, machines	-14.42	-17.53	-21.96	-38.17	-38.08	-38.50	-37.51
Wood and products of wood	6.39	9.12	12.65	13.14	11.43	13.17	11.41
Residues and waste from the food industry	5.38	8.06	10.95	11.57	10.41	10.70	12.82
Steel or iron products	12.50	7.97	5.07	1.24	2.22	3.14	1.35

Table 2. Index of Ukrainian services export competitiveness (according to the data [18, 19]))

Categories	Years						
	2013	2014	2015	2016	2017	2018	2019
Transportation	152.10	143.85	153,21	160.01	145.73	123.37	151.94
Computer and information services	13.16	30.32	35.80	41.05	42.00	50.68	38.19
Business services	-20.03	-21.75	-21.40	-33.06	-29.19	-58.70	-49.92
Tourism	-29.45	-39.93	-40.45	-42.10	-54.84	-59.86	-73.72
Communication services	-8.11	-0.47	-6.44	-1.13	-3.09	-1.96	-3.03
Construction services	-8.64	-1.42	10.47	12.71	-4.45	1.68	0.41
Financial services	-49.99	-48.75	-64.10	-44.13	-31.60	-30.26	-26.16
Insurance services	-7.30	-4.27	-4.00	-7.69	-8.84	-3.90	-3.53
Fees for royalty and licenses	-47.38	-28.54	-22.80	-26.27	-31.99	-33.96	-33.59
Personal, cultural and health services	-0.91	0.52	0.06	-0.07	-1.07	-0.11	-0.09

The basis of Ukrainian innovative competitiveness is human capital and research, as well as knowledge and research results. Their effective implementation is the main competitive advantage. However, compared to 2017, Ukraine lost 2 positions in the sub-index "Human Capital and Research", decreasing from 41st to 43rd place due to a reduction in education costs as a percentage of GDP and research and development costs as a percentage of GDP. At the same time, in the sub-index "Knowledge and results of scientific research", Ukraine is in a high 27th place in the overall ranking. The main strengths of this sub-index are the following: knowledge creation (15th place), the ratio of patents to GDP at purchasing power parity (19th place), the ratio of utility models to GDP at purchasing power parity (1st place), the cost of computer software as a percentage of GDP (17th place), exports of ICT services as a percentage of total trade (15th place).

In the Bloomberg Index of Innovative Development (BIID 2019) Ukraine obtained the 53rd place in 2019 with a total score of 48.09, deteriorating its position by a total of 7, due to the weakening of Ukraine's position on 6 of the seven components of this index (table 3), authors' calculations according to the data [23].

Table 3. Ukraine's position in the Bloomberg Innovation Index

Index	Years	
	2018	2019
General index	46	53
Intensity of research and development (R&D expenditures / GDP)	47	54
Productivity	50	60
Penetration of high technologies (share of innovative companies in the total number of enterprises)	32	37
Concentration of researchers (number of scientists per 1 million inhabitants)	46	46
Value added production (value added of production / GDP)	48	58
Efficiency of higher education (share of freelance graduates in the total number of graduates of educational institutions)	21	28
Patent activity	27	35

According to the latest published report of the World Economic Forum "The Global Competitiveness Report 2019" [21], Ukraine ranked 83rd among 140 countries. The Dominican Republic is in the 82nd place, and Macedonia is in the 84th place. Compared to the data of the previous report, Ukraine lost 2 positions, but such a comparison is incorrect, as the report for 2018 changed the method of calculating the index and its components. According to the new approach, the ranking is led by the United States, Singapore and Germany. Yemen and Chad have the lowest level of competitiveness – 139th and 140th respectively. Ukraine's geographical neighbours have the best positions: Poland – 37th place, Slovakia – 41st, the Russian Federation – 43rd, Hungary – 48th, Romania – 52nd. The authors of the rating (GCR 2019) noted that Ukraine occupies a fairly high position in the sub-indices: market size – 47th place, skills – 46th place. Instead, the most problematic are the sub-indices macroeconomic stability (131st), the financial system (117th), institutions (110th).

In the report "European Innovation Scoreboard 2019" [24], the leader in innovation is Sweden. The leader's group also includes Denmark, Finland, Germany and the Netherlands. Ukraine's position has decreased, and now its level corresponds to the group of European innovation outsiders (Romania, Bulgaria). The value of the aggregate index for Ukraine in 2019 was 27.8% (in 2016 – 28.9%), which indicates a decline of 1.1% compared to 2016. According to the methodology, Ukraine is in the group of slow innovators. Ukraine lags behind in all indicators, except for human resources (110.3%) and the impact of employment (77.5%). The weakest sides are communications and entrepreneurship (9.5%), the innovation environment (4.1% in 2017, no data in 2010 and 2016). According to the report, Ukraine is characterized by a low level of GDP per capita in 2017 – 6,600 USD (by PPP), while the average value of this indicator in the EU is 28,600 dollars. Ukraine's GDP growth rate and population growth rate are negative.

The export of goods consists of about 80% in the general structure of Ukraine's exports, and the strategic goal of increasing its manufacturability is important to strengthen the key characteristics of competitiveness in foreign markets. The production of iron and steel industry is stable at 20-25% of the total volume of industrial production in Ukraine during 2001-2019 years. The results of the calculating relationship between the share (percent) of innovative industrial products (x) in iron and steel industry, and the Ukrainian iron and steel industry competitiveness (y) shown in table 4.

Table 4. The share of innovative products and competitiveness of iron and steel industry (according to the [18] data)

Year	x	y
2000	9.4	163.67
2001	6.8	144.16
2002	7.0	140.11
2003	5.6	134.86
2004	5.8	150.20
2005	6.5	151.24
2006	6.7	152.79
2007	6.7	149.64
2008	5.9	149.88
2009	4.8	116.29
2010	3.8	125.51
2011	3.8	117.14
2012	3.3	96.93
2013	3.3	97.66
2014	2.5	107.78
2015	1.4	96.78
2016	1.0	89.35
2017	0.7	88.30
2018	0.8	91.58
2019	1.3	76.25

The research result proved that the growth of innovative products in the iron and steel industry increases its competitiveness in the world market. The values of the calculated parameters of the linear function are presented in the linear regression equation (2).

$$Y = 77.55618 - 10.2068x \quad (2)$$

The linear correlation coefficient for this model is 0.95. The coefficient of determination shows that the competitiveness of iron and steel industry products (y) by 89.35% depends on the share of innovative industrial products (x), and 10.65% is the influence of other factors. Thus, the increase in the share of innovative products in the exports of the Ukrainian iron and steel industry is the determining factor in its competitiveness on the world market. From 2013 to 2019, the level of manufacturability of Ukrainian merchandise exports and its competitiveness decreased, while the competitiveness of Ukrainian services, primarily IT, as well as financial and insurance services increased. The position of Ukrainian producers in global value chains remains unfavourable due to the export of low value-added products and imports of high-tech goods.

The level of state participation and regulation of economic and private sphere is another nowadays topical issue. According to Y. N. Harari [25], humanity is faced with two particularly important choices between totalitarian surveillance and the support and information of citizens; between nationalist isolation and world solidarity during the crisis. The experience of countries that slowed down the spread of the epidemic is essential. Some countries use the geolocation data actively for tracking the movement of sick persons or population at risk. In China, Taiwan, Singapore, South Korea, Israel, various technical solutions are used to monitor people, their movement and contacts. The so-called "digital strategy of human identification" is used by European countries for the period of the pandemic. Such decisions are not easy for democratic countries, because the threat to civil liberties is quite real and conscious in the current circumstances. Therefore, their governmental and interstate policies are aimed at combining the necessary restrictive measures and protection of confidentiality as much as possible. However, in autocratic states, there is no public discussion about the expediency of using additional tools for state control of private life. It is possible to assume a significant transformation of political models in some countries based on assessing the prospects of the post-crisis period of global development. However, the existing political tradition, maturity and responsibility of political elites and civil society for the implementation of priority worldview values will remain decisive.

3 Conclusions and suggestions

The global economy today faces serious and numerous challenges. Methods for overcoming the traditional problems of the cyclical downturn are complemented by the daily experience of overcoming the COVID-19 pandemic. The challenges for government anti-crisis policy in EMDE are reinforced by the need to establish long-term trends in sustainable economic growth.

The countries represented on the world market by transnational companies are gaining more and more competitive advantages in the international division of labour with the strengthening of global interdependence. Digital technology-based businesses have significantly improved their competitive position in terms of the pandemic. Namely, e-commerce has almost tripled during this time. The significant funding for the formation of a long-term national strategy for post-pandemic growth is the emphasis on high-tech and digital businesses.

Ukraine's economy has a high level of integration into world markets. At the same time, it has loses because of the efficiency and manufacturability of foreign economic activity. The established reduction of the identified comparative advantages index for the main branches of goods' exports reflects the presence of structural and technological lags in the modern economic structure of the country and requires economic policy measures aimed at long-term action. Instead, the Ukrainian economy demonstrates positive dynamics in the export of services and reflects the progressive trends of technological development, first of all in the export of IT services. Analysis of the Index of Ukrainian goods exports competitiveness, Index of Ukrainian services export competitiveness reflects a steady trend towards the loss of long-term competitiveness of Ukrainian industries for the production of goods, especially raw materials. At the same time, there is a positive trend to expand the competitive capabilities of modern high-tech services – Computer and information services, Communication, Construction services. Sustain the conditions for further foreign economic expansion of these Ukrainian businesses is one of the main tasks for the government's economic policy.

Governments are taking emergency measures to address global economic challenges and support national economies. The scale of financial support for citizens during the period of temporary loss of their solvency and business unprecedented, especially for small and medium businesses. The considering the growing global challenges including the COVID-19 pandemic, returning the crucial role of economic policy to national governments, also increasing the international coordination of crisis efforts and shifting the emphasis to stimulating economic activity of small and medium businesses would shape the economy policy shortly. The maximally support the above-mentioned high-tech businesses-locomotives of long-term growth and using of such experience is significantly important for Ukraine. Primarily, fiscal and monetary stimulus measures will be appropriate.

Despite the need for additional protection of national economies and, consequently, change of the existing practice of international integration, the objective reasons for its renewal will not disappear in the post-crisis period. The commonality of the current epidemiological, economic, and social challenges also determines the joint efforts to overcome them. The world economy will change as with any crisis, but with a new experience in countering global threats, the number of which isn't decreasing.

Ukraine's foreign economic policy has to be based on the possibility of effective integration into world value chains, logistics and communication networks due to the further consolidation of international efforts to solve global problems.

Further research will be devoted to elaborating the specific mechanisms to increase the manufacturability and competitiveness of both traditional and new industries of Ukrainian exports.

References

[1] S. Semerikov, S. Chukharev, S. Sakhno, A. Striuk, V. Osadchyi, V. Solovieva, T. Vakaliuk, P. Nechypurenko, O. Bondarenko, H. Danylchuk, E3S Web of Conferences **166** (2020)

[2] V.N. Soloviev, A.O. Bielinskyi, N.A. Kharadzjan, CEUR Workshop Proceedings **2832**, 24 (2020)

[3] M.E. Porter, *Competitive advantage of nations: creating and sustaining superior performance* (Simon and Schuster, 2011)

[4] H. Davies, P. Ellis, Journal of Management Studies **37**, 1189 (2000)

[5] A. Smith, Southern African Business Review **14** (2010)

[6] S. Chopra, P. Meindl, *Supply chain management: strategy, planning, and operation* (2001), `https://base-logistique-services.com/storage/app/media/Chopra_Meindl_SCM.pdf`

[7] J.A. Robinson, D. Acemoglu, *Why nations fail: The origins of power, prosperity and poverty* (Profile, London, 2012)

[8] J.R. Baldwin, B. Diverty, D. Sabourin, *Technology use and industrial transformation: Empirical perspective* (1995), `http://citeseerx.ist.psu.edu/viewdoc/download?doi=10.1.1.622.1001&rep=rep1&type=pdf`

[9] A. Isaksson, *Determinants of total factor productivity: a literature review* (2007), `https://open.unido.org/api/documents/4812034/download/UNIDO-Publication-2007-4812034`

[10] B.F. Filip, Journal of Public Administration, Finance and Law **10**, 125 (2016)

[11] B. Dettori, E. Marrocu, R. Paci, Regional Studies **46**, 1401 (2012)

[12] G. Muratoğlu, Y. Muratoğlu, Journal of Economics and Political Economy **3**, 111 (2016)

[13] L. Petkova, O. Berezina, I. Honcharenko, L. Berezhna, D. Marushchak, E3S Web of Conferences **166**, 13014 (2020)

[14] S.R. Semiv, *Research of competitiveness of branches of Ukraine economy in the conditions of European integration* (2007), `http://mev.lac.lviv.ua/downloads/vyklad/semiv/stattya_competitiveness.pdf`

[15] International Monetary Fund, *World Economic Outlook, October 2020: A Long and Difficult Ascent* (2020), `https://www.imf.org/en/`

Publications/WEO/Issues/2020/09/30/world-economic-outlook-october-2020

[16] European Comission, *Cohesion policy action against coronavirus* (2020), https://ec.europa.eu/regional_policy/en/newsroom/coronavirus-response/

[17] *A brief overview of corona-related national and regional policy measures taken to support SMEs* (2020), https://drive.google.com/file/d/13dJFoeyuOCMA6FPAnnfxM1n5rmvRSSqo/view?fbclid=IwAR3GrAy6u7np6HQ_i0dOHDI

[18] Ukrstat, *Official site of the State Statistics Service of Ukraine* (2020), http://www.ukrstat.gov.ua/

[19] United Nations, *UN Comtrade Database* (2020), https://comtrade.un.org/

[20] World Bank Group, *Doing Business 2020: Comparing Business Regulation in 190 Economies* (2020), https://tinyurl.com/3uh5bnjm

[21] K. Schwab, ed., *The Global Competitiveness Report 2019* (World Economic Forum, Cologny/Geneva, 2019), http://www3.weforum.org/docs/WEF_TheGlobalCompetitivenessReport2019.pdf

[22] *Global innovation index 2020* (2020), https://www.globalinnovationindex.org/

[23] P. Coy, *Bloomberg Index of Innovative Development* (2015), https://www.bloomberg.com/graphics/2015-innovative-countries/

[24] *Official site EU Commission* (2021), http://ec.europa.eu

[25] Y.N. Harari, *The world after coronavirus* (2020), https://www.ft.com/content/19d90308-6858-11ea-a3c9-1fe6fedcca75

Comparative analysis of the stock quotes dynamics for IT-sector and the entertainment industry companies based on the characteristics of memory depth

Nataliia Maksyshko[1,*] and *Oksana* Vasylieva[1,**]

[1]Zaporizhzhia National University, 66 Zhukovskoho Str., Zaporizhzhia, 69600, Ukraine

Abstract. The article is devoted to the study and comparative analysis of the stock quotes dynamics for the world's leading companies in the IT sector and the entertainment industry. Today, these areas are developing the fastest and most powerful, which attracts the attention of investors around the world. This is due to the rapid development of digital communication technologies, the growth of intellectualization and individualization of goods and services, and so on. These spheres have strong development potential, but the question to how their companies' stock quotes respond to the impact of such a natural but crisis phenomenon as the COVID-19 pandemic remains open. Based on the nonlinear paradigm of the financial markets dynamics, the paper considers and conducts a comprehensive fractal analysis of the quotations dynamics for six leading companies (Apple Inc., Tesla Inc., Alphabet Inc., The Walt Disney Company, Sony Corporation, Netflix) in this area before and during the COVID-19 pandemic. As a result of the application of the rescaled range analysis (R/S analysis), the presence of the persistence property and long-term memory in the stock quotes dynamics for all companies and its absence in their time series of profitability was confirmed. The application of the method of sequential R/S analysis made it possible to construct fuzzy sets of memory depths for the considered time series and to deepen the analysis of the dynamics due to the quantitative characteristics calculated on their basis. Taking into account the characteristics of memory depth in the dynamics of quotations made it possible to conduct a comparative analysis of the dynamics, both under the influence of the natural crisis situation and in terms of investing in different terms. The peculiarities of the delayed profitability dynamics of quotations for each of the companies are also taken into consideration and compared. The developed recommendations can be used in investment activities in the stock market.

1 Introduction

Trading in financial instruments on stock exchanges is increasingly becoming a source of income for various investors. At the same time, investors face the problem of choosing financial instruments in which to invest. For the investor, effective management of their financial resources now – means to get additional benefits in the future. But in order to get these benefits, they need to compare financial instruments and choose the most profitable and least risky among them.

The issue of analyzing the stock markets dynamics in order to develop practical recommendations for the investor is not new, but it remains relevant and extremely important. Development of investment strategy and awareness of prospects and risks of specific investment instruments is the key to successful investment activities.

For comparative analysis, traditionally, statistical characteristics of dynamics are used. Traditionally, statistical methods have been used to confirm the efficient market hypothesis [1]. For example, in [2]: Kolmogrov-Smirnov and Shapiro-Wilk tests; a run test and an auto-correlation test are used to check the randomness and the normality of the data. In [3] there is applied a rolling variance ratio test; in [4] European stock markets are tested by a runs test and joint variance ratio tests. In the study [3] the French stock market is considered to change its properties from efficient to adaptive.

An alternative theory of financial markets is the fractal market hypothesis [5], which considers markets as complex nonlinear systems in which randomness and hidden patterns interact, resulting in the dynamics of financial instruments has a fractal nature and the property of persistence (the presence of long-term memory). Methods of nonlinear dynamics are considered as relevant tools of its research. Therefore, recently more and more attention is paid to the study and application of these methods.

The most popular method of studying the fractal properties of dynamics is the Hurst exponent (R/S analysis) [5]. The paper [6] provides an excursion into the research of scientists on the application of the Hurst exponent to analyze the dynamics of capital markets.

Along with the Hurst exponent for the study of financial markets and, in particular, stock markets, the following indicators are used: Lyapunov exponent (indicator of nonlinear dynamics) to diagnose the crash of stock mar-

*e-mail: maxishko@ukr.net
**e-mail: oksanabay@ukr.net

kets [7], Shannon information entropy [8, 9], Renyi entropy [9], the Hurst-Holder exponent [10], local Whittle estimator [11]. The presence of fractal properties in the dynamics of financial markets is also investigated by calculating the Hausdorff dimension and applying the Mittag-Leffler functions [12].

In particular, in the works [13, 14], a method of sequential R/S analysis is proposed, which allows not only to establish the presence of long-term memory in the time series, but also to evaluate its quantitative and qualitative characteristics. The authors show that for the dynamics of different financial instruments that have long-term memory, memory characteristics may differ.

The authors also consider the use of different tools of nonlinear dynamics in terms of different stock markets and their segments. For example, in [15, 16] – the stock markets of individual countries are considered, in [17] – quotations of shares of certain companies, in [18] – multifractality the autocorrelations in stock portfolio returns is studied and other.

The research and comparison object of this work is the dynamics of shares quotations of leading companies in the IT sector and the entertainment industry. Today, these areas are developing the fastest and most powerful, which attracts the attention of investors around the world. This is due to the rapid development of digital communication technologies, the growth of intellectualization and individualization of goods and services, and so on. These areas have strong development potential, but the question of how stock quotes in this area respond, for example, to the impact of such a natural but crisis phenomenon as the COVID-19 pandemic remains open.

The hypothesis of this work is that the characteristics of long-term memory can also be used for comparative analysis and development of recommendations for investors operating in the stock market. The purpose of this work is to study the characteristics of the memory depth for the stock quotes dynamics of selected companies, their stability under the influence of such a crisis as the global COVID-19 pandemic, conducting a comparative analysis based on them and developing recommendations for investors.

2 Materials, methods and results

2.1 Materials and methods

Since the behavior of stock prices in the stock markets is mostly not normally distributed or close to normal distribution [5], after testing for deterministic chaos and proof of its fractal nature, it is advisable to use fractal analysis methods for its study. Such methods include the method of the Rescaled range or R/S analysis of Hurst (denote it A_1) [5] and the method of sequential R/S analysis (A_2) [13].

Let the time series (TS) X be considered. The result of applying method A_1 is to determine the Hurst exponent (H) and check its significance based on the application of the mixing test.

The value of the Hurst exponent $H \in [0; 1]$ determines the presence of certain properties of the dynamics: the value of $H \in (0; 0.5)$ corresponds to the antipersistent or ergodic TS; the value of $H = 0.5$ (and in its vicinity) indicates a random TS, in which events are random and uncorrelated, the present does not affect the future; the value of $H \in (0.5; 1]$ indicates that the TS is persistent or trend-resistant, characterized by the presence of long-term memory. The closer the value of H to 1, the more correlated the levels of the series.

Recall that the essence of the method of the Rescaled range exponent of Hurst A_1 is to construct the R/S trajectory for the TS X and determine the angle of the linear trend, built on its (R/S trajectory) starting points. At some value of $k = k_0$ R/S trajectory changes its slope quite sharply, i.e. at the point (x_{k_0}, y_{k_0}) the trajectory receives a significant in absolute value negative gain $\delta_k = y_{(k+1)} - y_k$ – there is a break from the trend and there is no return to the previous trend. It is assumed that at the point k_0 the effect of "long-term memory of the beginning of the series" dissipates. That is, the "breakdown of the trend" demonstrates the loss of memory about the initial conditions, and also signals (possibly with a lag, ie with some delay) the exhaustion of the cycle or quasi-cycle contained in the initial segment of this TS. According to [5], we adhere to the statement that after the end of the cycle (quasi-cycle) the memory about the initial conditions is lost, ie the long-term correlation of the following observations with respect to the initial one is lost. However, based on the peculiarities of the construction of the R/S-trajectory [5], the method of the R/S analysis of Hurst A_1 provides only (statistically) the average characteristic of the trend stability of the TS X as a whole and does not take into account the changing dynamics during the whole observation period.

However, the method of sequential R/S analysis A_2 [13] by modifying the scheme of construction of R/S-trajectory, allows to take into account the changing nature of the dynamics, to identify many cycles (quasi-cycles) that are characteristic of the studied TS, and ensure a more detailed assessment of the memory depth from the beginning of this TS.

Performing an iterative procedure (method A_3) using method A_2 and detecting the point of memory loss at the beginning of the time series for a set of nested segments $X = X^0 \supset X^1 \supset \ldots \supset X^r \supset X^{n-3}$(a family of time series differing by the starting point) allows to estimate the memory depth as a fuzzy set "memory depth of the TS as a whole" [14]. Note that the transition from a "clear" estimate (based on probability and requiring statistical significance) to a fuzzy estimate (non-additive measure) is due to the availability of data for which these requirements are not met.

According to [13, 14] the concept of "memory depth of TS X" is defined as a discrete fuzzy set (FS)

$$L(X) = \{(l_i, \mu_L(l_i)); l_i \in N\}, \mu_L : N \longrightarrow [0; 1], \quad (1)$$

where $N = \{l_i, i = 1, 2, ...\}$ – natural numbers set – set of basic values for memory depth,

https://doi.org/10.1051/shsconf/202110701003

$\mu_L(l_i) = \mu(l_i)$ – the value of the membership function, which determines the degree of belonging of a natural number l_i ("depth l_i") to a fuzzy set $L(X)$. The function $\mu_L(l_i)$ displays the base value l_i in the interval [0; 1] and displays the degree of possibility (confidence measure) in relation to the membership of the element l_i fuzzy set of memory depth $L(X)$.

The carrier of the fuzzy set $L(X)$ is the set "supp" $L(X) = L^0 = \{l_i \in N, i = 1, 2, ... : \mu_L(l_i) > 0\}$. Therefore, we finally consider the fuzzy set of memory depth of the time series X as a whole in the form

$$L(X) = \left\{(l, \mu(l)), l \in L^0\right\}. \tag{2}$$

Important characteristics based on the use of fuzzy memory set (2) are given in [14]. In this paper, for a comparative analysis of price dynamics, the following characteristics are used: l_{max} – the greatest value of the memory depth encountered:

$$l_{max} = \max_{l \in L^0} l, \tag{3}$$

l_{mc} – the memory depth value that has the largest value of the membership function $\mu(l)$:

$$\mu(l_{mc}) = \max_{l \in L^0} \mu(l), \tag{4}$$

l_{gc} – the gravity center of the fuzzy set $L(X)$ as a whole – is obtained using the defuzzification procedure:

$$l_{gc} = \frac{\sum_{l \in L^0} [l \cdot \mu(l)]}{\sum_{l \in L^0} \mu(l)}, \tag{5}$$

H_{entr_L} – information entropy indicator for the fuzzy set of memory depth $L(X)$ – used to assess the uncertainty degree regarding the variety of TS's behavior variants:

$$H_{entr_L} = -\sum_{l \in L^0} \mu(l) * \log_2 \mu(l). \tag{6}$$

Note that the scale for estimating the degree of uncertainty is considered to be ordinal by type.

Thus, the use of the selected considered characteristics of the memory depth for time series allows to deepen the comparative analysis of the stock prices dynamics for companies in the IT sector and the entertainment industry.

2.2 Stock market overview and input data

Today in the modern stock market there are separate sectors, which on the basis of information from sites [19, 20], are distributed according to the level of volatility.

The names of the companies for reduction are given according to their stock tickers.

Very high volatility: energy sector – oil (PZE, EC, HOM), gas (GAS, WGR, SPH), coal (BTU, ACI, MEE), alternative fuel (PTOI, GRPEF, BLDP); Industrial sector – industrial electronics (EDIG, SIRC, RGSE), industrial products (BOOM, PLPC, TPIC), machinery (SOHVY, CTEQF, PERT), aviation and space (Boeing, Generaldynamics, Tesla).

High volatility: Basic resources and materials – gold (HGLC, RMRK, NSRPF), metals (ACH, USNZY, GGIFF), chemical products (SKVI, LPAD, NL), forest resources and paper production (AZFL, NKSJF, FBI); Financial services – brokers (GS, LEH, BSC), banks (C, USB, NYB), insurance companies (AIG, XL, JH), management companies (WM, JNS, JNC), stock exchanges (CME, NYX, ICE), mortgage (FNM, FRE, AHM); Media / Entertainment – Worldflix Inc (WRFX), Dolby Laboratories (DLB), Disney (DIS), New-York times (NYT), Netflix (NFLX).

Average volatility: Retail and wholesale – clothing (DEST, BGI, ANF), food (WDRP, CHEF, GPDB), medicines (THCBF, HEWA), household goods (UPPR, RH); Medicine, pharmaceuticals, health care – biotechnology (VKTXW, DMTX, GNLKQ), pharmaceuticals (GMBL, MNZO), honey equipment (ARYC, SMLR, PLSE); Technology sector – Apple (APPL), Nvidia (NVDA), China Intelligence Information Systems (IICN), Genesis Electronics Group (GEGI), Microsoft (MSFT), Google (GOOG); Leisure / Restaurants / Tourism – casino (ERI, NNSR, SGMS), hotel and restaurant business (WTBDY, BCCI, UPZC), travel agencies (ACGX, AIOM, BDGN); Automotive sector – General motors (GM), Tesla Inc (TSLA), Harley Davidson (HDL).

Below average volatility: Telecommunication sector – wire communication (LICT, NULM, OTEL), wireless communication (TALK, NTL, MFOYY).

In 2020, the media and entertainment sector, especially HBO, Disney+ and Netflix, the pharmaceutical and medical sector, and the technology sector, thanks to Tesla, became the most popular in the world among investors.

Among investors, the most relevant financial instrument are shares, which are a security that certifies the participation of its owner in the formation of the authorized capital of the company and gives the right to receive a share of its profits in the form of dividends and accumulated capital.

Stock investing strategies are based on the purpose:

- receive a fixed income immediately. To do this, shares with the maximum dividends should be bought. In most cases, these are preferred shares;
- buy shares on the electronic exchange platform and wait until their value increases to sell profitably. In this case, you need to choose stocks with maximum growth potential. Experienced financiers prefer such an investment strategy, as after a steady rise in shares, the proceeds from the sale will significantly exceed the size of any dividends.

However, investing in stocks has both advantages and disadvantages. The advantages include reliability, protection against fraud, small amounts at the start of the investment, acceptable liquidity (quite high) and reliability. The disadvantages of stocks can be described as the complexity of investing, because any investment requires an understanding of the process and nuances.

Another disadvantage is the inaccessibility of investing for most Ukrainian investors, despite the fact that although almost everyone can open a brokerage account technically,

https://doi.org/10.1051/shsconf/202110701003

not everyone will be able to put a serious amount on it. The requirements of financial monitoring and the low level of tax culture lead to shadowing of revenues. Because of this, many Ukrainians simply cannot invest in this way.

And the last disadvantage, but very significant – is the risk of losing the value of the asset. Securities can lose quite sharply in price. If, except in case of force majeure, this is not typical for real estate and individual business, then for stocks it is a reality. Unsuccessful financial statements and securities lose in value. Unsuccessful actions of the government and the price of the financial instrument falls. At the same time, the value of securities may not be restored.

Therefore, to choose a financial instrument, both an experienced investor and a beginner need to conduct a thorough analysis and choose the best instruments with minimal risk to themselves.

Today, the IT industry in quarantine is developing the fastest, modernizing and finding ways to move production to remote work. In second place in terms of development is the field of cinema. Many cinema companies are creating online broadcasting services for the library of films and TV programs to compete between companies and attract more consumers.

Financial instruments were selected for comparative analysis – shares of the six most famous companies in the world in the IT sphere and entertainment industry.

Apple Inc. — American technology company with an office in Cupertino (California), which designs and develops consumer electronics, software and online services [21].

Tesla Inc. – American car company – startup from Silicon Valley. Focused on the design, manufacture and sale of electric vehicles and their components. The main production facility is the Tesla plant [22].

Alphabet Inc. — an international conglomerate of companies created on October 2, 2015 by American programmers and entrepreneurs Larry Page and Sergey Brin, which includes Google and other companies they owned directly or through Google [23].

The Walt Disney Company — one of the largest corporations in the entertainment industry in the world. Founded on October 16, 1923 by brothers Walter and Roy Disney as a small animation studio, as of June 2015 it is one of the largest Hollywood studios, the owner of 11 theme parks and two water parks, as well as several television and radio networks, including American Television and Radio ABC [24].

Netflix Inc. — an American media service provider and production company. As of April 2020, Netflix has 182 million subscribers worldwide, of which 69 million are in the United States. Netflix is available in all countries and regions except mainland China (due to local restrictions), Iran, Syria, North Korea and the Autonomous Republic of Crimea (due to US sanctions). The company also has offices in Brazil, the Netherlands, India, Japan and South Korea. Netflix is a member of the American Film Association [25].

Sony Corporation is one of the world's largest media companies. Sony manufactures consumer and professional electronics and other high-tech products. In addition, Sony is one of the world's largest media companies, with the record label Sony BMG (jointly with Bertelsmann), Columbia Pictures and TriStar Pictures, as well as a complete archive of MGM films (jointly with Comcast) [26].

Stock prices dynamics for the period from September 2017 to September 2020 (daily values) for selected companies can be seen in the figure 1.

The input data for the study are daily, weekly and monthly prices for the period from 09/11/2017 to 09/08/2020 obtained from the site Investing.com [19].

For the selected time series (TS), let's define the following notation:

$$Z^i = \langle z_j^i \rangle; i \in (\overline{1,6}), \tag{7}$$

where $j \in \{d, w, m\}$, index d denotes daily, w – weekly, m – monthly prices;

i = 1 – corresponds to the time series of Apple Inc., i = 2 – time series of Tesla Inc., i = 3 – time series of Alphabet Inc., i = 4 – time series of The Walt Disney Company, i = 5 – time series of Netflix Inc., i = 6 – time series of Sony Corporation.

2.3 Results

For a general understanding of the series dynamics based on the input data, historical volatility is calculated for each selected investment instrument (table 1).

Table 1: Historical volatility of the stock prices

	Apple	Tesla	Alphabet	Disney	Netflix	Sony
Historical volatility (%)	40.6	107.4	39.7	40.3	43.9	81.5

Let's move on to the comparative analysis of the stock markets using the methodology of fractal analysis and calculation of memory depth characteristics. Note that February and March 2020 were the worst months for global stock markets since 2008. Stock indexes lost tens of percent, and experts said that the 11-year growth cycle since the last financial crisis has come to an end.

The cause of the fall – an outbreak of coronavirus, which grew into a pandemic. Against the background of the COVID-19 outbreak, investors have reconsidered their views on the future of the global economy [27]. The restrictive measures introduced in different countries have negatively affected almost all areas related to consumer activity: tourism, trade, catering, entertainment and others. Under quarantine, people spend less and move less. Bidders began to get rid of shares of airlines, oil companies, consumer electronics manufacturers and other companies, waiting for falling income and retvenue. Indices of the world's leading stock exchanges collapsed. For example, the Italian FTSE MIB index alone lost 29.8% from February 19 to March 11.

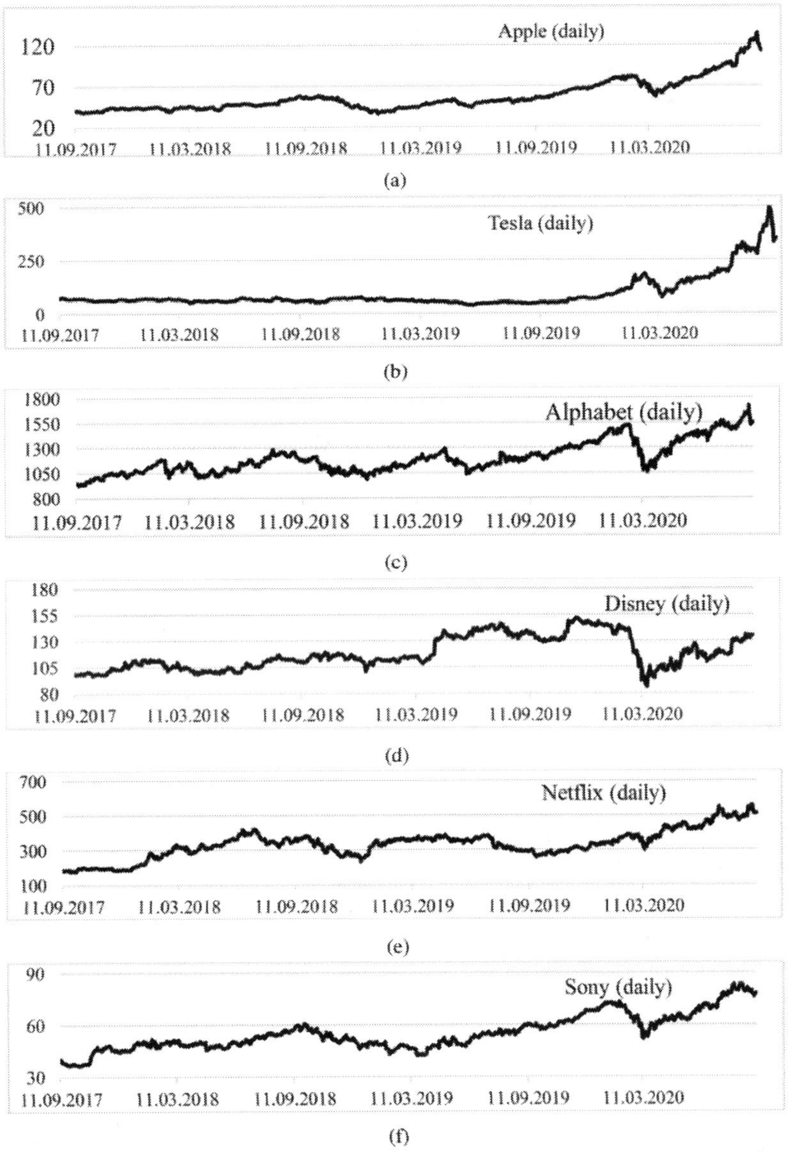

Figure 1: Stock prices dynamics for the period from September 2017 to September 2020 (daily values) for: a) Apple Inc.; b) Tesla Inc.; c) Alphabet Inc.; d) The Walt Disney Company; e) Netflix Inc.; f) Sony Corporation

In connection with the consequences of the pandemic, the dynamics of stock prices for six corporations before and after the pandemic were analyzed. According to figure 1, we can see a fairly stable situation of all corporations until 2020, fluctuations in company share prices are quite small, almost imperceptible, but from February to March 2020, this situation is changing.

Consider the obtained values of the Hurst exponent (table 2).

They indicate that all time series throughout the study period from September 2017 to February 2020 are persistent, i.e. have a long-term memory. The value of the Hurst exponent for all time series is in the range $H \in [0.89; 0.95]$. For mixed values, the Hurst exponent is in the range $H \in [0.56; 0.66]$. The results of fractal analysis TS during a pandemic show that the dynamics of all financial instruments for the entire study period from February 2020 to September 2020 is persistent, i.e. have long-term memory. The value of the Hurst exponent for all time series is in the range $H \in [0.90; 0.92]$. For mixed values, the Hurst exponent is in the range $H \in [0.61; 0.71]$.

The method of sequential R/S analysis [13] was used to study the dynamics and determine such a characteristic of fractal dynamics as long-term memory (i.e. its depth). As a result, a fuzzy set of memory depth for each TS is constructed. A visual representation of the fuzzy set of

Table 2: The results of the application of The Hurst's rescaled range method for time series daily prices before the pandemic and during the pandemic

TS	H	H_{mixed}	TS	H	H_{mixed}
before the pandemic					
Z_d^1	0.92	0.64	Z_d^4	0.92	0.66
Z_d^2	0.89	0.57	Z_d^5	0.95	0.64
Z_d^3	0.91	0.63	Z_d^6	0.92	0.56
during the pandemic					
Z_d^1	0.92	0.65	Z_d^4	0.92	0.65
Z_d^2	0.90	0.64	Z_d^5	0.91	0.65
Z_d^3	0.92	0.71	Z_d^6	0.90	0.61

Table 3: The results of the application of The Hurst's rescaled range method for time series weekly and monthly prices before the pandemic and during the pandemic

TS	H before	H during	TS	H before	H during
Z_w^1	0.87	0.82	Z_w^4	0.87	0.86
Z_w^2	0.83	0.80	Z_w^5	0.91	0.85
Z_w^3	0.86	0.83	Z_w^6	0.87	0.84
Z_m^1	0.73	0.60	Z_m^4	0.80	0.56
Z_m^2	0.74	0.58	Z_m^5	0.80	0.66
Z_m^3	0.72	0.60	Z_m^6	0.74	0.60

memory depth on the example of the time series of Apple and Tesla shares is shown in table 4.

Based on the fuzzy memory depth using formulas (3) - (6), the following characteristics of the time series dynamics were calculated: the greatest value of the memory depth l_{max}; the gravity center for the fuzzy set of memory depth l_{gc}; the most common memory depth l_{mc} and information entropy (H_{entrL}) [14]. The results of the calculations are shown in table 5.

When studying the dynamics of stock prices before the pandemic, it was found that for TS Z_d^1 the most often memory is stored for 5 and 6 days, where the gravity center to the fuzzy set of memory depth is 21.8, and the most common memory depth is 7, for TS Z_d^2 – during 4 and 5 days, where $l_{mc} = 8$ and $l_{gc} = 21.7$, for TS Z_d^3 – during 5, 8 days, where the center of gravity is 18.7, and the memory depth, which is most common from the fuzzy set of memory depths – 6, for TS Z_d^4 – during 6 and 7 days, where the center of gravity is 17.4, and the memory depth, which is most common from the fuzzy set of memory depth – 5, for TS Z_d^5 – during 6 and 8 days, where $l_{mc} = 5$ and $l_{gc} = 19.7$, and for TS Z_d^6 – during 5 and 9 days, where the center of gravity is 17.8 and the depth of memory is often 8, respectively (the number days is given in ascending order of the value of the membership function μ, $\mu \geq 0.60$).

Let's analyze the results of a consistent R/S analysis for the dynamics of stock prices during the pandemic. Long-term memory changed for all tested TS: for TS Z_d^1 the memory is stored for 7 and 11 days, for TS Z_d^2 – 7 and 9 days, for TS Z_d^3 – 5 and 8 days, for TS Z_d^4 – 4, 5 and 10 days, for TS Z_d^5 – 8 and 9 days, and for TS Z_d^6 – 5 and 7 days (the number of days is given in ascending order of the value of the membership function μ, $\mu \geq 0.60$). According to the results, during the pandemic period, long-term memory in some TS is shifted.

The following conclusions can be drawn from the analysis of the fuzzy set of memory depth.

The most stable and trend-resistant series were TS Z_d^2, Z_d^3 and Z_d^6. For these time series, the L_{mc} indicator is at a relatively high level and does not decrease during the pandemic, which indicates the trend stability of the time series. Shannon's entropy decreases (for TS Z_d^2, Z_d^3), which shows a decrease in uncertainty, or remains unchanged (for TS Z_d^6). It should be noted that for TS Z_d^2 the entropy was at the highest level, however, during the pandemic this indicator improved. That is, the crisis in the economy did not affect the increasing uncertainty of the time series.

For TS Z_d^1 negative is a significant decrease in L_{mc} from 7 to 3. TS Z_d^4 was marked by an increase in the uncertainty (entropy), and for TS Z_d^5, despite the decrease in entropy, l_{gc} and l_{max} decreased, which showed instability of these series to external risks.

It should be noted that the Hurst exponent of all series is at a high level, which indicates the persistence of the series.

In addition to the fractal analysis of corporate stock prices, we conduct a corresponding analysis of stock profitability.

In this regard, the TSs of stock profitability are built and studied:

$$P_s^i = \langle p_s^i \rangle \tag{8}$$

where $i = 1$ – corresponds to the time series of Apple Inc., $i = 2$ – time series of Tesla Inc., $i = 3$ – time series of Alphabet Inc., $i = 4$ – time series of The Walt Disney Company, $i = 5$ – time series of Netflix Inc., $i = 6$ – time series of Sony Corporation;

$p_s = \frac{z_d - z_{(d-s)}}{z_{(d-s)}} * 100\%$ – price profitability on day d relative to price on day $(d - s)$, i.e. profitability with lag $s = 1, 7, 10, 14, 21, 28, 30, 37, 42$.

The calculation results of the Hurst exponent for time series of profitability with the lag 1 ($s = 1$) indicate the randomness of TSs (table 6).

As a result of the study of the time series of the delayed profitability, it was found that the nature of the dynamics of profitability varies from stochastic to persistent depending on the magnitude of the time lag and acquires the characteristics inherent in the "parent" TS of price. But the lag for each time series is different:

TS P^1 (Apple), P^6 (Sony) acquire persistence at a lag(s) value of 14 days, and the Hurst exponent is already 0.70;

TS P^3 (Alphabet), P^5 (Netflix) acquire persistence at a lag(s) value of 10 days, and the Hurst exponent is already 0.74;

TS P^4 (Disney) acquires persistence at a lag(s) value of 7 days, and the Hurst exponent is already equal to 0.71;

TS P^2 (Tesla) acquires persistence at a lag(s) value of 21 days, and the Hearst index is already 0.70.

Table 4: The fuzzy set of memory depth for Apple Inc. and Tesla Inc. stock prices

Before the pandemic	During the pandemic
The fuzzy set of memory depth for TS Z_d^1	
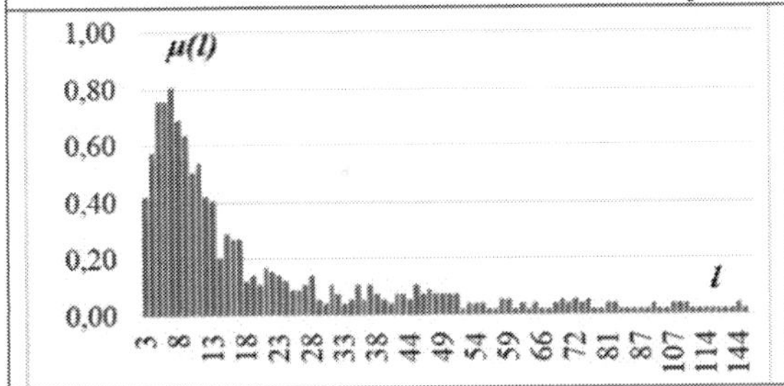	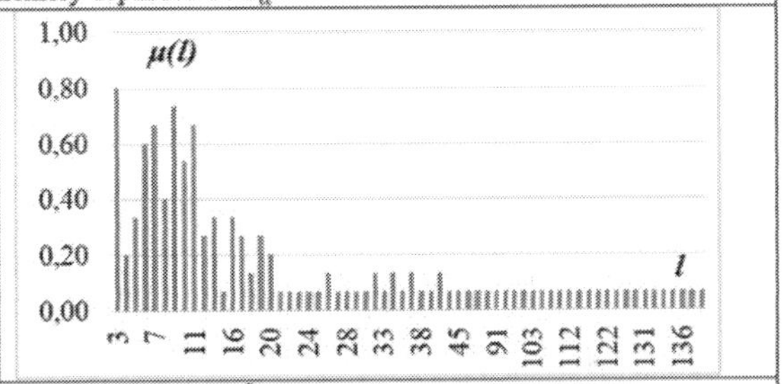
The fuzzy set of memory depth for Z_d^2	
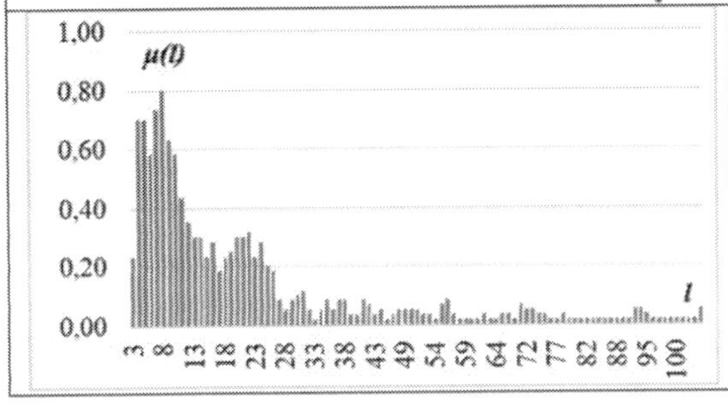	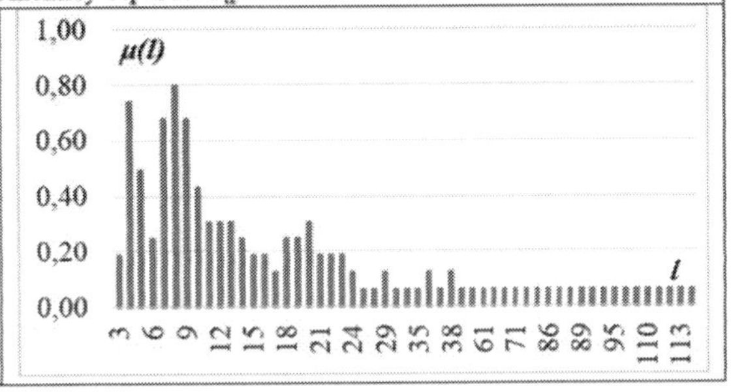

Table 5: The main system characteristics of the fuzzy set of memory depth

TS	Z_d^1		Z_d^2		Z_d^3		Z_d^4		Z_d^5		Z_d^6	
Characteristics	Before	During	Before	During	Before	During	Before	During	Before	During	Before	During
l_{max}	145	154	104	114	153	153	152	138	122	119	154	135
l_{mc}	7	3	8	8	6	7	5	4,5,10	5	7	8	8
l_{gc}	21.8	30.8	21.7	23.8	18.8	28.1	17.6	21.4	19.7	18.8	17.9	28.3
H_{entrL}	22.1	20.1	23.1	18.0	19.5	17.6	18.1	20.3	20.2	16.1	18.8	18.8

Table 6: The results of the application of The Hurst's rescaled range method for profitability time series before the pandemic and during the pandemic

TS	*H* before	*H* after	TS	*H* before	*H* after
Pd1	0.62	0.59	Pd4	0.56	0.63
Pd2	0.57	0.70	Pd5	0.64	0.61
Pd3	0.57	0.60	Pd6	0.59	0.61

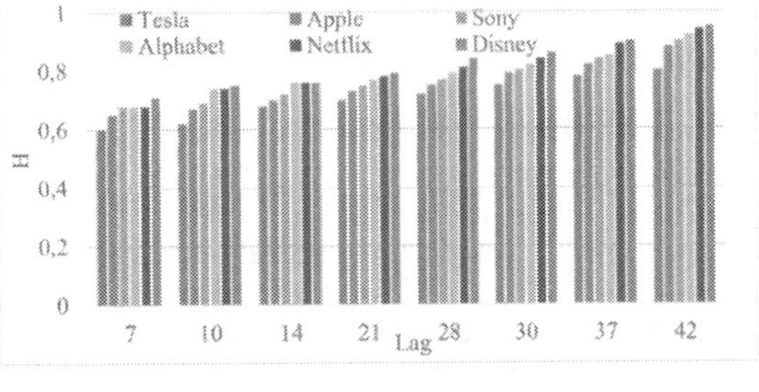

Figure 2: The value of the Hurst exponent for the profitability of shares of Apple, Tesla, Alphabet, Disney, Netflix and Sony with different lags

A graphical representation of the values of the Hurst exponent for profitability TS with different lag is shown in figure 2.

Therefore, the time series P^4 (Disney) acquires persistence the fastest. We observe that with increasing lag, the Hurst exponent increases and becomes suitable for study by fractal analysis. This allows the investor to unify the analysis tools and forecast profitability in accordance with the characteristics of the price dynamics of a particular investment instrument.

3 Conclusions

Lack of investment strategy can cause loss of funds and complete frustration in the choice of a particular stock market by the investor. To reduce the level of risk, it is necessary to choose how long the wherewithal will be invested before making the investment. According to the term of investment, there are three strategies: short-term, medium-term and long-term strategy.

Short-term investment can last from a few days to 3 months. The medium term lasts from 3 months to 3 years. Long-term – from 3 years.

The results of the research allow us to offer the following (certain) recommendations for choosing an investment strategy for each selected security.

The first factor for choosing an appropriate strategy for investing in stocks is usually the price and volatility of the security and the sector as a whole. The media and entertainment sectors, automotive and technology have average volatility, ie price change, the rate of change is average. However, the selected financial instruments are characterized by a fairly high historical volatility (table 1). It follows that for all selected financial instruments, we can recommend a short-term investment strategy (table 7) with the possibility of investing for a small period.

Table 7: Choosing an investment strategy

	Short-term strategy	Medium-term strategy	Long-term strategy
Type of analysis	Volatility	Delayed profitability	Memory depth
Apple	+	++	
Tesla	+	++	++
Alphabet	++	+++	++++
Disney	+++		
Netflix	+++		
Sony			++

Netflix and Disney remain the best option for a short-term strategy, because due to the pandemic, which will cover the world for another one or two years, the stock prices of these companies will be constantly volatile. The price will fluctuate depending on the growth of demand for remote viewing of movies, series and shows, or the return of demand again for watching movies in regular cinemas. As a result, there is a constant risk of losing money if you invest them for more than 3 months.

Consider the second investment strategy – medium-term. To do this, we turn to the analysis of the results of complex fractal analysis. All series are persistent both before the pandemic and now, ie persistent (tables 2, 3). However, the profitability of instruments is not persistent, and the time series of delayed profitability with different time lag was used to study the level of trend stability (figure 2). Due to which it was determined that Tesla Inc., Apple Inc. and Sony gained the fastest profitability, ie you can invest in these shares for more than a week or two. Because in 7–14 days, the yield becomes persistent and it can be predicted for more than three months. These shares will generate passive income and can be reinvested again. In addition, these corporations are quite ambitious and have grand plans for the future, such as space exploration and shuttle construction, the creation of an electric car, etc., so the prices of their securities will only increase.

Alphabet Inc. has the most stable results, this corporation is often chosen for long-term investment, the results of volatility are the lowest, because the price does not fluctuate strongly enough, the dynamics of the time series is persistent, deferred profitability becomes persistent with a lag of 10 days. The results of the memory depth study show the stability of the corporation even under the influence of natural external factors such as the COVID-19. Stable memory depth characteristics during the pandemic were also demonstrated by shares of Tesla Inc. and Sony (table 7).

Note that these recommendations are formed only on the basis of this study and they may change depending on changes in the dynamics of financial instruments under the influence of various external and internal factors.

References

[1] E.F. Fama, The Journal of Finance **46**, 1575 (1991)

[2] R. Singh, N. Leepsa, N. Kushwaha, Iranian Journal of Management Studies **9**, 627 (2016)

[3] C.M. Boya, Research in International Business and Finance **49**, 156 (2019)

[4] M.R. Borges, The European Journal of Finance **16**, 711 (2010), `https://doi.org/10.1080/1351847X.2010.495477`

[5] E.E. Peters, *Fractal Market Analysis. Applying Chaos Theory to Investment and Analysis* (John Wiley & Sons, Inc., New York, NY, 1994)

[6] L.O. Dittrich, P. Srbek, Atlantic Economic Journal **48**, 491 (2020)

[7] V. Soloviev, A. Bielinskyi, O. Serdyuk, V. Solovieva, S. Semerikov, CEUR Workshop Proceedings **2732**, 455 (2020)

[8] L.M. Calcagnile, F. Corsi, S. Marmi, Computational Economics **55**, 143 (2020)

[9] Y. Karaca, Y.D. Zhang, K. Muhammad, Expert Systems with Applications **144**, 113098 (2020)

[10] S. Bianchi, A. Pianese, Chaos, Solitons & Fractals **109**, 64 (2018)

[11] Z. Katarzyna, S. Mikolaj, *Analysis of the phenomenon of long-term memory in financial time series*, in *36th International Conference on Mathematical Methods in Economics* (České Budějovice, Czech Republic, 2018), pp. 216–221

[12] S. Erokhin, O. Roshka, MATEC Web Conf. **170**, 01074 (2018)

[13] V. Perepelitsa, N.K.Maksyshko, *Analysis and forecasting of the economic systems evolution: problems of data structuring in conditions of uncertainty and pre-forecast analysis* (Lambert Academic Publishing GmbH & Co. KG, Saarbrucken, 2012)

[14] N. Maksyshko, Bulletin of Zaporizhzhia National University **2**, 119 (2011)

[15] Z. Tan, Y. Fu, H. Cheng, J. Liu, International Journal of Emerging Markets **23** (2020)

[16] X. Wang, T. Lei, Z. Liu, Z. Wang, *Long-memory behavior analysis of China stock market based on*

Hurst exponent, in *2017 29th Chinese Control And Decision Conference (CCDC)* (2017), pp. 1710–1712

[17] H. Tebyaniyan, A. Jahanshad, F. Heidarpoor, International Journal of Nonlinear Analysis and Applications **11**, 161 (2020)

[18] C. Chen, Y. Wang, Physica A: Statistical Mechanics and its Applications **466**, 346 (2017)

[19] Fusion Media Limited, *Investing.com - Stock Market Quotes & Financial News* (2021), `https://www.investing.com/`

[20] Investfunds, *Investfunds website about tools for private investors* (2020), `https://investfunds.ru/`

[21] Apple Inc., *Apple* (2021), `https://www.apple.com/`

[22] Tesla, *Electric Cars, Solar & Clean Energy* (2021), `https://www.tesla.com/`

[23] Alphabet, *Alphabet* (2021), `https://abc.xyz/`

[24] Disney, *The Walt Disney Company* (2021), `https://thewaltdisneycompany.com/`

[25] Netflix, *Watch TV Shows Online, Watch Movies Online* (2021), `https://www.netflix.com/`

[26] Sony Corporation, *Portal site* (2021), `https://www.sony.net/`

[27] S. Semerikov, S. Chukharev, S. Sakhno, A. Striuk, V. Osadchyi, V. Solovieva, T. Vakaliuk, P. Nechypurenko, O. Bondarenko, H. Danylchuk, E3S Web of Conferences **166** (2020)

Application of blockchain technology in accounting and audit: international and domestic experience

Mariia Nezhyva[1,*], *Olha* Zaremba[1,**], and *Viktoria* Nehodenko[1,***]

[1]Financial analysis and audit department, Kyiv National University of Trade and Economics, 19 Kyoto Str., Kyiv, 02156, Ukraine

Abstract. The article is devoted to the study of blockchain technology in the financial sphere of Ukraine, as well as accounting and audit of their application in domestic and international practice. The article contains data on the prospects of blockchain development in the world and in Ukraine, as well as information on the peculiarities of the use of such technologies in the perspective areas – accounting and audit. The article argues that the development of blockchain technology will enable the transition to more modern methods and techniques, which in turn will allow to modernize the accounting and auditing system. The principle of triple entry as a major component of blockchain technology in accounting is emphasized. One of the most promising areas of application of blockchain is accounting for accounts payable and receivable, tax accrual and payment, record keeping and documentation. The article highlights the advantages and disadvantages of using blockchain technology in the financial sphere of Ukraine.

1 Introduction

Given the current development of the information environment, the question of improving information technologies and their application in the enterprise management system to increase the competitiveness of economic entities is particularly acute. This necessitates an increase in the cost of IT technology, the introduction of innovation and the broad informatization of the business environment.

Among the advanced technologies of the last decade, the most important is the blockchain technology, which is implemented in various fields, including financial. The experience of scientists and practitioners suggests that the use of blockchain technology in accounting and audit will directly improve these processes, in particular, change the work of an accountant. This will happen through the introduction of modified methods and techniques of information processing, its distribution, as well as dramatically change the very basis of the accounting system. However, there are a number of unresolved problems, as blockchain is currently in the experimental stage of development. This determines the need to improve this issue and accordingly determines the relevance of the topic under study.

The term "blockchain" has been introduced not so long ago in the economic literature. It was only after 2014 that blockchain publications began to appear in the economic literature. We study the peculiarities of the use of blockchain technology in the works [1–19]. Ukrainian enterprises have been slow to adapt their accounting and auditing systems to modern information technologies. Therefore, modern blockchain technology is in its infancy. This is explained by the limited scientific work of a methodological nature, which deals with the prospects of blockchain development, its advantages and disadvantages.

The purpose of the article is to determine the role of modern blockchain technology in financial areas, in particular in the field of accounting and audit, to study the results of implementation of this technology in foreign countries and to identify promising areas of development for Ukraine.

2 Results

A positive moment for the development of the blockchain was the approval of the Concept of Digital Economy Development [20]. This Concept is designed to encourage the digitization of all areas of economic activity, stimulate the use of digital technologies in enterprises and indicates the priority of the development of these technologies at the state level. Consider the key strategic goals facing Ukraine in the run up to 2030:

- increase the nominal GDP of Ukraine by 8 times – up to 1 trillion dollars in 2030;
- become an influential regional entity, independent, independent in making economic and geopolitical decisions, as far as the global situation will make it possible;
- ensure higher-than-average incomes and quality of life of Ukrainians in Europe and neighboring countries (nominal average monthly salary in 2030 is not lower than 2700 dollars; average monthly pension is 775 dollars);
- create conditions for the growth of capitalization of Ukrainian business;

*e-mail: marijka@ukr.net
**e-mail: olga.zaremba@midland-development.com.ua
***e-mail: vikla1977@ukr.net

- create conditions for an 8-10-fold increase in the economy over the next 10 years;
- significantly increase labor productivity (according to UIF estimates, to reach a GDP of 1 trillion dollars, productivity should increase 8.5 times to 71,000 dollars per employee. This is in line with the current level of productivity in high-tech sectors such as Belgium, Denmark, Finland, Germany, Sweden);
- make a technological leap in the Ukrainian economy;
- significantly increase the competitiveness of the Ukrainian economy in global markets;
- create new business models, new segments and sectors, opportunities to quickly, conveniently, cheaply develop, scale and capitalize any business;
- create new jobs, demand for new professions.

The main effect of digitalization is to change the value chains [21, 22]. Technology and digitalization will displace people from their usual processes – production, services, entertainment, trade, education and even medicine [23]. At the same time, the return on investment in business will dramatically increase. According to Digital Spillover [24], the ROI for digital technologies is 6.7 times higher than for traditional ones. According to research in the Digital Spillover report, every dollar invested in digital technology causes GDP growth of 20 dollars, ie the investment multiplier is 20. For comparison: the multiplier of traditional investment (non-digital) is 2-8 (depending on the industry).

Data becomes an asset. Collection, description, storage and processing of data allow to obtain valuable information for use in business processes, public life, the work of the state. The ability to work with data and analyze it is an opportunity to be the first to receive valuable market "insights", ie to be more competitive. Data is accessed via the Internet and other networks. Much of the world's data is (or has become) open.

Digital technologies have become the basis for the creation of new products, values, properties and, consequently, the basis for gaining competitive advantage in most markets. Digitization has given small companies and project teams the opportunity to create new products and quickly bring them to market along with the large companies present there. This led to the shift of "innovation centers" from large companies to small ones (startups, etc.). Digital transformation leads to the emergence of new unique systems and processes that make up their new value essence (e.g., Uber, Airbnb, digital banking, etc. [25–30]).

The world economy is significantly affected by business models of the common economy, or the sharing economy, which operates on the basis of digital technologies. The sharing economy includes living together (sharing), car sharing, offices (coworking), gadgets, clothes and even food (food sharing – "saving" products intended for disposal in the interests of the needy). Part of this trend is cloud technology [31].

Virtualization makes it possible to significantly reduce the initial capital cost of deploying the necessary digital infrastructure through the use of cloud technologies and software-defined architecture. Technology allows you to rent computing capabilities and services based on the needs of a particular business process. The user has the ability to quickly access the service and rent for the required time the required capacity on secure and technological platforms.

The use of artificial intelligence [32] is spreading to more and more areas and sectors of the economy. The number of companies that use it to one degree or another is growing exponentially. Countries are developing national artificial intelligence strategies and trying to compete with each other for talent. Ethical norms and limits of use are defined. Today, artificial intelligence combines many technologies, the main of which are: machine learning [33], computer vision [34], deep learning [35] and natural language processing (NLP) [36]. Artificial intelligence based on the processing of large data sets allows you to optimize processes and improve the quality of digital products and services.

Over the last 10 years, digitalization has become a key driver of growth in the Ukrainian economy and the well-being of citizens. Today, every Ukrainian can easily capitalize on themselves, their skills, knowledge through the use of digital technologies. With a smartphone and Internet access, everyone was able to sell their services through online services without intermediaries and employers; create a business from scratch without start-up capital and assets; receive income from the realization of their creative potential and creative ideas. Ukraine has become an intellectual hub, where all conditions are created for the development of people's potential.

In Ukraine, the principle of digital by default applies – all spheres of public life services are permeated by digitalization. The maintenance and development of any physical system occurs only in the absence of a digital alternative. 95% of all companies have changed their business models and goals under the influence of new technologies. Those who did not have time to do so, remained outsiders or disappeared altogether.

Digitization has made it possible to: optimize business processes; create new products and services thanks to the Internet of Things technology, virtual reality, cloud services, artificial intelligence; reduce costs (savings on people through automation and robotics, accounting automation, "smart" supply and fleet management, control over fuel consumption); offer completely new business solutions: new insurance models, P2P systems, alternative banking services, mobile training applications, individual approach to the online buyer, personal advertising [37–41].

Today, companies work at the intersection of industries and segments, becoming platforms and platforms, where they offer their customers not just a product, but the solution of any problems, gaining experience, values, emotions and impressions.

One of the most important values for the development of the digital economy is data [42]. Access to public, professional, industrial and other data is an important tool for the emergence of new products and services, i.e. directly affects economic growth.

Data is the basis of many digital services and products, working with large arrays helps to make research and development work more efficient and effective, more dynamically with the availability of data, develop technologies of artificial intelligence and predictive analytics. To take full advantage of the analysis, Big Data must be available [43].

Open non-personal data should be of better quality – in a form suitable for processing, indicating the main indicators (data legends).

The government also needs to encourage and create conditions for the exchange of data between private sector companies, while ensuring the security and confidentiality of such data. For example, to introduce tax benefits or make priority partners of the company, ready to share non-personal data with other companies through government digital platforms. The idea of paying taxes with data should be considered as an incentive.

You also need to keep in mind that data sharing can be costly for its owners. This can reduce incentives to ensure data transmission and raise the question of who should bear the costs of developing and maintaining transmission mechanisms. Therefore, government initiatives that facilitate data transfer may be needed. In the EU, for example, data exchange between companies began with the development of codes of conduct for suppliers.

The state, in turn, can create secure digital platforms for storing, transmitting, and administering citizens' data. Such platforms will allow citizens to manage their data and configure access rights to it by public institutions and private companies. In addition, the state needs to develop rules for managing personal data for users on such platforms. An example is the joint project of Amsterdam and Barcelona, funded by the European Union – DECODE. It is a digital blockchain platform that allows the user to set rules for the use of his data.

An important issue in working with data is the degree of their protection. This is what the state should focus on when creating digital platforms and providing access to personal but not personalized data. The higher trust of citizens, the greater their willingness to share their information.

The state needs to open access to any of its data that does not relate to national security issues and does not violate the right of citizens to privacy. Another function of the state in working with data should be training programs for entrepreneurs and researchers on what data from government agencies are, where to find them, how to work with them. Public authorities and companies themselves should also increase their competencies in data science and artificial intelligence.

Large-scale digitalization of spheres of life will lead to the fact that citizens of Ukraine and business will increasingly suffer from the growth of cybercrime. For example, when shopping or doing banking online.

The most dangerous for the economy and citizens are cyber attacks on critical infrastructure (energy, transport management, banking and telecommunications sectors, health care, water supply, etc.).

Finally, increasingly sophisticated and interconnected digital technologies will lead to new forms of cyberattacks (zero-day vulnerabilities).

Ukraine and its specialized organizations should work with all stakeholders, especially at the European level, to ensure national cybersecurity and cyber defense. The development of technical and organizational solutions should, on the one hand, not limit the opportunities offered by the country's transition to digital technologies, and on the other hand - to ensure the appropriate level of trust and security for each user.

In 2008, Satoshi Nakamoto highlighted the principles of blockchain technology in his work called "Bitcoin: A Peer-to-Peer Electronic Cash System" [44]. The basis of the blockchain is the principle of information exchange in peer network (Peer-to-Peer, P2P), which does not need for the functioning of the central server [2].

It is impossible to study the peculiarities of the implementation of blockchain technology, including accounting and audit, without understanding the very essence of this latest technology. There are different approaches to interpreting blockchain. For example, the founder of the Blockchain Research Institute in England, M. Swan interprets blockchain technology as a multifunctional and multilevel information technology designed to securely account for various assets; it is a decentralized transparent transaction record book – a database updated and controlled by participants [4].

In a general sense, a blockchain is a chain consisting of a large number of blocks, each containing a variety of information. Each subsequent block contains information about the previous one in encrypted form. The uniqueness of this technology is that the entire data register is decentralized (figure 1), and its copies are stored simultaneously on tens of thousands of computers around the world. This provides the highest level of protection against external influences. The data contained in the system cannot be deleted or replaced. Such a database is characterized by anonymity, an agreed mechanism, it is not owned by a specific entity, is not controlled or regulated by third parties. All functions in a blockchain system are distributed among its members, who pre-approve changes that may occur in the system. Thus, it is virtually impossible to make changes and edit after the transaction. The above information proves the ideal of blockchain technology for accounting purposes, because it monitors all transactions and changes in the system, prevents manipulation and distortion.

There are different types of blockchain, the classification of which is shown in figure 2. The level of access is open, closed and consortium blockchain. The peculiarity of open blockchain is that absolutely anyone can join the public network. The participant has the opportunity to read, add new entries and participate in blockchain processes. There is no external control in the open type. Closed blockchain, in turn, requires permissions for user access, imposes various restrictions, there is control over who exactly has the right to become a member of the network and in what transactions can participate.

https://doi.org/10.1051/shsconf/202110702001

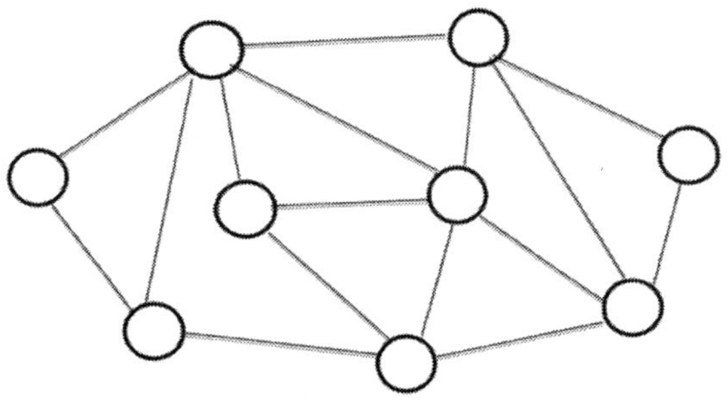

Figure 1. Decentralized principle of blockchain technology

Permissionless blockchain does not have a supervisory authority that validates transactions. This includes all known cryptocurrency platforms. The meaning of such data processing is the lack of benefits for any participant in the process of the operation. The use of open blockchain technology can serve to create a truly democratic system.

Permissioned blockchain is the confirmation of transactions by certain persons, in which the employee, the enterprise, the state can act. Participants have the ability to read system data, but some information may not be available. Regular users cannot enter data.

There are different categories of Blockchain 1.0-5.0 (the last two currently have only theoretical character). Blockchain 1.0 is a cryptocurrency that is used in various applications (such as wire transfer, digital payments). In turn, Blockchain 2.0 is contracts. Entire classes of economic, market, and financial applications based on blockchain work with different types of financial instruments (stocks, bonds, futures, smart assets, smart contracts). Blockchain 3.0 is an application whose scope goes beyond cash calculations. They extend to the spheres of public administration, health care, science, education, culture and the arts.

According to the Deloitte Digital-funded Coming of Age Digitally report, as well as PwC research, in 2018, 84% of respondents (600 business executives) are somehow involved in the blockchain process, and 30 percent of respondents (over 4,300 managers, executives) and analysts around the world point to the maturity of digital technology in their organizations, with the adoption of such technologies changing the speed, creativity, learning, risk, collaboration, quality of decision making, transparency, etc. for the better. In turn, 53% of companies say that blockchain technology is one of their top five strategic priorities in 2019, up 23.3% from the previous year [45].

There are some issues that adversely affect the competitiveness of companies that use blockchain technology. Among them, the expected rate of return on digital technology, which is 3–5 years, which is considered a long enough business.

The blockchain that recently started to be actively used in finance and accounting is still in its infancy, as stated in the 2018 KPMG report. Statistics show that only 1% of businesses use it at the mature level, while 74% of businesses say full maturity of blockchain. However, by 2021, at least 25% of the world's largest public companies are projected to use blockchain, with a value added to 176 billion dollars USA by 2025 [46].

According to research by one of the leading experts in blockchain technology in the world, Tapscott, Ukraine belongs to the 14 leading countries in the implementation of blockchain technology, among which the first are the USA, Canada, Brazil, Australia and Israel. Currently, the US is the leader, but China is forecast to be the leader in 2021-2023 [47].

Table 1. Opportunities for the impact of blockchain technology on accounting and audit

Characteristics of blockchain technology	Impact on accounting and audit
Distributed Book	The parties to the transaction share information about its implementation, by confirming one or another transaction increases the confidence that it will take place on the terms approved in the contract, facilitates interaction between the parties without paper workflow
Real-time updates	Accounting transactions are recorded in real time, information about them appears immediately on all sides of the contract, which allows for operational accounting and control
Confirmed, protected and digitally signed agreements	Since business transactions, once validated by the parties, are recorded on the system by means of a cryptographic hash, it is practically impossible to make irregular adjustments and changes to the accounting records after the fact
Consensus	Transactions are updated only by agreement between the parties, which minimizes the fact of non-fulfillment of contractual obligations
Irreversibility	Increased security of accounting data, reliability of records of performed business transactions
Transparency	Availability of accounting information including auditors, continuity and ease of audit
Triple entry	The method of accounting for the part of the double entry changes, which is supplemented by the third one – confirmation of the transaction in the blockchain

In Ukraine, there is no regulatory framework governing blockchain technology. The Law of Ukraine "On the Circulation of Cryptocurrency in Ukraine" is at the stage of the project, and the National Bank of Ukraine in its explanation considers bitcoin as a money surrogate forbidden to use in the territory of Ukraine [48]. There is ongoing debate in the scientific space as to whether it is nec-

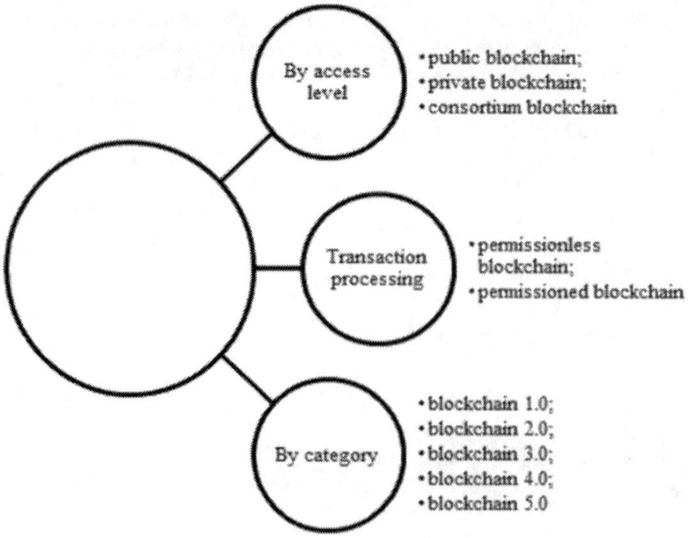

Figure 2. Classification of Blockchain technology

essary to legislate on this topic. On the one hand, lawlessness persists, which in turn can generate fraudulent schemes. On the other hand, state regulation of blockchain technology is able to establish uniform rules that must be observed by all market participants. However, additional burdens may be imposed on states. That is, there is a debatable nature of uncertainty.

Today, in Ukraine, obsolescence of the methods and techniques on which accounting and auditing is based is obsolete, the inconsistency of these methods with the digital economy of today, the basis of which is occupied by information technologies. Upgrading requires technology for processing, storing, transmitting and accumulating information. Therefore, blockchain technology is the main motive for the development and modernization of the entire accounting and audit methodology.

Blockchain technology allows participants to share data for each transaction that requires accounting records. By using blockchain, parties can set read-only mode specifically for external users, such as auditors and government agencies. Such access enables the transaction to be confirmed instantly for the purpose of reporting and other regulatory purposes. As a result, the work of the auditor will be automated, uninterrupted, more accurate and analytically disclosed. The introduction of blockchain technology in accounting and audit will reveal their processes in real time, meaning users will not have to wait for the end of the reporting period.

Creating a system of stable accounting – a common registry, its cryptographic coding virtually impossible to falsify, destroy data or hide activity. Blockchain technology can be combined with an existing accounting system in the form of a local blockchain within a single enterprise or group of companies. Today, some companies are implementing blockchain for certain types of transactions. Currently there are no enterprises with complete transformation of processes into blockchain technology.

There are features of blockchain technology such as: distributed ledger, triple entry (the main feature, while accounting uses dual entry), real-time updates, transparency, irreversibility, and more. All of them have a direct impact on accounting and auditing. The possibilities of such influence are shown in table 1 [5].

The application of blockchain technology in accounting is in the most promising areas, the main of which are payments with counterparties (accounts receivable and payables, including international purchases and transactions). There is also the use of this technology in operational accounting and reporting, when handling documents and storing them. The movement of assets within the enterprise, the calculation of taxes and their payment are increasingly accounted for by enterprises using blockchain technology.

Today, blockchain is increasingly popular in the audit field. In particular, the Big Four companies Deloitte and PwC are actively introducing blockchain technology into accounting practices.

In the field of audit Redchenko [3] identifies two areas of blockchain technology development: application in accounting systems to provide assurance or fulfillment of financial reporting procedures agreed with the client; providing services to companies that use blockchain in their own activities. The implementation of blockchain technology necessitates the continuous training of auditors and accountants, the continuous improvement of their knowledge in the field of information technology. The result of applying new technologies, like any phenomenon, has its advantages and disadvantages.

In the business environment, opinions are also divided on the feasibility of implementing blockchain technology in accounting and auditing. Many members of the business community are convinced that at this embryonic stage, it is difficult to understand the full meaning and power of new technologies. Note that for the effective and easy introduction of blockchain technology, you also need to improve the skills of employees in accounting and auditing, as well as to change business models in enterprises, which requires additional costs.

3 Conclusions

Thus, the implementation of blockchain technology in accounting and audit, according to most scientists, is a promising direction for improving this field. Investigating this issue, we can conclude that blockchain technology is developing at a rapid pace in various manifestations. This technology is transforming some of the outdated accounting and audit methods. Advantages of its introduction at the enterprise are reduction of expenses for keeping of accounting, carrying out of audit, increase of access to the information, its more reliable protection, unnecessary keeping of documentation in a manual way, in paper form. Blockchain improves relationships with contractors and extends enterprise access to finance. However, there are some doubts, including the overload of carriers with a wealth of information, the possibility of cyber attacks, the negative impact on the environment. Only through a comprehensive analysis, evaluation of all factors influencing management personnel can determine the feasibility of introducing blockchain technology in accounting and audit at the enterprise.

References

[1] J.L. Zhao, S. Fan, J. Yan, Financial Innovation **2**, 28 (2016)

[2] A. Porytko, Debit Credit **37** (2017)

[3] K. Redchenko, Statistics of Ukraine **1**, 85 (2018)

[4] M. Swan, *Blockchain: Blueprint for a new economy* (O'Reilly Media, Inc., 2015)

[5] L. Rao, S. Pandurangiah, The Management Accountant **53**, 42 (2018)

[6] M. Iansiti, K.R. Lakhani, *The truth about blockchain* (2017), https://hbr.org/2017/01/the-truth-about-blockchain

[7] R. Wattenhofer, *The science of the blockchain* (2016), https://lib.hpu.edu.vn/handle/123456789/28113

[8] N. Yushchenko, Economy and Management of National Economy **19**, 269 (2018)

[9] D. Tapscott, A. Tapscott, Business and the World pp. 269–275 (2016)

[10] P. Boucher, S. Nascimento, M. Kritikos, *How blockchain technology could change our lives* (Scientific Foresight Unit, Brussels, 2017), https://www.europarl.europa.eu/RegData/etudes/IDAN/2017/581948/EPRS_IDA(2017)581948_EN.pdf

[11] B. Marr, *How blockchain technology could change the world* (2016), https://tinyurl.com/9d3drn35

[12] J. Buchan, *Blockchain technology review – industry applications* (2017), http://web.archive.org/web/20210225021349/https://zudu.co.uk/blog/blockchain-technology-industry-applications/

[13] O. Boyko, Global and national problems of economy **22**, 31 (2018)

[14] A. Litoshenko, Economic analysis **23**, 88 (2016)

[15] A. Litoshenko, Economy and state **8**, 77 (2017)

[16] D. Efanov, P. Roschin, Procedia Computer Science **123**, 116 (2018)

[17] N. Kshetri, Telecommunications Policy **41**, 1027 (2017)

[18] M. Nezhyva, V. Miniailo, Herald of Kyiv National University of Trade and Economics **3**, 123 (2020)

[19] V. Soloviev, A. Belinskiy, Communications in Computer and Information Science **1007**, 276 (2019)

[20] *On approval of the concept of development of the digital economy and society of ukraine for 2018-2020 and approval of the action plan for its implementation* (2018), https://zakon.rada.gov.ua/laws/show/67-2018-%D1%80#Text

[21] O. Kravchenko, M. Leshchenko, D. Marushchak, Y. Vdovychenko, CEUR Workshop Proceedings **2422**, 434 (2019)

[22] H. Kucherova, D. Ocheretin, V. Los, N. Venherska, CEUR Workshop Proceedings **2732**, 385 (2020)

[23] I.K. Pokulyta, M.O. Kolotylo, Journal of Physics: Conference Series **1840**, 012055 (2021)

[24] *Digital Spillover: Measuring the true impact of the digital economy* (2017), https://www.huawei.com/minisite/gci/en/digital-spillover/files/gci_digital_spillover.pdf

[25] D. Trcek, CEUR Workshop Proceedings **2422**, 263 (2019)

[26] O. Kuzminska, M. Mazorchuk, N. Morze, O. Kobylin, CEUR Workshop Proceedings **2393**, 53 (2019)

[27] O. Pinchuk, O. Sokolyuk, O. Burov, M. Shyshkina, CEUR Workshop Proceedings **2433**, 90 (2019)

[28] A. Iatsyshyn, V. Kovach, Y. Romanenko, A. Iatsyshyn, CEUR Workshop Proceedings **2433**, 197 (2019)

[29] M.P. Leshchenko, A.M. Kolomiiets, A.V. Iatsyshyn, V.V. Kovalenko, A.V. Dakal, O.O. Radchenko, Journal of Physics: Conference Series **1840**, 012057 (2021)

[30] N.V. Morze, O.V. Strutynska, Journal of Physics: Conference Series (2021)

[31] O. Markova, S. Semerikov, A. Striuk, H. Shalatska, P. Nechypurenko, V. Tron, CEUR Workshop Proceedings **2433**, 499 (2019)

[32] O. Haranin, N. Moiseienko, CEUR Workshop Proceedings **2292**, 143 (2018)

[33] A. Kiv, V. Soloviev, S. Semerikov, H. Danylchuk, L. Kibalnyk, A. Matviychuk, CEUR Workshop Proceedings **2422**, 1 (2019)

[34] M. Bogdanovskyi, A. Tkachuk, O. Dobrzhanskyi, A. Humeniuk, E3S Web of Conferences **166**, 05004 (2020)

[35] V. Derbentsev, A. Matviychuk, N. Datsenko, V. Bezkorovainyi, A. Azaryan, CEUR Workshop Proceedings **2713**, 434 (2020)

[36] K. Osadcha, V. Osadchyi, S. Semerikov, H. Chemerys, A. Chorna, CEUR Workshop Proceedings **2732**, 547 (2020)

[37] I. Mintii, V. Soloviev, CEUR Workshop Proceedings **2257**, 227 (2018)

[38] S. Symonenko, N. Zaitseva, V. Osadchyi, K. Osadcha, E. Shmeltser, CEUR Workshop Proceedings **2547**, 37 (2020)

[39] M. Volikova, T. Armash, Y. Yechkalo, V. Zaselskiy, CEUR Workshop Proceedings **2433**, 486 (2019)

[40] V. Tkachuk, S. Semerikov, Y. Yechkalo, S. Khotskina, V. Soloviev, CEUR Workshop Proceedings **2732**, 1058 (2020)

[41] A. Vinkovska, A. Kiv, T. Koycheva, L. Bodnar, I. Donchev, CEUR Workshop Proceedings **2422**, 274 (2019)

[42] I. Gernego, L. Petrenko, M. Dyba, V. Tsarov, E3S Web of Conferences **166** (2020)

[43] S. Semerikov, I. Teplytskyi, Y. Yechkalo, O. Markova, V. Soloviev, A. Kiv, CEUR Workshop Proceedings **2393**, 833 (2019)

[44] S. Nakamoto, Tech. rep. (2008), https://downloads.coindesk.com/research/whitepapers/bitcoin.pdf

[45] *Deloitte's 2019 global blockchain survey: Blockchain gets down to business* (2019), https://www2.deloitte.com/content/dam/Deloitte/se/Documents/risk/DI_2019-global-blockchain-survey.pdf

[46] KPMG, *Blockchain and the Future of Finance: A Potential New World for CFOs – and How to Prepare* (2018), https://assets.kpmg/content/dam/kpmg/ca/pdf/2019/05/blockchain-and-the-future-of-finance.pdf

[47] PwC, *Blockchain is here. What's your next move? PwC's Global Blockchain Survey 2018* (2018), https://www.pwc.com/jp/ja/knowledge/thoughtleadership/2018/assets/pdf/blockchain-in-business-en.pdf

[48] National Bank of Ukraine, *Clarification on the lawfulness of the use of "virtual currency / cryptocurrency" bitcoin in ukraine* (2014), https://zakon.rada.gov.ua/laws/show/n0435500-14

Management of financial and economic security of critical infrastructure objects in the conditions of risks of quarantine restrictions: strategic and personnel aspects

Nataliia Zachosova[1,*], *Dmytro* Kutsenko[1], *Oleksii* Koval[1], and *Andrii* Kovalenko[1]

[1]Department of Management and Economic Security, The Bohdan Khmelnytsky National University of Cherkasy, 81 Shevchenko Blvd., Cherkasy, 18031, Ukraine

Abstract. The purpose of the study was to develop theoretical and methodological background and to make practical offers for solving the problem of financial and economic security management of critical infrastructure in terms of personnel and strategic aspects. The list of reasons and circumstances of necessity of maintenance of financial and economic security of objects of critical infrastructure and its management is established. The definition of financial and economic security of a critical infrastructure object is offered. The peculiarities of the interpretation of the concept of financial and economic security for the needs of critical infrastructure are summarized. Two levels of financial and economic security management of critical infrastructure objects are distinguished – state (external) level and internal level – and an explanation of the differences in financial and economic security management at each of them is provided. The risks caused by quarantine restrictions for the state of financial and economic security of critical infrastructure objects have been identified. As a main result of the study the TARGET model for strategic management of financial and economic security of critical infrastructure objects is proposed. Practical value of the proposed in the study TARGET model for strategic management of financial and economic security of critical infrastructure objects – it will be the basis for developing strategic documents for stabilizing and further developing and maintaining high functionality of critical infrastructure of Ukraine. As an additional result of the study proposals on the basics of personnel policy formation for the need to ensure financial and economic security of critical infrastructure objects in a global pandemic and quarantine restrictions are made. The novelty of proposed personnel policy is in a fact that it contains such modern components as staff training; employee incentives; overcoming resistance to change; staff involvement in solving various issues; staff discipline; leadership, partnership and teamwork; time management and self-management; staff safety, staff digital literacy.

1 Financial and economic security of critical infrastructure as a management object

An important problem of national security and defense of the state is to ensure a high functional state of the critical infrastructure of the country, which is based on efficient and continuous operation of economic entities. Vectors of research of modern scientists and practitioners should be directed to the development of innovative management tools, new technologies of management solutions to ensure economic security of business structures to stabilize the critical infrastructure of the state in the transformation of classical business practices caused by Industry 4.0 and globalization. It is also necessary to take into account in this process the latest non-traditional risks posed by the epidemiological situation in Ukraine and in the world, in particular, and those of their species that are provoked by quarantine restrictions.

*e-mail: natazachosova@gmail.com

1.1 The need to manage the financial and economic security of critical infrastructure

According to the draft regulation acts, "the object of critical infrastructure defined in the manner prescribed by law, a component of critical infrastructure, functionality, continuity, integrity and stability of which ensure the realization of vital national interests" [1]. In the context of exacerbation of the epidemiological situation, the continuity of action of the critical infrastructure objects was threatened, which created risks not only for their financial and economic security, but also for the security of the state and the population as a whole

In order to target efforts to develop modern directions and concepts of financial and economic security management of critical infrastructure objects, in particular, at the state level, to establish opportunities for their financial support, etc., it is important to determine their list. At the legislative level, it is proposed to consider the objects of critical infrastructure as "enterprises, institutions, organizations, regardless of ownership, which:

https://doi.org/10.1051/shsconf/202110702002

1) carry out activities and provide services in the fields of energy, chemical industry, defense industry, transport, information and communication technologies, electronic communications, in the banking and financial sectors;
2) provide services in the spheres of life support of the population, in particular in the spheres of centralized water supply, centralized drainage, supply of thermal energy, hot water, electricity and gas, food production, health care;
3) included in the list of enterprises of strategic importance for the economy and security of the state;
4) are subject to protection and defense in a state of emergency and a special period;
5) are objects of increased danger;
6) are objects of national importance, extensive connections and significant impact on other infrastructure;
7) are objects, the dysfunction of which will lead to a crisis situation of regional importance" [1].

Thus, the list includes business structures of various sectors of the economy, those objects, the shutdown of which may actually destabilize the state of the national economic system or lead to increased social tensions, humanitarian or man-made disaster. Obviously, the quarantine restrictions, in particular regarding the temporary suspension of activities, will not directly apply to such facilities. At the same time, the efficiency of their operation is directly affected by the suspension of the rest of the business structures, the transport system, which ensure and mediate the performance of critical functions by objects of critical infrastructure. It is in this context that the issue of ensuring financial and economic security should be addressed, because it is the system tools that characterize risks and minimize the consequences of the impact of threats that should be used in such conditions in the first place.

Continuity of financial transactions and financial services plays an important role in maintaining the normal state of the financial system of the country, the functioning of business and the well-being of the population. Therefore, the regulatory acts specifically state that "critical infrastructure in the banking system of Ukraine includes banks, the sustainable operation of which ensures the stability of the banking system, is essential for the economy and security of the state, the functioning of society, and which are of significant public interest. In particular, these are: National Bank; banks that meet at least one of the following criteria: a bank that is included in the list of systemically important banks; a bank that is included in the list of banks of Ukraine authorized to operate (carry out operations) during a special period; a bank in which the state directly or indirectly owns a share of more than 75 percent of the authorized capital of the bank" [2].

Awareness of the importance of maintaining an adequate level of security of critical infrastructure objects is reflected in the specification of the definition of this concept at the level of government as such a state of "security of critical infrastructure, which ensures the functionality and continuity of its work and / or ability to provide basic services" [3].

Given the importance of continuous and efficient operation of economic entities belonging to the list of critical infrastructure objects, the state of their financial and economic security is a characteristic that requires close control, monitoring and maintenance at a high level through rational and timely management actions by the management of such companies, and by public authorities through the tools of legal regulation.

The need to manage the financial and economic security of critical infrastructure objects is explained by the following circumstances and reasons:

- influence – economic and financial stability of economic entities that form the structural framework of the national economic, industrial, defense systems determines the functional capacity of the national economy as a whole;
- strategic importance – the state of financial and economic security of the subject is the starting point, the zero hypothesis for determining the strategic guidelines of its development in the future; at the same time, the condition of critical infrastructure is crucial for economic foresight within the development of the national strategy of the state;
- autonomy – critical infrastructure must be economically and financially independent if not from public authorities, given their special status, but at least from market business structures in order for their activities to be fully aimed at meeting national interests and not the desires and needs of certain groups of stakeholders;
- specifics – critical infrastructure objects must demonstrate the peculiarities of building their own systems of financial and economic security, for example, in the context of organizing the protection of information resources, which often have the status of state secrets or information with limited access; for such objects, it may be inadmissible to disclose information about their own financial condition or sources of income, existing contracts, which may be necessary to obtain investments or loans; such specific aspects significantly reduce the capacity of some critical infrastructure facilities to attract additional financial resources, leaving them the only opportunity to rely on public funding for their activities;
- time constraints – given the need for continuous operation of critical infrastructure objects, they need to have a sufficient supply of resources, especially material and financial, to ensure the permanence of operational business processes, which in a low level of financial and economic security will be difficult guarantee.

Concretization at the normative level of the concept of security of a critical infrastructure object indicates the need to clarify the content and essence of the category of financial and economic security of such an object.

1.2 Principles of financial and economic security management of critical infrastructure objects

Scientists have long been discussing the problems of supporting the continuous operation of critical infrastructure. Thus, we find interest in the outlined issues in the publications of researchers such as Cristina Alcaraz and Sherali Zeadally [4], Razieh Mosadeghi, Russell Richards and Rodger Tomlinson [5], Georgios Giannopoulos, Roberto Filippini and Muriel Schimmer [6], David Rehak, Pavel Senovsky and Simona Slivkova [7], Grangeat Amélie, Bony Aurélia, Lapebie Emmanuel, Eid Mohamed and Dusserre Gilles [8], Christer Pursiainen [9]. Problems of financial and economic security management are solved by such scientists as Alina Ianioglo and Tatjana Polajeva [10], Lars Osberg and Andrew Sharpe [11], Valeriu Ioan-Franc and Marius Andrei Diamescu [12], Kurt Schimmel, Sifeng Liu, Jeananne Nicholls, Nicholas A Nechval and Jeffrey Yi-Lin Forrest [13], Emanuel Kopp, Lincoln Kaffenberger and Christopher Wilson [14], Marek Kośny and Maria Piotrowska [15], Jacob S. Hacker, Gregory A. Huber, Austin Nichols, Philipp Rehm, Mark Schlesinger, Rob Valletta and Stuart Craig [16], Jacob S. Hacker, Gregory A. Huber, Philipp Rhem, Mark Schlesinger and Rob Valetta [17], Stefano Bistarelli, Fabio Fioravanti, Pamela Peretti and Francesco Santini [18], Jacob S. Hacker, Gregory A. Huber, Austin Nichols, Philipp Rehm, Mark Schlesinger, Rob Valletta and Stuart Craig [19], M. Eric Johnson and Eric Goetz [20], Miles Kahler [21]. Ukrainian researchers are also making successful attempts to study the country's critical infrastructure management issues, in particular, [22]. As a separate area of research on financial and economic security at the micro and macro levels has become issues of risk management, which can be traced in the publications of such scientists as Brian J. Allen and Rachelle Loyear [23], Alison Wakefield [24], Ryan McCormack and Albert Lord [25], Matthias Dehmer, Silja Meyer-Nieberg, Goran Mihelcic, Stefan Pickl and Martin Zsifkovits [26].

If from the functional plane – the study of financial and economic security subsystems and their condition – to move to the species parameter, the field of view of scientists in the field of security often includes such critical infrastructure as financial institutions, especially banks. It should be noted that this study also began with the scientific interest of its authors to the problems of economic security of various types of financial institutions [27, 28].

Since the objects of critical infrastructure at the state level include strategically important enterprises of different types of economic activity, banking institutions, etc., the definition of financial and economic security, proposed for many years by researchers for different types of economic entities, may be more or less suitable for use for these specific objects. However, it is worth considering the question of clarifying the concept of financial and economic security through the prism of the definitions proposed for business structures, because for the latter the purpose of the economic security system is to form such an internal environment that can provide its competitive advantages with minimal risk and threats to the loss of economic viability in the long run, while the priority of critical infrastructure objects financial and economic security should be to ensure a high level of protection of resources of such structures and support their effective use to achieve the necessary social effect, ensuring the viability and activity of related industries, finance, social spheres, national defense. Making a profit for critical infrastructure should be a secondary task compared to meeting the needs of stakeholders related to their work.

Even those definitions of financial and economic security that we have found in the scientific sources and which can be used to assist in performing the tasks of this study have a large number of shortcomings. That's why we can not call them universal and give them the status of classical definitions and use them to explain the core of financial and economic security of critical infrastructure. For example, interpretations that explain the core of financial and economic security through a category of "state" are criticized because of the fact that in a changing, risky environment, maintaining a certain state of resources or economic opportunities or characteristics requires a number of dynamic action by company management, which makes such a state volatile, changeable, but not "fix" it. Therefore, some parameters should be added to the concept of financial and economic security in a "state", plane which characterize it at that particular moment in time, when the level of financial and economic security is assessed and experts conclude whether in the state of security or, conversely – danger – the business entity is.

We believe that the financial and economic security of critical infrastructure should be considered a set of financial and economic conditions that are created by making management decisions on the formation of economic resources of the entity, areas of their preservation and use within the main, financial, investment and other activities in order to ensure the continuity of the functioning of the structure and the timely implementation of the tasks of its existence in the architecture of the critical infrastructure of the state.

Given that the objects of critical infrastructure differ in the so-called level of criticality – that is, fall under a certain gradation of importance for the state of national security and sustainability of economic development, we propose to highlight certain characteristics of their financial and economic security table 1.

In the course of any enterprise, institution or organization, financial and economic security should be recognized as an object of management. Only in this case, the management of the entity will cease to consider financial and economic security as a result of external and internal environmental factors that partly determine and partly affect the results of its activities, and realize the need and ability to influence the results of economic and financial risks. Bobro offers to put the task of the object's security protection on the government forces or on the top-management of the business structure – depending on the level of "criticality" of this object for the national security [29]. Thus, it is advisable to distinguish two levels of management of financial and economic security of critical infrastructure objects – public, state, governmental (external) and internal.

Table 1. Features of the concept of financial and economic security of critical infrastructure

Level of importance	Core of financial and economic security essence
National level	Financial and economic security as a mandatory condition for stable functioning of the economic system of the country and for keeping of the proper state of national security
Industry level	Financial and economic security as a situation of supporting the proper level of financial and economic capabilities and the level of functionality of the object of critical infrastructure, giving the possibility to contribute to the stabilization and development of the industry through the mechanisms of efficient use of available resources in the face of external and internal risks, threats and challenges to this process performance
Regional level	Financial and economic security as a set of conditions for the continuous functioning of the economic system for achiving the goal of receiving economic or social effect in the surroundings, full of risks, for the purpose to ensure the sustainability of economic entities relations in the region
Local economic or social system level	Financial and economic security as an availability of the necessary amounts of financial and economic resources and opportunities for their effective use, taking into account the existing risks and threats and the possible consequences of their manifestation to maintain the viability and ensure the proper conditions for the future development of the system

Internal level – a set of management decisions aimed at maintaining financial and economic security through planning, organization, coordination and control of actions to form the resource provision of the business entity and prevent threats to the continuity of this process. External (state) level – a set of administrative, control and supervisory measures to maintain the state of financial and economic security, formed on the object of critical infrastructure or to initiate actions to improve it in the short and long term through advisory, support or enforcement actions in the direction of operation of the critical infrastructure object.

2 Risks caused by quarantine restrictions for the state of financial and economic security of critical infrastructure objects

The main risk posed by quarantine restrictions for all businesses without exception is the forced cessation of their operation, for example, during a period of lockdown announced in the country. At the same time, we should not forget that the concept of risk is dualistic, ie, in addition to a purely negative result that may have a risk to the enterprise, organization and institution, risk is also an opportunity for qualitative change, transformation, modernization of their organizational and business processes. Thus, the transformation of risks in the system of financial and economic security of economic entities into indicators and catalysts for qualitative change – is an important modern task of security professionals.

Ivanenko offers several stages of risk management that can be recommended for use in the context of financial and economic security management of critical infrastructure objects. These are such stages as: "identification of risks as a process of their recognition and description; risk analysis, which involves understanding the nature of risk and determining its level; risk assessment, which involves comparing the results of risk analysis with the criteria for determining whether the risk is acceptable or not acceptable" [30]. At the stage of this scientific study, we will try to identify the risks caused by quarantine restrictions, and currently have a negative impact, or in the future may cause it for the state of financial and economic security of critical infrastructure objects:

- diseases of employees (lack of replacement, downtime of equipment, shutdown of business processes, the need to involve outsiders to ensure the continuity of the work process);
- transport and logistics problems (supply disruptions, delayed delivery of necessary materials, personnel delays to the workplace, the need for additional costs for the organization of transportation of personnel to the place of work);
- non-fulfillment of contractual obligations (the need to cease operations during the lockdown causes a breach of contract by companies that must comply with government decisions and stop work; for critical infrastructure objects this may mean an increase in receivables and payables, breaks in chains supply, loss of customers and partners, the need to find financial and economic alternatives to maintain their own economic viability);
- reduction of the number of orders, refusal of previous agreements on the purchase of products or services due to their loss of relevance during the quarantine period;
- difficulty or inability to work with foreign partners and counterparties due to quarantine restrictions by both (or more) countries whose entities are involved in the process of such cooperation.

New risks, threats and challenges for critical infrastructure objects activities update the search for new approaches to managing their financial and economic security.

3 Features of financial and economic security management of critical infrastructure objects during quarantine restrictions

As already noted, the biggest risk of quarantine restrictions for critical infrastructure objects and for the state as

https://doi.org/10.1051/shsconf/202110702002

a whole is the termination of their activities, not due to the decision of the authorities, given the strategic importance of such facilities, but the impossibility of its continuation due to downtime of partners, contractors, etc. Given this, an important vector of financial and economic security management in such conditions should be strategic management, as to determine the potential goals and benchmarks of economic activity in force majeure, is a difficult but necessary management task. Management also needs close attention to the organization of the personnel security system, as most of the risks associated with the epidemiological situation or the source of its origin are connected with a person who can get sick, and this fact will affect the level of intellectual and, ultimately, economic security of the business entity, or the impact of other possible threats, intensified by quarantine and lockdown, will reduce staff efficiency (e.g., lower family budget incomes, higher health care costs, the need to spend time caring for children and elderly relatives, etc.).

3.1 Strategic management of financial and economic security of critical infrastructure objects during quarantine restrictions

Traditional approaches to strategic management of financial and economic security of economic entities are found in numerous publications, for example, [31, 32]. Deeply respecting the opinions of reputable scientists, however, we believe that the paradigms of strategic management in general and financial and economic security in particular in today's economy, given the trends of Industry 4.0 and the epidemiological situation in the world, require revision and modernization. In our opinion, strategic management should be organized according to the TARGET model figure 1.

We will provide some explanations on the essence of the proposed model. Thus, in our opinion, it is necessary to reconsider the very essence of the strategic management process in the context of the time parameter. In the current conditions in Ukraine, to anticipate the company's actions even for one year is a strategy! Strategic management must be transferred from the plane of time, when strategy means one thing – a long-term vision of the company's development, into the plane of goals. That is, the strategy is the goal or goals to be achieved, the means and tools to achieve them, and the time interval for which it must be done may change depending on the factors and circumstances that accompany the activities of the enterprise. Thus, the strategic management of financial and economic security of critical infrastructure should be implemented not in terms of time, but in terms of objectives. However, if the factors and risks of the external and internal environment of the enterprise and their impact are sufficiently predictable, then the time parameter can be divided into short-term, long-term, medium-term. Foresight means the importance of trying to predict developments as one of the tasks of strategic management of financial and economic security.

The goals set by the financial and economic security system of the enterprise manager must be agreed with the

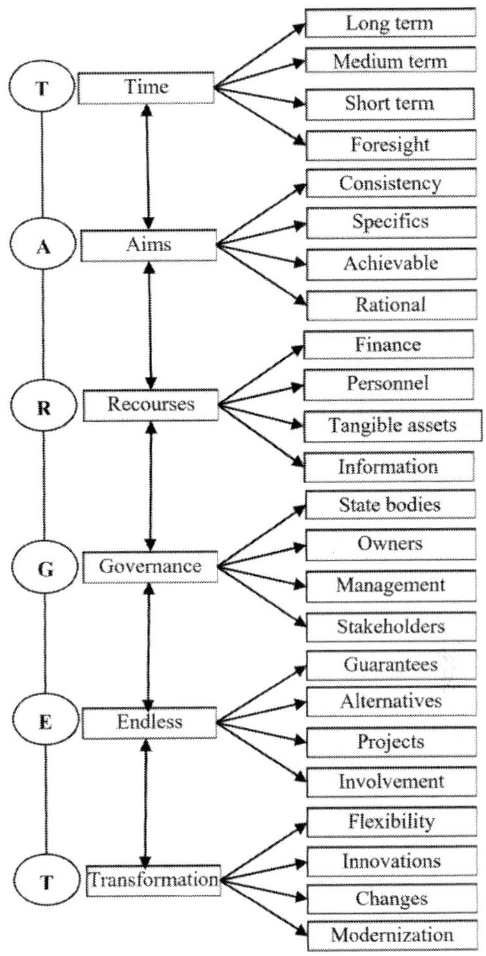

Figure 1. TARGET model for strategic management of financial and economic security of critical infrastructure objects

interests of different categories of stakeholders, and also should be measurable, achievable and rational from the standpoint of available resources (primarily financial, human, material, informational). Governance (management (supervision, leadership, subordination)), as already mentioned, should be carried out at two levels – internal and external, but the proposed model is not a level one, it used a subjective approach – and identifies the main actors influencing the process of strategic management of financial economic security. For critical infrastructure objects, their list includes: state bodies (supervision, control, monitoring), owners (in the case of state-owned enterprises, these two groups coincide – coordination of strategies), enterprise management (planning, organization, implementation and control of strategic tasks), stakeholders (multifaceted influence on the formation and implementation of strategies for financial and economic security to meet their own interests and within available capabilities).

Endless (continuity, infinity) is a specific feature of financial and economic security management of critical infrastructure objects. We believe that the strategies of such entities should be based on the assumption that their activities will continue for at least a few more decades. To do this, they need to obtain guarantees from public authorities, have several alternatives for each existing partner or counterparty in the event of the possibility of termination of the partnership (such as during quarantine), make the transition to project management within individual business processes – that the completion of one project initiates the beginning of a new one – to expand, improve the results, maximize their effect on the company and introduce a policy of involvement of employees (if not all, then representatives of the workforce) in strategic planning and financial and economic security management.

The last parameter – transformation – means the desire and readiness to upgrade, modernize production, management and other processes in accordance with the requirements of the time and new challenges (eg, global pandemic, quarantine restrictions, lockdown). Characteristic features of such management will be flexibility in making management decisions, even at the strategic level, innovation, change management and modernization of business processes as needed.

3.2 Personnel security of critical infrastructure objects during the quarantine period

Personnel of critical infrastructure object are the most vulnerable to the risks posed by a global pandemic [33]. Stress factors can lead to the fact that the actions or inaction of employees, their temporary absence from work, inattention or concern about the situation in the country, will threaten the state of personnel security of the entity, and as a consequence – will have negative effect and for the state of financial and economic security. The main vectors of work with employees, given the possibility of the need to transfer them to part-time work, leave on schedule, savings on bonuses and other forms of financial incentives, etc., should be: constant channels of communication, maximum information openness, flexibility in the organization work process (where possible) and in the distribution of working time (also if possible), involvement of staff in making various decisions on the organization of the business entity in a pandemic, stimulating leadership and supporting the initiative, etc. table 2.

Thus, the rational management of personnel, support and assistance of people who face new challenges of the global pandemic and quarantine restrictions, are under constant stress, worry about their own well-being, and feel additional responsibility for employment in a strategically important for state needs entity, is important for maintaining the continuous operation of critical infrastructure and the state of their financial and economic security.

4 Conclusions

The updated approach to managing the economic security of business processes in various modifications, taking into account the risks of quarantine restrictions, will be used in all sectors of the national economy, will modernize the management systems of various economic activities, and thus have a positive effect on strategic development of financial system, national economy and achieving a high level of national security and financial and political sovereignty.

The practical importance of the result lies in the next plane. It is expected that the proposed in the study TARGET model for strategic management of financial and economic security of critical infrastructure objects will be the basis for developing strategic documents for stabilizing and further developing and maintaining high functionality of critical infrastructure of Ukraine. In particular, testing the objectives of critical infrastructure development strategies in terms of consistency, specificity, achievable and rationality will allow to adjust them in a timely manner in accordance with existing challenges and time requirements. Compliance of strategic guidelines with the parameters of flexibility, innovation, readiness for change and modernization at the application level will allow to make quickly and effectively restructure of the strategies of critical infrastructure, to change their direction in accordance with new opportunities and needs.

The implementation of the proposals developed within this study at the applied level will be impossible without supply this process with a proper staff. In fact, the personnel component is important for achieving and maintaining the level of financial and economic security in both tactical and strategic areas. Of practical importance is the proposed approach to the formation and implementation of personnel policy for critical infrastructure objects, which, in contrast to existing ones, contains such modern components as staff training; employee incentives; overcoming resistance to change; staff involvement in solving various issues; staff discipline; leadership, partnership and teamwork; time management and self-management; staff safety, staff digital literacy. The presence of personnel policy at the enterprise in the form of a document with the proposed structure will allow top-management to organize personnel work effectively and efficiently and to increase significantly the level of personnel security of critical infrastructure.

The need to update personnel policy appeared due to the risks to staff and to the state of enterprises financial and economic security, which were evident during the epidemic, quarantine measures and lockdown. In particular, it is the inability to organize and work remotely, low level of digital technology using skills, poor communication and other social skills, dishonest performance of duties, failure to meet deadlines for work and projects, and, at the same time, overload, excessive time at the computer, the need to acquire new skills, emotional and professional burnout, declining income, constant stress, etc.

Making timely management decisions will minimize the losses of critical infrastructure objects, prevent their financial instability, reduce the risk of bankruptcy and liquidation in the strategic perspective.

Prospects for further research lie in the plane of possibility of developing methods for assessing the level of

Table 2. Features of the concept of financial and economic security of critical infrastructure

Directions (sections) of personnel policy	Description (details)
Studing	Encouraging staff to take online courses, payment or partial payment for the required higher education, encouragement of various forms of distance learning, providing opportunities for training during the working day (in case of flexible schedule), organization of field training, inviting lecturers, mentors, organizing with them online conferences
Promotion	Performance bonuses, various forms of financial assistance, payment for babysitting services, tutors for children, bonuses for saving and rational use of enterprise resources, compensation for taxi services or fuel for own car, provided if it is used for work purposes
Overcoming resistance to change	Explanatory work on the need for certain changes in production processes or work organization, organization of psychological consultations, involvement of employees in the process of development and implementation of changes, demonstration of intermediate results from the introduction of changes, punishment for active and aggressive resistance to change
Involvement	Striving to achieve the maximum level of employee involvement in all business processes that fall within their field of responsibility
Discipline and organization	Control of discipline, quality of performance of the set tasks, efficiency of labor activity, level of the organization of working process (in case of use of the form of remote work in the conditions of quarantine)
Leadership, partnership and teamwork	Stimulating the manifestation of leadership qualities, periodic rotation of personnel, organization of permanent and temporary teams for the implementation of projects and current tasks of the enterprise
Time management and self-management	Explaining the need and methods of combating chronophages, incentives for efficient and rational use of time, its savings when participating in various business processes
Security	Establishing the priority of occupational safety, life and health of the employee over other production, financial, economic goals of the enterprise; popularization of safety norms and rules, penalties for their violation, introduction of a social package and health insurance
Digital literacy	Assistance to work with digital technologies, gadgets, assistance in organizing the workflow online, organizing and conducting group chats, consulting support for staff working remotely

financial and economic security of critical infrastructure objects and creating strategies for their further development in terms of extending quarantine restrictions or new lockdowns.

References

[1] *Proekt zakonu Ukrayiny Pro krytychnu infrastrukturu ta yiyi zakhyst (Draft Law of Ukraine on Critical Infrastructure and its Protection)* (2019), http://search.ligazakon.ua/l_doc2.nsf/link1/JH7YW00A.html

[2] National Bank of Ukraine, *Vyznacheno kryteriyi vidnesennya obyektiv bankivskoyi systemy do krytychnoyi infrastruktury (The criteria for assigning the objects of the banking system to the critical infrastructure are defined)* (2020), https://bit.ly/3doFp5A

[3] Verkhovna Rada of Ukraine, *Poryadok vnesennya obyektiv krytychnoyi informatsiynoyi infrastruktury do derzhavnoho reyestru obyektiv krytychnoyi informatsiynoyi infrastruktury, yoho formuvannya ta zabezpechennya funktsionuvannya (The procedure for entering critical information infrastructure facilities in the state register of critical information infrastructure facilities, its formation and operation)* (2020), https://cutt.ly/2vf0xTE

[4] C. Alcaraz, S. Zeadally, International Journal of Critical Infrastructure Protection **8**, 53 (2015)

[5] R. Mosadeghi, R. Richards, R. Tomlinson, *Critical Infrastructure Protection and Uncertainty Analysis* (2017), chap. Chapter 8, pp. 193–223, https://www.worldscientific.com/doi/abs/10.1142/9789813207950_0008

[6] G. Giannopoulos, R. Filippini, M. Schimmer, *Risk Assessment Methodologies for Critical Infrastructure Protection. Part I: A state of the Art*, JRC Technical Notes (Publications Office of the European Union, Luxembourg, 2012), ISBN 978-92-79-23839-0, https://bit.ly/32jzHvu

[7] D. Rehak, P. Senovsky, S. Slivkova, Systems **6**, 21 (2018)

[8] G. Amélie, B. Aurélia, L. Emmanuel, E. Mohamed, D. Gilles, *The Challenge of Critical Infrastructure Dependency Modelling and Simulation for Emergency Management and Decision Making by the Civil Security Authorities*, in *Critical Information Infrastructures Security*, edited by E. Rome, M. Theocharidou, S. Wolthusen (Springer International Publishing, Cham, 2016), pp. 255–258, ISBN 978-3-319-33331-1

[9] C. Pursiainen, International Journal of Disaster Risk Reduction **27**, 632 (2018)

[10] A. Ianioglo, T. Polajeva, Journal of Business Economics and Management **18**, 1228 (2017)

[11] L. Osberg, A. Sharpe, *Economic Security in Nova Scotia* (2008), http://csls.ca/reports/csls2008-05.pdf

[12] V. Ioan-Franc, M.A. Diamescu, Revista Romana de Economie **31** (2010)

[13] K. Schimmel, S. Liu, J. Nicholls, N. Nechval, J. Forrest, Advances in Systems Science and Applications **17**, 14 (2017)

[14] E. Kopp, L. Kaffenberger, C. Wilson, *Cyber risk, market failures, and financial stability*, https://www.imf.org/-/media/Files/Publications/WP/2017/wp17185.ashx

[15] M. Kośny, M. Piotrowska, *Economic security of households: Assessment based on scenario analysis* (2019), http://www.ecineq.org/ecineq_bari13/FILESxBari13/CR2/p183.pdf

[16] J.S. Hacker, G.A. Huber, A. Nichols, P. Rehm, M. Schlesinger, R. Valletta, S. Craig, Review of Income and Wealth **60**, S5 (2014)

[17] J.S. Hacker, G.A. Huber, P. Rhem, M. Schlesinger, R. Valetta, *Economic Security at Risk: Findings from the Economic Security Index* (The Rockefeller Foundation, New York, NY, 2010), http://economicsecurityindex.org/upload/media/Economic_Security_Index_Full_Report.pdf

[18] S. Bistarelli, F. Fioravanti, P. Peretti, F. Santini, Journal of Experimental & Theoretical Artificial Intelligence **24**, 161 (2012)

[19] J.S. Hacker, G.A. Huber, A. Nichols, P. Rehm, M. Schlesinger, R. Valletta, S. Craig, Review of Income and Wealth **60**, S5 (2014)

[20] M.E. Johnson, E. Goetz, IEEE Security Privacy **5**, 16 (2007)

[21] M. Kahler, The Pacific Review **17**, 485 (2004)

[22] O. Sukhodolia, S. Kondratov, D. Bobro, V. Horbulin, *Developing The Critical Infrastructure Protection System in Ukraine* (NISS, Kyiv, 2017)

[23] R.L. B. J. Allen, *Enterprise Security Risk Management: Concepts and Applications* (Rothstein Publishing, Brookfield, Connecticut, 2017)

[24] A. Wakefield, *Corporate Security and Enterprise Risk Management*, in *Corporate Security in the 21st Century: Theory and Practice in International Perspective*, edited by K. Walby, R.K. Lippert (Palgrave Macmillan UK, London, 2014), pp. 235–253, ISBN 978-1-137-34607-0, https://doi.org/10.1057/9781137346070_13

[25] R. McCormack, A. Lord, *Risky business: Implementing risk management in national security decision-making* (2019)

[26] M. Dehmer, S. Meyer-Nieberg, G. Mihelcic, S. Pickl, M. Zsifkovits, EURO Journal on Decision Processes **3**, 305 (2015)

[27] N.V. Zachosova, Modern econ. pp. 14–25 (2011)

[28] N.V. Zachosova, Econ. Finan. Law pp. 7–13 (2010)

[29] D.B. Bobro, Strat. priorit. pp. 77–85 (2016)

[30] O.I. Ivanenko, Bul. of KNTU pp. 9–22 (2020)

[31] O.V. Cherevko, Efektyvna ekonomika (2014)

[32] I.O. Dotsenko, O.P. Melnychuk, SR pp. 79–84 (2018)

[33] M. Velykodna, Psychodynamic Practice **27**, 10 (2021)

Econophysics of cryptocurrency crashes: an overview

Andrii Bielinskyi[1,*], *Oleksandr* Serdyuk[2,**], *Serhiy* Semerikov[1,3,4,5,***], and *Vladimir* Soloviev[1,2,****]

[1]Kryvyi Rih State Pedagogical University, 54 Gagarin Ave., Kryvyi Rih, 50086, Ukraine
[2]The Bohdan Khmelnytsky National University of Cherkasy, 81 Shevchenko Blvd., Cherkasy, 18031, Ukraine
[3]Kryvyi Rih National University, 11 Vitalii Matusevych Str., Kryvyi Rih, 50027, Ukraine
[4]Institute of Information Technologies and Learning Tools of the NAES of Ukraine, 9 M. Berlynskoho Str., Kyiv, 04060, Ukraine
[5]University of Educational Management, 52-A Sichovykh Striltsiv Str., Kyiv, 04053, Ukraine

Abstract. Cryptocurrencies refer to a type of digital asset that uses distributed ledger, or blockchain technology to enable a secure transaction. Like other financial assets, they show signs of complex systems built from a large number of nonlinearly interacting constituents, which exhibits collective behavior and, due to an exchange of energy or information with the environment, can easily modify its internal structure and patterns of activity. We review the econophysics analysis methods and models adopted in or invented for financial time series and their subtle properties, which are applicable to time series in other disciplines. Quantitative measures of complexity have been proposed, classified, and adapted to the cryptocurrency market. Their behavior in the face of critical events and known cryptocurrency market crashes has been analyzed. It has been shown that most of these measures behave characteristically in the periods preceding the critical event. Therefore, it is possible to build indicators-precursors of crisis phenomena in the cryptocurrency market.

1 Introduction

The instability of global financial systems concerning normal and natural disturbances of the modern market and the presence of poorly foreseeable financial crashes indicate, first of all, the crisis of the methodology of modeling, forecasting, and interpretation of modern socio-economic realities. The doctrine of the unity of the scientific method states that for the study of events in socio-economic systems, the same methods and criteria as those used in the study of natural phenomena are applicable. Rapidly evolving coronavirus pandemic brings a devastating effect on the entire world and its economy as a whole [1–7]. Further instability related to COVID-19 will negatively affect not only on companies and financial markets, but also on traders and investors that have been interested in saving their investment, minimizing risks, and making decisions such as how to manage their resources, how much to consume and save, when to buy or sell stocks, etc., and these decisions depend on the expectation of when to expect next critical change [8–21]. Despite the complexity of the problem, the results of recent studies indicate that significant success has been achieved within the framework of interdisciplinary approaches, and the theory of self-organization – synergetics [22, 23]. The modern paradigm of synergetics is a complex paradigm associated with the possibility of direct numerical simulation of the processes of complex systems evolution, most of which have a network structure, or one way or another can be reduced to the network. The theory of complex networks studies the characteristics of networks, taking into account not only their topology but also statistical properties, the distribution of weights of individual nodes and edges, the effects of dissemination of information, robustness, etc. [1–4, 24–26].

Complex systems consist of a plurality of interacting agents possessing the ability to generate new qualities at the level of macroscopic collective behavior, the manifestation of which is the spontaneous formation of noticeable temporal, spatial, or functional structures [27–32]. As simulation processes, the application of quantitative methods involves measurement procedures, where importance is given to complexity measures. Prigogine notes that the concepts of simplicity and complexity are relativized in the pluralism of the descriptions of languages, which also determines the plurality of approaches to the quantitative description of the complexity phenomenon [5].

Financial markets have been attracting the attention of many scientists like engineers, mathematicians, physicists, and others for the last two decades. Such vast interest transformed into a branch of statistical mechanics — econophysics [30–34]. Physics, economics, finance, sociology, mathematics, engineering, and computer science are fields which, as a result of cross-fertilization, have created the multi-, cross-, and interdisciplinary areas of science and research such as econophysics and sociophysics, thriving in the last two and a half decades. These mixed research fields use knowledge, methodologies, methods, and tools of physics for modeling, explaining and forecasting

*e-mail: krivogame@gmail.com
**e-mail: serdyuk@ukr.net
***e-mail: semerikov@gmail.com
****e-mail: vnsoloviev2016@gmail.com

economic, social phenomena, and processes. Accordingly, econophysics is an interdisciplinary research field, applying theories and methods originally developed by physicists to solve problems in economics, usually those including uncertainty or stochastic processes, nonlinear dynamics, and evolutionary games.

There are deep relationships (as well as crucial differences) between physics and finance [35] that have inspired generations of physicists as well as economists. In general, physicists apprehend financial markets as complex systems and, as such, they conducted numerous scientific investigations [36].

Though statistical physics cannot get along without quantum-mechanical ideas and notions in its fundamentals, the main sphere of its interest is the macroscopic description of systems with a large number of particles, the dynamic behavior of which can't be brought to microscopic dynamical equations of quantum mechanics figured out for separate particles without the use of respective statistical postulates [37]. During last years an increasing flow of works was observed, in which detailed models of market process participants interactions and quantum-mechanical analogies, notions, and terminology based on methods of describing socio-economic systems are drawn to explain both particular peculiarities of modern market dynamics and economic functioning in whole [38–40]. In papers [41, 42], Soloviev and Saptsin have suggested a new paradigm of complex systems modeling based on the ideas of quantum as well as relativistic mechanics. It has been revealed that the use of quantum-mechanical analogies (such as the uncertainty principle, the notion of the operator, and quantum measurement interpretation) can be applied for describing socio-economic processes.

In this review, we will continue to study Prigogine's manifestations of the system complexity, using the current methods of quantitative analysis to determine the appropriate measures of complexity. The proposed measures of complexity, depending on the methodology and construction methods, can be divided into the following classes:

(1) informational,

(2) (multi-)fractal,

(3) chaos-dynamic,

(4) recurrent,

(5) irreversible,

(6) based on complex networks,

(7) quantum.

Econophysics, based on a rich arsenal of research on critical phenomena [43], very successfully copes with the description of similar events in economics and finance. These are crises and crashes that are constantly shaking the world economy. The introduced measures of complexity should, to one degree or another, respond to such phenomena.

The key idea here is the hypothesis that the complexity of the system before the crashes and the actual periods of crashes must change. This should signal the corresponding degree of complexity if they are able to quantify certain patterns of a complex system. A significant advantage of the introduced measures is their dynamism, that is, the ability to monitor the change in time of the chosen measure and compare it with the corresponding dynamics of the output time series. This allowed us to compare the critical changes in the dynamics of the system, which is described by the time series, with the characteristic changes of concrete measures of complexity. It turned out that quantitative measures of complexity respond to critical changes in the dynamics of a complex system, which allows them to be used in the diagnostic process and prediction of future changes.

The cryptocurrency market is a complex, self-organized system, which in most cases can be considered either as a complex network of market agents or as an integrated output signal of this network – a time series, for example, prices of individual cryptocurrency. The research on cryptocurrency price fluctuations being carried out internationally is complicated due to the interplay of many factors – including market supply and demand, the US dollar exchange rate, stock market state, the influence of crime, shadow market, and fiat money regulator pressure that introduces a high level of noise into the cryptocurrency data. Moreover, in the cryptocurrency market, to some extent, blockchain technology is tested in general. Hence, the cryptocurrency prices exhibit such complex volatility characteristics as nonlinearity and uncertainty, which are difficult to forecast, and any obtained results are uncertain. Therefore, cryptocurrency price prediction remains a huge challenge [44–58].

As can be seen, markets have seen significant numbers of investors selling off and rebalancing their portfolios with less risky assets. That has been leading to large losses and high volatility, typical of crisis periods. The economy key for preventing such activity may lie in cryptocurrency and constructing effective indicators of possible critical states that will help investors and traders fill in safety. Bitcoin, which is associated with the whole crypto market, has such properties as detachment and independence from the standard financial market and the proclaimed properties that should make it serve as the digital gold [59]. As was shown by Kristoufek [60], Bitcoin promises to be a safe-haven asset with its low correlation with gold, S&P 500, Dow Jones Industrial Average, and other authoritative stock indices even in the extreme events. But authors please not overestimate the cryptocurrency since according to their calculations and, obviously, the current structure of the system, gold remains more significant. But for ten years, this token has been discussed by many people, it has experienced a lot in such a short period, many people believe in it, and it has managed to form a fairly complex and self-organized system. The integrated actions from real-world merge in such dynamics and relevant information that is encoded in Bitcoin's time series can be extracted [61–63]. In the context of volatile financial markets, it is important to select such measures of complexity that will be able to notify us of upcoming abnormal events in the form of crises at an early stage.

In this review we:

- present such measures;
- study critical and crash phenomena that have taken place in the cryptocurrency market;
- try to understand whether crashes and critical events could be identified and predicted by such informative indicators or not.

This review is dedicated to the construction of such indicators based on the theory of complexity. According to our goals and actions, the paper is structured as follows. In Section 2, we present our classification of Bitcoin's crises for the period from July 16, 2010 to January 21, 2021. In Section 3, we describe the information measures of complexity. In Section 4, we describe the multifractal analysis methodology and its results for the crypto market. Section 5 defines what is chaos-based measures of complexity. In section 6, we deal with the recurrence quantification analysis of critical and crisis phenomena in the cryptocurrency market. Irreversible measure based on permutation patterns is defined in Section 7. Section 8 presents the theory and empirical results on network and multiplex measures of complexity and their robustness for digital currencies. Section 9 defines quantum complexity measures, the features of their manifestation on the crypto market are discussed. Section 10 contains conclusions and some recommendations for further research.

2 Data and classification

Bitcoin, being the most capitalized cryptocurrency, as a rule, sets and determines the main trends of the crypto market as a whole. Therefore, except for the part of the work where the study of collective processes in the market is carried out, we will use the time series of Bitcoin [64]. From figure 1 it can be seen that at the beginning of its existence, Bitcoin's dynamic was determined mainly by the processes of the formation of the market as a whole and characterized by high volatility, which, however, was not associated with critical phenomena. Bariviera et al. [65] find that the Hurt exponent changes significantly during the first years of existence of Bitcoin, and now it is less unstable than before. Moreover, they detect more evidence of information since 2014 [66].

Being historically proven, popular, and widely used cryptocurrency for the whole existence of cryptocurrencies in general, Bitcoin began to produce a lot of news and speculation, which began to determine its future life. Similar discussions began to lead to different kinds of crashes, critical events, and bubbles, which professional investors and inexperienced users began to fear. Thus, we advanced into action and set the tasks:

- classification of such critical events and crashes;
- construction of such indicators that will predict crashes, critical events in order to allow investors and ordinary users to work in this market.

Accordingly, during this period in the Bitcoin market, many crashes and critical events shook it. At the moment,

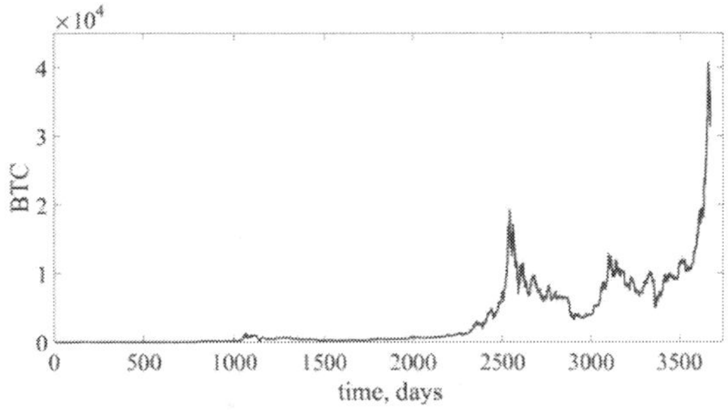

Figure 1: Bitcoin price development from July 16, 2010 to January 21, 2021

there are various research works on what crashes are and how to classify such risk events in the market. The definition of these events still has been debatable. Nevertheless, the proposals of most authors have common elements that allow us to arrive at a consensus. Generally, the market crash is a sudden drastic decline in the prices of financial assets in a specific market [67]. Additionally, the applied model for a specific market takes an important place in the definition of "drastic decline". For instance, Wang et al. [67] take into account events with a minimum one-day decrease of 5% in the stock returns. These authors [26] identify financial crashes as a decrease of 25% or less of multi-year financial returns. Lleo and Ziembda [68] define a crash as a specific event of a day, which decreasing closing price exceeds a fall of 10% between the highest and the lowest value of the stock index in a year. Hong and Stein [69] postulate that the market crash is an unexpected event in which appearance was not accompanied by any financial news. Moreover, the price change during this event is rather negative. Also, it is worth mentioning the study of Min Shu and Wei Zhu [70] where their classification of crashes included almost 50 crashes. It remains a little unclear which factors influence their choice of such an enormous amount of crashes in such a short period. Researchers emphasize these drops as such, with a fall of more than 15% and a duration of fewer than three weeks. Nevertheless, regarding this classification, we are going to emphasize the most relevant, where the complexity of the index started to decrease and whose initial deviation from regular behavior was noticeable in advance. Nowadays some people proclaim Bitcoin as a "digital gold". Gold as a material has served for jewelry and art as well as electronic or medical components. Limited supply and current acceptance of Bitcoin as a "digital gold" may erect it to the same level as gold. While some people back up Bitcoin's advantage, demonstrating its similarities with those of gold and silver [71], others argue that it is the new digital coin [72] due to its high volatility and unclear future development. However, researchers find its potential benefits during extreme market periods and provide a set of stylized factors that claim to be successful long-short strategies that generate sizable and statistically significant excess returns [73]. Despite volatile swings and many critics, Bitcoin has emerged and attracted much more confidence. These studies [74, 75] consider that measures of financial and macroeconomic activity can be drivers of Bitcoin returns. Reviewing papers of the researches above, the experience of others and our own [76–85], we have revised our classification of such leaps and falls, relying on Bitcoin time series during the entire period (01.01.2011-21.01.2021) of verifiable fixed daily values of the Bitcoin price (BTC) (https://finance.yahoo.com/cryptocurrencies). We emphasize that

- *crashes* are short and time-localized drops that last approximately two weeks, with the weighty losing of price each day. Their volatility is high. In percentage term, their decline exceeds 30 percent, and normalized returns proceed $\pm 3\sigma$ or near to it;
- *critical events* are those falls that, during their existence, have not had such massive changes in price as crashes.

Relying on these considerations, we emphasize 29 periods on Bitcoin time series, relying on normalized returns and volatility, where returns are calculated as

$$G(t) = \ln x(t + \Delta t) - \ln x(t) \cong [x(t + \Delta t) - x(t)]/x(t) \quad (1)$$

and normalized (standardized) returns as

$$g(t) \cong [G(t) - \langle G \rangle] / \sigma, \quad (2)$$

where $\sigma \equiv \sqrt{\langle G^2 \rangle - \langle G \rangle^2}$ is the standard deviation of G, Δt is time lag (in our case $\Delta t = 1$), and $\langle \dots \rangle$ denotes the average over the time period under study and volatility as

$$V_T(t) = \frac{1}{n} \sum_{t'=t}^{t+n-1} |g(t')|$$

From the mentioned stylized facts on BTC dynamics, it was noticed how considerably it started to change near 2014. To gain a deeper understanding of its existence in the starting period, we divided the BTC time series into two periods: (01.01.2011-31.08.2016) and (01.09.2016-21.01.2021). More detailed information about crises, crashes, and their classification under these definitions is given in table 1 and table 2.

Therefore, according to our classification crisis periods with numbers (1, 2, 4-6, 8, 9, 13, 18, 23-25, 27, 29) are *crashes*, all the rest – *critical events*.

Figure 2 confirms the importance of dividing the BTC time series in order to observe its dynamics in more detail. However, as it can be seen, we could separate time series in much deeper time scales.

In figure 3 output Bitcoin time series for the first and the second periods, their normalized returns $g(t)$, and volatility $V_T(t)$ calculated for the window of size 100 are presented.

From figure 3 we can see that during periods of crashes and critical events normalized returns g increases considerably in some cases beyond the limits $\pm 3\sigma$. This indicates deviation from the normal law of distribution, the presence of the "heavy tails" in the distribution g, which are characteristics of abnormal phenomena in the market. At the same time volatility also grows. Such qualities are the foundation of our classification for crashes, as it has been mentioned already. All the rest events are critical. These characteristics serve as indicators of crashes and critical events as they react only at the moment of the above-mentioned phenomena and do not allow identifying the corresponding abnormal phenomena in advance. In contrast, most of the indicators described below will respond to critical changes and crashes in advance. It enables them to be used as indicators – precursors of such phenomena.

Calculations were carried out within the framework of the algorithm of a rolling (sliding, moving) window. For this purpose, the part of the time series (window), for which there were calculated measures of complexity, was selected, then the window was displaced along with the time series in a predefined value, and the procedure repeated until all the studied series had exhausted. Further, comparing the dynamics of the actual time series and the corresponding measures of complexity, we can judge the characteristic changes in the dynamics of the behavior of complexity with changes in the cryptocurrency. If this or that measure of complexity behaves in a definite way for all periods of crashes, for example, decreases or increases during the pre-crashes or pre-critical period, then it can serve as their indicator or precursor.

Calculations of measures of complexity were carried out both for the entire time series, and for a fragment of the time series localizing some of the emphasized crashes and critical events. In the latter case, fragments of time series of the same length with fixed points of the onset of crashes or critical events were selected and the results of calculations of complexity measures were compared to verify the universality of the indicators. Following some described below procedures such time localization as, example, of length 100 or 200, either won't make any sense, or won't be possible as some of them are sensitive to time localization, or require a longer length of the time series as it is required by the procedure for better accuracy of further calculations

3 Informational measures of complexity

Complexity is a multifaceted concept, related to the degree of organization of systems. Patterns of complex organization and behavior are identified in all kinds of systems in nature and technology. Essential for the characterization of complexity is its quantification, the introduction of complexity measures, or descriptors [86].

We may speak of the complexity of a structure, meaning the amount of information (number of bits) of the structure; this is the minimum space we need to store enough information about the structure that allows us its reconstruction. We may also speak of the algorithmic complexity of a certain task: this is the minimum time (or other computational resources) needed to carry out this task on a computer. And we may also speak of the communication complexity of tasks involving more than one processor: this is the number of bits that have to be transmitted in solving this task [87–89].

Table 1: List of Bitcoin major crashes and critical events since June 2011 till July 2016

№	Name	Days in correction	Bitcoin High Price, $	Bitcoin Low Price, $	Decline, %	Decline, $
1	07.06.2011-10.06.2011	3	29.60	14.65	50	14.95
2	11.10.2011-18.10.2011	7	4.15	2.27	45	1.88
3	15.01.2012-16.02.2012	32	7.00	4.27	39	2.73
4	15.08.2012-18.08.2012	3	13.50	8.00	40	5.50
5	08.04.2013-15.04.2013	7	230.00	68.36	70	161.64
6	28.04.2013-02.05.2013	4	144.00	98.09	32	45.91
7	19.06.2013-04.07.2013	15	111.29	68.50	38	42.79
8	04.12.2013-07.12.2013	3	1237.66	697.02	44	540.64
9	05.02.2014-21.02.2014	16	904.52	111.55	88	792.97
10	24.03.2014-09.04.2014	17	567.56	384.63	32	182.93
11	09.08.2014-17.08.2014	8	592.06	462.18	22	129.88
12	22.09.2014-04.10.2014	12	436.86	322.86	26	114.00
13	12.01.2015-14.01.2015	2	269.33	164.92	39	104.41
14	27.07.2015-23.08.2015	27	293.70	211.43	28	82.27
15	09.11.2015-11.11.2015	2	380.22	304.71	28	75.51
16	18.06.2016-21.06.2016	3	761.04	590.56	22	170.48
17	29.07.2016-01.08.2016	3	654.74	513.43	24	141.31

Table 2: List of Bitcoin major crashes and critical events since January 2017 till March 2020

№	Name	Days in correction	Bitcoin High Price, $	Bitcoin Low Price, $	Decline, %	Decline, $
18	04.01.2017-11.01.2017	7	1135.41	785.42	30	349.99
19	15.03.2017-18.03.2017	3	1253.43	971.38	23	282.05
20	10.06.2017-15.07.2017	35	2973.44	1914.09	36	1059.35
21	31.08.2017-13.09.2017	13	4921.85	3243.08	34	1678.77
22	08.11.2017-12.11.2017	4	7444.36	5878.13	21	1566.23
23	16.12.2017-30.12.2017	14	19345.49	12531.52	35	6813.97
24	06.01.2018-05.02.2018	30	17172.30	6937.08	60	10235.22
25	04.03.2018-05.04.2018	33	11504.42	6634.86	42	4869.56
26	04.05.2018-27.05.2018	23	9845.90	7118.88	28	2727.02
27	18.11.2018-15.12.2018	27	5615.26	3232.51	42	2382.75
28	12.07.2019-16.07.2019	4	11797.37	9423.44	20	2373.93
29	06.03.2020-16.03.2020	10	9122.55	5014.48	45	4108.07

Historically, the first attempt to quantify complexity was based on Shannon's information theory [90] and Kolmogorov complexity [91].

3.1 Lempel-Ziv complexity

Lempel-Ziv complexity (LZC) is a classical measure that, for ergodic sources, relates the concepts of complexity (in the Kolmogorov-Chaitin sense), and entropy rate [92, 93]. For an ergodic dynamical process, the amount of new information gained per unit of time (entropy rate) can be estimated by measuring the capacity of this source to generate new patterns (LZC). Because of the simplicity of the LZC method, the entropy rate can be estimated from a single discrete sequence of measurements with a low computational cost [94].

In this paper, we show that the LZC measure can be just such a measure of complexity, which is an early precursor of crisis phenomena in the cryptocurrency market [2, 79, 95, 96].

Historically, the first LZC measure system studies for financial time series were conducted by S Da Silva et al. [96–99]. They considered the deviation of LZC from that value for a random time series as a measure of actual market efficiency in absolute [95, 96, 98, 99] or relative [97] terms. Using this approach authors were able to detect decreases in efficiency rates of the major stocks listed on the Sao Paulo Stock Exchange in the aftermath of the 2008 financial crisis [98]. In [100], authors have surveyed the principal applications of algorithmic (Kolmogorov) complexity to the problem of financial price motions and showed the relevance of the algorithmic framework to structure tracking in finance. Some empirical results are also provided to illustrate the power of the proposed estimators to take into account patterns in stock returns. In paper [101] was proposed a generic methodology to esti-

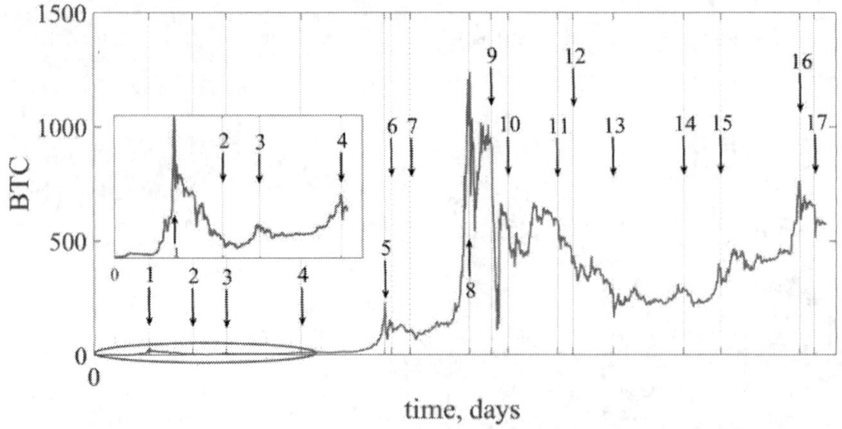

Figure 2: The dynamics of the daily values of the BTC price. The inset shows 1-4 crisis periods presented in table 1.

mate the Kolmogorov complexity of financial returns. Examples are given with simulated data that illustrate the advantages of our algorithmic method: among others, some regularities that cannot be detected with statistical methods can be revealed by compression tools. Applying compression algorithms to daily returns of the Dow Jones Industrial Average (DJIA), the authors concluded on an extremely high Kolmogorov complexity and by doing so, proposed another empirical observation supporting the impossibility to outperform the market. The structural complexity of time series describing returns on New York's and Warsaw's stock exchanges was studied using two estimates of the Shannon entropy rate based on the Lepel-Ziv and Context Tree Weighting algorithms [102]. Such structural complexity of the time series can be used as a measure of the internal (modelless) predictability of the main pricing processes and testing the hypothesis of an efficient market. Somewhat surprisingly, the results of [103], in which the authors computed the LZC from two composite stock indices, the Shanghai stock exchange composite index (SSE) and the DJIA, for both low-frequency (daily) and high-frequency (minute-to-minute) stock index data. The calculation results indicate that that the US market is basically fully random and consistent with the efficient market hypothesis (EMH), irrespective of whether low- or high-frequency stock index data are used. The Chinese market is also largely consistent with the EMH when low-frequency data are used. However, a completely different picture emerges when the high-frequency stock index data are used. H. Cao and Y. Li [104] presents a novel method for measuring the complexity of a time series by unraveling a chaotic attractor modeled on complex networks. The complexity index, which can potentially be exploited for prediction, has a similar meaning to the LZC and is an appropriate measure of a series' complexity. The proposed method is used to research the complexity of the world's major capital markets. The almost absent sensitivity of the LZC to fluctuations in the time series indicates most likely errors in the calculation algorithm during the transformation of the time series. The complexity–entropy causality plane is employed in order to explore disorder and complexity in the space of cryptocurrencies [104]. They are found to exist in distinct planar locations in the representation space, ranging from structured to stochastic-like behavior.

A brief analysis of the problem indicates that so far, the Lempel-Ziv informational measure of the complexity has not been used to study the stability and behavior of the cryptocurrency market in a crisis. In this section, we use the Lempel-Ziv complexity measure to study the cryptocurrency market. Using the example of the most capitalized cryptocurrency – Bitcoin – we demonstrate the ability to identify the dynamics of varying complexity. Particularly relevant is the identification of the characteristic behavior of Bitcoin during the crisis phases of market behavior. By observing the dynamics of the Lempel-Ziv measure, precursors of crisis phenomena can be constructed [105].

3.1.1 The concept of Kolmogorov complexity

Let us begin with the well-known degree of complexity proposed by Kolmogorov [106]. The concept of Kolmogorov complexity (or, as they say, algorithmic entropy) emerged in the 1960s at the intersection of algorithm theory, information theory, and probability theory. A. Kolmogorov's idea was to measure the amount of information contained in individual finite objects (rather than random variables, as in the Shannon theory of information). It turned out to be possible (though only to a limited extent). A. Kolmogorov proposed to measure the amount of information in finite objects using algorithm theory, defining the complexity of an object as the minimum length of the program that generates that object. This definition is the basis of algorithmic information theory as well as algorithmic probability theory: an object is considered random if its complexity is close to maximum.

What is the Kolmogorov complexity and how to measure it? In practice, we often encounter programs that compress files (to save space in the archive). The most

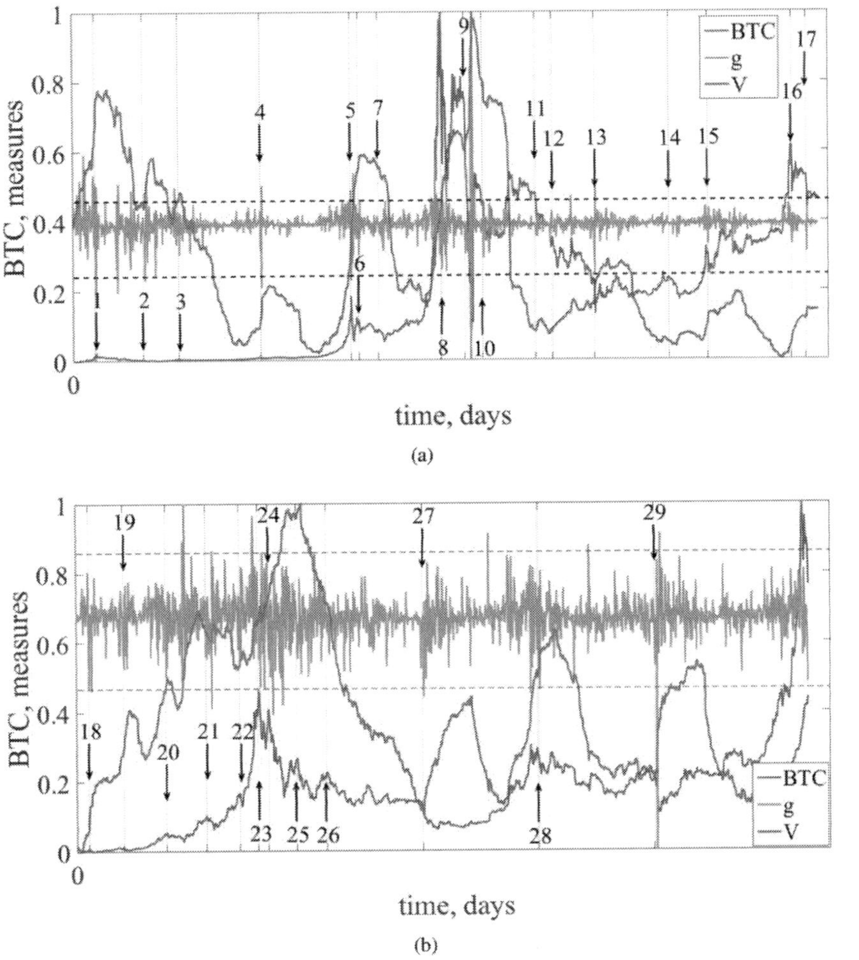

Figure 3: The standardized dynamics, returns $g(t)$, and volatility $V_T(t)$ of BTC/USD daily values for first (a) and second (b) periods. Horizontal dotted lines indicate the $\pm 3\sigma$ borders. The arrows indicate the beginning of one of the crashes or the critical events

common are called zip, gzip, compress, rar, arj, and others. Applying such a program to some file (with text, data, program), we get its compressed version (which is usually shorter than the original file). After that, you can restore the original file using the paired program "decompressor". Therefore, approximately, the Kolmogorov complexity of a file can be described as the length of its compressed version. Thus, a file that has a regular structure and is well compressed has a small Kolmogorov complexity (compared to its length). On the contrary, a badly compressed file has a complexity close to its length.

Suppose we have a fixed method of description (decompressor) D. For this word x, we consider all its descriptions, i.e., all words y for which $D(y)$ it is defined and equal to x. The length of the shortest of them is called the Kolmogorov complexity of the word x in this way of description D:

$$KS_D(x) = \min\{l(y) \mid D(y) = x\}$$

where $l(y)$ denotes the length of the word. The index D emphasizes that the definition depends on the chosen method D. It can be shown that there are optimal methods of description. The better the description method, the shorter it is. Therefore, it is natural to make the following definition: the method D_1 is no worse than the method D_2 if

$$KS_{D_1} \leq KS_{D_2}(x) + c$$

for some c and all x.

Thus, according to Kolmogorov, the complexity of an object (for example, the text is a sequence of characters) is the length of the minimum program that outputs the text, and entropy is the complexity that is divided by the length of the text. Unfortunately, this definition is purely speculative. There is no reliable way of identifying this program uniquely, but there are algorithms that are actually just trying to calculate the Kolmogorov complexity of text and entropy.

A universal (in the sense of applicability to different language systems) measure of the complexity of the finite character sequence was suggested by Lempel and Ziv [100]. As part of their approach, the complexity of a sequence is estimated by the number of steps in the process that gives rise to it.

Acceptable (editorial) operations are: a) character generation (required at least for the synthesis of alphabet elements) and b) copying the "finished" fragment from the prehistory (i.e. from the already synthesized part of the text).

Let be Σ a complete alphabet, S – text (a sequence of characters) composed of elements Σ; $S[i]$ – i^{th} text symbol; $S[i:j]$ – a snippet of text from the ith to j^{th} character inclusive ($i < j$); $N = |S|$ – length of text S. Then the sequence synthesis scheme can be represented as a concatenation

$$H(S) = S[1:i_1]S[i_1+1:i_2]\ldots$$
$$\ldots S[i_{k-1}+1:i_k]\ldots S[i_{m-1}+1:N],$$

where $S[i_{k-1}+1:i_k]$ is the fragment S generated at the kth step, and $m = m_H(S)$ is the number of process steps. Of all the schemes of generation is chosen the minimum number of steps. Thus, the Lempel-Ziv complexity of the sequence S is

$$c_{LZ}(S) = \min_H\{m_H(S)\}.$$

The minimum number of steps is provided by the choice to copy at each step the longest prototype from the prehistory. If you mark by the position number $j(k)$ from which the copying begins in step k the length of the copy fragment

$$l_{j(k)} = i_k - i_{k-1} - 1 = \max_{j\le i_{k-1}}\{l_j : S[i_{k-1}+1:i_{k-1}+l_j] =$$
$$= S[j:j+l_j-1]\}$$

and the kth component of these complex decomposition can be written in the form

$$S[i_{k-1}+1:i_k] = \begin{cases} S[j(k):j(k)+l_{j(k)}-1] & \text{if } j(k)\neq 0, \\ S[i_{k-1}+1] & \text{if } j(k)=0. \end{cases}$$

The case $j(k) = 0$ corresponds to a situation where a symbol is in the position $i_{k-1}+1$ that has not been encountered previously. In doing so, we use a character generation operation.

Complex text analysis can be performed in two regimes – segmentation and fragmentation. The first regime is discussed above. It gives an integrated view of the structure of the sequence as a whole and reduces it to disjoint but interconnected segments (without spaces). The other regime is to search for individual fragments characterized by an abnormally low complexity which means that they characterized by a sufficiently high degree of structure. Such fragments are detected by calculating local complexity within variable-length windows that slide along a sequence. Curves of change of local complexity along a sequence are called complex profiles. A set of profiles for different window sizes reveals the boundaries of anomalous fragments and their relationship.

We will find the LZC complexity for the time series, which is, for example, the daily values of the cryptocurrency price $x(t)$. To investigate the dynamics of LZC and compare it with cryptocurrency prices, we will find this measure of complexity for a fixed length (window) contract. To do this, we calculate the logarithmic returns accordingly to equation (1) and turn them into a sequence of bits.

You can specify the number of states that are differentiated (calculus system). Yes, for two different states we have 0, 1, for three – 0, 1, 2, etc. In the case of three states, unlike the binary coding system, a certain threshold σ is set and the states g are coded as follows [96, 98, 99]:

$$g = \begin{cases} 0 & \text{if } g < -\sigma, \\ 1 & \text{if } -b \le g \le b, \\ 2 & \text{if } g > b. \end{cases}$$

The algorithm performs two operations: (1) adds a new bit to an already existing sequence; (2) copies the already formed sequence. Algorithmic complexity is the number of such operations required to form a given sequence.

For a random sequence of lengths n, the algorithmic complexity is calculated by expression $LZC_r = n/\log n$. Then, the relative algorithmic complexity is the ratio of the obtained complexity to the complexity of the random sequence $LZC = LZC/LZC_r$.

Obviously, the classical indicators of algorithmic complexity are unacceptable and lead to erroneous conclusions. To overcome such difficulties, multiscale methods are used.

The idea of this group of methods includes two consecutive procedures: 1) coarse-graining ("granulation") of the initial time series – the averaging of data on non-intersecting segments, the size of which (the window of averaging) increased by one when switching to the next largest scale; 2) computing at each of the scales a definite (still mono scale) complexity indicator. The process of "rough splitting" consists in the averaging of series sequences in a series of non-intersecting windows, and the size of which – increases in the transition from scale to scale [107]. Each element of the "granular" time series follows the expression:

$$y_j^{\tau} = \frac{1}{\tau}\sum_{i=(j-1)\tau+1}^{j\tau} g(i), \quad \text{for } 1 \le j \le N/\tau, \tag{3}$$

with corresponding scale factor τ. The length of each "granular" row depends on the length of the window and is even N/τ. For a scale of 1, the "granular" series is exactly identical to the original one.

The coarse graining procedure for scales 2 and 3 is shown in figure 4.

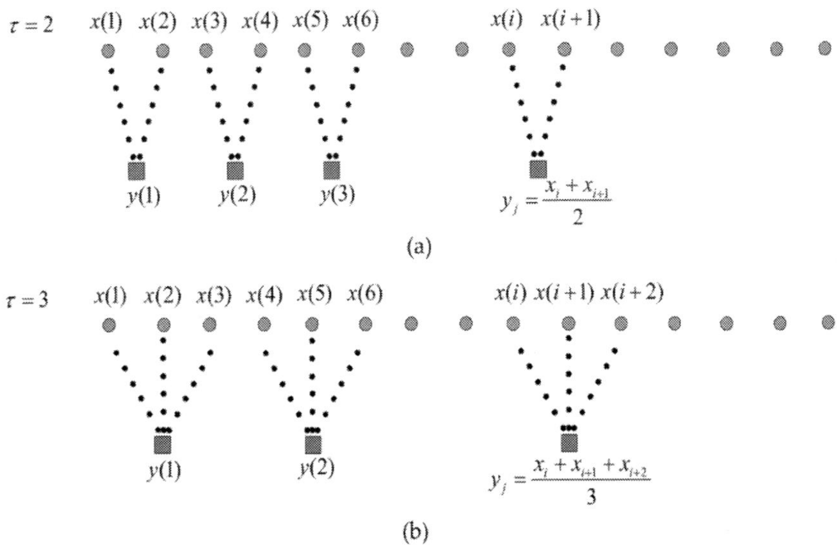

Figure 4: Coarse-graining procedure diagram: (a) scale factor $\tau = 2$; (b) scale factor $\tau = 3$

To find the LZC measure of the time series, the rolling time windows were considered; the index for every window was calculated, and then the average was obtained.

Obviously, the crisis in the cryptocurrency market responds to noticeable fluctuations in standardized returns. Therefore, it is logical to choose σ as the value for the threshold value b.

Figure 5 shows the dependence of the LZC on the scale. The absence of LZC fluctuations at scales exceeding 40 allows us to confine ourselves to this magnitude of the scale when calculating the multiscale measure.

Calculations of measures of complexity were carried for the two periods of BTC. Figure 6 presents the results of calculations of mono-(LZC_{m1}) and multi-(LZC_{m40}) scaling LZC measures. The calculations were performed for a rolling window of 100 days and an increment of 1 day.

The data in figure 6 indicate that the LZC measure is noticeably reduced both in the case of mono-scale (m_1) and averaged over the scales from 1 to 40 (m_{40}) for all 29 crashes and critical events in the immediate vicinity of the crisis point.

As the results of calculations showed, the choice of the size of a rolling window is important: in the case of large windows, points of crises of different times can fall into the window, distorting the influence of each of the crises. When choosing small windows, the results fluctuate greatly, which makes it difficult to determine the actual point of the crisis. The used window length of 100 days turned out to be optimal for the separation of crises and fixing the LZC measure as an indicator.

Since the LZC measure begins to decrease even before the actual crisis point, it can be called an indicator-precursor of crisis phenomena in the cryptocurrency market.

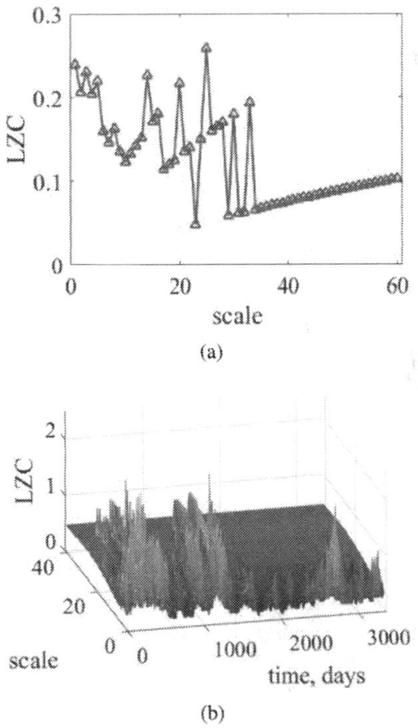

Figure 5: Scale-dependent LZC (a) and its version with the rolling window approach (b)

3.2 Entropy as a measure of complexity

Nowadays, the most important quantity that allows us to parameterize complexity in deterministic or random processes is entropy. Originally, it was introduced by R. Clau-

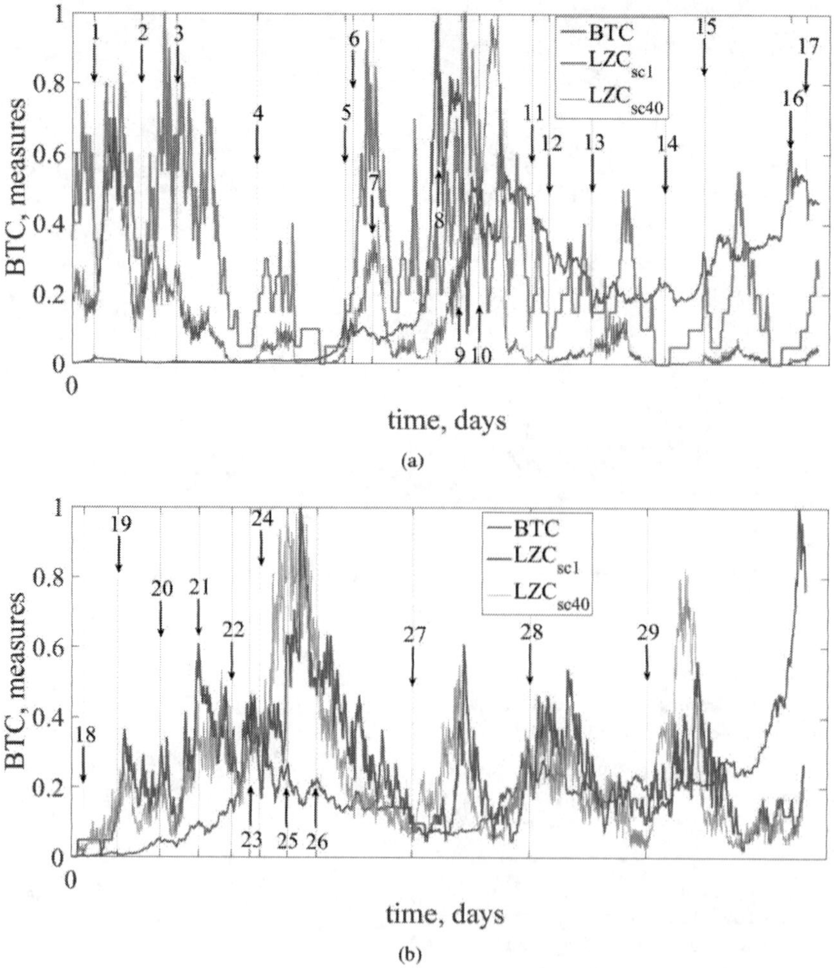

Figure 6: Comparative dynamics of BTC price fluctuations and mono- and multi-scaling LZC measures for first (a) and second (b) periods of the entire time series of Bitcoin

sius [108], in the context of classical thermodynamics, where according to his definition, entropy tends to increase within an isolated system, forming the generalized second law of thermodynamics. Then, the definition of entropy was extended by Boltzmann and Gibbs [109, 110], linking it to molecular disorder and chaos to make it suitable for statistical mechanics, where they combined the notion of entropy and probability [111].

After the fundamental paper of Shannon [90] in the context of information theory, where entropy denoted the average amount of information contained in the message, its notion was significantly redefined. After this, it has been evolved along with different ways and successful enough used for the research of economic systems [112–115].

A huge amount of different methods, as an example, from the theory of complexity, the purpose of which is to quantify the degree of complexity of systems obtained from various sources of nature, can be applied in our study. Such applications have been studied intensively for an economic behavior system.

The existence of patterns within the series is the core in the definition of randomness, so it is appropriate to establish such methods that will be based on the different patterns and their repetition [116]. In this regard, Pincus described the methodology *Approximate entropy* (ApEn) [117] to gain more detail analysis of relatively short and noisy time series, particularly, of clinical and psychological. Its development was motivated by the length constraints of biological data. Since then it has been used in different fields such as psychology [118], psychiatry [119], and finance [120–124]. Pincus co-authored with Kalman [124], considering both empirical data and models, including composite indices, individual stock prices, the random-walk hypothesis, Black-Sholes, and fractional Brownian motion models to demonstrate the benefits of

 https://doi.org/10.1051/shsconf/202110703001

ApEn applied to the classical econometric modeling apparatus. This research the usefulness of ApEn on the example of three major events of the stock market crash in the US, Japan, and India. During the major crashes, there is significant evidence of a decline of ApEn during and pre-crash periods. Based on the presented results, their research concludes that ApEn can serve as a base for a good trading system. Duan and Stanley [125] showed that it is possible to effectively distinguish the real-world financial time series from random-walk processes by examining changing patterns of volatility, approximate entropy, and the Hurst exponent. The empirical results prove that financial time series are predictable to some extent and ApEn is a good indicator to characterize the predictable degree of financial time series. Alfonso Delgado-Bonal [126] gives evidence of the usefulness of ApEn. The researcher quantifies the existence of patterns in evolving data series. In general, his results present that degree of predictability increases in times of crisis.

Permutation entropy (PEn), according to the previous approach, is a complexity measure that is related to the original *Shannon entropy* (ShEn) that applied to the distribution of ordinal patterns in time series. Such a tool was proposed by Bandt and Pompe [127], which is characterized by its simplicity, computational speed that does not require some prior knowledge about the system, strongly describes nonlinear chaotic regimes. Also, it is characterized by its robustness to noise [128, 129] and invariance to nonlinear monotonous transformations [130]. The combination of entropy and symbolic dynamics turned out to be fruitful for analyzing the disorder for the time series of any nature without losing their temporal information.

As an example, Henry and Judge [131] applied PEn to the Dow Jones Industrial Average (DJIA) to extract information from this complex economic system. The result demonstrates the ability of the PEn method to detect the degree of disorder and uncertainty for the specific time that is explored. Higor et al. [132] applied PEn and statistical complexity over sliding time-window of daily closing price log-returns to quantify the dynamic efficiency of more than four hundred cryptocurrencies. Authors address to the efficient market hypothesis when the values of two statistical measures within a time-window cannot be distinguished from those obtained by chance. They found that 37% of the cryptocurrencies in their study stayed efficient over 80% of the time, whereas 20% were informationally inefficient in less than 20% of the time. Moreover, the market capitalization was not correlated with their efficiency. Performed analysis of information efficiency over time reveals that different currencies with similar temporal patterns formed four clusters, and it was seen that more young currencies tend to follow the trend of the most leading currencies. Sensoy [133] compared the time-varying weak-form efficiency of Bitcoin prices in terms of US dollars (BTC/USD) and euro (BTC/EUR) at a high-frequency level by PEn. He noticed that BTC/USD and BTCEUR have become more informationally useful since the beginning of 2016, namely Bitcoin in the same period. Researcher also found that with higher frequency in the Bitcoin market, we had lower price efficiency. Moreover, cryptocurrency liquidity (volatility) had a significant positive (negative) effect on the informational efficiency of its price.

Also, Ayşe Metin Karakaş dedicated [134] both to Bitcoin and Ethereum. Here, the concept of entropy was applied for characterizing the nonlinear properties of the cryptocurrencies. For her goal, Shannon, Tsallis, Rényi, and Approximate entropies were estimated. From her empirical results, it was obtained that all entropies were positive. Of great interest was the results of ApEn which demonstrated larger value for Ethereum than for Bitcoin. In this case, it concluded that Ethereum had higher volatility.

Pele [135] investigated the ability of several econometrical models to forecast value at risk for a sample of daily time series of cryptocurrency returns. Using high-frequency data for Bitcoin, they estimated the entropy of the intraday distribution of log-returns through the symbolic time series analysis STSA, producing low-resolution data from high-resolution data. Their results showed that entropy had strong explanatory power for the quantiles of the distribution of the daily returns. They confirmed the hypothesis that there was a strong correlation between the daily logarithmic price of Bitcoin and the entropy of intraday returns based on Christoffersen's tests for Value at Risk (VaR) backtesting, they concluded that the VaR forecast built upon the entropy of intraday returns was the best, compared to the forecasts provided by the classical GARCH models.

3.2.1 Time delay method

The state of the system can be described by the set of variables. Its observational state can be expressed through a d-dimensional vector or matrix, where each of its components refers to a single variable that represents a property of the system. After a while, the variables change, resulting in different system states.

Usually, not all relevant variables can be captured from our observations. Often, only a single variable may be observed. *Thakens' theorem* [136] that was mentioned in previous sections ensures that it's possible to reconstruct the topological structure of the trajectory formed by the state vectors as the data collected for this single variable contains information about the dynamics of the whole system.

For an approximate reconstruction of the original dynamics of the observed system, we project the time series onto a Reconstructed Phase Space [130, 137, 138] with the commonly used time delay method [130] which relied on the *embedding dimension* and *time delay*.

The embedding dimension is being the dimensionality of the reconstructed system (corresponds to the number of relevant variables that may differ from one system to another. The time delay parameter specifies the temporal components of the vector components. As an example, in recurrence analysis that will be described in section 6, Webber and Zbilut [139] recommend setting the embedding dimension between 10 and 20. Regarding the analysis of financial systems, values between 1 and 20 for the embedding dimension are considered to be reasonable as well as the time delay.

3.2.2 Shannon entropy

The general approach can be described as follows. Formally, we represent the underlying dynamic state of the system in probability distribution form P and then the Shannon entropy S with an arbitrary base (i.e. 2, e, 10) is defined as:

$$S[P] = -\sum_{i=1}^{N} p_i \log p_i. \quad (4)$$

Here, in equation 4, p_i represents the probability that price i occurs in the sample's distribution of the Bitcoin time series, and N is the total amount of data in our system. Dealing with continuous probability distributions with a density function $f(x)$, we can define the entropy as:

$$H(f) = -\int_{-\infty}^{+\infty} f(x) \log f(x) dx. \quad (5)$$

According to the approach, the negative log increases with rarer events due to the information that is encoded in them (i.e., they surprise when they occur). Thus, when all p_i have the same value, i.e. where all values are equally probable, and $S[P]$ reaches its minimum for more structured time series (events that are more certain). Equation 5 is obeyed to the same rules as 4. In figure 7 are the empirical results for Shannon entropy and Bitcoin time series.

It can be seen from the figure that Shannon's entropy is rapidly increasing at the very moment of the crisis itself and is an excellent indicator of crisis phenomena.

3.2.3 Approximate entropy

To gain more detail analysis of the complex financial systems, it is known other entropy methods have become known, particularly, ApEn developed by Pincus [117] for measuring regularity in a time series.

When calculating it, given N data points $\{x(i) \mid i = 1, \ldots, N\}$ are transformed into subvectors $\vec{X}(i) \in \Re^{d_E}$, where each of those subvectors has $[x(i), x(i+1), \ldots, x(i+d_E-1)]$ for each $i, 1 \le i \le N-m+1$. Correspondingly, for further estnimations, we would like to calculate a probability of finding such patterns whose Chebyshev distance $d[\vec{X}(i), \vec{X}(j)]$ does not exceed a positive real number r:

$$C_i^{d_E}(r) = (N - d_E + 1)^{-1} \sum_{j=1}^{N-d_E+1} \mathcal{H}(r - d[\vec{X}(i), \vec{X}(j)])$$

where $\mathcal{H}(\cdot)$ is the Heviside function which count the number of instances $d[\vec{X}(i), \vec{X}(j)] \le r$.

Next, we estimate

$$F^{d_E}(r) = (N - d_E + 1)^{-1} \sum_{i=1}^{N-d_E+1} \ln(C_i^{d_E}(r)),$$

and ApEn of a corresponding time series (for fixed d_E and r) measures the logarithmic likelihood that patterns that are close for d_E adjacent observations remain close on the next comparison:

$$ApEn(d_E, r, N) = F^{d_E}(r) - F^{d_E+1}(r), \quad (6)$$

i.e., equation (6) measures the logarithmic likelihood that sequences of patterns that are close for d_E observations will remain close after further comparisons. Therefore, if the patterns in the sequence remain close to each other (high regularity), the ApEn becomes small, and hence, the time series data has a lower degree of randomness. High values of ApEn indicate randomness and unpredictability. But it should be considered that ApEn results are not always consistent, thus it depends on the value of r and the length of the data series. However, it remains insensitive to noise of magnitude if the values of r and d_E are sufficiently good, and it is robust to artifacts and outliers. Although ApEn remains usable without any models, it also fits naturally into a classical probability and statistics frameworks, and, generally, despite its shortcomings, it is still the applicable indicator of system stability, which significantly increased values may prognosticate the upcoming changes in the dynamics of the data.

The empirical results for the corresponding measure of entropy along with two periods of BTC are presented in figure 8.

Long before the crisis, the value of this type of entropy begins to decrease, the complexity of the system decreases. This measure, in our opinion, is one of the earliest precursors of the crisis.

3.2.4 Permutation entropy

PEn, according to the previous approach, is a complexity measure that is related to the fundamental Information theory and entropy proposed by Shannon. Such a tool was proposed by Bandt and Pompe [127], which is characterized by its simplicity, computational speed that does not require some prior knowledge about the system, strongly describes nonlinear chaotic regimes. Also, it is characterized by its robustness to noise [128, 129] and invariance to nonlinear monotonous transformations [130]. The combination of entropy and symbolic dynamics turned out to be fruitful for analyzing the disorder for the time series of any nature without losing their temporal information. According to this method, we need to consider "ordinal patterns" that consider the order among time series and relative amplitude of values instead of individual values. For evaluating PEn, at first, we need to consider a time series $\{x(i) \mid i = 1, \ldots, N\}$ which relevant details can be "revealed" in d_E-dimensional vector

$$\vec{X}(i) = [x(i), x(i+\tau), \ldots, x(i+(d_E-1)\tau)],$$

where $i = 1, 2, \ldots, N-(d_E-1)\tau$, and τ is an embedding delay of our time delayed vector. After it, we consider $d_E!$ permutation patterns $\pi = (k_0, k_1, \ldots, k_{d_E-1})$ of symbols $(0, 1, \ldots, d_E-1)$ if the following condition for each $\vec{X}(i)$

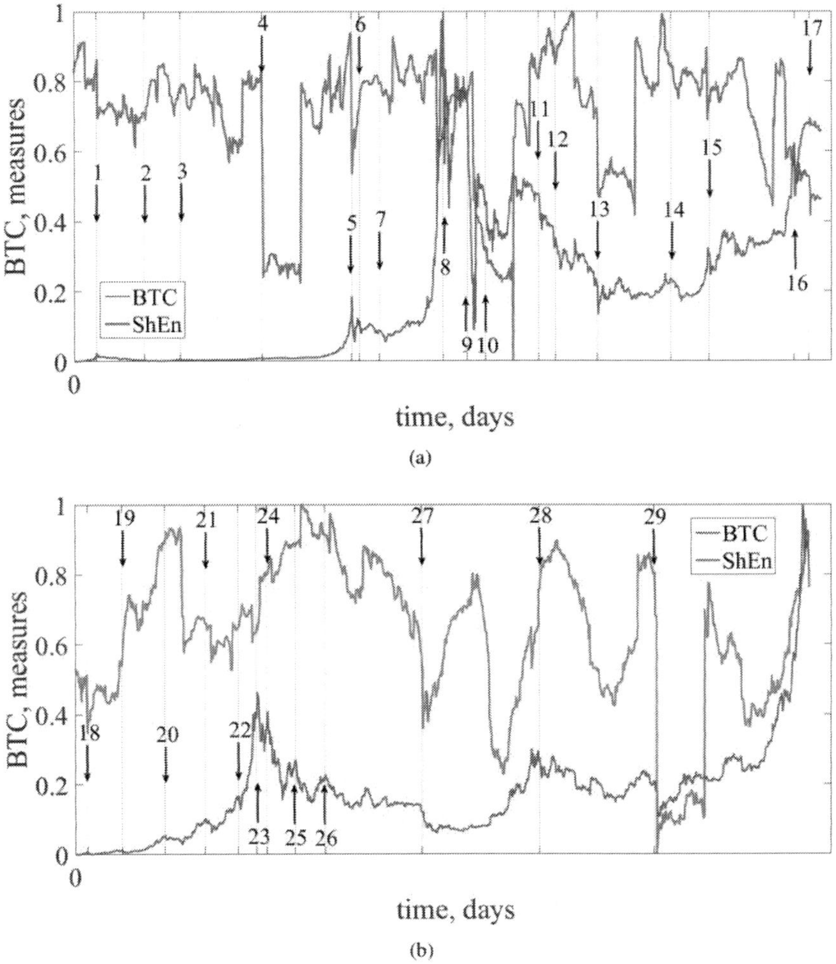

Figure 7: ShEn dynamics along with the first (a) and the second (b) periods of the entire time series of Bitcoin

is satisfied:

$$x(i + k_0) \leq x(i + k_1) \leq \ldots \leq x(i + k_{d_E-1}).$$

We will use ordinal pattern probability distribution as a basis for entropy estimation. Further, let us denote $f(\pi_l)$ as the frequency of occurrence of the pattern π_l. Then, the relative frequencies of permutations in the time series are defined as

$$p(\pi_l) = \frac{f(\pi_l)}{N - (d_E - 1)\tau},$$

where the ordinal pattern probability distribution is given by $P = \{p(\pi_l) \,|\, l = 1, \ldots, d_E!\}$. Finally, permutation entropy (denoted by $S[P]$) of the corresponding time series presented in the following form:

$$S[P] = -\sum_{l=1}^{d_E!} p(\pi_l) \log p(\pi_l).$$

Then, to get more convenient values, we calculate *Normalized permutation entropy* as:

$$E_s[P] = \frac{S[P]}{S_{max}},$$

whose $S_{max} = \ln d_E!$ represents the maximum value of $E_S[P]$ (a normalization constant), and normalized entropy has a range $0 \leq PEn \leq 1$. Here, the maximal entropy possible value is realized when all $d_E!$ possible permutations have an equal probability of occurrence (more noise and random data). With the much lower entropy value, we get a more predictable and regular sequence of the data. Therefore, the *PEn* gives a measure of the departure of the time series from a complete noise and stochastic time series.

There must be predefined appropriate parameters on which *PEn* relying, namely, the embedding dimension d_E is paramount of importance because it determines $d_E!$ possible states for the appropriate probability distribution. With small values such as 1 or 2, parameter d_E will not

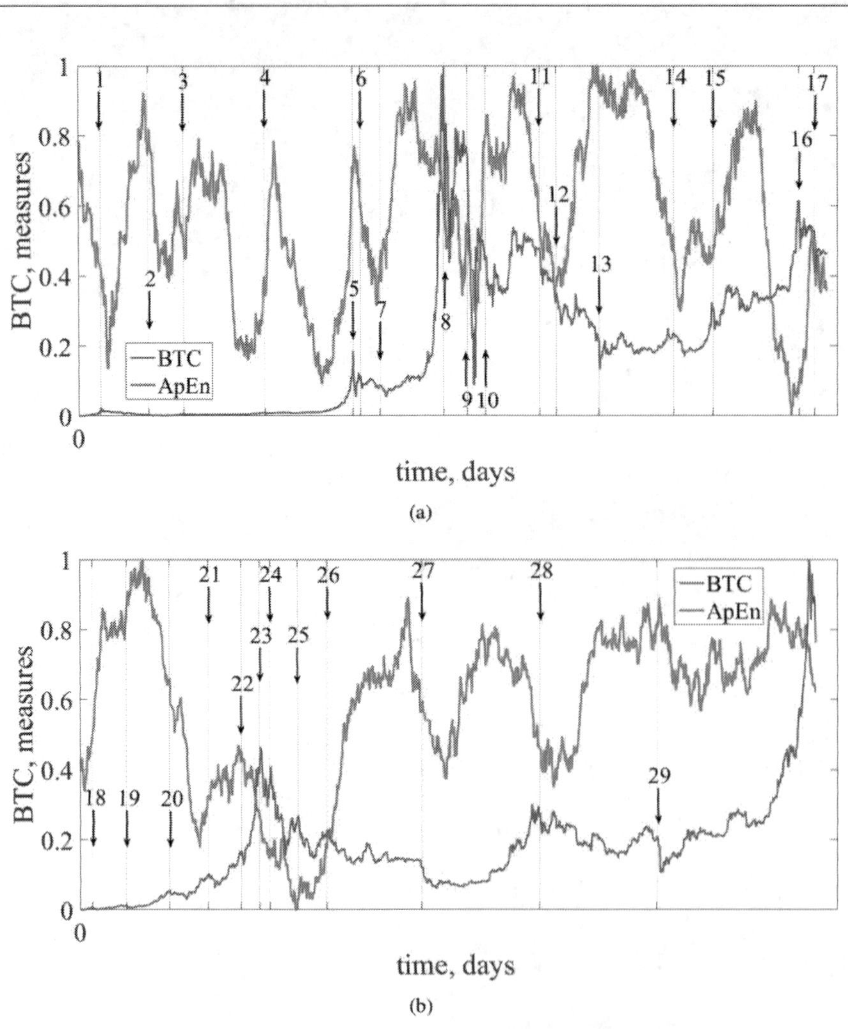

Figure 8: ApEn dynamics along with the first (a) and the second (b) periods of the entire time series of Bitcoin calculated with rolling window of 100 days and the step size of 1 day

work because there are only few distinct states. Furthermore, for obtaining reliable statistics and better detecting the dynamic structure of data, d_E should be relevant to the length of the time series or less [140]. For our experiments, $d_E \in \{3,4\}$ and $\tau \in \{2,3\}$ indicate the best results. Hence, in figure 9 we can observe the empirical results for permutation entropy, where it serves as indicator-precursor of the possible crashes and critical events.

Information measures of complexity due to their initial validity and transparency, ease of implementation and interpretation of the results occupy a prominent place among the tools for the quantitative analysis of complex systems.

4 Fractal and multifractal measures of complexity

The economic phenomena that cannot be explained by the traditional efficient market hypothesis can be explained by the fractal theory proposed by Mandelbrot [141]. Before, fractal studies focus on the Rescaled Range (R/S) analysis were proposed by Hurst [142, 143] in the field of hydrology. However, Lo [144] discovered that the R/S method is sensitive to short-term autocorrelation, which may lead to a bias error of nonstationary time series. To solve this problem, Penget et al. [145] proposed a widely used detrended fluctuation analysis (DFA) that uses a long-range power law to avoid significant long-term autocorrelation false detection [146]. As a multifractal extension (MF) of the DFA approach, Kantelhardt et al. [147] introduced the MF-DFA method that for a long time has been successfully applied for a variety of financial markets, such as stock [148–156], commodity [154, 157–160], tanker [161], derivative [162], foreign exchange rates [163–167], and electricity markets [168]. An especially interesting application of multifractal analysis is measuring the degree

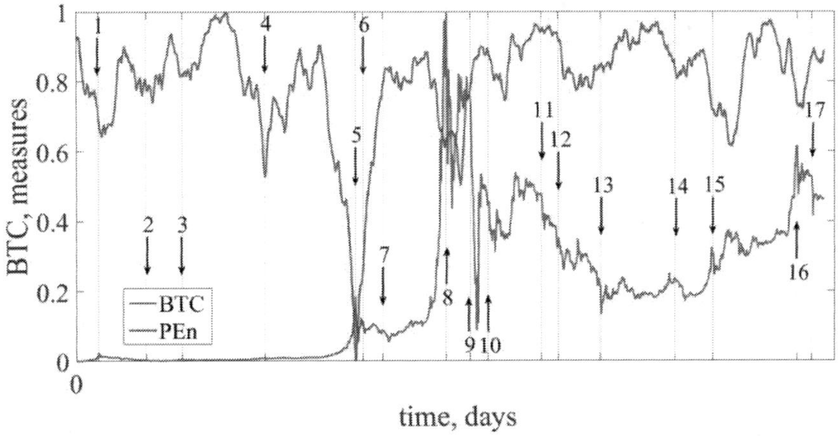

(a)

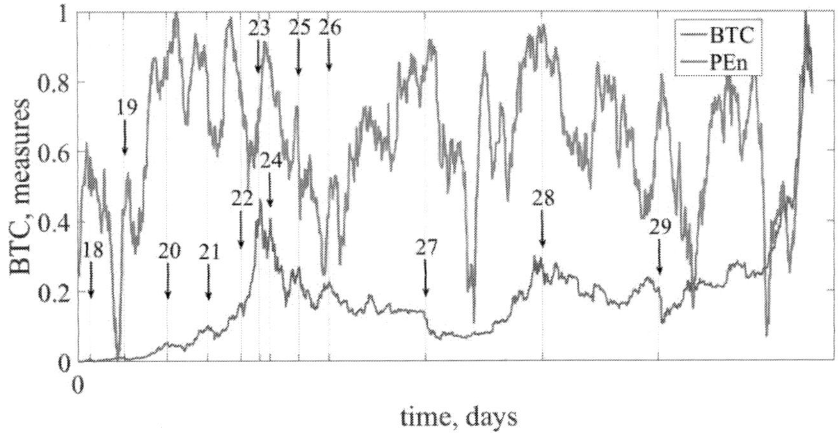

(b)

Figure 9: PEn dynamics along with the first (a) and the second (b) periods of the entire time series of Bitcoin

of multifractality of time series, which can be related to the degree of efficiency of financial markets [169–172].

Podobnik and Stanley [173] extended DFA by introducing a detrended cross-correlation analysis (DCCA) approach that can be used to study long-range cross-correlation between two non-stationary time series. Guided by ready-made approaches, Zhou [174] proposed a multifractal detrended cross-correlation analysis (MF-DCCA) [175], which is a combination of MF-DFA and DCCA. Then the number of interesting methods has been proposed, such as the method of MF-PX-DFA and MF-PX-DMA [176], MF-X-DMA [177], MF-HXA [178], MF-X-PF [179], etc. These increase the efficiency of some applications of the MF-DCCA method. The MF-DCCA method has been widely applied to describe the multifractal characteristics of two cross-correlated nonstationary time series in the financial field such as the foreign exchange market [180, 181], the stock market [182–184], the crude oil market [185–187], carbon market [188, 189], and the commodity market [154, 190]. Zhang et al. also employ MF-DCCA to examine the relationship between mass media and new media [191] and to quantify cross-correlation between investor sentiment proxies [192], i.e., fears [193] and Twitter happiness sentiment [194].

Along with common multifractal methods, Sattarhoff and Gronwald [195] applied an intermittency coefficient for the evaluation of financial market efficiency. While the random walk corresponds to the most genuine form of market efficiency, the larger the value of the intermittency coefficient is, the more inefficient a market would be. In an empirical application using data from the largest current market for tradable pollution permits, the European Union Emissions Trading Scheme, they show that this market becomes more efficient over time. Besides, the degree of market efficiency is overall similar to that for the US stock market; for one sub-period, the market efficiency is found to be higher. While the first finding is anticipated, the second finding is noteworthy, as various observers expressed regarding the information efficiency of this newly established artificial market.

Since Bitcoin was born, it has attracted the considerable attention of researchers from different fields of science that apply modern methods and models of analysis of the peculiarities of the dynamics of the popular digital currency, namely, the methods of multifractal analysis to gain a deeper understanding of its inherent nonlinear statistical properties.

Using 1-min returns of Bitcoin price, Takaishi investigated statistical properties and MF of Bitcoin time series [196]. His results present that 1-min returns distribution is fat-tailed, and kurtosis largely deviates from the Gaussian expectation. Although with large time scales, kurtosis is anticipated to approach the Gaussian expectation, he found that convergence to that is very slow. Skewness is found to be negative at short time scales and becomes consistent with zero at large time scales. Also, he analyzed daily volatility-asymmetry by using GARCH, GJR, and RGARCH models and found no evidence of volatility asymmetry. On exploring MF using MF-DFA, it was confirmed that the Bitcoin time series exhibits MF. The sources of MF are also investigated, and it is confirmed that both temporal correlation and the fat-tailed distribution contribute to the MF, and the degree of MF for the temporal correlation is stronger than that for the fat-tailed distribution.

Generally, regarding these papers [65, 66], Bariviera et al. investigated the long memory of the Bitcoin market using the Hurst exponent. Their research proves the advantages of the DFA methods, basically, because it is more robust and less sensitive to departures from conditions of stationarity. They find that daily returns suffered a regime switch. From 2011 until 2014 the Hurst exponent was showing persistence behavior, whereas after 2014, it equals to white noise, while daily volatility exhibits persistent behavior during the period under study. Also, daily volatility presents stronger fluctuations than in daily returns. In particular, that volatility characteristic is the main peculiarity of the Bitcoin market.

Kirichenko et al. [197] conducted a comparative correlation and fractal analysis to time series of the Bitcoin cryptocurrency rate and community activities in social networks associated with famous cryptocurrency. The results of their study show a significant correlation and similar multifractal structure between the Bitcoin rate and the community activities. Time series fractal analysis indicated the presence of self-similar and multifractal properties.

As an example, Hong et al. [198] attempt to investigate the time-varying long-term memory in the Bitcoin market through a sliding window approach and by employing a new efficiency index [199]. The daily dataset for the period from 2010 to 2017 is utilized, and some interesting findings emerge that:

- generalized Hurst exponents in the Bitcoin market are above 0.5;
- long-term memory exists in the Bitcoin market;
- high degree of inefficiency ratio;
- Bitcoin market does not become more efficient over time;
- rolling window approach can help to obtain more reliable results. Some conclusions for those who deal with the cryptocurrency market were made.

The Al-Yahyaee et al. [200] paper provides the results on the efficiency of the Bitcoin market compared to gold, stock, and foreign exchange markets. By applying the MF-DFA approach, researchers found that the long-memory feature and MF of the Bitcoin market were more robust and, therefore, more inefficient than the gold, stock, and currency markets.

Gajardo et al. [201] applied MF-ADCCA to analyze the presence and asymmetry of the cross-correlations between the major currency rates, Bitcoin, the DJIA, gold price, and the oil crude market. They found that multifractality existed in every cross-correlation studied, and there was an asymmetry in the cross-correlation exponents under the different trends of the WTI, Gold, and the DJIA. Bitcoin showed greater multifractal spectra than the other currencies on its cross-correlation with the WTI, the Gold, and the DJIA. Bitcoin presented a different relationship between commodities and stock market indices, which had to be taken into consideration when investing. The reason is that over the years the currency was traded and over time, it has earned the trust of the community.

The nonlinear patterns of the volatility of the seven Bitcoin markets were investigated by Lahmiri et al. [202]. Using four diverse distribution inferences: Normal, Student-t, Generalized Error, and t-Skewed distribution, they explored the fractional long-range dependence in conjunction with the potential inherent stochasticity of volatility time series. Their results testify to the existence of long-range memory in Bitcoin market volatility, irrespectively of distributional inference. The entropy measurement, which indicates a high degree of randomness in the estimated series, shows the same. As Bitcoin markets are highly disordered and risky, they cannot be considered suitable for hedging purposes. Their exploration provides strong evidence against the efficient market hypothesis.

Wei Zhang et al. [203] investigated the cross-correlations of the return-volume relationship of the Bitcoin market. In particular, they selected eight exchange rates whose trading volume accounts for more than 98% market shared to synthesize Bitcoin indexes. The empirical results based on MF-DCCA revealed that: (1) return-volume relationship exhibited the nonlinear dependencies and power-law cross-correlations; (2) all cross-correlations were multifractal, and there were anti-persistent behaviors of cross-correlation for $q = 2$; (3) the price of small fluctuations was more persistent than that of the volume, while the volume of larger fluctuations was more anti-persistent; (4) the sliding window approach showed that the cross-correlations of return-volume were anti-persistent in the entire sample period.

Similarly to our article [204] where we applied the MF-DFA method to Ukrainian and Russian stock markets, we use it here to explore the multifractal property of Bitcoin and construct reliable indicator for it.

4.1 Multifractal detrended fluctuation analysis (MF-DFA)

As an extension to the original DFA [145, 205, 206], the multifractal approach [147, 207] estimates the Hurst exponent of a time series at different scales. Based on a given time series $\{x(i) \mid i = 1, \ldots, N\}$, the MF-DFA is described as follows:

(i) The profile $Y(i)$ (accumulation) is defined as:

$$Y(i) = \sum_{j=1}^{i} (g(j) - \langle g \rangle), \tag{7}$$

where $\langle g \rangle$ stands for the average of returns.

(ii) The profile $Y(i)$ is then divided into $N_s \equiv \text{int}(N/s)$ non-overlapping time segments of equal length s and the local trend Y_v^{fit} for each segment is calculated by the least-square fit. Since the length of the time series is not always a multiple of s, a short period at the end of the profile, which is less than the window size, may be removed. For taking into account the rejected part and, therefore, to use all the elements of the sequence, the above procedure is repeated starting from the end of the profile. Therefore, the total $2N_s$ segments are obtained together, and the variance is computed as

$$F^2(v, s) = \frac{1}{s} \sum_{i=1}^{s} \left[Y((v-1)s + i) - Y_v^{fit}(i) \right]^2, \quad \text{for } v = 1, \ldots, N_s \tag{8}$$

and

$$F^2(v, s) = \frac{1}{s} \sum_{i=1}^{s} \left[Y(N - (v - N_s)s + i) - Y_v^{fit}(i) \right]^2, \quad \text{for } v = N_s + 1, \ldots, 2N_s \tag{9}$$

Various types of MF-DFA such as linear, quadratic, or higher order polynomials can be used for eliminating local trend in segment v; we use a cubic order polynomial.

(iii) Considering the variability of time series and the possible multiple scaling properties, we obtain the qth order fluctuation function by averaging over all segments:

$$F_q(s) = \left[\frac{1}{2N_s} \sum_{v=1}^{2N_s} \left[F^2(v, s) \right]^{q/2} \right]^{1/q}. \tag{10}$$

The index q can take any non-zero value. For $q = 0$, $F_q(s)$ is divergent and can be replaced by an exponential of a logarithmic sum

$$F_0(s) = \exp \left[\frac{1}{4N_s} \sum_{v=1}^{2N_s} \ln F^2(v, s) \right].$$

(iv) At least, we determine the scaling behavior of the fluctuation function by analyzing $\log F_q(s)$ vs $\log s$ graphs for each value of q. Here, $F(s)$ is expected to reveal power-law scaling

$$F_q(s) \sim s^{h(q)} \tag{11}$$

for large n. The scaling exponent $h(q)$ can be considered as generalized Hurst exponent. With $q = 2$ MF-DFA transforms into standard DFA, and $h(2) = H$, where H is the well-known Hurst exponent.

- $h(2) \in [0.0, 0.5) \to$ anti-persistency. The process under study tends to decrease (increase) after a previous increasing (decreasing);
- $h(2) = 0.5 \to$ uncorrelated process. The fluctuations that depend on q tend to a random walk behavior [131];
- $h(2) \in (0.5, 1.0] \to$ persistency. If a process tends to increase (decrease) for a some period T, then it expected to continue to increase (decrease) for a similar period of time;
- $h(2) > 1.0 \to$ nonstationary process, stronger long-range correlations are present.

Those intervals with time intervals v will dominate which variance $F^2(v, s)$ is large and q values are positive. Therefore, for positive values of q , $h(q)$ describes the scaling behavior of time intervals with large fluctuations. Large fluctuations are usually characterized by smaller scaling coefficients of $h(q)$ for multifractal series. On the contrary, for negative values of q , time intervals with a small variance $F^2(v, s)$ will dominate. Thus, $h(q)$ will describe the scaling behavior of time intervals with small fluctuations.

(v) Another way of characterizing multifractality of a time series is in terms of the multifractal scaling exponent $\tau(q)$ which is related to the generalized Hurst exponent $h(q)$ from the standard multifractal formalism and given by [146]:

$$\tau(q) = qh(q) - 1. \tag{12}$$

Equation (12) reflects temporal structure of the time series as a function of moments q, i.e., it represents the scaling dependence of small fluctuations for negative values of and large fluctuations for positives values. If (12) represents linear dependence of q, the time series is said to be monofractal. Otherwise, if (12) has a nonlinear dependence on q, then the series is multifractal.

(vi) The different scalings are better described by the singularity spectrum $f(\alpha)$ which can be defined as:

$$\alpha = \frac{d\tau(q)}{dq} = h(q) + q\frac{dh(q)}{dq},$$

$$f(\alpha) = q\alpha(q) - \tau(q) = q^2 \frac{dh(q)}{dq} + 1,$$

with α – the Hölder exponent or singularity strength. Following the methods described above, we present results that reflect multifractal behavior of the Bitcoin time series.

Figure 10a presents $F_q(s)$ in the log-log plot. The slope changes dependently on q , which indicates the multifractal property of a time series. As it was pointed out, multifractality emerges not only because of temporal correlation, but also because the Bitcoin returns distribution turns out to be broad (fat-tailed) [147], and this distribution could contribute to the multifractality of the time series. The same dependence can be observed in the remaining plots. The scaling exponent $\tau(q)$ remains nonlinear, as well as generalized Hurst exponents that can serve as evidence that Bitcoin exhibit multifractal property.

In the case of multifractals, the shape of the singularity spectrum typically resembles an inverted parabola (see figure 10d); furthermore, the degree of complexity is straightforwardly quantified by the width of $f(\alpha)$, simply defined as $\Delta\alpha = \alpha_{max} - \alpha_{min}$, where α_{max} and α_{min} correspond to the opposite ends of the α values as projected out by different q-moments (equation (10)).

In the figure below we present the width of the spectrum of multifractality that changes over time accordingly to the sliding window approach. The whole figure consists of both a three-dimensional plot (singularity spectrum) and two-dimensional representation of its surface.

If the series exhibited a simple monofractal scaling behavior, the value of singularity spectrum $f(\alpha)$ would be a constant. As can be observed, here our series exhibits a simple multifractal scaling behavior, as the value of singularity spectrum $f(\alpha)$ changes dependently on α , i.e., it exhibits different scalings at different scales. Moreover, with the sliding window of the corresponding length, we understand that at different time periods Bitcoin becomes more or less complex. The value of $\Delta\alpha$ gives a shred of additional evidence on it.

As we can see from the presented results, the width of the singularity spectrum after the crisis starts to increase, which tells us that more violent price fluctuations are usually expected. With the decreasing width of the singularity spectrum, the series is expected to hold the trend. As the rule, it reaches its minimum before the collapse of the BTC price.

5 Chaos-dynamical measures

Apparently random fluctuations in financial systems often tend to exhibit varying levels of complexity and chaos. Regarding limited data, it becomes hard to define the boundaries of predictability of them. The analysis of such systems, the processes driving their dynamics chaos theory has been considered in various fields such as economics, finance, physics, and others [208–213]. Regarding the analysis of Bitcoin dynamics, the knowledge about its completely random and, at the same time, deterministic processes can potentially explain fluctuations in time series of different nature. Considering the financial sector, evidence on deterministic chaos, the knowledge of such moments when two initially close trajectories start to diverge, and of the periods for which they will stay close to each other would have important implications for regulators and traders, who will develop effective short-term trading strategies. During the years, chaos theory has been providing approaches to study some interesting properties of time series. The most widespread are: correlation dimension, the BDS test, Kolmogorov entropy, Lyapunov exponent, close returns test, etc. [214–217].

Endowing Bitcoin time series with the sliding window approach and efficient methods of Lyapunov exponents, and Levy alpha-stable distribution, we are going to reflect its transition between chaotic and non-chaotic behavior. Also, as it has been observed, such unstable events as market crashes correspond to fat tails. Thus, the analysis of such extreme events can be understood throughout Levy alpha-stable distribution.

5.1 Lyapunov exponents

The evolution of the system exhibits *sensitive dependence* on *initial conditions*. It means that initially close trajectories that evolve may rapidly diverge from each other and have totally different outcomes. Accordingly, with small uncertainties that amplify enormously quickly, long-term predictions turn out to be impossible. On the other hand, in a system with *attraction points* or *stable points*, the distance between them decreases asymptotically in time or with the number of points, which tend to converge [218].

To present the idea more precisely, let's consider two consecutive points presented as $x(t)$ and its initially close neighbor as $x(t) + \delta(t)$, where $\delta(t)$ represents a tiny deviation in time t, as presented in figure 13

As two initially close points disturbed by some event, they start to diverge, and the distance between them grows following exponential law:

$$\|\delta(t)\| \approx \|\delta(0)\| \exp(\lambda t) \tag{13}$$

with λ that denotes the *Lyapunov exponent* (LE); $\delta(t)$ is the distance between the reference point and its nearest neighbor after t iterations; $\delta(0)$ is the initial distance between the reference point and its nearest neighbor perturbed with a small error at $t = 0$.

LE is a measure of the exponential rate of nearby trajectories in the phase-space of a dynamical system. In other words, it quantifies how fast converge or diverge trajectories that start close to each other, quantifying the strength of chaos in the system.

In such cases when our system n-dimensional, we have as many LEs as the dimensions in it. To define them, we consider the evolution of an infinitesimal sphere of perturbed initial conditions. During its evolution, the sphere will become distorted into an infinitesimal ellipsoid. Defining the length of the ith principal axis as $\delta_i(t)$, there are n-Lyapunov exponents given by:

$$\|\delta_i(t)\| \approx \|\delta_i(0)\| \exp(\lambda_i t), \quad \text{for } i = 1, \ldots, n. \tag{14}$$

To identify whether the motion is periodic or chaotic, especially, for large t it is recommended to contribute to

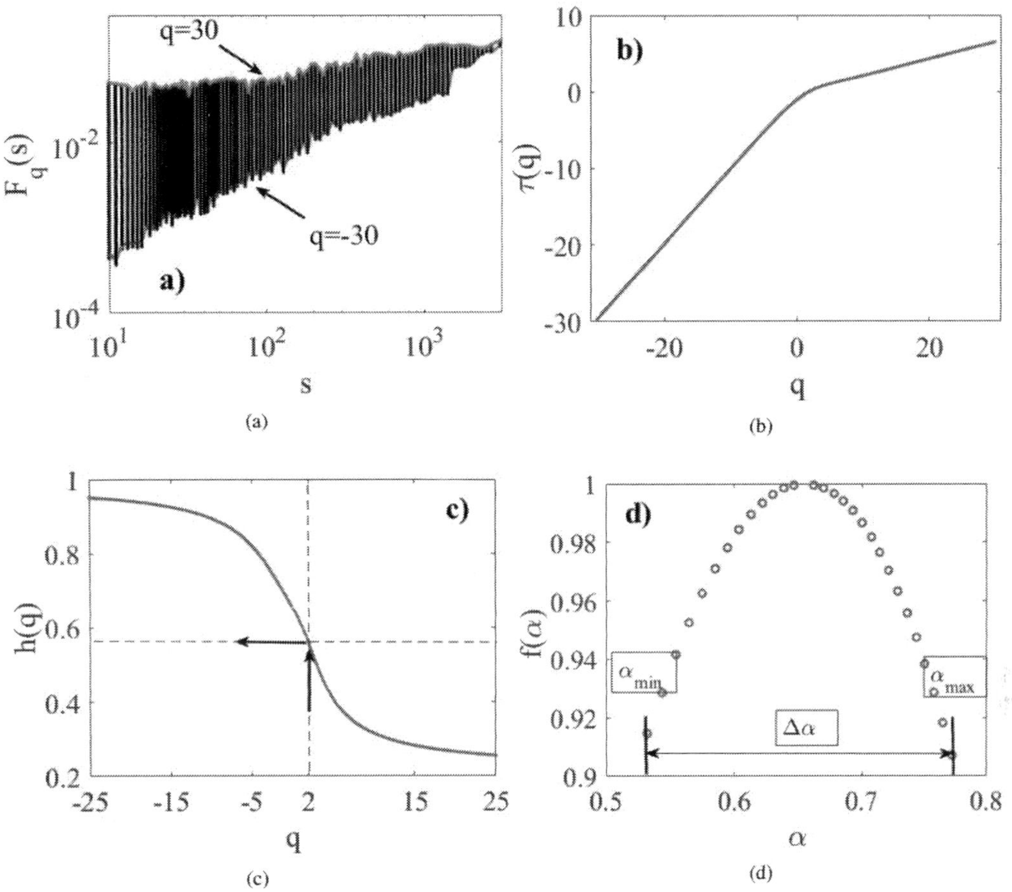

(a) (b) (c) (d)

Figure 10: The fluctuation function $F_q(s)$ (a), multifractal scaling exponent $\tau(q)$ (b), $h(q)$ versus q of the BTC return series (c), and singularity spectrum $f(\alpha)$ (d) obtained from MF-DFA for BTC time series.

the *largest Lyapunov exponent* (LLE) among the others of the n-dimensional dynamical system [219], as the diameter of ellipsoid starts to be controlled by it. Exactly the LLE is used to quantify the predictability of the systems, since exponential divergence means that in the system where the initial difference was infinitesimally small, start to rapidly lose its predictability behaving differently. However, it should be noted that other exponents also contain important information about the stability of the system, including the directions of convergence and divergence of the trajectories [220].

The existence of at least one positive LE is generally seen as a strong indicator of chaos. Positive LE means that initially similar, phase space trajectories that are sensitive to initial conditions and diverge exponentially fast, characterize chaotic behavior of the system. Negative LE responds to the cases when trajectories remain close to each other, but it is not necessarily implied stability, and we have to examine our system in more detail. Zero or very close to zero exponents indicate that perturbations made along the trajectory neither diverge nor converge.

With the great interest in LE, more and more methods and proposals for their calculating have appeared. Unfortunately, there has not been obtained accepted and universal method for estimating the whole spectrum of Lyapunov exponents from a time series data. One of the most common and popular algorithms have been applied by Wolf et al. [221], Sano and Sawada [222], and later improved by Eckmann et al. [223], Rosenstein et al. [224], Parlitz [225], and Balcerzak et al. [226]. Here, we followed the methods proposed by Eckmann and Gao et al. [81, 227–230] to compute the spectrum of Lyapunov exponents. With Rosenstein's algorithm, we compute only the LLE from an experimental time series. As again suggested by Eckmann et al. [231] one of the measures from recurrence quantification analysis can be considered for estimation of the LLE since it detects in a similar way highly non-monotonic behavior.

With the high growth in computer science, computer simulations of complex and chaotic systems become increasingly appreciated. For at least two decades, with development in numerical computations and quantitative

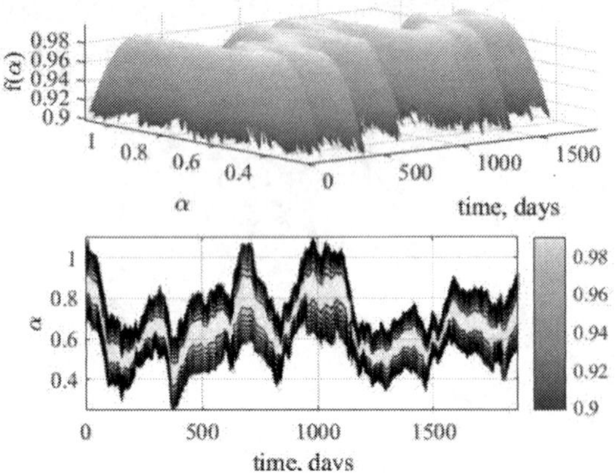

Figure 11: Changes in the spectrum of multifractality in time

analysis, no doubt left that chaos theory suggests the same unstable fluctuation that may be as common as the extreme events and critical transitions in financial markets. For instance, Scheinkman and LeBaron [232] explored several indications of nonlinear dynamic structure in stock market returns. In their opinion, the weaknesses of such studies are based on time series that are not long enough to reveal the strange (fractal) attractors. On the other hand, the reason may be chaos that comprises a class of signals intermediate between regular periodic or quasiperiodic motions and unpredictable, truly stochastic behavior [233]. Kulkarni [234] denotes that, probably, random financial fluctuations often exhibit varying levels of fluctuations, chaos. Kulkarni's paper represents the efficiency of LE for the complexity analysis of shortly limited data. The analysis constitutes weakly chaotic behavior which alternates with non-chaotic over the entire period of analysis.

Lyapunov exponents are a natural first choice in exploring and indicating such chaotic behaviors that occur in it. They do not only classify the system but also tell us the limits of predictability of the chaotic system [233]. During the last few decades, there was plenty of scientific research that was related to chaos systems, chaos behavior and, namely, to the LE. The earliest papers, in which authors [235, 236] try to use LE to detect chaos dynamics in financial time series, it is determined that linear, deterministic processes are characterized with negative LE from nonlinear, chaotic processes with the largest exponent (where it is positive). Besides, there is an article [237] in which Gençay presents a methodology to compute the empirical distributions of LE's using a blockwise bootstrap technique. This method provides a formal test of the hypothesis that the LLE equals some hypothesizes value, and can be used to test the system for the presence of chaotic dynamics. Such methodology is particularly useful in those cases where the largest exponent is positive but very close to zero.

Sarkar and Chadha [238] in their paper investigated the local fractal and chaotic properties of financial time series by calculating two exponents, the Local Hurst Exponent and LE. As was seen in their research, all calculations were made with the algorithm of the sliding window where they had considered two major financial indices of the US: the DJIA and the S&P 500. Regarding the considered measures, they attempted to predict the major crashes that took place in these markets.

Srinivasan et al. [239] have provided an explanation and motivation for reconstructed phase spaces using the methods of time delay and SVD embedding. They explained the meaning of LE and an algorithm for its estimation for the corresponding chaotic, deterministic, and periodic time series. From their presented results it is seen that estimated positive and zero exponents converge to the expected, documented values. Mastroeni and Vellucci [240] obtained empirical results with the help of the LLE and a determinism test that shows that commodity and futures prices are representatives of a nonlinear deterministic, rather than stochastic systems. Similarly to [238], Plakandaras et al. [241] measured the Hurst exponent and LE in the sliding window to focus on persistence and chaotic behavior — two prime characteristics of uncertainty indices. For such purpose, they analyzed 72 popular indices constructed by forecasting models, text mining from news articles, and data mining from monetary variables. More specifically, researchers found that almost all uncertainty indices are persistent, while the chaotic dynamics are detected only sporadically and for certain indices during recessions of economic turbulence. Authors of empirical analysis [242] in one of their chapters explored whether the global markets are intrinsically unstable where unpredictability, disorder, and discontinuities are inherent and not aberrations. They investigated a huge amount of literature and examine the possible nonlinear, particularly chaotic nature of the global stock markets. Their study explores the possible presence of chaos in two phases: over the period for 1998-2005 and from 2006 to 2011. Over 30 indices had been investigated. Empirical results showed that for the first phase, 29 indices are deterministic. But 10 of them are found to be non-chaotic. Estimated determinism factors for all the indices are quite high, but Lyapunov exponent is presented to be non-positive for at least 6 of them, where others are chaotic.

As it is seen, chaos theory and its tools remain a huge challenge for researchers of different fields of science and, namely, in the financial industry, and, as it was suggested in [241], the examination of persistent and chaos should be a prerequisite step before using financial indices in economic policy model. The world of Lyapunov exponents remains a growing interest in their definition, numerical methods, and application to various complex systems. In summary, LLE allows us to establish [219]:

- transition region between stable and unstable;
- stability region;
- unstable region;
- chaotic region, including a possible transition between unstable and chaotic.

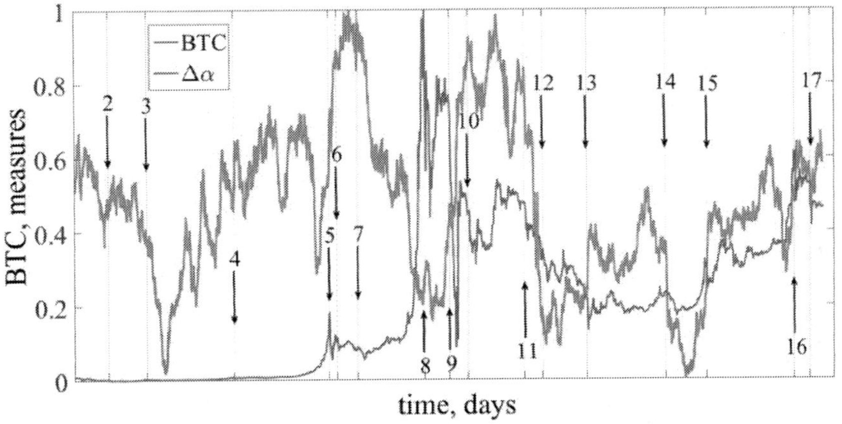

(a)

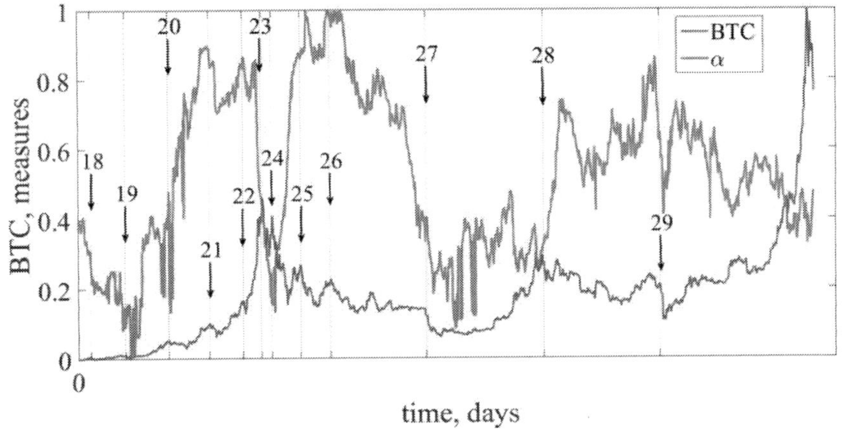

(b)

Figure 12: The comparison of the corresponding first (a) and second (b) Bitcoin time intervals with the width of the multifractality spectrum measure

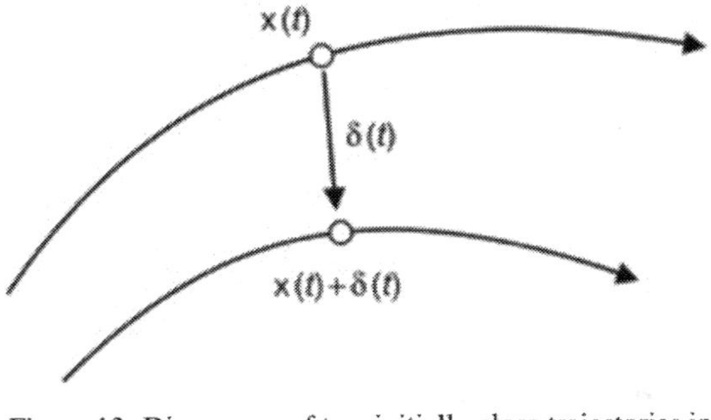

Figure 13: Divergence of two initially close trajectories in a dynamical system [219]

5.1.1 *Eckmann et al. method*

Firstly, according to the approach [223], we need to reconstruct attractor dynamics from a single time series $\{x(i) \mid i = 1, \ldots, N\}$ with the embedding dimension d_E , and after this, we construct d_E-dimensional orbit representing the time evolution

$$\vec{X}(i) = [x(i), x(i+1), \ldots, x(i+(d_E-1))], \quad \text{for } i = 1, \ldots, N - d_E + 1.$$

Then, we have to determine the most neighboring trajectories with $\vec{X}(i)$:

$$\left\|\vec{X}(i) - \vec{X}(j)\right\| = \max_{0 \le \alpha \le d_E - 1} \{|x(i+\alpha) - x(j+\alpha)|\}. \quad (15)$$

We sort the $x(i)$ so that $x(\Pi(1)) \le x(\Pi(2)) \le \cdots \le x(\Pi(N))$ and store the permutation Π, and its inverse Π^{-1}. Then, we try to find the neighbors of $x(i)$ in dimension 1 by looking at $k = \Pi^{-1}(i)$ and scan the $x(\Pi(s))$ for $s = k+1, k+2, \ldots$ and $k-1, k-2, \ldots$ until $x(\Pi(s)) - x(i) > r$. For chosen embedding dimension $d_E > 1$, we select the value of s for which further condition is true

$$|x(\Pi(s)+\alpha) - x(j+\alpha)| \le r, \quad \text{for } \alpha = 0, 1, \ldots, d_E - 1.$$

After we embedded our system in d_E dimensions, we need to determine the $d_E \times d_E$ matrix M_i that will describe time evolution of the vectors that surround trajectory $\vec{X}(i)$, and how they map onto $\vec{X}(i+1)$ state. The matrix M_i is obtained by looking for neighbors

$$M_i(\vec{X}(i) - \vec{X}(j)) \approx \vec{X}(i+1) - \vec{X}(j+1). \quad (16)$$

Nevertheless, the vectors $\vec{X}(i) - \vec{X}(j)$ may not span $\mathfrak{R}^{d_E}$. In this case, such indeterminacy may lead to spurious exponents which confuse the analysis. To overcome such obstacles, the projection of the trajectories is determined on a subspace of dimension $d_M \le d_E$. Thus, the manifold on which the dynamics takes place corresponds to the local dimension d_M, where d_E should be somewhere larger than d_M to avoid the presence of false neighbors [130, 233]. Hence, the trajectory $\vec{X}(i)$ is associated with a d_M-dimensional vector

$$\begin{aligned}\vec{X}(i) &= [x(i), x(i+\tau), \dots, x(i+(d_M-1)\tau)] = \\ &= [x(i), x(i+\tau), \dots, x(i+d_E-1)],\end{aligned} \quad (17)$$

where $\tau = (d_E - 1)/(d_M - 1)$. When $\tau > 1$, the condition (16) is replaced by

$$M_i(\vec{X}(i) - \vec{X}(j)) \approx \vec{X}(i+\tau) - \vec{X}(j+\tau). \quad (18)$$

The M_i is then defined by the linear least-square method [243]. The last step of the algorithm is the classical QR matrix decomposition to find orthogonal matrices Q_i and upper-triangular matrices R_i with non-negative diagonal elements such that

$$M_{1+i\tau}Q_i = Q_{i+1}R_{i+1}, \quad \text{for } i = 0, 1, 2, \dots .$$

As it was proposed by Eckmann et al. [137, 223, 244], in order to calculate d_M Lyapunov exponents, the equation for the kth Lyapunov exponent with K number of points on the attractor, for which the Jacobian has been estimated, the diagonal eigenvalues of the matrix R_i and the sampling step Δt is given by:

$$\lambda_k = \frac{1}{\Delta t}\frac{1}{\tau}\frac{1}{K}\sum_{i=0}^{K-1} \ln (R_i)_{kk}.$$

Thus, with linearizations by using the diagonal elements from the QR decomposition, we can calculate Lyapunov exponents.

The calculation results for the LLE on the example of BTC are presented in figure 14.

Let us pay attention to the behavior of λ_{max} at moments of the known failures noted in the list of crashes and critical events. Definitely, we can see that in the pre-crisis period, the value of LLE decreases markedly, then increases in the post-crisis period.

5.1.2 Rosenstein's et al. method

Rosenstein's algorithm [224] uses the delay embedding method that reconstructs the most important features of a multi-dimensional attractor into a single one-dimensional time series of some finite size N. For the time series $\{x(i) \mid i = 1, \dots, N\}$, each delay embedded vector $\vec{X}(i)$ will be presented similarly to the vector (17) with embedding dimension d_E and time delay τ. Then, in the reconstructed trajectory, we initialize searching for in the state space for the nearest neighbor $\vec{X}(j)$ of the trajectory $\vec{X}(i)$:

$$\delta_i(0) = \min_{\vec{X}(i)} \left\|\vec{X}(i) - \vec{X}(j)\right\|, \quad \text{for } |i-j| > \text{mean period},$$

where $\|\,\|$ is the Euclidian norm, $\vec{X}(j)$ is the nearest neighbor, and $\vec{X}(i)$ is the reference point.

From equation (13) we have already known that the distance between states $\vec{X}(i)$ and $\vec{X}(j)$ will grow in time accordingly to a power law, where λ is a good approximation of the LLE. For further estimations, we look at the logarithm of the distance trajectory $\ln \delta_i(k) \approx \lambda(k \cdot \Delta t) + \ln c_i$, where $\delta_i(k)$ is the distance between i^{th} pair of the nearest neighbors defined in equation (18) after k time steps, c_i is the initial separation of them, and Δt is the time interval between measurements (sampling period of the time series).

Further result of this algorithm is not a numerical value, but a function of time

$$y(k, \Delta t) = \frac{1}{\Delta t}\frac{1}{M}\sum_{i=1}^{M} \ln \delta_i(k)$$

with the size of the reconstructed time series $M = N - (d_E - 1)\tau$, and a set of approximately parallel lines $\delta_i(k)$ whose slope roughly proportional to the LLE. Then, it is proposed to be calculated as the angle of inclination of its most linear section. Finding such a section turns out to be a non-trivial task, and sometimes it is impossible to specify such a section at all. Despite this problem, Rosenstein's method is easy for implementing and computing.

The LLE behavior for a window procedure with the length of 250 days and the step size of 1 day is shown in figure 15.

It can be seen that, as before, the LLE is also sensitive to the crisis conditions of BTC.

5.1.3 Scale-dependent Lyapunov exponent (SDLE)

We briefly describe the idea and the formal foundations of the method SDLE, introduce new measures of complexity, and illustrate their effectiveness with the example of the BTC index. Let us have a single observation conducted at a discrete time interval Δt in the form of a time series $\{x_i \mid i = 1, \dots, N\}$ where $t = i \cdot \Delta t$. After reconstructing the phase space, let us consider the ensemble of trajectories.

Let us denote the initial distance between two close trajectories $\delta(0)$, and their average distance at time t and $t + \Delta t$ as $\delta(t)$ and $\delta(t + \Delta t)$ respectively. Note that the classical algorithm for calculating the LLE (λ_{max}) is based on the assumption (13) and its estimation as $(\ln(\delta(t) - \delta(0)))/t$. Depending on $\delta(0)$, this property may not be true even for truly chaotic systems. To calculate the SDLE, we check whether the following inequality holds for a pair of vectors $\left(\vec{X}(i), \vec{X}(j)\right)$:

$$\delta(k) \le \left\|\vec{X}(i) - \vec{X}(j)\right\| \le \delta(k) + \Delta\delta(k), \quad \text{for } k = 1, 2, 3, \dots ,$$

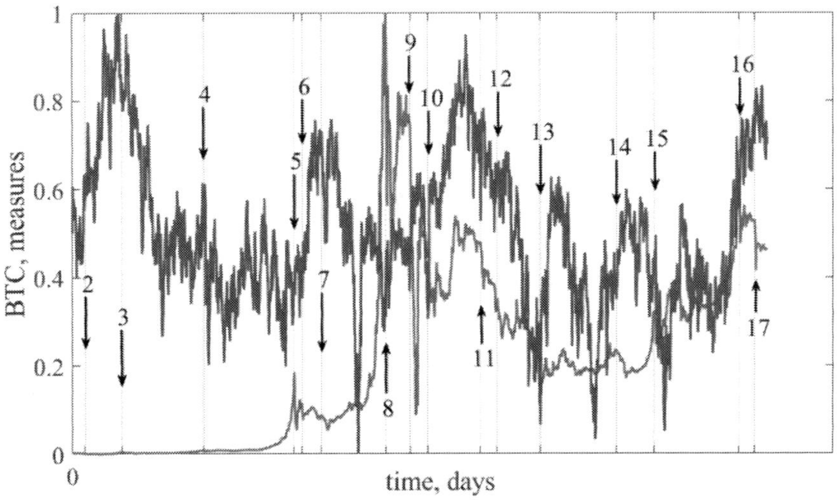

(a)

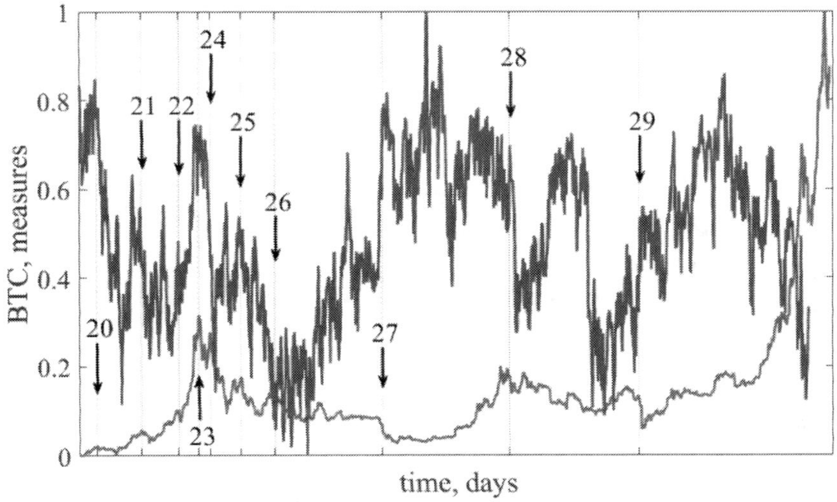

(b)

Figure 14: The dynamics of the LLE (λ_{max}) with Eckmann et al. method within the time window of the length 250 days and the time step of 1 day for the first (a) and second (b) periods. Exponents are calculated with $d_E = 3$ and $d_M = 2$

where $\delta(k)$ and $\Delta\delta(k)$ are arbitrarily chosen small values of distances, and

$$\left\|\vec{X}(i) - \vec{X}(j)\right\| = \sqrt{\sum_{w=1}^{d_E} (x(i + (w-1)\tau) - x(j + (w-1)\tau))^2}.$$

Geometrically, the last inequality defines a shell in high-dimensional space. Next, we investigate the dynamics of the same pairs of vectors $\left(\vec{X}(i), \vec{X}(j)\right)$ in the middle of the shell and perform averaging over the ensemble by indices i, j. Since the exponential or power functions are of the greatest interest, we assume that logging and averaging can be reversed. Finally, the following equation looks like:

$$\lambda(\delta(t)) = \left\langle \ln\left\|\vec{X}(i + t + \Delta t) - \vec{X}(j + t + \Delta t)\right\| - \right.$$
$$\left. - \ln\left\|\vec{X}(i + t) - \vec{X}(j + t)\right\|\right\rangle / \Delta t,$$

with the sampling intervals t and Δt; the angle brackets correspond to the averaging over the indices i, j inside the shell and

$$\delta(t) = \left\|\vec{X}(i + t) - \vec{X}(j + t)\right\| =$$
$$= \sqrt{\sum_{w=1}^{d_E} (x(i + (w-1)\tau + t) - x(j + (w-1)\tau + t))^2}.$$

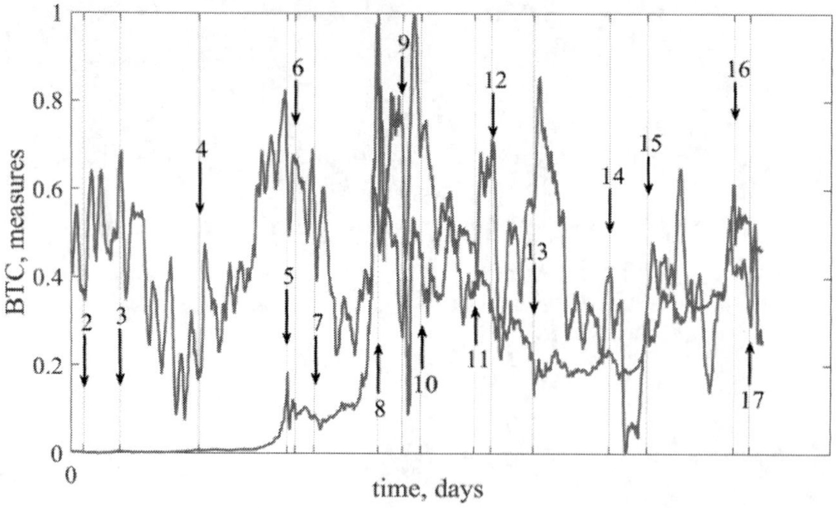

(a)

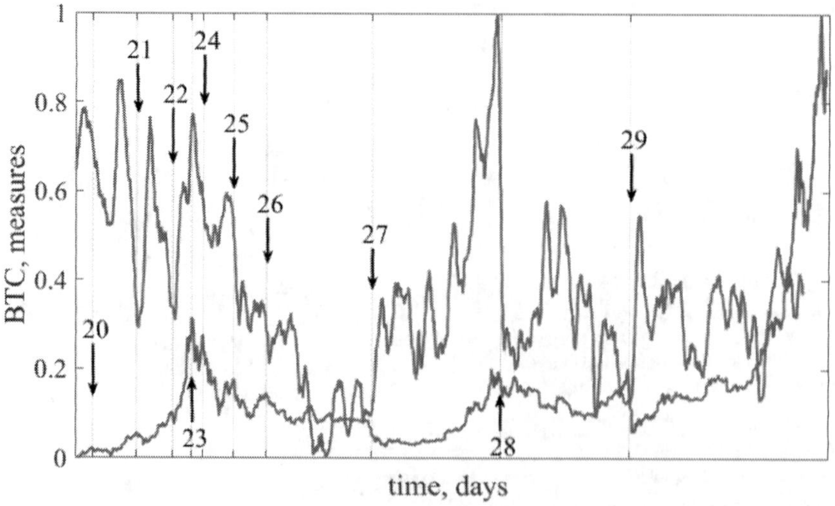

(b)

Figure 15: The dynamics of the LLE calculated with Rosenstein et. al. method within the time window length of 250 days and the time step of 1 day for first (a) and second (b) periods. Exponents are calculated with $d_E = 3$ and $\tau = 2$

Finally, note that

$$I = \ln \delta(t) = \ln \delta(0) + \int_0^t \lambda(\delta(t))dt.$$

Time series, characterizing economic systems of varying degrees of complexity, differ in magnitudes $\Delta\lambda = \lambda_{max} - \lambda_{min}$, $\Delta\delta = \delta_{max} - \delta_{min}$, and I. As an example, the integral measure I is calculated for a sliding window of 400 days and step size of 1 day for the daily values of the BTC index (see figure 16).

Like the previous LLE indicators, SLDE is also a leading indicator. However, its disadvantage is the impossibility of accurate calculations for small window sizes which limits the possibilities of its usage.

5.2 Lévy alpha-stable distribution

Financial crises that regularly shake the world economy are characterized by noticeable fluctuations in stock indices, thereby causing noticeable changes in the statistical distributions of empirical data [111, 245].

In 1900, Bachelier proposed the first model for the stochastic process of returns – an uncorrelated random walk with independent, identically Gaussian distributed (i.i.d) random variables [246]. This model is natural if one considers the return over a time scale Δt to be the result of many independent "shocks", which then lead by the central limit theorem to a Gaussian distribution of returns [246]. Some stylized facts of daily returns [247–249] re-

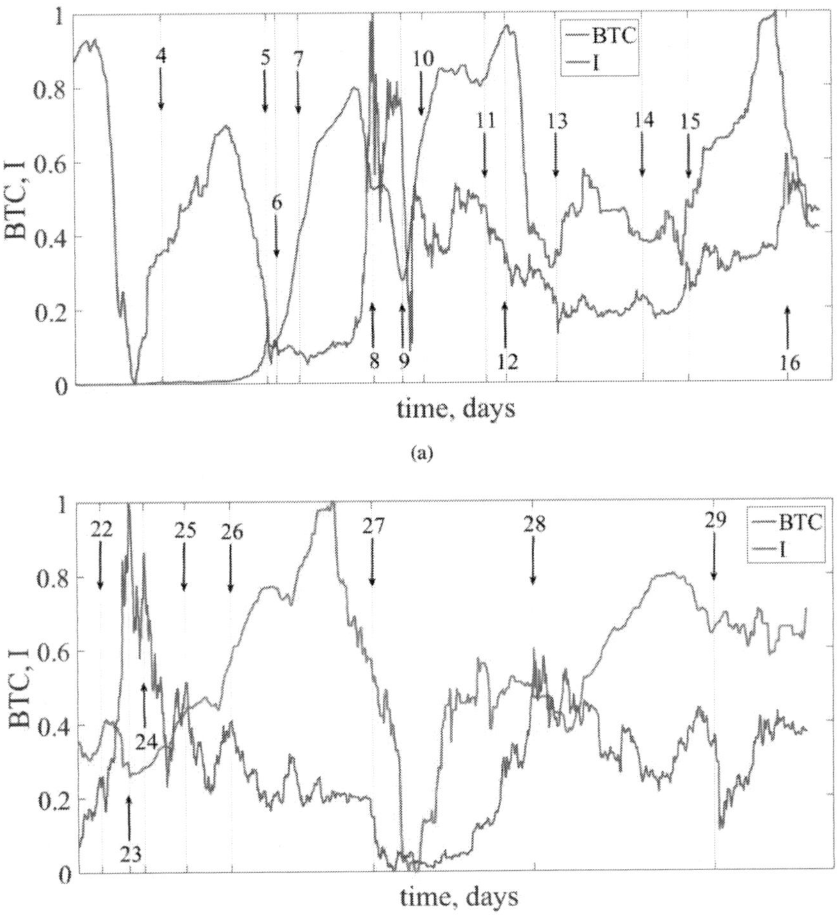

(a)

(b)

Figure 16: The dynamics of the SDLE exponents within the time window of length 400 days and the time step of 1 day for first (a) and second (b) periods

veal that distributions are leptokurtic and, therefore, Gaussian distribution does not fit well to the data. Sornette and Lux [250] pronounce that the distribution of such data may be not only leptokurtic, but it can also be characterized by fat tails [251–254]. Thus, it should belong to the class of fat-tailed distributions. Formally, it is said that they follow a *power law*. The emergence of power-law behavior in price fluctuations is argued to be a consequence of underlying complex mechanisms, such as feedback effects and correlations in financial markets [255–258]. Some theories associate this phenomenon with market impact and the distribution of large investors [247, 259], while other studies model the power-law behavior as a consequence of limited information and the true value of companies [260]. Such property is a symptom of self-organization and complexity which are prominent for economic systems. In Chakraborty et al. [261] paper it was established that currencies of several frontiers that are outside of *inverse cubic law* (with an exponent of $\alpha \simeq 3$) belong to the Levy-stable regime and are expected to be yet emerging and having sudden large changes such as crashes and critical events, while those of most developed exhibited *inverse cubic law*.

Recently, it has been reported that Bitcoin is becoming more mature, following *inverse cubic law* [262–264]. Besides, Stjepan Begušić et al. [265], motivated by the rise of novel assets based on blockchain technology, presented a detailed analysis of trade-level data of the BTC/USD pairs from five large Bitcoin exchanges: Mt. Gox, BTC-e, Bitstamp, Bitfinex, and Kraken. They applied two estimation methods and a resampling-based technique to statistically validate if the main cryptocurrency follows power-law behavior or not. Their study presented that the exponent α lie within the range $2 < \alpha < 2.5$ that gives the evidence that the cryptocurrencies market is much more volatile than the stock market, and that Bitcoin returns exhibit much heavier tails. Moreover, they find that such a phenomenon is universal as such behavior holds across multiple exchanges and tiny intervals. Their results imply that Bitcoin lies outside of the Levy-stable region which provides the existence of a finite second moment, and a basis for the usage of standard financial theories for portfolio optimization and risk management.

In figure 17 the daily returns of Bitcoin and contrast it with a sequence of i.i.d. Gaussian random variables are presented.

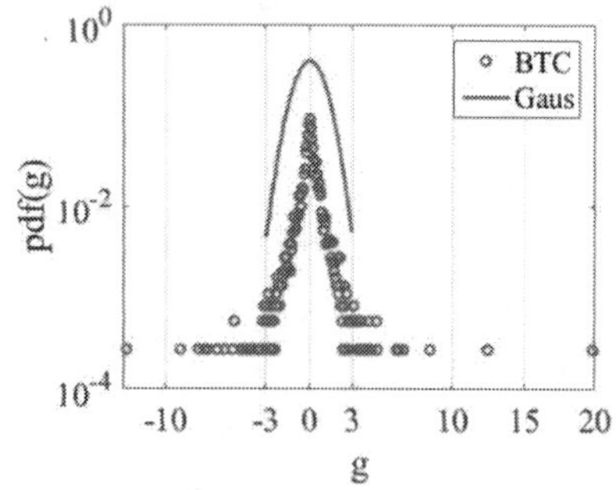

(a)

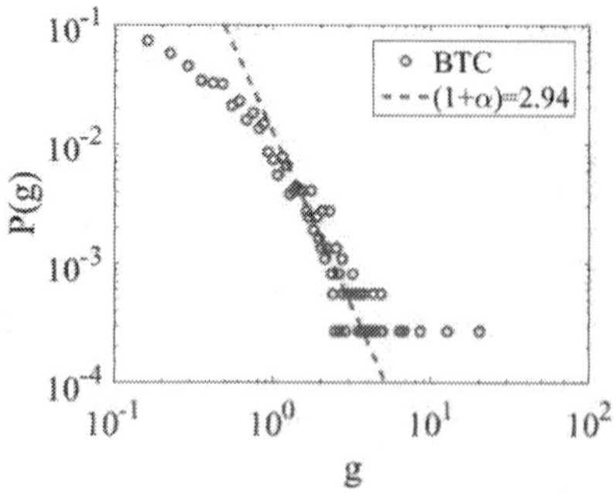

(b)

Figure 17: Probability density function of BTC daily normalized returns for the whole period (a). Cumulative distribution of the normalized BTC daily returns (b). Fits yield values $\alpha = 1.94 \pm 0.02$

It is obvious that the distribution of returns has heavy tails and in general case can be described as

$$P(g > x) \sim x^{-(1+\alpha)}, \quad \alpha \in (0, 2]. \tag{19}$$

Figure 17 b it can be seen that Bitcoin exhibits the inverse cubic law.

In the analysis of cotton prices, Mandelbrot observed that in addition to being non-Gaussian, the process of returns shows another interesting property: "time scaling" – that is, the distributions of returns for various choices of Δt, ranging from 1 day up to 1 month have similar functional forms [266]. Motivated by (i) pronounced tails, and (ii) a stable functional form for different time scales, Mandelbrot proposed that the distribution of returns is consistent with a Levy stable distribution [266–268] – that is, the returns can be modeled as a Levy stable process. Levy stable distributions arise from the generalization of the Central Limit Theorem (CLT) to random variables that do not have a finite second moment.

The CLT [269], which offers the fundamental justification for approximate normality, points to the importance of alpha-stable distribution: they are the only limiting laws of normalized sums of independent, identically distributed random variables. Gaussian distributions, the best-known member of the stable family, have long been well understood and widely used in all sorts of problems. However, they do not allow for large fluctuations and are thus inadequate for modeling high variability. Non-Gaussian stable models, on the other hand, do not share such limitations. In general, the upper and lower tails of their distributions decrease like a power function. In literature, this is often characterized as heavy or long tails. In the last two or three decades, data that seem to fit the stable model has been collected in fields as diverse as economics, telecommunications, hydrology, and physics [248].

Consequently, a probability model with a power tail can be suitable for identifying processes with extreme events. It was discovered that alpha-stable distributions fit better than the Gaussian distribution to financial and spot markets. It is still debatable whether Lévy stable distribution is applicable since there is not enough theoretical material, and there is not a universal analyzing method for estimating parameters of Lévy stable distribution.

5.2.1 Lévy's stable distribution properties

Lévy stable distribution being the generalization of the CLT, became an addition to a wide class of distributions. The term *stable* is such characteristic of distribution where the shape (up to scale and shift) retains under addition: if $X, X_1, X_2, \ldots, X_n$ are independent, identically distributed random variables, then for every n

$$X_1 + X_2 + \cdots + X_n \stackrel{d}{=} c_n X + d_n \tag{20}$$

for some constants $c_n > 0$ and $d_n \in \mathbb{R}$, where $X_1, \ldots, X_n$ are independent, identical copies of X.

The class of all laws that satisfy condition (20) is presented by 4 parameters: $\alpha \in (0, 2]$ is the *index of stability* or *characteristic exponent* where a smaller value of α corresponds to more severe tails of the distribution. The parameter $\beta \in [-1, 1]$ is called the *skewness* parameter of the law. If $\beta = 0$, the distribution is symmetric. In the case when $\beta > 0$, it is skewed toward the right, otherwise to the left. The last two parameters stand for the *scale* $\gamma \in [0, \infty)$ and $\delta \in (-\infty, \infty)$ the *location* parameters of the distribution. Since random variables X is characterized by four parameters, we will denote α-stable distribution by $S(\alpha, \beta, \gamma, \delta)$ and write

$$X \sim S(\alpha, \beta, \gamma, \delta). \tag{21}$$

Lévy stable distributions cannot be defined in closed form expression except few cases: the case of $(\alpha, \beta) = (2, 0)$ corresponds to the Gaussian distribution, $(\alpha, \beta) = (1, 0)$ to the Cauchy distribution. Instead, it is expressed in terms of their Fourier transforms or characteristic functions (CF). If the density $f(x)$ exists, CF of that density can be expressed as

$$\lambda(k) = E\exp(ikX) = \int_{-\infty}^{\infty} \exp(ikx) f(x) dx,$$

where k denotes the Fourier transformed variable. Thus, the inverse Fourier transform

$$f(k) = \frac{1}{2\pi}\int_{-\infty}^{\infty} \exp(-ikx)\lambda(k) dk$$

allows us to reconstruct probability density function with a known characteristic function.

As we do not have any analytical expression for the probability density of a random variable X_i, for Lévy stable distribution, if the variable x_i follows $S(\alpha, \beta, \gamma, \delta)$, the CF can be expressed as [270]:

$$\lambda(k) = \begin{cases} \exp i\delta k - \gamma^\alpha |k|^\alpha \left[1 - i\beta \mathrm{sgn}(k)\tan(\frac{\pi\alpha}{2})\right], & (\alpha \neq 1), \\ \exp i\delta k - \gamma|k| \left[1 + i\beta \mathrm{sgn}(k)\frac{2}{\pi}\ln|k|\right], & (\alpha = 1). \end{cases} \tag{22}$$

5.2.2 Methods for estimation of stable law parameters

There are numerous approaches that can estimate stable distribution parameters. Since the probability density functions are not always expressed in a closed form, there are some challenges to overcome the analytic difficulties. Thus, there have been constructed a variety of methods: the approximate maximum likelihood (ML) estimation [271, 272], quantiles method [273, 274], fractional lower order moment method [275, 276], method of log-cumulant [277], the logarithmic moment method [278] and more. Unfortunately, some of those methods cannot be applied due to computational problems associated with a limited range of estimation, restricted range of parameters, high computational costs, or requiring a large number of data. However, several of them should be mentioned.

5.2.3 Maximum likelihood method

DuMouchel was the first to obtain approximate ML estimates of α and γ (assuming $\delta = 0$) [279]. A multinomial approximation to the likelihood function is used in his approach. Under some additional assumptions on $\hat{\alpha}$ and the likelihood function, DuMouchel has shown the obtained estimates to be consistent and asymptotically normal. However, the computational effort involved seems considerable.

A direct method can be formulated, after Brorsen and Yang [271], as follows. The standard symmetric probability density functions defined by Zolotariev [280] is presented as:

$$f_\alpha(x) = \frac{\alpha}{\pi|1-\alpha|} x^{1/(\alpha-1)} \times$$
$$\times \int_0^{\pi/2} U_\alpha(\eta, 0) \exp -x^{\alpha/(\alpha-1)} U(\eta, 0) d\eta, \tag{23}$$

for $\alpha \neq 1$, $x > 0$, where U_α is defined by:

$$U_\alpha(\eta, \eta_0) = \left(\frac{\sin\alpha(\eta - \eta_0)}{\cos\eta}\right)^{\alpha/(1-\alpha)} \left(\frac{\cos\eta - \alpha(\eta - \eta_0)}{\cos\eta}\right) \tag{24}$$

and η_0 is explained here [281]. Therefore, the parameters α, γ, and δ can be estimated from the observations $\{x_i \mid i = 1, 2, \ldots, N\}$ by maximizing the log likelihood function:

$$\sum_{i=1}^{N} \log f_\alpha(z_i) = n\log\alpha - n\log(\alpha - 1) +$$
$$+ \sum_{i=1}^{N} \frac{\log z_i}{\alpha - 1} +$$
$$+ \sum_{i=1}^{N} \log \int_0^{\pi/2} U_\alpha(\eta, 0) \exp -z_i^{\alpha/(\alpha-1)} U_\alpha(\eta, 0) d\eta, \tag{25}$$

where $z_i = |x_i - \delta|/\gamma$.

To avoid the discontinuity and non-differentiability of the symmetric α-stable density function at $\alpha = 1$, *alpha* is restricted to be greater than one. Caution must be used when evaluating the integrals in equations (23) and (25), since the integrals are singular at $\eta = 0$.

An obvious disadvantage of this method is that it is a highly nonlinear optimization problem and no initialization and convergence analysis is available.

5.2.4 Quantiles methods

This method is focused on empirical quantiles, which has been introduced by Fama and Roll, with the assumptions that $\alpha > 1, \beta = 0$, and $\delta = 0$ [273]. However, it was much more appreciated through McCulloch [274] after its extension to include asymmetric distribution for $\alpha \in [0.6, 2]$.

In order to quantify the four parameters of the stable distribution, we consider N independent variables x_i that follow alpha-stable distribution (21) [274, 282]. Then, we need to define five empirical quantiles of probability levels 5%, 25%, 50%, 75%, and 95%. Then we have to obtain two intermediate quantities:

$$\begin{cases} \hat{\upsilon}_\alpha = \frac{\hat{Q}_{0.95} - \hat{Q}_{0.05}}{\hat{Q}_{0.75} - \hat{Q}_{0.25}}, & \hat{\upsilon}_\beta = \frac{\hat{Q}_{0.95} + \hat{Q}_{0.05} - 2\hat{Q}_{0.05}}{\hat{Q}_{0.95} - \hat{Q}_{0.05}}, \\ \hat{\alpha} = \Psi_1(\hat{\upsilon}_\alpha, \hat{\upsilon}_\beta), & \hat{\beta} = \Psi_2(\hat{\upsilon}_\alpha, \hat{\upsilon}_\beta), \end{cases}$$

where $\hat{Q}_p (p = 0.05, 0.25, 0.5, 0.75, 0.95)$ is the corresponding sample data with which quantile is calculated, and Ψ_1 with Ψ are the interpolating functions the values of which can be found in Table I-IV by McCulloch [274]. Further, the scale parameter is given by:

$$\hat{\gamma} = \frac{\hat{Q}_{0.75} - \hat{Q}_{0.25}}{\Psi_3(\hat{\alpha}, \hat{\beta})},$$

where $Psi_3(\hat{\alpha}, \hat{\beta})$ is given in Table V [274]. For simplicity of the location parameter, we can define the variable which is predefined in the following form:

$$\xi = \begin{cases} \delta + \beta\gamma\tan\frac{\pi\alpha}{2}, & \text{if } \alpha \neq 1, \\ \delta, & \text{if } \alpha = 1. \end{cases} \tag{26}$$

In consistence with the corresponding parameter ξ which can be estimated by $\hat{\xi} = \hat{Q}_{0.5} + \hat{\gamma}\Psi_5(\hat{\alpha}, \hat{\beta})$ (Ψ_5 can be obtained through linear interpolation according to Table VII [274]), the location parameter δ is given by:

$$\hat{\delta} = \hat{\xi} + \hat{\beta}\hat{\gamma}tan\frac{\pi\hat{\alpha}}{2}.$$

5.2.5 Empirical characteristic function method

Analyzing data, we often assume that they are ergodic [283]. In general, if random variables are ergodic with the integrable function $f(x)$, and the measure $\rho(x)dx$ in the space M, then the following equation holds [284]:

$$\lim_{N\to\infty} \frac{1}{N}\sum_{i=1}^{N} \exp(ikx_i) = \int_{-\infty}^{\infty} \exp(ikx)\rho(x)dx. \quad (27)$$

Then, to consider characteristic functions, equation (27) comes out to be the following ergodic equality [284]:

$$\lim_{N\to\infty} \frac{1}{N}\sum_{i=1}^{N} \exp(ikx_i) = \int_{-\infty}^{\infty} \exp(ikx)f(x)dx \quad (28)$$

for which we have

$$\hat{\lambda}(k) = \lim_{N\to\infty} \frac{1}{N}\sum_{i=1}^{N} \exp(ikx_i). \quad (29)$$

This assumption allows us to empirically obtain the probability distribution. Hence, the empirical characteristic function $\lambda_N(k)$ can be calculated as

$$\hat{\lambda}_N(k) = \frac{1}{N}\sum_{i=1}^{N} \exp(ikx_i). \quad (30)$$

Then, according to Koutrouvelis' [270, 285] regression type from (22) it can be derived that

$$\log(-\log(|\lambda(k)|^2)) = \log(2\gamma^\alpha) + \alpha\log(k). \quad (31)$$

The imaginary and real parts of $\lambda(k)$ are given by

$$\begin{cases} \lambda_I(k) = \exp(-|\gamma k|^\alpha)\cdot i\sin[\delta k - |\gamma k|^\alpha \beta \operatorname{sgn}(k)\omega(k,\alpha)], \\ \lambda_R(k) = \exp(-|\gamma k|^\alpha)\cdot \cos[\delta k - |\gamma k|^\alpha \beta \operatorname{sgn}(k)\omega(k,\alpha)], \end{cases} \quad (32)$$

where

$$\omega(k,\alpha) = \begin{cases} \tan\frac{\pi\alpha}{2}, & \alpha \neq 1, \\ \frac{2}{\pi}\ln|k|, & \alpha = 1. \end{cases}$$

Suppose $\Upsilon := \arctan(\lambda_I(k)/\lambda_R(k))$. Then, in the condition $\alpha \neq 1$, the last two equations lead, apart from considerations of principal values, to

$$\Upsilon(k) = \delta k - \beta\gamma^\alpha \tan(\frac{\pi\alpha}{2})\operatorname{sgn}(k)|k|^\alpha. \quad (33)$$

Equation (31) depends only on α and γ, and it suggests that we estimate these parameters by regressing

$$y = \log(-\log|\lambda_N(k)|^2)$$

on $\omega = \log(k)$ in the model

$$y_l = m + \alpha\theta_l + \epsilon_l, \quad \text{for } l = 1,\ldots,L, \quad (34)$$

where $y_l = \log(-\log(\hat{\lambda}(k_l))^2)$, $m = \log(2\gamma^\alpha)$, $\theta_l = \log(k_l)$, and ϵ_l responds for an error term. The proposed real data set for L (see Koutrouvelis [270], Table I) is given by $k_l = \pi l/25$ ($l = 1,\ldots,L$).

With estimated and fixed parameters α and γ, the symmetric parameter β and location parameter δ can be obtained by linear regression estimation

$$z_q = \delta k_q - \beta\gamma^\alpha \tan(\frac{\pi\alpha}{2})\operatorname{sgn}(k_q)|k_q|^\alpha + v_q, \quad \text{for } q = 1,\ldots,Q, \quad (35)$$

where $z_q = \Upsilon_N(k_q) + \pi l_N(k_q)$, v_l denotes an error term, and the proposed real data set for Q (see Koutrouvelis [270], Table II) is $k_q = \pi q/50$ ($q = 1,\ldots,Q$).

5.2.6 Related studies and corresponding results

Recently, the use of dynamic indicators, precursors of crashes in stock markets using the parameters of a α-stable distribution was proposed by us in the papers [76, 77, 286] and later repeated in a recent paper [30]. From the data above, we estimate the parameters α and β of the stable distribution that the best describes the empirical returns. Figure 18 shows the dynamics of the parameter α as a more informative indicator.

From the figure 18 we can see that our parameters start to decrease in crisis states. Such abnormal behavior can serve as an indicator or precursor of crashes and critical states.

6 Recurrence analysis

In 1890 the mathematical foundations of recurrence were introduced by Henri Poincaré, resulting in the *Poincaré recurrence theorem* [287]. This theorem states that certain systems will return to their arbitrarily close, or exactly the same initial states after a sufficiently long but finite time. Such property in the case of deterministic behavior of the system allows us to make conclusions regarding its future development.

6.1 Recurrence plot

Recurrence plot (RP) have been introduced to study dynamics and recurrence states of complex systems. When we create RP, at first, from recorded time series we reconstruct phase-space trajectory. Then, according to Eckmann et al. [231], we consider a trajectory $\vec{X}(i)$ on the reconstructed trajectory. The recurrence plot is an array of dots in a $N \times N$ matrix, where dot is placed at (i, j) whenever $\vec{X}(j)$ is sufficiently close to $\vec{X}(i)$, and both axes are time axes which mathematically can be expressed as

$$R_{ij} = \mathcal{H}(\epsilon - \left\|\vec{X}(i) - \vec{X}(j)\right\|), \quad \text{for } i,j = 1,\ldots,N, \quad (36)$$

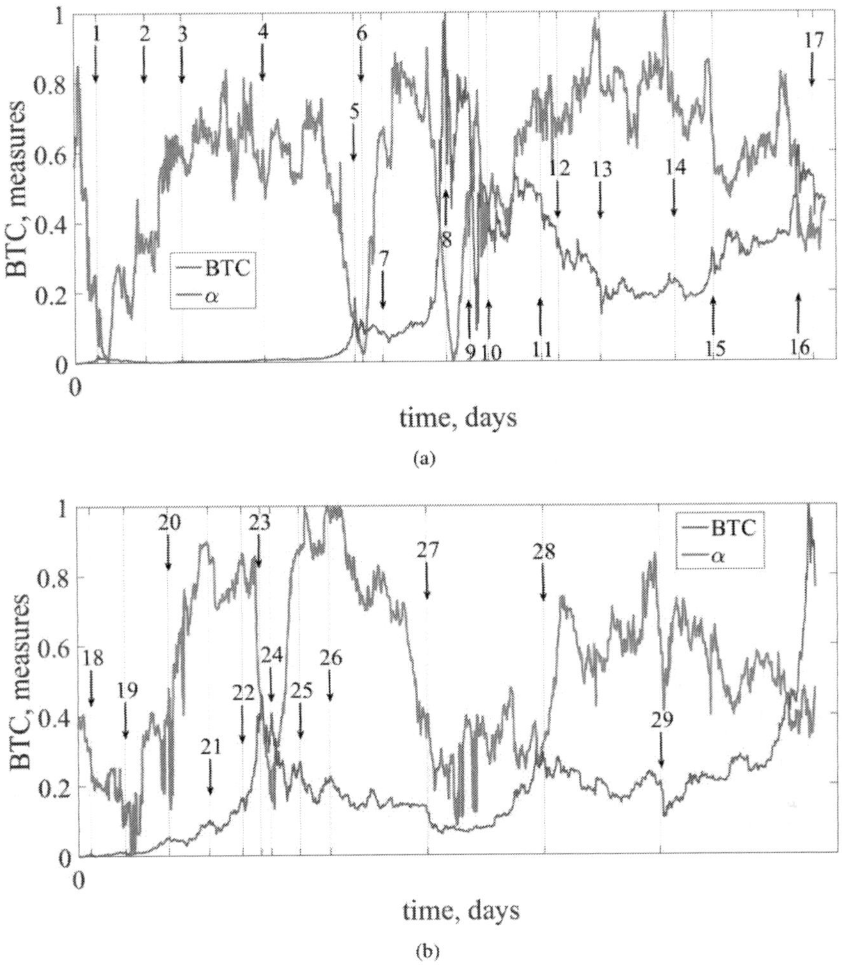

Figure 18: The Bitcoin time series and estimated for them parameter α. Vertical arrows indicate crashes and critical events

where $\| \|$ is a norm (representing the spatial distance between the states at times i and j); ϵ is a predefined recurrence threshold, and $\mathcal{H}$ is the Heaviside function. As a result, the matrix captures a total of N^2 binary similarity values. A synthetic example is presented in figure 19.

Typically an L_p-norm is applied to determine the pairwise similarity between two vectors. Accordingly to Webber and Zbilut [139], there are such candidates that can serve as a distance measure:

- the L_1-norm (Manhattan distance);
- the L_2-norm (Manhattan distance);
- the L_∞-norm (Manhattan distance);

In accordance with our results, Maximum distance seems to be a suitable choice. It is often used as it is independent of the phase space dimension, easy to calculate, and allows some analytical expression [288–290].

Also, as it can be seen from equation (36), the similarity between vectors is determined by a threshold ϵ. The choice of $\epsilon > 0$ ensures that all vectors that lie within this radius are similar to each other, and that dissimilarity up to a certain error is permitted [287].

The fixed radius for recurrent states is the commonly used condition, which leads to equally sized ϵ-neighborhoods. The shape in which neighborhoods lie is determined by the distance metric. Applying the fixed threshold with the distance metric, we define recurrence matrices that are symmetric along the middle diagonal. The self-similarity of the multi-dimensional vectors reflects in the middle diagonal which is commonly referred to as *line of identity* (LOI). In contrast, it is not guaranteed that a recurrence matrix is symmetric, if the condition of fixed number of nearest neighbors is applied. For specific purposes (e.g., quantification of recurrences), it can be useful to exclude the LOI from the RP, as the trivial recurrence of a state with itself might not be of interest [291].

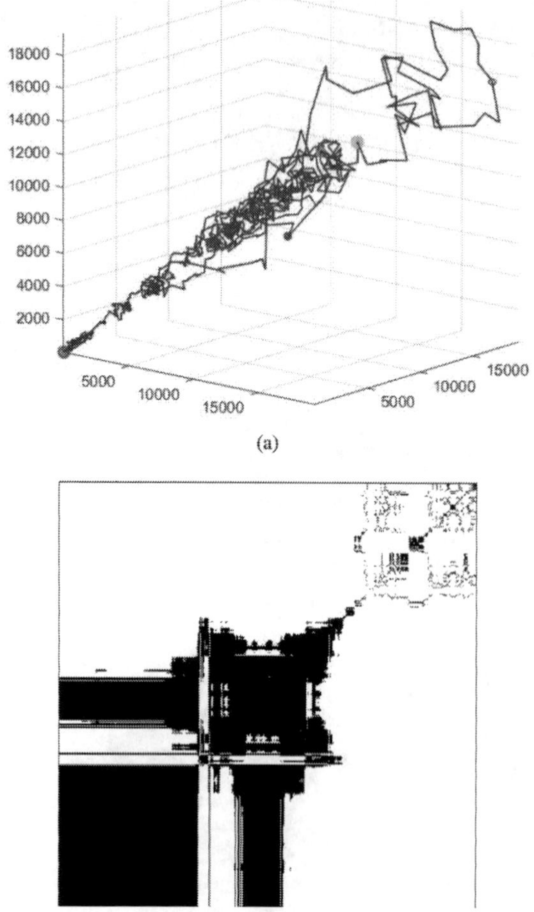

(a)

(b)

Figure 19: Phase portrait (a) and corresponding RP (b) for BTC

6.1.1 Recurrence plots and their structures

The visualization of trajectories and hidden patterns of the systems is the "destiny" of RP [291, 292].

The dots within RP, representing the time evolution of the trajectories, exhibit characteristic large-scale and small-scale patterns. Large-scale patterns of RP can be classified as *homogeneous*, *periodic*, *drift*, and *disrupted* [289, 293, 294]:

- *Homogeneous* typify behavior of autonomous and stationary systems, which consist of a large number of recurrence points that are homogeneously distributed (relaxation times are short).
- *Periodic* represents long, uninterrupted, and diagonally oriented structures that represent which indicate periodic behavior. These lines are usually distributed regularly.
- Paling or darkening from the LOI to the outer corners in RP is characteristic of *drift*.
- *Disrupted* distribution of recurrence points may serve as an indicator of drastic changes as well as extreme events in the system dynamics. In these cases, RP can be used to find and assess extreme and rare events by scoring the frequency of their repeats.

6.1.2 Recurrence quantification analysis

For a qualitative description of the system, the graphic representation of the system suits perfectly. However, the main disadvantage of graphical representation is that it forces users to subjectively intuit and interpret patterns and structures presented within the recurrence plot. Also, with plots' increasing size, they can be hardly depicted on graphical display as a whole. As a result, we need to work with separated parts of the original plot. Analysis in such a way may create new defects that which should distort objectivity of the observed patterns and lead to incorrect interpretations. To overcome such limitation and spread an objective assessment among observers, in the early 1990s by Webber and Zbilut were introduced definitions and procedures to quantify RP's complexity, and later, it has been extended by Marwan et al. [294–296].

The small-scale clusters can represent a combination of *isolated dots* (chance recurrences). Similar evolution at different periods in time or in reverse temporal order will present *diagonal lines* (deterministic structures) as well as *vertical/horizontal lines* to inscribe laminar states (intermittency) or systems that paused at singularities. For the quantitative description of the system, such small-scale clusters serve the base of the recurrence quantification analysis (RQA).

Usually, first acquaintance with classical RQA starts with recurrence point density, or, as it is known, recurrence rate (RR):

$$RR = \frac{1}{N^2}\sum_{i,j=1}^{N} R_{i,j}.$$

It enumerates the probability that any state of the system will recur. It is the simplest measure that is computed by taking the number of nearest points forming short, spanning row and columns of the recurrent plot. It summarizes them and divides by the number of possible points in the recurrence matrix of size N^2.

The remaining measure relies on the frequency distribution of line structures in the RP. First, we consider the histogram of the length of the diagonal structures in the RP

$$P(l) = \sum_{i,j=1}^{N}\left\{(1 - R_{i-1,j-1})\cdot(1 - R_{i+l,j+l})\cdot\prod_{k=0}^{l-1} R_{i+k,j+k}\right\}.$$

The fraction recurrence points on the recurrence plots that form line segments of minimal length μ parallel to the matrix diagonal is the measure of determinism (DET):

$$DET^{(\mu)} = \frac{\sum_{l=\mu}^{N} l\cdot P(l)}{\sum_{i,j=1}^{N} R_{i,j}} = \frac{\sum_{l=\mu}^{N} l\cdot P(l)}{\sum_{l=1}^{N} l\cdot P(l)}.$$

Systems that exhibit deterministic dynamics are mainly characterized by diagonal lines. Long diagonal lines indicate periodic signals, but short diagonal lines stand for chaotic behavior. Regarding the quantitative analysis, typically, only the lines with minimal length $\mu = 2$ are considered. If $\mu = 1$, then DET and RR are identical. For some systems, DET becomes more reliable if $\mu > 2$. Here, μ serves as a filter, excluding the shorter lines. However, it should be noted that too large μ may spoil the histogram $P(l)$ and thus the reliability of DET.

The results of calculations of window dynamics of the considered recurrence measures are presented in Figure 20. RR and DET are calculated for local time series of 50 days and a step of 1 day. In this case, the beginning of a crash or critical event is at point 100.

It is evident that the two recurrent measures during abnormal periods decrease long before the actual anomaly. The complex system becomes less recurrent and less deterministic which is logical in the periods approaching critical phenomena. And, consequently, RR and DET can be used as precursors of critical and crash phenomena.

6.1.3 Chaos recurrent measures

The corresponding measure of entropy is related to the recurrence properties that may be peculiar for the nonlinear complex system and important class of recurrence quantifiers are those that try to capture the level of complexity of a signal [78, 84, 297]. In accordance with this study, the entropy diagonal line histogram (DLEn) is of the greatest interest which uses the Shannon entropy of the distribution of diagonal lines $P(l)$ to determine the complexity of the diagonal structures within the recurrence plot. One of the most know quantitative indicators of the recurrence analysis can be defined as:

$$DLEn = -\sum_{l=l_{min}}^{l=l_{max}} p(l) \ln p(l)$$

and

$$p(l) = \frac{P(l)}{\sum_{l=l_{min}}^{N} P(l)},$$

where $p(l)$ captures the probability that a diagonal line has exactly length l, and DLEn reflects the complexity of deterministic structure in the system. Further calculations were provided and presented in figure 21 for both Bitcoin time series.

However, as follows from the analysis of the entropy indicators, the results may differ for different data preparation. Further, we take into account two types of Shannon entropy-based approaches: recurrence period density entropy (RPDEn) and recurrence entropy (RecEn).

The RPDEn is the quantitative measure of the recurrence analysis that is useful for characterizing the periodicity or absolutely random processes in the time series. It is useful for quantifying the degree of repetitiveness [298, 299]. Considering embedded data point $\vec{X}(i)$ from the phase space and suitable threshold ϵ in d_E-dimensional space. Then the trajectory is followed forward in time until it has left the corresponding threshold ϵ. Subsequently, the time j at which the trajectory first returns to this ball and the period T of previous and current states is recorded. The procedure is repeated for all states of the RPs, forming a histogram of recurrence times $R(T)$. The histogram is then normalized to give the recurrence time probability density:

$$P(T_i) = \frac{R(T_i)}{\sum_{i=1}^{T_{max}} R(T_i)},$$

where $T_{max} = \max\{T_i\}$. The normalized entropy of the obtained density can be defined as:

$$RPDEn = \frac{-\sum_{i=1}^{T_{max}} P(T_i) \ln P(T_i)}{\ln T_{max}}. \quad (37)$$

In fact, based on the length of the sequences of neighboring points in the phase space: the more points are neighborhoods, the lower the value of the entropy according to equation (37). The comparing of RPDEn and Bitcoin's critical states can be seen in figure 22.

However, recent articles [297, 300] present a slightly different technique for calculating recurrent entropy using a novel way to extract information from the recurrence matrix. To properly define it, we need to define the microstates $F(\epsilon)$ for the RP that are associated with features of the dynamics of the time series. Selecting the appropriate metric and using the Heaviside function, we evaluate the matrices of dimensions $N \times N$ that are sampled from the RP. The total number of microstates for a given N is $N_{ms} = 2^{N^2}$. The microstates are populated by $\overline{N}$ random samples obtained from the recurrence matrix such that:

$$\overline{N} = \sum_{i=1}^{N_{ms}} n_i$$

with n_i representing the number of times that a microstate i is observed. The probability of occurrence of the related microstate i can be obtained as:

$$p_i = n_i \cdot \left(\overline{N}\right)^{-1}.$$

The RecEn of the RP associated with the probability distribution of the corresponding microstates is given by the following equation:

$$RecEn = \sum_{i=1}^{N_{ms}} p_i \ln p_i. \quad (38)$$

In figure 23 we can see the performance of *RecEn* accordingly to the described above method.

A vertical line of length l starting from a dot (i, j) means that the trajectory starting from $\vec{X}(j)$ remains close to $\vec{X}(i)$ during $l-1$ time steps. A diagonal black line of length l starting from a dot i, j means that trajectories starting from $\vec{X}(i)$ and $\vec{X}(j)$ remain close during $l-1$ time steps, thus these lines are related to the divergence of the trajectory segments. The average diagonal line length

$$L_{mean} = \frac{\sum_{l=l_{min}}^{N} l \cdot P(l)}{\sum_{l=l_{min}}^{N} P(l)}$$

is the average time that two segments of the trajectory are close to each other and can be interpreted as the mean prediction time. Here, $P(l)$ is a histogram of diagonal lines of length l.

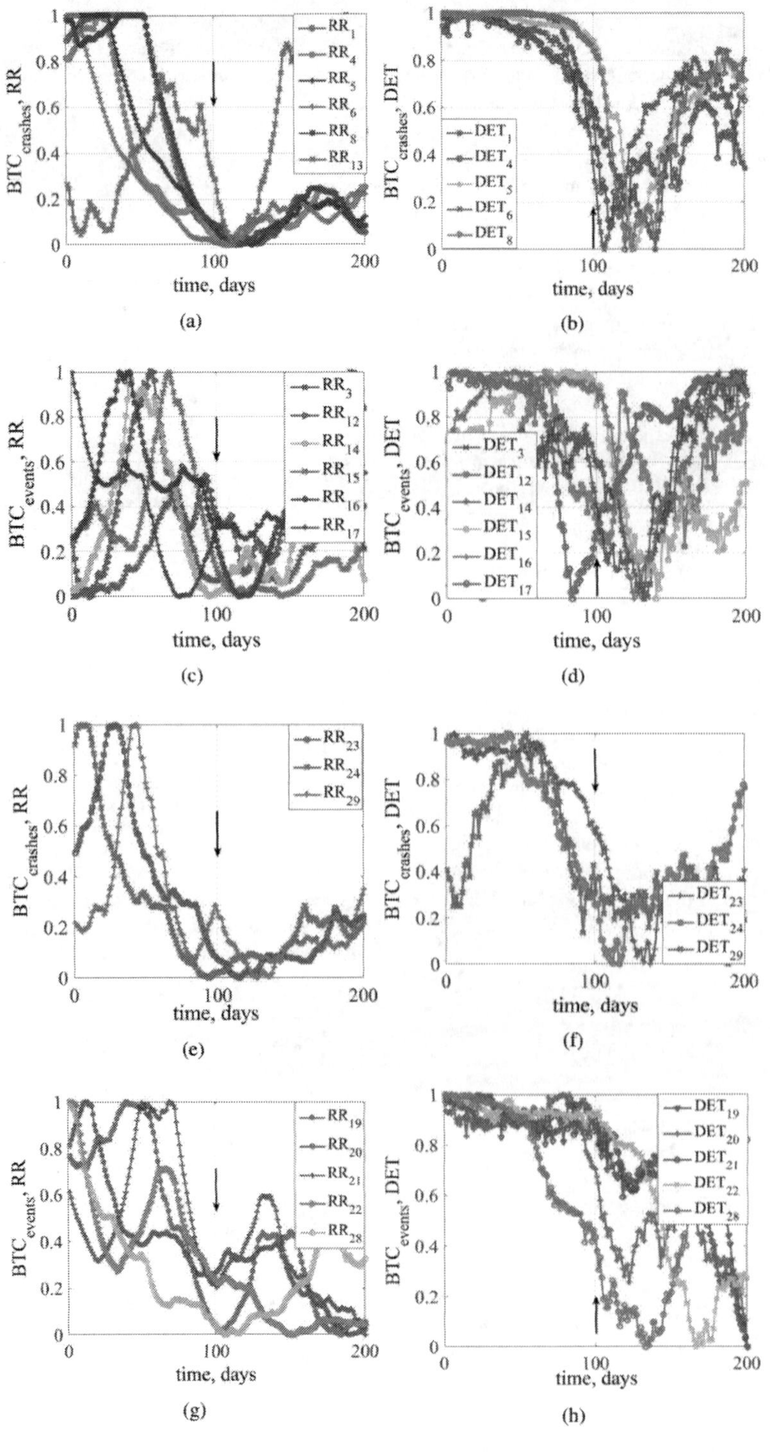

Figure 20: Dynamics of RR and DET for crashes (a, b, e, f) and critical events (c, d, g, h)

Another measure (L_{max}) considers the length of the longest diagonal line found in the RP. In other words, it means the maximum time that two segments of the trajectory are close to each other, and the following equation can

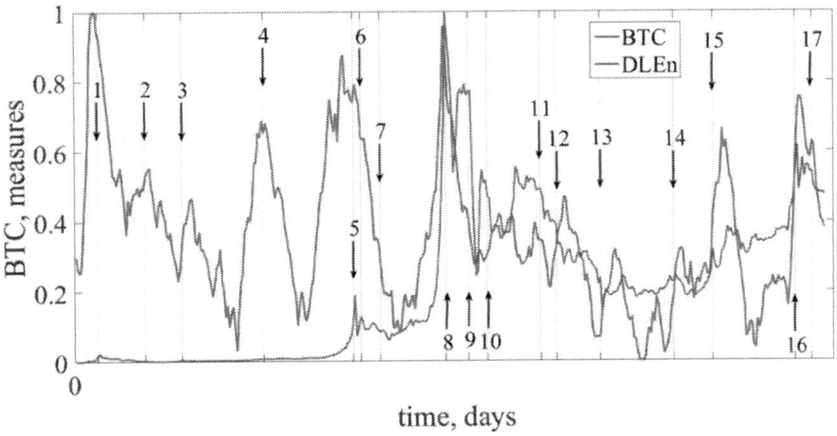

(a)

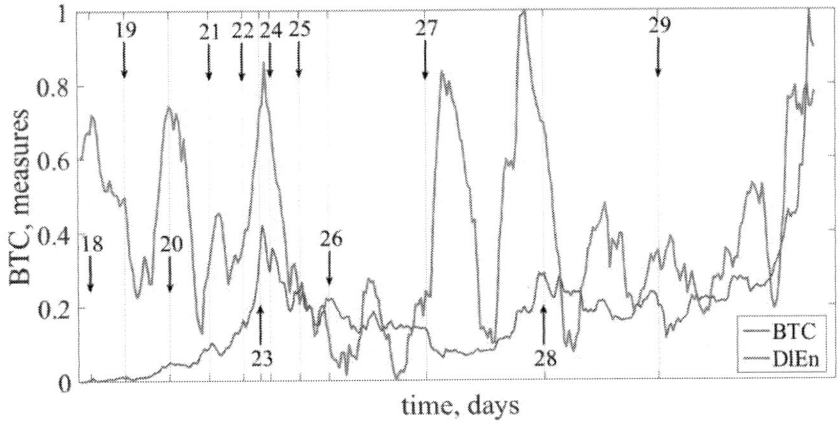

(b)

Figure 21: *DLEn* dynamics along with first (a) and second (b) periods of the entire time series of Bitcoin

be defined as

$$L_{max} = \max\left(\{l_i \mid i = 1, \ldots, N_l\}\right),$$

where $N_l = \sum_{l \geqslant l_{min}} P(l)$ is the total number of diagonal lines.

Respectively, the inverse of L_{max} characterizes the exponential divergence of the phase space trajectory [301, 302]. Faster the trajectory segments diverge, shorter are the diagonal lines and higher is the measure of *divergence* (*DIV*). It is given by the following equation:

$$DIV = 1/L_{max}.$$

Therefore, the measure of *DIV*, according to Eckmann [231], can be used to estimate the largest positive Lyapunov exponent. The comparative dynamics of the measure of divergence and BTC time series are presented in figure 24.

A comparative analysis of the measures under consideration revealed an obvious advantage of the recursive measure. In addition to the smoothness of the measure itself, it can be calculated for windows of small sizes, which leads to inaccurate or incorrect results for other methods.

7 Irreversibility

Complex systems are open systems that exchange energy, matter, and information with the environment. Investigating complex systems in the natural sciences, Prigogine made a fundamental generalization, indicating the need for consideration of the phenomena of irreversibility and non-equilibrium as principles of selection of space-time structures that are implemented in practice [303]. Later it became clear that this generalization extends to complex systems of another nature: social, economic, biomedical, etc. [304]. Prigogine believed that the most important changes in the modern scientific revolution are related to the removal of previous restrictions in the scientific understanding of time. The nonlinear world is characterized by features of temporality, i.e., irreversibility and transience

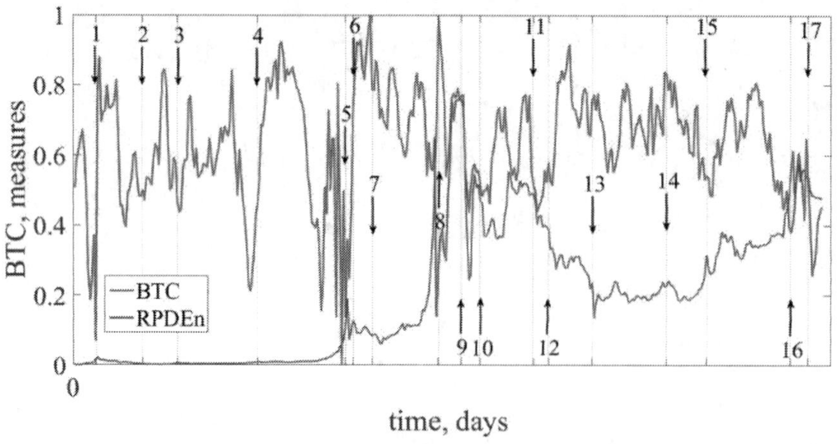

(a)

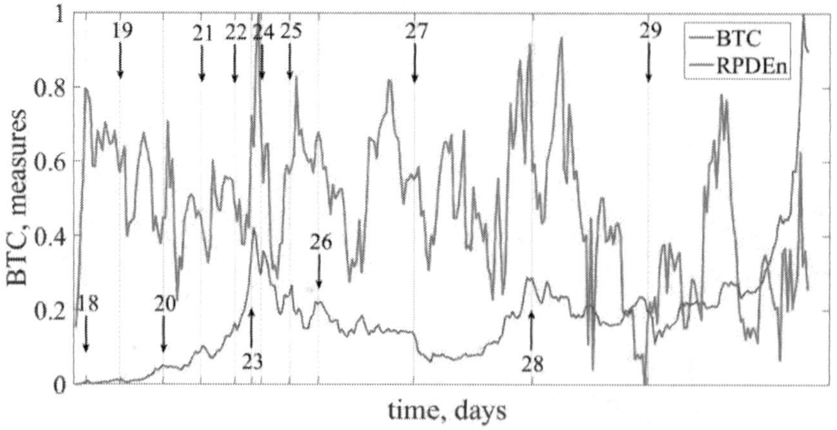

(b)

Figure 22: *RPDEn* dynamics along with first (a) and second (b) periods of the entire time series of Bitcoin

of processes and phenomena. Self-organization is considered as a spontaneous process of formation of integrating complex systems. It is due to the ambiguity of choice at bifurcation points that time in theories of self-organization becomes truly irreversible. In contrast to linear dynamic theories – classical, relativistic, quantum (where time is reversed), in the thermodynamics of dissipative structures created by Prigogine, time ceases to be a simple parameter and becomes a concept that expresses the pace and direction of events.

Thus, the irreversibility of time is a fundamental property of non-equilibrium dissipative systems, and its loss may indicate the development of destructive processes [107, 304].

A stationary process X is called statistically inverse in time, if for any N, the series $\{x(i) \mid i = 1, \dots, N\}$ and $\{x(i) \mid i = N, \dots, 1\}$ will have the same compatible probability distributions [305]. The irreversibility of time series indicates the presence of nonlinearities in the dynamics of a system far from equilibrium, including non-Gaussian random processes and dissipative chaos. Since the definition of the irreversibility of the time series is formal, there is no a priori optimal algorithm for its quantification. Several methods for measuring the irreversibility of time have been proposed [107, 304, 306–312].

In the first group of methods, the symbolization of time series is performed, and then the analysis is performed by statistical comparison of the appearance of a string of symbols in the forward and reverse directions [307].

Sometimes additional compression algorithms are used [306]. An important step for this group is the symbolization – the conversion of the time series into a character series requires additional special information (e.g., division of the range or size of the alphabet) and, therefore, contains the problem of the algorithm's dependence on these additional parameters. The second problem arises when considering the large-scale invariance of complex signals. Since the procedures of typical symbolizations are local, taking into account different scales can cause some difficulties [107].

https://doi.org/10.1051/shsconf/202110703001

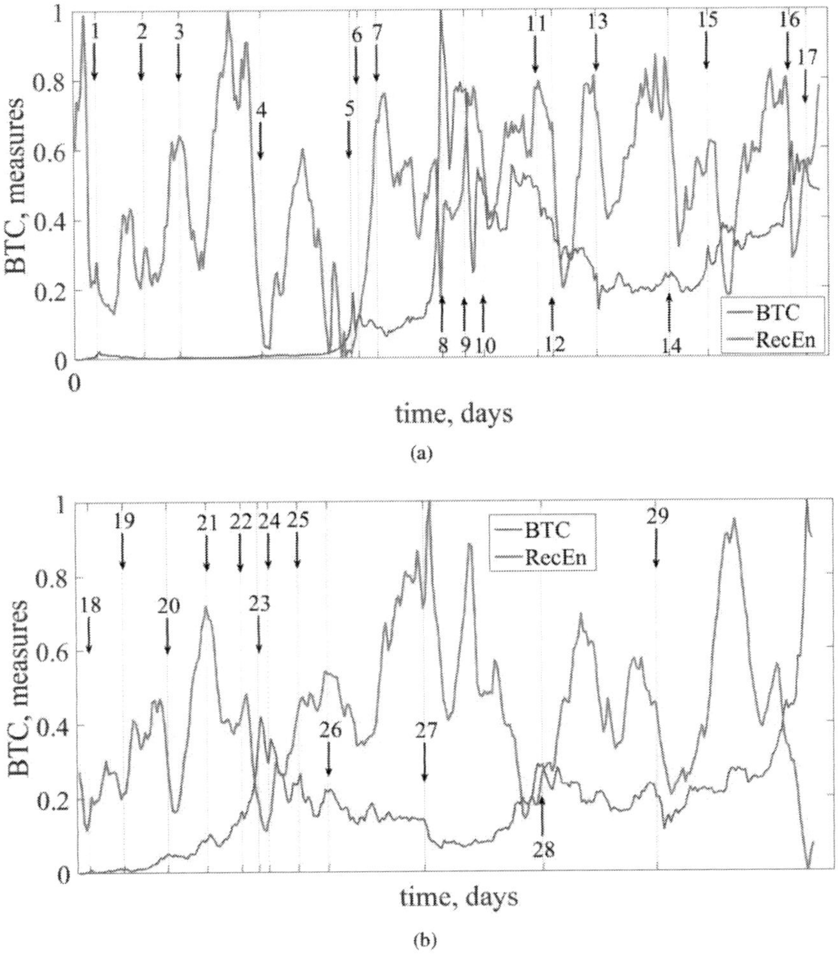

Figure 23: *RecEn* dynamics along with first (a) and second (b) periods of the entire time series of Bitcoin

Another group of methods in formalizing the index of irreversibility does not use the symbolization procedure but is based on the use of real values of the time series or returns.

One of such approaches is based on the asymmetry of the distribution of points in the Poincare map, built on the basis of the values of the analyzed time series [309, 312].

Recently, a fundamentally new approach to measuring the irreversibility of time series has been proposed, which uses the methods of complex network theory [308, 311] and which combines two tools: the algorithm for visibility of time series recovery into a complex network and the Kullbak-Leibler divergence algorithm [311]. The first forms a directional network according to the geometric criterion. The degree of irreversibility of the series is then estimated by the Kullbak-Leibler divergence (i.e., the resolution) between the distribution of the input and output stages of the associated count. This method is computationally efficient, does not require any special symbolization of the process, and, according to the authors, naturally takes into account multiscale.

In this study, we consider the irreversibility of time as a measure of the complexity of the system.

Let us consider non-reversible measures of complexity based on the construction and analysis of ordinal permutation patterns.

7.1 Time series irreversibility measure based on permutation patterns

The concept of permutation patterns (PP) was introduced by Bandt and Pompe [127]. PP is based on the idea of finding the order patterns that result in sorted (ascending) sub-sequences, and of then studying the probability distribution of these patterns. Authors of the work [313] introduce a new method, based on permutation entropy [127, 314, 315], to evaluate irreversibility of time series at various temporal scales. The proposed one presents various advantages: (1) it has no free parameters other than the

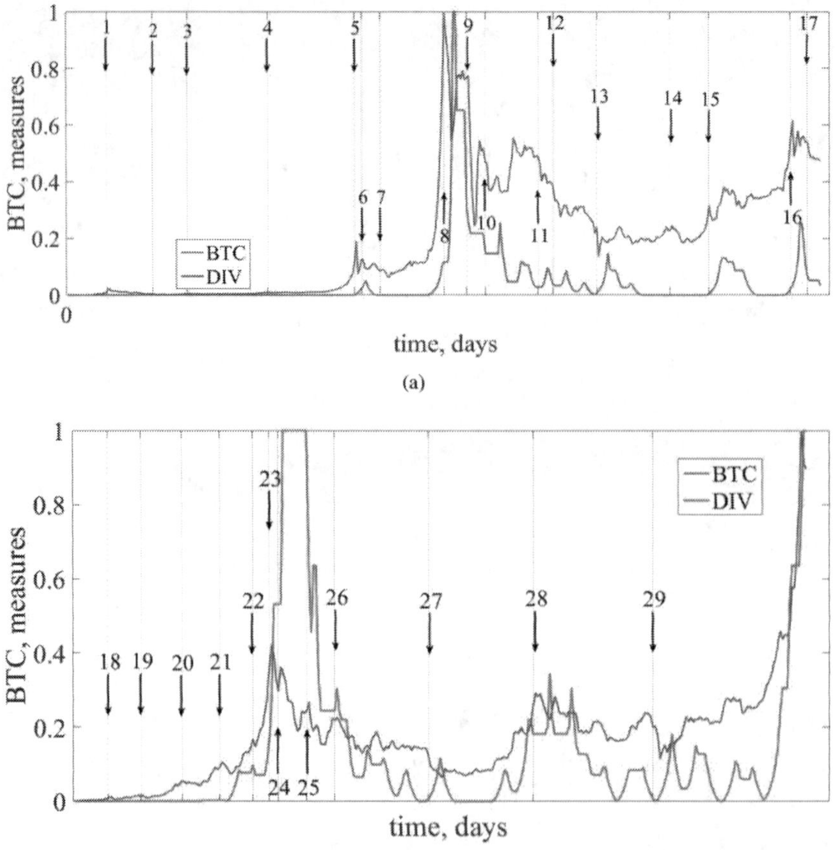

Figure 24: The measure of divergence along with first (a) and second (b) periods for BTC time series

embedding dimension of the permutation entropy; (2) similar to visibility graph methods [311, 316], it is temporally local, and therefore allows assessing fluctuations; (3) assessing significance is straightforward, and does not rely on scaling arguments as in visibility graph methods; and (4) it has a convergence speed advantage over visibility graph methods graph methods. They also demonstrated how the proposed approach can help elucidate the complex irreversibility dynamics of financial time series, representing 30 major European stocks and 12 world indices [313] and gait analysis on late stages of neurodegenerative dementias [317]. This technique was also used to analyze the irreversibility of time series in ecology (the time series of lynx abundance), epidemiology (dengue prevalence), economy (the S&P price-index series), neuroscience (electroencephalographic data from an epileptic patient) [318, 319], and heart rate [320].

The idea of irreversibility analysis implies that regarding each permutation pattern that can be obtained following procedure from section 3.2.4, there will be reversed one under the operation of time reversal. Example, if for an embedded vector $\vec{X}$, a pattern π_i is found, reversing time series will necessary imply π_i^r. As an example, let us consider a fragment of BTC time series for the period 21.01.2021-31.01.2021:

$$X^d = [30825.70, 33005.76, 32067.64, 32289.38, 32366.39]$$

and

$$X^r = [32366.39, 32289.38, 32067.64, 33005.76, 30825.70].$$

According to mentioned steps, we will construct embedded matrix of overlapping column vectors with $d_E = 3$ and $\tau = 1$. Our sampled data is partitioned as follows:

$$X_t^d(d_E, \tau) = \begin{bmatrix} 30825.70 & 33005.76 & 32067.64 \\ 33005.76 & 32067.64 & 32289.38 \\ 32067.64 & 32289.38 & 32366.39 \end{bmatrix} \tag{39}$$

and for $X_t^r(d_E, \tau)$:

$$\begin{bmatrix} 32366.39 & 32289.38 & 32067.64 \\ 32289.38 & 32067.64 & 33005.76 \\ 32067.64 & 33005.76 & 30825.70 \end{bmatrix}. \tag{40}$$

After it, our time-delayed vectors are mapped to *permutations* or *ordinal patterns* of the same size. Our example consists $3! = 6$ different ordinal patterns in total. They can be paired together, such that each pattern composing a pair is the time reversal of the other. For instance:

$$\{0,1,2\} \leftrightarrow \{2,1,0\}$$
$$\{1,0,2\} \leftrightarrow \{2,0,1\}$$
$$\{1,2,0\} \leftrightarrow \{0,2,1\}$$

with $\leftrightarrow$ representing a time reversal transformation.

As an example, the corresponding permutation of the first column from (39) would be $\phi(30825.70, 33005.76, 32067.64]) = 021$ since we arrange values in ascending order and replace them by their ordinal ranking from original placement. Therefore, after mapping from the time-series data into a series of permutations ($\phi : \mathbb{R}^{d_E} \to S_{d_E}$), we obtain the ordinal matrices first of all for initial time series:

$$\begin{bmatrix} 0 & 1 & 0 \\ 2 & 2 & 1 \\ 1 & 0 & 2 \end{bmatrix}. \tag{41}$$

and its reversed version:

$$\begin{bmatrix} 2 & 1 & 1 \\ 1 & 0 & 2 \\ 0 & 2 & 0 \end{bmatrix}. \tag{42}$$

Finally, the probability of each pattern in initial and reversed time series is calculated as

$$p(\pi) = \frac{\#\{t \le N - (d_E - 1)\tau, \phi(X_t^{d_E,\tau}) = \pi\}}{N - (d_E - 1)\tau}, \tag{43}$$

forming the probability distributions $P^d = [p_{0,1,2}, p_{2,1,0}, p_{1,0,2}, p_{2,0,1}, p_{1,2,0}, p_{0,2,1}]$ and $P^r = [p_{2,1,0}, p_{0,1,2}, p_{2,0,1}, p_{1,0,2}, p_{0,2,1}, p_{1,2,0}]$. Using the Kullback-Leibler divergence, we can define the degree of irreversibility in a time series:

$$D_{KL} = \sum_{i=1}^{d_E!} P^d(i) \log \frac{P^d(i)}{P^r(i)}. \tag{44}$$

If $P^d \approxeq P^r$, the time series is presented to be reversible, thus yielding a $D_{KL} \approxeq 0$ and vise versa. Estimating varying bitcoin's irreversibility according to (44) with time window of 100 days and time step of 1 day, we obtain following results.

The figures show that time series are significantly irreversible. When moving the original rows of their irreversible disappears. Draws attention and noticeable unevenness introduced measures, which correlate with the fluctuations of the input time series. Identifying significant changes in the time series and comparing them with the corresponding changes of non-reversible measures of complexity, it is possible to construct the corresponding indicators.

8 Single and multiplex networks

The new interdisciplinary study of complex systems, known as the complex networks theory, laid the foundation for a new network paradigm of synergetics [321]. In the framework of the complexity paradigm, it became apparent that we should move from well-studied systems and processes, taking into account the minimal number of new entities that are characteristic of the social sciences or the humanities. Apparently, one of these entities is the bonds, that is, what characterizes the interaction of the elements that are part of the system, that makes parts of the whole. The set of these links is called the network. Investigating networks, we take into account their topology, statistical properties, the distribution of weights of individual nodes and edges, the effects of information dissemination, robustness, etc. [1, 4, 322, 323]. Complex networks include electrical, transport, information, social, economic, biological, neural, and other networks [3, 324, 325]. The network paradigm has become dominant in the study of complex systems since it allows you to enter new quantitative measures of complexity not existing for the time series [326]. Moreover, the network paradigm provides adequate support for the core concepts of Industry 4.0 [321].

Previously, we introduced various quantitative measures of complexity for individual time series [57, 327–329]. However, except a graph for an individual time series, it is necessary to take into account the interconnection interaction, which can be realized within the framework of different models [330]. We will consider it by simulating so-called multiplex networks, the features of which are reduced to a fixed number of nodes in each layer, but they are linked by different bonds [330].

Recently, the first papers using the spectral and topological characteristics of dynamic systems presented as networks have appeared. Thus, in [331], it has been investigated universal and non-universal allometric scaling behaviors in the visibility graphs of 30 world stock market indices. It has been established that the nature of such behavior is due to the returns distribution that is characterized by fat-tails, the nonlinear long-term correlation, and a coupling effect between the set of influential factors.

Birch [332] compared the mean degree value and clustering coefficient for a group of companies included in the DAX 30 index basket. He observed the companies from the DAX 30 index for two time periods: the first from the beginning of 2008 through the end of 2009 and the second from the beginning of 2010 up to the end of 2011 as these include the dates – a period of crisis (7th October 2008 – 31st December 2008) and a period of recovery (7th May 2010 – 3rd August 2010). Contrary to expectations, the results differed little from the relatively low accuracy of the horizontal visibility graph procedure compared to visibility graph.

Wand and Wei [333] collected from the Chinese stock market the data of 2571 stock companies in 2012 and the data of 2578 stock companies in 2013. Every year, data of these stock companies are randomly arranged. These data are then converted into some complex networks based on the visibility graph method. For these complex networks, degree distribution and clustering coefficient are considered. These results show that complex networks have power-law distribution and small-world properties.

Yan and Serooskerken [334] construct an indicator to measure the magnitude of the super-exponential growth of

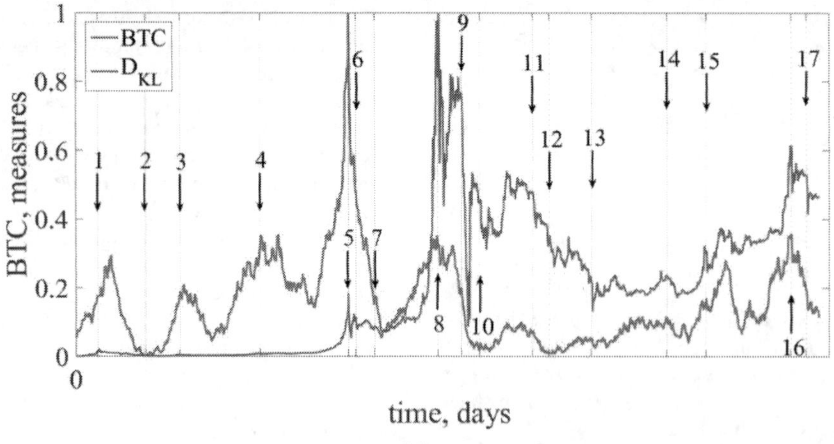

(a)

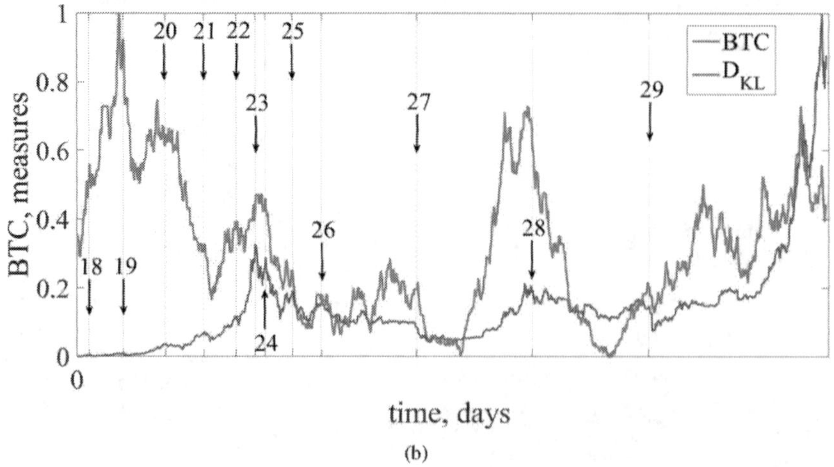

(b)

Figure 25: Dynamics of permutation-based time irreversibility measure for the first (a) and second (b) periods

stock prices, by measuring the degree of the price network, generated from the price time series. Twelve major international stock indices have been investigated. The work results show that this new indicator has strong predictive power for financial extremes, both peaks, and troughs. By varying the model parameters, the authors show the predictive power is very robust. The new indicator has a better performance than the indicator based on a well-known model of log-periodic oscillations of Sornette [335].

Authors of another paper [336] analyze high-frequency data from the S&P 500 via the horizontal visibility graph method and find that all major crises that have taken place worldwide for the last twenty years, affected significantly the behavior of the price-index. Nevertheless, they observe that each of those crises impacted the index in a different way and magnitude. These results suggest that the predictability of the price-index series increases during periods of crisis.

In the paper [337] Serafino et al. studied the visibility graphs built from the time series of several stock market indices. They proposed a validation procedure for each link of these graphs against a null hypothesis derived from ARCH-type modeling of such series. Building on this framework made it possible to devise a market indicator that turned out to be highly correlated and even predictive of financial instability periods.

Francés and Carles [25] examined the characteristics of the daily price series of 16 different cryptocurrencies between July 2017 and February 2018. Using Minimum Spanning Tree (MST) and hierarchical analysis by dendrograms that were obtained from Pearson correlation between daily returns, they visualized and identified a high correlation between price movements of all the currencies. From the obtained results, it is seen that the most interconnected with all cryptocurrencies is Ethereum, while Bitcoin standed for one of its branches. They concluded that Ethereum might be a benchmark currency in the cryptocurrency market, rather than Bitcoin.

Coquide et al. [338] constructed the Google matrices of bitcoin transaction for all year quarters from the very start of it till April 10, 2013. From PageRank and CheiRank probabilities that serve as analogues to trade import and export, they determined the dimensionless trade balance of each user and analyzed the direct and hidden (indirect) links between top PageRank users of BCN using the recently developed reduced Google matrix algorithm. They modeled the contagion propagation of the transactions assuming that a user goes bankrupt if its dimensional balance exceeds a certain bankruptcy threshold k. Their results present that the phase transition neighboring with the critical threshold $k = k_c \approx 0.1$ below which almost all users remain safe. For $k > 0.55$ almost all users remain safe, and for $0.1 < k > 0.55$ more than a half of users go bankrupt. Moreover, their result present that even being not very close to the critical threshold $k_c \approx 0.1$, almost all top PageRank and CheiRank users rapidly become bankrupts that give the evidence about their strong interconnectivity. With the reduced Google matrix algorithm, they presented the most preferable interlinks of the most valuable users.

Multiplex networks are actively used to simulate complex networks of different nature: from financial (stock market [336, 337, 339, 340], banks [341], guarantee market [342]) to social [343]. Particular attention should be paid to the work [339], in which the above multiplex measures are analyzed for the subject of correlations with known stock market crises.

8.1 Methods of converting time series into graphs

In recent years, interesting algorithms for the transformation of time series into a network have been developed, which allows extending the range of known characteristics of time series even to network ones. Recently, several approaches have been proposed to transform time sequences into complex network-like mappings. These methods can be conventionally divided into three classes [344].

The first is based on the study of the convexity of successive values of the time series and is called visibility graph (VG) [344, 345]. The second analyzes the mutual approximation of different segments of the time sequence and uses the technique of recurrent analysis [344]. The recurrent diagram reflects the existing repetition of phase trajectories in the form of a binary matrix whose elements are units or zeros, depending on whether they are close (recurrent) with given accuracy or not, the selected points of the phase space of the dynamic system. The recurrence diagram is easily transformed into an adjacency matrix, on which the spectral and topological characteristics of the graph are calculated. Finally, if the basis of forming the links of the elements of the graph is to put correlation relations between them, we obtain a correlation graph [327, 344]. To construct and analyze the properties of a correlation graph, we must form an adjacency matrix from the correlation matrix. To do this, you need to enter a value that for the correlation field will serve as the distance between the correlated agents. Such a distance may be dependent on the ratio of the correlation C_{ij} value $d_{ij} = \sqrt{2(1 - C_{ij})}$. So, if the correlation coefficient between the two assets is significant, the distance between them is small, and, starting from a certain critical value x_{crit}, assets can be considered bound on the graph. For an adjacency matrix, this means that they are adjacent to the graph. Otherwise, the assets are not contiguous. In this case, the binding condition of the graph is a prerequisite.

The main purpose of such methods is to accurately reproduce the information stored in the time series in an alternative mathematical structure, so that powerful graph theory tools could eventually be used to characterize the time series from a different point of view in order to overcome the gap between nonlinear analysis of time series, dynamic systems and the graphs theory.

The usage of the complexity of recurrent networks to prevent critical and crisis phenomena in stock markets has been considered by us in a recent paper [346]. Therefore, we will focus on algorithms of the VG and multiplex VG (MVG).

The recurrence diagrams for the visualization of phase space recurrences are based on Henri Poincare's idea of the phase space recurrence of dynamical systems. According to Takens' theorem [136], an equivalent phase trajectory that preserves the structure of the original phase trajectory can be recovered from a single observation or time series by the time delay method: the recurrence diagram is easily transformed into an adjacency matrix, by which the spectral and topological characteristics of the graph are calculated [327].

The algorithm of the VG is realized as follows [345].

Take a time series $\{x(t_i) \mid i = 1, \ldots, N\}$ of length N. Each point in the time series data can be considered as a vertex in an associative network, and the edge connects two vertices if two corresponding data points can "see" each other from the corresponding point of the time series (see Figure 26). Formally, two values $x(t_a)$ (at a point in time t_a) and $x(t_b)$ (at a point in time t_b) are connected, if, for any other value $(x(t_c), t_c)$, which is placed between them (i.e., $t_a < t_c < t_b$), the condition is satisfied:

$$x(t_c) < x(t_a) + (x(t_b) - x(t_a))\frac{t_c - t_a}{t_b - t_a}.$$

Note that the visibility graph is always connected by definition and also is invariant under affine transformations, due to the mapping method.

An alternative (and much simpler) algorithm is the horizontal visibility graph (HVG) [344], in which a connection can be established between two data points a and b if one can draw a horizontal line in the time series joining them that does not intersect any intermediate data by the following geometrical criterion: $x(t_a), x(t_b) > x(t_c)$ for all c such that $t_a < t_c < t_b$ (see figure 26).

In multiplex networks, there are two tasks [346]: (1) turn separate time series into the network for each layer; (2) connect the intra-loop networks to each other. The first problem is solved within the framework of the standard algorithms described above. For multiplex networks, the algorithm of the MVG for the three layers is presented in figure 27.

The cross-recurrent multiplex network (MCRP) is formed from recursive diagrams of individual layers.

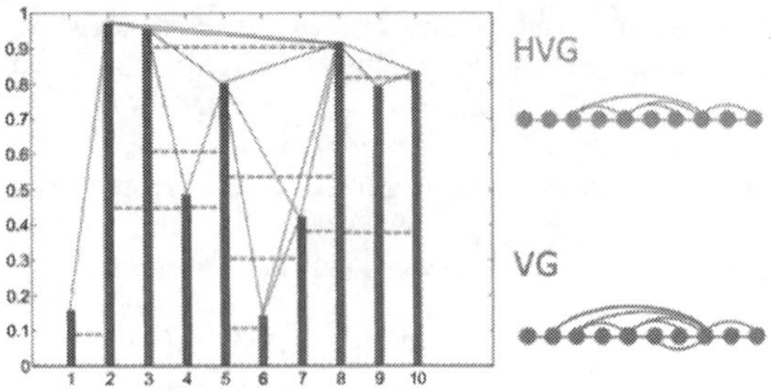

Figure 26: Illustration of constructing the visibility graph (red lines) and the horizontal visibility graph (green lines) [347]

8.2 Spectral and topological graph properties

Spectral theory of graphs is based on algebraic invariants of a graph – its spectra [27]. The spectrum of graph G is the set of eigenvalues $S_p(G)$ of a matrix corresponding to a given graph. For adjacency matrix A of a graph, there exists a characteristic polynomial $|\lambda I - A|$, which is called the characteristic polynomial of a graph $P_G(\lambda)$. The eigenvalues of the matrix A (the zeros of the polynomial $|\lambda I - A|$) and the spectrum of the matrix A (the set of eigenvalues) are called respectively their eigenvalues.

Another common type of graph spectrum is the spectrum of the Laplace matrix L. The Laplace matrix is used to calculate the tree graphs, as well as to obtain some important spectral characteristics of the graph. It can be defined as $L = D - A$ where D – diagonal matrix of order n:

$$d_{ij} = \begin{cases} d_i, & i = j, \\ 0, & i \neq j, \end{cases} \tag{45}$$

with d_i – the degree of corresponding vertex in the graph.

The spectrum $S_{p_L}(G)$ of the matrix L is the root of the characteristic equation:

$$|\lambda I - L| = |\lambda I - D + A| = 0. \tag{46}$$

Comparing the spectra of S_p with S_{p_L}, it is easy to establish that

$$S_p(G) = [\lambda_1, \lambda_2, \ldots, \lambda_n],$$
$$S_{p_L}(G) = [r - \lambda_n, \lambda_{n-1}, \ldots, r - \lambda_1],$$

where $\lambda_1 = r$.

The number zero is the eigenvalue of the matrix L, which corresponds to an eigenvector whose coordinates are equal to unity. The multiplicity of the null eigenvalue is equal to the number of connected components of the graph. The rest of eigenvalues L are positive. The least of the positive eigenvalues λ_2 is called the index of *algebraic connectivity* of the graph. This value represents the "force" of the connectivity of the graph component and is used in the analysis of reliability and synchronization of the graph.

Important derivative characteristics are spectral gap, graph energy, spectral moments, and spectral radius. The spectral gap is the difference between the largest and the next eigenvalues of the adjacency matrix and characterizes the rate of return of the system to the equilibrium state. The *graph energy* is the sum of the modules of the eigenvalues of the graph adjacency matrix:

$$E(G) = \sum_{i=1}^{n} |\lambda_i|.$$

The spectral radius is the largest modulus of the eigenvalue of the adjacency matrix. Denote by N_c the value, which corresponds to an *average eigenvalue* of the graph adjacency matrix:

$$N_c = \ln\left(\frac{1}{n}\sum_{i=1}^{n} \exp(\lambda_i)\right)$$

and is called *natural connectivity*.

The kth spectral moment of the adjacency matrix is determined by the expression

$$m_k(A) = \frac{1}{n}\sum_{i=1}^{n} \lambda_i^k,$$

with λ_i that represents eigenvalues of the adjacency matrix, and n is the vertex of G.

Among the topological measures, one of the most important is the *node degree* k – the number of links attached to this node. For non-directed networks, the node's degree k_i is determined by the sum $k_i = \sum_j a_{ij}$, where the elements a_{ij} of the adjacency matrix.

For characterizing the *linear size* of the network, there are useful concepts of average $\langle l \rangle$ and maximum l_{max} shortest paths. For a connected network of n nodes, the *average path length* (ApLen) is equal to

$$\langle l \rangle = \frac{2}{n(n-1)}\sum_{i>j} l_{ij}, \tag{47}$$

where l_{ij} – the length of the shortest path between the nodes. The *diameter* of the connected graph is the maximum possible distance between its two vertices, while the minimum possible is the radius of the graph.

If the average length of the shortest path gives an idea of the whole network and is a global characteristic, the next parameter – the clustering coefficient – is a local value and characterizes a separate node. For a given node i, the *clustering coefficient* C_i is defined as the ratio of the existing number of links between its closest neighbors to the maximum possible number of such relationships:

$$C_i = \frac{2E_i}{k_i(k_i - 1)}. \tag{48}$$

In equation (48), $k_i(k_i-1)/2$ is the maximum number of links between the closest neighbors. The clustering coefficient of the entire network is defined as the average value of C_i for all its nodes. The clustering coefficient shows how many of the nearest neighbors of the given node are also the closest neighbors to each other. It characterizes the tendency to form groups of interconnected nodes – clusters. For the real-life networks, the high values of the clustering coefficient are high.

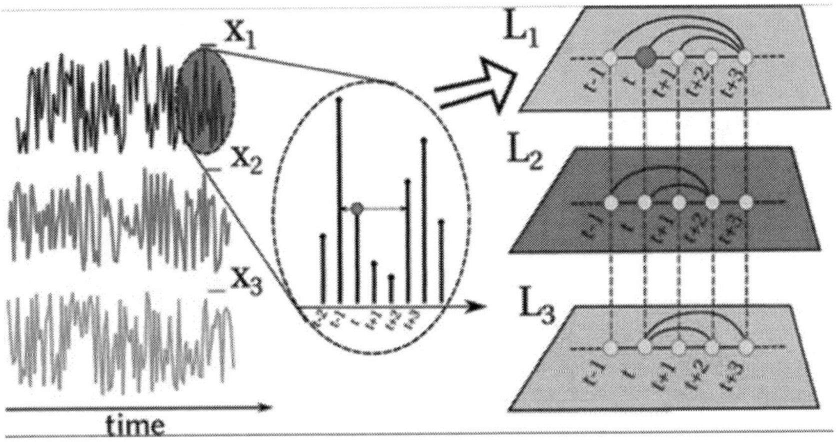

Figure 27: Scheme for forming bonds between three layers of the multiplex network [346]

Another feature of the node is *betweenness*. It reflects the role of the node in establishing network connections and shows how many shortest paths pass through this node. Node betweenness $c(v)$) is defined as

$$c(v) = \sum_{i \neq j} \frac{\sigma(i, j \mid v)}{\sigma(i, j)}, \tag{49}$$

where $\sigma(i, j)$ – the total number of shortest paths between nodes i and j; $\sigma(i, j \mid v)$ – the number of those shortest paths between i, j passing through v. The value of (49) is also called the *load* or *betweenness centrality*.

One of the main characteristics of the network is the degree distribution $P(d)$, which is defined as the probability that the node i has a degree $d_i = d$. Most natural and actual artificial networks follow a power law distribution

$$P(d) \sim 1/d^{\gamma},\ d \neq 0,\ \gamma > 0. \tag{50}$$

Also, important topological characteristic is the *vertex eccentricity* – the largest distance between i and any other vertex, that is, how far the vertex is far from the other vertices of the graph. The centrality of the vertex measures its relative importance in the graph. At the same time, the *farness* of a node is defined as the sum of its distances to all other nodes, and its *closeness* is defined as the backward distance. Thus, the centrality of the node is lower than its total distance to all other nodes.

Another important measure is the *link density* in the graph, which is defined as the number of links n_e, divided by the expression $n_n(n_n - 1)/2$, where n_n is the number of nodes of the graph.

From spectral measures, we consider the maximal node degree (d_{max} – figure 28a and figure 28b). From the topological measures, the average path length (APLen – figure 28c and 28d) is found, which is in accordance with equation (47).

Figure 28 demonstrates the asymmetric response of the spectral and topological measures of network complexity. For the complete series, the calculation parameters are as follows: window length of 100 days, step is of 1 day.

Figures above show that all of the above spectral measures have maximum values in pre-crisis periods. The complex system has the greatest complexity. With the approach of the crisis, the complexity of the system decreases, recovering after the crisis. Some of the topological, in particular, APLen, shows an opposite relationship. Indeed, in more complex systems you can always find shorter paths that connect any nodes. During the crisis (reducing complexity, increasing the chaotic component), the length of the corresponding path increases.

8.3 Multiplex complexity measures

A multilayer/multiplex network [83, 348] is a pair $M=(G,C)$ where $\{G_\alpha \mid \alpha \in 1,\ldots,M\}$ is a family of graphs (whether directed or not, weighed or not) $G_\alpha = (X_\alpha, E_\alpha)$ that called layers and

$$C = \left\{E_{\alpha\beta} \subseteq X_\alpha \times X_\beta \mid \alpha,\beta \in 1,\ldots,M, \alpha \neq \beta\right\}$$

is a set of links between nodes of layers G_α and G_β where $\alpha \neq \beta$. The elements of each E_α are intralayer edges in M in contrast to the elements of each $E_{\alpha\beta}$ that called interlayer edges.

A set of nodes in a layer G_α is denoted as $X_\alpha=\left\{x_1^\alpha,\ldots,x_{N_\alpha}^\alpha\right\}$, and an intralayer adjacency matrix as $A^{[\alpha]} = (a_{ij}^\alpha) \in \mathrm{Re}^{N_\alpha \times N_\alpha}$, where

$$a_{ij}^\alpha = \begin{cases} 1, & (x_i^\alpha, x_j^\alpha) \in E_\alpha, \\ 0, & \text{otherwise.} \end{cases} \tag{51}$$

for $1 \le i \le N_\alpha$, $1 \le j \le N_\beta$ and $1 \le \alpha \le M$. For an interlayer adjacency matrix, we have $A^{[\alpha,\beta]}(a_{ij}^{\alpha\beta}) \in \mathrm{Re}^{N_\alpha \times N_\beta}$, where

$$a_{ij}^{\alpha\beta} = \begin{cases} 1, & (x_i^\alpha, x_j^\beta) \in E_{\alpha\beta}, \\ 0, & \text{otherwise.} \end{cases} \tag{52}$$

A multiplex network is a partial case of interlayer networks, and it contains a fixed number of nodes connected

https://doi.org/10.1051/shsconf/202110703001

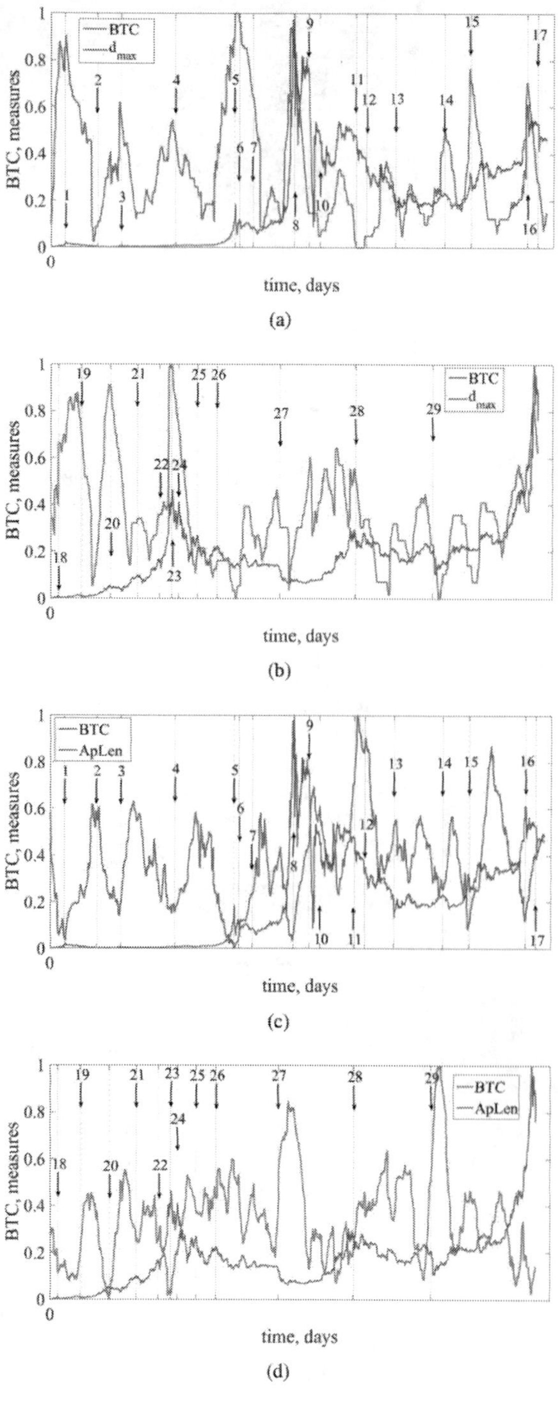

Figure 28: Maximal node degree d_{max} (a, b) and average path length $\langle l \rangle$ (c, d) for two periods of Bitcoin time series

by different types of links. Multiplex networks are characterized by correlations of different nature [330], which enable the introduction of additional multiplexes.

Let's evaluate the quantitative overlap between the various layers. The average edge overlap is defined as [340]

$$\omega = \frac{\sum_i \sum_{j>i} \sum_\alpha a_{ij}^{[\alpha]}}{M \sum_i \sum_j 1 - \delta_{0,\sum_\alpha a_{ij}^{[\alpha]}}} \tag{53}$$

and determines the number of layers in which this edge is presented. Its value lies on the interval $[1/M, 1]$ and equals $1/M$ if the connection (i, j) exists only in one layer. In other words, if there is a layer α such that $a_{ij}^{[\beta]} = 0 \, \forall \beta \neq \alpha$. If all layers are identical, then $\omega = 1$. Consequently, this measure can serve as a measure of the coherence of the output time series: high values of ω indicate a noticeable correlation in the structure of the time series.

The *total overlap* $O^{\alpha\beta}$ between the two layers α and β is defined as the total number of edges that are shared between the layers α and β:

$$O^{\alpha\beta} = \sum a_{ij}^{\alpha} a_{ij}^{\beta}, \tag{54}$$

where $\alpha \neq \beta$.

For a multiplex network, the node degree k is already a vector

$$k_i = (k_i^{[1]}, \ldots, k_i^{[M]}), \tag{55}$$

with the degree $k_i^{[\alpha]}$ of the node i in the layer α, namely

$$k_i^{[\alpha]} = \sum_j a_{ij}^{[\alpha]},$$

while $a_{ij}^{[\alpha]}$ is the element of the adjacency matrix of the layer α. Specificity of the node degree in vector form allows to describe additional quantities. One of them is the *ovelapping degree* of node i

$$o_i = \sum_{\alpha=1}^{M} k_i^{[\alpha]}. \tag{56}$$

The next measure quantitatively describes the interlayer degree correlations of the selected node in two different layers. If a chosen pair (α, β) from M layers is characterized by the distributions $P(k^{[\alpha]}), P(k^{[\beta]})$, the so-called *interlayer mutual information* is determined as:

$$I_{\alpha\beta} = \sum \sum P(k^{[\alpha]}, k^{[\beta]}) \log \frac{P(k^{[\alpha]}, k^{[\beta]})}{P(k^{[\alpha]}), P(k^{[\beta]})}, \tag{57}$$

where $P(k^{[\alpha]}, k^{[\beta]})$ is the probability of finding a node degree $k^{[\alpha]}$ in a layer α and a degree $k^{[\beta]}$ in a layer β. The higher the value of $I_{\alpha\beta}$, the more correlated (or anti-correlated) is the degree distribution of the two layers and, consequently, the structure of a time series associated with them. We also find the mean value of $I_{\alpha\beta}$ for all possible pairs of layers – the scalar $\langle I_{\alpha\beta} \rangle$ that quantifies the information flow in the system.

The quantity that quantitatively describes the distribution of a node degree i between different layers is the *entropy of a multiplexed degree*:

$$S_i = -\sum_{\alpha=1}^{M} \frac{k_i^{[\alpha]}}{o_i} \ln \frac{k_i^{[\alpha]}}{o_i}. \tag{58}$$

Entropy is zero if all connected to i edges are in the same layer, and has the maximum value when they are evenly distributed between different layers. So, the higher value of S_i, the more evenly distributed edges connected to i between different layers.

A similar indicator is the *multiplex participation coefficient*:

$$P_i = \frac{M}{M-1}\left[1 - \sum_{\alpha=1}^{M}\left(\frac{k_i^{\alpha}}{o_i}\right)^2\right]. \quad (59)$$

The P_i takes values on the interval $[0, 1]$ and determines how homogeneously are distributed the links of a node i among M layers. If a node is only active on one layer, $P_i = 0$; $P_i = 1$ if a node has an precisely defined number of incident links that are equally distributed across M layers.

Obviously, measures S_i and P_i are very similar.

We will show that some of these spectral and topological indicators serve as the measures of system complexity, and the dynamic of their changes allow us to build precursors of crashes and critical events in the cryptocurrency market.

As far as multiplex measures are concerned, they are very similar in their dynamic to the spectral and topological representations above (see figure 28). In the case of a shorter sample of a base of three layers – Bitcoin, Ethereum, Litecoin (see figure 29-31), we have the asymmetric behavior of the multiplex measures I, O, o (Equations (57), (54), (56) and S, P (Equations (58) and (59)) for different methods of building multiplex networks.

Figures 29-31 show that the mentioned multiplex measures are excellent indicators that warn about the approaching crisis phenomenon, that is, are indicator-precursors.

9 Quantum precursors

Quantum econophysics, a direction distinguished by the use of mathematical apparatus of quantum mechanics as well as its fundamental conceptual ideas and relativistic aspects, developed within its boundaries just a couple of years later, in the first decade of the 21st century [41, 349–351].

According to classical physics, immediate values of physical quantities, which describe the system status, not only exist but can also be exactly measured. Although non-relativistic quantum mechanics does not reject the existence of immediate values of classic physical quantities, it postulates that not all of them can be measured simultaneously (Heisenberg uncertainty ratio). Relativistic quantum mechanics denies the existence of immediate values for all kinds of physical quantities, and, therefore, the notion of system status seizes to be algoristic.

In this section, we will demonstrate the possibilities of quantum econophysics on the example of the application of the Heisenberg uncertainty principle and the Random Matrices Theory to the actual and debatable now market of cryptocurrencies [80, 352].

9.1 Heisenberg uncertainty principle and economic analogues of basic physical quantities

In our paper [42], we have suggested a new paradigm of complex systems modeling based on the ideas of quantum as well as relativistic mechanics. It has been revealed that the use of quantum-mechanical analogies (such as the uncertainty principle, the notion of the operator, and quantum measurement interpretation) can be applied to describing socio-economic processes. Methodological and philosophical analysis of fundamental physical notions and constants, such as time, space, and spatial coordinates, mass, Planck's constant, light velocity from modern theoretical physics provides an opportunity to search for adequate and useful analogs in socio-economic phenomena and processes.

The Heisenberg uncertainty principle is one of the cornerstones of quantum mechanics. The modern version of the uncertainty principle, deals not with the precision of a measurement and the disturbance it introduces, but with the intrinsic uncertainty any quantum state must possess, regardless of what measurement is performed [353, 354]. Recently, the study of uncertainty relations, in general, has been a topic of growing interest, specifically in the setting of quantum information and quantum cryptography, where it is fundamental to the security of certain protocols [355, 356].

To demonstrate it, let us use the known Heisenberg's uncertainty ratio which is the fundamental consequence of non-relativistic quantum mechanics axioms and appears to be (e.g., [357]):

$$\Delta x \cdot \Delta \upsilon \geqslant \frac{\hbar}{2m_0}, \quad (60)$$

where Δx and $\Delta \upsilon$ are mean square deviations of x coordinate and velocity υ corresponding to the particle with (rest) mass m_0, $\hbar$ – Planck's constant. Considering values Δx and $\Delta \upsilon$ to be measurable when their product reaches their minimum, according to equation (60) we derive:

$$m_0 = \frac{\hbar}{2 \cdot \Delta x \cdot \Delta \upsilon}, \quad (61)$$

i.e., the mass of the particle is conveyed via uncertainties of its coordinate and velocity – time derivative of the same coordinate.

Economic measurements are fundamentally relative, local in time, space and other socio-economic coordinates, and can be carried out via consequent and/or parallel comparisons "here and now", "here and there", "yesterday and today", "a year ago and now", etc.

Due to these reasons constant monitoring, analysis, and time series prediction (time series imply data derived from the dynamics of stock indices, exchange rates, cryptocurrency prices, spot prices, and other socio-economic indicators) become relevant for the evaluation of the state, tendencies, and perspectives of global, regional, and national economies.

Suppose there is a set of K time series, each of N samples, that correspond to the single distance T, with an

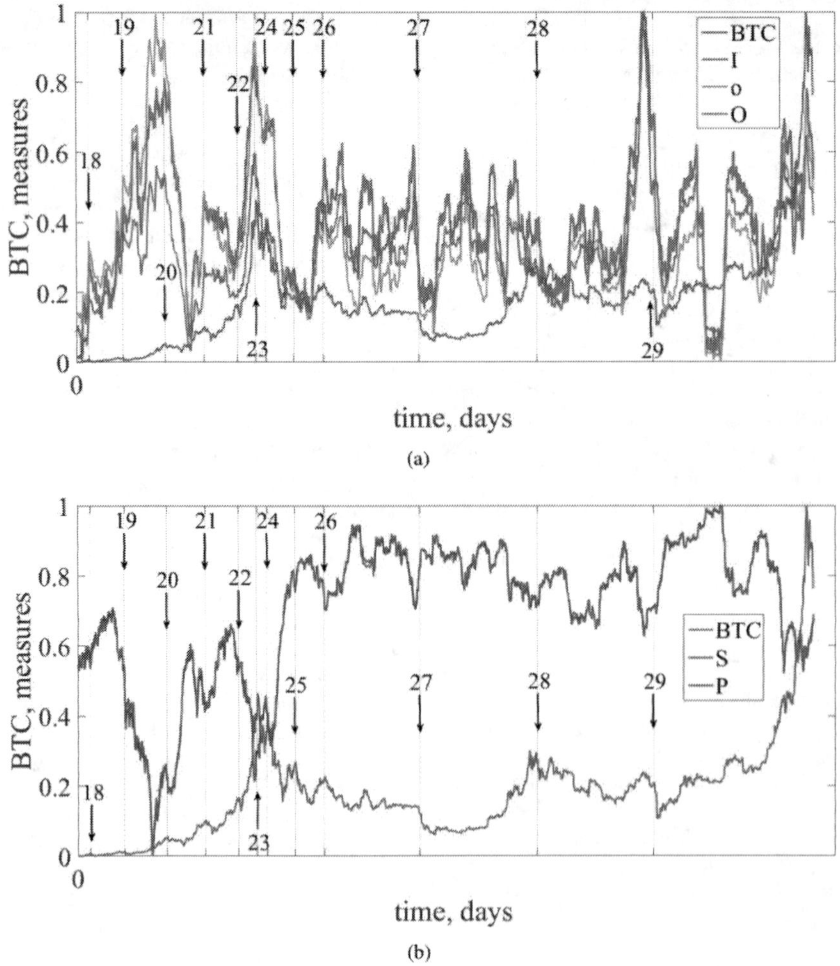

Figure 29: Dynamics of the second BTC period and multiplex measures for a base of three layers. The graph was built using the multiplex visibility graph

equally minimal time step Δt_{min}:

$$X_i(t_n), \quad \text{for } n = 0, 1, \ldots, N-1; \text{ for } i = 1, 2, \ldots, K. \quad (62)$$

To bring all series to the unified and non-dimensional representation, accurate to the additive constant, we normalize them, have taken a natural logarithm of each term of the series. Then, consider that every new series $X_i(t_n)$ is a one-dimensional trajectory of a certain fictitious or abstract particle numbered i, while its coordinate is registered after every time span Δt_{min}, and evaluate mean square deviations of its coordinate and speed in some time window $\Delta T = \Delta N \cdot \Delta t_{min} = \Delta N$, $1 \ll \Delta N \ll N$. The "immediate" speed of ith particle at the moment t_n is defined by the ratio:

$$v_i(t_n) = \frac{x_i(t_{n+1}) - x_i(t_n)}{\Delta t_{min}} = \frac{1}{\Delta t_{min}} \ln \frac{X_i(t_{n+1})}{X_i(t_n)} \quad (63)$$

with variance D_{v_i} and standard deviation Δv_i.

Keeping an analogy with equation (1), after some transformations, we can write an uncertainty ratio for this trajectory [42]:

$$\frac{1}{\Delta t_{min}} \left(\langle \ln^2 \frac{X_i(t_{n+1})}{X_i(t_n)} \rangle_{n,\Delta N} - \left(\langle \ln^2 \frac{X_i(t_{n+1})}{X_i(t_n)} \rangle_{n,\Delta N} \right)^2 \right) \sim \frac{h}{m_i}, \quad (64)$$

where m_i – economic "mass" of a X_i series, h – value which comes as an economic Planck's constant.

Since the analogy with physical particle trajectory is merely formal, h value, unlike the physical Planck's constant $\hbar$, can, generally speaking, depend on the historical period, for which the series are taken, and the length of the averaging interval (e.g., economical processes are different in the time of crisis and recession), on the series number i etc. Whether this analogy is correct or not depends on the particular series' properties.

In recent research [30, 358], we tested the economic mass as an indicator of crisis phenomena on stock index

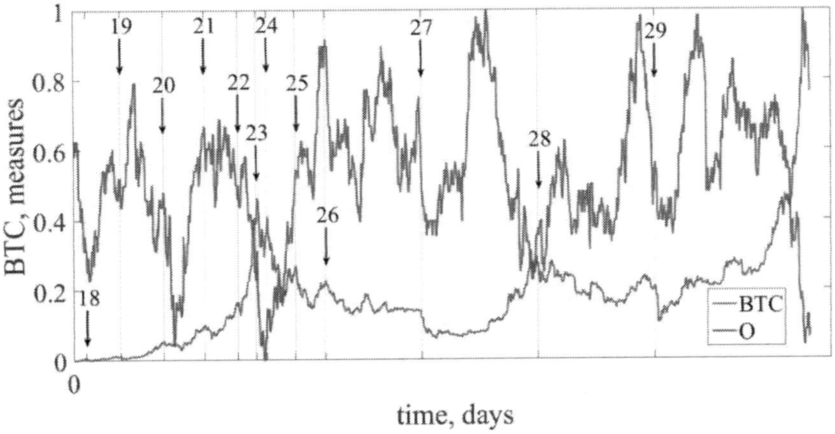

(a)

Figure 30: Dynamics of BTC daily prices and multiplex measures O. The graph was built using the MHVG

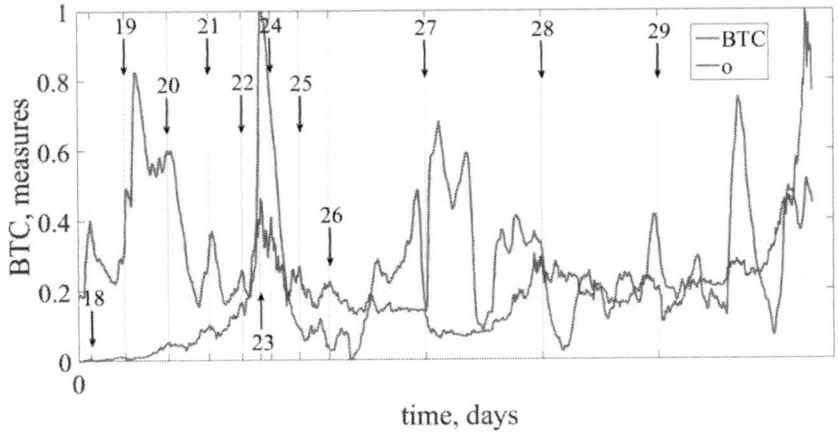

(a)

Figure 31: Dynamics of BTC daily prices and multiplex measures o. The graph was built using the MCRP

data. Here, we test the model for the cryptocurrency market on the example of Bitcoin [79, 80, 352, 359].

Obviously, there is a dynamic characteristic value m, depending on the internal dynamics of the market. In times of crashes and critical events marked by arrows, mass m is significantly reduced in the pre-crash and pre-critical periods.

Obviously, m remains a good indicator-precursor even in this case. Value m is considerably reduced before a special market condition. The market becomes more volatile and prone to changes.

Next method of quantum econophysics is borrowed from nuclear physicists and is called Random Matrix Theory.

9.2 Random matrix theory and quantum indicators-precursors

Random Matrix Theory (RMT) developed in this context the energy levels of complex nuclei, which the existing models failed to explain (Wigner, Dyson, Mehta, and others [353, 354, 356]). Deviations from the universal predictions of RMT identify specific, nonrandom properties of the system under consideration, providing clues about the underlying interactions.

Unlike most physical systems, where one relates correlations between subunits to basic interactions, the underlying "interactions" for the stock market problem are not known. Here, we analyze cross-correlations between stocks by applying concepts and methods of random matrix theory, developed in the context of complex quantum

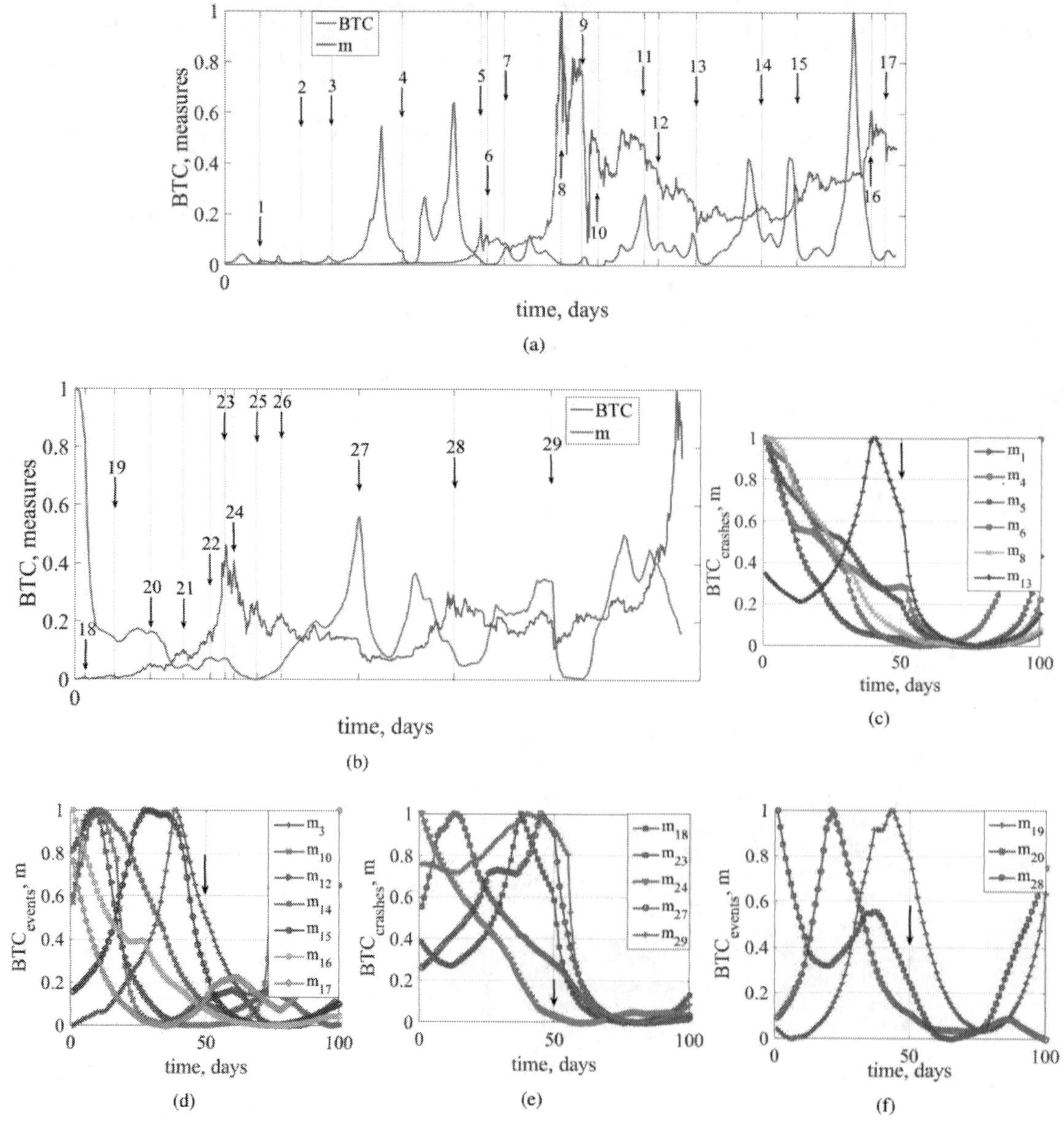

Figure 32: Dynamics of measure m for two entire periods, and its dynamics for their local crashes (c, e) and critical events (d, f) with the window size of 50 days and step of 1 day

systems where the precise nature of the interactions between subunits are not known.

RMT has been applied extensively in studying multiple financial time series [42, 352, 355, 357–362].

Special databases have been prepared, consisting of cryptocurrency time series for a certain time. The largest number of cryptocurrencies 689 contained a base of 456 days from 31.12.2017 to 31.05.2020, and the smallest (22 cryptocurrencies) contained a base, respectively, from 11.09.2013 to 31.05.2020 (https://coinmarketcap.com/all/views/all/). In order to quantify correlations, we first calculate the logarithmic returns of the cryptocurrencies price series over a time scale $\Delta t = 1$ day. We calculate the pairwise cross-correlation coefficients between any two cryptocurrency time series returns. for the largest database, a graphical representation of the pair correlation field is shown in figure 33a. For comparison, a map of correlations of randomly mixed time series of the same length shown in figure 33 b.

For the correlation matrix C we can calculate its eigenvalues, $C = U\Lambda U^T$, where U denotes the eigenvectors, Λ is the eigenvalues of the correlation matrix, whose density

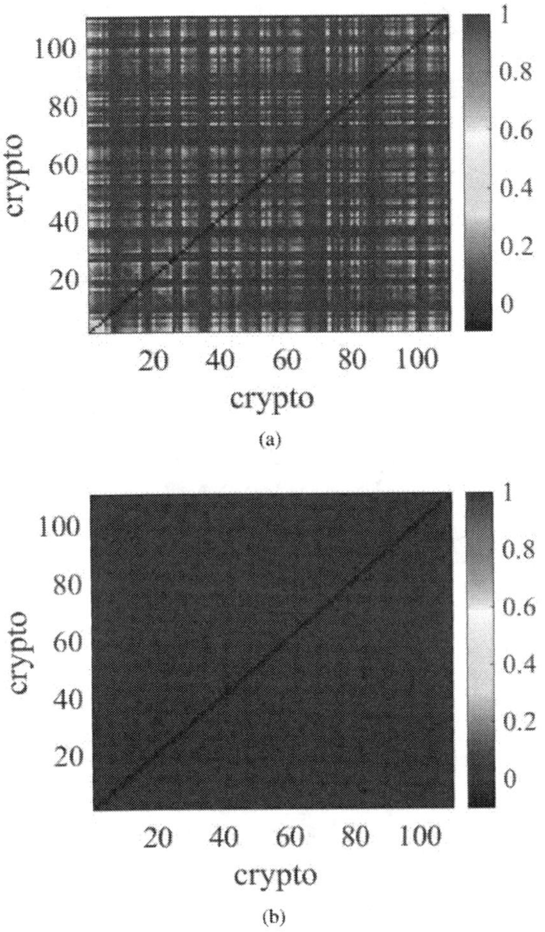

Figure 33: Visualization of the field of correlations for the initial (a, c) and mixed (b, d) matrix cryptocurrency

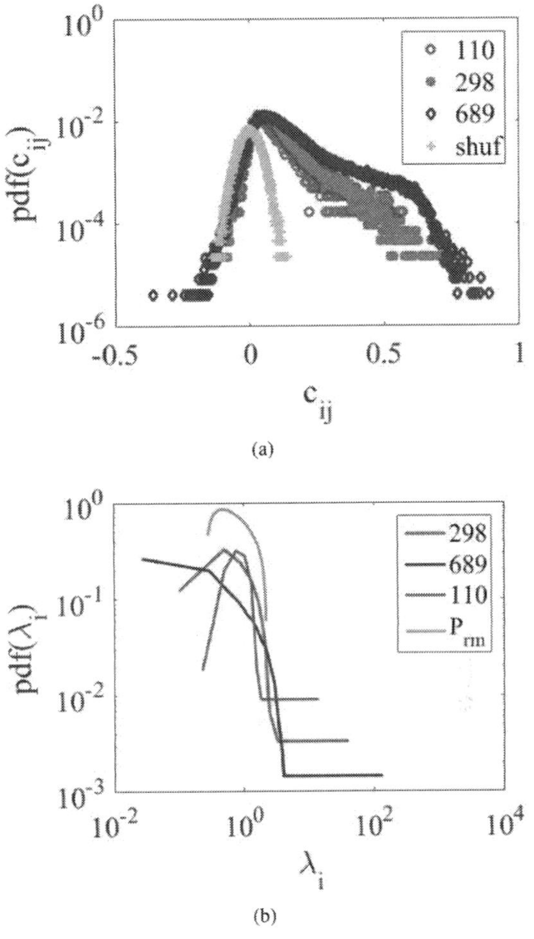

Figure 34: Comparison of distributions of the pair correlation coefficients (a) and eigenvalues of the correlation matrix (b) obtained for initial cryptocurrencies and their shuffled version

$f_c(\lambda)$ is defined as $f_c(\lambda) = (1/N)dn(\lambda)/d\lambda$, where $n(\lambda)$ is the number of eigenvalues of C that are less than λ. In the limit $N \to \infty, T \to \infty$ and $Q = T/N \geqslant 1$ fixed, the probability density function $f_c(\lambda)$ of eigenvalues λ of the random correlation matrix M has a closed form:

$$f_c(\lambda) = \frac{Q}{2\pi\sigma^2}\frac{\sqrt{(\lambda_{max}-\lambda)(\lambda-\lambda_{min})}}{\lambda}, \qquad (65)$$

with $\lambda \in [\lambda_{min}, \lambda_{max}]$, where λ_{min}^{max} is given by $\lambda_{min}^{max}=\sigma^2(1 + 1/Q \pm 2\sqrt{1/Q})$ and σ^2 is equal to the variance of the elements of matrix M.

We compute the eigenvalues of the correlation matrix $C, \lambda_{max} = \lambda_1 > \lambda_2 > \cdots > \lambda_{15} = \lambda_{min}$. The probability density functions of paired correlation coefficients c_{ij} and eigenvalues λ_i for matrices of 110, 298, and 689 cryptocurrencies are presented in figure 34.

From figure 34 it can be seen that the distribution functions for the paired correlation coefficients of the selected matrices differ significantly from the distribution function described by the RMT. It can be seen that the crypto market has a significantly correlated, self-organized system, and the difference from the RMT of the case, the correlation coefficients exceed the value of 0.6-0.8 on "thick tails". The distribution of eigenvalues of the correlation matrix also differs markedly from the case of RMT. In our case, only one-third of its values refer to the RMT region.

Eigenvectors correspond to the participation ratio PR and its inverse participation ratio IPR

$$I^k = \sum_{l=1}^{N}\left[u_l^l\right]^4, \qquad (66)$$

where $u_k^l, l = 1,\ldots,N$, are the components of the eigenvector u^k. So, PR indicates the number of eigenvector components that contribute significantly to that eigenvector. More specifically, a low IPR indicates that cryptocurrencies contribute more equally. In contrast, a large IPR would imply that the factor is driven by the dynamics of a small number of cryptocurrencies.

The irregularity of the influence of eigenvalues of the correlation matrix is determined by the absorption ratio (AR)

$$AR_n = \frac{\sum_{k=1}^{n} \lambda_l}{\sum_{k=1}^{N} \lambda_k}, \tag{67}$$

which is a cumulative risk measure and indicates which part of the overall variation is described from the total number of N eigenvalues.

The difference in dynamics is due to the peculiarities of non-random correlations between the time series of individual assets. Under the framework of RMT, if the eigenvalues of the real-time series differ from the prediction of RMT, there must exist hidden economic information in those deviating eigenvalues. For cryptocurrencies markets, there are several deviating eigenvalues in which the largest eigenvalue λ_{max} reflects a collective effect of the whole market. As for PR, the differences from RMT appear at large and small λ values and are similar to the Anderson quantum effect of localization [363]. Under crash and critical event conditions, the states at the edges of the distributions of eigenvalues are delocalized, thus identifying the beginning of one of these events. This is evidenced by the results presented in figure 35b.

We find that both λ_{max} and $PR\lambda_{max}$ have large values for periods containing the crypto market crashes and critical events. At the same time, their growth begins in the pre-crashes periods. As well as the economic mass, they are quantum precursors of crashes and critical events phenomena.

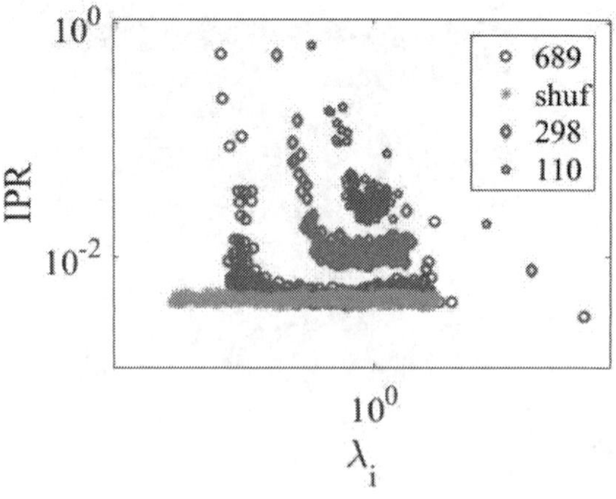

(a)

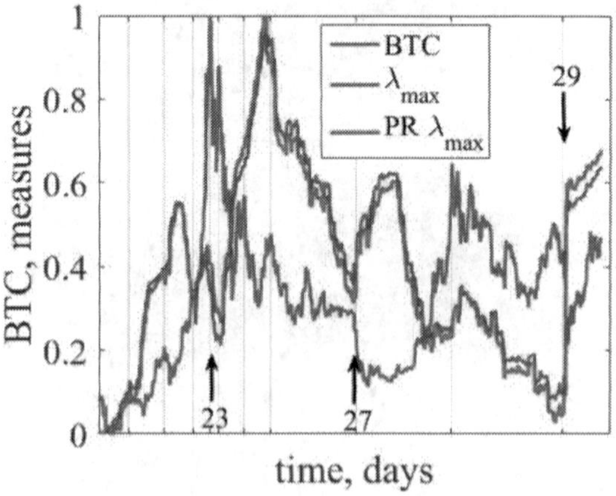

(b)

Figure 35: Inverse participation ratio (a) and (b) quantum measures of complexity λ_{max} and its participation ratio

10 Conclusions

Definitely, the factors from within and outside of the cryptocurrencies universe are going to evolve all of them. The great influence will go from incumbents and policymakers, as well as challengers and users. Current mistrust on the part of the government may lead to the introduction of specific licensing requirements that may make these digital currencies less attractive. Similarly, the adaptation to them and acceptance of cryptocurrencies may lead to increasing demand for them. The current situation with coronavirus is of paramount importance and is of significant danger.

From the literature overview, we have understood that crashes and critical events do not disappear without a trace, but will also affect the fate of individuals. On the other hand, in the future, the influence of these events may attract users to alternative forms of currency such as cryptocurrencies. Increased trading activity on cryptocurrency exchanges could positively affect the popularity of stablecoins. Lastly, the overall monetary system may be fundamentally changed through the introduction of a central bank digital currency, potentially upstaging stablecoins [364].

In order to give reliable, powerful, and simple indicators-precursors that are able to minimize further losses as a result of changes, we addressed the reach arsenal of the theory of complexity and the methods of nonlinear dynamics that can identify special trajectories in the complex dynamics and classify them. Following our research, we obtain informational, (multi-)fractal, chaos-dynamical, recurrent, irreversible, based on complex networks, and quantum measures of complexity.

The obtained quantitative methods were applied to classified crashes of the Bitcoin market, where it was seen that these indicators can be used in order to protect yourself from the upcoming critical change. To draw some conclusions about its evolution and factors that influence it, we pointed out the most influential critical changes in this market. Relying on different research articles and our previous experience, such changes were classified as crashes and critical events. Moreover, we assume crash as strong, time localized drop with high volatility whose percentage decline exceeds 30%. Critical events are those falls with less percentage decline and volatility. The analysis of the crypto market with the sliding (rolling or moving) window approach allowed us to draw some conclusions about its evolution and factors that influence it. Regarding empirical results, we have shown that some of the measures are very sensitive to the length of the sliding window and its time step. For example, if we consider two closest to each other events, a previous event that had much more volatility can have a great influence on the corresponding measure of complexity and spoil the identification of the next less influential, but important event. Thus, time localization is significant while calculating the measure of complexity. The less time localization and time step, the more corresponding changes are taken into account. For a much larger time window and its step, we can have less accurate estimations.

It turned out that most of the chosen measures of complexity respond in advance to the corresponding changes of complexity in the cryptocurrency market and can be used in the diagnostic processes. Such measures can be presented as indicators or even indicators-precursors of the approaching crashes and critical events.

Relying on the information theory and its powerful toolkit, we emphasized four measures of complexity, such as the measure of Lempel-Ziv, classical Shannon entropy, and its two modifications (Approximate entropy and Permutation entropy). We referred to the complexity of the systems, how it was described in different studies, and what methods were applied to quantify its degree. Our results show that in the pre-crash or pre-critical period, the complexity of Bitcoin starts to change that is it starts to decrease, indicating that such events presented to be more predictable and corresponding patterns are more structured. Thus, the degree of predictability increases in times of such events.

Along with information theory, we referred to the multifractal properties of the cryptocurrency market. As it was obtained with multifractal detrended fluctuation analysis, the scaling exponents remain non-linear, and the width of singularity spectrum changes in time that gives evidence that at different times (scales) BTC time series exhibits more or less complex behavior, indicating that cryptocurrency exhibits multifractal properties. Applying the width of multifractality as an indicator of possible critical states we found that before crash or critical event, this measure starts to decrease that tell us that the series is expected to be more predictable and stable, while its dynamics after such events is increasing that present system to be more susceptible to fluctuations.

Chaos-dynamical measures were applied to study bitcoin from the perspective of Chaos theory. Transitions between chaotic and non-chaotic behavior were identified with Lyapunov exponents where there was selected the largest Lyapunov exponent. The beginning of each crash or critical event can be characterized by the convergence of the two considered, initially close trajectories where the Largest Lyapunov exponent reflects the rate of such convergence, while the high-price level regimes are characterized by the divergence of the initially close trajectories. Whereas decreasing largest Lyapunov exponent indicates that patterns in the system tend to be predictable (trajectories converge), increasing indicates that the system goes into an unstable regime. Our empirical results obtained with the sliding window presented the largest Lyapunov exponent to be an indicator of such regimes. Moreover, we considered that such extreme events can be related to the fat-tails and better described with non-Gaussian distributions, particularly, described by Lévy alpha-stable distribution and its four parameters. As it is still debatable whether the stable distribution is completely applicable or not, we addressed to its group of stable parameters, and during tests, we emphasized that the characteristic exponent that serve to describe the thickness of tails is the best for serving as an indicator-precursor of possible crashes and critical events. Thus, is shown that such a complex system as the bitcoin market, with growth and preferential attachments, is characterized by power-laws.

The analysis of the crypto market with the measures from the recurrence quantification analysis revealed that its toolkit is suitable for distinguishing diverse market periods. Such measures as recurrence rate, determinism, the entropy diagonal line histogram, recurrence periodic density entropy, recurrence entropy, and divergence, which methodology is based on clusters of isolated points, vertical/diagonal lines, etc., are presented to be great for detection of the periods of instability or relaxation.

Also, applying the concept of time irreversibility (asymmetry), we found it changes for the period of crashes and critical events, and corresponding measures detect such phenomena. Thus, time irreversibility based on permutation patterns can serve as a good base for further models of financial control. However, much more measures can be added to this list [107, 316, 365–367].

Moreover, we have demonstrated the possibility of studying complex cryptosystems within the network paradigm. The time series can be presented as an economic network (visibility graph) and multiplex network with a set of both spectral and topological characteristics, which are sensitive to the critical changes in the BTC market.

Addressing to quantum econophysics and its apparatus where appropriate measures of complexity were obtained. Such quantitative methods as Heisenberg uncertainty and the Random matrix approach have confirmed their effectiveness for studying the cryptocurrency market. We found that economic "mass" along with λ_{max} and $PR\ \lambda_{max}$ are presented to be effective due to their robustness, computational efficiency, and simplicity.

Apparently, the impact of different crashes and critical events was reflected in the cryptocurrency market, as well as the coronavirus pandemic and therefore, the dynamics of past events, as well as of the subsequent could be identified in advance using the appropriate indicators of the theory of complexity. In our further studies, we are going to aim our view on exploring and analyzing of other methods from the theory of complexity. Being emerging currency that still needs to become trustworthy among as many people as possible, bitcoin's dynamics is the subject not only to high fluctuations but also to various attacks [368–372] of blockchain [373] and Proof-of-Work protocol [374]. Thus, particular interest presents research on implementing the theory of complex systems and its enormous toolkit for identifying such abnormal activity when transactions between users happen to make appropriate actions in advance. Equally interesting is the development of the theory of quantum computation and quantum information [375–378] where it would be interesting to see their influence on blockchain and cryptocurrencies. Moreover, the research in the field of artificial intelligence, machine, and deep learning does not remain without attention [51, 379–390].

References

[1] R. Albert, A.L. Barabási, Rev. Mod. Phys. **74**, 47 (2002)

[2] C. Reuven, H. Shlomo, *Complex Networks: Structure, Robustness and Function* (Cambridge University Press, 2010), ISBN 9780521841566

[3] M.E.J. Newman, SIAM Review **45**, 167–256 (2003)

[4] M. Newman, A.L. Barabasi, D.J. Watts, eds., *The Structure and Dynamics of Networks* (Princeton University Press, Princeton, NJ, USA, 2006)

[5] G. Nicolis, I. Prigogine, *Exploring complexity: an introduction* (W.H. Freeman, 1989), ISBN 9780716718598

[6] A. Rai, A. Mahata, M. Nurujjaman, O. Prakash, *Statistical properties of the aftershocks of stock market crashes: evidence based on the 1987 crash, 2008 financial crisis and COVID-19 pandemic* (2020), `arXiv:2012.03012 [q-fin.ST]`

[7] E. Mnif, A. Jarboui, K. Mouakhar, Finance Research Letters **36**, 101647 (2020)

[8] A. Ammy-Driss, M. Garcin, *Efficiency of the financial markets during the COVID-19 crisis: time-varying parameters of fractional stable dynamics* (2020), `arXiv:2007.10727 [q-fin.ST]`

[9] R. Cerqueti, V. Ficcadenti, *Anxiety for the pandemic and trust in financial markets* (2020), `arXiv:2008.01649 [q-fin.ST]`

[10] M. Costola, M. Iacopini, C.R.M.A. Santagiustina, *Public Concern and the Financial Markets during the COVID-19 outbreak* (2020), `arXiv:2005.06796 [q-fin.ST]`

[11] M. Feldkircher, F. Huber, M. Pfarrhofer, *Measuring the Effectiveness of US Monetary Policy during the COVID-19 Recession* (2020), `arXiv:2007.15419 [econ.EM]`

[12] M. Garcin, J. Klein, S. Laaribi, *Estimation of time-varying kernel densities and chronology of the impact of COVID-19 on financial markets* (2020), `arXiv:2007.09043 [q-fin.ST]`

[13] M. Pagano, C. Wagner, J. Zechner, *Disaster resilience and asset prices* (2020), `arXiv:2005.08929 [q-fin.GN]`

[14] A.A. Toda, *Susceptible-Infected-Recovered (SIR) Dynamics of COVID-19 and Economic Impact* (2020), `arXiv:2003.11221 [q-bio.PE]`

[15] S. Drożdż, J. Kwapień, P. Oświę, T. Stanisz, M. Wątorek, Entropy **22** (2020)

[16] R.K.K. Pang, O. Granados, H. Chhajer, E.F. Legara, *An analysis of network filtering methods to sovereign bond yields during COVID-19* (2021), `arXiv:2009.13390 [q-fin.ST]`

[17] S. Semerikov, S. Chukharev, S. Sakhno, A. Striuk, V. Osadchyi, V. Solovieva, T. Vakaliuk, P. Nechypurenko, O. Bondarenko, H. Danylchuk, E3S Web of Conferences **166** (2020)

[18] H. Danylchuk, L. Kibalnyk, O. Kovtun, A. Kiv, O. Pursky, G. Berezhna, CEUR Workshop Proceedings **2713**, 352 (2020)

[19] A. Kaminskyi, M. Nehrey, N. Rizun, CEUR Workshop Proceedings **2713**, 204 (2020)

[20] N. Maksyshko, O. Vasylieva, I. Kozin, V. Perepelitsa, CEUR Workshop Proceedings **2713**, 219 (2020)

[21] S. Semerikov, H. Kucherova, V. Los, D. Ocheretin, CEUR Workshop Proceedings **2845**, 22 (2021)

[22] G. Malinetsky, Modeling and Analysis of Information Systems **19**, 5 (2015)

[23] V. Soloviev, N. Moiseienko, O. Tarasova, CEUR Workshop Proceedings **2393**, 905 (2019)

[24] S. Somin, Y. Altshuler, G. Gordon, A. Pentland, E. Shmueli, Scientific Reports **10**, 4587 (2020)

[25] P. Grau, C. Jaureguizar, D. Jaureguizar, Esic Market Economics and Business Journal **49**, 569 (2018)

[26] J. Liang, L. Li, D. Zeng, PLOS ONE **13**, 1 (2018)

[27] S. Thurner, P. Klimek, R. Hanel, *Introduction to the theory of complex systems* (Oxford University Press, Oxford, 2018)

[28] S. Drożdż, L. Minati, P. Oświęcimka, M. Stanuszek, M. Wątorek, Chaos: An Interdisciplinary Journal of Nonlinear Science **30**, 023122 (2020)

[29] M. Wątorek, S. Drożdż, J. Kwapień, L. Minati, P. Oświęcimka, M. Stanuszek, Physics Reports **901**, 1 (2021), multiscale characteristics of the emerging global cryptocurrency market

[30] A.O. Bielinskyi, I. Khvostina, A. Mamanazarov, A. Matviychuk, S. Semerikov, O. Serdyuk, V. Solovieva, V.N. Soloviev, IOP Conference Series: Earth and Environmental Science **628**, 012019 (2021)

[31] M. Ausloos, D. Grech, T.D. Matteo, R. Kutner, C. Schinckus, H.E. Stanley, Physica A: Statistical Mechanics and its Applications **531**, 136 (2020)

[32] W.B. Arthur, Nature Reviews Physics **3**, 136 (2021)

[33] R. Kutner, M. Ausloos, D. Grech, T. Di Matteo, C. Schinckus, H. Eugene Stanley, Physica A: Statistical Mechanics and its Applications **516**, 240 (2019)

[34] R. Mantegna, H. Stanley, *An Introduction to Econophysics: Correlations and Complexity in Finance*, Vol. 53 (2000)

[35] Z.Q. Jiang, W.J. Xie, W.X. Zhou, D. Sornette, Reports on Progress in Physics **82**, 125901 (2019)

[36] J. Kwapień, S. Drożdż, Physics Reports **515**, 115 (2012), physical approach to complex systems

[37] B.E. Baaquie, *Quantum Finance: Path Integrals and Hamiltonians for Options and Interest Rates*

(Cambridge University Press, 2004)

[38] M. Schaden, Physica A: Statistical Mechanics and its Applications **316**, 511–538 (2002)

[39] V.P. Maslov, V.E. Nazaikinskii, *Mathematics underlying the 2008 financial crisis, and a possible remedy* (2009), `arXiv:0811.4678 [q-fin.GN]`

[40] C. Schinckus, *A Methodological Call for a Quantum Econophysics*, in *Selected Papers of the 7th International Conference on Quantum Interaction - Volume 8369* (Springer-Verlag, Berlin, Heidelberg, 2013), QI 2013, p. 308–316

[41] V. Saptsin, V. Soloviev, *Relativistic quantum econophysics - new paradigms in complex systems modelling* (2009), `arXiv:0907.1142 [physics.soc-ph]`

[42] V. Soloviev, V. Saptsin, *Heisenberg uncertainty principle and economic analogues of basic physical quantities* (2011), `arXiv:1111.5289 [physics.gen-ph]`

[43] D. Sornette, *Why Stock Markets Crash: Critical Events in Complex Financial Systems* (Princeton University Press, 2003)

[44] Y. Peng, P. Albuquerque, J. Camboim de Sá, A.J. Padula, M. Montenegro, Expert Systems with Applications **97** (2017)

[45] Q. Zhao, *A deep learning framework for predicting digital asset price movement from trade-by-trade data* (2020), `arXiv:2010.07404 [q-fin.ST]`

[46] M. Amjad, D. Shah, *Trading Bitcoin and Online Time Series Prediction*, in *Proceedings of the Time Series Workshop at NIPS 2016*, edited by O. Anava, A. Khaleghi, M. Cuturi, V. Kuznetsov, A. Rakhlin (PMLR, Barcelona, Spain, 2017), Vol. 55 of *Proceedings of Machine Learning Research*

[47] T. Chen, C. Guestrin, Proceedings of the 22nd ACM SIGKDD International Conference on Knowledge Discovery and Data Mining (2016)

[48] V. Derbentsev, A. Matviychuk, V. Soloviev, *Forecasting of Cryptocurrency Prices Using Machine Learning* (Springer, 2020), pp. 211–231, 1st edn., ISBN 978-981-15-4498-9

[49] A.H. Dyhrberg, Finance Research Letters **16**, 85 (2016)

[50] M. Ortu, N. Uras, C. Conversano, G. Destefanis, S. Bartolucci, *On technical trading and social media indicators in cryptocurrencies' price classification through deep learning* (2021), `arXiv:2102.08189 [q-fin.ST]`

[51] N. Uras, L. Marchesi, M. Marchesi, R. Tonelli, *Forecasting bitcoin closing price series using linear regression and neural networks models* (2020), `arXiv:2001.01127 [q-fin.ST]`

[52] A. Hachicha, F. Hachicha, Review of Quantitative Finance and Accounting **56**, 647 (2021)

[53] J. Kaminski, *Nowcasting the bitcoin market with twitter signals* (2016), `arXiv:1406.7577 [cs.SI]`

[54] M.A. Kennis, *Multi-channel discourse as an indicator for bitcoin price and volume movements* (2018), `arXiv:1811.03146 [q-fin.ST]`

[55] O. Kodama, L. Pichl, T. Kaizoji, CBU International Conference Proceedings **5**, 384 (2017)

[56] D. Shah, K. Zhang, *Bayesian regression and bitcoin* (2014), `arXiv:1410.1231 [cs.AI]`

[57] V. Soloviev, S. V., in *Modern problems of forecasting socio-economic processes: concepts, models, applied aspects* (Tkachuk O. V., 2012), pp. 174–188

[58] M. Tarnopolski, *Modeling the price of bitcoin with geometric fractional brownian motion: a monte carlo approach* (2017), `arXiv:1707.03746 [q-fin.CP]`

[59] N. Courtois, M. Grajek, R. Naik, *Optimizing SHA256 in Bitcoin Mining* (2014), Vol. 448 of *Communications in Computer and Information Science*, pp. 131–144, ISBN 978-3-662-44892-2

[60] L. Kristoufek, Frontiers in Physics **8** (2020)

[61] D. Broomhead, G.P. King, Physica D: Nonlinear Phenomena **20**, 217 (1986)

[62] V.I. Ponomarenko, M.D. Prokhorov, Phys. Rev. E **66**, 026215 (2002)

[63] M. Rajković, Physica A: Statistical Mechanics and its Applications **287**, 383 (2000)

[64] G.M. Caporale, L.A. Gil-Alana, A. Plastun, CESifo Working Paper Series 6811, CESifo (2017), `https://ideas.repec.org/p/ces/ceswps/_6811.html`

[65] A.F. Bariviera, M.J. Basgall, W. Hasperué, M. Naiouf, Physica A: Statistical Mechanics and its Applications **484**, 82–90 (2017)

[66] A.F. Bariviera, Economics Letters **161**, 1–4 (2017)

[67] J. Wang, G. Meric, Z. Liu, I. Meric, Journal of Banking & Finance **33**, 1563 (2009)

[68] S. Lleo, W.T. Ziemba, Financial Markets, Institutions & Instruments **26**, 61 (2017)

[69] H. Hong, J. Stein, Review of Financial Studies **16**, 487 (2003)

[70] M. Shu, W. Zhu, Physica A: Statistical Mechanics and its Applications **548**, 124477 (2020)

[71] T. Klein, H. Pham Thu, T. Walther, International Review of Financial Analysis **59**, 105 (2018)

[72] K. Gkillas, F. Longin, SSRN Electronic Journal pp. 1–85 (2019)

[73] Y. Liu, A. Tsyvinski, X. Wu, NBER Working Papers 25882, National Bureau of Economic Research, Inc (2019)

[74] L. Kristoufek, PLOS ONE **10**, 1 (2015)

[75] X. Li, C. Wang, Decision Support Systems **95**, 49 (2017)

[76] A. Bielinskyi, V. Soloviev, S. Semerikov, V. Solovieva, CEUR Workshop Proceedings **2422**, 420 (2019)

[77] A. Bielinskyi, S. Semerikov, V. Solovieva, V. Soloviev, SHS Web Conf. **65**, 06006 (2019)

[78] V. Derbentsev, S. Semerikov, O. Serdyuk, V. Solovieva, V. Soloviev, E3S Web Conf. **166**, 13031 (2020)

[79] V. Soloviev, A. Belinskiy, *Complex Systems Theory and Crashes of Cryptocurrency Market* (Springer, Cham, 2019), Vol. 1007 of *ICTERI 2018*, pp. 276–297, ISBN 978-3-030-13928-5

[80] V. Soloviev, A. Belinskiy, CEUR Workshop Proceedings **2104**, 116 (2018)

[81] V. Soloviev, A. Bielinskyi, O. Serdyuk, V. Solovieva, S. Semerikov, CEUR Workshop Proceedings **2732**, 455 (2020)

[82] V. Soloviev, S. Yevtushenko, V. Batareyev, CEUR Workshop Proceedings **2393**, 434 (2019)

[83] V. Soloviev, V. Solovieva, A. Tuliakova, M. Ivanova, *Construction of crisis precursors in multiplex networks*, in *Proceedings of the 2019 7th International Conference on Modeling, Development and Strategic Management of Economic System (MDSMES 2019)* (Atlantis Press, 2019/10), pp. 361–366, ISBN 978-94-6252-800-0, ISSN 2352-5428, `https://doi.org/10.2991/mdsmes-19.2019.68`

[84] V. Soloviev, O. Serdiuk, S. Semerikov, O. Kohut-Ferens, *Recurrence entropy and financial crashes*, in *Proceedings of the 2019 7th International Conference on Modeling, Development and Strategic Management of Economic System (MDSMES 2019)* (Atlantis Press, 2019/10), pp. 385–388, ISBN 978-94-6252-800-0, ISSN 2352-5428, `https://doi.org/10.2991/mdsmes-19.2019.73`

[85] V. Soloviev, A. Bielinskyi, N. Kharadzjan, CEUR Workshop Proceedings **2832**, 24 (2020)

[86] M.S. Kanwal, J.A. Grochow, N. Ay, Entropy **19** (2017)

[87] D. Bonchev, *Information Theoretic Complexity Measures* (Springer, 2009), pp. 4820–4838

[88] L. Lovász, The Emergence of Complexity in Mathematics, Physics, Chemistry and Biology, Pontifical Academy of Sciences p. 12 (1996)

[89] L.T. Lui, G. Terrazas, H. Zenil, C. Alexander, N. Krasnogor, Artificial Life **21**, 205 (2015)

[90] C.E. Shannon, Bell System Technical Journal **27**, 379 (1948)

[91] G. Sommazzi, Ph.D. thesis (2016)

[92] J.L. Blanc, L. Pezard, A. Lesne, Phys. Rev. E **84**, 036214 (2011)

[93] S. Zozor, P. Ravier, O. Buttelli, Physica A: Statistical Mechanics and its Applications **345**, 285 (2005)

[94] E. Estevez-Rams, R. Lora Serrano, B. Aragón Fernández, I. Brito Reyes, Chaos: An Interdisciplinary Journal of Nonlinear Science **23**, 023118 (2013)

[95] S.D. Silva, R. Matsushita, R. Giglio, Economics Bulletin **7**, 1 (2008)

[96] R. Giglio, R. Matsushita, A. Figueiredo, I. Gleria, S.D. Silva, EPL (Europhysics Letters) **84**, 48005 (2008)

[97] S. Da Silva, MPRA Paper 64497, University Library of Munich, Germany (2015)

[98] S. Da Silva, C. Taufemback, R. Giglio, Economics Bulletin **31**, 1631 (2011)

[99] R. Giglio, S. Da Silva, MPRA Paper 22720, University Library of Munich, Germany (2009)

[100] A. Lempel, J. Ziv, IEEE Transactions on Information Theory **22**, 75 (1976)

[101] O. Brandouy, J.P. Delahaye, L. Ma, H. Zenil, Research in International Business and Finance **30**, 336 (2014)

[102] P. Fiedor, Risks **3**, 219 (2015)

[103] J. Gao, Y. Hou, F. Fan, F. Liu, Entropy **22** (2020)

[104] H. Cao, Y. Li, Chaos: An Interdisciplinary Journal of Nonlinear Science **24**, 013134 (2014)

[105] V. Soloviev, S. Semerikov, V. Solovieva, *Lempel-Ziv Complexity and Crises of Cryptocurrency Market*, in *Proceedings of the III International Scientific Congress Society of Ambient Intelligence 2020 (ISC-SAI 2020)* (Atlantis Press, 2020), pp. 299–306, ISBN 978-94-6252-933-5, ISSN 2352-5428, `https://doi.org/10.2991/aebmr.k.200318.037`

[106] A.N. Kolmogorov, International Journal of Computer Mathematics **2**, 157 (1968)

[107] M. Costa, C.K. Peng, A. Goldberger, Cardiovascular engineering (Dordrecht, Netherlands) **8**, 88 (2008)

[108] R. Clausius, T. Hirst, *The Mechanical Theory of Heat: With Its Applications to the Steam-Engine and to the Physical Properties of Bodies* (Creative Media Partners, LLC, 2017)

[109] L. Boltzmann, *Weitere Studien über das Wärmegleichgewicht unter Gasmolekülen* (Vieweg+Teubner Verlag, 1970), Vol. 67, pp. 115–225

[110] J.W. Gibbs, *Elementary principles in statistical mechanics developed with especial reference to the rational foundation of thermodynamics* (C. Scribner, New York, 1902)

[111] M. Vosvrda, J. Barunik, L. Vacha, M. Vošvrda, Czech Economic Review **4**, 281 (2010)

[112] J.S. Richman, J.R. Moorman, American Journal of Physiology-Heart and Circulatory Physiology **278**, H2039 (2000)

[113] R. Sole, S. Valverde, *Information Theory of Complex Networks: On Evolution and Architectural Constraints* (Springer, 2004), Vol. 207, pp. 189–207, ISBN 978-3-540-22354-2

[114] V. Soloviev, O. Serdiuk, in *Information technologies and modeling in economics: on the way to interdisciplinarity* (Gate-Ukraine, 2013), pp. 115–129

[115] C. Tsallis, *Introduction to nonextensive statistical mechanics: Approaching a complex world* (Springer, 2009), ISBN 978-0-387-85358-1

[116] A. Delgado-Bonal, A. Marshak, Entropy **21** (2019)

[117] S. Pincus, Proceedings of the National Academy of Sciences of the United States of America **88**,

2297—2301 (1991)
[118] S.M. Pincus, A.L. Goldberger, American Journal of Physiology-Heart and Circulatory Physiology **266**, H1643 (1994)
[119] K. Yun, H.K. Park, D.H. Kwon, Y.T. Kim, S.N. Cho, H.J. Cho, B.S. Peterson, J. Jeong, Psychiatry Research: Neuroimaging **201**, 226 (2012)
[120] S.N. Bhaduri, Journal of Emerging Market Finance **13**, 43 (2014)
[121] C. Eom, G. Oh, W.S. Jung, Physica A: Statistical Mechanics and its Applications **387**, 5511 (2008)
[122] S. Lahmiri, S. Bekiros, Chaos, Solitons & Fractals **138**, 109936 (2020)
[123] I. Mahmoud, K. Naoui, H. Jemmali, International Journal of Economics and Financial Issues **3**, 683 (2013)
[124] S. Pincus, R.E. Kalman, Proceedings of the National Academy of Sciences **101**, 13709 (2004)
[125] W.Q. Duan, H.E. Stanley, Phys. Rev. E **81**, 066116 (2010)
[126] A. Delgado-Bonal, Sci. Rep. **9**, 2045 (2019)
[127] C. Bandt, B. Pompe, Phys. Rev. Lett. **88**, 174102 (2002)
[128] J. Amigó, *Permutation Complexity in Dynamical Systems* (Springer-Verlag Berlin Heidelberg, 2010)
[129] M. Zanin, L. Zunino, O.A. Rosso, D. Papo, Entropy **14**, 1553 (2012)
[130] H. Kantz, T. Schreiber, *Nonlinear Time Series Analysis*, 2nd edn. (Cambridge University Press, 2003)
[131] M. Henry, G. Judge, Econometrics **7** (2019)
[132] H. Sigaki, M. Perc, H. Ribeiro, Scientific Reports **9** (2019)
[133] A. Sensoy, Finance Research Letters **28**, 68 (2019)
[134] A. Metin Karakaş, Asian Journal of Business and Management **7**, 10 (2019)
[135] D.T. Pele, M. Mazurencu-Marinescu-Pele, Entropy **21** (2019)
[136] F. Takens, in *Dynamical Systems and Turbulence, Warwick 1980*, edited by D. Rand, L.S. Young (Springer, Berlin, 1981), Vol. 898 of *Lecture Notes in Mathematics*, chap. 21, pp. 366–381
[137] J.P. Eckmann, D. Ruelle, Rev. Mod. Phys. **57**, 617 (1985)
[138] E. Ott, T. Sauer, J. Yorke, *Coping with Chaos*, Wiley Series in Nonlinear Science (Wiley, 1994)
[139] C. Webber, Jr., J. Zbilut, in *Tutorials in Contemporary Nonlinear Methods for the Behavioral Sciences*, edited by M.A. Riley, G.C.V. Orden (National Science Foundation (NSF), 2005), chap. 2
[140] R. Gu, Physica A: Statistical Mechanics and its Applications **484**, 215 (2017)
[141] B.B. Mandelbrot, J.A. Wheeler, American Journal of Physics **51**, 286 (1983)
[142] H.E. Hurst, Transactions of the American Society of Civil Engineers **116**, 770 (1951)
[143] H.E. Hurst, Nature **180**, 494 (1957)
[144] A.W. Lo, Working Paper 2984, National Bureau of Economic Research (1989)
[145] C.K. Peng, S.V. Buldyrev, S. Havlin, M. Simons, H.E. Stanley, A.L. Goldberger, Phys. Rev. E **49**, 1685 (1994)
[146] Z.Q. Jiang, W.J. Xie, W.X. Zhou, Physica A: Statistical Mechanics and its Applications **405**, 235 (2014)
[147] J.W. Kantelhardt, S.A. Zschiegner, E. Koscielny-Bunde, S. Havlin, A. Bunde, H. Stanley, Physica A: Statistical Mechanics and its Applications **316**, 87–114 (2002)
[148] F. Aslam, W. Mohti, P. Ferreira, International Journal of Financial Studies **8** (2020)
[149] R. Hasan, S.M. Mohammad, Physica A: Statistical Mechanics and its Applications **419**, 746 (2015)
[150] S. Kumar, N. Deo, Physica A: Statistical Mechanics and its Applications **388**, 1593 (2009)
[151] J. Kwapień, P. Oświęcimka, S. Drożdż, Physica A: Statistical Mechanics and its Applications **350**, 466 (2005)
[152] S. Lahmiri, Physica A: Statistical Mechanics and its Applications **486**, 183 (2017)
[153] J.W. Lee, K. Eun Lee, P. Arne Rikvold, Physica A: Statistical Mechanics and its Applications **364**, 355 (2006)
[154] K. Matia, Y. Ashkenazy, H.E. Stanley, Europhysics Letters (EPL) **61**, 422 (2003)
[155] P. Suárez-García, D. Gómez-Ullate, Physica A: Statistical Mechanics and its Applications **394**, 226 (2014)
[156] L. Zunino, A. Figliola, B.M. Tabak, D.G. Pérez, M. Garavaglia, O.A. Rosso, Chaos, Solitons & Fractals **41**, 2331 (2009)
[157] F. Delbianco, F. Tohmé, T. Stosic, B. Stosic, Physica A: Statistical Mechanics and its Applications **457**, 573 (2016)
[158] R. Gu, H. Chen, Y. Wang, Physica A: Statistical Mechanics and its Applications **389**, 2805 (2010)
[159] Z. Li, X. Lu, Energy Procedia **5**, 1920 (2011), 2010 International Conference on Energy, Environment and Development - ICEED2010
[160] P. Mali, A. Mukhopadhyay, Physica A: Statistical Mechanics and its Applications **413**, 361–372 (2014)
[161] S. Zheng, X. Lan, Physica A: Statistical Mechanics and its Applications **444**, 547 (2016)
[162] G. Lim, S. Kim, H. Lee, K. Kim, D.I. Lee, Physica A: Statistical Mechanics and its Applications **386**, 259 (2007)
[163] P. Caraiani, E. Haven, Physica A Statistical Mechanics and its Applications **419**, 395 (2015)
[164] P. Norouzzadeh, B. Rahmani, Physica A: Statistical Mechanics and its Applications **367**, 328 (2006)
[165] G. Oh, C. Eom, S. Havlin, W.S. Jung, F. Wang, H. Stanley, S. Kim, Eur. Phys. J. B **85**, 214 (2012)
[166] J. Qin, X. Lu, Y. Zhou, L. Qu, Physica A: Statistical Mechanics and its Applications **421**, 443 (2015)

[167] D.H. Wang, X.W. Yu, Y.Y. Suo, Physica A: Statistical Mechanics and its Applications **391**, 3503 (2012)
[168] P. Norouzzadeh, W. Dullaert, B. Rahmani, Physica A: Statistical Mechanics and its Applications **380**, 333 (2007)
[169] W. Mensi, A.K. Tiwari, S.M. Yoon, Physica A: Statistical Mechanics and its Applications **471**, 135 (2017)
[170] A.K. Tiwari, C.T. Albulescu, S.M. Yoon, Physica A: Statistical Mechanics and its Applications **483**, 182 (2017)
[171] Y. Wang, Y. Wei, C. Wu, Physica A: Statistical Mechanics and its Applications **390**, 817 (2011)
[172] L. Zunino, B. Tabak, A. Figliola, D. Pérez, M. Garavaglia, O. Rosso, Physica A: Statistical Mechanics and its Applications **387**, 6558 (2008)
[173] B. Podobnik, H.E. Stanley, Phys. Rev. Lett. **100**, 084102 (2008)
[174] W.X. Zhou, Physical Review E **77** (2008)
[175] N. Costa, C. Silva, P. Ferreira, International Journal of Financial Studies **7** (2019)
[176] X.Y. Qian, Y.M. Liu, Z.Q. Jiang, B. Podobnik, W.X. Zhou, H.E. Stanley, Phys. Rev. E **91**, 062816 (2015)
[177] Z.Q. Jiang, W.X. Zhou, Phys. Rev. E **84**, 016106 (2011)
[178] L. Kristoufek, EPL (Europhysics Letters) **95**, 68001 (2011)
[179] J. Wang, P. Shang, W. Ge, Fractals **20**, 271 (2012)
[180] J. Li, X. Lu, Y. Zhou, Physica A: Statistical Mechanics and its Applications **453**, 131 (2016)
[181] C. Xie, Y. Zhou, G. Wang, X. Yan, Fluctuation and Noise Letters **16**, 1750004-226 (2017)
[182] F. Ma, Y. Wei, D. Huang, Physica A: Statistical Mechanics and its Applications **392**, 1659 (2013)
[183] Y. Wang, Y. Wei, C. Wu, Physica A: Statistical Mechanics and its Applications **389**, 5468 (2010)
[184] P. Yue, H.C. Xu, W. Chen, X. Xiong, W.X. Zhou, Fractals **25**, 1750041 (2017)
[185] F. Ma, Y. Wei, D. Huang, L. Zhao, Physica A: Statistical Mechanics and its Applications **392**, 5356 (2013)
[186] F. Ma, Q. Zhang, C. Peng, Y. Wei, Physica A: Statistical Mechanics and its Applications **410**, 154 (2014)
[187] Y. Wang, Y. Wei, C. Wu, Physica A: Statistical Mechanics and its Applications **390**, 864 (2011)
[188] X. Zhuang, Y. Wei, F. Ma, Physica A: Statistical Mechanics and its Applications **430**, 101 (2015)
[189] X. Zhuang, Y. Wei, B. Zhang, Physica A: Statistical Mechanics and its Applications **399**, 113 (2014)
[190] L. Xinsheng, L. Jianfeng, Z. Ying, Q. Yubo, Physica A: Statistical Mechanics and its Applications **486**, 168 (2017)
[191] Z. Zhang, Y. Zhang, D. Shen, W. Zhang, Complexity **2018**, 1 (2018)
[192] Z. Zhang, Y. Zhang, D. Shen, W. Zhang, Physica A: Statistical Mechanics and its Applications **508**, 67 (2018)
[193] Z. Da, J. Engelberg, P. Gao, Review of Financial Studies **28**, 1 (2015)
[194] W. Zhang, P. Wang, X. Li, D. Shen, Journal of Behavioral and Experimental Finance **18**, 50 (2018)
[195] M. Gronwald, C. Sattarhof, Working Paper 5, University of Aberdeen Business School (2018)
[196] T. Takaishi, Physica A: Statistical Mechanics and its Applications **506**, 507–519 (2018)
[197] L. Kirichenko, V. Bulakh, T. Radivilova, 2017 4th International Scientific-Practical Conference Problems of Infocommunications. Science and Technology (PIC S&T) (2017)
[198] Y. Jiang, H. Nie, W. Ruan, Finance Research Letters **25**, 280 (2018)
[199] A. Sensoy, E. Hacihasanoglu, Energy Economics **46**, 318 (2014)
[200] K.H. Al-Yahyaee, W. Mensi, S.M. Yoon, Finance Research Letters **27**, 228 (2018)
[201] G. Gajardo, W.D. Kristjanpoller, M. Minutolo, Chaos, Solitons & Fractals **109**, 195 (2018)
[202] S. Lahmiri, S. Bekiros, A. Salvi, Chaos, Solitons & Fractals **107**, 43 (2018)
[203] W. Zhang, P. Wang, X. Li, D. Shen, Complexity **2018**, 1 (2018)
[204] A. Ganchuk, V. Derbentsev, V. Soloviev, *Multifractal Properties of the Ukraine Stock Market* (2006), `arXiv:physics/0608009v1 [physics.data-an]`
[205] K. Hu, P.C. Ivanov, Z. Chen, P. Carpena, H. Eugene Stanley, Phys. Rev. E **64**, 011114 (2001)
[206] Z. Chen, P.C. Ivanov, K. Hu, H.E. Stanley, Phys. Rev. E **65**, 041107 (2002)
[207] J.R. Thompson, J.R. Wilson, Mathematics and Computers in Simulation **126**, 63 (2016)
[208] D. Clark, L. Tarra, A. Berera, Phys. Rev. Fluids **5**, 064608 (2020)
[209] R. Engelken, F. Wolf, L.F. Abbott, *Lyapunov spectra of chaotic recurrent neural networks* (2020), `arXiv:2006.02427 [nlin.CD]`
[210] K. Krishnamurthy, T. Can, D.J. Schwab, *Theory of gating in recurrent neural networks* (2021), `arXiv:2007.14823 [cond-mat.dis-nn]`
[211] W.S. Lee, S. Flach, *Deep learning of chaos classification* (2020), `arXiv:2004.10980 [cs.LG]`
[212] M.B. Tayel, E.I. AlSaba, *Robust and sensitive method of lyapunov exponent for heart rate variability* (2015), `arXiv:1508.00996 [cs.SY]`
[213] H. De Thélin, T. Gauthier, G. Vigny, Bulletin of the London Mathematical Society (2020)
[214] M. Cencini, F. Cecconi, A. Vulpiani, *Chaos: From Simple Models to Complex Systems*, Vol. 17 of *Series on Advances in Statistical Mechanics* (2010)
[215] P. Grassberger, I. Procaccia, Phys. Rev. Lett. **50**, 346 (1983)

[216] J.C. Sprott, *Chaos and Time-Series Analysis* (Princeton University Press, 2001), ISBN 978-0198508403

[217] L.S. Young, Journal of Physics A: Mathematical and Theoretical **46**, 254001 (2013)

[218] C.J. Gavilán-Moreno, G. Espinosa-Paredes, Nuclear Engineering and Technology **48**, 434 (2016)

[219] A. Prieto-Guerrero, G. Espinosa-Paredes, *Dynamics of BWRs and mathematical models* (2019), pp. 193–268

[220] D. Nychka, S. Ellner, A.R. Gallant, D. McCaffrey, Journal of the Royal Statistical Society: Series B (Methodological) **54**, 399 (1992)

[221] A. Wolf, J. Swift, H.L. Swinney, J. Vastano, Physica D: Nonlinear Phenomena **16**, 285 (1985)

[222] M. Sano, Y. Sawada, Phys. Rev. Lett. **55**, 1082 (1985)

[223] J.P. Eckmann, S.O. Kamphorst, D. Ruelle, S. Ciliberto, Phys. Rev. A **34**, 4971 (1986)

[224] M.T. Rosenstein, J.J. Collins, C.J. De Luca, Physica D: Nonlinear Phenomena **65**, 117 (1993)

[225] U. Parlitz, International Journal of Bifurcation and Chaos **02**, 155 (1992)

[226] M. Balcerzak, D. Pikunov, A. Dabrowski, Nonlinear Dynamics **94**, 3053 (2018)

[227] J. Gao, Y. Cao, W.W. Tung, J. Hu, *Multiscale Analysis of Complex Time Series: Integration of Chaos and Random Fractal Theory, and Beyond* (Wiley, 2007), ISBN 978-0-470-19164-4

[228] J. Gao, J. Hu, W.W. Tung, Y. Zheng, Quantitative Finance **13**, 265 (2013)

[229] V. Soloviev, D. Chabanenko, S. I., *Using the scale-dependent Lyapunov exponent as a measure of complexity*, in *The banking system of Ukraine in the context of globalization of financial markets: proceedings of VII International scientific and practical conference*, edited by M. Gedz (CHIBS UBS NBU, 2012), pp. 469–471

[230] V. Soloviev, I. Stratiychuk, The problems of economics **2**, 279 (2013)

[231] J.P. Eckmann, S.O. Kamphorst, D. Ruelle, Europhysics Letters (EPL) **4**, 973 (1987)

[232] J. Scheinkman, B. Lebaron, The Journal of Business **62**, 311 (1989)

[233] H.D.I. Abarbanel, R. Brown, J.J. Sidorowich, L.S. Tsimring, Rev. Mod. Phys. **65**, 1331 (1993)

[234] V.S. Kulkarni, *Complexity, chaos, and the duffing-oscillator model: An analysis of inventory fluctuations in markets* (2013), `arXiv:1308.1616 [q-fin.GN]`

[235] O. Bajo-Rubio, F. Fernandez-Rodriguez, S. Sosvilla-Rivero, Economics Letters **39**, 207 (1992)

[236] W.D. Dechert, R. Gencay, Journal of Applied Econometrics **7**, S41 (1992)

[237] R. Gençay, Physica D: Nonlinear Phenomena **89**, 261 (1996)

[238] S. Shreemoyee, C. Vikhyat, Engineering and Technology International Journal of Economics and Management Engineering **13** (2019)

[239] S. Srinivasan, S. Prasad, S. Patil, G. Lazarou, J. Picone, *Estimation of Lyapunov Spectra from a Time Series*, in *Proceedings of the IEEE SoutheastCon 2006* (2006), pp. 192–195

[240] L. Mastroeni, P. Vellucci, *"Chaos" in energy and commodity markets: a controversial matter* (2017), `arXiv:1611.07432 [q-fin.ST]`

[241] V. Plakandaras, R. Gupta, M.E. Wohar, Applied Economics **51**, 4477 (2019)

[242] G. Chakrabarti, C. Sen, *Anatomy of Global Stock Market Crashes*, Number 978-81-322-0463-3 in SpringerBriefs in Economics (Springer, 2012)

[243] J. Liesen, V. Mehrmann, *Linear Algebra*, Springer Undergraduate Mathematics Series, 1st edn. (Springer, 2015)

[244] C. Touzé, A. Chaigne, Acustica **86**, 1 (2000)

[245] B. Podobnik, A. Valentinčič, D. Horvatić, H.E. Stanley, Proceedings of the National Academy of Sciences of the United States of America **108**, 17883—17888 (2011)

[246] L. Bachelier, Annales scientifiques de l'École Normale Supérieure **3e série, 17**, 21 (1900)

[247] X. Gabaix, P. Gopikrishnan, V. Plerou, H. Stanley, Nature **423**, 267 (2003)

[248] M. Kateregga, S. Mataramvura, D. Taylor, Cogent Economics & Finance **5**, 1318813 (2017)

[249] D. Krężołek, Studia Ekonomiczne **247**, 57–68 (2015)

[250] T. Lux, D. Sornette, Journal of Money, Credit and Banking **34**, 589 (2002)

[251] Y. Malevergne, V. Pisarenko, D. Sornette, Quantitative Finance **5**, 379 (2005)

[252] Y. Malevergne, V. Pisarenko, D. Sornette, Phys. Rev. E **83**, 036111 (2011)

[253] N.N. Taleb, International Journal of Forecasting **36**, 1228–1240 (2020)

[254] N.N. Taleb, Y. Bar-Yam, P. Cirillo, *On single point forecasts for fat-tailed variables* (2020), `arXiv:2007.16096 [physics.soc-ph]`

[255] P. Gopikrishnan, M. Meyer, L. Amaral, H. Stanley, The European Physical Journal B **3**, 139–140 (1998)

[256] P. Gopikrishnan, V. Plerou, L.A. Nunes Amaral, M. Meyer, H.E. Stanley, Phys. Rev. E **60**, 5305 (1999)

[257] B. Podobnik, D. Horvatic, A.M. Petersen, H.E. Stanley, Proceedings of the National Academy of Sciences of the United States of America **106**, 22079—22084 (2009)

[258] B. Podobnik, K. Matia, A. Chessa, P.C. Ivanov, Y. Lee, H.E. Stanley, Physica A Statistical Mechanics and its Applications **300**, 300 (2001)

[259] X. Gabaix, Annual Review of Economics **1**, 255 (2009)

[260] Z. Kostanjčar, B. Jeren, Advances in Complex Systems **16**, 1350008 (2013)

[261] A. Chakraborty, S. Easwaran, S. Sinha, *Deviations from universality in the fluctuation behavior of a heterogeneous complex system reveal intrinsic properties of components: The case of the international currency market* (2018), `arXiv:1606.06111 [q-fin.ST]`

[262] T. Takaishi, *Recent scaling properties of bitcoin price returns* (2020), `arXiv:2009.06874 [q-fin.ST]`

[263] S. Drożdż, R. Gębarowski, L. Minati, P. Oświęcimka, M. Wątorek, Chaos: An Interdisciplinary Journal of Nonlinear Science **28**, 071101 (2018)

[264] T. Takaishi, PLOS ONE **16**, e0246209 (2021)

[265] S. Begušić, Z. Kostanjčar, H. Eugene Stanley, B. Podobnik, Physica A: Statistical Mechanics and its Applications **510**, 400 (2018)

[266] B. Mandelbrot, The Journal of Business **36** (1963)

[267] P. Levy, Bulletin de la Société Mathématique de France **52**, 49 (1924)

[268] B. Mandelbrot, International Economic Review **1**, 79 (1960)

[269] E.S. Andersen, Mathematica Scandinavica **3**, 185 (1955)

[270] I.A. Koutrouvelis, Journal of the American Statistical Association **75**, 918 (1980)

[271] B. Brorsen, S. Yang, Communications in Statistics Part B: Simulation and Computation **19**, 1459 (1990)

[272] J. Nolan, *Maximum Likelihood Estimation and Diagnostics for Stable Distributions* (Birkhäuser, Boston, MA, 2001), pp. 379–400

[273] E.F. Fama, R. Roll, Journal of the American Statistical Association **66**, 331 (1971)

[274] J.H. McCulloch, Communications in Statistics - Simulation and Computation **15**, 1109 (1986)

[275] X. Ma, C.L. Nikias, IEEE Transactions on Signal Processing **43**, 2884 (1995)

[276] M. Shao, C.L. Nikias, Proceedings of the IEEE **81**, 986 (1993)

[277] J.M. Nicolas, Traitement Du Signal **19**, 139 (2002)

[278] E.E. Kuruoglu, IEEE Transactions on Signal Processing **49**, 2192 (2001)

[279] W.H. DuMouchel, The Annals of Statistics **1**, 948 (1973)

[280] V.M. Zolotarev, *One-dimensional stable distributions*, Vol. 65 of *Translations of Mathematical Monographs* (American Mathematical Society, Providence, RI, 1986), translated from the Russian by H. H. McFaden, Translation edited by Ben Silver

[281] J.M. Chambers, C.L. Mallows, B.W. Stuck, Journal of the American Statistical Association **71**, 340 (1976)

[282] S. Bianchi, A. Pantanella, International Journal of Trade, Economics and Finance **2**, 52 (2011)

[283] V.I. Arnold, A. Avez, ZAMM - Journal of Applied Mathematics and Mechanics / Zeitschrift für Angewandte Mathematik und Mechanik **50**, 506 (1970)

[284] K. Umeno, Nonlinear Theory and Its Applications, IEICE **7**, 14 (2016)

[285] I.A. Koutrouvelis, Communications in Statistics - Simulation and Computation **10**, 17 (1981)

[286] V. Soloviev, V. Solovieva, D. Chabanenko, in *Contemporary concepts of forecasting the development of complex socio-economic systems*, edited by O. Chernyak, P. Zakharchenko (FO-P Tkachuk O V, Berdyansk, 2014), pp. 257–264

[287] H. Poincaré, *The Three-Body Problem and the Equations of Dynamics*, Astrophysics and Space Science Library, 1st edn. (Springer, Cham, 2017)

[288] P. Faure, H. Korn, Physica D: Nonlinear Phenomena **122**, 265 (1998)

[289] M. Thiel, M. Romano, J. Kurths, *Analytical Description of Recurrence Plots of White Noise and Chaotic Processes*, in *Applied Nonlinear Dynamics* (2003), Vol. 11

[290] M. Thiel, M. Romano, J. Kurths, R. Meucci, E. Allaria, F. Arecchi, Physica D: Nonlinear Phenomena **171**, 138 (2002)

[291] L. Charles, J. Webber, I. Cornel, M. Norbert, eds., *Recurrence Plots and Their Quantifications: Expanding Horizons*, Vol. 180 of *Springer Proceedings in Physics* (Springer, 2015)

[292] N. Marwan, M. Carmen Romano, M. Thiel, J. Kurths, Physics Reports **438**, 237 (2007)

[293] G. Jianbo, C. Huaqing, Physics Letters A **270**, 75 (2000)

[294] N. Marwan, N. Wessel, U. Meyerfeldt, A. Schirdewan, J. Kurths, Phys. Rev. E **66**, 026702 (2002)

[295] C.L. Webber, J.P. Zbilut, Journal of Applied Physiology **76**, 965 (1994)

[296] J.P. Zbilut, C.L. Webber, Physics Letters A **171**, 199 (1992)

[297] G. Corso, T.L. Prado, G.Z. dos S. Lima, S.R. Lopes, *A novel entropy recurrence quantification analysis* (2017), `arXiv:1707.00944 [stat.OT]`

[298] M.A. Little, P.E. McSharry, S.J. Roberts, D.A. Costello, I.M. Moroz, BioMedical Engineering On-Line **6**, 23 (2007)

[299] H. Rabarimanantsoa, L. Achour, C. Letellier, A. Cuvelier, J.F. Muir, Chaos: An Interdisciplinary Journal of Nonlinear Science **17**, 013115 (2007)

[300] S. Lopes, T. Prado, G. Corso, G. dos S. Lima, J. Kurths, Chaos, Solitons & Fractals **133**, 109616 (2020)

[301] A.L. Goldberger, L.A.N. Amaral, L. Glass, J.M. Hausdorff, P.C. Ivanov, R.G. Mark, J.E. Mietus, G.B. Moody, C.K. Peng, H.E. Stanley, Circulation **101**, e215 (2000)

[302] M. Kirchner, P. Schubert, M. Liebherr, C.T. Haas, PLOS ONE **9**, 1 (2014)

[303] I. Prigogine, *From Being to Becoming Time and Complexity in the Physical Sciences*, 1st edn. (W.H.

Freeman & Co, 1981)
[304] M. Costa, A.L. Goldberger, C.K. Peng, Phys. Rev. E **71**, 021906 (2005)
[305] V. Soloviev, O. Rybchinska, Bulletin of the Kryvyi Rih Economic Institute KEI **2**, 52 (2010)
[306] S. Daw, C. Finney, M. Kennel, Physical review. E, Statistical physics, plasmas, fluids, and related interdisciplinary topics **62**, 1912 (2000)
[307] C. Diks, J. van Houwelingen, F. Takens, J. DeGoede, Physics Letters A **201**, 221 (1995)
[308] J. F.Donges, R. V. Donner, J. Kurths, EPL (Europhysics Letters) **102**, 10004 (2013)
[309] P. Guzik, J. Piskorski, T. Krauze, A. Wykretowicz, H. Wysocki, Biomedizinische Technik. Biomedical engineering **51**, 272 (2006)
[310] M.B. Kennel, Phys. Rev. E **69**, 056208 (2004)
[311] L. Lacasa, A. Nuñez, E. Roldán, J.M.R. Parrondo, B. Luque, The European Physical Journal B **85** (2012)
[312] A. Porta, S. Guzzetti, N. Montano, T. Gnecchi-Ruscone, R. Furlan, A. Malliani, *Time reversibility in short-term heart period variability*, in *Computers in Cardiology* (2006), Vol. 33, pp. 77–80, 2006 Computers in Cardiology, CIC ; Conference date: 17-09-2006 Through 20-09-2006
[313] M. Zanin, A. Rodríguez-González, E. Menasalvas Ruiz, D. Papo, Entropy **20** (2018)
[314] W. Yao, W. Yao, D. Yao, D. Guo, J. Wang, Applied Physics Letters **116**, 014101 (2020)
[315] J. Li, P. Shang, X. Zhang, Nonlinear Dynamics **96**, 2637 (2019)
[316] R. Flanagan, L. Lacasa, Physics Letters A **380**, 1689 (2016)
[317] J.A. Martín-Gonzalo, I. Pulido-Valdeolivas, Y. Wang, T. Wang, G. Chiclana-Actis, M.d.C. Algarra-Lucas, I. Palmí-Cortés, J. Fernández Travieso, M.D. Torrecillas-Narváez, A.A. Miralles-Martinez et al., Entropy **21**, 868 (2019)
[318] J.H. Martínez, J.L. Herrera-Diestra, M. Chavez, Chaos: An Interdisciplinary Journal of Nonlinear Science **28**, 123111 (2018)
[319] Y. Wenpo, J. Dai, M. Perc, J. Wang, D. Yao, D. Guo, Nonlinear Dynamics **100**, 907 (2020)
[320] W. Yao, W. Yao, J. Wang, Physics Letters A **383**, 1764 (2019)
[321] G.G. Malinetsky, Computer research and modeling **5**, 315 (2013)
[322] T.U. Grund, The Journal of Mathematical Sociology **37**, 131 (2013)
[323] R. Cohen, S. Havlin, *Complex Networks: Structure, Robustness and Function* (Cambridge University Press, 2010)
[324] G. Bianconi, EPL (Europhysics Letters) **111**, 56001 (2015)
[325] S. Boccaletti, V. Latora, Y. Moreno, M. Chavez, D.U. Hwang, Physics Reports **424**, 175 (2006)
[326] *Dynamics of graph spectral entropy in financial crisis*, in *Socio-Economic Aspects of Economics and Managment* (Aspekt Publishing of Budget Printing Cente, 2015), pp. 227–234
[327] V. Soloviev, Bulletin of Cherkasy University **38**, 67 (2015)
[328] V. Soloviev, V. Solovieva, A. Tuliakova, Neuro-Fuzzy Modeling Techniques in Economics **8**, 3 (2019)
[329] V. Soloviev, V. Solovieva, A. Tuliakova, A. Hostryk, L. Pichl, CEUR Workshop Proceedings **2713**, 53 (2020)
[330] S. Boccaletti, G. Bianconi, R. Criado, C. del Genio, J. Gómez-Gardeñes, M. Romance, I. Sendiña-Nadal, Z. Wang, M. Zanin, Physics Reports **544**, 1 (2014), the structure and dynamics of multilayer networks
[331] M.C. Qian, Z.Q. Jiang, W.X. Zhou, Journal of Physics A: Mathematical and Theoretical **43**, 335002 (2010)
[332] J. Birch, Ph.D. thesis (2015)
[333] F. Liu, N. Wang, D. Wei, The Open Cybernetics & Systemics Journal **11**, 36 (2017)
[334] W. Yan, E. van Tuyll van Serooskerken, PLOS ONE **10**, 1 (2015)
[335] A. Johansen, O. Ledoit, D. Sornette, International Journal of Theoretical and Applied Finance **03**, 219 (2000)
[336] M.D. Vamvakaris, A.A. Pantelous, K.M. Zuev, Physica A: Statistical Mechanics and its Applications **497**, 41 (2018)
[337] M. Serafino, A. Gabrielli, G. Caldarelli, G. Cimini, *Statistical validation of financial time series via visibility graph* (2017), `arXiv:1710.10980 [q-fin.RM]`
[338] C. Coquidé, J. Lages, D.L. Shepelyansky, Lecture Notes in Business Information Processing p. 208–219 (2019)
[339] T. Squartini, A. Gabrielli, D. Garlaschelli, T. Gili, A. Bifone, F. Caccioli, Complexity **2018**, 1 (2018)
[340] L. Lacasa, V. Nicosia, V. Latora, Scientific Reports **5** (2015)
[341] L. Bargigli, G. di Iasio, L. Infante, F. Lillo, F. Pierobon, Quantitative Finance **15**, 673 (2015)
[342] S. Li, S. Wen, Complexity **2017**, 1 (2017)
[343] C. Stephen, Procedia Computer Science **53**, 410 (2015)
[344] R.V. Donner, M. Small, J.F. Donges, N. Marwan, Y. Zou, R. Xiang, J. Kurths, International Journal of Bifurcation and Chaos **21**, 1019 (2011)
[345] L. Lacasa, B. Luque, F. Ballesteros, J. Luque, J.C. Nuño, Proceedings of the National Academy of Sciences **105**, 4972–4975 (2008)
[346] V. Soloviev, A. Tuliakova, Neuro-fuzzy modeling technologies in economics **5**, 152 (2016)
[347] J. Iacovacci, L. Lacasa, Physical Review E **94** (2016)
[348] A. de la Concha, S. Martinez-Jaramillo, C. Carmona, in *Complex Networks & Their Applications VI* (Springer International Publishing, 2018), pp.

1135–1148

[349] G. Colangelo, F.M. Ciurana, L.C. Bianchet, R.J. Sewell, M.W. Mitchell, Nature **543**, 525–528 (2017)

[350] E.G. Hidalgo, *Quantum econophysics* (2006), `arXiv:physics/0609245 [physics.soc-ph]`

[351] V. Maslov, Math. Notes **72**, 811 (2002)

[352] V. Soloviev, O. Serdiuk, Bulletin of Cherkasy University **1**, 3 (2009)

[353] E. Benítez Rodríguez, L. Aguilar, Scientific Reports **8**, 4010 (2018)

[354] L.A. Rozema, A. Darabi, D.H. Mahler, A. Hayat, Y. Soudagar, A.M. Steinberg, Phys. Rev. Lett. **109**, 100404 (2012)

[355] M. Berta, M. Christandl, R. Colbeck, J.M. Renes, R. Renner, Nature Physics **6**, 659–662 (2010)

[356] R. Prevedel, D.R. Hamel, R. Colbeck, K. Fisher, K.J. Resch, Nature Physics **7**, 757–761 (2011)

[357] L. Landau, E. Lifshitz, *Quantum Mechanics: Non-Relativistic Theory*, Course of Theoretical Physics, 3rd edn. (Butterworth-Heinemann, 1981)

[358] V. Soloviev, Y. Romanenko, *Economic analog of Heisenberg uncertainly principle and financial crisis*, in *System analysis and information technology : 19-th International conference* (ESC "IASA" NTUU "Igor Sikorsky Kyiv Polytechnic Institute", Kyiv, Ukraine, 2017), SAIT 2017, pp. 32–33

[359] V. Soloviev, S. Yevtushenko, V. Batareyev, CEUR Workshop Proceedings **2546**, 87 (2019)

[360] S. Drozdz, J. Kwapien, P. Oswiecimka, *Empirics versus rmt in financial cross-correlations* (2007), `arXiv:0711.0644 [q-fin.ST]`

[361] F.J. Dyson, Journal of Mathematical Physics **3**, 140 (1962)

[362] E.P. Wigner, Annals of Mathematics **53**, 36 (1951)

[363] P.W. Anderson, Phys. Rev. **109**, 1492 (1958)

[364] A. Lipton, A. Sardon, F. Schär, C. Schüpbach, *From tether to libra: Stablecoins, digital currency and the future of money* (2020), `arXiv:2005.12949 [cs.CY]`

[365] C. Karmakar, A. Khandoker, J. Gubbi, M. Palaniswami, *Modified Ehlers' index for improved detection of heart rate asymmetry in Poincaré plot*, in *36th Annual Computers in Cardiology Conference (CinC)* (IEEE, 2009), Vol. 36, pp. 169 – 172

[366] A. Porta, S. Guzzetti, N. Montano, T. Gnecchi-Ruscone, R. Furlan, A. Malliani, *Time reversibility in short-term heart period variability*, in *2006 Computers in Cardiology* (IEEE, 2006), Vol. 2006, pp. 77 – 80

[367] A. Puglisi, D. Villamaina, EPL **88**, 30004 (2009)

[368] A. Abhishta, R. Joosten, S. Dragomiretskiy, L. Nieuwenhuis, *Impact of Successful DDoS Attacks on a Major Crypto-Currency Exchange*, in *2019 27th Euromicro International Conference on Parallel, Distributed and Network-based Processing (PDP)* (IEEE, United States, 2019), pp. 379–384

[369] A.A. Kochkarov, S.D. Osipovich, R.A. Kochkarov, Revista ESPACIOS **41**, 29 (2020)

[370] M. Conti, E. Sandeep Kumar, C. Lal, S. Ruj, IEEE Communications Surveys & Tutorials **20**, 3416–3452 (2018)

[371] M. Mirkin, Y. Ji, J. Pang, A. Klages-Mundt, I. Eyal, A. Juels, *Bdos: Blockchain denial of service* (2020), `arXiv:1912.07497 [cs.CR]`

[372] M. Vasek, M. Thornton, T. Moore, *Empirical Analysis of Denial-of-Service Attacks in the Bitcoin Ecosystem*, in *Financial Cryptography and Data Security. FC 2014. Lecture Notes in Computer Science*, edited by R. Böhme, M. Brenner, T. Moore, M. Smith (Springer, 2014), Vol. 8438 of *FC 2014*, pp. 57–71

[373] U. Hacioglu, ed., *Blockchain Economics and Financial Market Innovation*, Contributions to Economics, 1st edn. (Springer, 2019), ISBN 978-3-030-25275-5

[374] S. Nakamoto, *Bitcoin: A peer-to-peer electronic cash system* (2009), `http://www.bitcoin.org/bitcoin.pdf`

[375] D. Aggarwal, G. Brennen, T. Lee, M. Santha, M. Tomamichel, Ledger **3** (2018)

[376] D. Sapaev, D. Bulychkov, F. Ablayev, A. Vasiliev, M. Ziatdinov, *Quantum-assisted blockchain* (2018), `arXiv:1802.06763 [quant-ph]`

[377] O. Sattath, International Journal of Information Security **19**, 291–302 (2020)

[378] L. Tessler, T. Byrnes, *Bitcoin and quantum computing* (2018), `arXiv:1711.04235 [quant-ph]`

[379] L. Alessandretti, A. ElBahrawy, L.M. Aiello, A. Baronchelli, Complexity **2018**, 1–16 (2018)

[380] N. Gandal, H. Halaburda, Games **7** (2016)

[381] T. Guo, A. Bifet, N. Antulov-Fantulin, 2018 IEEE International Conference on Data Mining (ICDM) (2018)

[382] H. Jang, J. Lee, IEEE Access **6**, 5427 (2018)

[383] O. Sattarov, A. Muminov, C.W. Lee, H.K. Kang, R. Oh, J. Ahn, H.J. Oh, H.S. Jeon, Applied Sciences **10** (2020)

[384] D. Zhao, A. Rinaldo, C. Brookins, *Cryptocurrency price prediction and trading strategies using support vector machines* (2019), `arXiv:1911.11819 [q-fin.TR]`

[385] T.R. Li, A.S. Chamrajnagar, X.R. Fong, N.R. Rizik, F. Fu, Frontiers in Physics **7**, 98 (2019)

[386] W. Wei, Q. Zhang, L. Liu, *Bitcoin transaction forecasting with deep network representation learning* (2020), `arXiv:2007.07993 [cs.SI]`

[387] A.H.A. Othman, S. Kassim, R.B. Rosman, N.H.B. Redzuan, Journal of Revenue and Pricing Management **19**, 314 (2020)

[388] S. McNally, J. Roche, S. Caton, *Predicting the Price of Bitcoin Using Machine Learning*, in *2018 26th Euromicro International Conference on Parallel, Distributed and Network-based Processing*

(PDP) (2018), pp. 339–343

[389] S.M. Raju, A.M. Tarif, *Real-time prediction of bitcoin price using machine learning techniques and public sentiment analysis* (2020), arXiv:2006.14473 [q-fin.ST]

[390] Hua, Yiqing, E3S Web Conf. **218**, 01050 (2020)

The lack of reversibility during financial crisis and its identification

Andrii Bielinskyi[1,*], *Serhii* Hushko[2,**], *Andriy* Matviychuk[3,***], *Oleksandr* Serdyuk[4,****], *Serhiy* Semerikov[1,5,6,7,†], and *Vladimir* Soloviev[1,‡]

[1]Kryvyi Rih State Pedagogical University, 54 Gagarin Ave., Kryvyi Rih, 50086, Ukraine
[2]State University of Economics and Technology, 16 Medychna Str., Kryvyi Rih, 50005, Ukraine
[3]Kyiv National Economic University named after Vadym Hetman, 54/1 Peremogy Ave., Kyiv, 03680, Ukraine
[4]The Bohdan Khmelnytsky National University of Cherkasy, 81 Shevchenko Blvd., Cherkasy, 18031, Ukraine
[5]Kryvyi Rih National University, 11 Vitalii Matusevych Str., Kryvyi Rih, 50027, Ukraine
[6]Institute of Information Technologies and Learning Tools of NAES of Ukraine, 9 M. Berlynskoho Str., Kyiv, 04060, Ukraine
[7]University of Educational Management, 52-A Sichovykh Striltsiv Str., Kyiv, 04053, Ukraine

Abstract. The focus of this study to measure the varying irreversibility of stock markets. A fundamental idea of this study is that financial systems are complex and nonlinear systems that are presented to be non-Gaussian fractal and chaotic. Their complexity and different aspects of nonlinear properties, such as time irreversibility, vary over time and for a long-range of scales. Therefore, our work presents approaches to measure the complexity and irreversibility of the time series. To the presented methods we include Guzik's index, Porta's index, Costa's index, based on complex networks measures, Multiscale time irreversibility index and based on permutation patterns measures. Our study presents that the corresponding measures can be used as indicators or indicator-precursors of crisis states in stock markets.

1 Introduction

Complex systems are open systems that exchange energy, matter, and information with the environment. Investigating complex systems in the natural sciences, Prigogine made a fundamental generalization, indicating the need for consideration of the phenomena of irreversibility and non-equilibrium as principles of selection of space-time structures that are implemented in practice [1]. Later it became clear that this generalization extends to complex systems of another nature: social, economic, biomedical, etc. [2]. Prigogine believed that the most important changes in the modern scientific revolution are related to the removal of previous restrictions in the scientific understanding of time. The nonlinear world is characterized by features of temporality, i.e., irreversibility and transience of processes and phenomena. Self-organization is considered as a spontaneous process of formation of integrating complex systems. It is due to the ambiguity of choice at bifurcation points that time in theories of self-organization becomes truly irreversible. In contrast to linear dynamic theories – classical, relativistic, quantum (where time is reversed), in the thermodynamics of dissipative structures created by Prigogine, time ceases to be a simple parameter and becomes a concept that expresses the pace and direction of events.

Thus, the irreversibility of time is a fundamental property of non-equilibrium dissipative systems, and its loss may indicate the development of destructive processes [2, 3].

Considering the statistical properties of a signal under study, its evolution could be called irreversible if there is the lack of invariance, i.e., the same signal would have been obtained if we measured it in the opposite direction. The function f could be applied to find characteristics that differ forward and backward versions, i.e., time series would be irreversible if $f(X^d) \neq f(X^r)$. The main idea of this definition there is no any restrictions on f.

Our study implies that a stationary process X is called statistically inverse in time if the probability distributions of the forward and backward in time systems are approximately the same [4–6]. The irreversibility of time series indicates the presence of nonlinear dependencies (memory) [7] in the dynamics of a system far from equilibrium, including non-Gaussian random processes and dissipative chaos. Since the definition of the irreversibility of the time series is formal, there is no a priori optimal algorithm for its quantification. Several methods for measuring the irreversibility of time have been proposed [2–4, 8–13]. Such methods significant as their purpose to deal with signals that exclude linear Gaussian random processes and, there by, allow to quantify the degree of predictability in the system.

*e-mail: krivogame@gmail.com
**e-mail: dep.director\protect_edu@kneu.edu.ua
***e-mail: editor@nfmte.com
****e-mail: serdyuk@ukr.net
†e-mail: semerikov@gmail.com
‡e-mail: vnsoloviev2016@gmail.com

In the first group of methods, the symbolization of time series is performed, and then the analysis is performed by statistical comparison of the appearance of a string of symbols in the forward and reverse directions [9]. Sometimes additional compression algorithms are used [8]. An important step for this group is the symbolization – the conversion of the time series into a character series requires additional special information (e.g., division of the range or size of the alphabet) and, therefore, contains the problem of the algorithm's dependence on these additional parameters. The second problem arises when considering the large-scale invariance of complex signals. Since the procedures of typical symbolizations are local, taking into account different scales can cause some difficulties [3].

Another group of methods in formalizing the index of irreversibility does not use the symbolization procedure but is based on the use of real values of the time series or returns.

One such approaches is based on the asymmetry of the distribution of points of the Poincare map, built on the basis of the values of the analyzed time series [10, 13].

Recently, a fundamentally new approach to measuring the irreversibility of time series has been proposed, which uses the methods of complex network theory [4, 12] and which combines two tools: the algorithm for visibility of time series recovery into a complex network and the Kullbak-Leibler divergence algorithm [12]. The first forms a directional network according to the geometric criterion. The degree of irreversibility of the series is then estimated by the Kullbak-Leibler divergence (i.e., the resolution) between the distribution of the input and output stages of the associated count. This method is computationally efficient, does not require any special symbolization of the process, and, according to the authors, naturally takes into account multiscale.

In this study, we apply irreversibility analysis and construct indicators or indicators-precursors of crashes and critical events, which dynamics is associated with luck of irreversibility the system. To these measures we include Guzik's index, Porta's index, Costa's index, based on complex networks, multiscale time irreversibility index with measure based on ordinal patterns.

For analyzing and explaining basic characteristics of stock market with time irreversibility measures, we have chosen Dow Jones Industrial Average index (DJIA) as the most quoted financial barometer in the world. In order to have better look on its intraday dynamics, we have separated its time series into two parts: from 2 January 1920 to 3 January 1983 and second part from 4 January 1983 to 3 March 2021. Both periods of daily values have been obtained through Yahoo Finance (http://finance.yahoo.com/) and Investing.com (https://www.investing.com/).

Regarding our previous studies [14–23], we have emphasized 30 crisis events that were classified as *crashes* and *critical events*. According to classification:

- Crashes are short, time-localized drops, with strong losing of price each day.
- Critical events are those falls that, during their existence, have not had such serious changes in price as crashes.

Table 1 shows the major crashes and critical events related to our classification.

Table 1: Major Historical Corrections of the DJIA price since 1920

№	Interval	Days in correction	Decline, %
1	03.09.1929-29.10.1929	41	39.64
2	01.03.1938-31.03.1938	23	24.15
3	08.04.1940-05.06.1940	42	25.1
4	21.08.1946-10.09.1946	14	16.35
5	30.07.1957-22.10.1957	60	17.51
6	19.03.1962-28.05.1962	50	19.91
7	18.07.1966-07.10.1966	59	12.84
8	09.04.1970-26.05.1970	34	20.35
9	24.10.1974-04.10.1974	52	27.45
10	02.10.1987-19.10.1987	12	34.16
11	17.07.1990-23.08.1990	28	17.21
12	01.10.1997-21.10.1997	15	12.43
13	17.08.1998-31.08.1998	11	18.44
14	14.08.2002-01.10.2002	34	19.52
15	16.10.2008-15.12.2008	42	30.21
16	09.08.2011-22.09.2011	32	11.94
17	18.08.2015-25.08.2015	6	10.53
18	29.12.2015-20.01.2016	16	11.02
19	03.12.2018-24.12.2018	15	15.62
20	04.03.2020-23.03.2020	13	31.38

As it is seen from the Table, during DJIA existence, many crashes and critical events shook it. According to our classification, events with number (1, 10, 13, 15, 20) are crashes, all the rest – critical events.

The calculations of indicators for them will be carried out within the sliding window approach. According to the procedure, we emphasize the frame of a predefined length in which the calculation of the corresponding measure is obtained. For this fragment measure of irreversibility is obtained regarding normalized returns, where returns are calculated as

$$G(t) = \ln x(t + \Delta t) - \ln x(t) \cong [x(t + \Delta t) - x(t)] / x(t) \quad (1)$$

and normalized (standardized) returns as

$$g(t) \cong [G(t) - \langle G \rangle] / \sigma, \quad (2)$$

where $\sigma \equiv \sqrt{\langle G^2 \rangle - \langle G \rangle^2}$ is the standard deviation of G, Δt is the time shift (in our case $\Delta t = 1$), and $\langle \ldots \rangle$ is the average over studied time period.

Then, the time window is shifted along the time by a predefined value, and the procedure is repeated until the entire series is exhausted. Comparing the calculated measure of irreversibility (asymmetry) and the actual time series of DJIA, we can analyze changes of complexity in the system. Our measures can be called indicators or precursors if they behave in a definite way for all periods of crashes, for example, decreases or increases during the pre-crash or pre-critical period. For our calculations time frame with the length 500 and step 1 are seemed to be the most reasonable parameters.

2 Assessing financial crises throughout irreversibility analysis

2.1 Irreversible complexity measures based on Poincaré diagrams

The Poincaré diagram for the time series is a graph on the x axis of which the normalized returns for current time $g(t)$ are plotted, and subsequent values $g(t+1)$ on the y axis. In Figure 1 the Poincaré diagram for the initial and shuffled series of the DJIA is shown.

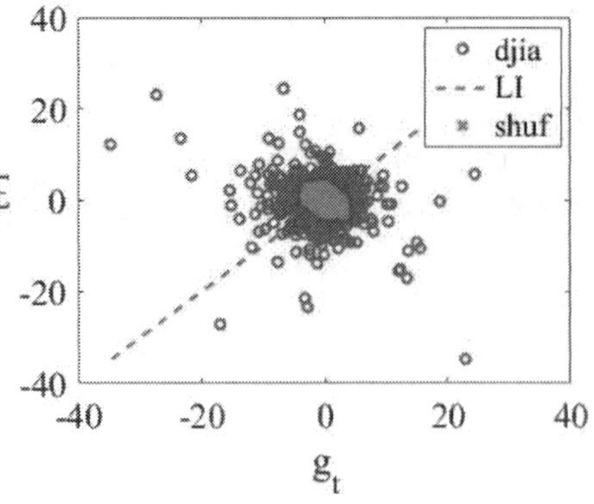

(a)

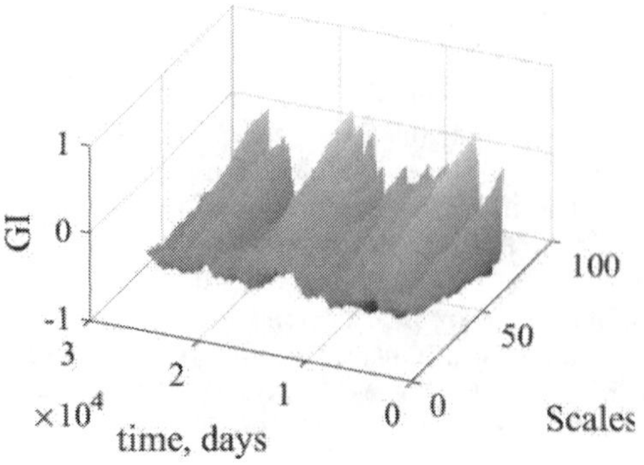

(b)

Figure 1: The Poincaré diagram (a) and the dependence of the Costa's index that will be described further on time and scale (b)

All consequent values that are equal to each other ($g(t) = g(t+1)$) are located on the line of identity (LI). Intervals, representing increasing in returns, above LI ($g(t) < g(t+1)$), whereas shortenings of two succeeding returns represent points below this line ($g(t) > g(t+1)$). By assessing the asymmetry of points in the diagram, further, we will present quantitative measures for varying degree of irreversibility in the DJIA.

2.1.1 Guzik's index

Guzik's index (GI) was defined as the distance of points above LI to LI divided by the distance of all points in Poincaré plot except those that are located on LI [10, 24]. Specifically,

$$GI = \frac{\sum_{i=1}^{a}\left(D_i^+\right)^2}{\sum_{i=1}^{m}(D_i)^2}, \tag{3}$$

where $a = C(P_i^+)$ means the number of points above LI; $m = C(P_i^+) + C(P_i^-)$ means the number of points in Poincaré plot except those which are not on LI; D_i^+ is the distance of points above the line to itself, and D_i is the distance of point $P_i\,(g(i), g(i+1))$ to LI which can be defined as

$$D_i = \frac{|g(i+1) - g(i)|}{\sqrt{2}}. \tag{4}$$

In fugure 2 is illustrated GI for two periods of the DJIA.

As we can see from illustration above, GI for crashes and critical events noticeably falling before deviant event and rising during emerging crises, which makes it as an excellent indicator-precursor of abnormal events.

2.1.2 Porta's index

Porta's index (PI) [13] was defined as the number of points below LI divided by the total number of points in Poincaré plot except those that are located on LI, specifically

$$PI = \frac{b}{m}, \tag{5}$$

where $b = C(P_i^-)$ is the number of points below LI, and $m = C(P_i^+) + C(P_i^-)$ is the total number of points below and above LI.

In figure 3 is illustrated PI for two periods of DJIA.

As we can see, according to Porta's index, irreversibility decreases during crash and critical events similarly to previous index which makes it appropriate indicator.

2.1.3 Costa's index

Costa's index represents a simplified version of [24] where number of increments ($x(i+1) - x(i) > 0$) and decrements ($x(i+1) - x(i) < 0$) are taken into account. They are presented to be symmetric if equal to each other. The procedure is implemented for coarse-grained time series. For scale τ, we consider the time series $G_\tau = \{g(i)\}, g(i) = x(i+\tau) - x(i), 1 \le i \le N - \tau$. The Costa's index [3], which displays the asymmetry of the probability distribution of positive and negative returns, is calculated by the formula:

$$CI_\tau = \frac{\sum_{i=1}^{N-\tau}\mathcal{H}[-g(i)] - \sum_{i=1}^{N-\tau}\mathcal{H}[g(i)]}{N-\tau}. \tag{6}$$

The generelized Costa's index according to can be defined as

$$CI = \frac{1}{L}\sum_{\tau=1}^{L}|CI_\tau|, \tag{7}$$

where L is the maximal scale.

In figure 4 CI presents the similar pehavior for the two periods of DJIA as in previous two measures.

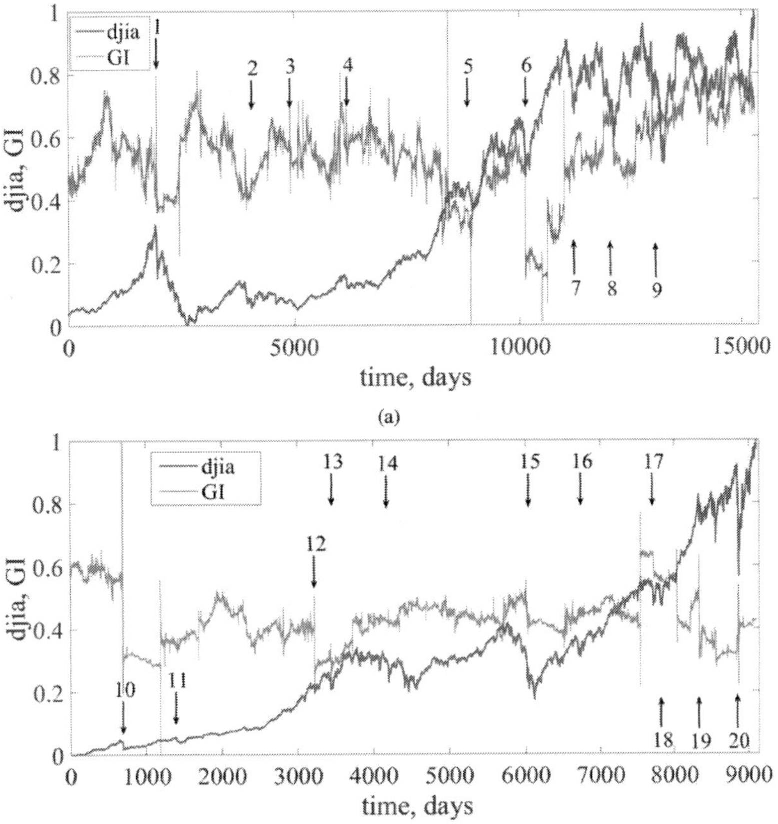

Figure 2: Guzik's index with corresponding first (a) and second (b) periods of the DJIA time series

2.2 Complex network methods

Visibility graphs (VGs) are based on a simple mapping from the time series to the network domain exploiting the local convexity of scalar-valued time series $\{x_i \mid i = 1, \ldots, N\}$ where each observation x_i is a vertex in a complex network. Two vertices i and j are linked by an edge (i, j) if for all vertices k with $t_i < t_k < t_j$ the following condition is applied [25]:

$$x_k < x_j + (x_i - x_j)\frac{t_j - t_k}{t_j - t_i}. \quad (8)$$

This is, the adjacency matrix (A_{ij}) of the following undirected and unweighted VG is presented as:

$$A_{ij}^{(VG)} = A_{ji}^{(VG)} = \prod_{k=i+1}^{j-1} \mathcal{H}\left(x_k < x_j + (x_i - x_j)\frac{t_j - t_k}{t_j - t_i}\right), \quad (9)$$

where $\mathcal{H}(\cdot)$ is the Heaviside function.

Horizontal visibility graphs (HVGs) provide a simplified version of this algorithm [26]. For a given time series, the vertex sets of VG and HVG are the same, whereas the edge set of the HVG maps the mutual horizontal visibility of two observations x_i and x_j, i.e., there is an edge (i, j) if $x_k < \min(x_i, x_j)$ for all k with $t_i < t_k < t_j$, so that

$$A_{ij}^{(VG)} = A_{ji}^{(VG)} = \prod_{k=i+1}^{j-1} \mathcal{H}(x_i - x_k)\mathcal{H}(x_j - x_k). \quad (10)$$

VG and HVG capture essentially the same properties of the system under study (e.g., regarding fractal properties of a time series), since the HVG is a subgraph of the VG with the same vertex set, but possessing only a subset of the VG's edges. Note that the VG is invariant under a superposition of linear trends, whereas the HVG is not.

Since the definition of VGs and HVGs takes the timing (or at least time-ordering) of observations explicitly into account, the direction of time is intrinsically interwoven with the resulting network structure. To account for this fact, we define a set of novel statistical network quantifiers based on two simple vertex characteristics:

(i) As the number of edges incident to a given vertex i can be defined as $k_i^r = \sum_j A_{ij}$, for a (H)VG, we rewrite this quantity for a vertex of time t_i, regarding

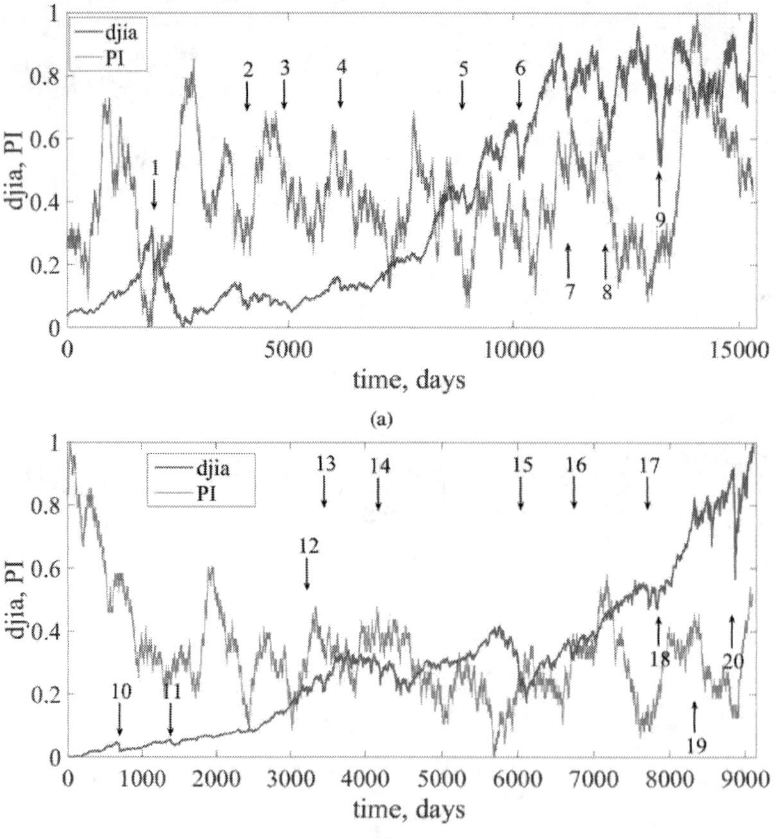

(a)

(b)

Figure 3: Dynamics of Porta's index for first (a) and second (b) periods of DJIA time series

its past and future vertices (prices):

$$k_i^r = \sum\nolimits_{j<i} A_{ij}, \tag{11}$$

$$k_i^a = \sum\nolimits_{j>i} A_{ij}, \tag{12}$$

where $k_i = k_i^r + k_i^a$, and k_i^r with k_i^a referred to as the *retarded* and *advanced* degrees. As it is defined in [12], following measures correspond to the in- and out-degrees of time-directed (H)VGs.

(ii) The local clustering coefficient $C_i = \binom{k_i}{2}^{-1} \sum_{j,k} A_{ij} A_{jk} A_{ki}$ is another vertex property of higher order characterizing the neighborhood structure of vertex i [27]. Similarly to (11) and (12), for studying the connectivity due to past and future prices, we rewrite the standard coefficient as the *retarded* and *advanced local clustering coefficients*

$$C_i^r = \binom{k_i^r}{2}^{-1} \sum\nolimits_{j<i,k<i} A_{ij} A_{jk} A_{ki}, \tag{13}$$

$$C_i^a = \binom{k_i^a}{2}^{-1} \sum\nolimits_{j>i,k>i} A_{ij} A_{jk} A_{ki}, \tag{14}$$

According to graph-based method, we will utilize the probability density functions (PDFs) of (11)-(14). If our system is presented to be time-reversible, we conjecture that probability distributions of forward and backward in time characteristics should be the same. For irreversible processes, we expect to find statistical non-equivalence. According to [12], this deviation will be defined through Kullback-Leibler divergence:

$$D_{KL}(p \parallel q) = \sum_{i=1}^{N} p(x_i) \cdot \log \frac{p(x_i)}{q(x_i)}, \tag{15}$$

where, in our case, p responds to a distribution of the *retarded* characteristics and q is of the *advanced*.

Figure 5 presents D_{KL} measure for the distribution of degrees and local clustering coefficients.

As it can be seen for figure 5 and b, both irreversibility measures for degrees and local clustering decrease during crashes and critical events which tells about luck of irreversibility during them. Also, it is shown in figure 4 that the first period of the DJIA is presented to be more reversible as the distance between distribution of degrees is close to zero for almost the entire period. Local clustering coefficient is seemed to be more robust and informative comparing to degree.

https://doi.org/10.1051/shsconf/202110703002

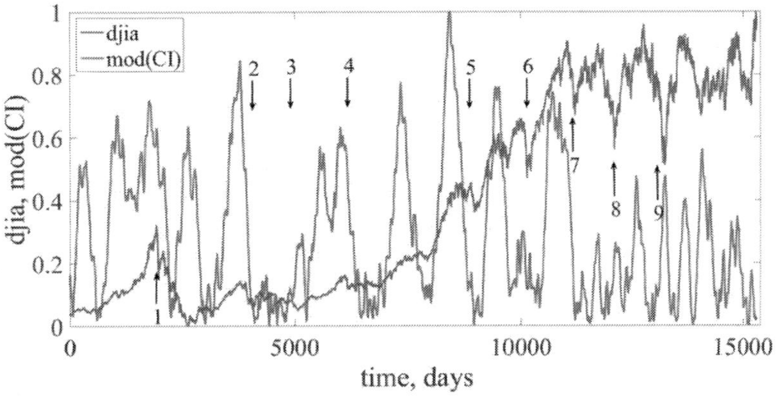

(a)

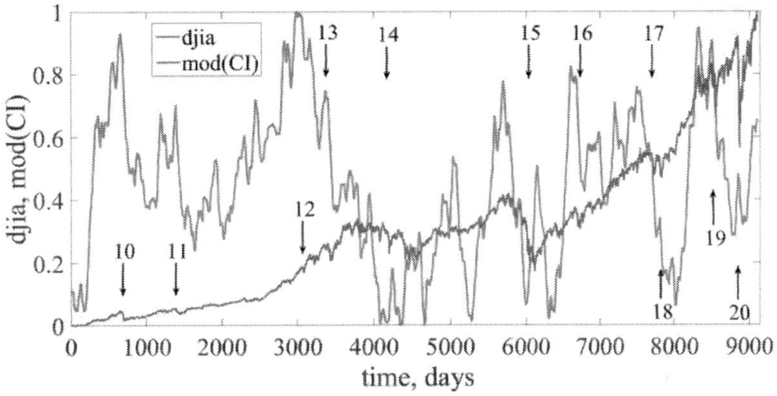

(b)

Figure 4: Dynamics of Costa's index for first (a) and second (b) periods of DJIA time series

2.3 Multiscale time irreversibility index

For the following procedure [24], first of all, we need to construct goarse-grained time series which can be defined as

$$y_\tau(j) = \frac{1}{\tau} \sum_{i=(j-1)\tau+1}^{j\tau} g(i), \quad \text{for } 1 \le j \le \frac{N}{\tau}. \tag{16}$$

Then, using a statistical physics approach, we make the simplifying assumptions that each transition (increase or decrease of $y_\tau(j)$) is independent and requires a specific amount of "energy" E. The probability density function of this class of system [28] can be assumed to follow $\rho \propto \exp(-\beta E - \gamma Q)$ where Q represents the non-equilibrium heat flux across the boundary of the system, and β and γ are the Lagrange multipliers derived from the constraints on the average value of the energy E per transition and the average contribution of each transition to the heat flux Q.

Since the time reversal operation on the original financial index time series inverts an increase to a decrease and vice versa, the difference between the average energy for the *activation* of information rate, i.e., $\langle \beta E + \gamma Q \rangle_{y_\tau > 0}$, and the *relaxation* of information rate, i.e., $\langle \beta E + \gamma Q \rangle_{y_\tau < 0}$, can be used as measurement of time reversal asymmetry.

Taking into consideration that the assumption of the distribution function ρ links the energy to the empirical distribution, we, following , define the next measure of temporal irreversibility:

$$a(\tau) = \frac{\int_0^\infty \left[\rho(y_\tau) \ln \rho(y_\tau) - \rho(-y_\tau) \ln \rho(-y_\tau)\right]^2 dy_\tau}{\int_{-\infty}^\infty \rho(y_\tau) \ln \rho(y_\tau) dy_\tau} \tag{17}$$

The time series is called reversivle if $a(\tau) = 0$.

Sometimes it is important for us to know not only the degree of irreversibility but also whether it reversed in time or not. For this purpose, we will replace equation (17) by the following one:

$$A(\tau) = \frac{\int_0^\infty \left[\rho(y_\tau) \ln \rho(y_\tau) - \rho(-y_\tau) \ln \rho(-y_\tau)\right] dy_\tau}{\int_{-\infty}^\infty \rho(y_\tau) \ln \rho(y_\tau) dy_\tau} \tag{18}$$

The time series is said to be irreversible for all scale τ if $A(\tau) > 0$. In case when $A(\tau) = 0$, the time series may be reversible or not for scale τ.

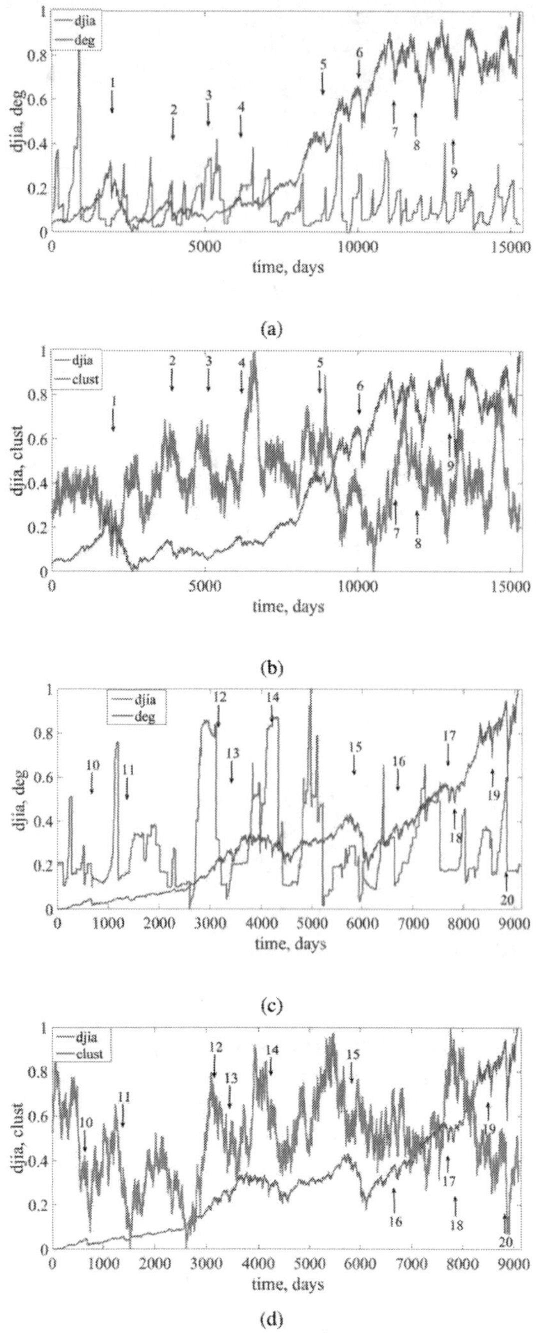

Figure 5: Dynamics of graph-based time irreversibility measures for the first (a, b) and second (c, d) periods of DJIA time series

For the analysis of discrete values, equation (18) can be presented as:

$$\hat{A}(\tau) = \frac{\sum_{y_\tau>0} \Pr(y_\tau) \ln [\Pr(y_\tau)]}{\sum_{y_\tau} \Pr(y_\tau) \ln [\Pr(y_\tau)]} - \frac{\sum_{y_\tau<0} \Pr(y_\tau) \ln [\Pr(y_\tau)]}{\sum_{y_\tau} \Pr(y_\tau) \ln [\Pr(y_\tau)]}. \quad (19)$$

The generalized multiscale asymmetry index (A_I) is defined as the summation of $\hat{A}(\tau)$ obtained for a predefined range of scales, i.e.,

$$A_I = \sum_{\tau=1}^{L} \hat{A}(\tau). \quad (20)$$

The figures illustrate that time series are significantly irreversible. For initial time series (for approximately 5-10 scales), the *transition* of prices is presented to be reversible (symmetric). After it, *transitions* presented to be asymmetric. Draws attention and noticeable unevenness introduced measures, which correlate with the fluctuations of the input time series. Identifying significant changes in the time series and comparing them with the corresponding changes of non-reversible measures of complexity, it is possible to construct the corresponding indicators.

2.4 Time series irreversibility measure based on permutation patterns

The idea of analyzing the permutation patterns (PP) was initially introduced by Bandt and Pompe [29] to provide researchers with a simple and efficient tool to characterize the complexity of the real systems dynamics. With respect to other approaches, as entropies, fractal dimensions, or Lyapunov exponents, it avoids amplitude threshold and instead dealing with casual values inhereted from time series dynamics, deals with ordinal permutation patterns [30]. Their frequencies allow us to distinguish deterministic processes from completely random.

The calculations of PP assume that the time series is partitioned with the *embedding dimension* d_E (number of elements to be compared) and the *embedding delay* τ (time separation between elements). In our opinion, $d_E \in \{3, 4\}$ and $\tau \in \{2, 3\}$ are the best parameters that encapsulate all the necessary quantitative information.

Further, all embedded patterns are assigned to their ordinal *rankings*. As an example, let us consider a fragment of the DJIA time series for period 18.08.2015-26.08.2015:

$$X = \{17511.34, 17348.73, 16990.69, 16459.75, 15871.35, 15666.44, 16285.51\}.$$

According to mentioned steps, we will construct embedded matrix of overlapping column vectors with $d_E = 3$ and $\tau = 2$. Our sampled data is partitioned as follows:

$$X_t^{d_E,\tau} = \begin{bmatrix} 17511.34 & 16990.69 & 15871.35 \\ 17348.73 & 16459.75 & 15666.44 \\ 16990.69 & 15871.35 & 16285.51 \end{bmatrix}. \quad (21)$$

After it, our time-delayed vectors are mapped to *permutations* or *ordinal patterns* of the same size. Our example consists $3! = 6$ different ordinal patterns in total:

$$\pi_1 = \{0, 1, 2\}$$
$$\pi_2 = \{0, 2, 1\}$$
$$\pi_3 = \{1, 0, 2\}$$
$$\pi_4 = \{1, 2, 0\}$$
$$\pi_5 = \{2, 0, 1\}$$
$$\pi_6 = \{2, 1, 0\}$$

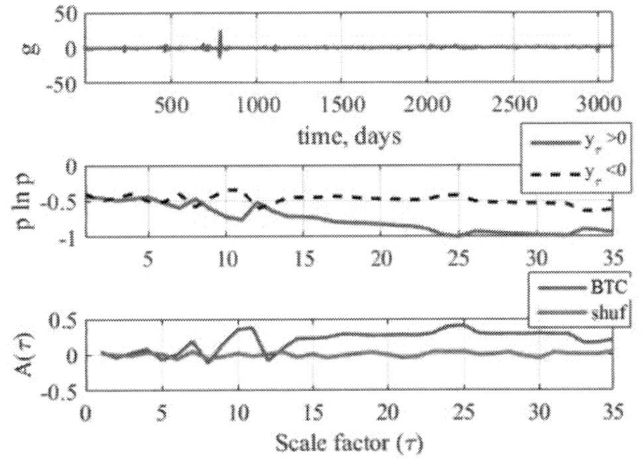

(a)

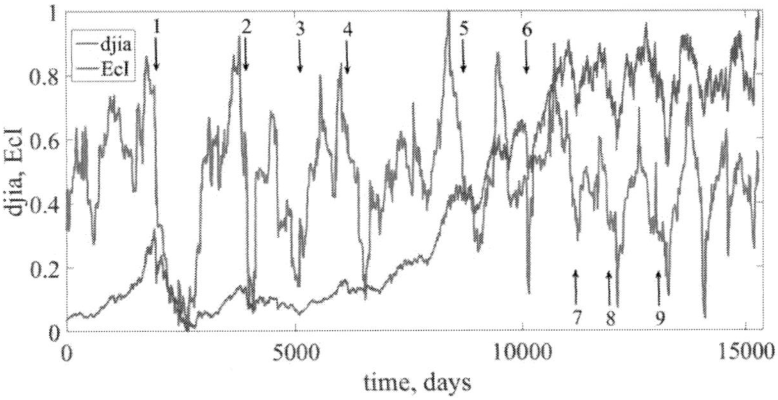

(b)

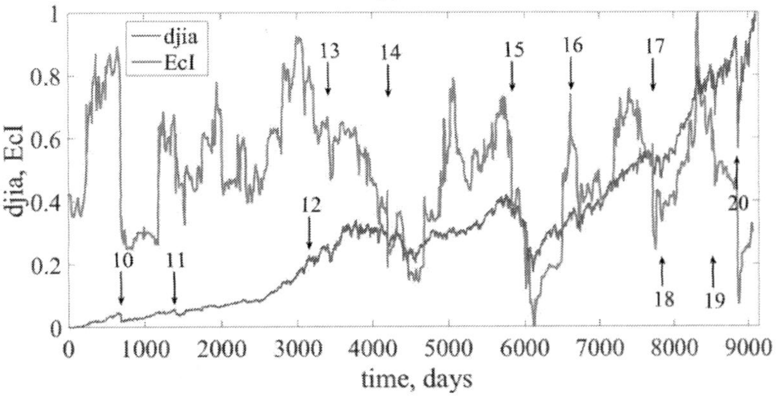

(c)

Figure 6: Dynamics of asymmetry index for first (a) and second (b) periods.

As an example, the corresponding permutation of the first column from (21) would be $\phi([17511.34, 17348.73, 16990.69]) = 210$ since $x(3) \leq x(2) \leq x(1)$. Therefore, after mapping from the time-series data into a series of permutations ($\phi : \mathbb{R}^{d_E} \rightarrow S_{d_E}$), we obtain the ordinal matrix:

$$\begin{bmatrix} 2 & 2 & 1 \\ 1 & 1 & 0 \\ 0 & 0 & 2 \end{bmatrix}. \tag{22}$$

https://doi.org/10.1051/shsconf/202110703002

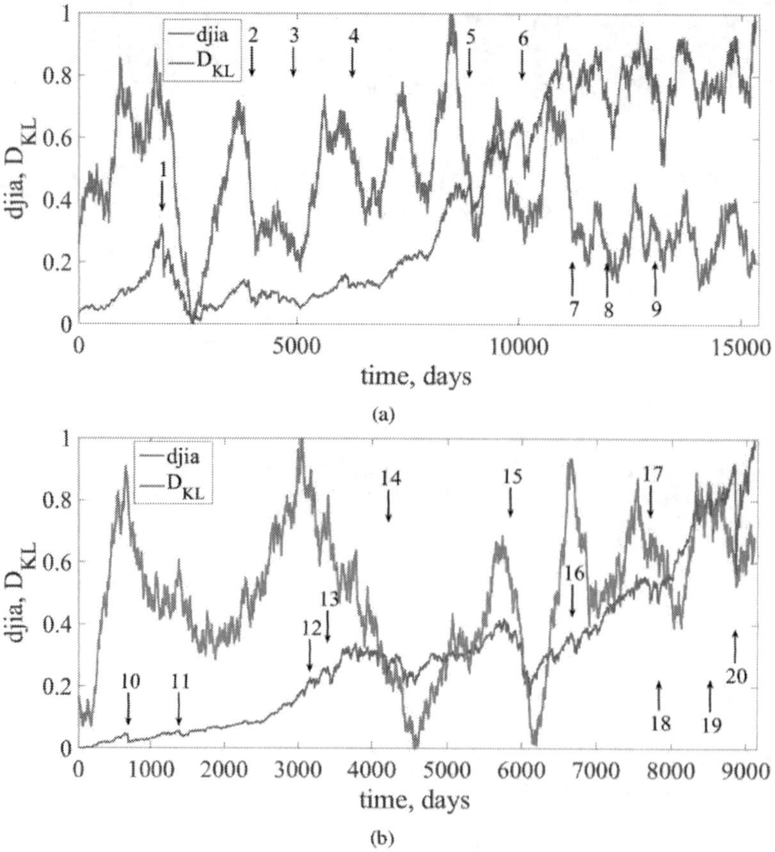

Figure 7: Dynamics of permutation-based time irreversibility measure for first (a) and second (b) periods.

Finally, the probability of each pattern is calculated as

$$p(\pi) = \frac{\#\{t \leq N - (d_E - 1)\tau, \phi(X_t^{d_E,\tau}) = \pi\}}{N - (d_E - 1)\tau}, \quad (23)$$

where $\#\{\cdot\}$ denotes the cardinality of a set, and *permutation entropy* is calculated regarding a probability distribution P, whose elements $p_i \equiv p(\pi_i)$ are the probabilities associated with the ith permutation pattern, $i = 1, \ldots, d_E!$:

$$S[P] = -\sum_{i=1}^{d_E!} p_i \log_2 p_i. \quad (24)$$

Interesting for us time irreversibility of permutation patterns is not related on (24), but on the probability distribution of ordinal patterns. That is, we find probabilities of finding corresponding ordinal patterns for both initial and reversed times series. Correspondingly, if both types have approximately the same probability distributions of their patterns, time series is presented to be reversible and the opposite conclusion for the other case.

The difference between distributions of direct time series (P^d) and reversed (P^r) can be estimated with equation (15).

From the presented figures it can be seen that as financial crisis comes, the distance between two distributions becomes more close to zero, denoting that those period is less irreversible and efficient. Moreover, in this case we see that D_{KL} for permutaiton patterns acts as a measure of complexity. The dynamics before crisis events starts do decrease, presenting trend to be more predictable, and after them it increases, demonstrating the increasing complexity.

3 Conclusions

Financial systems does not always evolve with precisely the same values. Instead, their prices increase or decrease over time due to different market conditions, political, and economical situations in concrete countries or in the word.

In this work we have presented how to deal with (statistical) time irreversibility, varying over time. Using the time series of Dow Jones Industrial Average index and the sliding window procedure, first of all, we have presented our classification of crisis events in DJIA index, and we have constructed econophysical and econometrical indicators of financial crashes and critical events. Our study affirms ranging degrees of irreversibility in DJIA stock index. Some of its periods of existence are presented to be more irreversible comparing to others. Namely, periods of financial stress are characterized by higher irreversibility and, thus, by increasing predictability and less efficiency.

References

[1] I. Prigogine, *From Being to Becoming: Time and Complexity in the Physical Sciences* (W. H. Freeman, 1980)
[2] M. Costa, A.L. Goldberger, C.K. Peng, Phys. Rev. E **71**, 021906 (2005)
[3] M. Costa, C.K. Peng, A. Goldberger, Cardiovascular Engineering **8**, 88 (2008)
[4] J. Donges, R. Donner, J. Kurths, EPL **102**, 10004 (2013)
[5] M. Zanin, A. Rodríguez-González, E. Menasalvas Ruiz, D. Papo, Entropy **20** (2018)
[6] R. Flanagan, L. Lacasa, Phys. Lett. A **380**, 1689 (2016)
[7] A. Puglisi, D. Villamaina, EPL **88**, 30004 (2009)
[8] C.S. Daw, C.E.A. Finney, M.B. Kennel, Phys. Rev. E **62**, 1912 (2000)
[9] C. Diks, J. van Houwelingen, F. Takens, J. DeGoede, Phys. Lett. A **201**, 221 (1995)
[10] P. Guzik, J. Piskorski, T. Krauze, A. Wykretowicz, H. Wysocki, Biomedizinische Technik. Biomedical engineering **51**, 272 (2006)
[11] M.B. Kennel, Phys. Rev. E **69**, 056208 (2004)
[12] L. Lacasa, A. Nuñez, E. Roldán, J. Parrondo, B. Luque, Eur. Phys. J. B **85**, 217 (2012)
[13] A. Porta, S. Guzzetti, N. Montano, T. Gnecchi-Ruscone, R. Furlan, A. Malliani, *Time reversibility in short-term heart period variability*, in *2006 Computers in Cardiology* (IEEE, 2006), Vol. 2006, pp. 77–80, ISBN 978-1-4244-2532-7
[14] V. Soloviev, V. Solovieva, A. Tuliakova, Neuro-Fuzzy Modeling Techniques in Economics **8**, 3 (2019)
[15] V. Soloviev, V. Solovieva, A. Tuliakova, A. Hostryk, L. Pichl, CEUR Workshop Proceedings **2713**, 53 (2020)
[16] V. Soloviev, A. Bielinskyi, O. Serdyuk, V. Solovieva, S. Semerikov, CEUR Workshop Proceedings **2732**, 455 (2020)
[17] A. Bielinskyi, S. Semerikov, V. Solovieva, V. Soloviev, SHS Web of Conferences **65**, 06006 (2019)
[18] V. Soloviev, A. Bielinskyi, V. Solovieva, CEUR Workshop Proceedings **2393**, 434 (2019)
[19] V. Soloviev, A. Belinskiy, Communications in Computer and Information Science **1007**, 276 (2019)
[20] V. Soloviev, S. Yevtushenko, V. Batareyev, CEUR Workshop Proceedings **2546**, 87 (2019)
[21] V. Soloviev, A. Belinskij, CEUR Workshop Proceedings **2104**, 116 (2018)
[22] A.O. Bielinskyi, I. Khvostina, A. Mamanazarov, A. Matviychuk, S. Semerikov, O. Serdyuk, V. Solovieva, V.N. Soloviev, IOP Conference Series: Earth and Environmental Science **628**, 012019 (2021)
[23] V.N. Soloviev, A.O. Bielinskyi, N.A. Kharadzjan, CEUR Workshop Proceedings **2832**, 24 (2020)
[24] M. Costa, A.L. Goldberger, C.K. Peng, Phys. Rev. Lett. **95**, 198102 (2005)
[25] L. Lacasa, B. Luque, F. Ballesteros, J. Luque, J.C. Nuño, Proceedings of the National Academy of Sciences **105**, 4972 (2008)
[26] B. Luque, L. Lacasa, F. Ballesteros, J. Luque, Physical Rev. E **80**, 046103 (2009)
[27] M.E.J. Newman, SIAM Rev. **45**, 167–256 (2003)
[28] D. Jou, J. Casas-Vázquez, G. Lebon, Reports on Progress in Physics **51**, 1105 (1999)
[29] C. Bandt, B. Pompe, Phys. Rev. Lett. **88**, 174102 (2002)
[30] M. Zanin, L. Zunino, O.A. Rosso, D. Papo, Entropy **14**, 1553 (2012)

Ukraine's foreign trade: responses to global challenges

Svitlana Radzivivska[1,*] and *Ivan* Us[2,**]

[1]National Academy of Management, 15 Ushynskoho Str., Kyiv, 03151, Ukraine
[2]The National Institute for Strategic Studies, 7-a Pyrogova Str., Kyiv, 01030, Ukraine

Abstract. The overview of the country's trade with all the continents during the period of 2000-2019 is given. The description of the commodity structure of exports/imports of goods, with particular attention to the industrial products, is followed by the detailed analysis of Ukraine's foreign trade in 2020. The decrease in Ukraine's foreign trade in 2020 by 6.4% is fully consistent with the projected WTO reduction of world trade in 2020 by 9.2%. Although COVID-19 had negative impact on Ukraine's trade with the EU and the EAEU, it contributed to closer trade ties with Asia, improving Ukraine's trade balance. The government and the national business elite should aim at solving the problems of increasing the volume and improving the commodity structure of Ukraine's foreign trade with emphasis on the development of transport system for exporting agricultural and food products to the developed countries of the West and to the prospective economies of the East and the South. It is essential, on the one hand, to focus on the inflows of FDI and their appropriate use, and, on the other hand, on Ukraine's participation in the formation of GVCs, global production networks. In the conditions of the Fourth industrial revolution, the economy finds itself transformed due to the fundamental changes. The optimization of foreign trade relations of Ukraine will not only improve the economy, but also enable the country to become a better functioning element of the global economic system.

1 Introduction

The globalization of the world economy intensifies the processes of international division of labor, causing transformations in the directions and intensity of trade flows between countries. The severe competition for resources, the need to preserve the environment, and the Fourth Industrial Revolution, as well as the global challenges, e.g. the Crisis of 2008-2009, sanctions policy, COVID-19 [1, 2], require the states to adapt their production and, consequently, foreign economic activities to meet the new requirements, created by the specific development of the global processes [3]. Timely understanding of the new requirements and making appropriate decisions are the imperatives for any country to survive and function in the global economic system.

The conclusions of the classical, neo-classical, as well as modern international trade theories are gaining new insights nowadays since foreign trade is the key factor influencing economic growth not only of separate states, but regional blocks (e.g., USMCA, EU, ASEAN, EAEU, etc.), transcontinental coalitions (e.g., Transatlantic Trade and Investment Partnership, Regional Comprehensive Economic Partnership, etc.), as well as the global economy as a whole [4–7].

That is why, in our opinion, the optimization of foreign trade relations of Ukraine is of vital importance since it will not only improve the country's economy, but also enable Ukraine to become a better functioning element of the global economic system [8]. Our hope is that Ukraine will be able to adapt to the new realities and become a more confident and sufficient player of the modern world economy – an open complex stochastic nonlinear system, the elements of which are highly interconnected and interdependent [9–20].

The methodological basis of the research is the systemic approach combined with the historical and logical methods, as well as the method of comparative statistics analysis, based on the data provided by the State Statistics Service of Ukraine and the State Customs Service of Ukraine.

The paper comprehensively examines three interrelated aspects of Ukraine's foreign trade: 1) the geographical expansion, i.e. its globalization; 2) the development of its commodity structure, i.e. its specialization; 3) the impact of COVID-19 on it.

Ukraine's economy is shrinking, but more developed countries have it worse. The country's biggest stroke of luck is that industries that suffered most from COVID-19 – tourism, petroleum production, services, and manufacturing – don't dominate the economy [21]. Ukraine's main sectors like agriculture (40% of export) and metallurgy either escaped unscathed or even grew [22]. COVID-19 contributed to solving the topical problem of the large trade deficit of Ukraine [23] since in 2020 the negative balance of trade is more than twice less than in 2019. At the same time COVID-19 has had negative impact on Ukraine's eco-

*e-mail: svitrad98@ukr.net
**e-mail: ivanusus@gmail.com

nomic relations with both – the EU and the EAEU – and positive impact on those with the Asian economies.

The objective of the paper is to demonstrate the peculiarities of Ukraine's adaptation to global challenges by providing the detailed analysis of Ukraine's merchandise trade with the purpose of improving the country's foreign economic activity.

2 Overview of Ukraine's intensity of trade flows of goods with continents in the 21st century

Over the past 20 years Ukraine has made some progress in terms of expanding the geography of its trade. Since one of the key peculiarities of the economy is that the share of trade in goods in all trade in goods and services makes up for more than 80% [23], we will focus on the analysis of exports and imports of goods.

The official site of the State Statistics Service provides data for the analysis of the dynamics of geographical structure of Ukraine's exports of goods (table 1) and imports of goods (table 2).The dynamics of Ukraine's exports/imports of goods is calculated as the change in the ratio of volume of goods taken at the particular year – 2005, 2010, 2015, 2019 to 2000 since the benchmark year adopted for our analysis is 2000. For example, table 1 indicates that Ukraine's exports of goods to Europe in 2005 increased 2.32 times compared to 2000, in 2010 – 2.95, in 2015 – 2.83, and in 2019 – 5.71 times. In 2019 the share of exports of the Ukrainian goods to Europe was 53.6% of total merchandise exports of Ukraine. Thus, *the European countries are the main consumers of the Ukrainian products*.

The exports of Ukraine's goods to Asia are characterized by the following dynamics: in 2005 they grew 2.47 times compared to 2000, in 2010 – 3.95, in 2015 – 3.56, in 2019 – 4.79 times (table 1). In 2019 the share of Ukraine's exports of goods to the Asian countries reached 33.37% of all the Ukrainian merchandise exports. Moreover, *in 2010 the share of Ukraine's exports of products to Europe and Asia was nearly the same – 42%* of the Ukrainian goods to each continent, but in 2019 Europe took the leading role with nearly 56% while Asia received 33% of Ukraine's commodities. Still, the analysis of table 1 allows to conclude that *the Asian countries are important partners for Ukraine*.

It is interesting to note that Ukraine's exports of products to the African countries in 2005 increased 3.27 times compared to 2000, in 2010 – 4.13, in 2015 – 5.2 times while the country's exports of goods to the American countries in 2005 rose 1.5 times compared to 2000, in 2010 – 1.64, 2015 – 0.65 times. In 2005 Ukraine's merchandise exports to Australia and Oceania grew 1.96 times compared to 2000, in 2010 – 4.06 times. Considering Ukraine's trade with Africa, it is obvious that Ukraine's potential is to be fulfilled yet. At the same time trade relations of Ukraine with the American countries demonstrate that their role as consumers of the Ukrainian goods is declining. To sum up, Europe and Asia consumed 86.97% of Ukrainian goods in 2019, Africa – 9.97%.

Table 1. Dynamics of geographical structure of Ukraine's exports of goods, 2000–2019, based on [24]

	Total, including the undistributed goods	**CIS**	**Other countries of the world**	**EU**
2000	14572.5	4459.7	10112.8	4576.3
2005	34228.4	10531.1	23697.3	10293.3
2010	51405.2	18740.6	32664.6	13085.3
2015	38127.1	7806.1	30321.0	13015.2
2019	50054.6	—	—	20750.7
2015 /2000, %	261.6	175.0	299.8	284.4

	TOTAL		**EUROPE**		**ASIA**	
	mln USD	%	mln USD	%	mln USD	%
2000	10111.9	100	4680.2	**46.28**	3475.7	**34.37**
2005	23696.2	100	10881.4	**45.92**	8576.0	**36.19**
2010	32592.1	100	13829.6	**42.43**	13715.4	**42.08**
2015	30229.7	100	13248.3	**43.83**	12378.9	**40.95**
2019	49884.0	100	26739.5	**53.60**	16646.1	**33.37**
2015/ 2000, %	299.0		283.1		356.2	
2019/ 2000, %	493.3		571.3		478,9	

	AFRICA		**AMERICA**		**AUSTRALIA and OCEANIA**	
	mln USD	%	mln USD	%	mln USD	%
2000	731.5	**7.23**	1217.5	12.04	7.0	0.08
2005	2393.9	**10.10**	1831.2	7.73	13.7	0.06
2010	3018.7	**9.26**	2000.0	6.14	28.4	0.09
2015	3803.3	**12.58**	785.6	2.60	13.6	0.04
2019	4971.5	**9.97**	1467.9	2.94	59.0	0.12
2015/ 2000, %	519.9		64.5		194.3	
2019/ 2000, %	679.6		120.6		842.9	

Let's have a closer look at the dynamics of geographical structure of Ukraine's imports of goods, presented in table 2. The analysis of Ukraine's imports of products from Europe indicates that the country's imports from the European states in 2005 increased 2.94 times compared to 2000, in 2010 – 4.64 times, in 2015 – 3.87 times, in 2019 – 8.81 times. Merchandise imports from Asia to Ukraine in 2005 grew 5.56 times compared to 2000, in 2010 – 11.9, in 2015 – 8.59, in 2019 – 20.65 times.

As table 2 shows, *the share of imports commodities from Europe to Ukraine in 2000 was 72.75% while in 2019 it was 62.58% of total merchandise imports of Ukraine.*

Table 2. Dynamics of geographical structure of Ukraine's imports of goods, 2000–2019, based on [24]

	Total, including the undistributed goods	**CIS**	**Other countries of the world**	**EU**
2000	13956.0	8029.6	5926.4	4047.7
2005	36136.3	16988.3	19148.0	12211.3
2010	60742.2	26697.4	34044.8	19151.4
2015	37516.4	10485.5	27030.9	15330.2
2019	60800.2	—	—	25012.2
2015 /2000, %	268.8	130.6	456.1	378.7

	TOTAL		**EUROPE**		**ASIA**	
	mln USD	%	mln USD	%	mln USD	%
2000	5926.3	100	4311.5	**72.75**	842.3	**14.21**
2005	19147.5	100	12666.4	**66.15**	4685.5	**24.47**
2010	34043.0	100	20004.5	**58.76**	10023.3	**29.44**
2015	27008.9	100	16665.3	**61.70**	7235.7	**26.79**
2019	60730.1	100	38001.9	**62.58**	17394.5	**28.64**
2015/ 2000, %	455.7		386.5		859.0	
2019/ 2000, %	1024.8		881.4		2065.1	

	AFRICA		**AMERICA**		**AUSTRALIA and OCEANIA**	
	mln USD	%	mln USD	%	mln USD	%
2000	136.4	2.30	581.4	9.81	54.7	0.93
2005	426.2	2.23	1265.5	6.61	103.9	0.54
2010	874.4	5.57	2879.4	8.46	261.4	0.77
2015	601.7	2.23	2336.6	8.65	169.6	0.63
e **2019**	819.2	1.35	4397.7	7.24	116.8	0.19
2015/ 2000, %	441.1		401.9		310.1	
2019/ 2000, %	600.6		756.4		213.5	

At the same time, the share of imports of goods from the Asian countries to Ukraine in 2000 was 14.21%, and in 2019 reached 28.64% of total imports of products. Thus, the analysis of trade relations of Ukraine with the Asian countries suggests that their role as suppliers of goods for Ukraine is becoming more important. However, *it should be noted that Ukraine has been having negative balance of trade in goods with Europe since 2005 (–1785 mln USD), which has grown to – 11262.4 mln USD in 2019, while the trade balance of Ukraine with Asia had been positive throughout the analyzed period up until 2019 when it became negative(–748.4 mln USD).* So, up until 2019 Ukraine's trade in goods with both – Europe and Asia didn't contribute to solving the problem of decreasing the large trade deficit of the country [23].

Imports of products from America to Ukraine in 2005 increased 2.18 times, in 2010 – 5 times, in 2015 – 4 times, in 2019 – 7.56 times. The share of imports from the American countries in 2019 was 7.24% of total imports of goods of Ukraine. Nevertheless, goods imports from Africa to Ukraine in 2005 rose 3.12 times, in 2010 – 6.41 times, in 2015 – 4.41, in 2019 – 6 times.

It should be underlined that the trade balance of goods of Ukraine with the African countries is positive, while with those of America – negative. Merchandise imports from Australia and Oceania to Ukraine in 2005 grew 1.9 times compared to 2000, in 2010 – 4.8, in 2015 – 3.1, in 2019 – 2.1 times. Ukraine's trade with Africa and especially Australia and Oceania has not been developed – the turnover of goods between Ukraine and those continents is not significant. To sum up, the European countries remain the main suppliers of goods for Ukraine; however, the Asian ones steadily improve their positions.

3 Ukraine's trade specialization and its deindustrialization

Let's have a closer look at the commodity structure of exports during the period of 2016-2019 (table 3).

As table 3 shows, the exports of Ukraine are formed by the following main product groups: I. Live animals and livestock products; II. Plant products; III. Animal or plant fats and oils; IV. Finished food industry products; V. Mineral products; VI. Products of chemical and allied industries; XV. Base metals and preparations thereof; XVI. Machines, equipment and mechanisms, electric and technical equipment; XVII. Ground, air and water transport facilities. The linear regression trend model of the relationship between the two variables – the time and the size of goods is calculated by the method of least squares. The results presented in the tables 3 and 4 are based on the Microsoft Excel program. The pair linear regression is applied in the research. The models enable the economic interpretation and the analysis of the current economic situation, as well as the suggestions on the ways for its improvement.

During the period under review, the share of "I. Live animals and livestock products" grew from 2.13% to 2.55% of Ukraine's merchandise exports (table 3). There was also an increase in the share of "II. Plant products" from 22.26% to 25.8%. However, the share of merchandise exports of "III. Animal or plant fats and oils" decreased from 10.9% to 9.45% and "IV. Finished food industry products" – from 6.74% to 6.43%.

Thus, in 2019 the share of food and agricultural products was 44.23% of all exports of goods of Ukraine. Moreover, as table 3 indicates "V. Mineral products", "VI. Products of chemical and allied industries"; "XV. Base metals and preparations thereof" accounted for 34.07% of all merchandise exports in 2019.

It should be emphasized that all the above described product groups (I, II, III, IV, as well as V, VI, XV) are

Table 3. Dynamics of commodity structure of volume (mln USD) and share (%) of Ukraine's exports of goods, 2016–2019, based on [24]

MAIN COMMODITY GROUPS	2016	2017	2018	2019
I. Live animals and livestock products	775.0 2.13	1108.8 2.56	1210.6 2.55	1277.0 2.55
Volume trend	$y = 160.78x + 690.9$			
II. Plant products	8093.7 22.26	9215.7 21.3	9886.1 20.89	12914.5 25.8
Volume trend	$y = 1513.3x + 6244.3$			
III. Animal or plant fats and oils	3963.0 10.9	4605.7 10.65	4496.5 9.5	4732.2 9.45
Volume trend	$y = 219.84x + 3899.8$			
IV. Finished food industry products	2450.1 6.74	2826.7 6.53	3018.6 6.38	3220.4 6.43
Volume trend	$y = 250.28x + 2253.3$			
V. Mineral products	2728.8 7.5	3947.7 9.12	4339.6 9.16	4866.5 9.72
Volume trend	$y = 680.5x + 2269.4$			
VI. Products of chemical and allied industries	1558.2 4.29	1660.6 3.84	1871.3 3.95	1930.8 3.86
Volume trend	$y = 132.85x + 1423.1$			
XV. Base metals and preparations thereof	8338.9 22.93	10124.6 23.4	11632.7 24.58	10255.7 20.49
Volume trend	$y = 725.85x + 8273.4$			
XVI. Machines, equipment and mechanisms, electric and technical equipment	3637.9 10.0	4276.8 9.89	4654.7 9.83	4464.4 8.92
Volume trend	$y = 285.74x + 3544.1$			
XVII. Ground, air and water transport facilities	555.7 1.53	625.9 1.45	669.3 1.41	882.3 1.76
Volume trend	$y = 102.32x + 427.5$			
TOTAL EXPORTS	36361.7 100	43264.7 100	47335.0 100	50054.6 100

considered to be low value-added and their total share in 2019 was 78.3% of all Ukraine's goods for exports.

The analysis of the Ukrainian exports and imports of the selected industrial goods – "XVI. Machines, equipment and mechanisms, electric and technical equipment" and "XVII. Ground, air and water transport facilities" – demonstrates that in exports of Ukraine the share of those groups of goods in all the exports of goods is decreasing (in 2016 – 11.53%, in 2019 – 10.68%) while the share of those groups of goods in all the imports of goods is increasing (in 2016 – 27.64%, in 2019 – 32.04%).

As the analysis of tables 3 and 4 suggests, Ukraine is in a state of economic decline and deindustrialization. According to Viktor Halasiuk, the country has to use the available resources to become a world leader in certain niches. So, it is necessary to develop the sectors Ukraine can quickly scale by several times [25]. In this regard, we believe that the first and foremost factor of catching-up development is the state itself interested in improving competitiveness of its economy in the global market.

Thus, the reindustrialization should be reviewed with the objective of identifying the nature of commodity imports of Ukraine and manifest itself in the increase in its commodity exports of industrial goods. Finally, Ukraine's quick and careless participation in globalization, organized according to the recommendations of the Washington Consensus, has brought negative results, i.e. the large trade deficit, the drastic decline of its high tech production, orientation of the country's economy development primarily on agricultural and food production, on the extraction of raw materials, their primary processing. Consequently, the great number of highly skilled professionals could not find job corresponding to their qualifications, and the majority of them was forced either to work in trade or to emigrate abroad. Overcoming negative tendencies in the economy of Ukraine requires the refusal of the recommendations of the Washington Consensus, above all of the neoliberalism ideology, forming their basis and contradicting the main postulate of adhering to justice in the world.

G. Gereffi noted that the Washington Consensus model of development, which held away from the mid-1980s through the mid-2000s, is a nation-state-centered view of the global economy, in which countries are the primary units of analysis in international production and trade. Then, the scholar emphasized that "the GVC framework fundamentally challenges this view of the global economy and it provides a different interpretation of the key drivers of change over the past four decades [26]. GVCs account for around half of world trade today. On the one hand, GVC participation is determined by factor endowments,

 https://doi.org/10.1051/shsconf/202110704001

Table 4. Dynamics of commodity structure of volume (mln USD) and share (%) of Ukraine's imports of goods, 2016–2019, based on [24]

MAIN COMMODITY GROUPS	2016	2017	2018	2019
I. Live animals and livestock products	626.3 1.6	731.5 1.47	918.0 1.61	1071.5 1.76
Volume trend	$y = 152.21x + 456.3$			
II. Plant products	1284.8 3.27	1368.0 2.76	1529.2 2.67	1794.6 2.95
Volume trend	$y = 169.06x + 1071.5$			
III. Animal or plant fats and oils	246.0 0.63	266.6 0.54	267.4 0.47	253.3 0.42
Volume trend	$y = 2.27x + 252.65$			
IV. Finished food industry products	1734.0 4.42	1935.0 3.9	2340.9 4.09	2616.6 4.3
Volume trend	$y = 305.37x + 1393.2$			
V. Mineral products	8495.0 21.64	12504.7 25.21	14191.9 24.82	12984.6 21.36
Volume trend	$y = 1515.6x + 8255.1$			
VI. Products of chemical and allied industries	5619.5 14.32	6545.9 13.2	7058.3 12.34	7483.4 12.3
Volume trend	$y = 610.41x + 5150.8$			
XV. Base metals and preparations thereof	2306.5 5.88	3012.8 6.07	3575.2 6.25	3650.7 6.0
Volume trend	$y = 459.5x + 1987.6$			
XVI. Machines, equipment and mechanisms, electric and technical equipment	7889.4 20.1	9902.6 19.96	11955.2 20.91	13312.8 21.9
Volume trend	$y = 1832.3x + 6184.3$			
XVII. Ground, air and water transport facilities	2959.5 7.54	4182.1 8.43	4554.0 7.96	6162.8 10.14
Volume trend	$y = 998.18x + 1969.2$			
TOTAL EXPORTS	39249.8 100	49607.2 100	57187.6 100	60800.2 100

geography, market size, and institution; on the other hand, these fundamentals alone need not dictate destiny since policies to attract FDI can remedy the scarcity of capital, technology, and management skills [27]. In our opinion, the development of industry requires the expansion of demand for products, the improvement of the rules to access the markets, the diversification of the geographical directions and commodity structure of the industrial exports.

In order to solve the problems of increasing the volume and improving the commodity structure of Ukraine's foreign trade, firstly, it is essential to focus on the inflows of FDI and their appropriate use, and, secondly, on the active involvement in global value chains and production networks on regional and global levels. These actions will promote the formation of the socio-market economy in Ukraine, the knowledge economy, approval of justice and high technology, the economy which enables to keep the environment in the parameters required for the functioning of the ecosystem and for preventing climatic disasters on regional and global levels.

4 World trade in 2020

World trade in 2020 marked a significant decline as the result of countries' measures to combat the spread of COVID-19. The most significant reduction occurred in the second quarter of 2020, when compared to the second quarter of 2019, the value of world trade decreased by 21% [28]. For comparison, the decline in the value of trade in goods during the financial crisis was deeper (in the second quarter of 2009 there was a decline of 33%).

In terms of monthly indicators, the value of world trade in goods decreased by 23% in April 2020 compared to April 2019 and by 26% in May (monthly estimates are based on data from 72 countries, which accounted for 92% of world trade in goods). In the third quarter of 2020, the volume of trade in goods increased by 11.6% compared to the previous quarter. Despite this growth, the value of trade in July-September was 5.6% lower than in the same period of 2019. This improvement has taken place largely as the European countries began to ease restrictive measures. At the same time some of the Asian states saw a slight increase, reflecting different pandemic trajectories in different regions.

The actual dynamics has led the World Trade Organization to revise its own forecast of the world trade in 2020 for the better. If in April 2020, according to the optimistic scenario, a reduction of 12.9% was forecast, and according to the pessimistic scenario by 31.9%, in the October forecast the WTO estimates a decrease in world trade in goods at 9.2% [29]. Thus, the updated forecast is better than the optimistic scenario, which is evidence for better adaptation of the world trade processes in the conditions of pandemic caused by the COVID-19 than it was expected in April.

According to WTO forecasts, Asian countries will come out of the crisis faster than countries in the other regions. At the end of 2020, their exports are projected to decline by 4.5% and imports by 4.4%, while European exports will decline by 11.7% and North American countries by 14.7%. Imports from European countries will decrease by 10.3% and from North America by 8.7%. It should be noted that the WTO made its forecast before the data for the three quarters of 2020 were obtained, according to which the volume of world trade in goods was lower by 8.2% comparing with the same period of 2019. The sooner the effects of the COVID-19 can be overcome, the greater will be the resumption of world trade and the return of the world to the track of globalization, which in the context of shifting the emphasis from globalization to regionalization is important [30].

5 Foreign trade in goods of Ukraine in 2020

Following the results of 2020, the value of foreign trade in goods of Ukraine decreased by 6.4% compared to the same period of 2019 [31]. The decline was mainly due to reduction in imports of goods by 10.3%, while their exports decreased by 1.7%. Thus, shrinking Ukraine's foreign trade is in a better state in comparison with the WTO's projected reduction for the world trade in 2020.

As in the rest of the world, the main reduction during 2020 occurred in April and May, when the turnover decreased compared to the corresponding months of 2019 by 17.4% and 27.3%, while for the period from January to March the reduction was 1%, and for the period from June to December, turnover dropped by 4% (figure 1). Significantly, the reduction in the value of trade was due to lower prices, as the physical volume of foreign trade in 2020 decreased by 4.7%.

Turnover in April 2020 was 1.6 bln USD less than in March and amounted to 7.3 bln USD. It was even lower in May – 6.9 bln USD. In the second half of 2020 it began to show steady growth. In June it rose to 7.4 bln USD, in July – to 8.3 bln USD, in August – to 8.7 bln USD, in September – to 8.9 bln USD, in October – to 9.6 bln USD, in November – to 10.1 bln USD and in December – to 10.7 bln USD (figure 2). Higher rates of decline in imports of goods in comparison with exports of goods resulted in more than two times reduction of the negative balance in foreign trade. If the negative balance for 2019 was 10.4 bln USD, then in 2020 it decreased to 5 bln USD. Exports of goods for 2020 compared to 2019 dropped by 0.8 bln USD and amounted to 49.2 bln USD.

Moreover, during the fall of 2020, we observed the tendency of exports growth compared to the fall of 2019. In September the exports increased by 2.8%, in October – by 1.4% and in November – by 9.1%. In December, export growth was more significant and amounted to 18.5% (figure 1) – the fact demonstrates the gradual resumption of economic activity in the main markets for domestic products.

It should be noted that the reduction in 2020 was formed by lower revenues from exports of metals and metal products; goods of this group decreased by 1.2 bln USD, which was 11.7% lower than at the same period last year. At the same time exports of fuel and energy products dropped (by 307 mln USD or 35.6%), as well as machinery, equipment and transport (by 98 mln USD or 1.8%). However, exports of mineral products increased by 772 mln USD, or 19.3%. Exports of food products and agricultural products also increased by 49 mln USD or 0.2%. Despite a slight increase, the share of this sector in the commodity structure of exports increased more significantly. If in 2019 the share of food products and agricultural products in the commodity structure was 44.3%, in 2020 it increased to 45.1%. At the same time during this period the share of metals and metal products decreased from 20.5% to 18.4%. These two product groups account for almost two thirds of total exports of goods from Ukraine (table 5).

Table 5. Commodity structure of Ukraine's foreign trade in 2020, bln USD

Industry	Export	Import
Food and agricultural products	22,2	6,5
Mineral products	4,8	0,6
Fuel and energy products	0,6	8,0
Chemical industry products, rubber	2,7	10,8
Textiles, footwear	0,9	2,7
Metals and articles thereof	9,1	3,2
Machinery, equipment and transport	5,4	18,6
Other goods	3,5	3,8

The analysis indicates that the level of reduction in the value of exports was similar to the level of reduction in its physical volumes (decreased by 1.5%). This is largely due to the fluctuations in prices for main product groups of Ukraine's exports. It's worth to underline that price trends had different directions.

If prices for the agricultural exports increased (as a result, when the value of exports of goods of this group increased by 0.2%, their physical volumes decreased by 9%), while prices for the other goods mostly fell. It is due to the decline in prices for chemical products that the increase in the physical volume of their exports by 24.9% in value brought much lower growth (by 2.3%).

Although the decline in world metal prices resulted in the reduction of the value of exports of this product group by 14.5%, their physical volumes remained at the level of the previous year. Prices for machine-building products were more stable. With the decrease in the physical volume of exports of goods in this group by 5,8%, value receipts from them fell by 1.8%.

Imports of goods in 2020 compared to 2019 decreased by 6.2 bln USD to 54.2 bln USD. The decline in imports was largely due to reduction in Ukraine's spending on fuel

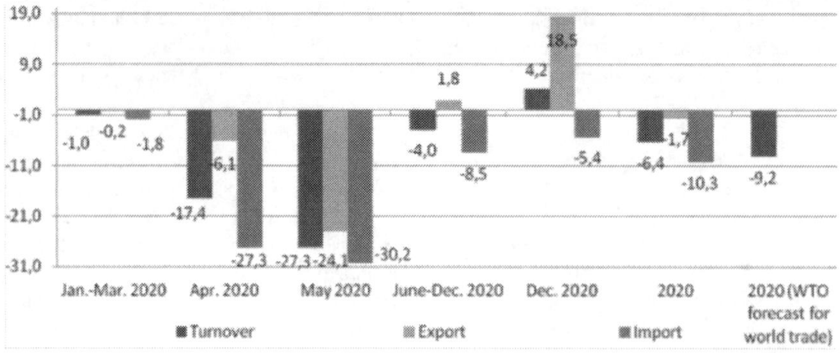

Figure 1. Dynamics of commodity foreign trade of Ukraine in 2020, %

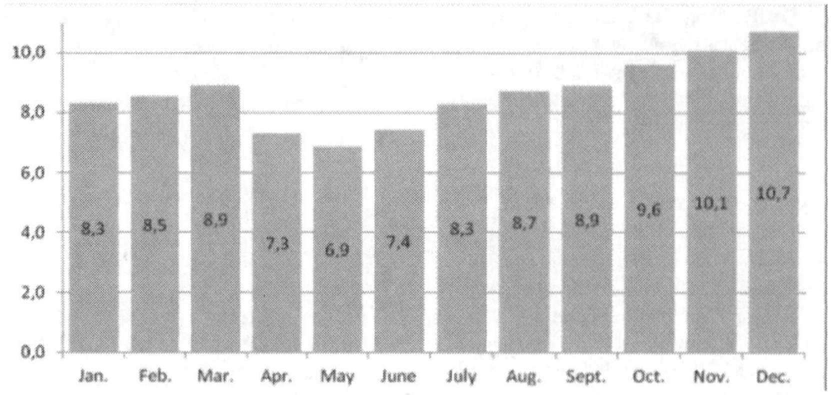

Figure 2. Foreign trade commodity turnover of Ukraine in 2020, bln USD

and energy products by 4.2 bln USD, as well as machinery, equipment, transport (by 2 bln USD).

It's necessary to mention that during this period, imports of food and agricultural products increased by 759 mln USD. These changes resulted in the growth of the share of food and agricultural products in the commodity structure of imports by 2.5 percentage points with the decrease in the share of fuel and energy products by 5.5 percentage points, while the share of other industries remained unchanged.

The leading role in the commodity structure of imports is played by machinery, equipment and transport (34.3% of total imports), chemical products (20%) and fuel and energy products (14.7%). These three product groups account for almost 70% of total imports of goods to Ukraine.

In contrast to the physical volumes of exports, which remained almost unchanged, the physical volumes of imports of goods decreased by 11.8%. The largest decrease in physical volumes took place in imports of hides, furs and fur products (by 24.4%), but due to the relatively small share of goods in this group in total imports, this did not have a significant impact. Greater impact on the overall indicator had decrease in imports of machinery, equipment and transport (by 15.8%) and chemical products (by 15.7%).

The physical volumes of imports of fuel and energy products also decreased by 15.1%, while the value of imports of this commodity group decreased by 34.6%. This reduction was caused by significantly lower prices for goods in this group. Thus, the price of natural gas imports for 2019 was 282 USD per ton, and for 2020 it was just 219 USD per ton. The cost of a ton of crude oil imports in 2019 was 513 USD, and for the same period in 2020 – 328 USD.

In the geographical structure of Ukraine's commodity foreign trade in 2020, the leading partner of Ukraine was the EU (calculations include the UK). However, the EU countries' share decreased from 41.9% in 2019 to 41% in 2020. This change was due to the reduction in the value of trade by 8.4%. As table 6 indicates, the EU is Ukraine's largest trading partner in both exports of goods (37.8% of total exports) and imports (43.8% of total imports).

Contrary to the general trend, the decline in exports to the EU countries was faster than in imports. According to the results of 2020, exports to them decreased by 10.2%, or 2.1 bln USD up to 18.6 bln USD, while imports fell by 1.8 bln USD or 6.9% to 21.3 bln USD. The reduction in exports to the EU was based on the decline in the supply of vehicles and spare parts to them by 1.4 bln USD (or 81.6%) and industrial goods by 653 mln USD (or 5.7%).

Table 6. Geographical structure of Ukraine's commodity exports and imports in 2020, %

	Export	Import
EU	37,8	43,8
China	14,4	15,3
EAEU	9,2	14,6
Turkey	5,0	4,4
Other countries	33,6	21,9

At the same time exports of food products and beverages showed the increase of 76 mln USD (or 1.8%). It should be noted that in December 2020, compared to December 2019, exports of goods from Ukraine to the EU increased by a quarter.

Somewhat similar situation was observed in the dynamics of imports from the EU countries. The reduction of imports of fuels and lubricants by 1.4 bln USD or 42.7%, vehicles by 542 mln USD or by 15.5% and capital equipment (except for transport) by 326 mln USD or 6.4% occurred with the parallel increase in imports of food and beverages by 354 mln USD or 16.4%.

The share of the Eurasian Economic Union (EAEU) countries in Ukraine's trade turnover also decreased from 14.9% in 2019 to 12% during 2020. The rate of reduction in trade was higher than in trade with the EU and amounted to 24.5%. This was largely due to the 28.7% reduction in trade with the Russian Federation. At the heart of this reduction is the decrease of 1.4 bln USD (or 35.3%) in imports from Russia of fuels and lubricants. Despite this reduction, the goods of this group account for 56.7% of total imports from the Russian Federation. In general, imports from the EAEU countries decreased by 29.1%.

The rate of decline in exports to the EAEU countries in the 2020 was lower than imports from them and amounted to 15% or 798 mln USD. This was largely due to lower (by 576 mln USD) volumes of industrial supplies.

If during 2020 Ukraine's trade with the EU countries, as well as with the EAEU countries, decreased, the dynamics of trade in goods with the other countries increased. The trade turnover with them increased by 875 mln USD (or by 1.8%) due to the growth in exports by 2.1 bln USD (or 8.6%), while imports fell by 1.2 bln USD (or 5%). It is important to stress that exports from Ukraine to China almost doubled (growth was 97.7%), which contributed to the increase in the share of this country in the geographical structure of Ukraine's foreign trade to 14.9% (14.4% in exports and 15.3% in imports). Higher export growth rates, while declining imports to countries other than the EU and the EAEU, contributed to the increase in the trade surplus. If in 2019 the positive trade balance with the countries, except EU and EAEU, was 0.2 bln USD, then in 2020 it was 3.5 bln USD.

The increase in the trade surplus with the countries other than the EU and the EAEU was due to the improving negative trade balance with China – from -5.6 bln USD to -1.2 bln USD – due to this fact the negative balance in trade with the Asian countries of 0.8 bln USD in 2019 was changed to the positive balance in 2020 in the amount of 3.1 bln USD, while maintaining positive trade balance with the African countries of 3,2 bln USD. Considering the negative trade balances with both unions – the EU with 5.1 bln USD and the EAEU with 3,4 bln USD – Ukraine's focusing on the improving the dynamics of increasing trade with the other countries can be viewed as positive factor that contributes to the partial compensation of the negative balance in trade with both the EU and the EAEU. It is also essential to develop the transport system for increasing the exports of agricultural and food products. So, COVID-19 has become a new challenge for the structure of Ukraine's foreign trade, and the national economy's prospects will depend largely on the level of the international technological competitiveness of the country [32, 33].

6 Conclusions

Europe was 53.6% of total merchandise exports of Ukraine. The European countries are the main consumers of the Ukrainian products.

Europe and Asia consumed 86.97% of Ukrainian goods in 2019, Africa – 9.97%. The role of the Asian countries as suppliers of goods for Ukraine is becoming more important. Ukraine has been having the negative balance of trade in goods with Europe since 2005 (–1785 mln USD), which has grown to – 11262.4 mln USD in 2019, while the trade balance of Ukraine with Asia had been positive throughout the analyzed period up until 2019 (–748.4 mln USD). The trade balance of goods of Ukraine with the African countries is positive, and with those of America – negative. The share of imports from the American countries in 2019 was 7.24% of total imports of goods of Ukraine.

The analysis of the Ukrainian exports and imports of the selected industrial goods – "XVI. Machines, equipment and mechanisms, electric and technical equipment" and "XVII. Ground, air and water transport facilities" – demonstrates that in exports of Ukraine the share of those groups of goods in all the exports of goods is decreasing (in 2016 – 11.53%, in 2019 – 10.68%) while the share of those groups of goods in all the imports of goods is increasing (in 2016 – 27.64%, in 2019 – 32.04%).

The reduction in the value of world trade in the 2020 due to the COVID-19 pandemic was not as significant as after the global crisis of 2008–2009. During 2020 the WTO improved its forecast for the dynamics of world trade in 2020. The decrease in Ukraine's foreign trade in 2020 by 6.4% is fully consistent with the projected WTO reduction of world trade in 2020 by 9.2%. During the last two months of 2020, for the first time since the beginning of COVID-19, the growth in the value of trade turnover of Ukraine was recorded compared to the same months of the previous year, which indicates a gradual resumption of foreign trade in the context of easing quarantine restrictions. The reduction in the value of foreign trade in 2020 by 6.4% with the fall in their physical volume by 4.7% indicates the prevailing trend of the decline in the prices. The world oil prices fell which resulted in the decrease of 34.6% in the cost of imports of fuel and energy products, while reducing purchases of their physical volumes by 15.1%.

The prices for agricultural and food products increased, which resulted in the growth in the value of trade in these goods by 2.9% while reducing their physical volume by 8.3%. So, one of the prospects is the development of the country's transport system for exporting agricultural and food products. Despite the general trend of declining trade, exports from Ukraine to China have doubled, leading to the increase in China's role in Ukraine's economy. The reduction of trade with the countries of the EU and the EAEU is observed against the background of the increased trade with the other countries, which, provided that the trade surplus is maintained, contributes to improving Ukraine's balance of payments.

Finally, COVID-19 demonstrated that agricultural and food products are of primary importance during the challenging times. In 2020 Ukraine improved trade with the Asian countries, especially China; however, the rise of the prices for the agricultural and food products makes it logical for the future of Ukraine to pay more attention to the development of the transport system in order to increase their exports to the developed countries of the West and to the prospective economies of the East and the South.

References

[1] A. Bielinskyi, I. Khvostina, A. Mamanazarov, A. Matviychuk, S. Semerikov, O. Serdyuk, V. Solovieva, V. Soloviev, IOP Conference Series: Earth and Environmental Science **628**, 012019 (2021)

[2] S. Semerikov, H. Kucherova, V. Los, D. Ocheretin, CEUR Workshop Proceedings **2845**, 22 (2021)

[3] J. Baylis, S. Smith, P. Owens, *The globalization of world politics. An introduction to International relations*, 8th edn. (Oxford University Press, 2020)

[4] V. Jeníček, V. Krepl, Agricultural Economics (Zemědělská Ekonomika) **55**, 211 (2009)

[5] S. Sen, Economics Working Paper Archive 635, Levy Economics Institute (2010)

[6] S.K. Gnangnon, The Journal of International Trade & Economic Development **29**, 211 (2020)

[7] A.U. Santos-Paulino, The Journal of International Trade & Economic Development **26**, 693 (2017)

[8] W. Leontief, A. Strout, *Multiregional input-output analysis*, in *Structural interdependence and economic development*, edited by T. Barna (Palgrave Macmillan, 1961), pp. 119–150, ISBN 978-1-349-81634-7

[9] L. von Bertalanffy, *General System Theory. Foundations, Development, Applications* (Braziller, New York, 1969), ISBN 0807604534

[10] G.J. Klir, *An Approach to General Systems Theory* (Van Nostrand Reinhold Co, 1969)

[11] L. Friendshuh, L. Troncale, *Identifying Fundamental Systems Processes for a General Theory of Systems (GTS)*, in *Proceedings of the 56th Annual conference, International Society for the Systems Sciences (ISSS)* (2012), p. 23

[12] V. Soloviev, A. Belinskiy, Communications in Computer and Information Science **1007**, 276 (2019)

[13] V. Soloviev, A. Bielinskyi, V. Solovieva, CEUR Workshop Proceedings **2393**, 434 (2019)

[14] V. Soloviev, N. Moiseienko, O. Tarasova, CEUR Workshop Proceedings **2393**, 905 (2019)

[15] V. Soloviev, N. Moiseienko, O. Tarasova, Communications in Computer and Information Science **1175 CCIS**, 231 (2020)

[16] V. Soloviev, A. Bielinskyi, O. Serdyuk, V. Solovieva, S. Semerikov, CEUR Workshop Proceedings **2732**, 455 (2020)

[17] V. Soloviev, V. Solovieva, A. Tuliakova, A. Hostryk, L. Pichl, CEUR Workshop Proceedings **2713**, 53 (2020)

[18] V. Soloviev, O. Serdiuk, S. Semerikov, A. Kiv, CEUR Workshop Proceedings **2713**, 21 (2020)

[19] A. Bielinskyi, S. Semerikov, O. Serdyuk, V. Solovieva, V. Soloviev, L. Pichl, CEUR Workshop Proceedings **2713**, 372 (2020)

[20] V. Derbentsev, S. Semerikov, O. Serdyuk, V. Solovieva, V. Soloviev, E3S Web of Conferences **166**, 13031 (2020)

[21] S. Semerikov, S. Chukharev, S. Sakhno, A. Striuk, V. Osadchyi, V. Solovieva, T. Vakaliuk, P. Nechypurenko, O. Bondarenko, H. Danylchuk, E3S Web of Conferences **166**, 00001 (2020)

[22] N. Datskevych, *Netherlands 'will pay the price' for blocking Turkish visit – Erdoğan* (2020), `https://tinyurl.com/z74dz8xt`

[23] S. Sardak, S. Radziyevska, I. Us, CEUR Workshop Proceedings **2422**, 84 (2019)

[24] *The State Statistics Service of Ukraine database* (2020), `http://www.ukrstat.gov.ua`

[25] V. Halasiuk, *Ukraine on the eve of the fourth industrial revolution* (2016)

[26] G. Gereffi, Review of International Political Economy **21**, 9 (2014)

[27] *World Development Report 2020: trading for development in the age of global value chains* (World Bank Group, Washington, D.C., 2020), `https://tinyurl.com/252wpfby`

[28] *World merchandise trade fell 14% in volume, 21% in value in q2 amid global lockdown* (2020), `https://www.wto.org/english/news_e/news20_e/stat_23sep20_e.htm`

[29] *Trade shows signs of rebound from covid-19, recovery still uncertain* (2020), `https://www.wto.org/english/news_e/pres20_e/pr862_e.htm`

[30] A.J. Morrison, D.A. Ricks, K. Roth, Organizational Dynamics **19**, 17 (1991)

[31] *The State Customs Service of Ukraine database* (2021), `https://customs.gov.ua/`

[32] D. Pokryshka, European Science **3**, 32 (2019)

[33] N. Cherkas, Ph.D. thesis, Kyiv, Ukraine (2018)

Assessment of factors influencing the stability of Ukrainian export based on the fractal analysis

Dmytro Kainara[1,*] and *Iuliia* Kaliuzhna[1,**]

[1]Zaporizhzhia National University, 66 Zhukovsky Str., Zaporizhzhia, 69063, Ukraine

Abstract. The article considered the assessment of the sustainability of the export flow of Ukraine. It was determined that the markets of Slovakia, the Russian Federation, Germany, Poland and Lithuania possess long-term memory. Also, the main factors that affect the sustainability of export have been analyzed. Among cultural factors, a weak correlation was demonstrated by Long Term Orientation, Indulgence and Power Distance according to the Geert Hofstede method. At the same time, the cultural patterns according to the S. H. Schwartz method did not show a correlation with the long-term memory of product export. On the stability of the export flow, the role of the linguistic distance and the factor of the diaspora presence have been assessed. It was proved that the gravitational effect has the greatest impact on the stability of export; Power Distance, linguistic distance and the factor of the diaspora presence have the negligible impact as well.

1 Introduction

International trade in its nature is a self-organizing system under the influence of international politics, the politics of national states, the politics of international organizations, the politics of individual market entities, etc.

The influence of the international trade entities forms the cycles and trends in the market. Thus, even at first sight, the unstable markets, with a high level of price volatility and a wide range of trade volumes, in fact, may have a clear trade structure, which is subject to the seasonality factors, economic cycles within the country and international policies of their states.

To estimate the factor of a long-term memory, the market uses fractal analysis, which allows one to find patterns that cannot be estimated by the traditional statistical methods. And so that, to understand the trends that actually exist on the market, as well as factors that have a latent effect on the export of the products, which forms the relevance of this work.

2 Methodology

Fractal analysis is used to assess the stability of export flows taking into account the influence of cyclical factors. This method was founded by H. E. Hurst [1] and based by R/S analysis technique. R/S analysis reveals patterns to detect regularities and long-term memory in conditions where traditional methods do not allow it.

Using a temporal analysis approach, B. Mandelbrot [2], and later E. Peters [3] discovered the effect of long-term memory in the stock markets. The stock market has been proven to have a long-term memory.

E. Peters identified 3 types of markets, depending on the indicator that the Hurst coefficient took [3]:

- $0 \leq H < 0.5$ – fractal market with an antipersistent series – these markets are characterized by unstable dynamics and the tendency of the indicator to the "average" value;
- $H = 0.5$ – efficient market (EMH), where prices are set according to the theory of E. F. Fama;
- $0.5 > H \geq 1$ – fractal market with a persistent series. These markets are characterized by a stable trend and long-term memory.

Since the volume of sales depends on the price of the product, this indicator can be used to analyze the level of stability in the market.

Data for R/S analysis was taken from UN Comtrade Database [4]. The sampling interval for the analysis was from January 01, 2010 to December 31, 2020: monthly indicators. This sample was determined due to the fact that the average economic cycle in the world lasts 10 years. However, since at the time of submitting the work to the editorial office, there are no data for October 01, 2020 – December 31, 2020, they were predicted using a simple moving average. Upon that, the deviation of the Hurst coefficient from the real one will not exceed 1%, which will insignificantly affect the results of the work.

For R/S analysis the following algorithm was used [5]:

1. Conversion of the time series in the logarithmic number:
$$t_i = \ln \frac{x_i}{x_{i-1}}, \ i = 2 \ldots n \qquad (1)$$
2. Finding the divisor for dividing a logarithmic series into k intervals with b number of elements in each interval.

*e-mail: Kainara1992@gmail.com
**e-mail: kalyuzhnaya.ju@gmail.com

Table 1. The results of assessing the export flow to Ukraine's main trading partners

Country	Hurst coefficient
Fractal market with persistence of trends	
Slovak Republic	0,9427
Russian Federation	0,9422
Germany	0,8939
Poland	0,8593
Lithuania	0,8537
Belarus	0,8316
United States	0,8303
Romania	0,8067
Georgia	0,7641
Moldova	0,7371
Bulgaria	0,7327
Hungary	0,7327
Chile	0,709
Brazil	0,6968
Azerbaijan	0,6949
Iran, Islamic Republic	0,6906
Sweden	0,6828
Italy	0,6795
Egypt, Arab Republic	0,6762
Spain	0,6761
France	0,6739
Turkey	0,6725
Armenia	0,6587
China	0,6479
Algeria	0,6271
Canada	0,6266
Austria	0,6238
Kazakhstan	0,612
Effective market (EMH)	
Jordan	0,5984
Israel	0,5979
United Kingdom	0,5927
Albania	0,5909
Ireland	0,5809
Japan	0,5793
Mexico	0,5742
Denmark	0,5705
Greece	0,5683
Estonia	0,5672
Finland	0,5619
Portugal	0,552
New Zealand	0,535
Latvia	0,4946
Fractal market with anti-persistence of trends	
Argentina	0,4348
Australia	0,4197
Peru	0,4106
South Africa	0,4078

3. Finding the average value of the interval:

$$\overline{S}_i = \frac{1}{b}\sum_{i=1}^{b} t_i \quad (2)$$

Table 2. The gravity effect between Ukraine and main trading partners

Country	Gravity coefficient
Albania	34,8575
Algeria	35,8521
Argentina	33,654
Armenia	34,3709
Australia	34,5056
Austria	38,6596
Azerbaijan	35,2912
Belarus	38,4896
Brazil	35,469
Bulgaria	36,8507
Canada	36,1735
Chile	33,0968
China	38,5091
Denmark	37,9589
Egypt, Arab Republic	36,7419
Estonia	35,9915
Finland	38,009
France	39,1631
Georgia	34,7596
Germany	40,554
Greece	37,2181
Hungary	37,9761
Ireland	36,7848
Israel	37,1875
Italy	39,2376
Japan	36,9915
Jordan	34,9293
Kazakhstan	35,7818
Latvia	36,5566
Lithuania	37,7291
Mexico	35,0504
Moldova	36,9715
New Zealand	32,3209
Peru	33,0756
Poland	39,5154
Portugal	35,2591
Romania	37,9941
Russian Federation	40,5115
Slovak Republic	37,3133
South Africa	34,0342
Spain	37,8024
Sweden	38,4725
Turkey	38,9623
United Kingdom	39,0992
United States	38,522

4. Finding the accumulated deviation from the average value for each value and series:

$$X_{i,k} = \sum_{i=1}^{b}\left(t_i - \overline{t_i}\right) \quad (3)$$

5. Calculating the range for each interval:

$$R_k = \max(X_{i,k}) - \min(X_{i,k}) \quad (4)$$

Table 3. Correlation analysis of the impact of cultural factors on export sustainability

	Hurst coefficient	Power Distance	Individualism	Masculitinity	Uncertainty Avoidanance	Long Term Orientation	Indulgence	Gravity Effect
Hurst coefficient	1							
Power Distance	-0,333	1						
Individualism	-0,111	0,75	1					
Masculitinity	0,039	0,1	0,3	1				
Uncertainty Avoidanance	-0,207	0,66	0,6	0,2	1			
Long Term Orientation	-0,451	0,24	0,1	-0,1	0,1	1		
Indulgence	-0,394	0,5	0,4	0,16	0,5	0,6	1	
Gravity Effect	0,594	0,1	0,3	0,1	0,1	-0,4	-0,2	1

6. Calculating the standard deviation for each interval:

$$S_k = \sqrt{\frac{\sum_{i=1}^{b} (t_i - \overline{t_i})^2}{b-1}} \quad (5)$$

7. Finding the normalized range for the interval:

$$R/S_k = \frac{R_k}{S_k} \quad (6)$$

8. Finding the average value of the normalized range over all intervals:

$$R/S = \frac{1}{k} \sum_{i=1}^{k} R_i \quad (7)$$

9. Calculating the standard error:

$$V = \frac{R/S}{\sqrt{b}} \quad (8)$$

10. Doing subsequent iterations of the algorithm increasing the number of b values in intervals.

    Selecting the b values was carried out according to the principle: 3 months (quarterly cycle of companies), 6 months (half-year cycle of companies), 12 months (annual cycle of companies), 24 months (two-year cycle), 40 months (average duration of the economic cycle in the Ukrainian economy).

3 Literature review

Today, the fractal analysis is used mainly for analyzing the characteristics of the stock market and predicting price trends [6–9] and less often for forecasting the market [10–14]. First of all, it is necessary to highlight the work of T. Ikeda [6], who conducted the fractal analysis of the main stock markets of the world and proved the presence of fractality in them. This is the opinion of R. F. Mulligan in his work [7].

In the work of M. Frezza, S. Bianchi and A. Pianese [15], the authors use the fractal analysis capabilities to track propagation waves of the COVID-19 crisis. In the work of M. Karpiarz, P. Fronczak and A. Fronczak [16], the fractal analysis was used to interpret the distance coefficient in the distance of the gravity model.

However, scientists practically did not use the fractal analysis to assess the export flow, which significantly limits the understanding of the foreign trade trends.

4 Impact and sustainability of export flows

The analysis of trends in the international trade in Ukraine shows that in 2019 the main partner in the export of products is the European Union, China, Russia, Turkey. The share of the Russian Federation in the export of products has been steadily decreasing over the past 6 years, which may indicate, at first sight, the instability of trade between the countries. Herewith, the trade with the EU and the PRC in the initial analysis of the data indicates stable dynamics [4].

The results of assessing the export flow using the R/S analysis are shown in table 1. The source of information is the UN Comtrade Database [4].

Also, despite a significant decrease in trade between the Russian Federation and Ukraine, the trade relations between the countries remain significant. The reason for this is the preservation of economic relationships that have existed since the USSR, cultural closeness, high gravitational influence of the Russian economy and low language barriers.

Further, it is worth noting the almost identical Hurst coefficient for Poland and Lithuania. This may indicate the preservation of the relationships between the countries, which has existed since the times of the Polish-Lithuanian Commonwealth. The high Hurst coefficient for Germany can be explained by the gravitational influence of the economy on Ukraine.

In the international trade, the gravitational effect lies in the hypothesis that trade volumes depend on the level of their GDP and are inversely proportional to the distance between the countries. To assess the gravitational effect between the given countries, a modified J. Tinbergen formula was taken [17]:

$$M_{ij} = \log \frac{Y_j \cdot Y_i}{D_{ij}^2} \quad (9)$$

where M_{ij} — attraction of two countries,

Table 4. Assessment of linguistic distance and diaspora factor

Country	Level of English knowledge [23]	Diaspora factor [24]	Total
Australia	1	0	1
Austria	0,623	0	0,623
Azerbaijan	0,432	0,5	1,932
Argentina	0,566	0	0,566
Armenia	0,494	0,5	1,994
Belarus	0,513	0,5	2,013
Bulgaria	0,579	0	1,079
Brazil	0,49	0,5	0,99
United Kingdom	1	0	1
Hungary	0,598	0,5	1,098
Germany	0,616	0,5	1,116
Greece	0,578	0	0,578
Georgia	0,503	0,5	1,003
Egypt, Arab Republic	0,437	0	0,437
Ireland	1	0	1
Spain	0,537	0,5	1,037
Italy	0,547	0,5	1,047
Kazakhstan	0,412	0,5	1,912
Canada	1	0,5	1,5
China	0,52	0	0,52
Latvia	0,555	0,5	2,055
Lithuania	0,57	0,5	2,07
Moldova	0,589	0,75	2,339
Poland	0,596	0,75	1,846
Portugal	0,618	0,5	1,118
Russian Federation	0,512	0,75	2,262
Romania	0,589	0,5	1,089
Slovak Republic	0,577	0,5	1,577
United States	1	0,5	1,5
Turkey	0,465	0	0,465
France	0,559	0	0,559
Sweden	0,625	0	0,625
Estonia	0,566	0,5	2,066
South Africa	0,607	0	0,607
Japan	0,487	0	0,487
Finland	0,613	0	0,613
Albania	0,445	0	0,445
Algeria	0,442	0	0,442
Denmark	0,632	0	0,632
Chile	0,532	0	0,532
Mexico	0,440	0	0,440
Peru	0,482	0	0,482
New Zealand	1	0	1
Jordan	0,456	0	0,456

Y_j, Y_i — GDP of the countries i and j,

D_{ij}^2 — distance between the capitals of the countries.

The results of the gravitational effect assessment are shown in table 2. The source of information is the World bank database [18] and the Distance database [19].

Germany and the Russian Federation have the greatest gravitational effect on the Ukrainian economy. It is also worth noting the high gravitational influence of other EU countries on the economy of Ukraine, primarily France, Poland and Sweden. At the same time, the United States of America, although locates at a considerable distance from Ukraine, has an equal comparable to neighboring Belarus.

The second factor that affects the long-term memory of product exports is cultural factors. These include both cultural patterns and language barriers, as well as the presence of a diaspora in the partner country.

G. Hofstede [20] and S. H. Schwartz [21] indicators will be used to assess the cultural barriers. To assess the cultural difference between the countries, it is necessary to calculate the difference between the corresponding indicator for Ukraine and the partner country:

$$CD_{ij} = |CH_i - CH_j|, \qquad (10)$$

where CD_{ij} – cultural distance between the countries i and j;

CH_i and CH_j – G. Hofstede or S. H. Schwartz indicator for countries i and j.

The correlation analysis for indicators of gravity effect and cultural distance is shown in table 3.

The average correlation effect between the stability of export and the gravitational effect of 0,5939, as well as a weak influence between the Hurst coefficient and Long Term Orientation, Indulgence and Power Distance is preserved.

The presence of a negative correlation between individual cultural patterns confirms the thesis about the influence of cultural distance on the long-term memory of product exports. At the same time, the pattern of long-term "Long Term Orientation and Indulgence" has the greatest influence, which indicates the best interaction in export transactions between people of similar nature. As for the weak inverse correlation of Power Distance and Hurst coefficient, we can say that company employees prefer to work more with countries with similar governments and distance from society.

At this, there is no correlation between the Hurst coefficient and S. H. Schwartz indicators. Linguistic factors and factors of the presence of a common diaspora also affect the stability of export. Let's consider this factor. As a methodology for assessing the impact of linguistic distance and the presence of a diaspora, we use the approach proposed by L. Vlasenko [22].

The factor of knowledge of the English language was taken as a factor in assessing the linguistic distance. The EF EPI rating was used to assess the level of knowledge of the English language [23].

Table 5. Linguistic distance and diaspora factor correlation analysis

	Hurst coefficient	Linguistic distance and diaspora factor
Hurst coefficient	1	
Linguistic distance and diaspora factor	0,460836	1

Table 6. The results of the regression model

	Coefficient	Standard error	t-Statistic	P-Value	Lower 95%	Upper 95%
Hurst coefficient	-0,67428	0,26512	-2,54331	0,014951	0,033937	0,228792
Gravity coefficient	0,034611	0,00742	4,661742	$3.46 \cdot 10^{-55}$	-1,21011	-0,13845
Power distance	-0,00152	0,000632	-2,41061	0,020613	0,019606	0,049617
Linguistic distance and diaspora factor	0,131365	0,048206	-0,0028	-0,00025		

Moreover, in the countries where knowledge of the language is native, the indicator equals to 1. As for the diaspora, based on Ethnologue data [24] the indicator of the diaspora presence was assigned a value of 0 (in the absence of a significant common diaspora), in the presence of a significant common diaspora a value of 0,5, and in the presence of very significant common diaspora a value of 0,75.

The results of assessing the indicator of linguistic distance and the diaspora factor are shown in table 4.

The results of the correlation analysis are shown in table 5.

Thus, a correlation index of 0,46 indicates the presence of an average correlation between the linguistic distance and diaspora factor and with the Hurst coefficient. Consequently, the level of knowledge of the English language in a society affects the size of the cultural barriers that arise in trade. And in the context of globalization, an increase in the level of English proficiency in society will allow the formation of stable ties in trade, which will positively affect the trade flow.

Further, based on the factor of culture, gravitational impact, linguistic distance and diaspora, let's conduct a regression analysis to identify the level of influence of each of the indicators on the level of stability.

Thus, there is an average correlation between the gravity coefficient and the Hurst coefficient, the weak correlation between the Hurst coefficient and the Power Distance, and the linguistic distance and the diaspora factor. The results of a regression model building are shown in table 6.

Thus, all indicators are statistically significant. And the given t-statistic values are above the critical value of 2,039.

The indicators of the regression model quality are shown in table 7.

Table 7. Quality indicators of the regression model

Indicator	Value
Multiple R	0,723433
R Square	0,523355
Adjusted R Square	0,487607
Standard Error	0,095665
Fisher criterion	14,63998263
Approximation index	14,64%

As a result, the data presented indicate the statistical significance of the regression model. In fact, the Fisher's criterion is higher than the critical value (2,911). At the same time, the approximation indicator is 14,64%, which indicates the high quality of the model. To assess the level of influence of factors on the Hurst coefficient, an elasticity analysis has been performed.

The results are shown in table 8.

Table 8. The elasticity analysis

Indicator	Level of 1% variation
Gravity factor	1,846%
Power distance	1,843%
Linguistic distance and diaspora factor	0,194%
Other factors	0,979%

Thus, the analysis showed that cultural factors have the same influence as the factors of gravitational influence on long-term memory in export shipments. At the same time, the linguistic distance and the factor of the presence of a diaspora in the country have an insignificant effect on long-term memory.

This suggests that Ukraine's export instead of economic factors and economic benefits is more susceptible primarily to personal connections and preferences of the management of firms and managers who are engaged in the foreign economic activity.

5 Conclusion

As a result of the conducted study, it was established, first of all, that Ukraine's export supplies to the main partner countries are subject to fractality, which indicates the presence of a long-term memory in the market. Exports to the Russian Federation are indicative, the share of which in the total export structure was significantly reduced after 2014. However, despite this, a large number of companies still supply their products to the Russian market and depend on the economic processes that are taking place in the Russian Federation. At the same time, a high level of a long-term memory of export trade is also observed with the EU countries and the USA.

In this case, the size of the country's GDP and the proximity of its location have the greatest influence on the level of fractality, which was calculated using the Gravity factor. An interesting result is the fact that cultural affinity between countries plays an almost identical role to that of the Gravity factor. At the same time, the presence of a diaspora, as well as the level of knowledge of the English language in society, has little effect on the level of a long-term memory.

This trend is more negative than positive. The high level of dependence on key partner countries with a similar culture indicates the inability of Ukrainian management to quickly navigate the constantly changing economic environment and get significant benefits from trading with new partners, giving preference to the old ones. As a result, Ukraine loses significant export benefits.

To improve this situation, it is necessary to train to increase the level of cross-cultural management of the companies and management of the foreign economic activity. This will help lower cultural barriers and, consequently, the level of fractality of markets, which will increase the efficiency of export trade. As for the knowledge of the language, as practice shows, the majority of foreign economic activity managers in Ukraine know English, which significantly reduces the level of linguistic barriers in the export deliveries.

References

[1] H.E. Hurst, International Association of Scientific Hydrology. Bulletin **1**, 13 (1956), `https://doi.org/10.1080/02626665609493644`

[2] B. Mandelbrot, R.L. Hudson, *The (Mis)Behaviour Of Markets : A Fractal View of Risk, Ruin, and Reward* (Profile Books, London, 2004)

[3] E.E. Peters, *Chaos and order in the capital markets: a new view of cycles, prices, and market volatility*, 2nd edn. (Wiley, New York, 1996), ISBN 978-0-471-13938-6

[4] United Nations, *Comtrade database* (2020), `https://comtrade.un.org/`

[5] E.E. Peters, *Fractal market analysis: applying chaos theory to investment and economics* (Wiley, Chichester, 1994), ISBN 978-0-471-58524-4

[6] T. Ikeda, Economics Bulletin **37**, 1514 (2017)

[7] R.F. Mulligan, The Quarterly Review of Economics and Finance **44**, 155 (2004)

[8] N. Maksyshko, O. Vasylieva, CEUR Workshop Proceedings **2422**, 467 (2019)

[9] H. Danylchuk, O. Kovtun, L. Kibalnyk, O. Sysoiev, E3S Web of Conferences **166** (2020)

[10] A. Matviychuk, O. Novoseletskyy, S. Vashchaiev, H. Velykoivanenko, I. Zubenko, CEUR Workshop Proceedings **2422**, 455 (2019)

[11] H. Danylchuk, L. Kibalnyk, O. Kovtun, A. Kiv, O. Pursky, G. Berezhna, CEUR Workshop Proceedings **2713**, 352 (2020)

[12] N. Maksyshko, O. Vasylieva, I. Kozin, V. Perepelitsa, CEUR Workshop Proceedings **2713**, 219 (2020)

[13] A. Bielinskyi, S. Semerikov, O. Serdyuk, V. Solovieva, V. Soloviev, L. Pichl, CEUR Workshop Proceedings **2713**, 372 (2020)

[14] A. Bielinskyi, I. Khvostina, A. Mamanazarov, A. Matviychuk, S. Semerikov, O. Serdyuk, V. Solovieva, V. Soloviev, IOP Conference Series: Earth and Environmental Science **628** (2021)

[15] M. Frezza, S. Bianchi, A. Pianese, Finance Research Letters **38**, 101851 (2021)

[16] M. Karpiarz, P. Fronczak, A. Fronczak, Physical Review Letters **113**, 248701 (2014)

[17] J. Tinbergen, The International Executive **5**, 27 (1963)

[18] *DataBank* (2021), `https://databank.worldbank.org/indicator/NY.GDP.MKTP.CD/1ff4a498/Popular-Indicators`

[19] *Distance calculator* (2021), `https://www.distance.to/`

[20] H. Insights, *Hofstede insights organisational culture consulting* (2021), `https://www.hofstede-insights.com`

[21] S.H. Schwartz, *The 7 schwartz cultural value orientation scores for 80 countries* (2008), `10.13140/RG.2.1.3313.3040`

[22] L. Vlasenko, Problemy systemnoho pidkhodu v ekonomitsi **1** (2017)

[23] *EF English Proficiency Index* (2020), `https://www.ef.com/assetscdn/WIBIwq6RdJvcD9bc8RMd/legacy/__/~/media/centralefcom/epi/downloads/full-reports/v10/ef-epi-2020-russian.pdf`

[24] *Ethnologue: Languages of the World* (2021), `https://www.ethnologue.com`

Decision support information and analytical technology in discharge military personnel employment

Mykhailo Medvid[1,*], *Peter* Ivashchenko[2,**], *Igor* Britchenko[3,***], *Iryna* Trubavina[1,****], and *Volodymyr* Liutyi[1,†]

[1]National Academy of the National Guard of Ukraine, 3 Zakhysnykiv Ukrainy Sq., Kharkiv, 61001, Ukraine
[2]V. N. Karazin Kharkiv National University, 4 Svobody Sq., Kharkiv, 61022, Ukraine
[3]State Higher Vocational School Memorial of Prof. Stanisław Tarnowski in Tarnobrzeg, 50 Henryka Sienkiewicza Str., 39-400 Tarnobrzeg, Poland

Abstract. The research material proposes the use of decision support information-analytical technology in discharge military personnel employment, which, in contrast to the usual processing of survey results, makes it possible to obtain more information for decision-making. Adherence to such an approach in the development of public administration mechanisms increases the likelihood that in the case of their implementation in the country there will be positive changes, as they will indirectly take into account the availability of necessary resources. Information and analytical technology to support decisions in the field of labor economics substantiates the development of the example of discharged military personnel social welfare. According to the content, the developed decision support information-analytical technology in the field of labor economics provides for the following components: formation of social protection mechanisms; development of a questionnaire on the use of the appropriate mechanism; statistical analysis of survey results (generation of empirical distributions of answers to questions relative frequencies; comparison of answers medians; correspondences analysis; factor analysis; analysis of hierarchies); meaningful interpretation; decision making.

1 Introduction

Labor is a complex and multifaceted phenomenon that plays a particularly important role both in the life of society and the individual, because this very concept is inseparable from human life. By changing the environment in the process of work and adapting it to their needs, people not only ensure their existence, but also create conditions for the development and progress of society. Labor is one of the most important forms of self-expression, self-actualization and human selfimprovement.

Military service is also a type of labor, the problems of which are devoted to many scientific studies. For example, the conscription forces young men to serve in the military, irrespective of their opportunity costs. As Paul Bingley, Petter Lundborg and Stéphanie Vincent Lyk-Jensen say "The opportunity costs of conscription are borne by men with the best labor market prospects" [1].

During hostilities, the number of servicemen who retire and have to apply to public employment service centers to find a new job. A significant part of this category of people has the desire and ability to create their own business and do business. However, in most of them the relevant competencies are not developed.

*e-mail: medvidmm@ukr.net
**e-mail: ipaplin7@gmail.com
***e-mail: trubavina@gmail.com
****e-mail: ibritchenko@gmail.com
†e-mail: liutyy@ukr.net

In order to prevent the growth of social tension in society, to employ of this category of people, and to ensure the creation of new jobs for them, mechanisms for public management of human resources with military competencies, including training, retraining and advanced training need to be defined. According to Henry J. Meyer and Erwin O. Smigel "fully understand this joblessness, it is necessary to study the types of unemployed veteran" [2].

Ignoring this problem, increasing the level of poverty of the Ukraine population, deteriorating socio-economic conditions of human development at the end of hostilities may increase migration, in particular, the part of the population with developed military competencies, as well as lead to creation of private military companies in Ukraine that deliver combat services to other states.

2 Intent of the research justification

Based on the needs of society, the development (improvement) of mechanisms of retired military personnel social welfare is substantiated, as well as determined their advantages and disadvantages. These include [3, 4]:

- the state management mechanism of development of the military instructor social institution;
- the state management mechanism of ex-servicemen retraining and their support in creation their own business and running business;

- the state management mechanism of social welfare of ex-servicemen through rehabilitation and employment abroad.

Decision-making on the allocation of resources for the functioning of the above mechanisms is a difficult task and it requires the development of information and analytical technology to support decisions on social welfare of military personnel. Among the publications there are many devoted to decisions support information and analytical technologies in the field of labor economics [5, 6]. However, their use to solve this problem is unacceptable.

3 Results and discussion

In order to determine the advantages in the functioning of one or another of the mechanisms suggested above, a questionnaire for military personnel was developed. The servicemen were invited to act as an expert and evaluate the options for the formation (development) of social welfare mechanisms for retired servicemen.

We developed a military personnel questionnaire to determine the advantages in the functioning of one or another of the mechanisms suggested above. And we invited the acting servicemen to expert and evaluate the options for the formation (development) of social welfare mechanisms for the retired military personnel.

Scenario № 1, The type of mechanism is *"the state management mechanism for development of the military instructor social institution"*. It is characterized by the influence on the formation of the public good "military security" through the human resources for combat activities formation and development (military-patriotic educated youth, the formation in another state population the military competencies (participation in the officers' training, including reserve officers' training program, NCO training program, in the developing of primary military competencies within "The Motherland Defense" subject in secondary schools and other subjects in military lyceums and lyceums with enhanced military and physical training)). The advantage of this type is that it requires minimum costs.

Scenario № 2, The type of mechanism is *"the state management mechanism for ex-servicemen retraining and their support in creation their own business and running business"*. It is characterized by affecting the state socio-economic development, creating new enterprises that leads to the creation of new jobs that can be allocated to ex-servicemen, including those who became disabled while participating in hostilities and affecting the formation of the public good "military security" with the deductions. But there is and disadvantage of this type as it requires significant funding.

Scenario № 3, The type of mechanism is *"the state management mechanism for social welfare of ex-servicemen through rehabilitation and employment abroad"*. It is characterized by affecting the reduction of social tensions in society by reducing the cost of social welfare, reducing unemployment (used in the absence of sufficient conditions for the functioning of mechanisms № 1 and № 2, or when it is impossible cover all persons in need of such social welfare). But there is and disadvantage of this type as there is a threat to the State secrecy and loss of human resources by the state, which can be used in the future in combat operations.

Scenario № 4, The type of mechanism is *"there is no need to develop any of these mechanisms"*. It is characterized in, that the shortcomings of social adaptation of servicemen traditional institutional mechanisms are explained by foreign scientists mainly for two main reasons. First, they respond to challenges rather than anticipate them, and second, the traditional system turns a serviceman into a passive recipient of benefits and services, a kind of "gift set" from the State.

The servicemen evaluated the proposed options on a ten-point scale from 1 – "not important" to 10 – "very important" in response to questions:

1) provided that all the proposed opportunities exist - what exactly did you use when you were released to the reserve (according to priorities)?;
2) evaluate each of the proposed options in terms of the need to form a public good "military security";
3) evaluate each of the proposed options in terms of the need for economic development of the state;
4) evaluate each of the proposed options in terms of the need to ensure the security of the state as a whole (social, economic, military, demographic, State secrecy);
5) provided that all the proposed options exist, predict the quality of implementation of each in our country;
6) estimate the cost of the state for each of the options (10 – the least, 1 – the most).

$$X = (x_{nkl}), \tag{1}$$

where x_{nkl} – the number of points awarded by the serviceman by number n (n = 1, ..., 40); focusing on the criterion k (k = 1, ..., 6); in relation to the choice of specialty l.

The following criteria were used:

$C1$ – the ability to choose any of the four mechanisms;

$C2$ – the need for the formation of the public good "military security";

$C3$ – economic development of the state;

$C4$ – ensuring state security;

$C5$ – an idea of the quality of each of the four mechanisms implementation;

$C6$ – government spending on the implementation of each mechanism.

The purpose of the application of the criteria is to link the solution of the problem of choosing the mechanism to the state and professional self-determination for further employment. The following professions were chosen: instructor; entrepreneur; mercenary; also, the fourth variant is considered – not professionally defined.

Let's perform a statistical analysis of the criteria. To do this, we regroup the data of the matrix X by combining the columns of occupations into one for each criterion. We obtain a matrix

$$Y = (y_{ij}),\ i = 1, ..., 160;\ j = 1, ..., 6. \qquad (2)$$

Diagrams (figures 1-6) of empirical distribution of relative frequencies of responses according to the criteria related to the mechanisms of state management for military personnel social welfare on a ten-point scale.

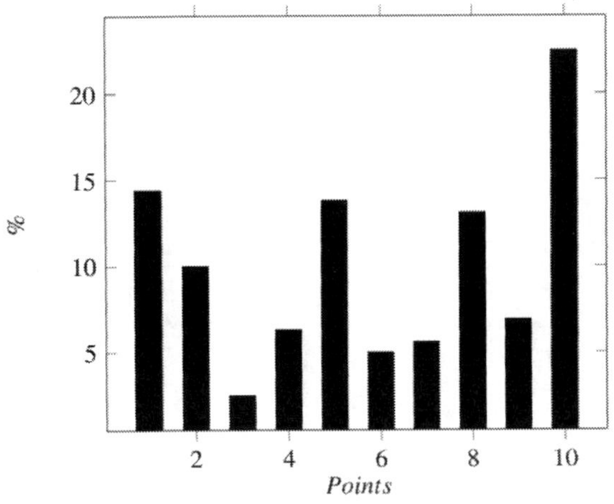

Figure 1. Criteria C1

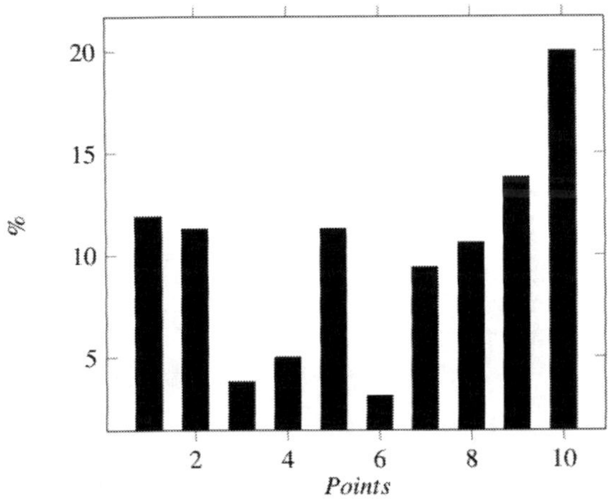

Figure 2. Criteria C2

In figures 1-6 it is presented the empirical distribution of the results of a survey of 40 retired servicemen (sample size is equal to 160), according to 6 criteria provided that the specialties are combined on a ten-point scale.

The following can be said about the peculiarities of the six given empirical criteria allocating.

Despite the presence of significant differences in the answers, they are characterized by categorical of answers for the lowest and highest scores. For example, when it is asked about the possibility of choosing any of the four specialties, points 1 and 10 were asked by 14.4% and 22.5%

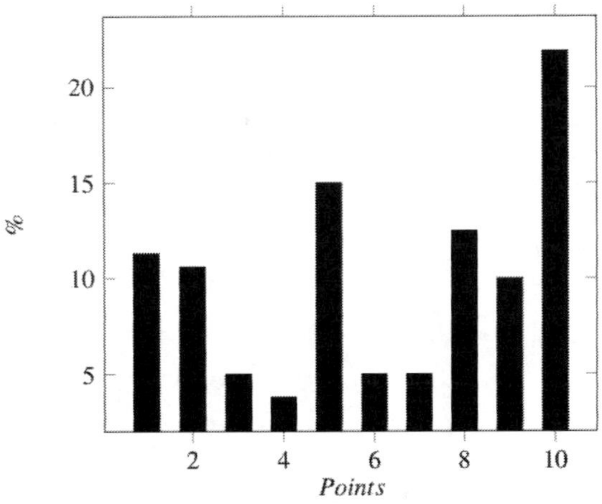

Figure 3. Criteria C3

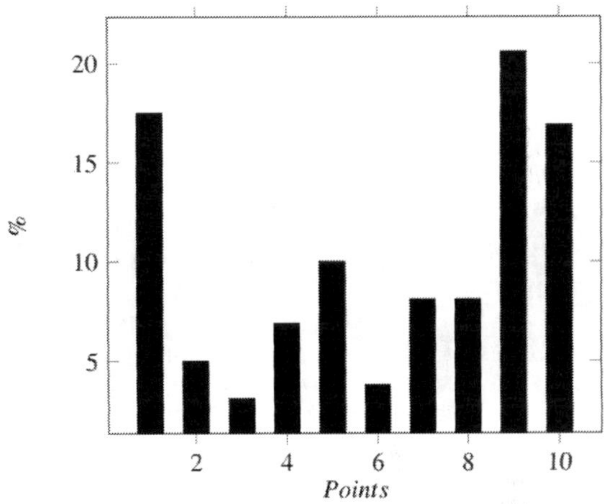

Figure 4. Criteria C4

of respondents (figure 1). Points 1 and 10 were given by 11.9% and 20% of respondents, realizing the need to form a public good "military security" (figure 2). Points 1 and 9 were given by 17.5% and 20.6% of respondents, in answers to questions about state security (figure 4).

It should be noted that there was a significant percentage of respondents with neutral responses (5 points). So, 5 points were given by 15% of respondents by criterion $C3$ (figure 3), 14.4% of respondents by criteria $C5$, $C6$. The significance of the neutral nature of the answers to the criteria can be verified using Student's t-test or by its non-parametric alternative – the Kolmogorov-Smirnov criterion [7, 8].

A complete description of the median responses of servicemen comparison is given in table 1.

We advance a number of hypotheses that the medians $Me(y_{*,j1})$, $Me(y_{*,j2})$ do not differ in pairs by 5% (p=0.05).

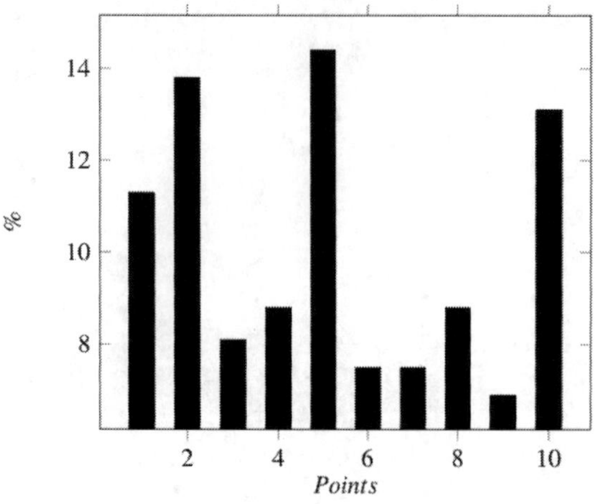

Figure 5. Criteria $C5$

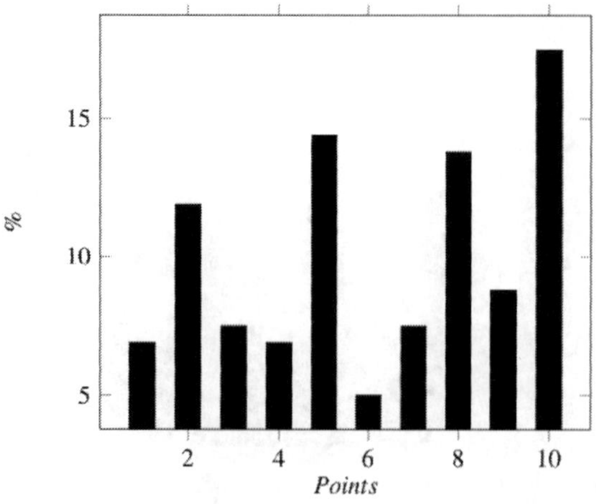

Figure 6. Criteria $C6$

That is:

$$H_0 : Me(y_{*,j1}) - Me(y_{*,j2}) = 0, \quad (3)$$

$$H_1 : Me(y_{*,j1}) - Me(y_{*,j2}) \neq 0 \quad (4)$$

where $j1 \neq j2$, $j1$, $j2 = 1, ..., 6$.

Since the variables C are measured on an ordinal scale, to test hypothesis (3-4) we use the nonparametric Wilcoxon test [8, 9].

The averages and the magnitude of variation for the vectors of the statistical population given by the matrix Y are calculated and constructed in the program STATISTICA. The results are presented graphically in figure 7.

We apply correspondence analysis [9].

In the table 2 it is shown a fragment of the frequency matrix obtained by the transition from frequencies (results of questionnaire processing) to particulars.

According to the table, using the tools of associative (corresponding, relevant) analysis [9], we establish associative links between criteria and specialties. The results are obtained by analyzing figure 8.

Table 1. The results of variable check $C1, \ldots, C6$ by Wilcoxon test (calculated by the authors)

Pairs variates	*W*	*p-value*
$C1$ & $C2$	2032.5	0.4474
$C1$ & $C3$	3013	0.5503
$C1$ & $C4$	3371.5	0.4956
$C1$ & $C5$	6001.5	**0.0262**
$C1$ & $C6$	5700	0.9445
$C2$ & $C3$	2287.5	0.6950
$C2$ & $C4$	2815	0.7784
$C2$ & $C5$	5749.5	**0.0020**
$C2$ & $C6$	5544.5	0.6186
$C3$ & $C4$	2999	0.7270
$C3$ & $C5$	4528	**0.0185**
$C3$ & $C6$	5156	0.7570
$C4$ & $C5$	5922	**0.0059**
$C4$ & $C6$	5487.5	0.7004
$C5$ & $C6$	3398.5	**0.0332**

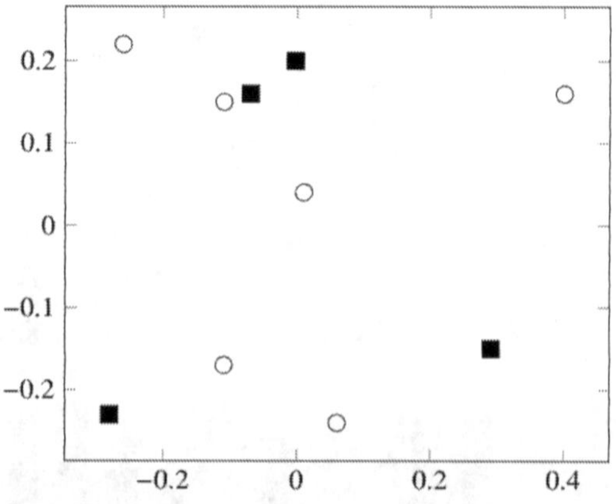

Figure 7. Position of rows (criteria, C_l) and columns frequency coordinates (professions, J_l)

Table 2. Fragment of two-input frequency matrix (C_i - criteria, $i = 1, \ldots, 6$; J_l - profession; $l = 1, \ldots, 4$),%

	*J*1	*J*2	*J*3	*J*4
$C1$	3.6	2.7	3.6	3.6
$C2$	1.8	3.6	5.4	4.5
$C3$	6.3	0.9	5.4	3.6
$C4$	2.7	2.7	3.6	4.5
$C5$	7.2	5.4	3.6	4.5
$C6$	5.4	6.3	5.4	3.6

The extended correlation matrix for variables $C1$-$C6$ has the form given in table 3.

The close (significant) positive relationship between the variables $C1$-$C6$ (the values of the pairwise correlation coefficients are equal to or greater than 0.7) is indicated in

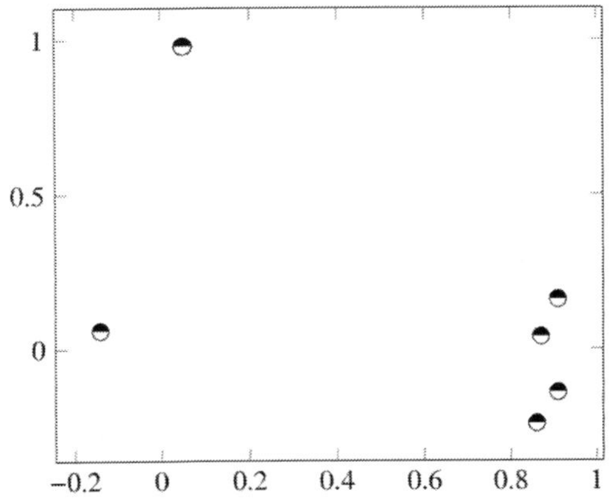

Figure 8. Representation of criteria $C1$, ..., $C6$ in the space of factors $F1$ and $F2$ after rotation) and columns frequency coordinates (professions, J_i)

Table 3. Extended correlation matrix for variables (C_1–C_6)

Criteria	*C***1**	*C***2**	*C***3**	*C*4	*C***5**	*C***6**
*C*1	1	**0.7**	0.6	0.6	-0.1	-0.2
*C*2	**0.7**	1	**0.7**	**0.8**	0.2	-0.2
*C*3	0.6	**0.7**	1	**0.8**	0.1	-0.2
*C*4	0.6	**0.8**	**0.8**	1	0.2	-0.3
*C*5	-.1	0.2	0.07	0.2	1	0.1
*C*6	-.2	-0.2	-0.2	-0.2	0.1	1

bold. The set of close relationships gives rise to the following structure (figure 9), that need to be studied further.

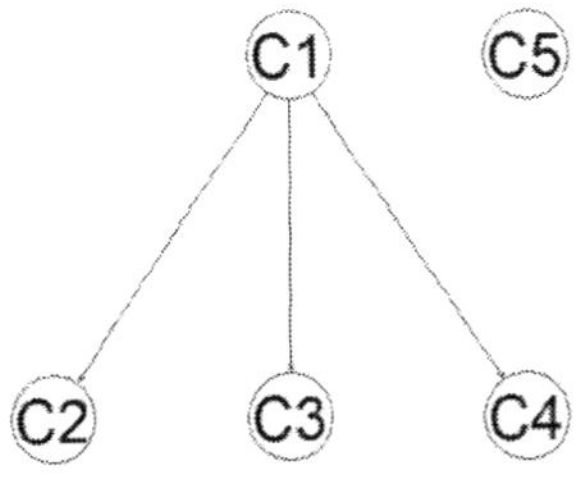

Figure 9. The structure of significant relationships between variables $C1, \ldots, C6$

Data analysis table 3 and figure 9 suggests the existence of three or four generalizing factors, which largely cover the percentage of the total variables $C1, \ldots, C6$ variance.

Using the method of factor analysis and taking into account the presence of a close correlation between the variables $C1$, $C2$, $C3$, $C4$ (see table 3), we build table 4.

Table 4 presents the factor loads (pairwise correlation coefficients between C_i and F_j), as well as the total variances of the factors and their fate. Sets of significant loads are highlighted in bold (>0.7).

Table 4. Factor loads and their total variances with corresponding fractions

Variable	*Factor (F1)*	*Factor (F2)*	*Factor (F3)*
*C*1	**0.86**	-0.24	0.08
*C*2	**0.91**	0.14	0.06
*C*3	**0.87**	0.04	0.12
*C*4	**0.91**	0.16	0.13
*C*5	0.05	**0.98**	-0.05
*C*6	-.14	-0.06	**-0.99**

Factor loads are obtained by the method of principal components with subsequent oblique rotation (Varimax normalized) [10].

Based on the comparative analysis method of public administration experience [3] and using the method of analysis of Saati hierarchies [11], we build a scheme of hierarchies for choosing the option of social welfare of servicemen discharged to reserve (figures 10, 11).

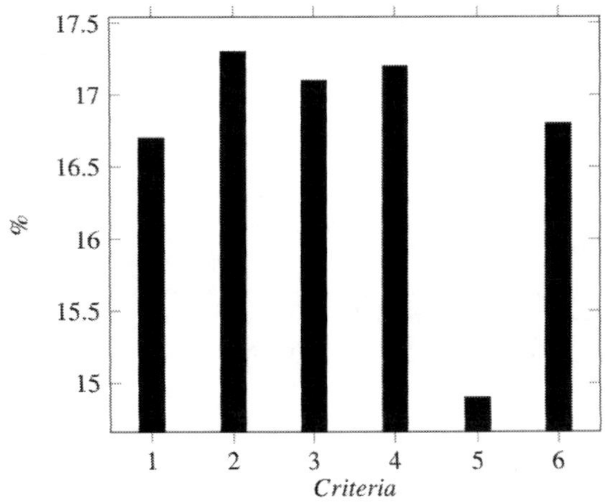

Figure 10. The structure of military personnel preferences of by criteria

The developed decision support information-analytical technology in the field of labor economics within its content provides the following components:

1. Establishment of social welfare mechanisms.
2. Development of a questionnaire regarding the use of the appropriate mechanism.
3. Statistical analysis of survey results:
    - generation of empirical distributions of answers to questions relative frequencies;
    - comparison of answers medians;
    - correspondences analysis;
    - factor analysis;
    - analysis of hierarchies.
4. Meaningful interpretation.
5. Decision making.

 https://doi.org/10.1051/shsconf/202110705001

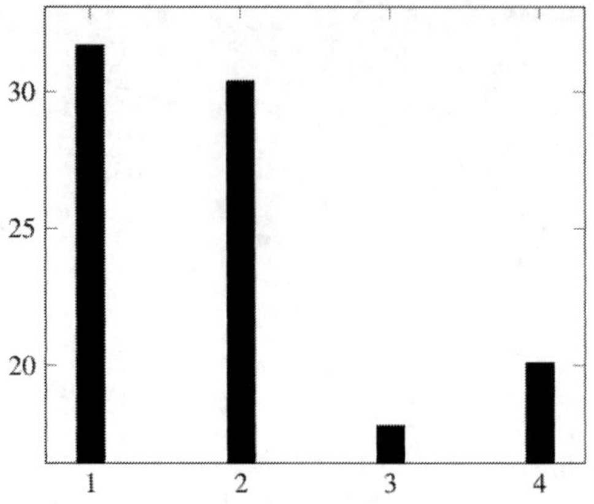

Figure 11. The structure of military personnel preferences by alternatives (1 – instructor; 2 – entrepreneur; 3 – mercenary; 4 – not professionally defined)

According to the results of data processing it is established that:

1. Respondents prefer such mechanisms as the state management mechanism for development of the military instructor social institution; and the state management mechanism of ex-servicemen retraining and their support in creation their own business and running business (figure 11). This is important information for the resources' allocation for the operation of these mechanisms.

2. Despite the fact that according to figure 10 criteria $C2$, $C3$ and $C4$ are equivalent, among them the most important in the opinion of respondents is $C2$ criteria. This criterion is the need to form a public good "military security", because it has the largest number of links with others (figure 9). This is important information that shows that the servicemen who took part in the survey are worried about the fate of the state and they are patriotic.

3. The relatively low value of $C5$ criterion that means the idea of the quality of implementation of each of the mechanisms indicates distrust of the military in the implementation of socio-economic policy in the state (figure 10). This is important information for further research in this area.

Thus, the implementation of the developed decision support information-analytical technology in the field of labor economics makes it possible to obtain more information for decision-making in comparison with the usual processing of survey results.

4 Conclusion

The research material suggests the use of decision support information-analytical technology in discharge military personnel employment developed by its authors. And, in contrast to the usual processing of survey results, it makes possible to obtain more information for decision-making. Adherence to such approach in the development of public administration mechanisms, it increases the likelihood that in the case of their implementation in the country the positive changes will be expected, as they will indirectly take into account the availability of necessary resources. Information and analytical technology to support decisions in the field of labor economics substantiates the development of the example of discharged servicemen social welfare.

According to the content, the developed decision support information-analytical technology in the field of labor economics provides for the following components: formation of social protection mechanisms; development of a questionnaire on the use of the appropriate mechanism; statistical analysis of survey results (generation of empirical distributions of answers to questions relative frequencies; comparison of answers medians; correspondences analysis; factor analysis; analysis of hierarchies); meaningful interpretation; decision making.

According to the results of data processing it is established that:

1. Respondents prefer such mechanisms as the state management mechanism for development of the military instructor social institution; and the state management mechanism for ex-servicemen retraining and their support in creation their own business and running business. This is important information for the resources' allocation for the operation of these mechanisms.

2. The servicemen who took part in the survey are worried about the fate of the state and they are patriotic.

3. There is a distrust of the military personnel in the implementation of socio-economic policy in the state. This is an important information for further research in this field.

References

[1] P. Bingley, P. Lundborg, S.V. Lyk-Jensen, Journal of Labor Economics **38**, 39 (2020)
[2] H.J. Meyer, E.O. Smigel, American Journal of Sociology **56**, 341 (1951)
[3] M.M. Medvid, Law and Honor **2**, 107 (2020)
[4] M.V. Zverev, Public administration studies **3** (2019)
[5] Y. Smith, Social Service Review **88**, 407 (2014)
[6] L. Garicano, P. Heaton (2010)
[7] T. Cleff, *Applied Statistics and Multivariate Data Analysis for Business and Economics* (Springer, 2019)
[8] NIST/SEMATECH, *e-Handbook of Statistical Methods* (2013), `https://www.itl.nist.gov/div898/handbook/`
[9] E.J. Beh, *Correspondence Analysis: Theory, Practice and New Strategies* (John Wiley & Sons Ltd, 2014)

https://doi.org/10.1051/shsconf/202110705001

[10] A.C. Weide, A. Beauducel, Frontiers in Psychology **10**, 645 (2019)

[11] T.L. Saati, *Fundamentals of Decision Making and Priority Theory with the Analytic Hierarchy Process* (RWS, 1994)

Marketing forecasting based on Big Data information

Sergey Ivanov[1,*] and *Mykola* Ivanov[1,**]

[1]Zaporizhzhia National University, 66 Zhukovsky Str., Zaporizhzhia, 69063, Ukraine

Abstract. In the paper discusses the use of big data as a tool to increase data transfer speed while providing access to multidimensional data in the process of forecasting product sales in the market. In this paper discusses modern big data tools that use the MapReduce model. The big data presented in this article is a single, centralized source of information across your entire domain. In the paper also proposes the structure of a marketing analytics system that includes many databases in which transactions are processed in real time. For marketing forecasting of multidimensional data in Matlab, a neural network is considered and built. For training and building a network, it is proposed to construct a matrix of input data for presentation in a neural network and a matrix of target data that determine the output statistical information. Input and output data in the neural network is presented in the form of a 5x10 matrix, which represents static information about 10 products for five days of the week. The application of the Levenberg-Marquardt algorithm for training a neural network is considered. The results of the neural network training process in Matlab are also presented. The obtained forecasting results are given, which allows us to conclude about the advantages of a neural network in multivariate forecasting in real time.

1 Introduction

As Laurent Flores explained [1], the success of digital marketing is determined by the success of digital marketing, thus how they are measured and used. However, attention is paid to marketing forecasting in the digital economy, taking into account intelligent systems. Intelligent systems and the use of multidimensional communication determined the emergence of a new concept by Klaus Schwab [2]. According to this concept, it is argued that we are in the era of the fourth industrial revolution (Industry 4.0), when the virtual world is combined with the physical world using information technology. The fourth industrial revolution is characterized by a change in economic relations and the widespread use of intelligent technologies (big data [3], artificial neural networks [4], and others).

It should be noted that with the use of digital marketing, D2C models have come to be used. The D2C (Direct to Consumer) model represents a direct selling system, where companies themselves manufacture, promote, sell and deliver their product without the involvement of intermediaries. So, the article provides an analysis that shows that, unlike traditional promotion through retail chains, companies using the D2C model develop faster with their own distribution channels [5]. Thanks to their good positioning, these companies not only have a competitive advantage in the market, but also have their own structure on the Internet. These companies have changed the producer-consumer relationship and are reducing the distance between them. Today, any customer can contact the manufacturer directly, ask their question and make a purchase, avoiding extra charges and saving time. Renowned manufacturers have recognized the need to develop their own D2C strategies based on marketing analytics. The authors of the article acknowledge that the use of D2C opens up additional opportunities for companies. According to the authors, Nike is a prime example, with D2C sales accounting for a third of total revenues by the end of 2020 based on its Consumer Direct strategy.

Given the widespread use of digital marketing, Rimma Katz in article [6] explored social commerce, which fostered development along the D2C model. Today, social commerce is used to increase the reach of consumers, those who may know about direct contact with the manufacturer.

So in the presented report by Jasmine Enberg [7], a summary of the main events, their analysis for marketing management, solving problems strategic development of companies. Report author Jasmine Enberg determines that the global forecast for monthly social media users in 2020 has increased due to the effects of the pandemic. However, no platform will be able to maintain the growth it picked up at the beginning of the year. Therefore, in 2021, the growth rate will begin to normalize. Recent product launches, including Facebook live shopping and Instagram shopping tags, show that e-commerce continues to be a priority for the two platforms. Snapchat and Twitter were more focused on effective marketing, namely the release of new sets of promotional offers with direct consumer response and others.

*e-mail: flydaiver@gmail.com
**e-mail: nn_iva@ukr.net

Therefore, the modern market is based on direct selling (D2C) models. allow you to get images of buyers and segment them. In addition to these tasks, it is necessary to solve forecasting the market, which changes every year.

It should be noted that worldwide targeted statistics for the entire sales system. This information is stored in cloud storage (Big Data). The information used includes data from the time of attracting a new consumer to the required resource information about the number, including repeated ones.

In this article, we propose a marketing analytics method based on artificial neural networks.

Marketing forecasting is part of any firm's overall marketing analytics. An important role in the forecasting method is played by the multidimensionality of information and methods of their processing. The use of Big Data with OLAP technologies requires new approaches to processing and applying large amounts of data. This is due to the wide range of communication systems used in the e-commerce market. Therefore, marketing forecasting based on Big Data is an urgent task, which is discussed in this article.

The article is devoted to marketing forecasting based on the use of neural networks. The information base of marketing forecasting is Big Data. The process of constructing training matrices, training a neural network and making the predictions are presented in the result of paper.

2 Literature review

Modern companies use many available forecasting methods, they not only improve the quality of their products, but also get information about the needs of customers. Marketing forecasting models are a great way to predict customer preferences and apply new ways to stand out from the competition. Using practical forecasting models today is the best way to get the most effective and complete data to improve marketing decisions. In this case, forecasting methods in digital marketing should include not only customer surveys, their age, interests and price, but also the characteristics of the product, brand, logistics and others.

In the work of Yonathan Bard [8], the problem of fitting mathematical models to numerical data was considered. Such a fit is often performed by the least squares method, regardless of previous knowledge of parameter values or the statistical nature of measurement errors. According to the results of a study of the opinions of experts, Alison Hubbard Ashton [9] considered a scenario in which the results of the opinions of experts differ significantly from the polls of intentions. For this case, the author considered the problem of predicting market behavior.

Further development of the principle of intentions was found in the work of Vicki G. Morwitz [10]. The author proposed the principles of using intentions when solving the forecasting problem. These tasks included researching people and how they would behave in different situations. For this, the method of polling the intentions of people was used. Intent surveys are widely used in marketing when sales data is unknown, for example, to forecast new products.

A continuation of these works found themselves in the work of J. Scott Armstrong [11], where the role of a person as a dominant factor was considered. This task was solved as a role-playing game for making predictions of the behavior of people who interact with others. A key tenet of this approach is to provide realistic simulation of interactions.

This forecasting method is currently rarely used. Gene Rowe and George Wright [12] considered the application of the Delphi method as a procedure. The authors found that the accuracy of expert predictions can be improved through the use of Delphi structured methods. One of the principles of the method is that experts' forecasts should not depend on each other. Expert groups sometimes violate this principle; as a result, the data should not be used in forecasting.

The task "Intentions" was considered in their work by Dick R. Wittink and Trond Bergestuen [13]. This paper examines the intention as an indicator of a consumer to purchase a product under the influence of various factors. The consumer can declare his intentions to make a purchase of various goods. This method is based on the following principles, namely, using a new design to create an acceptable situation.

The formation of a digital marketing strategy was considered in the work of Pinaki Mandal and Nitin Joshi [14]. The authors of the article emphasize that digital technologies make marketing more effective, since they allow to identify individual consumer interests, better manage campaigns and improve the product. In this article, the authors propose a flowchart for developing marketing strategies.

The analysis of the state of the digital economy and digital marketing is discussed in the article by Mykola Ivanov [15]. The author in the article shows that the dynamics of processes in the economy is quite high and requires a quick analysis of multidimensional data. The author proposes a conceptual model and a method for assessing consumer demand in the target market, aimed at the prospective management of trading floors using Big Data.

The analysis of the overview of macroeconomic forecasting are discussed in the [16–19]. The authors take the focus on a wide range of theories as well as empirical methods: business cycle analysis, time series methods, macroeconomic models, medium and long-run projections, fiscal and financial forecasts, and sectoral forecasting.

The research of the technological sources of the next long wave of growth is made in [20, 21]. The authors study the impacts of national innovation systems (NIS) and economic complexity index (ECI) on economic growth.

The use of Exponential smoothing (ES) forecasting methods is discussed in [22, 23]. The authors research revolution of exponential smoothing, which has been improved with the introduction of a complete modeling framework incorporating innovations state space models, likelihood calculation, prediction intervals and procedures for model selection.

3 Marketing forecasting modeling

The modern economy is characterized by rapid dynamics of economic processes. Under these conditions, marketing forecasting models acquire new meanings in managerial decision-making. The process of the importance of making management decisions in digital marketing systems is shown in figure 1.

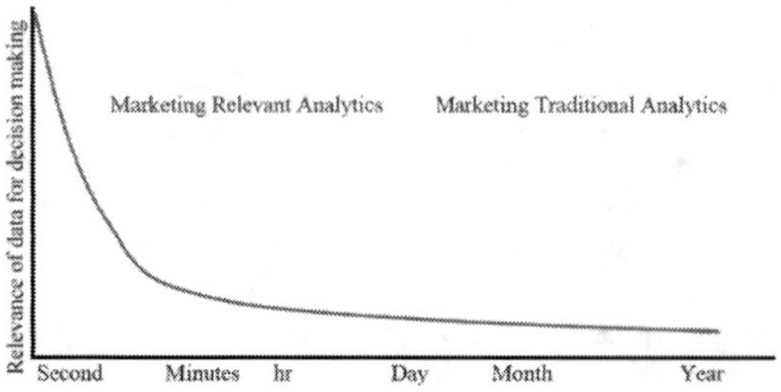

Figure 1. The importance of management decision making in digital marketing

Traditional Marketing Analytics (MTA) is based on the use of classical approaches and forecasting methods. MRA is aimed at solving economic important forecasting problems from the moment information appears to several hours.

These tasks are solved using neural networks and marketing robots.

Traditional Marketing Analytics is based on the use of classical approaches and forecasting methods. Marketing Actual Analytics (MRA) is aimed at solving economic important forecasting problems from the moment information appears to several hours. These tasks are solved using neural networks and marketing robots. Today the number of information sources of data in the world is growing rapidly. Therefore, storage technologies and their processing of information are becoming more and more in demand. By storing information, one can single out the use of Big Data for which the basic principles of work can be formulated:

1. Horizontal scalability, which takes into account that the data can be arbitrarily large from any system. They have the ability to handle big data.

2. Tolerance to failures, which use the principles of horizontal scalability and apply methods of clustering systems; locality of data, which allows in large distributed systems to separate data from a large number of data centers. All modern tools for working with big data, one way or another, follow these three principles. The first principle is based on the MapReduce model. The MapReduce model provides for distributed data processing proposed by Google and is shown in figure 2.

MapReduce provides that data is organized as relational or multidimensional data (OLAP). The data processing method takes place in three stages. The first stage is

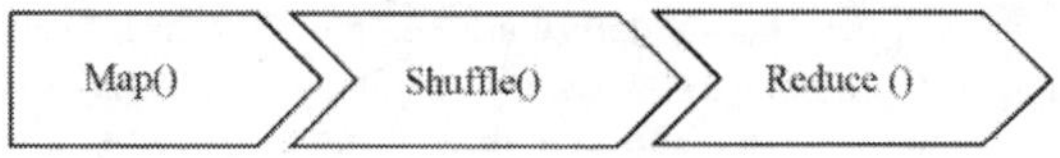

Figure 2. Data processing according to the MapReduce model

aimed at executing the Map() function. At this stage, the data is preprocessed using the Map() function, which is defined by the user. The work of this stage is to preprocess and filter the data. The second stage of the model is performed by the Shuffle() function. This stage goes unnoticed by the user. At this stage, the Map() function performs the data immersion procedure similarly to the formation of data marts (Data Mart), that is, one Map() data output corresponds to each mart. In the future, these showcases will serve as an input for the Reduce() function. The third stage of the model is aimed at executing the Reduce() function. Each data mart, which is formed in the second stage, transfers information to the input of the Reduce() function. The Reduce() function is user defined and calculates the result for individual storefronts. The set of all values returned by Reduce() is the result in this method. Therefore, Big Data technology is consider as a tool that allows you to increase the speed of data transfer while providing a large capacity of information carriers. In addition, this technology can improve the availability of cloud applications and data services. Thus, digital marketing is shape around the mainstream e-commerce models. The interconnection of the main models (B2B, B2A, D2C, C2A and C2C) of e-commerce systems based on the systems for collecting, storing and analyzing information in real time. Which based on subsequent storage in historical data layers. For the implementation of systems that perform Marketing Relevant Analytical tasks using data, OLAP data systems are used, which are structured according to the principle of multidimensional information presentation [15]. Reducing the cost of creating multidimensional warehouses can be achieve by using Data Mart. A data mart can only contain thematically aggregated data. Big Data is today a single, centralized source of information for the entire subject area. The structure of the marketing analytical system can be represent as follows (figure 3).

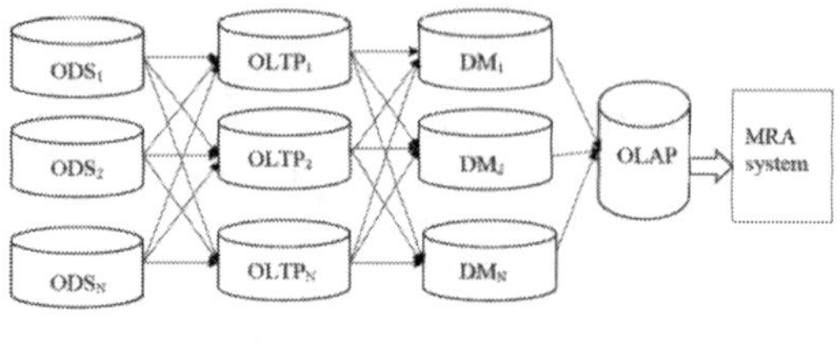

Figure 3. The structure of the marketing analytical system

In a marketing analytical system, there are many databases. where transaction processing is done in real time. Therefore, online data source systems (ODS) provide information for processing in OLTP. OLTP systems provide storage and processing of information in real time. The processed data in OLTP is transferred to the Data Mart systems with the subsequent construction of multidimensional OLAP data cubes.

This multidimensional data is aimed at presenting information on thematic sections both on marketing information and other information from different areas of the economy. The marketer has the ability to access multidimensional data in the repository, as well as complete economic information for conducting an MRA. The advantages of this approach are:

- simplicity of creating and filling OLAP, since filling comes from reliable sources of data marts;
- reducing the load on working with multidimensional data, namely, one multidimensional query processes multiple OLAP layers.

The data coming from the OID is transferred to the OLTP and the data marts are moved. OLAP stores data as multidimensional layers of measures and dimensions [15]. For marketing forecasting, a multidimensional query is formed to multidimensional data, which allows obtaining the following information input stream (Inflow) – formed by data from OLTP and DMN subsystems:

$$I = \{p_j = (g_j, in_j, mi_j)\}, \ j = 1, .., N, \tag{1}$$

where g_j is the product included in the analyzed sets of N – the object of research,

in_j – indicator of income j of the product,

mi_j – product marketing indicators j.

Datasets from set I stored in OLAP and on demand allow you to obtain and conduct marketing analysis with the subsequent storage of data, which are called transactions. Description of a transaction to set I as follows:

$$T = \{in_j | in_N \in I\}. \tag{2}$$

Such transactions for retail outlets on the Internet correspond to the nomenclature of goods that the consumer buys and the data stored in OLAP as multidimensional data cubes (OLAP).

Then solving the problem of marketing forecasting based on a neural network, data arrays of its training are formed. The forecasting technique using a neural network is formalized through the problem of pattern recognition. Data on the predicted economic indicators of a product for a certain period of time form an image, the class of which is determined by the values of the predicted indicators.

In the proposed methodology, the dimension of the multidimensional array will determine both the forecasting interval and the number of predicted indicators. Each next line of the array is formed as a result of a shift by one interval equal to the prediction interval.

The neural network is trained on the generated training array of product indicators and adjusts its weights accordingly. As a result, the neural network is trained to solve the forecasting problem for a certain forecasting horizon. It should be noted that two forecasting approaches are used: one-step and multi-step. One-step forecasting is used for short-term forecasts and multistep forecasting is used for long-term forecasting.

Let the time interval $[t_0, t_k]$ be given, the indicators g_j, in_j, pr_j of the product are defined, where t_0 is the initial time value, t_k is the current time value. To find the predicted values on the prediction interval Δ, a method is proposed that includes the following stages:

1. Analytical analysis of marketing indicators and the formation of a learning matrix from selected values from historical slices of multidimensional databases (OLAP technology).

A learning matrix (ML - matrix learning) can be written as input data for representation in a neural network (equation (3)).

The input data in the neural network in Matlab is presented in the form of a 5x10 matrix, which represents static information on 10 products for five days of the week and has the following form (figure 4).

Target data that determine the output statistical information for the neural output can be represented in the form of learning value matrix (TVM_{Output}, equation (4)).

The target data, which determines the output statistical information in Matlab for neural output, is presented in figure 5.

2. Neural network training. The process of training a neural network is to match to each ML_Input element the value of the TVM_Output matrix corresponding to the mapping in the value of the elements of the weight matrix w_j:

$$w_j : ML_{Input} \rightarrow TVM_{Output}. \tag{5}$$

In the process of training the neural network, the task of minimizing the objective function is solved. With this approach, an algorithm is used for training, which is the most efficient not only in terms of errors, but also in time. The neural network in Matlab is trained using the Levenberg-Marquardt error backpropagation algorithm. The Levenberg-Marquardt algorithm uses a scalable conjugate gradient backpropagation. Therefore, the training of the neural network is represented in time, and the network is tuned in accordance with its error. The magnitude parameter is used to measure the generalization of the neural network and stop learning when the generalization stops improving. The test score itself does not affect training and provides an independent assessment of the performance of the neural network during and after training. The choice of the algorithm, as well as the learning process of the neural network, is shown in figure 6 and 7.

The number of neural network training epochs can be written as follows:

$$epochs = \left| \frac{t_k - t_0}{\Delta} \right|. \tag{6}$$

In the process of forecasting by a neural network, it is necessary to take into account the forecasting horizon. In the Matlab system, the sim (net,[;;]) function is implemented, which allows you to supply a variety of input values and get a solution at the output of a neural network. The forecast results for the sale of 10 goods are shown in figure 8.

$$ML_{Input} = \begin{bmatrix} in_{01} = f_{01}(t_0) & in_{02} = f_{02}(t_0 + \Delta) & \dots & in_{0m} = f_0 m(t_0 + (k-1)\Delta) \\ in_{11} = f_{11}(t_0) & in_{12} = f_{12}(t_0 + \Delta) & \dots & in_{1m} = f_1 m(t_0 + (k-1)\Delta) \\ \dots & \dots & \dots & \dots \\ in_{N1} = f_{N1}(t_0) & in_{N2} = f_{N2}(t_0 + \Delta) & \dots & in_{Nm} = f_N m(t_0 + (k-1)\Delta) \end{bmatrix}, m = 1, ..., k \quad (3)$$

where Δ - is the horizon (time interval) of forecasting.

$$TVM_{Output} = \begin{bmatrix} in_{01} = f_{01}(t_0 + \Delta) & in_{02} = f_{02}(t_0 + 2\Delta) & \dots & in_{0m} = f_{0m}(t_0 + k\Delta) \\ in_{11} = f_{11}(t_0 + \Delta) & in_{12} = f_{12}(t_0 + 2\Delta) & \dots & in_{1m} = f_{1m}(t_0 + k\Delta) \\ \dots & \dots & \dots & \dots \\ in_{N1} = f_{N1}(t_0 + \Delta) & in_{N2} = f_{N2}(t_0 + 2\Delta) & \dots & in_{Nm} = f_{Nm}(t_0 + k\Delta) \end{bmatrix}, m = 1, ..., k \quad (4)$$

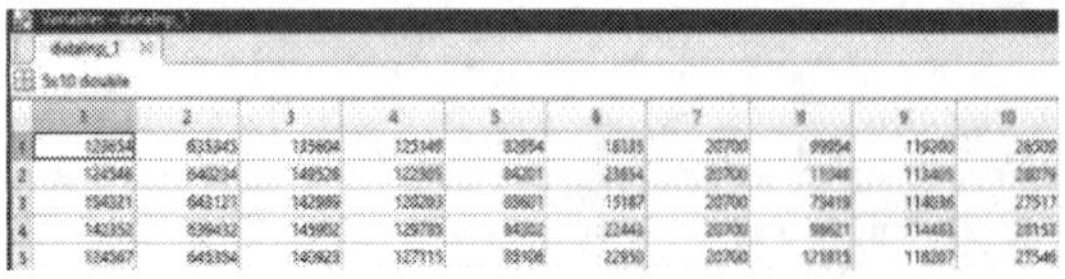

Figure 4. DataInp matrix view

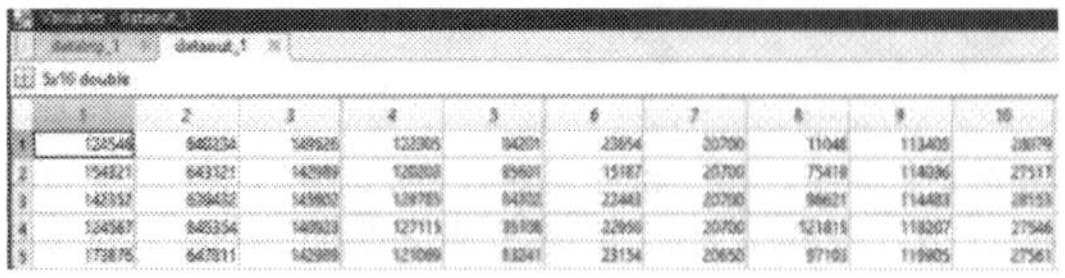

Figure 5. DataOut matrix view

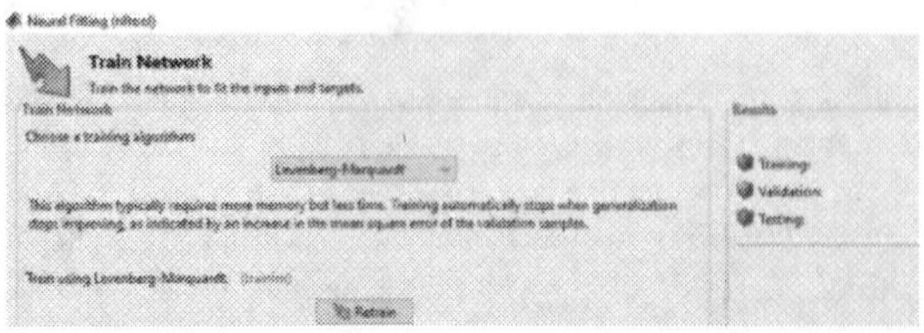

Figure 6. The choice of the learning algorithm

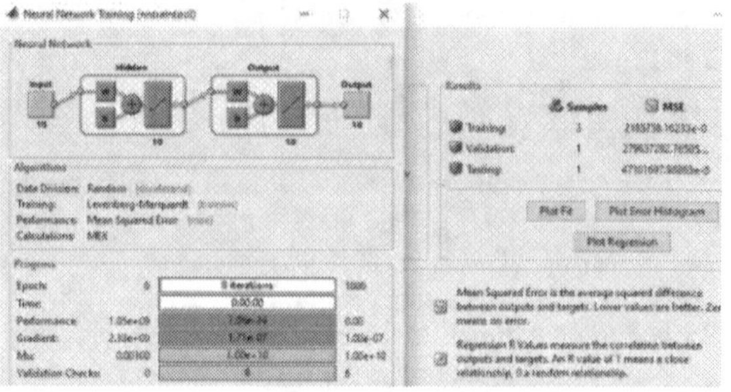

Figure 7. The process of learning the neural network

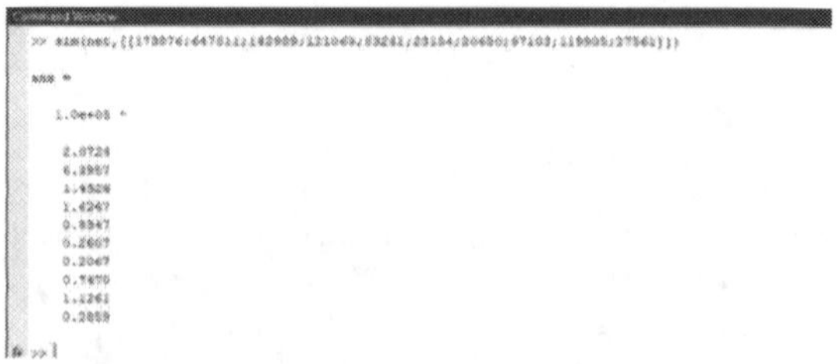

Figure 8. The result of forecasting the sale of 10 products

Therefore, the created neural network does indeed make multiple predictive decisions. It allows you to solve the sales function of marketing and consider the dynamics of the sale of many products in real time. Marketing forecasting looks at the number of future periods that the forecast will cover. That is, you may need a forecast 7 days ahead, with data for every day. In this case, the period is a day, and the horizon is 7 days. Finally, the prediction interval is the frequency with which a new prediction is made. Often the prediction interval coincides with the prediction period. The choice of the forecasting period and horizon is usually dictated by the conditions for making marketing decisions. Choosing these two parameters is one of the hardest parts of marketing forecasting. For forecasting to be meaningful, the forecasting horizon must be no less than the time required to implement the decision made on the basis of the forecast.

Thus, forecasting is highly dependent on the nature of the decision being made.

4 Conclusion

It is known that making management decisions in digital marketing, taking into account the time of receipt and a large amount of information are an urgent task today.

In addition, in the article, the authors consider the use of Big Data as a tool to increase the data transfer speed while providing access to multidimensional data (OLAP).

The article proposes the structure of the marketing analytics system. It includes many databases, where transactions are processed in real time. Consequently, online data source systems (ODS) provide information for processing in OLTP. This ensures prompt processing of information in real time. For marketing forecasting of multidimensional data, a neural network in Matlab is built. To solve the problem of improving forecasts, the authors have proposed building input data matrices for presentation in a neural network and target data matrices that determine the output statistical information.

Also in the paper, the results of the neural network training process in Matlab are represented on a chosen learning algorithm. The obtained forecasting results allow us to draw a conclusion about the advantages of a neural network in multivariate forecasting in real time. Multidimensional data and the level of their detail are important to solving the problem of forecasting in real time.

The wider use of digital-marketing systems gives an opportunity for applying the proposed approach to marketing forecasting.

References

[1] L. Flors, *How to Measure Digital Marketing* (Springer, London, 2014)

[2] K. Schwab, *The Fourth Industrial Revolution* (World Economic Forum, Cologny/Geneva, 2017)

[3] M. Ivanov, S. Ivanov, O. Cherep, N. Terentieva, V. Maltiz, I. Kaliuzhna, V. Lyalyuk, CEUR Workshop Proceedings **2713**, 295 (2020)

[4] S. Semerikov, H. Kucherova, V. Los, D. Ocheretin, CEUR Workshop Proceedings **2845**, 22 (2021)

[5] Insider Intelligence Editors, *Nike's D2C sales will comprise a third of its total revenues* (2021), https://tinyurl.com/3dwxndf9

[6] R. Kats, *Mucinex eyes social commerce to help bolster D2C business* (2020), https://tinyurl.com/zzsm5zea

[7] J. Enberg, *US Digital Ad Spending 2019* (2019), https://www.emarketer.com/content/us-digital-ad-spending-2019

[8] Y. Bard, *Nonlinear Parameter Estimation*, 4th edn. (Academic Press, New York, NY, 1973)

[9] A.H. Ashton, The Accounting Review **60**, 173 (1985)

[10] V.G. Morwitz, *Methods for Forecasting from Intentions Data* (Springer US, Boston, MA, 2001), pp. 33–56, ISBN 978-0-306-47630-3, https://doi.org/10.1007/978-0-306-47630-3_3

[11] J.S. Armstrong, *Role Playing: A Method to Forecast Decisions* (Springer US, Boston, MA, 2001), pp. 15–30, ISBN 978-0-306-47630-3, https://doi.org/10.1007/978-0-306-47630-3_2

[12] G. Rowe, G. Wright, in *Principles of Forecasting: A Handbook for Researchers and Practitioners*, edited by J.S. Armstrong (Springer US, Boston, MA, 2001), pp. 125–144, ISBN 978-0-306-47630-3, https://doi.org/10.1007/978-0-306-47630-3_7

[13] D.R. Wittink, T. Bergestuen, *Forecasting with Conjoint Analysis* (Springer US, Boston, MA, 2001), pp. 147–167, ISBN 978-0-306-47630-3, https://doi.org/10.1007/978-0-306-47630-3_8

[14] P. Mandal, N. Joshi, International Journal of Scientific Research and Management **5**, 5428 (2017)

[15] M. Ivanov, CEUR Workshop Proceedings **2422**, 395 (2019)

[16] N. Carnot, V. Koen, B. Tissot, *Economic Forecasting* (Palgrave Macmillan UK, London, 2005)

[17] N. Carnot, V. Koen, B. Tissot, *Economic Forecasting and Policy* (Palgrave Macmillan UK, London, 2011)

[18] S. Ivanov, CEUR Workshop Proceedings **2422**, 385 (2019)

[19] C.P. Lim, ed., *Handbook on Decision Making*, Vol. 1: Techniques and Applications (Springer-Verlag, Berlin Heidelberg, 2010)

[20] D. Vázquez, Journal of Evolutionary Economics **30**, 949 (2020)

[21] K. Lee, J. Lee, Journal of Evolutionary Economics **30**, 897 (2020)

[22] C. Chatfield, A.B. Koehler, J.K. Ord, R.D. Snyder, Journal of the Royal Statistical Society: Series D (The Statistician) **50**, 147 (2001)

[23] R. Hyndman, A. Koehler, K. Ord, R. Snyder, *Forecasting with Exponential Smoothing: the State Space Approach* (Springer, New York, 2008)

Fuzzy arithmetic in economic research

Kateryna Gorbatiuk[1]*, *Tetiana* Zavhorodnia[1], *Oksana* Proskurovych[1], and *Olha* Mantalyuk[1]

[1]Khmelnytskyi National University, 11 Instytutska Str., Khmelnytskyi, 29016, Ukraine

Abstract. Many tasks in economic research are based on arithmetic calculations of indicators that reflect the state of economic development. The general incompleteness of publicly available data, designed to solve such problems, has led to the emergence of numerous decision support systems based on fuzzy arithmetic. The article presents a study on the approach aimed at integrating fuzzy information about economic indicators into economic models. The definition of arithmetic operations on fuzzy values is given, and the methods of obtaining the resulting fuzzy indicators with the help of some software tools are considered. Analytical and numerical methods of obtaining the resulting indicators in the form of fuzzy numbers are described and analyzed. A direct calculation algorithm for all arithmetic operations is proposed, utilized, and used for the evaluation of resulting indicators. Also, analytical and numerical methods for obtaining fuzzy results are considered in the article, and some of them are proposed for utilization. On the example of economic indicators used in the labor rationing, the results of an evaluation of some indicators for a technological operation in the form of fuzzy numbers were obtained by different methods and compared. The practical recommendations, given in the article, on the use of fuzzy arithmetic in decision support systems outline the directions of further research.

1 Introduction

To make an effective management decision under the conditions of uncertainty and risk, it is often required to take into account incomplete information about the indicators that affect the efficiency of processes in various sectors of the economy. So, the tasks of assessing economic indicators are often performed taking into account the uncertainty. One of the ways to describe the incompleteness of information in indicators of different nature is to formalize the uncertainty using the fuzzy set theory approach.

The presentation of economic indicators in the form of fuzzy numbers opens up ample opportunities for taking into account uncertainty and using all available information in calculations without significant losses due to assumptions about the determinism and clarity of specific values. Working with economic indicators presented as fuzzy values requires the use of appropriate calculation methods, in particular, using the software. This will automate the fuzzy assessment of indicators obtained by establishing the relationship between factors described as fuzzy numbers.

The fuzzy set theory, proposed in 1965 by L. A. Zadeh [1, 2] is becoming more and more widespread in the practice of modeling uncertainties in indicators of various natures. In particular, the methods and basic concepts of the fuzzy set theory have found their application in economic research tasks.

Thus, if we move away from the idea of considering the economic indicators as random values with unknown probabilistic characteristics and move on to the concepts of the fuzzy set theory, we can obtain full advantage of the fuzzy set theory technique over the technique of probability theory, which does not provide a suitable way to answer many practical questions.

At the same time, the similarity in the concepts of fuzziness and randomness allows us to draw correspondences between the membership functions in the fuzzy set theory and likelihood characteristics in probability theory.

In many works devoted to the mathematical modeling of various economic processes, it is noted that the practice of modeling economic indicators with only average characteristics is not always acceptable. Most indicators that take into account only the average values should also be described with a margin of error estimate. According to the authors of this article, most of the economic indicators should be described using fuzzy values (fuzzy numbers). Moreover, the construction of their membership functions can be based both on the use of expert information and on a statistical analysis of the distributions of the studied indicators.

The main purposes of this study are: considering the existing fuzzy arithmetic methods; applying fuzzy arithmetic methods to the problems of fuzzy estimation of economic indicators on the example of calculating indicators in the field of labor rationing; comparing the results of estimation obtained by different methods, in particular, implemented in the programming language R; providing recommendations for further use of fuzzy arithmetic for assessing economic indicators taking into account an uncertainty of fuzzy nature.

There are six sections in this article. The first one contains the description of the background of this research.

*e-mail: ecvgor@ukr.net

The second section presents an overview of the scientific sources on the topic of research. The theoretical details of the proposed methods are explained in the third section. Then, the stages of the study and their main results are shown in the fourth and fifth parts of this work. The last part of the article contains several conclusions based on the research results and covers a discussion on further research directions.

2 Literature review

Since 1965 [1], the methods and basic concepts of fuzzy set theory have become widespread in the practice of solving economic problems. The introduction of the fuzzy sets' concept, classes with inaccurate boundaries described by membership functions, provided a basis for the development of a more flexible approach to the analysis of judgments and modeling of complex systems, whose behavior is described by linguistic rather than numerical variables.

Working with fuzzy indicators requires the use of appropriate theoretically sound calculation methods, which is impossible without the use of modern computing tools. The basic concepts of fuzzy arithmetic are represented in different scientific works all over the world. In particular, the articles [3–9] contain the description and theoretical reasoning of the most widely used methods of fuzzy calculations. Some of the studies [10–13] are provided to improve the basic ideas by the new computational approaches. These works are only a small part of the total number of works in this area.

Recently, many software products have been created for fuzzy calculations, which allow taking into account the need to perform arithmetic operations on fuzzy values. One such software is the R programming language, which contains the corresponding function libraries, which are constantly supplemented by implementations of new methods. Work with these methods is described in detail in the relevant literature [14–16], which simplifies their practical application for fuzzy calculations.

Also, some implementations of fuzzy arithmetic are presented in the studies [17–20] related to several economic tasks. In [17] is proposed a method based on fuzzy arithmetic, designed to estimate the possible range of flow rates (and levels) of water, to make a forecast, based on possible rainfalls provided by the uncertainty model. The paper [18] outlines a computational method for decreasing the imperfection of the fuzzy arithmetic models output, where was suggested an approach aiming at integrating into the arithmetic models fuzzy information still not taken into account: the imprecise knowledge about the interaction potentially existing between two variables when performing an arithmetic operation (addition, subtraction, division or product). The article [19] represents a model of the power plant as a multivariable system using fuzzy arithmetic based on the Transformation Method. Finally, in cash flow analysis [20], instead of single-valued estimation, each cost data value is proposed to be designated as a trapezoidal fuzzy set number.

3 Research methodology

Fuzzy set theory tools allow to intuitively display inaccurate or fuzzy data. Fuzzy numbers form a subclass of fuzzy sets built on subsets of real numbers. By definition, a fuzzy set A with a membership function $\mu_A : R \rightarrow [0, 1]$ is a fuzzy number if it has at least the following properties:

1) it is a normalized fuzzy set, i.e. $\mu_A(x_0) = 1$ for a certain value $x_0 \in R$;

2) the membership function is convex, i.e. it is true that $\mu_A\big(\lambda x_1 + (1 - \lambda)x_2\big) \geq \mu_A(x_1) \wedge \mu_A(x_2)$ for any $x_1, x_2 \in R$ and $\lambda \in [0, 1]$;

3) the carrier set (universe of discourse) of the fuzzy set A is limited.

The specific types of membership functions are determined based on various additional assumptions about the properties of these functions (symmetry, monotonicity, continuity of the first derivative, etc.) taking into account the specifics of the existing uncertainty and the real situation [4].

In particular, fuzzy numbers with a triangular membership function $\mu(t)$ are called triangular fuzzy numbers, and they are denoted by $\tilde{t} = \big(t_{min}/t_c/t_{max}\big)$, where t_{min}, t_{max}, t_c – the minimum, maximum value, and some measure of the central value (mean, mode, median, etc.) of a particular parameter, respectively. The membership function of this triangular fuzzy number is:

$$\mu(t) = \begin{cases} \frac{t-t_{min}}{t_c-t_{min}}, \text{ if } t_{min} \leq t \leq t_c, \\ \dfrac{t - t_{max}}{t_c - t_{max}}, \text{ if } t_c \leq t \leq t_{max}. \end{cases} \quad (1)$$

To solve the problem of inflexibility and incompleteness of using single-valued estimates of economic indicators, fuzzy numbers can be employed to deal with the uncertainty in economic analysis.

There is no doubt about the need to automate calculations on determining the results of arithmetic operations on fuzzy numbers. Also, the implementation of the fuzzy arithmetic methods requires using of computer technologies, which can allow developing the corresponding algorithms. Some of the algorithms were developed by the authors using VBA in Excel, and utilized for fuzzy arithmetic calculations of economic indicators in the field of labor rationing.

Nowadays, the design of fuzzy systems is assisted by the general-purpose fuzzy systems software (FSS) that has been actively developed during recent years [15]. Most of this FSS is released to free and open-source libraries and toolboxes with the aim of being reusable and easy to integrate with software developed for specific application purposes [15].

One of such general-purpose fuzzy systems software is an R programming language (with RStudio environment). R is a free, open-source language that is already widely used in various fields (from computer science to social sciences) [14]. At present, the R programming language is gaining importance as a non-commercial solution that is competitive with Matlab [15]. Thus, the provision of freely available software for the use of non-classical fuzzy systems within the R packages allows practitioners and researchers from a wide range of fields to use in their research the developments of the entire research community [14].

Fuzzy numbers, which play an important role in many practical applications, have a software implementation in the R package. FuzzyNumbers is an open-source package for the R programming language, which deals with all major operating systems, i.e. Windows, Linux, and macOS [16]. FuzzyNumbers is designed for easy and efficient use of fuzzy numbers and various fuzzy operations.

A fuzzy number A can be specified by giving its carrier set (universe of discourse), or left/right side functions, or lower/upper limits of α-cuts [16]. In practice, most algorithms that work with fuzzy numbers involve the use of at least the latter, i.e. α-cuts.

A fuzzy number given by side functions (LR fuzzy number) has the following form of membership function [16]:

$$\mu_A(x) = \begin{cases} 0, \text{ if } x < a_1, \\ left\left(\frac{x-a_1}{a_2-a_1}\right), \text{ if } a_1 \le x < a_2, \\ 1, \text{ if } a_2 \le x \le a_3, \\ right\left(\frac{x-a_3}{a_4-a_3}\right), \text{ if } a_3 < x \le a_4, \\ 0, \text{ if } x > a_4. \end{cases} \quad (2)$$

where $a_1, a_2, a_3, a_4 \in R$, $a_1 \le a_2 \le a_3 \le a_4$, $left$ – is a non-decreasing function (left generator A), $right$ – is a non-increasing function (right generator A).

For example, a fuzzy number A with linear side functions (trapezoidal number) can be specified in R as: "A->FuzzyNumber (1, 2, 4, 7, left = function (x) x, right = function (x) 1-x)".

To graphically represent a given fuzzy number we can use the command: "plot (A)" (figure 1).

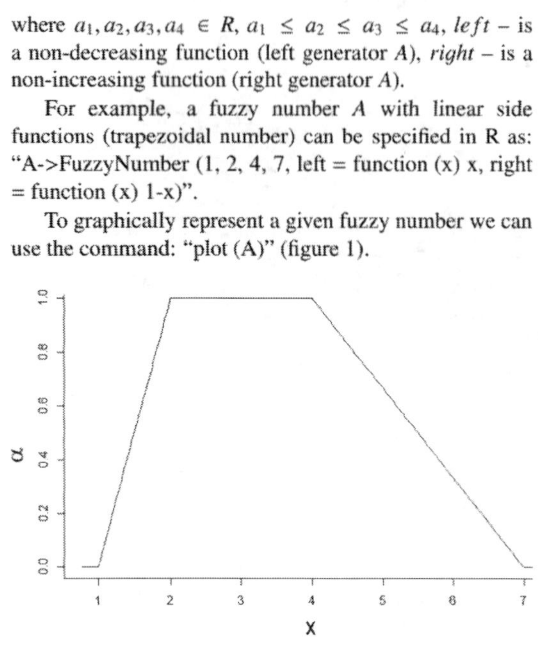

Figure 1. Membership function of a trapezoidal fuzzy number

A similar trapezoidal fuzzy number can also be specified using the function: "TrapezoidalFuzzyNumber(1,2,4,7)".

So, to specify such a number, we must provide four parameters: $a_1, a_2, a_3, a_4 \in R$.

Also, the FuzzyNumbers package contains the methods for performing arithmetic operations on different types of fuzzy numbers. But, while addition and subtraction are fairly simple operations, multiplication and division, unfortunately, remain unclosed and are more complex [8], and do not always have a solution.

4 Case study: fuzzy arithmetic methods in calculating the economic indicators

4.1 Direct and inverse methods of performing arithmetic operations on fuzzy numbers

If we are talking about the economic indicators in different areas of research, the shape and values of membership functions of fuzzy numbers can be obtained using expert information, as well as using probabilistic methods based on certain similarities between the concepts of fuzziness and probability. Without the ability to accurately determine the type of distribution, from which the samples are taken, we only can deal with the distribution frequency characteristics.

We propose the method of fuzzy representation of economic indicators based on statistical data. Thus, based on statistical data of selected economic indicators, appropriate distribution histograms or frequency polygons can be constructed with the number of intervals m in the range from 7 to 12 according to the recommendations for histogram building [21]. Each value of the membership function μ of the corresponding indicator is calculated for the middle of each interval as a relative frequency divided by the maximum value of all frequencies.

Then, it is possible to obtain a fuzzy number $A = \langle \mu_A; [a; b] \rangle$ with a discretized membership function μ_A given on the real interval $[a; b]$, which is divided into m equal parts: $[a; b] = [b_0; b_1] \cup [b_1; b_2] \cup \ldots [b_{m-1}; b_m]$, where $b_0 = a, b_m = b$. Each value $\mu_A(b_i)$ of the membership function corresponds to the value $b_i, i = \overline{0, m}$. Therefore, μ_A is a point membership function specified on the real interval $[a; b]$.

In probability theory, there are no general definitions or methods for determining the parameters of the distribution of an indicator, which is the result of performing arithmetic operations on random variables. And if we need to form our opinion on the shape of the indicator's resulting distribution, we need to build modeling (simulating) algorithms.

On the contrary, in fuzzy set theory, there are several methods designed to obtain the results of arithmetic operations on fuzzy numbers. Among them, there are analytical and numerical methods, as well as direct and inverse methods. Numerical methods are more universal and convenient than the analytical methods for use in practical problems of automated calculations.

The fuzzy extension principle, introduced by Zadeh in [1, 2], is the basic tool for fuzzy arithmetic. It extends functions of real numbers to functions of fuzzy numbers and it allows the extension of arithmetic operations to fuzzy arguments [4]. According to the extension principle, the following definition of a membership function for a fuzzy number, which is the result of an arithmetic operation on two fuzzy numbers A and B, is:

$$\mu_{A\circ B}(z) = \sup_U\{\mu_A(x) \wedge \mu_B(y)\} \quad (3)$$

where $U = \{(x,y) \in \sigma(A \times B) | x \circ y = z\}$, $\{\circ\}$ – is an arithmetic operation from the set $\{+,-,*,/\}$, $\sigma(A \times B)$ - is the Cartesian product of the partitions of the fuzzy set carriers A and B.

If we use this definition directly, and try to calculate the resulting membership function values as required in formula (3), then we obtain the method for determining the result of arithmetic operations on fuzzy numbers called a direct method.

In addition to direct methods, there are several inverse methods for solving the problem, based on so-called α-cuts representation. By the α-cuts approach, it is possible to define a parametric representation of fuzzy numbers that allows a large variety of possible shapes, and is very simple to implement [4]. Both analytical and numerical methods have been developed in the theory of fuzzy sets, in particular, there are matrix methods. One of them, described in [9], allows finding a solution for any arithmetic operation from the set $\{+,-,*,/\}$, provided that the membership functions of the numbers, combined by these operations, are convex. And also, the division operation requires compliance with the condition: $0 \notin [c;d]$, where $[c;d]$ - is the carrier of a fuzzy number representing the divisor.

The difference between the inverse methods and the direct ones is in the fact, that the discretization of the membership function in inverse methods is not based on the parameter X, but on the value of $\alpha \in [0;1]$, i.e. α-level sets (α-cuts) are constructed for the values of α (for example, $\alpha = 0.2, 0.4, 0.6, 0.8$ and 1). But in this case, there is an additional issue with determining of the value x_i for each level α. That is due to the fact that the membership function is built by frequencies based on the uniform partition of the parameter X, which does not guarantee the presence of exact values of α at the obtained partition points.

We assume, that if the value is given by the discretized membership function μ by the parameter X, then in the interval between the nodes of the partition, μ is approximated by a linear function $ax + b$, which parameters are calculated by the formulas:

$$a = \frac{\mu(x_1) - \mu(x_2)}{x_1 - x_2};\ b = \frac{\mu(x_2)x_1 - \mu(x_1)x_2}{x_1 - x_2} \quad (4)$$

where $(x_1, y_1), (x_2, y_2)$ – the partition nodes.

Then, the value x_i for a given α-level could be calculated by the formula: $x_i = \frac{\alpha - b}{a}$, having determined in advance on which section, relative to the nodes, the value of α is.

Thus, having the membership function discretized by α-levels, we will be able to apply the matrix method [9] for finding the results of arithmetic operations on fuzzy numbers. To do this, we construct a matrix of dimension 18×18 (there are nine values of α-levels = 0.2, 0.4, 0.6, 0.8, 1.0, 0.8, 0.6, 0.4, 0.2 for each fuzzy number). Even values of matrix indices both in rows and columns will correspond to the values of arithmetic operation results (performed on the particular values of the carriers of A and B). Odd indexes will indicate the level of the corresponding membership function, which is the result of the operation:

$$\mu_c(x_{ij}) = \min(\mu(x_{Ai}); \mu(x_{Bj})). \quad (5)$$

So, the results of the operation $x_{ij} = x_{Ai} \circ x_{Bj}$ are placed in the cells of the matrix with even indices i and j. All other cells of the matrix remain unfilled.

Studies show that the solution for the resulting function is located depending on the type of operation on one of the matrix diagonals: for addition and multiplication – on the diagonal that runs from the upper left corner to the lower right corner, and for subtraction and division – on another diagonal.

Then, we carry out the numerical experiment to find the result of the multiplication of two fuzzy numbers A and B by the method of direct application of the formula (3), and also by the method of a full scan of frequency characteristics with subsequent multiplication of corresponding data values.

The membership functions of fuzzy numbers are often based on expert information. However, there are also probabilistic methods for constructing membership functions based on the similarity of the concepts of fuzziness and probability.

Suppose, that there is statistical data on some economic indicators A and B. According to the obtained samples, the corresponding frequency histograms are constructed using the number of intervals in the range defined as recommended by Sturges' Rule for building histograms [21]. But, in our study, we have chosen the number of bins quite larger, to have more data points for calculations, and we set $m = 19$.

To obtain an unambiguous correspondence between the values of A and B to the frequencies, we replace the frequency histogram with another graphical characteristic used in probability theory and mathematical statistics – a frequency polygon. The values of the fuzzy number carriers for A and B and the corresponding frequencies N are represented in table 1.

So, we construct the membership function for each number A and B based on the available frequency polygons. To do this, we normalize the values by dividing the current frequencies by the maximum frequency value n_{max}. Thus, we obtain two fuzzy numbers: $A = \langle \mu_A; [a;b] \rangle$ and $B = \langle \mu_B; [c;d] \rangle$, where μ_A, μ_B – are the discretized membership functions given on the carriers $[a;b]$ and $[c;d]$ (real intervals) respectively.

Moreover, the real intervals $[a;b]$ and $[c;d]$ are divided into m equal parts: $[a;b] = [b_0;b_1] \cup [b_1;b_2] \cup \dots [b_{m-1};b_m]$, where $b_0 = a, b_m = b$; and $[c;d] = [d_0;d_1] \cup [d_1;d_2] \cup \dots [d_{m-1};d_m]$, where $d_0 = c, d_m = d$.

Thus, two point membership functions are set on the intervals $[a;b]$ and $[c;d]$ by the values $b_i, i = \overline{0,m}$ and $d_i, i = \overline{0,m}$, and corresponding values of the membership functions μ_A and μ_B.

Now, we need to construct the membership function of the fuzzy number AB directly using the definition (3).

Table 1. Membership functions for *A*, *B* and *AB*

A	N	μ_A	B	N	μ_B	AB	μ_{ab}
0.05	1	0.03	0.37	1	0.02	0.02	0.07
0.07	2	0.07	0.39	1	0.03	0.03	0.32
0.10	11	0.32	0.41	3	0.08	0.05	0.54
0.12	17	0.49	0.42	6	0.17	0.07	0.73
0.15	23	0.67	0.44	8	0.24	0.09	0.89
0.17	25	0.73	0.46	19	0.54	0.10	1.00
0.20	34	1.00	0.48	17	0.49	0.12	0.86
0.22	33	0.96	0.50	26	0.77	0.14	0.76
0.25	29	0.86	0.51	30	0.89	0.15	0.59
0.27	23	0.67	0.53	34	1.00	0.17	0.55
0.30	19	0.55	0.55	28	0.82	0.19	0.41
0.32	14	0.41	0.57	26	0.77	0.21	0.25
0.35	9	0.25	0.59	20	0.59	0.22	0.14
0.37	4	0.12	0.61	14	0.41	0.24	0.14
0.39	5	0.14	0.62	9	0.27	0.26	0.07
0.42	2	0.07	0.64	5	0.15	0.27	0.05
0.44	2	0.04	0.66	2	0.05	0.29	0.04
0.47	1	0.02	0.68	2	0.04	0.31	0.02
0.49	1	0.02	0.70	1	0.02	0.33	0.01

To do this, for each element of the Cartesian product of the partitions of the fuzzy set carriers *A* and *B*, we calculate the value $b_i{\cdot}d_j$ for all $i, j = \overline{0, m}$. Therefore, we obtain a matrix of dimension $(m + 1) \times (m + 1)$. For the obtained range of values, we find the minimum and maximum elements to determine the carrier for the resulting fuzzy number $[w; z]$. Then, we divide the obtained real interval by the points z_i into m equal parts ($i = \overline{0, m}$): $[w; z] = [z_0; z_1] \cup [z_1; z_2] \cup \ldots [z_{m-1}; z_m]$, where $z_0 = w$, $z_m = z$.

For each part, we determine its middle point:

$$z_k^* = \frac{z_k + z_{k+1}}{2}, \tag{6}$$

where $k = \overline{0, m-1}$.

From the set of Cartesian product elements, we select those pairs (b_i, d_j), for which $b_i \cdot d_j \in [z_k; z_{k+1}]$. For each such pair (b_i, d_j), we determine $\mu_C(b_i, d_j) = \mu_A(b_i) \wedge \mu_B(d_j)$ (minimum of μ_A and μ_B), and set the corresponding $\mu_c(z_k^*)$ to the maximum of $\mu_C(b_i, d_j)$. The process is repeated for each interval $[z_k; z_{k+1}]$, where $k = \overline{0, m-1}$.

So, using the developed algorithm it is possible to obtain a discrete membership function as the result of multiplying of two fuzzy numbers (table 1).

For comparison, we can simulate the distribution of the values *AB* directly using the data of the frequency characteristics of *A* and *B*. That is, we "unfold" the frequency distribution into a set of values according to the frequency of each value. For the obtained samples we perform the operation of pairwise multiplication of all possible sample values. As a result, we obtain a set of 62500 values, which can be used to construct a frequency polygon, and consequently determine the membership function μ_C.

Then, we compare the fuzzy numbers obtained by two different methods (figure 2).

The graphs show the absence of significant differences in the constructed membership functions. But it is quite obvious that the amount of calculations in the direct "multiplication" of the samples is always significantly larger.

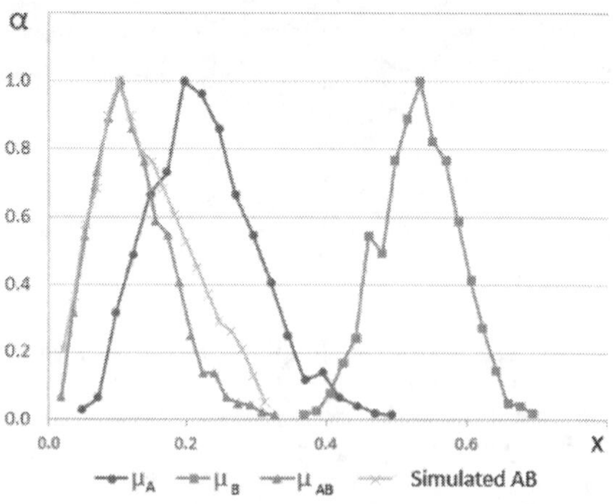

Figure 2. Membership functions of original and resulting fuzzy numbers

4.2 Using fuzzy arithmetic in labor rationing tasks

The proposed methods can be tested on examples of assessing indicators used in labor rationing. Consider the time of maintenance of workplace (T_{tech}), which is calculated by the formula:

$$T_{tech} = \frac{T_n}{T_f} \cdot T_m, \tag{7}$$

where T_f – the time of equipment failure at the workplace;
T_m – machine operation time;
T_n – the rate of time (or complexity) for the elimination of the failure, equal to the downtime of the workplace.

In this case, to describe the characteristics required for the calculations, we will use triangular fuzzy numbers.

So, let $\overline{T_n} = (10/15/20)$ (min.), $\overline{T_f} = (200/250/300)$ (min.). The value T_m is given as a crisp real number, and it is equal to 20 (min.).

Applying the division operation to the fuzzy numbers T_n and T_f, we obtain the result (after multiplying by T_m): $\overline{T_{tech}} = (0.625/1.132/1.905)$. The table 2 is filled with intermediate and resulting calculations.

Table 2. Calculation of workplace maintenance time

α	$\overline{T_n}$	$\overline{T_f}$	$\overline{T_n/T_f}$	$\overline{T_{tech}}$
0.00	10.00	200.00	0.031	0.625
0.33	11.67	216.67	0.039	0.773
0.67	13.33	233.33	0.047	0.941
1.00	15.00	250.00	0.057	1.132
0.67	16.67	266.67	0.068	1.351
0.33	18.33	283.33	0.080	1.606
0.00	20.00	300.00	0.095	1.905

A graphic representation of the obtained membership function of the workplace maintenance time is shown in figure 3.

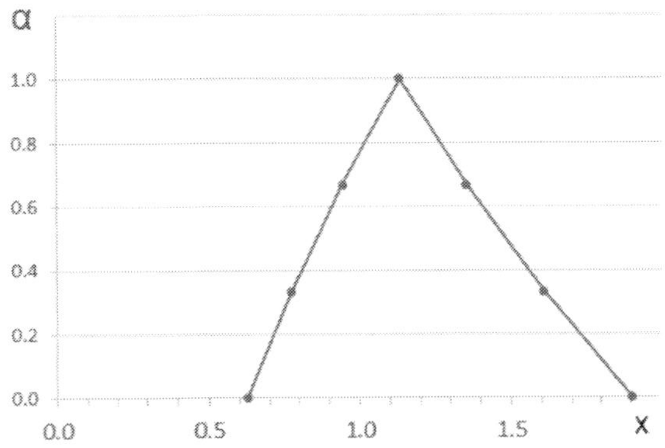

Figure 3. Membership function of T_{tech}

It is seen from the graph of the membership function, that the central value is shifted to the left, and the deviations from it are not quite symmetrical.

The proposed methods can be used when working with other economic indicators of a fuzzy nature to take into account the uncertainty associated with the fuzzy estimation of their components.

Consider the possibility of applying these methods on the example of estimating the fuzzy duration of a technological operation T, consisting of several elements, which duration is presented as fuzzy numbers.

The parameters for a fuzzy description of each operation element are calculated based on the statistical analysis of the observations on the duration of the operation elements at one workplace for one work shift (table 3). Using fuzzy numbers as the duration of each operation element, we can determine the fuzzy duration of the entire operation using some methods of performing arithmetic operations on fuzzy numbers (table 4). Observational data is considered stable because the coefficients of constancy K_{st} = max/min (used in labor rationing) do not exceed the allowable values for this type of production.

Table 3. Technological operation elements

Element description	Element name
Take workpiece, install and fasten	T_1
Switch on machine, withdraw tool	T_2
Sharpen the part in one pass	T_3
Withdraw tool, stop speed	T_4
Unfasten, remove, set aside	T_5

Both triangular and trapezoidal fuzzy numbers can be used to describe the duration of the technological operation elements. For trapezoidal numbers, we use the values of the boundaries of the entropy interval as the interval on which the membership function is equal to 1. Four parameters for each description of the number are given in table 5 and table 6.

Using the tools of programming language R, we consistently find the sums of the fuzzy numbers described above for triangular as well as trapezoidal membership functions [16]. The results of fuzzy calculations we summarize in table 5 and table 6, and then provide a graphical representation of the resulting membership function for two mentioned types: triangular (figure 4) and trapezoidal (figure 5).

It is seen from the graphs of membership functions of the resulting indicators, that when performing the addition operation, the shape of the resulting membership function coincides with the shape of the membership function of terms, and the real interval of the fuzzy number, which determines the duration of the entire technological operation, does not depend on the membership function type.

As we can see, the carriers of fuzzy numbers are the same for both forms of fuzzy numbers, and the resulting membership functions allow us to make a conclusion about the fuzzy duration of the technological operation. This will focus the production system on the stabiliza-

Table 4. Observations on the operation duration

Observation	Duration of operation element, sec.				
	T_1	T_2	T_3	T_4	T_5
1	9	3	103	4	5
2	10	5	102	3	6
3	7	4	115	5	7
4	11	3	118	4	5
5	14	4	121	6	9
6	8	5	136	4	7
7	12	7	139	6	6
8	11	6	110	5	10
9	13	4	128	4	8
10	8	5	125	6	6
K_{st}	2	2.33	1.36	2	2

Table 5. Fuzzy duration of technological operation elements (triangular fuzzy numbers)

Element name	a_1	a_2	a_3	a_4
T_1	7.0	10.47	10.47	14.0
T_2	3.0	4.74	4.74	7.0
T_3	102.0	121.61	121.61	139.0
T_4	3.0	4.78	4.78	6.0
T_5	5.0	7.05	7.05	10.0
$T_1 + T_2$	10.0	15.21	15.21	21.0
$T_3 + T_4$	105.0	126.39	126.39	145.0
$\sum_{i=1}^{4} T_i$	115.00	141.6	141.6	166.0
$\sum_{i=1}^{5} T_i$	120.00	148.65	148.65	176.0

Table 6. Fuzzy duration of technological operation elements (trapezoidal fuzzy numbers)

Element name	a_1	a_2	a_3	a_4
T_1	7.0	8.89	12.03	14.0
T_2	3.0	3.97	5.51	7.0
T_3	102.0	113.97	129.21	139.0
T_4	3.0	4.00	5.45	6.0
T_5	5.0	5.97	8.14	10.0
$T_1 + T_2$	10.0	12.86	17.54	21.0
$T_3 + T_4$	105.0	117.97	134.66	145.0
$\sum_{i=1}^{4} T_i$	115.0	130.83	152.2	166.0
$\sum_{i=1}^{5} T_i$	120.0	137.88	159.25	176.0

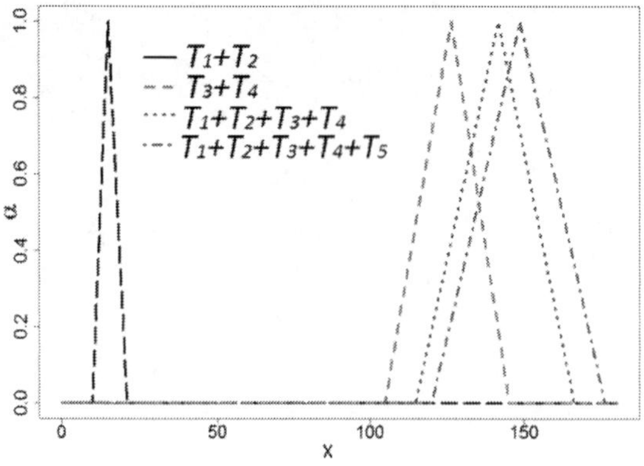

Figure 4. Membership functions of duration of technological operation and its elements (triangular fuzzy numbers)

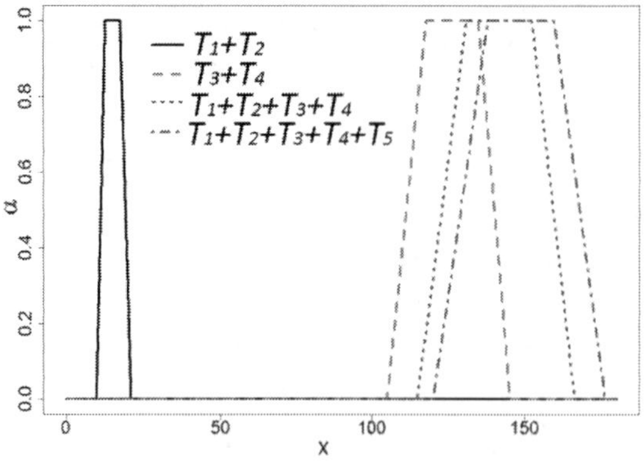

Figure 5. Membership functions of duration of technological operation and its elements (trapezoidal fuzzy numbers)

tion of labor processes by reducing the need for regulation within the fuzzy values of norms. In particular, such fuzzy estimators of the duration of different technological operations could be used in planning and forecasting the results of production processes on the enterprises. So, the decision-making systems can rely on more adequate and consistent information about labor and production processes, which allows determining the entire state of economic development on the enterprise.

5 Results and discussion

The stated approach to modeling uncertainty in economic indicators using fuzzy numbers provides considerable opportunities to obtain the most adequate estimates of these indicators using all available information on their structure and dynamics. Fuzzy arithmetic gives a powerful tool, which allows working with fuzzy economic indicators. We considered the appropriate theoretically sound calculation methods, and showed the examples of their use by modern computing tools.

In particular, the programming language R, as freely available and open-source fuzzy system software, was used for automating arithmetic calculations using a fuzzy representation of economic indicators. We have shown that this approach is quite objective in terms of a formalized description of the existing uncertainty in any economic indicators, and as an example in the labor rationing. R is an ideal choice of language for such a toolbox for fuzzy calculations, as it is freely available and used by a wide variety of researchers from plenty of research fields [14].

6 Conclusion

In this paper, we have analyzed the state of development of the fuzzy arithmetic approach in economic problems in order to provide samples of application fuzzy arithmetic methods in estimating different economic indicators related to a particular branch (for example, labor rationing), and to focus on significant further developments.

Further research should improve the methodology of calculating fuzzy values of indicators used in other economic branches, and it will allow achieving a more adequate description of uncertainties in economic indicators, and to some extent simplify the decision-making processes in economic tasks. With this approach, decision-makers will be able to have more space to define important economic indicators and obtain more valuable information from the final results.

References

[1] L.A. Zadeh, Information and Control **8**, 338 (1965)

[2] L. Zadeh, Information Sciences **8**, 199 (1975)

[3] A.M.S. Ramasamy, *Fuzzy arithmetic*, in *Proceedings of the National Seminar on Algebra and It's applications* (Annamalainagar, 1999), pp. 213–223, `https://www.researchgate.net/publication/279973774_FUZZY_ARITHMETIC`

[4] L. Stefanini, L. Sorini, M.L. Guerra, in *Handbook of Granular Computing* (John Wiley & Sons, Ltd, 2008), chap. 12, pp. 249–283, ISBN 9780470724163, `https://onlinelibrary.wiley.com/doi/abs/10.1002/9780470724163.ch12`

[5] L. Stefanini, *New Tools in Fuzzy Arithmetic with Fuzzy Numbers*, in *Information Processing and Management of Uncertainty in Knowledge-Based Systems. Applications*, edited by E. Hüllermeier, R. Kruse, F. Hoffmann (Springer Berlin Heidelberg, Berlin, Heidelberg, 2010), pp. 471–480, ISBN 978-3-642-14058-7

[6] Y. Wang, Journal of Advanced Mathematics and Applications **3**, 165 (2014)

[7] P. Dutta, H. Boruah, T. Ali, International Journal of Latest Trends in Computing **2**, 99 (2011)

[8] S. Gao, Z. Zhang, C. Cao, Journal of Software **4**, 331 (2009)

[9] D.A. Wismer (McGraw-Hill, New York, 1971), chap. 6. Distributed multilevel systems, p. 233–273

[10] N. Gerami Seresht, A.R. Fayek, International Journal of Approximate Reasoning **106**, 172 (2019)

[11] J.P. Mukeba, Journal of Fuzzy Set Valued Analysis **2019**, 1 (2019)

[12] V. Balopoulos, *Efficient Interval-Based Fuzzy Arithmetic*, in *1st International Conference "From Scientific Computing to Computational Engineering", Athens, Greece, 8-10 September 2004*, edited by D.T. Tsahalis (Patras University Press, 2004), Vol. 2, pp. 498–505, https://www.researchgate.net/publication/236019730_Efficient_Interval-Based_Fuzzy_Arithmetic

[13] L. Stefanini, L. Sorini, *Fuzzy Arithmetic with Parametric LR Fuzzy Numbers*, in *Proceedings of the Joint 2009 International Fuzzy Systems Association World Congress and 2009 European Society of Fuzzy Logic and Technology Conference, Lisbon, Portugal, July 20-24, 2009*, edited by J.P. Carvalho, D. Dubois, U. Kaymak, J.M. da Costa Sousa (2009), pp. 600–605, http://www.eusflat.org/proceedings/IFSA-EUSFLAT_2009/pdf/tema_0600.pdf

[14] C. Wagner, S. Miller, J.M. Garibaldi, *A fuzzy toolbox for the R programming language*, in *2011 IEEE International Conference on Fuzzy Systems (FUZZ-IEEE 2011)* (2011), pp. 1185–1192

[15] J. Alcalá-Fdez, J.M. Alonso, IEEE Transactions on Fuzzy Systems **24**, 40 (2016)

[16] M. Gagolewski, J. Caha, *A Guide to the FuzzyNumbers Package for R* (2015), https://cran.r-project.org/web/packages/FuzzyNumbers/vignettes/FuzzyNumbersTutorial.pdf.

[17] B. de Bruyn, L. Fayet, V. Laborie, E3S Web of Conferences **7**, 18002 (2016)

[18] Cédric Lesage, E.J.E.S.S. **15**, 49 (2001)

[19] W.M. Wan Mohamad, T. Ahmad, N.A. Ab Karim, A. Ashaari, Mathematical Modelling and Analyses **23**, 101 (2018)

[20] S.M. Lee, K.L. Lin, S. Gupta, *Economic decision making using fuzzy numbers*, in *Proceedings of 2nd World Conference of POM* (2003), Vol. 15, https://www.poms.org/archive/conferences/Meeting2004/POMS_CDDel/Browse%20This%20CD/PAPERS/002-0634.pdf

[21] H.A. Sturges, Journal of the American Statistical Association **21**, 65 (1926), https://doi.org/10.1080/01621459.1926.10502161

The development of energy consumption forecasting model for a metallurgical enterprise

Anna Bakurova[1,*], *Olesia* Yus'kiv[1,**], *Dima* Shyrokorad[1,***], *Elina* Tereschenko[1,****], and *Anton* Riabenko[1,†]

[1]Zaporizhzhia Polytechnic National University, 64 Zhukovskoho Str., Zaporizhzhia, 69063, Ukraine

Abstract. An up-to-date issue of a modern metallurgical enterprise is the increase of its energy efficiency, which is related, first of all, with energy saving. Therefore, the purpose of this paper is to develop a model for forecasting the metallurgical enterprise power system consumption and its experimental testing based on the PJSC "Electrometallurgical plant "Dniprospetsstal" named after A. M. Kuzmin data. In order to build a forecasting model, a neural network apparatus in the MATLAB system was used and it was done in two stages. At the first stage, as an experiments series result, the optimal architecture and algorithm of neural network training were determined. In the second stage, the dependence of the modeling graphs load error from the influence of daily consumption graphs is identified. The MATLAB software package has been adapted for the needs of "Dniprospetsstal" named after A. M. Kuzmin. Neural networks designed in this way can be used to solve applied issues of electrometallurgy, in particular, the long-term estimation of time series of hourly power for the 24 hours ahead.

1 Introduction

Black metallurgy is one of the most energy-intensive industrial sectors. It is characterized by a high level of electricity consumption that is a significant component of energy use. Thus, the share of electricity in the cost of production of large enterprises in the industry varies from 11% to 16%, and in some cases its share increases to 30% [1].

The most electricity-intensive are steel production (its share of consumption is 32.3%), production of hot (21.9%) and cold-rolled products (12%), as well as aglomerative production (14.5%), blast-furnace based steel production(6.9%) and coke production (5.2%) [2].

The maximum specific consumption of electricity that is in electric steel production is 727 kWh/t of steel. Under this condition the consumption of electricity per ton of steel depends on the power of transformers, the specific electric power of arc furnaces, the use of fuel and oxygen burners, preheating of scrap and out-of-furnace processing.

The power consumption forecast in metallurgy that is necessary for optimal control of loading modes of electric power plants, which includes regulation of active and reactive loads of metallurgical enterprise, minimization of losses from reactive power flows and maintenance of voltage within set limits in electric networks due to strict power quality requirements.

*e-mail: abaka11060@gmail.com
**e-mail: yuskivolesya@rambler.ru
***e-mail: hoveringphoenix@gmail.com
****e-mail: elina_vt@ukr.net
†e-mail: rjabenkoae@gmail.com

The power consumption forecast in metallurgy that is necessary for optimal control of loading modes of electric power plants, which includes regulation of active and reactive loads of metallurgical enterprise, minimization of losses from reactive power flows and maintenance of voltage within set limits in electric networks due to strict power quality requirements.

The peculiarities of electricity consumption by ferrous metallurgy enterprises include: a large number of electrical equipment that is used in the implementation of the technological process in each unit; great variety of types and capacities of electricity receivers; relatively weak links between the mutual influence of electricity receivers in the implementation of the technological process; a large number of electrical equipment that participates in ensuring the technological process in each unit and creates a conditionally constant load, also depends on the intensity of the technological process; factors those randomly affect the modes and volume of electricity consumption; a large number of hours of use of maximum electric power; large electrical intensity of the final product types; the possibility of changing the modes of operation and composition of equipment in the unit, product range and other systematically acting factors [3].

Therefore, the problem of forecasting electricity consumption in metallurgical production is one of the important scientific and practical tasks in the power industry today.

Planning is one of the main management functions. The deviation of actual consumption from the declared numbers by more than a certain percentage leads to the purchase of electricity from the balancing market at a higher price, deviation to a lesser value is also being punished by payment for undelivered electricity, which is determined by the difference between declared and actual consumption by the set fee rates.

The complexity of the energy consumption forecast caused by the need to consider many factors that affect electricity consumption (the level light; the day length; day of the week; transitions from winter to summer time and back; the presence of extraordinary events (disasters; mass actions); the state of other factors that affect the change in consumption in accordance to the data obtained from the processing of consumption statistics).

Improving the accuracy of forecasting saves energy resources, determines the efficiency of power supply management and, accordingly, increases the profits of energy companies. The need for accurate forecasting of electricity consumption is conditioned by technological and economic reasons. The total payment for the consumed electricity for the enterprise as a market participant consists of the payment for the actually consumed volume of electricity and the payment for the deviation of the actually consumed volume from the declared (forecasted) one. The higher possibility of the forecast error is, the greater is the deviation of the actual amount of electricity consumed from the declared (forecast) the greater is the additional costs of the enterprise to pay for electricity. Thus, reducing the forecast error will allow the enterprise to reduce the additional fee for electricity consumption caused by inaccurate forecasting.

In large energy-intensive industries, the share of electricity bills in the cost of production can reach tens of percent (for example, in metallurgical production, it is 11% – 30%) [4], and in a market economy, the prime cost of production will determine its competitiveness. Thus, in the conditions of modern energy market for large enterprises and mining industries it is important to create a system of hourly electricity consumption, which minimizes the deviation of power consumption from the declared.

Based on the done forecast, the actual and optimal modes of operation of power systems are calculated, as well as qualitative parameters (reliability, quality of electricity, etc.) are evaluated. Clear forecasting of electricity consumption provides optimal load distribution within the production process.

2 Literature review

Currently, forecasting of electricity consumption in most of industrial enterprises is made on the experts' evaluation method, which in most of cases is not able to provide the necessary forecast accuracy. Electricity consumption forecasting gives us primary information for planning normal operating modes in energy management.

The accuracy of forecasting directly depends on the calculation methods. There are a large number of models and methods of short-term forecasting of the load on the power system of a metallurgical enterprise.

The formation and development of methods of mathematical modeling, forecasting and planning of electricity consumption, in particular in metallurgy, is related with the work of scientists, an overview of the main ideas of those are given below.

The monograph by C. K. Belt [5] is a fundamental work in energy management in the metallurgical industry. The main idea that can be traced in it is that in the metallurgical industry there are unique processes and conditions those require a more individual approach. The author gives an explanation that the general methods of energy management in the very metallurgical industry, describes the advantages and reasons for implementing an energy management program, as well as reveal the basic indicators, tells us about data analysis, project identification and processes improvement.

In the article [6] a group of scientists presents a systematic review of existing publications on energy management in industry, that identifies five main elements of energy management, based on the following topics: strategy; planning, implementation; operation, control; organization and culture. The authors have developed a conceptual structure of energy management, which shows that a comprehensive approach is needed to effectively use the existing potential of energy efficiency.

The book [7] was written by the authors of the Neural Network Toolbox for MATLAB and provides a detailed description of the fundamental architectures of neural networks, learning rules, and applications for solving practical problems.

In [8] the authors offer the use of a deep neural network (DNN) for short-term electrical load prediction (STLF), which allowed to overcome the problems of non linearity and achieve higher prediction accuracy.

The manual [9] covers the latest trends and technologies in energy engineering and related industries, contains the latest materials on energy planning and policy.

The educational publication [10] explores a number of sustainable energy sources and tools used to analyze industry and future energy trends. The text contains an overview of energy economics with an integrated software package for research on energy use dynamics and forecasting; climatic and environmental factors of using energy are taken into consideration.

The main disadvantage of the existing methods is the need to build a load model and the constant refinement of the designed model. Another disadvantage of these methods is the inaccurate setup of the relationship between input and output variables, because the relationships between them are nonlinear.

Most of the existing power consumption forecasting algorithms are developed in the energy sector function as a combination of various statistical methods. However, due to the nonlinear relationship between the factors and the load on which it depends, qualitative modeling is a very complex process. The known methods of forecasting electricity consumption do not work with distorted or incomplete data, so we need new methods, approaches of forecasting electricity consumption, those could take into account different types of data.

In this paper, based on the analysis, the MATLAB complex with the Neural Network Toolbox was chosen to develop a model for forecasting the energy consumption of a metallurgical enterprise on the basis of artificial neural networks (ANN [11]).

3 Problem statement

When we are building a model of forecasting electricity consumption for a metallurgical enterprise, several tasks are being set. They are:

1. To construct a mathematical model by processing the data on the load for the past period of time.

2. To obtain a forecast based on the obtained model.

The task of this study is to get as much information as possible from the available data to do and to build the most accurate forecast possible. The specific actions are selected on the basis of available data and availability of informational and technical systems. The solution of this problem requires preliminary research and descriptive analysis.

To solve the problem of forecasting electricity consumption, a number of smaller subtasks are set, those at the final stage will achieve the goal – to build a forecast with maximum accuracy.

The general task is divided into a number of subtasks, those include:

- descriptive analysis of the time series, or graphical analysis. The graphical analysis gives us an ability to identify obvious trends and patterns in the original data;
- time series research, as well as identification of regular and permanent components;
- obtaining an accurate forecast of the time series, taking into account fluctuations;
- evaluation of the quality of the constructed forecast.

The study of forecasting approaches shows that there is no standard, single method for forecasting electricity consumption: any enterprise has its special technological cycles, those together form a process that is unique to every enterprise.

The electricity consumption processes have functional, cyclical and random tendencies. The cyclic dependencies (they are usually daily, weekly and annual) are the easiest to predict. The cyclical trends, according to preliminary estimates, are for about 70–80% of all changes in the process of electricity consumption [12]. For example, the most significant cyclic factors in almost all industrial processes are the time, day of the week and the length of daylight.

The second of the important factors those are studied when solving the problem of forecasting are functional nature patterns (10–15% of the total deviations) [13]. This group can include deviations, those are explained by known and predicted factors those are specific to the metallurgical enterprise they are: air temperature or coolant use. Data analysis helps us to identify these factors and calculate their weight share in the energy consumption process.

Operational forecast of electricity and power consumption is studied with the base on the current dynamics of power consumption at the last minute of each hour, but the main guideline for the technologist who forms the forecast of consumption for the planned period is the trajectory of power consumption, placed in short-term consumption forecast. Therefore, providing the high accuracy of short-term forecast of electricity demand and capacity in power systems is a priority task. The operational forecasting of electricity and power consumption is a component of short-term forecasting.

The objective of this paper is to develop a model for forecasting the power system consumption of a metallurgical enterprise and its experimental testing on the data of "Dniprospetsstal".

Solving the issue of improving the quality of operational consumption forecasting will be considered as a set of measures to improve the quality of short-term forecasting.

4 Materials and methods

The task of short-term forecasting of electricity consumption is a difficult task. The apparatus of artificial neural networks has the ability to generalize information that describes nonlinear dependencies in complex objects, processes and phenomena.

The shares of electricity consumption in the energy consumption system of "Dniprospetsstal" are determined by the production cycle, but in addition to the process of furnaces loading, they are affected by other needs those appear during the working hours. Therefore, as an array of input data for forecasting, statistical information was taken on the hourly consumption of electricity during the 24 hours period for the autumn months. Units of measurement of input information are MW/h.

Based on the training varieties and the formation of the forecast at the initial stage of the research, it is necessary for the ANN to select the weight coefficients in such a way that the standard deviation of the values of the outputs will be minimal. This requirement applies to the network of direct data extension and reverse error propagation (Feed-forward backpropagation – further FfB) [14]. Such network architecture requires its training at the expense of setting time targets, those should be the standard in the formation of the forecast and the input data that must be processed.

In order to set the input and target values of the function for the ANN model, it is necessary to use the architecture with nonlinear autoregression (NARX), which is presented in figure 1. It is a recurring dynamic network with reversed connection that has several layers and is based on the autoregressive model:

$$y(t) = f(y(t-1), .., y(t-n), x(t-1), .., x(t-n)) \quad (1)$$

The predicted value of $y(t)$ depends on the n previous values of the output and on the n previous values of the time series. Figure 1 shows the neural network scheme used in this research.

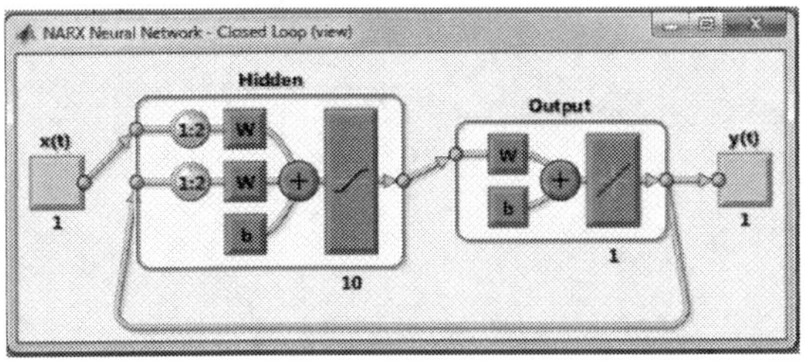

Figure 1. A closed-loop in a network of nonlinear autoregression

The NARX network in figure 1 is a two-layer reversed expansion neural network. The sigmoidal function acts as a transfer function in the hidden layer, and the linear function acts as a transfer function in the output layer.

The given network model uses delay lines with taps to store the previous values of $x(t)$ and $y(t)$. The original value of $y(t)$ is given back to the input of the network (after the delay), as $y(t)$ is a function dependent on the previous values: $y(t-1)$, $y(t-2)$, ..., $y(t-d)$.

There are two entrances in the network. One is external, and the other is connected to the output of the network, from that the input values are obtained. For each of the inputs there is a delay line with taps for storing the previous values. The number of neurons at the latent level will be established experimentally 10. The number of delays is 2.

A closed loop in nonlinear autoregression allows one-step predictions, that is, it predicts the value of $y(t)$ from the previous values of $x(t-1)$, $x(t-2)$, $y(t-1)$, and $y(t-2)$. A closed-loop circuit can be used to perform multi-step prediction. This is caused by the fact that the predicted values of $y(t)$ will be used instead of the actual future values of $y(t)$.

Before you can start training a network that contains delay lines with taps, you must fill them with the initial values of input and output.

The task is to predict the time series, so as the input data will be used the numbers of time intervals in the prescribed manner, and as the output – the value of the analyzed energy consumption.

For effective network training, all input data is divided into three subsets. The first will be needed for training of a network (correction of scales). 70% of the data was used for the first one.

The second subset is necessary to avoid network over fitting. The control of network learning is carried out by calculating the error on the data from this subset. At the initial stage of learning the neural network, the error should be reduced. The over fitting will be evidenced by an increase in error. Therefore, a trained system will have scales, at those the error on the test subset would be minimal.

The third subset is a test. It is useful when comparing different models as an independent test of a trained network. Also, the test subset may indicate poor separation of the output data, if the minimum errors on the data and the verifiable subset are achieved in substantially distant ones from one iteration. For the second and third subsets, 15% of the data were allocated.

Another approach to forecasting energy consumption is deep learning [15]. A deep neural network allows you to process a large amount of input information in a short time, has the ability to build dependencies on non-informative input information, to detect hidden dependencies between the inputs and outputs.

Neural networks with long short-term memory have been created as a modification of recurrent neural networks, those can be selected. Any recurrent neural network has the form of a chain of repeating modules of the neural network.

The core components of the LSTM network are the input sequence layer and the LSTM layer. The LSTM layer studies long-term relationships between time steps of given sequences.

The diagram in figure 2 illustrates the architecture of a simple LSTM network for forecasting. The network is started from the input layer of the Sequence Input sequence, which provides the LSTM layer. To predict class marks, it ends with a Fully Connected layer and regression of the source layer.

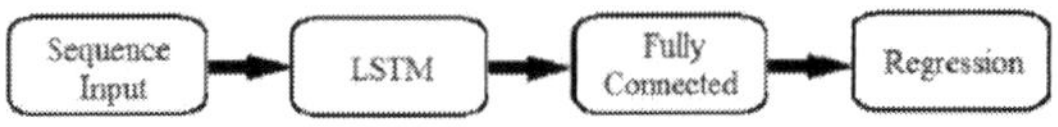

Figure 2. Simple LSTM network architecture

To predict the meaning of future sequence time steps, it is possible to train the LSTM chain from sequence to sequence, where the answers are training sequences with values shifted by one. Thus, at each time step of the input sequence, the LSTM network learns to predict the value of the next time step. The training of such a network is carried out using the ADAM method – adaptive moment estimation optimization algorithm [15].

Forecast accuracy is estimated as the mean absolute error in percentages (MAPE):

$$MAPE = \frac{1}{p} + \sum_{i=1}^{p} \frac{|y_i - \overline{y_i}|}{y_i} * 100 \quad (2)$$

where y_i – actual load values, forecast result,

p is the amount of data.

According to the analysis of publications [7, 14, 16], a forecast error of no more than 5% is acceptable.

5 Experiments results

In the Neural Network Toolbox environment, it is possible to use the following algorithms for learning neural networks: nonlinear Levenberg-Marquardt optimization, Bayesian Regularization methods and conjugate gradient methods (Scaled Conjugate Gradient) those are detailed described in sources [7, 14].

For objectivity of comparison of indicators of quality of forecasting at application of various algorithms of forecasting with use of the corresponding architecture of construction of ANN one was chosen exactly and on the same day that gave an opportunity to compare the results of received forecasts in the most accurate way.

When modeling the process of electricity consumption in the power system at the initial stage, a network that has an architecture of direct data propagation and reverse error propagation (FfB) and trained according to the Levenberg-Marquardt algorithm was chosen.

The training results of the neural network in Matlab are presented in the following way (figure 3).

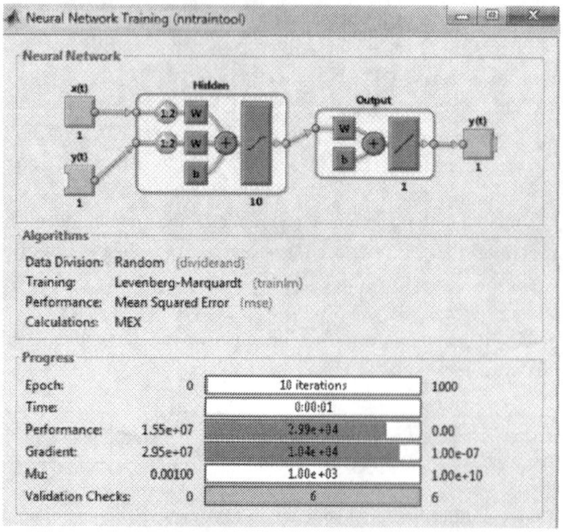

Figure 3. Neural network training results performed by Levenberg-Marquardt algorithm on the architecture of the FfB model

In the following stages of the training, the same type of architecture, FfB, was used, but the learning algorithm changed: first, the Bayesian regularization algorithm, then the combined gradient method.

According to the obtained results, the comparison of the generated forecast from the actual values using different training algorithms on the FfB architecture are presented in figure 4.

In contrast to the above ANN, a separate software solution is required for forecasting using the LSTM network. The sequence of software implementation of deep training in the Matlab package is:

1) to import data from Excel;

2) to divide the data into a training sample and a test;

3) to standardize the data to have a zero mean and modular deviation;

4) to create an LSTM regression network. Set the LSTM layer, which has 5 hidden modules. Further increase in the number of LSTM modules or layers doesn't improve the forecast quality due to insignificant quantity of training data in our case;

5) to set training options. Specify the 'adam' algorithm and train LSTM with the given training options using train network for 250 epochs. To prevent the gradients from exploding, set the gradient threshold to 1;

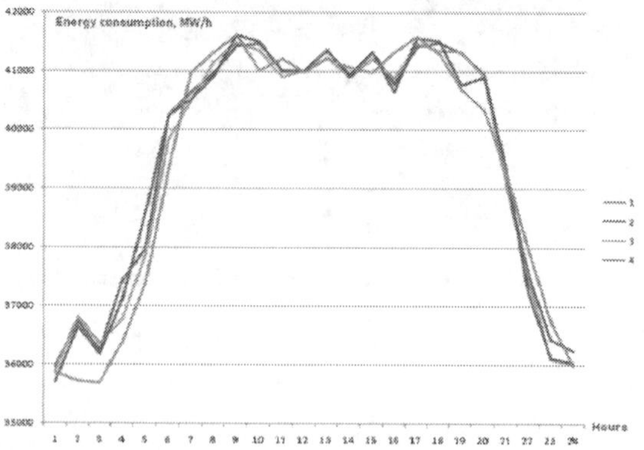

Figure 4. Results of energy consumption forecasting using ANN: 1 – actual value of energy consumption; 2 – forecast value of energy consumption according to Levenberg-Marquardt algorithm; 3 – forecast value of energy consumption according to Bayesian regularization algorithm; 4 – forecast value of energy consumption by the method of combined gradients

6) to predict the value of several time steps in the future, use the predict and Update State function;

7) to initiate a network state, first predict the training data XTrain. Then we make the first prediction using the last time step of the learning answer YTrain.

The results of forecasting with the LSTM network are shown in figure 5.

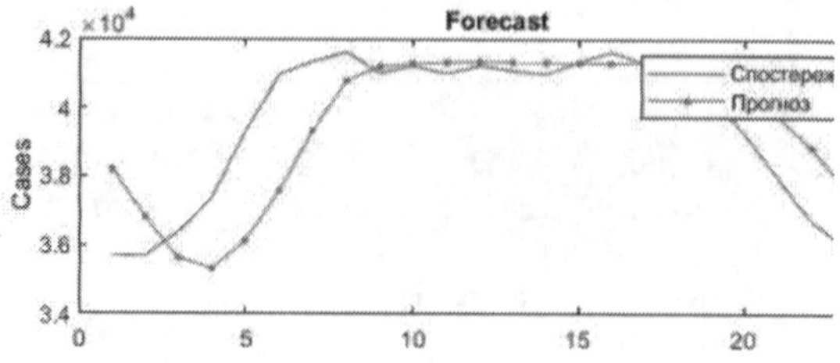

Figure 5. Display of actual and forecasted data values with the LSTM network

The results of the obtained forecasts and calculations of forecasting accuracy according to formula (2) on different training algorithms are summarized in table 1.

According to table 1, the best among the used ANN for this training is the LSTM network, the maximum prediction error is 2.59% modulus per operating 24 hours period. In absolute terms, this is 1,069.76 MW.

The use of artificial neural networks allows to increase the consumption forecasting efficiency, as well as to consider the dynamic processes those are occurring in the power system, which affect the nature of electricity consumption and power, which increases the accuracy of forecasts.

Table 1. Characteristics of ANN training

Parameters	Levenberg-Marquardt algorithm	Bayesian regularization algorithm	Combined gradients	LSTM network
Epoches number	10	211	7	250
Training time	0:00:01	0:00:05	0:00:01	0:00:05
MAPE, %	2.87	3.07	3.27	2.59

To implement this task, the optimal software and hardware complex MATLAB was selected, that has tools for building and realizing ANN. We had the wide range of possibilities and proposed ready-made solutions so an analysis of the choice of the ANN architecture and the algorithm of its training was carried out in this software package.

Using MATLAB software package, experiments on the basis of statistical data of the metallurgical enterprise by the method of artificial neural networks were conducted. It is revealed that the developed neural networks may be used to make a forecast of electricity consumption for the 24 hours' time period ahead with a maximum accuracy of 2.8% of the actual volume of electricity consumed.

6 Discussion

The use of artificial neural networks allows to increase the consumption forecasting efficiency, as well as to consider the dynamic processes those are occurring in the power system, which affect the nature of electricity consumption and power, which increases the accuracy of forecasts.

To implement this task, the optimal software and hardware complex Matlab was selected, that has tools for building and realizing ANN. We had the wide range of possibilities and proposed ready-made solutions so an analysis of the choice of the ANN architecture and the algorithm of its training was carried out in this software package.

Using Matlab software package, experiments on the basis of statistical data of the metallurgical enterprise by the method of artificial neural networks were conducted. It is revealed that the developed neural networks may be used to make a forecast of electricity consumption for the 24 hours' time period ahead with a maximum accuracy of 2.87% of the actual volume of electricity consumed.

Developed models for forecasting energy consumption of "Dniprospetsstal" power systems based on LSTM ANN meet the requirements for the quality of short-term forecasting of consumption in power systems. But they fail to include all energy consumption problems and factors that affect the energy system of the enterprise. Therefore, there are a number of ways for the further model's improvement.

An important aspect of ferrous metallurgy enterprises is the risk of downtime for various reasons, the prediction of those using neural networks is considered in the article [17]. This fact makes a challenge for further research in this area regarding uninterrupted power supply and the impact of risk factors on the random component of the time series.

The issue is being analyzed in the article [18] relate to forecasting energy consumption in non-ferrous metallurgy, those enterprises have their own characteristics. The method of hybrid regression of reference vectors for research, that inspires to compare the results obtained with the help of neural networks of different architecture with the results obtained by other methods is used.

In the monograph [5] shows a tendency to increase gas consumption in the winter months and emphasizes that reducing the energy used for the station heating may be a viable project. Therefore, the following studies we provide the possibility of introducing meteorological factors into the model of electricity consumption forecasting to increase the efficiency of ANN.

High accuracy of short-term hourly forecasting of electricity consumption is a necessary condition for normal operation in the wholesale market of electricity and capacity (WMEC), otherwise such market entry does not make sense. In such conditions, the financial position of the WMEC participant depends on the accuracy of forecasting, so there is a problem of estimation the economic efficiency of electricity consumption forecasting, but this requires additional research.

7 Conclusions

Management of a large electrometallurgical enterprise in terms of energy saving policy involves effective management of energy consumption by forecasting electricity consumption, that leads to new challenges and enhances the relevance of research in this area.

The forecasting results analysis done with the use of the developed models showed that the chosen approach with experimentally selected architectures and learning algorithms meets the necessary requirements that were set at the initial stage of creating a dynamic model of forecasting based on ANN. It can be concluded that the networks are built in the correct way, because the chosen learning algorithms make the model effective.

The practical value of the modeling carried out in this work is confirmed by a number of experiments performed on real and available for research data, those demonstrate the possibility of using the developed tool to solve forecasting problems.

The scientific novelty of this work is the formalization based on the neural networks theory of energy consumption forecasting models those include the daily volumes of electricity consumed.

The work was carried out as the part of the research work "Mathematical modeling of socioeconomic processes and systems", the registration number DB05038, at the Department of System Analysis and Computational Mathematics of Zaporizhzhia Polytechnic National University.

References

[1] S. Kiyko, Innovative Technologies and Scientific Solutions for Industries **4**, 56 (2020)

[2] H.M. Hnatiienko, V.I. Snytiuk, *Ekspertni tekhnolohii pryiniattia rishen* (Maklaut, Kyiv, 2008)

[3] V.M. Molokanova, O.P. Orliuk, V.O. Petrenko, O.B. Butnik-Syverskyi, V.L. Khomenko, Naukovyi Visnyk Natsionalnoho Hirnychoho Universytetu **2**, 131 (2020)

[4] S.G. Kiyko, Science and technology of the Air Force of the Armed Forces of Ukraine **4**, 133 (2020)

[5] C.K. Belt, *Energy Management for the Metals Industry* (CRC Press, New York, 2017)

[6] M. Schulze, H. Nehler, M. Ottosson, P. Thollander, Journal of Cleaner Production **112**, 3692 (2016)

[7] M.T. Hagan, H.B. Demuth, M.H. Beale, O.D. Jesús, *Neural Network Design*, 2nd edn. (2014), ISBN 978-0-9717321-1-7, `https://hagan.okstate.edu/NNDesign.pdf`

[8] P.P. Phyo, C. Jeenanunta, *Electricity load forecasting using a deep neural network* (2019), `https://ph01.tci-thaijo.org/index.php/easr/article/view/116025`

[9] D.Y. Goswami, F. Kreith, eds., *Energy Efficiency and Renewable Energy Handbook*, 2nd edn. (CRC Press, Boca Raton, 2015), ISBN 9780429103070

[10] C.F. Kutscher, J.B. Milford, F. Kreith, *Principles of Sustainable Energy Systems*, 3rd edn. (CRC Press, Boca Raton, 2018), ISBN 9780429485589

[11] S. Semerikov, I. Teplytskyi, Y. Yechkalo, A. Kiv, CEUR Workshop Proceedings **2257**, 122 (2018)

[12] I.M. Kirpichnikova, L.A. Saplin, K.L. Solomakho, Vestnik YuUrGU. Energetika **14**, 16 (2014)

[13] G.P. Shumilova, N.E. Gotman, T.B. Starczeva, *Prognozirovanie elektricheskikh nagruzok pri operativnom upravlenii elektroenergeticheskimi sistemami na osnove nejrosetevykh struktur* (URO RAN, Ekaterinburg, 2008)

[14] E.V. Bodyanskij, O.G. Rudenko, *Iskusstvenny'e nejronny'e seti: arkhitektury, obuchenie, primeneniya* (Teletekh, Khar'kov, 2004)

[15] V. Derbentsev, A. Matviychuk, N. Datsenko, V. Bezkorovainyi, A. Azaryan, CEUR Workshop Proceedings **2713**, 434 (2020)

[16] I.V. Brejdo, Y.F. Bulatbaeva, G.D. Orazgaleeva, *Algoritm sozdaniya modeli prognozirovaniya energopotrebleniya na osnove nejronnoj seti v Matlab* (Novacziya, Krasnodar, 2020)

[17] V.V. Vichuzhanin, N.D. Rudnichenko, Informatics and Mathematical Methods in Simulation **6**, 333 (2016)

[18] Z. Huang, C. Yang, X. Zhou, S. Yang, Cognitive Computation **12**, 357 (2020)

A flexible evolutionary model of machine learning of the most successful strategies of human capital development

Vasyl Porokhnya[1] and *Vladyslav* Penev[1,*]

[1]Classical private university, 70B Zhukovskoho Str., Zaporizhzhia, 69002, Ukraine

Abstract. As a result of research, the concept of a flexible evolutionary model is proposed, which with the help of machine learning allows obtaining the most successful strategy for the development of human capital. The proposed conceptual and methodological approach to machine learning of the process of assessing human capital of enterprises, taking into account the cognitive psychology of man and reflective attitudes in the human environment, can increase the effectiveness of decision-making in the field of human capital development management. The training involves indicators of return on investment in the individual, in the types of components of human capital, which are characterized by properties (creativity, competence, purposefulness, communication, motivation), where between their varieties there are appropriate reflective relationships. The main difficulty of this approach to the choice of alternative solutions for finding options for the use of human capital is the correct selection of indicators of significance (return) of contributions to the development of types of human capital, on the basis of which cycles occur of systemic learning. This approach can simplify the search for and developments of human capital development strategies, present alternative ways, and simplify management decisions.

1 Introduction

Human capital (HC) in the modern economy is one of the most important components of an enterprise or state for economic growth. The article [1] uses the decision tree model is used to analyze the performance appraisal of hospital staff to explore the several important factors that affect performances. The purpose of the article [2] is to analyze how objective factors influence the departure of employees in order to determine the main reasons contributing to the decision of an employee to leave the company, as well as the ability to predict whether a particular employee will be able to leave the company. Unfortunately, machine learning for assessing the properties of human capital is in its early stages. Human capital has a strong impact on the competitiveness of the assessed entity. Although the use of funds to improve the value and quality of human capital is still considered to be partly costly rather than an investment, the decision to increase human capital is a hot topic of research, because decision-making in this area is associated with significant uncertainty about the characteristics of properties. and relations that form the concept of human capital. And first of all, there are natural signs of human behavior and their measurement: reflexive, cognitive. Therefore, the main thing is that we need to know the characteristics of human behavior. To know to what extent the process of assessing the development of human capital of an individual, enterprise, or state personnel coincides with the knowledge about a man that humanity possesses. To do this, it will be useful, in our opinion, to build a model for assessing human capital, which would be reflected in the mechanisms of machine learning to find alternatives to increase human capital.

In the framework of this study, the expediency of using machine learning to increase the reliability of obtaining the most effective strategy for human capital development is substantiated.

The study simulated strategies for human capital development using machine learning as a set of possible actions that differently affect the values of parameters that change the amount of human capital to increase its value. As a result of the study, the concept of a flexible model was proposed, which using machine learning allows obtaining the most successful alternative to the development of human capital.

2 Results

If we consider the structure of human capital as a set of qualities of its properties, their ratios, which directly affect productivity, which increases income for staff, the company as a whole, society, nation, it is possible to cover all possible options for its evaluation. It should be said that human capital is a form of capital that is able to generate income that is valued with the influence of natural factors inherent only in man.

The power of the impact of human capital on the competitiveness of the enterprise exacerbates the interest of entrepreneurs and the state to increase the human capital of individuals, to make it more efficient. Thus, Simon Smith Kuznets, introducing the concept of GDP, noted that one

*e-mail: penev.vladislav@gmail.com

 https://doi.org/10.1051/shsconf/202110705005

of the necessary factors for the development of the country (including increasing its GDP) is the growth of human capital.

In order to assess the impact of the HC of a company, organization, enterprise on its financial results, it is necessary that it has a quantitative assessment. Since the calculation of financial indicators is mainly based on costs, it is optimal to take as a basis for the quantitative assessment of the human capital cost approach, taking into account the cognitive psychology of man and reflexive relations in the system of HC figure 1.

The advantage of the cost approach to the evaluation of human capital is that it allows you to get the amount of HC in the valuation, integrated with estimates of other sources, such as fixed, working capital, organizational, and consumer capital [3, 4].

The structure of HC costs is primarily related to staff remuneration. Of course, these are payments to employees, including wages, taxes, bonuses, incentive bonuses, etc. These payments themselves already contain a qualitative and quantitative assessment of the work of employees.

Under the cost of human capital to increase the level of the intellectual capital of the enterprise means the cost of training, retraining, training of employees, participation in thematic seminars, conferences, symposia; costs of research, information, and legal systems, software, literature, including periodicals, etc.

These costs, investments in HC are aimed at innovative reproduction of intellectual capital, to increase the intellectual level of employees, which in turn allows more efficient use of available limited human resources in accordance with the needs of the enterprise.

The cost of health capital is an investment in a person, made in order to form, maintain, and improve his health and efficiency, aimed at increasing the capital of the culture of possession of information. Such costs include health care costs (preventive examinations of employees, additional health insurance, spa treatment, and other disease prevention and control measures), as well as costs aimed at improving the safety of workers places, organization of sports leisure of employees, improvement of living conditions, etc. The main thing in this is the awareness of the individual that his health and culture of behavior is an opportunity to replenish the information potential of activity in making informed decisions. That is, investment in the personality mentality is an investment in the culture of the individual, which determines the efficiency and accelerates the growth of human capital.

In formal form, the expression for quantifying HumanCap's human capital will look like an additive model representing the sum of five components:

$$HumCap = WF + IntCap + HealCap + CultCap + InfCap, \quad (1)$$

where *WP* – remuneration of personnel of the enterprise to the general fund of remuneration;

IntCap – the share of costs in human capital to increase the level of the intellectual capital of the enterprise:

$$IntCap = InvestCap/TotalExp,$$

where: *InvestCap* – investment in the HC for innovation reproduction of the intellectual capital;

TotalExp – the total cost of the company to increase HC;

HealCap – the share of costs (investments) of the enterprise for health capital,

$$HealCap = HealthReprod/TotalExp,$$

where: *HealthReprod* – the cost (investment) to reproduce the health of staff;

CultCap – the company's cost of capital culture of information ownership:

$$CultureCap = CultureKnowH/TotalExp,$$

where: *CultureKnowH* – the share of personal spending on *KnowHow* and possession of information and the ability to use it;

InfCap – enterprise costs for information capital,

$$InformCap = InformConvert/TotalExp,$$

where: *InformConvert* – the cost of the individual's ability to turn information into the necessary knowledge of the company, which is of particular importance for its activities.

Keep in mind that each cost element has its impact.

For example, the return on investment in intellectual capital is higher than on investment in "health capital". Therefore, it is advisable to introduce weighting factors – for each type of cost.

$$HumCap = WF\,\alpha_1 + IntCap\,\alpha_2 + HealCap\,\alpha_3 + CultCap\,\alpha_4 + InfCap\,\alpha_5, \quad (2)$$

where α_1 – capital of competence "property-competence";

α_2 – educational capital "property of creativity";

α_3 – health capital "property-motivation";

α_4 – culture capital "property-purpose";

α_5 – the information capital "property-communicative".

The value of the coefficients is assigned based on the value of the return of each component in the HC of the enterprise. In determining the return on each type of cost must take into account the following, most important factors:

1. The industry in which the company operates.
2. The degree of "intellectualization" of labor, the level of knowledge-intensive products.
3. Stage of the life cycle of the enterprise.
4. The presence of a system for assessing the intellectual level of staff and its motivation.
5. Culture of information ownership.
6. Investments in human capital for innovations in the reproduction of intellectual capital.

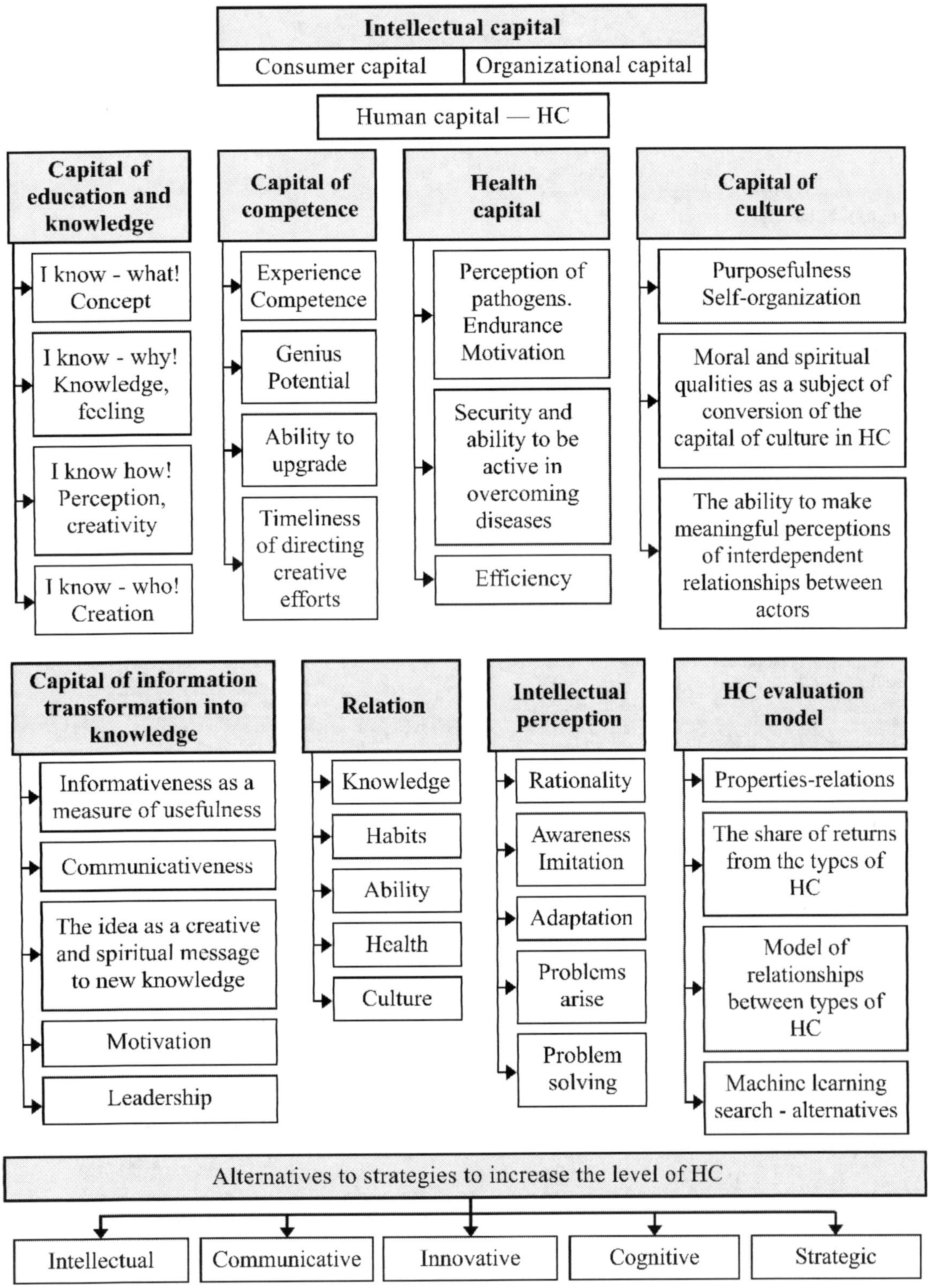

Figure 1. Human capital

7. The cost of the ability of the individual to turn information into the necessary knowledge of the company.

In addition, investment in human capital for innovation reproduction of intellectual capital and company health is influenced by concepts such as secondary education and their average age. The dependence of the efficiency of intellectual capital costs on the general level of education, the cost of health capital on the average age of staff can be taken into account through the parameters ε_1 and ε_2 respectively.

$$WF\,\alpha_1 = IntCap\,\alpha_2\varepsilon_1 + HealCap\,\alpha_3\varepsilon_2 + \\ + CultCap\,\alpha_4 + InfCap\,\alpha_5, \quad (3)$$

where ε_1, ε_2 - parameters that reflect the dependence of cost recovery on the level of education and age, respectively.

$$\varepsilon_1 = \frac{\sum_{i=1}^{n} r_i\, z_i}{R_S z}, \quad (4)$$

where r_i and z_i – the number of employees with i-level of education and their average salary;

n – the number of levels of education in a particular enterprise;

R_S – the total number of personnel of the enterprise;

z – the average salary of employees of the basic level of education.

The function that reflects the dependence of the return on health capital costs is as follows:

$$\varepsilon_2 = \gamma_m \sum_{j=1}^{n} k_j\, d_j + \gamma_w \sum_{j=1}^{n} k_j\, t_j, \quad (5)$$

where γ_m, γ_w – the share of men and women, respectively, in the enterprise;

d_j, t_j – the share of employees of men and women of the j-age group;

n is the number of age groups;

k_j – rate of return on investment in health capital depending on age groups, which are determined on the basis of retirement age. Appropriate coefficient values are assigned to each age group.

Coefficients of return on investment depending on the components of human capital are characterized by cognitive properties (creativity, competence, purposefulness, communication, motivation), between varieties (S) in which there are appropriate reflexive relations (α) of their interaction, figure 2 [5].

Thus, the model for assessing the level of a particular type of specific property of human capital and its quantitative assessment is determined by the formula:

$$HC = \alpha_i(\sum_{t} S^z_{ip} \times HumanCap_i), \quad (6)$$

where $S^z_{ip} = \sum_{j=1}^{n_{ip}} k^a_{jp} * \alpha^z_{ijp}$ – the level of the i variety of the p property;

α^z_{ijp} – the share of participation of the j relation in the formation of the i variety of the p property of the z individual;

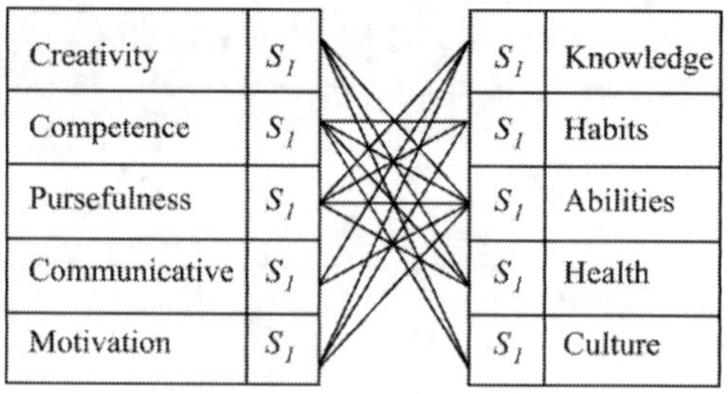

Figure 2. The properties and reflexive relations of human capital

k^a_{jp} – the basic coefficient of distribution of the influence of the j relation on the formation of the p property;

n_{ip} – the number of relations of the i variety of the p property;

S^z_{ip} – the level of the i variety of the p property;

$HumanCap_i$ – human capital of the species.

The level of significance of the criteria is determined by the methodology of the MAI (table 1).

Table 1. The level of the weight of the criteria

Criterion	Result
Creativity	0.223
Competence	0.244
Purposefulness	0.2
Communicative	0.189
Motivation	0.144

Analysis of the criteria, their weight, allows you to choose the best alternative for the development of the properties of human capital. To do this, use the method of hierarchical comparisons in assessing the level of priorities of alternatives, the results of which are given in the table 2 and in figure 3.

Table 2. Influence of criteria on a choice of alternatives (properties) of improvement of the level of human capital

Criterion	Properties				
	Intellectual	Communicative	Strategic	Cognitive	Innovative
Creativity	0.216	0.172	0.207	0.172	0.23
Competence	0.208	0.2	0.189	0.226	0.17
Purposefulness	0.176	0.177	0.27	0.19	0.18
Motivation	0.168	0.19	0.189	0.2	0.26
Communicative	0.164	0.227	0.182	0.23	0.2
General approach	0.189	0.195	0.208	0.202	0.21

The process of investing in human capital should be divided into several stages of investing: education costs; the cost of finding and hiring staff; staff costs during the training period; staff costs during the period of accumulation

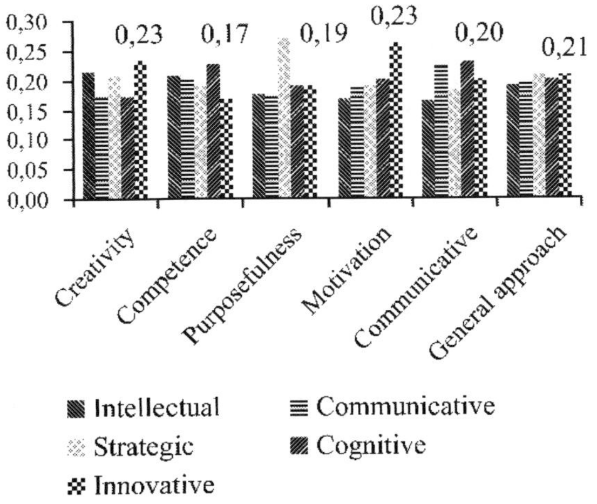

Figure 3. The influence of criteria on the choice of alternatives

of growth potential; staff costs during the period of professionalism; staff costs during training, advanced training; staff costs in the period of decline and "obsolescence" of professionalism; investment in human capital for innovations in the reproduction of intellectual capital; the cost of the individual's ability to turn information into the necessary knowledge companies; the cost of information culture.

The development entity itself can invest in its personal human capital, but these investments can be formed by the enterprise, to increase the human capital of the organization or the state, to increase the national human capital.

Based on the assessment of investment in human capital, the assessment of the direct economic effect is carried out. Such an assessment is possible, as a rule, only after the implementation of any investment project involving staff. To assess the economic effect is more suitable known ROI (Return on Investment):

$$ROI = (OutInv - PostInv) - Z)/Z\ 100\%, \qquad (7)$$

where $OutInv$ – the cost of the human capital of the individual before training;

$PostInv$ – the cost of the human capital of the individual after receiving educational services;

Z – direct and indirect costs of the investment project.

The value of ROI is crucial in measuring the effectiveness of human capital development strategy. Thus, if the value of ROI is less than 20%, the development strategy is considered ineffective, in the case of around 20%, we can say that the company is developing moderately, and sometimes completely stopped in development. A good indicator of development is a value of 150% to 200%, and anything between 20% and 150% can be called an effective strategy.

The relevance of the use of machine learning in the field of economics allows us to consider differently many aspects of human capital development strategies. Training to find the best strategy for human capital development can be represented as a continuous cycle, which will end only after reaching the specified conditions (figure 4).

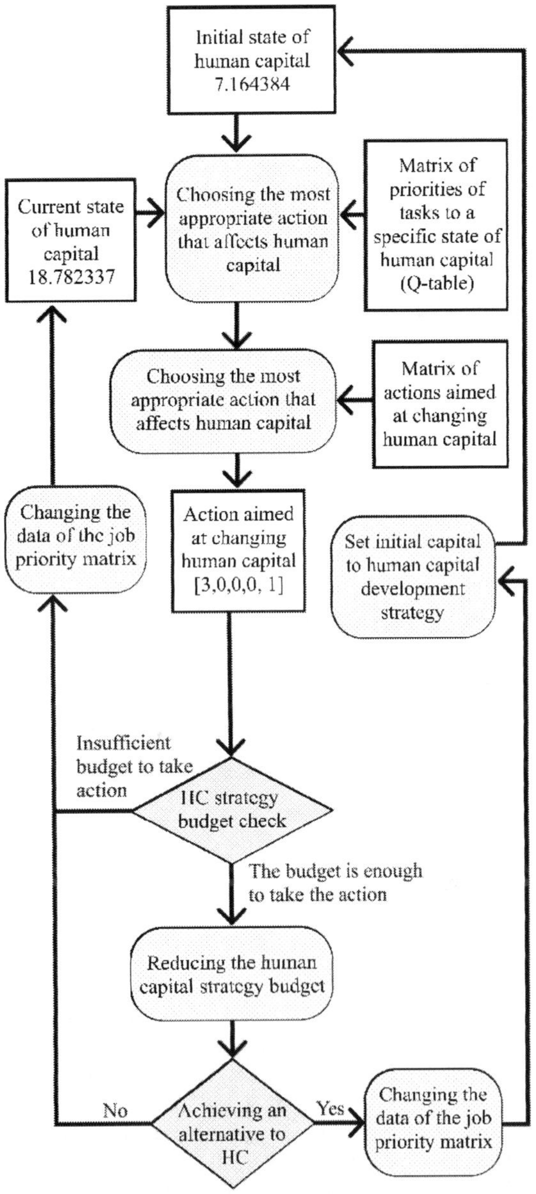

Figure 4. Machine learning of alternative development of human capital of the enterprise

In the reinforcement learning algorithm, the agent's actions are directed to the steps to achieve success with a reward estimate. After Δt steps into the next step, the human capital will decide some next step. The weight for this step is calculated as $\gamma^{\Delta t}$, where γ is the discount factor, which can take a value from 0 and 1, $0 \leq \gamma \leq 1$ and has the effect of evaluating actions that are aimed at achieving the human capital goal. γ can be called the level of success in achieving the desired state by human capital, when the investment data changes at the Δt step.

Thus, we can conclude that a function is required that will determine the quality of combinations of the state of human capital and the action aimed at it:

$$Q \div S \times A \rightarrow R, \quad (8)$$

At the beginning of training, Q is initialized, possibly with an arbitrary fixed value – 0. After initialization, at each moment of time t, the agent selects an action, observes a reward, enters a new state (that may depend on both the previous state and the selected action), and Q is updated. The core of the algorithm is a Bellman[6] equation as a simple value iteration update, using the weighted average of the old value and the new information [7]:

$$Q^{new}(s_t, a_t) \leftarrow Q(s_t, a_t) + \alpha \times (r_t + \gamma \times \max Q(s_{t+1}, a) - Q(s_t, a_t)), \quad (9)$$

where r_t is the reward received when moving from the state S_t to the state S_{t+1}, and $0 < \alpha \leq 1$.

Note that $S^{new}(s_t, \alpha_t)$ is the sum of three factors:

1) $(1 - \alpha)Q(s_t, \alpha_t)$: the current value weighted by the learning rate. Values of the learning rate near to 1 made faster the changes in Q;

2) αr_t: the reward $r_t = r(s_t, a_t)$ to obtain if action a_t is taken when in state s_t (weighted by learning rate);

3) $\alpha\gamma \max Q(s_{t+1}, \alpha)$: the maximum reward that can be obtained from state s_{t+1} (weighted by learning rate and discount factor).

The episode of the algorithm ends when the state S_{t+1} of human capital is the desired one or the financial investment opportunities end. However, Q-learning can also be trained on non-ad-hoc tasks. If the discount factor is below 1, the action values are finite, even if the problem may contain infinite loops.

For all final states s_f, $Q(s_f, \alpha)$ is never updated but is set to the reward valuer observed for state s_f. In most cases, $Q(s_f, \alpha)$ can be taken to equal zero.

Before starting training according to the Q-learning method to find the best development strategy, the parameters of the current human capital are determined. The result of the calculations will be the state of human capital in the q-table. At the beginning of training, the entire matrix of the priority of the choice of actions aimed at the development of human capital will be equal to zero. If all the priorities or some of them are equal, then the choice is made randomly.

After choosing an action and applying it, we recalculate the value of human capital to determine a new state. Actions, in this case, are aimed at increasing the costs of one or more categories of human capital, table 3.

The success of performing actions is a certain human capital to which the subject strives. However, it is worth considering financial constraints, which may vary depending on the situation. For example, health costs for a specific activity cannot be less than 20% of total costs. Then, even if the desired value of human capital is obtained and modeled, but the condition for health capital is not met, then the learning cycle is considered ineffective. Limitations may also be on the part of the budget if any, but in some cases, with a budget limitation, the specified goal of human capital may not be achieved.

Table 3. Q-Learning state table for actions aimed at changing the values of human capital criteria, through training, when each cell is updated through training

The state of human capital	Actions aimed at changing human capital		
	Enhancing creativity	Health Capital Gain	Boost motivation
13.435367	-11,435600	-84,758326	-54,23453
...	...	...	...
18.782337	-2,543467	-3,246473	-26,26434
...	...	...	...
180.28351	-8,546123	-30,245365	-2,234563

If you do not limit the total budget, then after going through the training cycle and receiving the optimal algorithm for the development of human capital, at same time you can find out the required budget for the implementation of this strategy

Having completed the training, we get a filled matrix, where the rows are the states of human capital, and the columns are possible actions (table 4) aimed at human capital.

Table 4. Matrix of actions aimed at changing the parameters of human capital with indicators of the impact on human capital and the required budget

Action number	Impact values on human capital				
	Creativity	Competence	Purposefulness	Communicativeness	Motivation
1	12	0	0	0	0
2	3	0	4	0	0
3	0	0	0	7	2
4	1	1	1	0	0
5	0	0	21	0	0
6	0	0	0	0	32

The optimality of using machine learning in this method lies in the speed of finding a sequence of actions that affect the change in human capital and its structure. With its help, knowing the initial data and the planned budget for the development of human capital, it is possible to optimize costs, taking into account the specifics of the subject, whose human capital is being considered. That is, having the same budget, using machine learning, you can calculate the most effective strategy for the development of human capital.

The main difficulty of this approach to the choice of alternative solutions for the search for options for using human capital is the correct selection of the coefficients of the significance of the contributions, on the basis of which the learning cycles of the system will take place. This approach can simplify the creation of a strategy for human capital development, present alternative ways, and simplify management decisions, table 5.

Table 5. Initialized data affecting machine learning training to find optimal investments in human capital without the limitations of the human capital development fund

Criteria Properties (One hundredth part)						
Creativity	Competence	Purposefulness	Communicativeness	Motivation	Minimum interval in learning state value	The sought-after value of human capital
197	225	158	182	212	0.0001	25

After making calculations with the initial data presented in table 6, results were obtained that describe the strategy of investing in human capital.

It should be borne in mind that the coefficients of the importance of each of the parameters do not allow a decrease in investment for each of the parameters to zero. The input data for training are: the coefficients of the importance of each of the categories and the required value of human capital.

Table 6. Initialized data affecting machine learning training to find optimal investments in human capital without the limitations of the human capital development fund.

	Impact values on human capital					
Action number	Salary	Education capital	Health capital	Cultural capital	Information capital	Total human capital costs
15637	219	188	630	260	22	1519
21352	114	155	121	95	128	621
71521	103	157	151	61	129	601
128868	110	196	115	72	103	596
2034991	158	156	87	50	136	587
11000000	106	197	90	51	141	585
13243567	136	187	89	50	122	584

The results of the strategy of investing in human capital are highly dependent on the learning cycle. So on very small training cycles, investment is very inconsistent, but in the next step, although the investment funds change their direction, it is more and more difficult to achieve a stronger human capital with less investment.

The data in table 6 and figure 5 show the optimization of the costs of human capital development to achieve a human capital value of 25 units. It can be concluded that in order to obtain better results, it is necessary to carry out a sufficient number of training cycles. You can see how, when setting the initial data, the system gives priority to the wrong criteria for investment, which have become priority in recent cycles.

Taking into account the dynamics of changes in results, it can be concluded that subsequent training cycles can bring more optimized costs.

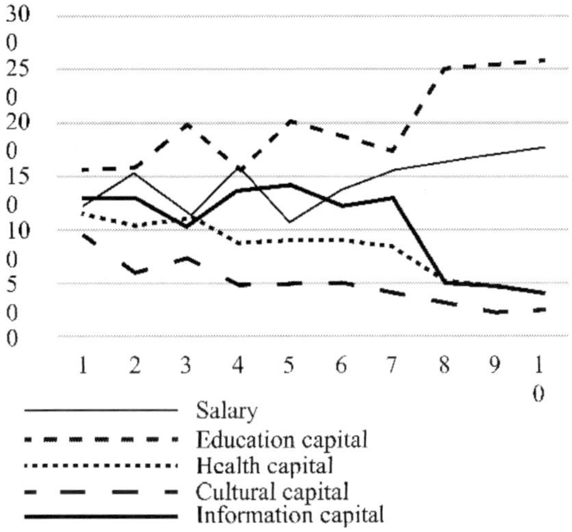

Figure 5. Machine learning of alternative development of human capital of the enterprise

Fig. 6 shows the optimization of the costs of human capital development, taking into account the same level of human capital development.

After training the parameters of the model, we determine the new values of the return coefficients for each type of human capital. As a result, the parameters of the alternatives that determine the planning of new investments in the HC structure change, which entails the search and recalculation of new behavior.

Thus, after each stage of training, new indicators of alternatives should be determined and calculations should be made that determine the subsequent investments in human capital. It should also be borne in mind that each alternative has its own characteristic features and properties of behavior and influence on the choice of options for investing in capital. Making decisions aimed at assessing the interaction of the relationship of these properties in the capital structure can change cognitive and reflexive attitudes and affect the level of human capital.

It should also be borne in mind that the definition of new properties in terms of alternatives can give us information about an insufficient assessment of the interaction of the relationships of these properties in the capital structure. That entails adjusting the choice of options and the redistribution of investment in the capital structure.

https://doi.org/10.1051/shsconf/202110705005

The results of changes in human capital allow us to determine the return of its structural elements of human capital. Redistribution of coefficients alternatives should be redistributed taking into account these data. The current situation requires development in the direction of choosing alternatives of a cognitive, strategic and innovative type.

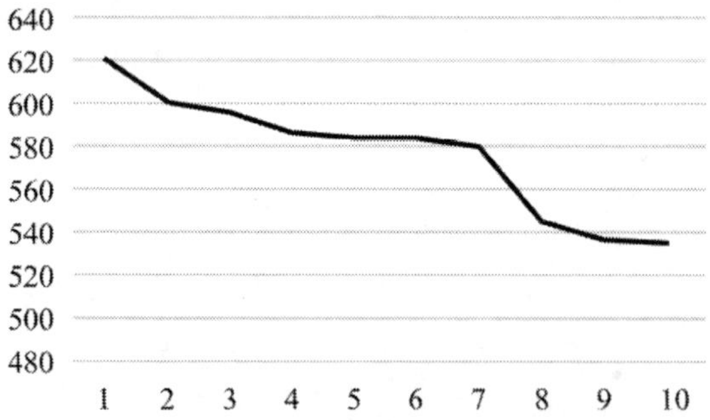

Figure 6. Machine learning of alternative development of human capital of the enterprise

Is a promising approach to modeling intellectual capital in general. It is also worth noting that when the input data changes, machine learning will be able to rebuild and perform calculations and optimize the result better and faster than a human.

3 Conclusions

As a result of research, the concept of a flexible evolutionary model is proposed, which with the help of machine learning allows obtaining the most successful strategy for the development of human capital. The proposed conceptual and methodological approach to machine learning of the process of assessing human capital of enterprises, taking into account the cognitive psychology of man and reflective attitudes in the human environment, can increase the effectiveness of decision-making in the field of human capital development management.

The training involves indicators of return on investment in the individual, in the types of components of human capital, which are characterized by properties (creativity, competence, purposefulness, communication, motivation), where between their varieties there are reflective relationships.

The main difficulty of this approach to the choice of alternative solutions for finding options for the use of human capital is the correct selection of indicators of significance (return) of contributions to the development of types of human capital, on the basis of which cycles occur of systemic learning.

This approach can simplify the search for and developments of human capital development strategies, present alternative ways, and simplify management decisions.

It is worth noting that setting up training with changing training parameters, namely the sum of the reward and the value of data optimization, can achieve better results by accelerating training, and therefore obtaining data on a more trained AI, which can give better results.

Using machine learning to optimize the cost of human capital development is a better method. Speed, lack of subjectivity, and the ability to quickly respond to external changes is an advantage over a person.

This method has a number of possible improvements. For example, it is worth considering the obsolescence of some of the capitals. Also, one of the criteria is wages and its impact on human capital depends on the return, but it should be borne in mind that low returns should not always affect the decrease in this parameter. Also, this method allows, using other input parameters and the limit in the human capital development fund, to find out the development limit and also optimize the costs for each of the human capital funds.

References

[1] L. Yang, H. Xu, Y. Jiang, Procedia Computer Science **131**, 1282 (2018)

[2] F. Fallucchi, M. Coladangelo, R. Giuliano, E. William De Luca, Computers **9** (2020)

[3] T.W. Schultz, *Human Capital: Policy Issues and Research Opportunities* (UMI, 1972), pp. 1–84, `https://core.ac.uk/download/pdf/6908045.pdf`

[4] G.S. Becker, *Human Capital: A Theoretical and Empirical Analysis, with Special Reference to Education*, 3rd edn. (University of Chicago Press, 1994)

[5] V.M. Porohnya, *Intellectual Capital of Economic Growth: A Manual* (CPU, 2012)

[6] R.E. Bellman, *Dynamic Programming* (Princeton University Press, New York, 1957)

[7] C.J.C.H. Watkins, P. Dayan, Machine Learning **8**, 279 (1992)

Optimization models for the use of agricultural aviation when performing chemical works

Abdizhapar Saparbayev[1,2,*], *Aiymzhan* Makulova[3,**], *Assel* Kaziyeva[4,***], *Zaure* Klyshbayeva[5,****], and *Elmira* Saparbayeva[6,†]

[1]Academy Kainar, 7a Satpayev street, Almaty, 050013, Kazakhstan
[2]International Academy of Innovative Technologies, 33 mkr. Samal-2, Almaty, 050059, Kazakhstan
[3]University Narxoz, 55 Zhandosov Str., Almaty, 050035, Kazakhstan
[4]Abai Kazakh National Pedagogical University, 13 Dostyk Ave., Almaty, 050010, Kazakhstan
[5]College of Catering and Service, 12 B. Mailin Str., Nur-Sultan, 010008, Kazakhstan
[6]Mukhtar Auezov South Kazakhstan University, 5 Tauke Khan Ave., Shymkent, 160012, Kazakhstan

Abstract. The article discusses the optimization models for the use of agricultural aviation when performing chemical works. The question of determining the need of vehicles for the maintenance of agricultural work is of great national economic importance. Aviachemical method is firmly established in the technology of cultivation of grain crops. First of all, this is due to the fact that, compared to ground-based methods, the aero-chemical method of chemization of agriculture has a number of technical and economic advantages, including a higher quality and uniformity of chemicals, eliminates mechanical damage to crops, and less dependence on the physical and geographical conditions of the area, allows to obtain significant savings of chemicals and fuel, reduces the time of work.

1 Introduction

Of particular importance is the use of aviation technology in the conditions of intensification of grain production. With the help of aviation equipment, mineral fertilizers are applied, aviation-chemical control of pests and diseases of field plants is carried out, weed vegetation is destroyed in grain crops, defoliation, desiccation of plants and a number of other works are carried out. Currently, with the help of aviation, more than a third of the chemical work in the grain industry is performed and more than 20 million hectares of land are processed annually [1].

Such a diverse use of aviation technology in grain farming is primarily due to the fact that the implementation of aviation chemical works on technical, economic and economic efficiency is not inferior to the ground-based method, and for a number of indicators, such as productivity, reduced labor costs, the possibility of processing by wet soil without over compaction and destruction of its structure, the exclusion of mechanical damage exceeds it. In addition, the use of aviation technology allows freeing up human resources: over a year, labor savings amount to more than 15 million man-days and about 6 million tractor shifts. According to scientific studies, on average, the aviation method of processing crops of grain crops, as compared with the ground, gives an additional yield increase of up to 2 c/ha. Additional net income from the use of aviation in grain farming exceeds 3 billion tenge [2].

*e-mail: sad171@mail.ru
**e-mail: aiymzhan.makulova@narxoz.kz
***e-mail: assel01@mail.ru
****e-mail: zaure1975.75@mail.ru
†e-mail: saparbaeva_e@list.ru

2 Material and methods

The use of an aircraft fleet (AF) when performing chemical works in the territory of the republic is characterized, on the one hand, by a large variety of types of these works, and on the other hand, by the uneven intensity of their performance in time and territory. In the transition from one type of work to another, both the productivity of the technical equipment that performs these works and the cost of their implementation changes. Therefore, with optimal distribution of the aircraft fleet at the aviation chemical works (AChW), conditions are created for the fullest possible satisfaction of applications for aviation service due to more intensive use of the aircraft.

AChW performance with deadlines, established by agrotechnology, can dramatically reduce the effect of the activities carried out, and sometimes completely unacceptable. Compressed and tightly fixed deadlines for conducting an emergency management dictate a particular area for a certain period of time, the need for such an aircraft fleet, which significantly exceeds the capacity of an airline located in a given area. "Peak" periods in different regions of the republic may partially or completely not coincide in time, which is explained by differences in climatic conditions and in "peak" periods in different regions of the republic may partially or completely not coincide in time, which is explained by differences in climatic conditions and in "peak" periods in different regions of the republic may partially or completely not coincide in time, which is explained by differences in climatic conditions and in structure of cultivated crops and land. Such a discrepancy in time of "peak" periods in various areas creates the necessary prerequisites for the temporary transfer of part of the aircraft fleet from home areas to other areas currently in need, i.e. for maneuvering fleet in the Republic.

The heterogeneity of the structure of cultivated crops and land in areas, mainly due to differences in soil and climatic conditions, is the main reason for the discrepancy between the "peaks" of AChW in these areas. Such a discrepancy in time of the periods of maximally intensive use of aircraft parks allows defining a simple logic scheme for redistributing the aircraft fleet between them at certain time intervals. Such redistribution, i.e. maneuvering allows you to perform a much larger amount of AChW at the agrotechnical periods most favorable for the regions without increasing the fleet in the aviation enterprises of each of them.

3 Results and discussion

To solve the problem of the optimal redistribution of agricultural aviation between areas (temporary secondment), taking into account the mismatch of their "peak" periods, the usual deterministic linear programming model can be used. Calculations on such a model provide a practically acceptable solution to the maneuvering problem, especially if it is solved as an operational management task, i.e. for a short time interval (decade, month). However, with an increase in the planning interval (quarter, year), apparently, the accuracy and, consequently, the practical suitability of calculations for the deterministic model can deteriorate sharply due to the fact that the influence of the random factor is not taken into account in such a model. So, in fact, the "peak" period in a region may not only not coincide in time with similar periods in other regions, but also under the influence of weather conditions can significantly mix with respect to the time interval in which its manifestation was previously (predicted). Therefore, a more accurate solution to the problem of using agricultural aviation in the performance of chemical work, taking into account the random factor, can be obtained by solving it in a stochastic formulation.

When constructing and studying mathematical models of the use of the fleet of aircraft for the performance of maintenance and repair, the following should be considered:

- the priority of the performance of the aircraft security personnel, for which the serviced territory is assigned;
- multipurpose use of aircraft due to the diversity of the types of work performed;
- the discrepancy of the situation during the year of the peaks of the performance of the AChW in different areas of the service area;
- the implementation of labor-intensive regulations is carried out at home enterprises.

Taking into account the above features, a mathematical model of optimal maneuvering of aircraft fleet on AChW in the context of current planning can be formulated as a multi-index production-transport task of resource allocation: find such a plan of maneuvering aircraft with which the expected amount of maintenance is performed with a minimum of total reduced costs.

It is necessary to find a solution that minimizes the reduced total costs [1, 3, 4]:

$$\sum_{i=1}^{m}\sum_{j=1}^{n}\sum_{l=1}^{L}\sum_{r=1}^{R}\sum_{t=1}^{T} C_{ijer}x_{ijert} \tag{1}$$

under restrictions on the amount of work performed

$$\sum_{i=1}^{m}\sum_{l=1}^{L} P_{ijr}x_{ijlrt} \geq d_{jrt}, \tag{2}$$

for the number of serviceable aircraft

$$\sum_{i=1}^{n}\sum_{r=1}^{R} x_{ijlrt} \geq a_{ilt}, i = \overline{1,m}, l = \overline{1,L}, t = \overline{1,T} \tag{3}$$

for consumption of fuels and lubricants

$$\sum_{i=1}^{m}\sum_{j=1}^{n}\sum_{l=1}^{L}\sum_{r=1}^{R} q_i x_{ijlrt} \leq Q_t, t = \overline{1,T}, \tag{4}$$

with the natural requirement of nonnegativity of variables

$$x_{ijlrt} \geq 0, i = \overline{1,m};$$
$$j = \overline{1,n}; r = \overline{1,R}; l = \overline{1,L}; t = \overline{1,T}; \tag{5}$$

where i – an index of type AF, $i = \overline{1,m}$;

j – an index of type AChW, $j = \overline{1,n}$;

l – airline index in which the aircraft is based AF, l=$\overline{1,L}$;

r – index of the area (region) of the AChW, $r = \overline{1,R}$;

t – the index of the subperiod, $t = \overline{1,T}$;

d_{irt} – the volume of chemical works of the j-th species in the r-th region, which must be performed in the subperiod t, ha;

a_{ilt} is the number of serviceable aircraft of the i-th type at enterprises l in the sub-period t;

C_{ijlr} – unit costs of the enterprise l for the execution of works of the j-th type by the i-th type of aircraft in the region of r, tenge;

P_{ijr} – specific productivity of the i-th aircraft type, performing the j-th type of work in the r-th region, ha/h;

q_i – specific fuel consumption of the i-th aircraft type, kg/h;

Q_t – disposable amount of fuel and lubricants in the subperiod t, t;

X_{ijlrt} – number of operated aircraft of the i-th type assigned to the aviation enterprise l in the j-th type of work in the r-th region in the sub-period t.

The task of maneuvering the fleet of aircraft on the AChW in the formulation (1) – (5) belongs to the class of linear programming problems. When solving the problem (1) – (5) by conventional linear programming methods, considerable difficulties arise associated with the high dimensionality of the problem (the number of variables is $mnlRT = 3 \cdot 4 \cdot 21 \cdot 28 \cdot 12$; the number of constraints is $nRT+mLT+T = 4{\cdot}12{\cdot}28+3{\cdot}12{\cdot}21+12$). Over the past two decades, many characteristic structures have been identified and a large number of algorithms have been created to solve them [5–7]. Two main approaches to solving problems of large dimensionality were defined: direct methods and methods based on ideas of decomposition (decomposition).

In the first case, for solving problems of a special type, general methods of mathematical programming are used. In this case, the main computational difficulties associated with the storage of large amounts of information and operations on them. Large sparse matrices (matrices having a small percentage of nonzero elements) are usually stored in computer equipment in packaged form, i.e. only nonzero elements are stored along with the necessary information about their position in the matrix. For example, when using the simplex method for solving linear problems of large dimension, the main computational difficulties are associated with the inversion of the basis matrix [8]. Therefore, when solving a large dimension problem, it is advisable to use a modified simplex method or one of the modifications of the generalized gradient descent method. A feature of the modified simplex method is the representation of the inverse matrix of the conditions of the problem in a multiplicative form, which provides a number of significant computational advantages. Most modern mathematical programming packages use the modified simplex method. If the matrix of conditions of the problem has a special structure, but the necessary calculations can be performed using a matrix of a lower dimension. Such an approach, for example, is implemented when applying the methods of taking into account bilateral restrictions or the method of reducing the basis to a triangular shape. Most modern mathematical programming packages use the modified simplex method. If the matrix of conditions of the problem has a special structure, but the necessary calculations can be performed using a matrix of a lower dimension. Such an approach, for example, is implemented when applying the methods of taking into account bilateral restrictions or the method of reducing the basis to a triangular shape. Most modern mathematical programming packages use the modified simplex method [9, chapter 3]. If the matrix of conditions of the problem has a special structure, but the necessary calculations can be performed using a matrix of a lower dimension. Such an approach, for example, is implemented when applying the methods of taking into account bilateral restrictions or the method of reducing the basis to a triangular shape.

The methods belonging to the second class are based on decomposing the system into subsystems of a lower dimension. Usually, the first and second level subsystems are distinguished in such a way that the second level subsystems determine the corresponding changes in the first level subsystems. The task of the second level is to coordinate the functioning of the elements of the first level in order to obtain a common solution to the original problem.

Two main decomposition approaches were put forward in the study of large linear programming problems [10–12]. The principles of Danzig-Wolfe entailed a numerous stream of works developing central ideas in various directions. If in the original version the solution of the coordinating problem is based on the method of improving the plan in linear programming, in subsequent works they proceeded from the method of simultaneous solution of the direct and dual problems, the generalized gradient descent, the use of modified Lagrange functions and other procedures.

Analysis of the existing use of aircraft and the seasonality of the AChW, the monthly maintenance of accounting and reporting documentation for airlines shows that a monthly maneuvering scheme is justified, with the aircraft returning to the base enterprise by the beginning of each month. Repeated flights in the adjacent two months, possible in a common solution for the entire planned period, can be eliminated by their subsequent analysis. Therefore, the task (1) – (5) will be solved for each "peak" month separately.

When implementing the task (1) – (5) it is advisable to consider the following.

1. In practice, it is necessary to perform calculations on the model for several objective functions.

2. In actual calculations, indexes of types of aircraft, types of maintenance organizations, airlines and serviced areas do not take all values, but only their various subsets. For example, for each airline, only a certain subset of the serviced areas can be defined, in which a certain subset of the types of maintenance should be performed. To describe this possibility, it is advisable to use in the model not an enumeration of index values from minimum to maximum, but an indication of its subset.

3. There is a need to solve the problem with a restriction on the fuel resource as a whole in terms of management, and sometimes even in the context of individual airlines.

4. The statement of the problem (1) – (5) is rigid, i.e. a situation of no valid solution is possible. However, this situation can be interpreted economically (as a deficit of the fleet of aircraft or an overestimated amount of work). Therefore, it is also of interest to the planner to analyze an admissible solution. In order to ensure the possibility of such an analysis, it is better to use instead of setting (1) – (5) a non-rigid statement with discrepancies in constraints that may become critical in the calculation process. But to match the solutions on the model (1) – (5) and the proposals of the following models by us, the sum of the residuals with the latter must be entered into the objective function with the corresponding penalty coefficient (large enough – positive) when solving the problem at minimum or very small (large negative) – to the maximum.

The need to take these four points into account caused the transition in the MANEVR dialogue complex from using the model (1) – (5) in the next model in the so-called technological formulation.

I. Calculated ratios

1. The calculation of the total costs associated with maneuvering the aircraft and the execution of works:

$$\sum_{l=1}^{L}\sum_{r\in R_l}\sum_{j\in N_r}\sum_{i\in M_{lrj}} \tilde{c}_r^{ij} x_{er}^{ij} - y_1 = 0 \qquad (6)$$

2. The calculation of the total value (tariffs) of work performed:

$$\sum_{l=1}^{L}\sum_{r\in R_1}\sum_{j\in M_{lrj}} \tilde{S}_r^{ij} x_{er}^{ij} - y_2 = 0 \qquad (7)$$

3. The calculation of the total amount of outstanding work:

$$\sum_{r\in R_e}\sum_{j\in N_r} V_i^j - y_0 = 0 \qquad (8)$$

II. Target functions

1. The minimization of the total cost of the AChW:

$$\mu y_0 + \epsilon y_1 \to min \qquad (9)$$

2. Maximizing profits from the performance of the AChW:

$$\epsilon(y_2 + y_1) - \mu y_0 \to max \qquad (10)$$

III. The main resource and technological limitations

1. Restrictions on the fleet of aircraft in the airline (in terms of aircraft days):

$$\sum_{r\in R_e}\sum_{j\in N_r} x_{er}^{ij} \le a_e^i, I = \overline{1,L}; i \in M_1 \qquad (11)$$

2. Restrictions on the volume of work performed:

$$\sum_{l=1}^{L}\sum_{i\in M_{lrj}} P_r^{ji} x_{lr}^{ij} + V_r^j = d_r^j, r \in R_1; j \in N_r \qquad (12)$$

3. Restrictions on fuel resources:

$$\sum_{r\in R_l}\sum_{j\in N_r}\sum_{i\in M_k} q_i^k x_{lr}^{ij} \le Q_l^k, l = \overline{1,L}; k = 1,2. \qquad (13)$$

or alternative

$$\sum_{L}^{l=1}\sum_{r\in R_1}\sum_{j\in N_r}\sum_{i\in M_l} q^k x_{lr}^{ij} \le Q^k, k = 1,2. \qquad (14)$$

IV. Nonnegativity conditions $x_{lr} \le 0, y_0 \le 0, Y_1 \le 0, y_2 \le 0,$

$$V_r \le 0, i \in M_1, j \in N_r, l = \overline{1,L}; r \in R_1 \qquad (15)$$

The designations of the model (6) – (15) are similar to the designations of the model (1) – (5) with the exception of the following: k - fuel type index; R_1 – a subset of the serviced areas allowed for the 1st airline; N_r – a subset of the work performed in the r-th region; M_{lrj} – a subset of aircraft types available in the l-st airline; $\tilde{C}_{lr}^{ij}$ – complex costs (specific) for the execution of works, taking into account maneuvering,

$$\tilde{C}_{er}^{ij} = C_{ie}^{l.ch.} \cdot h_e^i + C_{ie}^{l.ch.} \cdot \frac{2Ler}{V_i} \frac{h_e^i}{(T_i - \frac{2Ler}{V_i})}$$

where $C_{il}^{l.ch.}$ – cost of flight hours;

Ler – distances between airlines and serviced areas, km;

V_i – average ground speed;

T_I – time limit (monthly flying hours);

$\tilde{S}_r^{ij}$ – cost of works (calculated as a function of summation of tariff rates for a single treatment of 1 ha, established depending on the type of aircraft, type of work, the consumption rate of the substance and the rut length of the cultivated area and tariff rates per 1 hectare for approaching the aircraft from the aerodrome to the cultivated area depending on the type of aircraft and the approach distance);

a_1 – the fleet of aircraft (in terms of aircraft days);

μ and ϵ – weighting coefficients which establish the correct priorities for the objective function component ($\mu \sim 10^3$ and $\epsilon \sim 10^{-2}$);

x_{lr}^{ij} – amount of work j-th type AChWs Taken l-th airlines of a r-th region via i-th type aircraft;

V_r – the volume of outstanding work in each area;

At_0 – the amount of backlog;

I_1 is the sum of the costs;

At_2 – the total cost of the work.

In the dialogue complex "MANEVR" coefficients $\tilde{C}_{lr}^{ij}$ and $\tilde{S}_r^{ij}$ and models (6) – (15) are calculated by a specially developed program in the on-line mode.

So far, only the movement of the aircraft has been considered. But when performing chemical work, it is also practiced to move flight technical crews while maneuvering aircraft in the serviced territory of the republic. The relocation of the flight technical staff will make it possible to rationally combine the labor and material resources of airlines to obtain high economic indicators.

It is necessary to find such a maneuvering plan, in which the declared amounts of AChW will be performed with the maximum profit for the given resource constraints. Mathematically, this problem is written as follows.

Maximize

$$\sum_{i=1}^{m}\sum_{j=1}^{n}\sum_{l=1}^{L}\sum_{r=1}^{R}\sum_{t=1}^{T}(\tau_{ij}P_{ijr} - C_{ijl})X_{ijlrt}, \qquad (16)$$

under restrictions on the amount of chemical work

$$\sum_{i=1}^{m}\sum_{l=1}^{L} P_{ijr}X_{ijlrt} \le d_{jrt}, j = \overline{1,n}, l = \overline{(1,L)}; t = \overline{1,T} \qquad (17)$$

for the number of serviceable aircraft

$$\sum_{j=1}^{n}\sum_{r=1}^{R} X_{ijlrt} \leq a_{ilt}, i = \overline{1,m}, l = \overline{(1,L)}; t = \overline{1,T} \quad (18)$$

on the number of crews

$$\sum_{j=1}^{n}\sum_{r=1}^{R} F_i X_{ijlrt} \leq G_{ilt}, i = \overline{1,m}, l = \overline{(1,L)}; t = \overline{1,T} \quad (19)$$

on the number of aircraft technicians

$$\sum_{j=1}^{n}\sum_{r=1}^{R} \alpha_i X_{ijlrt} \leq V_{ilt}, i = \overline{1,m}, l = \overline{(1,L)}; t = \overline{1,T} \quad (20)$$

for consumption of fuels and lubricants

$$\sum_{j=1}^{n}\sum_{r=1}^{R}\sum_{t=1}^{T} q_{ij} X_{ijlrt} \leq Q_{lt}, l = \overline{1,L}, l = \overline{1,L}; \quad (21)$$

on the power of aviation technical bases

$$\sum_{j=1}^{n}\sum_{r=1}^{R}\sum_{t=1}^{T} \lambda_i X_{ijlrt} \leq M_{lt}, l = \overline{1,m}, l = \overline{1,L}; \quad (22)$$

on variables

$$X_{ijlrt} \geq 0, i = \overline{1,m}; j = \overline{1,n}; l = \overline{1,L}; r = \overline{1,R}; t = \overline{1,T}. \quad (23)$$

The indices and most of the designations of the model (16) – (23) coincide with the indices and designations adopted in (1) – (5). In the model (16) – (23) the following notation is added:

τ_{ij} – tariff for processing i ha of the j-th type of AChW, tenge;

G_{ilt} is the available number of crews of aircraft of type i at airline l in sub-period t;

δ_i – the number of people in the flight crew of the i-th type aircraft;

α_i – number of members of the technical personnel of the aircraft of the i-th type;

V_{ilt} is the available number of aircraft technicians of the i-th type aircraft at airline l in sub-period t;

λ_i – the specific volume of maintenance of the i-th type aircraft, g;

M_{il} is the total available capacity of aviation technical bases of airlines l for servicing aircraft of the i-th type, h.

Information support of the task is based on data on aircraft parameters, technological features of the performance of certain types of AChW, flight distance to the work site, etc.

As a result of solving the problems of linear programming (16) – (23), we will get not only the optimal maneuvering scheme $\{x^*_{ijlrt}\}$, but also a number of important matters for the user of technical and economic service indicators. So, thanks to the obtained maneuvering scheme, the optimal fuel reserves at the aviation enterprises, the deficit and excess of the fleet, the expedient redeployment of the fleet, etc. are easily determined. Experimental calculations using the model (6) – (15) were performed on real data from the Aviation Administration Department in the national economy of the Kazakh Civil Aviation Authority using the developed and specialized dialogue complex "MANEVR" [2]. Consider the specific results of calculations of optimal plans, made using the complex "MANEVR" for the "peak" period we give quantitative estimates of their effectiveness.

The results of the optimization calculations showed that the share of work performed by the aviation enterprises of the Kazakh Civil Aviation Authority due to maneuvering in the total volume of the AChW is rather high and amounts to 36.6%. Out of 18 airlines that service 19 regions of the republic, 17 airlines perform the work on the territory of other regions (maneuver) according to the optimal plan, and only the Atyrau airline carries out AChW exclusively in the area of its deployment. The amount of work performed by maneuvering increases significantly even as compared with the optimal plan calculated using fixed connections (36.6% vs. 31.7% in terms of fixed links). This allows us to conclude that, at least for the "peak" period, maneuvering the fleet is an important factor in increasing the efficiency of the performance of maintenance of operational personnel in the republic. Expanding maneuvering capabilities by taking into account the full links between airlines when servicing areas allows you to calculate more effective plan options, both in terms of the costs of performing maintenance, and the net cost of maneuvering. For example, for each month of a "peak" period, the costs of performing an AChW in the optimal plan calculated on full links are, on average, 2–5% lower than similar costs in the optimal plan calculated on fixed links. A similar comparison of the net cost of maneuvering results in 10-50% cost reduction. The effectiveness of the optimization calculations of the maneuvering plan is also visible in terms of its level of balance.

Complex "MANEVR" allows you to calculate the profitability ratios carried out AChW as a whole by default, and in terms of airline. A comparison of the achieved level of profitability in the basic calculation (for fixed connections) with the optimal plan (for full connections) shows that with optimal maneuvering of the fleet during the "peak" period, the profitability of management as a whole may increase by at least 10-20%, while an increase in the share of highly profitable airlines (from 80% profitability and above) on average 60%.

Since the optimization calculations were carried out according to two criteria – minimum costs and maximum profits, it should be clarified that the highest profitability rates of airlines are achieved in terms of maximum profits. Moreover, this criterion provides the most narrow range of differences in the level of profitability of AChW airlines. Comparison of calculations for the minimum cost and maximum profit showed that in the first case, the sun than in the second case.

In the previously implemented methodology, optimization calculations using the model (6) – (15) using the MANEVR dialogue complex were also carried out on real data of the Ukrainian and Kazakh Civil Aviation Authorities (GA) for 2010–2019.

Kazakhstan is one of the leading republics in the use of aviation in the grain industry. The Kazakh Civil Aviation Authority annually handles more than 12 million hectares of land, or about 20% of the total agricultural inventory performed in the CIS. Of these, fertilizer application accounts for 978.5 thousand hectares, plant protection from pests, diseases and weeds — 5629.2 thousand hectares, defoliation and desiccation of grain crops — 223.6 thousand hectares.

Currently, the AChWs performed by the Kazakh Department of Civil Aviation comprise several dozen species. Each type is characterized by the purpose, terms, conditions of performance.

On the territory of the republic in its various areas there are 18 airlines. Each of them has a certain fleet of one or more types of aircraft, designed to perform AChW. All types of aircraft are sufficiently universal, i.e. can be used to perform the entire spectrum of basic AChW. The main types of aircraft that are available in airline control companies include the An-2 aircraft and the Mi-2 and Ka-26 helicopters.

An-2 airplanes are more economical with the application of mineral fertilizers and with similar labor-intensive works. Therefore, they remain the main aviation technology for the production of chemical works in the grain industry. With their help, more than 90% of the total volume of AChW is performed annually. Recently, the possibility of mass production of An-3 aircraft, which are a deep modification of the An-2 aircraft, is being studied.

The use of Mi-2 and Ka-26 helicopters provides in some cases higher efficiency. However, the cost of work is also increasing, since helicopters are a relatively expensive vehicle.

Out of 18 airlines that serve 19 regions of the republic, 17 airlines, according to the optimal plan, perform work in the territory of other regions (maneuver) and in the area of their deployment. From table 1 it is clear that the largest amount of aviation work is carried out to combat weeds and pests and diseases.

The share of work performed by airlines due to maneuvering, to some extent, characterizes the efficiency (rationality) of locating the airlines themselves in the territory of the republic. Aviation enterprises, in which the percentage of work performed by maneuvering is high (50-80%), should be considered, apparently, as irrationally located (in the part of agricultural aviation), since they perform most of the work in other areas. And to perform work on their territory, even in the "peak" period, they have an excess of Armed Forces. Another group of airlines should include those that practically do not perform work in other areas. The fleet of these airlines is usually not enough to carry out work in their field during the entire "peak" period. According to the results of optimization calculations, such airlines should primarily include Burunday and Kyzylorda. The remaining airlines form a group with an average percentage of the amount of work performed by maneuvering. These enterprises, in our opinion, are most rationally located and optimal in size for the fleet of agricultural use. Their fleet provides more than half of the work in their "own" area on their own, and maneuvering avoids the irrational use of aircraft in certain months with lower volumes of declared work.

Table 1. Distribution of AChW between enterprises of the Kazakh State Aviation Administration according to the optimal maneuvering plan in the "peak" period

Aviation enterprise	Processed, thousand hectares			
	Mineral fertilizers	Weed control	Pest and disease control	Defoliation and desiccation
Burunday	21.2	21.7	8.3	-
West Kazakhstan	30.0	97.9	155.0	-
Aktobe	4.5	175.6	66.9	-
Karaganda	52.8	160.6	174.4	-
Kostanay	69.0	448.0	262.3	-
Atyrau	-	-	-	-
East Kazakhstan	3.5	84.8	53.3	7.8
Turkestan	281.6	256.8	154.2	35.0
Zhambyl	90.2	223.1	145.5	39.0
Akmola	100.6	551.4	255.2	57.7
Semipalatinsk	51.6	96.3	201.1	4.0
Kokshetau	45.7	262.1	168.4	18.3
Pavlodar	4.9	87.8	498.1	24.8
North Kazakhstan	29.7	193.1	71.5	-
Kzyl-Orda	50.3	15.8	0.7	35.3
Taldykorgan	96.2	79.6	129.4	-
Turgai	52.7	224.3	240.0	-
Zhezkazgan	-	11.4	54.6	-
Total management	978.5	2990.3	2638.9	223.5

Optimization calculations also make it possible to characterize the level of specialization of airlines in the performance of various maintenance companies and to evaluate the effectiveness (rationality) of their deployment in the republic at the present time.

4 Conclusion

The analysis of the calculations performed by the Kazakh State Aviation Administration show the high efficiency of using optimization methods and models for solving the problem of using agricultural aviation in the performance of chemical work. And the practical value of the optimization approach to solving this problem is directly provided by the territorial administration of the GA fully provided by the calculation technology implemented in the MANEVR dialogue complex.

References

[1] A. Saparbayev, *Transport logistics in grain processing* (Fortuna Polygraph LLP, 2019)

[2] A. Iztaev, T. Kulazhanov, A. Saparbayev, *Innovative technologies and logistics of processing businesses in AIC* (Fortuna Polygraph LLP, 2019)

[3] A. Bakaev, V. Revenko, V. Gritsenko, *Information technology in transport*, Vol. 3. Water transport (Naukova dumka, Kyiv, 2009)

[4] E. Vasilyeva, V. Levit, V. Livshits, *Nonlinear transport problems on networks* (Finansy i statistika, Moscow, 1981)

[5] E. Krylatykh, A. Strokov, *Food Security of Russia and the World: Theory and Methodology of Research, Analysis of Provision, Opportunities and Threats* (Infra-M, 2019)

[6] A. Ganicheva, *Mathematical methods and models in agroindustrial complex* (Tver State Agricultural Academy, Tver, 2019)

[7] A. Makulova, *Model assessment of agricultural production management* (Fortuna Polygraph LLP, 2019)

[8] L. Lesdon, *Optimization of large systems* (Nauka, Moscow, 1975)

[9] H.A. Taha, *Operations Research: An Introduction*, 10th edn. (Pearson, 2017)

[10] G.B. Dantzig, M.N. Thapa, *Linear programming 2: theory and extensions* (Springer Science & Business Media, 2006)

[11] M.R. Bartolacci, L.J. LeBlanc, Y. Kayikci, T.A. Grossman, Journal of Business Logistics **33**, 118 (2012)

[12] M. Savelsbergh, J.H. Song, Computers & Operations Research **35**, 2266 (2008), part Special Issue: Includes selected papers presented at the ECCO'04 European Conference on combinatorial Optimization

Improving the innovative development management of Zaporizhzhia region's industrial complex

Tetiana Pulina[1], *Nadiia* Shmygol[1,*], *Oleksii* Narivs'kyi[1], *Władysława* Łuczka[2], and *Yeugeniia* Shmygol[3]

[1]National University "Zaporizhzhia Polytechnic", 64 Zhukovsky Str., Zaporizhzhia, 69600, Ukraine
[2]University of Life Sciences in Poznań, 28 Polish Army Str., 60-637 Poznan, Poland
[3]Warsaw University of Technology, 1 Politechniki Sq., 00-661 Warsaw, Poland

Abstract. The paper carries out the strategic analysis of innovative possibilities of Zaporizhzhia region's industrial complex and investigates the indicators of innovative activity of industrial complexes of Zaporizhzhia region and Ukraine by Harrington's generalized desirability function. The evaluation results prove the "good" ability and readiness of the industrial complex of Zaporizhzhia region and Ukraine to carry out effective innovation activities. However, according to the comparative assessment of the innovative activity of Zaporizhzhia region's industrial complex and Ukraine's one, Zaporizhzhia region's industrial complex is found to lag 9.2% behind Ukraine's. A cluster development strategy is suggested to improve the innovative development management of Zaporizhzhia region's industrial complex.

1 Introduction

The modern model of sustainable development of the national economy as a whole and its regional components is based on the continuous conversion of modern achievements of scientific and technological progress into the latest technologies, quality goods and services. Therefore, the reproduction of productive forces and manufacturing relations at the regional level requires radical changes, both within the industrial complex of the region and its sub-complexes. These changes are determined by a combination of internal and external factors, namely innovative technological changes in the traditional real economy. Therefore, special attention is paid to defining strategic directions of innovative development of Zaporizhzhia region's industrial complex by assessing enterprises' innovative activity in the industrial complex. Cluster development models are used to improve the innovative development management of industrial complexes in most industrialized countries.

In this context, it is of particular importance to define long-term strategic directions of innovative development of the region's industrial complex.

Issues of innovative development of the regions are reflected in the works [1–23].

The conducted research offers basic guidelines for assessing the development of the industrial complex of the region, as well as develops scientific approaches and recommendations for the management of the region's industrial complex. However, in general, the problem of defining strategic directions of innovative development of the region's industrial complex remains open; moreover, the lack of innovative development strategy of the industrial complex slows down the development of projects, programs and scenarios of socio-economic development.

The *purpose* of the paper is to improve the innovative development management of Zaporizhzhia region's industrial complex.

The following *methods of economic research* were used in the research process: abstract-logical (for theoretical generalizations and formulation of conclusions), statistical-economic (diagnostics of the state and assessment of innovative development of the region's industrial complex).

2 The industrial potential of Zaporizhzhia region

Zaporizhzhia region is one of the most economically attractive regions of Ukraine due to great industrial and agricultural potential, natural, as well as own energy resources, high scientific and technical potential, developed transport infrastructure, developed banking system and accessibility to the markets of Ukraine, CIS, Europe and Asia. There are more than 160 powerful industrial enterprises in the region. States, particularly by the exchange of good practices [9].

The industrial complex of the region provides 8.2% of the national volume of sold industrial products (UAH 202.3 billion in 2019) – this is the 4th place among the regions (table 1).

The development of the region's industrial complex is a required condition to achieve long-term competitive advantages and, accordingly, the effective operation of enterprises in the long run.

Most enterprises in the region need implementing effective modern energy-saving production technologies, as

*e-mail: nadezdash@ua.fm

Table 1. The volume of sold industrial products by individual regions in 2019 [24]

Territory	Volume of sold industrial products in 2019	
	UAH mln.	% to total production sold
Ukraine	2480804.2	100.0
Dnipropetrovsk region	454124	18.3
Donetsk region	283946	11.4
Kyiv (city)	232979.5	9.4
Zaporizhzhia region	202366.2	8.2
Kharkiv region	185639.2	7.5
Others	1118842.7	45.19

well as introducing effective measures to support domestic producers in the local market, by creating regional clusters in promising areas of the region's development [25].

The basis of the region's industrial complex is metallurgical and energy complexes which produce 23.1% of total iron, 34.3% of steel, 8.9% of coke, 25.9% of electricity in Ukraine [24].

In the industrial complex, the leading place is occupied by metallurgy (35.7%), the second place in the structure of industry belongs to the power industry (24.2%), in the third place is mechanical engineering (11.4%) [26].

The industrial potential of Zaporizhzhia region is strengthened by a large number of branch scientific, research and designing institutes, which are fully capable of promoting the state policy of creating a strong industrial base that not only meets domestic needs, but is productively competitive on the world market

Recently, science, technology and innovation have become increasingly important for the economic development of both the region and the state as a whole, which provide a basis for effective development by generating new knowledge. It is innovation that becomes a prerequisite to create additional wealth and form the socio-economic paradigm of modern society.

In 2019, the region is not among the leading ones in terms of innovation processes activity, although this indicator is still higher than its average Ukrainian value. In 2019, the region ranked 4th among the regions of Ukraine in terms of the share of innovatively active enterprises, while in 2013 it was the leader of innovative activity in Ukraine and took the 1st place.

In recent years, in the region there has been a tendency to reduce the number of innovative enterprises (figure 1) [26].

Thus, in 2019 in Zaporizhzhia region, 47 enterprises and organizations of the region were engaged in innovation activities, or 19.36% of the total number of surveyed against 49 enterprises in 2015 and 56 in 1995. 30 enterprises (63.8% of total number of innovation- active enterprises) within the framework of innovation activities were engaged in purchasing equipment, 11 innovation-active

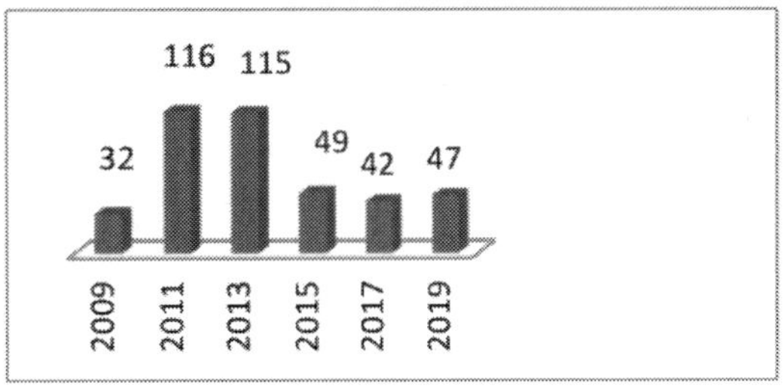

Figure 1. Number of enterprises and organizations that carry out scientific research and development

enterprises (23.4%) conducted their own internal research and development and only 5 enterprises (12.8%) were engaged in market innovation [26]. The number of industrial enterprises that have implemented innovations is given in table 2. The predominant number of innovation-active enterprises of Zaporizhzhia region is concentrated in the regional center – Zaporizhzhia (more than 90%), the second and third places in the number of innovation-active enterprises are occupied by Berdyansk and Melitopol.

Table 2. Number of industrial enterprises that implemented innovations

Year	2010	2015	2017	2019
Total, units	19	45	37	43
implemented new technological processes, units	11	20	21	32
per cent to the total number of companies implementing innovation	57.9	44.4	56.8	74.4
including low-waste, resource-saving and waste-free companies, units	6	11	7	8
per cent to the total number of enterprises,	31.6	24.4	18.9	18.6
mastered the production of innovative types of products, units	14	35	26	21
per cent to the total number of companies implementing innovation	73.7	77.8	70.3	48.8
– incl. types of equipment, units	3	21	16	13
per cent to the total number of companies implementing innovation	15.8	46.7	43.2	30.2
sold innovative production, units	18	42	31	25
per cent to the total number of companies implementing innovation	94.7	93.3	83.8	58.1

3 The main type of innovative activity of industrial enterprises

The main type of innovative activity of industrial enterprises in 2019 and so forth continues to be the development of innovative products manufacturing. 21 enterprises, or 48.8% of the total number of industrial enterprises that implemented innovations, are engaged in such activities. In 2013, there were 44 such enterprises. 32 enterprises introduced new technological processes (compared to 35 in 2013), but only every 3rd of these enterprises implemented resource-saving and waste-free technologies. The volume of sold innovative products decreased from UAH 3,162.3 mln. in 2017 to UAH 2,801.82 mln. in 2019, or by 11.4%. In terms of the volume of sold innovative products, the region in 2019 ranked 8th in Ukraine. The region accounted for 8.2% of the total volume of sold innovative products in Ukraine.

The volume of fundamentally new innovative products also remains insignificant. Industrial enterprises prove to sell innovative new products primarily for enterprises, not for the market.

The decrease in the output of innovative products in the region is largely due to innovation costs reduction. Thus, during 2017–2019 in the region the innovation activities costs were reduced from UAH 1,393.4 mln. to UAH 681.7 mln., or by 2.1 times. In 2019, the region ranked 9th in the total innovation funding in Ukraine. It accounted for 4.8% of its total volume.

Regarding the structure of innovation funding sources, in 2019, as in previous years, the key ones are the enterprises' own funds (98.05%). A number of large-scale innovative projects have been implemented at the region's enterprises, which are aimed at ecological and energy-efficient renovation, modernization of production and development of new types of products. The main innovation funding source over the past ten years are the enterprises' own funds (table 3).

Table 3. The industrial enterprises' innovation funding sources

Year	Total costs	Incl. at the expense of funds, UAH thous.			
		Own funds	Domestic investors	Foreign investors	Other sources
2009	163878.0	160788.5	–	–	623.8
2010	128595.1	127895.1	–	–	–
2011	800427.2	648835.4	12682.5	9463.5	130331.2
2012	242990.6	235426.2	–	–	6459.4
2013	298671.5	298667.5	–	–	4.0
2014	339943.9	335729.6	–	–	4214.3
2015	321051.3	321051.3	–	–	–
2017	1393360.2	1161569.9	–	–	231790.3
2019	681707.2	668400.3	–	3200	3400

The main direction of innovation expenditures during 2009–2019 in the region is still the purchase of machinery, equipment and software. In 2019, expenditures in this area of innovation amounted to UAH 514.4 mln., which was 75.5% of the total expenditures on innovation. Industrial enterprises spent the least on acquiring other external knowledge. In 2019, it amounted to UAH 0.3 mln., which is only 4.54% of the total innovation costs. Expenditures on purchasing machinery, equipment and software increased the most during this period.

It should be noted that most powerful industrial enterprises that are the main consumers of innovative products have their own research institutions or cooperate with leading industry research institutions of Ukraine which compete significantly with private innovative firms.

Due to the high cost of research and development (R&D) and designing, some companies refuse to develop and implement innovative products of their own production, explaining this by decrease in their competitiveness and insufficient financial support from the state.

Taking into account that there is an active market for innovation both within the country and abroad, in the region in most cases there is no mechanism of interaction between consumers and manufacturers of innovative products. The reason for this situation is the lack of an organizational system in the region, which allows to purposefully transform the results of scientific and technical activities into existing technologies and new enterprises that produce unique, competitive products on the world market.

Zaporizhzhia region has a very high innovation potential. However, in recent years, innovation and investment activities of the region's enterprises have not been developing effectively enough, despite the significant attention of the state to this field.

4 The integral indicators for assessing the innovative development of Zaporizhzhia region's industrial complex

To outline the innovation strategy of the region's industrial complex, it is necessary to determine the factors influencing the innovative development of Zaporizhzhia region's industrial complex and evaluate its innovative activity. To define the favorable and problematic factors influencing innovative development, it is relevant to evaluate the innovative activity of Zaporizhzhia region's industrial enterprises according to the methods proposed by O. B. Zhykhor and L. M. Matrosova [27, 28]. Data to calculate a comprehensive indicator of innovative development of Zaporizhzhia region's industrial complex are given in (table 4) [26].

Data to calculate a comprehensive indicator of innovative development of Ukraine industrial complex in 2009–2019 (table 5) [24].

To calculate innovative development indicators of Zaporizhzhia region's industrial complex in 2009-2019 and innovative development indicators of Ukraine's industrial complex in 2009-2019, there are such evaluation criteria:

1. The breadth of coverage of the region's enterprises with innovative activities:

Table 4. Data to calculate a comprehensive indicator of innovative development of Zaporizhzhia region's industrial complex in 2009–2019

Output data	Years								
	2009	2010	2011	2012	2013	2014	2015	2017	2019
Number of IAE N_{ia}	31	22	116	115	115	108	49	42	47
Number of surveyed enterprises (N)	441	415	421	386	399	470	403	384	397
Volumes of sold products UAH mln. (in actual prices), UAH mln. (Q_z)	53095.4	66768	81503.2	82505.7	78490	97797.6	124249.8	197495.3	202366.2
Manufacturing of new products types mastered, articles ($r_{3\ vid}$)	132	114	619	446	397	611	397	319	209
New technological processes implemented (H_m)	134	170	509	114	207	212	114	142	156
Total amount of innovation costs, UAH mln. (D_{ob})	163.88	128.6	800.4	243	298.7	339.9	321.1	1393.4	681.7
Number of employees by industrial activity type, thous. people (P)	176.5	175	173.6	170	163.1	151.2	149.6	138.8	137.3
Volume of shipped innovative products, UAH mln. (V_n)	2144.9	1979	2490.3	1671.6	1671.1	1530	3162.3	4041.2	2801.8

  - Share of IAE in their total number, % ($N_{ia}/N \cdot 100\%$);
  - Share of shipped innovative products in total industrial output, % ($V_z/Q_z \cdot 100\%$);

2. The range of distribution of types and objects of innovation among innovation-active enterprises:
   - Average number of new types of industrial products per IAE ($r_{3\ vid}/N_{ia}$);
   - Average number of new technologies per IAE (H_m/N_{ia}).
3. The degree of financial support for innovation activities of enterprises involved in developing and implementing innovations:
   - Average amount of innovation costs per one innovatively active enterprise, thous. UAH (D_{ob}/N_{ia});
   - Average number of employees per IAE (P/N_{ia}).

Economic and mathematical modeling based on a generalized desirability function (or Harrington's scale) was used to determine the complex indicator of innovative development of the industrial complex.

It should be noted that the integrated indicator of innovative development of machine-building enterprises based on the Harrington function varies in the range {0; 1}. Moreover, the closer it is to the unit, the greater the level of machine-building enterprises' innovative development is.

In general, the values of the Harrington function are interpreted as follows (table 6).

Table 5. Indicators of innovative development of Ukraine's industrial complex of Ukraine in 2009–2019

Output data	Years								
	2009	2010	2011	2012	2013	2014	2015	2017	2019
Number of IAE N_{ia}	1411	1462	1850	1758	1715	1609	1576	672	687
Number of surveyed enterprises (N)	11023	10594	11454	10089	10208	10010	10002	4978	4699
Volumes of sold products UAH mln. (in actual prices), UAH mln. (Q_z)	806550.6	1065851	1331887.6	1400680.2	1354130.1	1389140.5	1776603.7	2608027.7	2938830.8
Manufacturing of new products types mastered, articles ($r_{3\,vid}$)	806550.6	1065851	1331887.6	1400680.2	1354130.1	1389140.5	1776603.7	2608027.7	2938830.8
New technological processes implemented (H_m)	1893	2043	2510	2188	1576	1743	2117	1831	2318
Total amount of innovation costs, UAH mln. (D_{ob})	7949.9	8045.5	1433.9	11480.6	9562.6	7695.9	13813.7	9117.5	14220.9
Number of employees by industrial activity type, thous. people (P)	3031	2860	2828	2804	2673	2297	2040	1894	1867
Volume of shipped innovative products, UAH mln. (V_n)	31432.3	33697.6	50511.74	36157.8	35891.6	25669	23050	17714.2	34264.9

Table 6. Interpretation of estimates of the complex level of machine-building enterprises' innovative development based on the Harrington function

№	Values of the integrated indicator of machine-building enterprises' innovative development on the scale of the desirability function	Interpretation of the level of machine-building enterprises' innovative development
1	$D > 0.8$	High level of innovative development
2	$0.63 < D < 0.8$	Good level of innovative development
3	$0.37 < D < 0.63$	Satisfactory level of innovative development
4	$0.2 < D < 0.37$	Unsatisfactory level of innovative development
5	$D < 0.2$	Low level of innovative development

scale; y_{r_i} – an innovation development indicator in a dimensionless form.

Table 7. Natural values of indicators of enterprises' innovative development of the region's industrial complex

Year	r_1, %	r_2, %	r_3, units	r_4, units	r_5, UAH thous.	r_6, people
2009	7.03	4.04	4.26	4.32	5.29	5.69
2010	5.30	2.96	5.18	7.73	5.85	7.95
2011	27.55	3.06	5.34	4.39	6.90	1.50
2012	29.79	2.03	3.88	0.99	2.11	1.48
2013	28.82	2.13	3.45	1.80	2.60	1.42
2014	22.98	1.56	5.66	1.96	3.15	1.40
2015	12.16	2.55	8.10	2.33	6.55	3.05
2017	10.94	2.05	7.60	3.38	33.18	3.30
2019	11.84	1.38	4.45	3.32	14.50	2.92

Table 8. Determining the dimensionless type of innovative development indicators of the region's industrial complex

Year	y_{r_1}	y_{r_2}	y_{r_3}	y_{r_4}	y_{r_5}	y_{r_6}
2009	0.55	1.04	2.24	3.22	0.94	2.65
2010	0.38	0.94	3.15	5.53	1.06	4.07
2011	1.71	0.81	3.05	3.23	0.89	0.98
2012	1.71	0.78	2.00	0.80	0.32	0.93
2013	1.72	0.80	1.89	1.96	0.47	0.91
2014	1.43	0.85	2.49	1.81	0.66	0.98
2015	0.76	1.96	4.16	3.08	0.76	2.41
2017	0.81	3.01	2.14	1.24	2.45	1.17
2019	0.81	1.19	1.42	0.98	0.70	1.07

The natural values of enterprises' innovative activity indicators in the region are determined on the basis of enterprises' quantitative characteristics. For example, the indicator r_1 (share of IAE) is the ratio of the number of innovatively active enterprises to the total number of surveyed

The basis of this indicator is the idea of transforming the natural values of each innovation development indicator (table 7) into a dimensionless form in (table 8), followed by the definition of partial i functions by Harrington's scale d_{ri} (table 9) and a comprehensive indicator of innovative development of the region's industrial complex D (table 10) and Ukraine's (table 11):

$$D = \sqrt[r]{\prod_{i=1}^{r} d_{r_i}}, \quad d_{r_i} = \exp(-\exp(-y_{r_i})), \qquad (1)$$

where r – the number of indicators used to assess innovation development; d_{r_i} – a partial function by Harrington's

Table 9. Calculation of partial functions due to years

Year	d_{r_1}	d_{r_2}	d_{r_3}	d_{r_4}	d_{r_5}	y_{r_6}
2009	0.56	0.70	0.90	0.96	0.68	0.93
2010	0.51	0.68	0.96	1.00	0.71	0.98
2011	0.83	0.64	0.95	0.96	0.66	0.69
2012	0.83	0.63	0.87	0.64	0.49	0.67
2013	0.84	0.64	0.86	0.87	0.53	0.67
2014	0.79	0.65	0.92	0.85	0.60	0.69
2015	0.63	0.87	0.98	0.95	0.63	0.91
2017	0.64	0.95	0.89	0.75	0.92	0.73
2019	0.64	0.74	0.79	0.69	0.61	0.71

Table 10. Determining the dimensionless type of innovative development indicators of Ukraine's industrial complex for 2009–2019

Year	y_{r_1}	y_{r_2}	y_{r_3}	y_{r_4}	y_{r_5}	y_{r_6}
2009	0.84	1.66	0.86	0.85	0.64	1.14
2010	0.90	1.35	0.74	0.89	0.63	1.03
2011	1.06	1.62	0.79	0.86	0.89	0.81
2012	1.14	1.10	0.87	0.79	0.75	0.84
2013	1.10	1.13	0.82	0.58	0.64	0.82
2014	1.05	0.79	1.03	0.69	0.55	0.75
2015	1.05	0.55	0.88	0.48	0.98	0.67
2017	0.89	0.29	1.60	1.73	1.55	1.49
2019	0.96	0.50	1.41	2.14	2.37	1.44

Table 11. Calculation of partial functions due to years

Year	d_{r_1}	d_{r_2}	d_{r_3}	d_{r_4}	d_{r_5}	y_{r_6}
2009	0.65	0.83	0.65	0.65	0.59	0.73
2010	0.67	0.77	0.62	0.66	0.59	0.70
2011	0.71	0.82	0.63	0.65	0.66	0.64
2012	0.73	0.72	0.66	0.63	0.62	0.65
2013	0.72	0.72	0.65	0.87	0.59	0.64
2014	0.71	0.63	0.70	0.85	0.56	0.62
2015	0.71	0.56	0.66	0.95	0.69	0.60
2017	0.66	0.47	0.82	0.75	0.81	0.80
2019	0.68	0.54	0.78	0.69	0.91	0.79

enterprises, it is measured in % (2):

$$r_1 = \frac{N_{ia}}{N} \cdot 100\%. \quad (2)$$

The average value of innovative costs r_5 is the ratio of the total number of innovation costs to the number of innovatively active enterprises (3),

$$r_5 = \frac{D_{ob}}{N_{ia}}. \quad (3)$$

It is measured in monetary units (UAH thous.) (table 7).

In order to use an integrated efficiency indicator (in the work – the geometric mean), it is essential for the indicators used to have the same units of measurement. Therefore, the transition from natural values of enterprises' innovative activity indicators in the region to dimensionless. Each indicator refers to a similar one calculated for this period for Ukraine as a whole (4), (table 8):

$$y_{r_1} = \frac{region's}{Ukraine's}. \quad (4)$$

Table 12 provides the results of calculating the comprehensive indicator of innovative development of Zaporizhzhia region's and Ukraine's industrial complexes in 2009–2019.

Table 12. Integral indicators of innovative development of Zaporizhzhia region's industrial complex and Ukraine's industrial complex in 2009–2019

Year	Integral indicators of enterprises' innovative development			
	Zaporizhzhia region's industrial complex		Ukraine's industrial complex	
	Inx	Changes to the previous year	*Inn*	Changes to the previous year
2009	0.77	–	0.68	–
2010	0.78	0.01	0.77	-0.01
2011	0.78	0	0.68	0.01
2012	0.68	-0.10	0.67	-0.01
2013	0.72	0.04	0.65	-0.02
2014	0.74	0.02	0.64	-0.01
2015	0.81	0.07	0.62	-0.02
2017	0.81	0	0.72	0.10
2019	0.69	-0.12	0.76	0.04

According to the results of data analysis (table 12), it turns out that the integral indicators for assessing the innovative development of Zaporizhzhia region's industrial complex during 2009–2019 are unstable, varying from 0.68 to 0.81.

These indicators prove the "good" ability and readiness of Zaporizhzhia region's industrial complex to carry out effective innovation activities.

In particular, for Ukraine, the integrated indicators of innovative development of the industrial complex vary from 0.62 to 0.77. It was found that the indicator of innovative development of Zaporizhzhia region's industrial complex in 2019 is 9.2% less than Ukraine's industrial complex. This indicates that the innovative development of Zaporizhzhia region's industrial complex lags behind Ukraine's one.

Among the factors favorable for embodying the region's innovation potential, there should be noted the high potential of the region's enterprises to innovate, as well as the relatively high level of R&D expenditures. The most problematic factors in the region's innovative development are the quality of research institutions and the low correlation of foreign direct investment with new technologies transfer.

The assessment of the enterprises' innovative development in Zaporizhzhia region's industrial complex indicates the need to intensify innovation and optimize innovation potential. In order to further develop the region's industrial complex, it is necessary to move to its innovative development model. The optimal model of the innovation strategy of Zaporizhzhia region's industrial complex is a cluster development strategy [29].

5 Conclusions

In order to improve the innovative development management of Zaporizhzhia region's industrial complex, the innovative strategy of the region's industrial complex is substantiated by assessing the enterprises' innovative activity on an integrated indicator, which is a generalized desirability function. It is established that the comprehensive indicator of innovative development of Zaporizhzhia region's industrial complex in 2019 is 9.2% less than Ukraine's industrial complex. In addition, the volume of sold innovative products in the region in 2019 decreased by 11.4% compared to 2017. The decrease in the output of innovative products in the region is largely due to innovation costs reduction from UAH 1,393.4 mln. to UAH 681.7 mln., or by 2.1 times. This is due to the declining role of the state as a direct investor, the lack of private investors and the low level of innovation and investment infrastructure in Zaporizhzhia region.

It has been found necessary to intensify innovation and transition to an innovative development model. The cluster development strategy is substantiated by the innovation strategy model of Zaporizhzhia region's industrial complex.

References

[1] A. Granberg, *Osnovy regional'noj jekonomiki (Fundamentals of Regional Economics)* (GUVShJe, 2000)

[2] S. Ischuk, *Promyslovi kompleksy Ukrainy: Naukovi osnovy terytorial'noi orhanizatsii (Industrial complexes of Ukraine: Scientific bases of territorial organization)* (Palyvoda A.A., 2003)

[3] A. Marshall, *Principles of economics: unabridged eighth edition* (Cosimo, Inc., 2009)

[4] M.E. Porter, *On competition* (Harvard Business Press, 2008)

[5] J. Schumpeter, U. Backhaus, *The theory of economic development* (Springer, 2003)

[6] O. Prokopenko, O. Kudrina, V. Omelyanenko, CEUR Workshop Proceedings **2105**, 258 (2018)

[7] O. Prokopenko, V. Omelyanenko, T. Ponomarenko, O. Olshanska, Periodicals of Engineering and Natural Sciences **7**, 752 (2019)

[8] T. Ponomarenko, V. Khudolei, O. Prokopenko, J. Klisinski, Problems and Perspectives in Management **16**, 85 (2018)

[9] G. Savitskaya, *Methodology of a comprehensive analysis of economic activity (Metodika kompleksnogo analiza hozyaystvennoy deyatelnosti)* (INFRA-M, 2007)

[10] Yale Center for Environmental Law & Policy, *Environment performance index. methodology* (2018), `https://epi.envirocenter.yale.edu/2018-epi-report/methodology`

[11] V. Boiko, A. Kwilinski, M. Misiuk, L. Boiko, Economic Annals-XXI **175**, 68 (2019)

[12] T. Savchenko, N. Basiurkina, O. Rodina, A. Kwilinski, Manag. Th. St. Rur. Bus. Infr. Dev. **41**, 43 (2019)

[13] I. Perevozova, N. Shmygol, D. Tereshchenko, K. Kandahura, O. Katerna, J. Sec. Sust. **9**, 139 (2019)

[14] J. Polcyn, S. Stepień, A. Tosovic-Stevanovic, *Environmental Governance as an Element of Sustainable Development in the European Union Countries*, in *41st International Scientific Conference on Economic and Social Development* (2019), Economic and Social Development, pp. 111–121

[15] L. Zavidna, P. Makarenko, G. Chepurda, O. Lyzunova, N. Shmygol, Academy of Strategic Management Journal **18** (2019)

[16] V. Tkachenko, A. Kwilinski, O. Korystin, N. Svyrydiuk, I. Tkachenko, J. Sec. Sust. **8**, 379 (2019)

[17] A. Kwilinski, I. Ruzhytskyi, V. Patlachuk, O. Patlachuk, B. Kaminska, Journal of Legal, Ethical and Regulatory Issues **22**, 1 (2019)

[18] O. Baklanova, M. Petrova, V. Koval, Ikonomicheski Izsledvania **29**, 68 (2020)

[19] N. Shmygol, W. Luczka, O. Trokhymets, D. Pawliszczy, R. Zavgorodniy, E3S Web of Conferences **166**, 13005 (2020)

[20] V. Derbentsev, S. Semerikov, O. Serdyuk, V. Solovieva, V. Soloviev, E3S Web of Conferences **166** (2020)

[21] A. Lobanova, V. Bayura, Y. Viznytsia, L. Bratchenko, V. Karitka, E3S Web of Conferences **166**, 10020 (2020)

[22] D. Shiyan, I. Ostapchuk, O. Lakomova, E3S Web of Conferences **166**, 01012 (2020)

[23] A. Bielinskyi, I. Khvostina, A. Mamanazarov, A. Matviychuk, S. Semerikov, O. Serdyuk, V. Solovieva, V. Soloviev, IOP Conference Series: Earth and Environmental Science **628**, 012019 (2021)

[24] N. Shmygol, O. Cherniavska, T. Pulina, R. Zavgorodniy, Energy Policy Journal **23**, 135 (2020)

[25] *Official State Statistics Committee of Ukraine*, `http://www.ukrstat.gov.ua`

[26] *Official site of the Main Department of Statistics in Zaporizhzhia region*, `http://www.zp.ukrstat.gov.ua//`

[27] O. Zhykhor, Kommunal'noe khoziajstvo horodov. Nauchno-tekhnycheskyj sbornyk **73**, 56 (2006)

[28] L. Matrosova, *Formuvannia orhanizatsijno-ekonomichnoho mekhanizmu upravlinnia innovatsijnymy protsesamy u promyslovosti [Formation of organizational and economic mechanism of management of innovative processes in industry]* (SUDU, 2000)

[29] T. Pulina, T. But, O. Khrystenko, V. Zaytseva, Lecture Notes in Civil Engineering **73**, 709 (2020)

https://doi.org/10.1051/shsconf/202110706003

Modeling grain transportation in the system of grain processing industries

Abdizhapar Saparbayev[1,2,*], *Aiymzhan* Makulova[3,**], *Nesipkul* Bayboltaeva[4,***], *Nagima* Sarsenbieva[5,****], and *Aliya* Imatayeva[3,†]

[1]Academy Kainar, 7a Satpayev Str., Almaty, 050059, Kazakhstan
[2]International Academy of Innovative Technologies, 33 mkr. Samal-2, Almaty, 050059, Kazakhstan
[3]University Narxoz, 55 Zhandosov Str., Almaty, 050035, Kazakhstan
[4]NJSC "Kazakh National Agrarian University", 8 Abay Ave., Almaty, 050010, Kazakhstan
[5]Mukhtar Auezov South Kazakhstan University, 5 Tauke Khan Ave., Shymkent, 160012, Kazakhstan

Abstract. The article discusses the modeling of grain transportation in the system of grain processing industries. The question of determining the need of vehicles for the maintenance of agricultural work is of great national economic importance. It is associated with the problem of the effective use of material and cash resources, land and labor in grain farming. The resulting task is a difficult experimental task.

1 Introduction

To improve the efficiency of grain production, it is necessary to ensure a close connection of the grain industry with the branches engaged in the storage and processing of grain crops and the sale of their products.

Proper organization of the export of the grown grain crop plays an important role in improving the efficiency of the grain receiving enterprises (GRE) and the entire sector of the economy, as it contributes to a relative reduction in all transportation costs, improving the quality and safety of grain, reducing the cost of harvesting, improving the economic relations between the grain processing industries in the process of transportation and procurement works in a short time.

All sorts of oncoming, unnecessarily long-distance transportation of grain crops distract vehicles from transportation that is really necessary for the national economy, unreasonably increase the volume of transport and loading and unloading operations, which, in turn, causes an irrational expenditure of public funds allocated for reimbursement of expenses for the organization of transport systems. Consequently, the reduction of export costs can be achieved on the basis of rational transportation patterns by optimally linking grain farms with GRE [1].

A prerequisite for the procurement of grain is to deliver it to the GRE in the shortest possible time. At the same time, it is necessary to organize timely delivery of grain from the landlords to the GRE.

At present, a large number of economic and mathematical models in the grain processing industry have been developed [2–4], in which there are three main areas:

*e-mail: sad171@mail.ru
**e-mail: aiymzhan.makulova@narxoz.kz
***e-mail: nesipkul.bayboltaeva.53@mail.ru
****e-mail: sarsenbieva60@mail.ru
†e-mail: aliya.imatayeva@narxoz.kz

- development and solution of economic and mathematical problems of on-farm analysis;
- development and solution of economic and mathematical problems at the level of agro-industrial associations and individual units of the grain processing industry;
- development and solution of economic and mathematical problems of industry analysis.

At present, the tasks of the first direction are the most developed and implemented, since the information necessary for them is more accessible and reliable. The objectives of this direction include: optimization of the use of mineral and organic fertilizers; optimization of crop development plan; optimization of the production structure of the grain enterprise, etc.

The second direction, which has arisen in connection with the organization of agro-industrial associations, includes the tasks of optimizing not only the production of grain production, but also its industrial processing within the associations. The third direction is connected with the development and solution of problems of development of individual links of the grain processing industry at the level of the oblast, krai and republic. The main objective of this direction is the optimal placement and specialization of grain production by regions, as well as the optimization of purchases of grain products by farms, regions, regions and republics [2].

The main directions of improving the efficiency of transport in the grain processing industry include:

- the organization of the work of transport with the aim of the most efficient use of vehicles, labor, material and other resources in order to best meet the needs of the grain processing industries under consideration and reduce product losses;

- the distribution and use of resources within the transport industry in such a way as to ensure the implementation of transport work with minimal economic costs;
- improvement of the vehicle fleet by rationalization of its structure taking into account the specificity of grain products.

Until now, the above-mentioned tasks of optimal use of transport systems in the grain processing industry were solved separately: either products were distributed according to a given production nomenclature without taking into account the parameters of the vehicles used, or rolling stock between the clientele, without taking into account the specificity of the product being transported, or already established links.

Proceeding from this, the problem of optimizing the use of transport systems in the grain processing industry should be considered in an integrated approach in the field-procurement-processing-market production complex, taking into account market relations.

In the integrated field-harvesting-processing system, on the example of the Akmola region, optimal options for attaching grain farms to grain receiving enterprises are determined on the basis of effective use of their technical ability in operations of receiving, processing, placing, forming batches of grain and the optimal scheme for linking grain receiving enterprises to processing plants taking into account the distance of transportation for the intended purpose, as well as the annual demand. These model solutions can be used to justify the capacity, throughput of grain receiving elevators and managing grain sales.

The developed models for transporting grain and grain products in the production complex "field-harvesting-processing-market" to the interactive specialized software package "ASTYK" were introduced in the territorial departments of agriculture of the republic.

2 Material and methods

When optimizing the technology of grain procurement in the system "field-procurement", it is possible to determine the main parameters of the volume of procurement and the formation of optimal flows of grain transportation vehicles. These include: identifying the possible magnitude of the volume of workpieces and the proportion of grain crops, as well as determining the number of batches of grain to be formed for specific purposes, depending on the state and technological dignity.

The task of linking grain farms with GRE belongs to the class of operational planning tasks. The mathematical model of the task of linking grain farms with GRE is formulated as a multi-index transport-distributed linear programming problem of high dimensionality.

It is necessary to find in the area of permissible values such solutions, at which the work carried out by vehicles for the transportation of grain crops in the field-harvesting system is minimized.

Let us proceed to the consideration of the mathematical model of problems.

Required to minimize

$$F(x) = \sum_{i=1}^{m}\sum_{j=1}^{n}\sum_{l=1}^{L}\sum_{t=1}^{T} R_{il} X_{ijlt} \quad (1)$$

with restrictions:
per volume of crops grown

$$\sum_{l=1}^{L} X_{ijlt} \geq Q_{ijt}, i = \overline{1,m}; j = \overline{1,n}; t = \overline{1,T}; \quad (2)$$

on the volume of grain blanks of each GRE

$$\sum_{i=1}^{m} X_{ijlt} \geq P_{ljt}, l = \overline{1,L}; j = \overline{1,n}; t = \overline{1,T}; \quad (3)$$

per capita capacity

$$\alpha_1 \leq \sum_{i=1}^{m}\sum_{j=1}^{n}\sum_{t=1}^{T} X_{ijlt}, l = \overline{1,L}; \quad (4)$$

on technical capabilities

$$\sum_{i=1}^{m} U_{ijt}X_{ljt} \leq V_{ilt}P_{jlt}, j = \overline{1,m}; l = \overline{1,L}; t = \overline{1,T}; \quad (5)$$

with the natural requirement of nonnegativity of variables

$$X_{ijlt} \geq 0, i = \overline{1,m}; j = \overline{1,n}; l = \overline{1,L}; t = \overline{1,T}; \quad (6)$$

where i - the index of grain farming;

j – the index of the type of grain crops;

l – the index of GRE;

t – the index of the time subperiod;

Q_{ijt} – the volume of grain crops of the j-th species in the i-th farm, which must be transported to the subperiod t;

P_{jlt} – the volumes of preforms l-th GRE of j-th culture sub-period t;

α_l – the the minimum amount of grain required for the smooth operation of l-th GRE;

R_{il} – the distance between the i-th grain farms and l-th GRE;

U_{it} – the proportion of the j-th culture in the i-th economy sub-period t;

V_{ilt} – the proportion of j-th culture in l-th GRE and subperiod t;

X_{ijlt} – volumes of delivered grain crops of the j-th species from the grown i-th grain farming to the l-th in the subperiod t.

In the more general case, the model (1) - (6) should be supplemented with inequalities of the type (4), taking into account the grain drying capacities of the GRE. However, the mathematical content of the model will change.

When solving the problem (1) - (6) by conventional linear programming methods, considerable difficulties arise associated with the high dimensionality of the problem and the limited amount of high-speed memory of computer equipment [5].

The specificity of the constraint system (1) - (6) allows the use of the aggregation method: the process of solving the original problem is replaced by solving a number of problems of a much smaller dimension with the corresponding linking of the solutions obtained.

Information support of the task is based on data on the names of grain farms and GRE of the considered area, information about the capacity and throughput of GRE, distances of transportation of grain from producers to GRE, information about planned deliveries of grain from grain farms, about the availability of unloading mechanisms at GRE, etc.

3 Results and discussion

Experimental calculations using the model (1) - (6) were performed on real data from the Akmola regional department of agriculture and the Statistics Committee of the Ministries of the National Economy of the Republic of Kazakhstan in recent years using the developed specialized dialogue complex "ASTYK", which provides a real-time solution to the problem linking grain farms with GRE in the region. Intended for carrying out calculations on linear models of various contents of dimension and structure based on standard software, and the procedures allow: parametric generation of models, their additional calculation diagnostics, automatic setup of report generation programs, dialog management [3].

In accordance with this, the software performs:

- tabular input of information;
- control and correction of the initial information;
- calculation of the optimal linking plan;
- analysis and correction of the results;
- issue of output forms of documents.

Experimental calculations consist of three stages:

1) preparation of initial information for solving a specific variant of the problem;
2) problem solving;
3) the issuance of the results of the decision in a user-friendly form.

At the first stage, the MPS-format of data presentation was adopted for presenting initial information in modern packages for solving optimization problems of mathematical programming.

We have developed a set of programs that translate data into the MPS-format, which allows presenting input data in the form of tables, quickly and efficiently correcting the input information.

At the second stage, in order to establish the optimal variant of linking grain farms with GRE, the following tasks were solved:

- grouping of grain farms according to GRE, taking into account the existing list on the basis of contracting agreements;
- grouping of grain farms according to GRE, taking into account the shortest distance of grain delivery within the region;
- grouping of grain farms according to GRE, taking into account the shortest distance of grain delivery within the region.

The calculation was carried out in the context of 104 grain farms and 33 GRE area, where the power GRE correspond to the volume of produced grain. Minimum operating costs are obtained on the basis of a decrease in the distance of grain transportation by road from grain farms to GRE.

When evaluating the binding scheme of the studied variants, the analysis of the solution showed that the second option is considered the most optimal, i.e. linkage of grain farms for GRE, taking into account the shortest distance of grain delivery within the region.

As a result of work of farms, GRE and automobile transport in accordance with the optimal scheme of linking farms with GRE, the average distance of grain delivery to enterprises decreased against the existing one by 13.1 km, i.e. reduced the cost of transporting 1 ton of grain by 3024 tenge in the whole area. From optimizing the linkage of grain farms with GRE, the overall economic efficiency amounted (estimated) to 709,128 thousand tenge.

At the third stage, on the basis of the decision received and the initial information, the technical and economic indicators of interest to us are calculated in the form of reports.

Experimental calculations showed the need to take into account in the model (1) – (6) still a number of conditions and restrictions, and those that are usually unknown in advance and can only be updated by a planner directly in the process of forming a plan.

In the above model (1) – (6) it is assumed that all quantities are deterministic, but research shows that these quantities cannot be considered as predetermined, since they can vary significantly under different conditions. Therefore, the deterministic model (1) – (6) does not reflect the conditions associated with the random nature of the quantities, and it becomes necessary to consider models of stochastic programming that take into account the probabilistic nature of the initial information [4–7].

Suppose that all the data of the model (1) – (6), except for the right-hand sides of inequalities (3) and (5), are known in advance. Let the value of P_{ilt} – the volume of billet grain GRE – is a random variable with a given distribution law. Then problem (1) – (6) can be formulated as a one-stage stochastic problem in the form.

$$Z = \sum_{i=1}^{m}\sum_{j=1}^{n}\sum_{l=1}^{L}\sum_{t=1}^{T} R_{il}X_{ijlt}+$$
$$+M\left(\sum_{j=1}^{n}\sum_{l=1}^{L}\sum_{t=1}^{T}\alpha_{jlt}\left(\max\left(0;\sum_{i=1}^{m} X_{ijlt}-P_{jlt}(\Theta)\right)\right)\right)+$$
$$+M\left(\sum_{j=1}^{n}\sum_{l=1}^{L}\sum_{t=1}^{T}\beta_{jlt}\left(\max\left(0;\sum_{i=1}^{m} U_{ijt}X_{ijlt}-V_{jlt}P_{ljt}(\Theta)\right)\right)\right)$$
$$\rightarrow \min, \quad (7)$$

$$\sum_{l=1}^{L} X_{ijlt} \geq Q_{ijt}, i=\overline{1,m}; j=\overline{1,n}; t=\overline{1,T}; \quad (8)$$

$$\sum_{i=1}^{m}\sum_{j=1}^{n}\sum_{t=1}^{T} X_{ijlt} \geq \alpha_1, l=\overline{1,L}; \quad (9)$$

$$X_{ijlt} \geq 0, i=\overline{1,m}; j=\overline{1,n}; t=\overline{1,T}; \quad (10)$$

where α_{jlt}, β_{jlt} – penalty multipliers. Problem (7) – (10) can also be formulated as a two-stage stochastic linear problem in the form:

$$Z = \sum_{i=1}^{m}\sum_{j=1}^{n}\sum_{l=1}^{L}\sum_{t=1}^{T} R_{il}X_{ijlt} + My^{+}(\theta) + My^{-}(\theta)$$
$$\rightarrow \min, \quad (11)$$

$$\sum_{l=1}^{L} X_{ijlt} - Y^{+}_{jlt}(\Theta) \leq P_{jlt}(\Theta),$$
$$j=\overline{1,n}; l=\overline{1,L}; t=\overline{1,T}; \quad (12)$$

$$\sum_{i=1}^{m} U_{ijt}X_{ijlt} - Y^{-}_{jlt}(\Theta) \leq V_{jlt}P_{jlt}(\Theta),$$
$$j=\overline{1,n}; l=\overline{1,L}; t=\overline{1,T}; \quad (13)$$

$$\sum_{l=1}^{L} X_{ijlt} \geq Q_{ijt}, i=\overline{1,m}; j=\overline{1,n}; t=\overline{1,T}, \quad (14)$$

$$\sum_{i=1}^{m}\sum_{j=1}^{n}\sum_{t=1}^{T} X_{ijlt} \geq \alpha_1, l=\overline{1,L}; \quad (15)$$

$$X_{ijlt} \geq 0, i=\overline{1,m}; j=\overline{1,n}; l=\overline{1,L}; t=\overline{1,T}, \quad (16)$$

$$y^{+}(\Theta) = \sum_{j=1}^{n}\sum_{l=1}^{L}\sum_{t=1}^{T}\alpha_{jlt}y^{+}_{jlt}(\Theta);$$

$$y^{-}(\Theta) = \sum_{j=1}^{n}\sum_{l=1}^{L}\sum_{t=1}^{T}\beta_{jlt}y^{-}_{jlt}(\Theta);$$

$$y^{+}_{jlt}(\Theta) = \max\left(0;\sum_{i=1}^{m} X_{ijlt} - P_{jlt}(\Theta)\right),$$

$$y^{-}_{jlt}(\Theta) = \max\left(0;\sum_{i=1}^{m} U_{ijt}X_{ijlt} - X_{jlt}P_{jlt}(\Theta)\right)$$

The problem record in the form (11) – (16) is convenient for qualitative research. For the numerical solution, the record in the form (7) – (10) is adapted.

The resulting task is a difficult experimental task. Known linear programming methods are not applicable for its solution. It should be noted that the complexity of the solution is due to the fact that it is almost impossible to obtain an exact expression of the function Z and its derivatives. Therefore, to solve problem (7) – (10), it is advisable to apply direct probabilistic methods of stochastic programming, in particular, the method of stochastic linearization [8], which is defined by the following iterative procedures:

$$X^{S+1} = X^{S} + P_S(\bar{X}^{S} - X^{S}), \quad (17)$$
$$Z^{S+1} = Z^{S} + \delta_S(\xi^{S}(X^{S}) - Z^{S}), \quad (18)$$
$$(Z^{S}, (\bar{X}^{S}) = \max(Z^{S}, X^{S}) \quad (19)$$
$$x \in X$$

where $(X^0$, $(Z^0$ are arbitrary points, X – convex set:

$$M(\xi^{S}(X^{0},...X^{S})) = \nabla F(X^{S}), \quad 0 \leq P_S \leq 1;$$

$$\sum_{S=0}^{\infty} P_S^2 < \infty; \quad \sum_{S=0}^{\infty} P_S = \infty; \quad (20)$$

$$0 \leq \delta_S \leq 1, \quad \sum_{S=0}^{\infty}\delta_S^2 < \infty; \quad \sum_{S=0}^{\infty}\delta_S = \infty, \quad (21)$$

$$(P_S/\delta_S) \rightarrow 0, \quad \sum_{S=0}^{\infty} M(P_S^2 + \delta_S^2) < \infty \quad (22)$$

It should be noted that the algorithm of the method of stochastic linearization (17) – (19) can be applied to solve the problem (7) – (10) in the event that there exists $Z(X,\Theta)$. If it is difficult to guarantee the existence of $Z(X,\Theta)$, then one can apply the method of stochastic linearization with smoothing [9, 10]. In this case, one should calculate the stochastic generalized gradient of the function $Z(X,\Theta)$.

We now show the calculation of the stochastic generalization of the gradient of the function. Denote by $Z_1(X,\Theta)$ and $Z_2(X,\Theta)$ integrand expressions of objective function (7) i.e.

$$Z_1(X,\Theta) = \sum_{j=1}^{n}\sum_{l=1}^{L}\sum_{t=1}^{T}\alpha_{jlt}\max\left\{0;\sum_{i=1}^{m} X_{ijlt} - P_{jlt}(\Theta)\right\}$$

$$Z_2(X,\Theta) = \sum_{j=1}^{n}\sum_{l=1}^{L}\sum_{t=1}^{T}\beta_{jlt}\max\left\{0;\sum_{i=1}^{m} U_{ijt}X_{ijlt} - V_{jlt}P_{jlt}(\Theta)\right\}$$

according to the rules for calculating subgradients

$$\hat{Z}_{1x}(X,\Theta) = \begin{cases} a_{jlt}, & \text{if } \sum_{i=1}^{m} X_{ijlt} \geq P_{jlt}(\Theta) \\ 0, & \text{otherwise} \end{cases}$$

$$\hat{Z}_{2x}(X,\Theta) = \begin{cases} \beta_{jlt}U_{ijt}, & \text{if } \sum_{i=1}^{m} U_{ijt}X_{ijlt} \geq V_{jlt}P_{jlt}(\Theta) \\ 0, & \text{otherwise} \end{cases}$$

it follows that

$$\xi^S_{1j} = \begin{cases} 1, & \text{if } \sum_{i=1}^{m} X_{ijlt} \geq P_{jlt}(\Theta) \\ 0, & \text{otherwise} \end{cases} \quad (23)$$

$$\xi^S_{2j} = \begin{cases} U_{ijt}, & \text{if } \sum_{i=1}^{m} U_{ijt} X_{ijlt} \geq V_{jlt} P_{jlt}(\Theta) \\ 0, & \text{otherwise} \end{cases} \quad (24)$$

This means that the stochastic quasi-gradient of the objective function $Z(x, \Theta)$ of the problem (7) – (10) at any point can be determined by the formula of the following form:

$$xi^S_j = \sum_{i=1}^{m} \sum_{l=1}^{L} R_{il} - \sum_{j=1}^{n} \sum_{l=1}^{L} \sum_{t=1}^{T} \alpha_{jlt} \xi^S_{1j} - \sum_{j=1}^{n} \sum_{l=1}^{L} \sum_{t=1}^{T} \beta_{jlt} \xi^S_{2j}$$

where ξ^S_{1j} and ξ^S_{2j} determined according to (23) and (24).

Thus, according to the experimental calculation with the smallest distance of grain delivery, an economic prerequisite is created to reduce the cost of grain procurement in the grain processing industry.

Consider the model of transportation of grain in the integrated system "Field-preparation-processing".

Reducing transportation costs during transportation of grain in the integrated field-harvest-processing system, preserving their quality by accelerating transportation is achieved by attaching the most rational transportation patterns through optimal coordination and coordination of its component subsystems.

We have developed two-stage optimization models for managing the formation of stocks and the implementation of target batches of wheat in the production system "field-procurement-processing" [1, 3].

The mathematical model of grain transportation in the complex system "field-procurement-processing" is formulated as a production and transport problem of linear programming of large dimensionality.

It is necessary to find in the area of permissible values such solutions, at which the work performed by vehicles for the transportation of grain in this system is minimized.

We write the mathematical model in the form:
minimize

$$Z = \sum_{i=1}^{m} \sum_{l=1}^{L} \sum_{j=1}^{n} R^{(1)}_{ijl} X_{ijl} + \sum_{l=1}^{L} \sum_{K=1}^{K} \sum_{j=1}^{n} R^{(2)}_{lkj} Y_{lkj} \quad (25)$$

under restrictions
on the volume of grain production by an unattached grain manufacturer

$$\sum_{l=1}^{L} P_{lj} - \sum_{i=1}^{m} Q_{ij} = Q_{m+1,j}, j = \overline{1,n} \quad (26)$$

if $Q_{m+1,j} < 0$, then $Q_{m+1,j} = 0$;
on the volume of short delivery of grain at the enterprise for the storage of grain

$$P_{lj} - \sum_{i=1}^{m} X_{ijl} = H_{lj}, j = \overline{1,n}; l = \overline{1,L} \quad (27)$$

on the need for an unattached grain storage enterprise

$$\sum_{i=1}^{m} Q_{jl} - \sum_{l=1}^{L} P_{lj} = P_{m+1,j}, j = \overline{1,n} \quad (28)$$

if $P_{m+1,j} < 0$, then $P_{m+1,j} = 0$;
on the volume of grain to be sold on the free market

$$Q_{ij} - \sum_{l=1}^{L} X_{ijl} = S^{(1)}_{lj}, j = \overline{1,n}; i = \overline{1,m} \quad (29)$$

For the volume of grain supply to unattached grain storage enterprises

$$\sum_{K=1}^{K} F_{kj} - \sum_{l=1}^{L} (P_{lj} - H_{lj}) = A_{L+1,j}, j = \overline{1,n} \quad (30)$$

if $A_{L+1,j} < 0$, then $A_{L+1,j} = 0$;
on the volume of short delivery of grain to the processing plant

$$F_{kj} - \sum_{l=1}^{L} Y_{lkj} = B_{kj}, j = \overline{1,n}; k = \overline{1,K} \quad (31)$$

on the needs of an unattached grain processing plant

$$\sum_{l=1}^{L} (P_{lj} - H_{lj}) - \sum_{k=1}^{K} F_{kj} = C_{k+1,j}, j = \overline{1,n} \quad (32)$$

if $C_{k+1,j} < 0$, then $C_{K+1,j} = 0$;
for the volume of grain to be sold on the free market of a grain storage enterprise

$$(P_{lj} - H_{lj}) - \sum_{k=1}^{K} Y_{lkj} = S^{(2)}_{lj}, j = \overline{1,n}, l = \overline{1,L}; \quad (33)$$

on non-negativity of variables

$$X_{ijl} \geq 0, Y_{lkj} \geq 0, \quad i = \overline{1,m}; j = \overline{1,n}; \quad l = \overline{1,L}; k = \overline{1,K}; \quad (34)$$

where, i – the index of grain producers (collective and state farms);

j – the index of the batch of grain by purpose;

l – the index of enterprises for the storage of grain (GRE);

k – index of grain processing enterprises;

Q_{ij} – the volume of production of the j-th batch of grain at the i-th manufacturer;

P_{li} – the volume of storage of the j-th batch of grain at the l-th enterprise for the storage of grain;

$R^{(1)}_{il}$ – the distance from the i-th grain producer to the l-th enterprise grain storage;

$Q_{m+1,j}$ – the volume of production of the j-th batch of grain unattached by the grain producer;

H_{li} – the volume of short deliveries of the j-th batch of grain at the l-th enterprise for the storage of grain;

$P_{m+1,j}$ is the storage volume of the j-th grain batch of an unattached grain storage enterprise at the i-th grain producer to be sold on the free market;

$S_{ij}^{(1)}$ – j-th batch volume of grain from i-th grain producer to be sold on the free market;

F_{kj} – the need of the k-th grain processing enterprise in the j-th batch of grain;

$R_{lk}^{(2)}$ – the distance from the l-th enterprise for the storage of grain to the k-th enterprise for processing;

$A_{l+1,j}$ – the volume of delivery of the j-th batch of grain of an unattached enterprise for the storage of grain;

B_{kj} – the volume of short delivery of the j-th batch of grain to the k-th grain processing enterprise;

$C_{k+1,j}$ – the need for an unattached grain processing enterprise in the j-th batch of grain;

$S_{lj}^{(2)}$ – the amount of j-th shipment of grain, subject to the implementation of free market enterprise l-th grain storage;

X_{ijl} – volumes supplied j-th grain party i-th producer grain l-th grain storage entity;

Y_{lkj} – volume supplied by j-th batch of grain from the l-th enterprise for grain storage k-th company for processing.

The task of transporting grain in the complex field-harvesting-processing system was accomplished in two stages: a field-harvesting and a storage-processing, which determine the formation of optimal stocks and the implementation of target wheat lots.

When solving the problem in the system "field-procurement", the necessity was taken into account: timely delivery of products to their destinations with proper safety; effective use of vehicles, loading and unloading equipment of receiving and processing enterprises; rational use of technical capabilities of GRE for the operations of receiving, processing, placing the formation of batches of grain.

4 Conclusion

These solutions of the model (25) – (34) can be used to justify the capacity, throughput of the GRE and control the sale of grain. The developed models, in contrast to previously known ones, are built in accordance with the principles of targeting and taking into account the peculiarities of regional development. In the integrated field-harvesting-processing system, on the example of the Akmola region, optimal options for attaching grain farms to grain receiving enterprises are determined on the basis of effective use of their technical ability in operations of receiving, processing, placing, forming batches of grain and the optimal scheme for linking grain receiving enterprises to processing plants taking into account the distance of transportation for the intended purpose, as well as the annual demand. These model solutions can be used to justify the capacity, throughput of grain receiving elevators and managing grain sales. The digital (dialogue) technology of optimization calculations "ASTYK", complex, implemented on the basis of standard software tools, can be used for a wide class of tasks for optimizing transport and logistics systems for grain processing in real time planning process.

References

[1] A. Saparbayev, *Transport logistics in grain processing* (Fortuna Polygraph LLP, 2019)

[2] A. Iztaev, T. Kulazhanov, A. Saparbayev, *Innovative technologies and logistics of processing businesses in AIC* (Fortuna Polygraph LLP, 2019)

[3] A. Bakaev, V. Revenko, V. Gritsenko, *Information technology in transport*, Vol. 3. Water transport (Naukova dumka, Kyiv, 2009)

[4] E. Vasilyeva, V. Levit, V. Livshits, *Nonlinear transport problems on networks* (Finansy i statistika, Moscow, 1981)

[5] L. Lesdon, *Optimization of large systems* (Nauka, Moscow, 1975)

[6] E. Krylatykh, A. Strokov, *Food Security of Russia and the World: Theory and Methodology of Research, Analysis of Provision, Opportunities and Threats* (Infra-M, 2019)

[7] A. Ganicheva, *Mathematical methods and models in agroindustrial complex* (Tver State Agricultural Academy, Tver, 2019)

[8] A. Makulova, *Model assessment of agricultural production management* (Fortuna Polygraph LLP, 2019)

[9] H.A. Taha, *Operations Research: An Introduction*, 10th edn. (Pearson, 2017)

[10] G.B. Dantzig, M.N. Thapa, *Linear programming 2: theory and extensions* (Springer Science & Business Media, 2006)

https://doi.org/10.1051/shsconf/202110706004

Using a comprehensive index technology to analyze structural changes in the regions' economic development in a COVID-19 pandemic: the case of Ukraine

Pavlo Hryhoruk[1,*], *Nila* Khrushch[1,**], and *Svitlana* Grygoruk[1,***]

[1]Khmelnytskyi National University, 11 Instytutska Str., Khmelnytskyi, 29016, Ukraine

Abstract. The paper investigates the issues of evaluating structural changes in the regions' economic development based on the comprehensive index assessment technology. The impact of the COVID-19 pandemic on regional development and changes in the regional structure is considered. The authors propose the use of block convolution to design a comprehensive index based on a set of metric initial indicators that characterize the regions' economic development. Grouping the set of initial indicators is carried out based on the method of an extreme grouping of parameters and the method of principal components. A weighted linear additive convolution was used to develop partial composite indices and an economic development comprehensive index. The practical approbation was carried out for the regions of Ukraine according to the data of 9 months of 2019 and the same period of 2020. To establish the regions' structure, we used the division of the comprehensive index values into intervals and further distributing regions into classes according to the level of economic development. There is a general decrease in the value of the integrated indicator in 2020, caused by the impact of the COVID-19 pandemic. However, no significant changes in the structure of the regions were detected, which indicates an equally negative impact of the pandemic for all regions of Ukraine.

1 Introduction

One of the most significant problems of regional development is to ensure sustainable economic growth. The economic system of any country is a multifunctional regional entity, so the definition of long-term priorities of strategic planning of regional development should be based on comprehensive assessments of the level of their economic development. They allow tracking the dynamics and asymmetry of development, to establish inequalities and gaps in the region's structure, to provide an analytical basis for the preparation of strategic decisions on the transformation of socio-economic development policy of individual regions.

Global problems related to climate changes, financial crises, intensified competition in global and domestic markets, deepened in 2020 due to another global challenge – the COVID-19 pandemic [1, 2]. Its destructive impact has been reflected in all spheres of public life, destroying established socio-economic processes and relationships. Measures, severe restrictions, lockdowns aimed at curbing the spread of the pandemic, were reflected in the negative effects of slowing down the socio-economic development of both regions and the world economy as a whole. They were a prerequisite for a new financial and economic crisis. This is evidenced by the results of analytical studies and forecast estimates of basic macroeconomic indicators provided by global institutions, in particular, the World Bank (WB), the International Labor Organization (ILO), the World Health Organization (WHO), the United Nations (UN), the European Bank for Reconstruction and Development (EBRD) and others.

In particular, according to the ILO, the loss of labor income for the three quarters of 2020 compared to the corresponding period of 2019 is estimated at 10.7%, or 3.5 trillion USD [3]. The baseline forecast calculated by World Bank analysts [4] predicts a reduction in world GDP by 5.2% in 2020. And although the world economy is expected to grow by 4.3% in 2021, the pandemic may hold back economic activity and income growth for a long time [5]. UBS Chairman Axel Weber also made a cautious forecast about the pace of global economic recovery, noting that "it would be at least a year to go back to pre-crisis levels of GDP. It'll take another year or two to be anywhere near getting unemployment and pre-crisis growth back and so it would be quite a long recovery that we're facing" [6].

The consequences of the pandemic were especially acute in developing economies countries, particularly in Ukraine. Thus, according to the State Statistics Service of Ukraine [7], real GDP in the third quarter of 2020 compared to the third quarter of 2019 decreased by 3.5%. The financial result before taxation of large and medium-sized enterprises in the III quarter of 2020 amounted to 93.3 billion UAH of profit, while for the corresponding period of 2019 – UAH 342.8 billion in profit, which is 73% less. Exports of goods for the period under review decreased by 3.6%, and imports – by 14.3%.

*e-mail: violete@ukr.net
**e-mail: nila.ukr@gmail.com
***e-mail: grygoruk.svitlana@gmail.com

The main forecast macroeconomic indicators for the end of 2020, presented by the Cabinet of Ministers of Ukraine, envisage a fall in GDP by 4.8%, the inflation rate – 11.6%; unemployment rate – 9.4%; reduction of the average salary – 4.5%; decrease in exports – 5.5%, imports – 10% [8]. According to the EBRD, by the end of 2020 GDP was expected to decline by 5.5%, but in 2021 it is predicted to grow by 3% [9]. The most optimistic about the resumption of production are construction companies, the most pessimistic – service companies that have suffered the most from the introduction of quarantine restrictions.

The decline in macroeconomic indicators is directly caused by negative changes in regional development. To reduce the negative socio-economic consequences of the COVID-19 pandemic, it is necessary to identify trends, assess different scenarios of regional development, identify existing structural changes and develop a system of measures within regional development strategies to stabilize the situation. The presented macroeconomic forecasts necessitate research aimed at estimating the real losses from COVID-19 pandemic in terms of socio-economic development of regions, identifying areas of rational use of endogenous factors to ensure their sustainable economic growth, which will contribute to the achievement of the goals reflected in the State Strategy for Regional Development for 2021–2027 [8].

2 Literature review

Currently, there is a large number of different scholar's approaches to assess the economic development level and the establishment of regional differences and imbalances.

These studies are based mainly on the use of quantitatively measurable indicators that allow sound mathematical processing to shape conclusions. One of the most commonly used approaches is research based on the analysis of the GDP indicator and indicators derived from it like the Hoover Concentration Index, the Theil index, the Herfindahl index, etc. [10–14]. In particular, the authors also use the Klassen typology to track the dynamics and nature of changes in regional development.

Given the natural multidimensionality of regions' economic development description, widely used methods of multidimensional statistical analysis for their structuring by the level of this characteristic and determination of disparities between regions, in particular, cluster analysis, factor analysis, multidimensional scaling, structural equation method, Solow-Swan, and Mankiw-Romer–Weil growth models [12, 15–23], which allows grouping regions into homogeneous aggregates based on various quantifiable indicators, to identify gaps in the development of individual regions. Among the shortcomings of these approaches, in our opinion, it is worth noting the difficulty of taking into account the importance of individual indicators. The authors of the study, who used these tools, also noted that the grouping results are significantly influenced by clustering methods, which is also a disadvantage. The further development of multidimensional statistics' methods is reflected in the application of fuzzy clustering methods for structuring regions and identifying imbalances in their economic development, which is presented in [24–27].

Another way to take into account the multidimensionality for the description of regional development processes is to use the technology of comprehensive index assessment [28–33]. The vast majority of scientists' approaches in the presented studies are focused on designing a composite indicator of economic development by linear convolution of a set of quantitatively measured indicators. The differences are in the information base chosen for the study and how the results are interpreted. Among the shortcomings, it is worth noting the lack of consideration of the weight of the initial indicators or proper justification of the proposed weights, which in most cases it is proposed to determine the expert method. Besides, either a linear relationship between the values of the composite indicator and these levels, or a desirability scale without proper conversion of the original data is usually used to interpret the results and establish levels of economic development [31].

The study of issues related to assessing the impact of the COVID-19 pandemic on the economic development of economic systems both at the global level and at the level of individual national economies is currently one of the most relevant and is quite intensively studied by scientists. The vast majority of researchers are inclined to believe that overcoming the crisis is possible only after a few years, even with the total vaccination of the population, which should curb the spread of viral infection. Such conclusions are supported by the results of economic and mathematical modeling and evaluation of current and future trends in the economic system development. Issues related to the application of mathematical modeling to assess the impact of a pandemic on economic development are reflected, in particular, in publications [34–37]. However, it should be noted that the authors of these studies provide short-term forecast estimates of macroeconomic indicators at the level of national economies. with an emphasis on trends and potential scenarios for their development. The main attention is paid to the assessment of GDP change as one of the most important macroeconomic indicators. In our point of view, insufficient attention is currently paid to research to identify changes in the trends of economic development of certain regions of the country.

Our study aims to develop an approach to building an economic development comprehensive index for analyzing the impact of COVID-19 on Ukraine's regions development and identifying structural changes by combining the technology of comprehensive index assessment, multidimensional statistical analysis, and projection of results on the desirability scale.

3 Problem description and methodology

The economic development of the regions is characterized by a large number of indicators. They usually reflect the quantitative results of the activities of regional business entities and therefore have a metric origin, i.e., measured on one of the quantitative scales. This significantly simplifies their further analytical processing, because for indicators of this nature it is quite correct to use mathematical operations.

One of the difficulties that arise in the process of processing such data and interpretation of results is their internal inconsistency, diversity, and inequality of impact on the studied quality. To concentrate the information contained in the initial indicators and reduce the dimension of characteristics' space, various computing technologies are used. One of them is the technology of comprehensive index assessment, which allows reducing the description of the studied phenomenon, in this case, the economic development of the regions, to a single comprehensive indicator. This is usually done by weighted convolution of the initial units. At the same time, there are several methodological problems to realize this process. First, the economic development of regions, as a complex phenomenon, requires the use of a large number of baselines for their description. Thus, the relative impact of each indicator on the final result is reduced. Secondly, there is a problem of reasonable determination of the weight of each component when they are integrated into a composite indicator.

A possible solution to these obstacles is the use of block convolution. Under such conditions, the initial set of indicators is divided into subsets that don't intersect. A partial composite index is constructed for each subset. The final result is settled by convolution of the constructed partial composite indices taking into account the weight of each obtained subset.

One of the approaches that allow getting a solution to this problem is the method of an extreme grouping of parameters. It is based on the hypothesis that the set of initial characteristics can be divided into groups, each of which reflects the effect of a certain factor – the latent characteristics of the group. Therefore, the method focuses on the selection of groups of parameters such that the relationships between the parameters within the group are maximum under the assumption that the number of such groups is fixed. It is assumed that the relationships within the group are explained by the relationship between some generalized latent characteristic of the group (generalized index) and the initial indicators included in this group. Direct relationships between initial indicators are unknown and may be absent. Since the indicators within each of these groups must be more closely related than the indicators of different groups, the task is to identify highly correlated groups of indicators.

Denote by $G=X_1, X_2, \dots, X_n$ the set of initial indicators. The initial data for the method's computational procedure is the correlation matrix R of these indicators. Let $G_1, G_2, \dots, G_s$ be subsets into which the set of initial indicators is divided:

$$\bigcup_{i=1}^{s} G_i = G, \tag{1}$$

$$G_i \cap G_j = \varnothing, \tag{2}$$

$i \neq j, i, j = 1, 2, \dots, s.$

Denote by $H_1, H_2, \dots, H_s$ – the corresponding latent characteristics (indicators) of each group. The criterion that allows you to determine the best grouping of indicators has the form:

$$\sum_{i=1}^{s} \sum_{X_j \in G_i} |r_{X_j,H_i}|, \tag{3}$$

where r_{X_j,H_i} is the correlation coefficient between initial indicator X_j, which belongs to subset G_i, and common indicator H_i of subset G_i.

To obtain a division of the original set of indicators into subgroups, you can use the method of principal components. It is known that the model of transition from the system of initial indicators to the set of latent characteristics, which are the principal components, is reflected by the dependence:

$$Z^T = WF^T, \tag{4}$$

where Z^T – transposed matrix of standardized initial indicators' values, F^T – transposed matrix of principal components' values, W – matrix of principal components factor loadings:

$$Z = \begin{pmatrix} z_{11} & z_{12} & \dots & z_{1n} \\ z_{21} & z_{22} & \dots & z_{2n} \\ \vdots & \vdots & \ddots & \vdots \\ z_{m1} & z_{m2} & \dots & z_{mn} \end{pmatrix}, \tag{5}$$

$$F = \begin{pmatrix} f_{11} & f_{12} & \dots & f_{1n} \\ f_{21} & f_{22} & \dots & f_{2n} \\ \vdots & \vdots & \ddots & \vdots \\ f_{m1} & f_{m2} & \dots & f_{mn} \end{pmatrix}, \tag{6}$$

$$W = \begin{pmatrix} w_{11} & w_{12} & \dots & w_{1n} \\ w_{21} & w_{22} & \dots & w_{2n} \\ \vdots & \vdots & \ddots & \vdots \\ w_{n1} & w_{n2} & \dots & w_{nn} \end{pmatrix}, \tag{7}$$

where m – the volume of the sample, which is used to measure the initial set of indicators.

The relationship between the values of indicators and principal components (factors) can be written as follows:

$$z_{ji} = \sum_{k=1}^{n} w_{ik} f_{jk} \tag{8}$$

where z_{ji} – i-th component (value) of Z_j, w_{ik} – factor loadings for F_k, f_{jk} – j-th components of F_k, $i = 1, 2, \dots, n$, $j = 1, 2, \dots, m$.

Let us calculate the correlation coefficient between the initial indicator Xi and the principal component F_j, taking into account the fact that the principal components are non-correlated:

$$r_{F_i,F_j} = 0, \tag{9}$$

$i \neq j$.

As a result, we obtain:

$$r_{X_i,F_j} = r_{Z_i,F_j} = \frac{1}{m}\left(\sum_{k=1}^{n} W_{ik} F_k\right) F_j = w_{ij}. \tag{10}$$

Therefore, the correlation coefficient between the initial indicator and the principal component is equal to the factor load of this component for the corresponding indicator. This fact allows us to conclude that to get the desired grouping of indicators it is necessary to analyze the values of the factor loadings of the principal components for each initial indicator. In this case, as the latent characteristic H_j of the group G_j, we choose the corresponding principal component F_j. To avoid the formation of empty groups or all groups, each of which will contain only one initial indicator, for grouping, we choose not all the principal components, but only the first s most influential, which explain the given share of variance of initial indicators. The value of s is defined as the smallest value of the number of principal components for which the inequality is met:

$$\frac{\sum_{i=1}^{s} \lambda_i}{n} \geq \gamma \tag{11}$$

where λ_i – eigenvalues, ordered by decreasing their values, γ – a predetermined explanation fraction of the initial indicators' variance by the principal components. Typically, this value is selected from 0.70 to 0.80.

In the group of homogeneous indicators, it is expedient to include those initial indicators for which the corresponding values of factor loadings for the principal components on absolute value will have the greatest values. To construct a partial composite index I_j for each formed group G_j, we use one of the formulas for weighted convolution [38]:

$$I_j = \sum_{i \in K_j} \alpha_i^{(j)} U_i^{(j)} \tag{12}$$

$$I_j = \prod_{i \in K_j} \left(U_i^{(j)}\right)^{\alpha_i^{(j)}} \tag{13}$$

$$I_j = -1 + \prod_{i \in K_j} \left(1 + U_i^{(j)}\right)^{\alpha_i^{(j)}} \tag{14}$$

where $U_i^{(j)}$ – normalized values of those indicators X_i, that belong to subset G_j, $a_i^{(j)}$ – weight coefficients of appropriate indicators, K_j – set of indices for those indicators X_i, that belong to subset G_j, $j = 1, 2, \ldots, s$.

The initial indicators are transformed to normalize form according to the formula (15) or formula (16) [17]:

$$u_{ij} = 1 - \frac{|x_{ij} - x_j^*|}{x_{jmax} - x_{jmin}} \tag{15}$$

$$u_{ij} = \begin{cases} \frac{x_{ij}}{x_{jmax}}, & \text{when } X_j \text{ is an incentive;} \\ \frac{x_{jmin}}{x_{ij}}, & \text{when } X_j \text{ is a disincentive;} \end{cases} \tag{16}$$

where u_{ij} – normalized values of indicators, x_{ij} – initial values of indicators, $x_{jmin} = \min x_{ij}$, $x_{jmax} = \max x_{ij}$,

$$x_i^* = \begin{cases} x_{jmax}, & \text{when } X_j \text{ is an incentive;} \\ x_{jmin}, & \text{when } X_j \text{ is a disincentive;} \end{cases} \tag{17}$$

i=$1, 2, \ldots, m$, j=$1, 2, \ldots, n$, m – number of units under study, n – number of initial indicators.

The normalization procedure is necessary to extract the units of measurement of the original indicators and reducing their values to a scale from 0 to 1. This step is aimed at simplifying the further interpretation of the calculation result. To calculate the weight coefficients, we propose to use the components of eigenvector V_j:

$$\alpha_i^{(j)} = \frac{\left(v_i^{(j)}\right)^2}{\sum_{i \in K_j} \left(v_i^{(j)}\right)^2}, \tag{18}$$

where $a_i^{(j)}$ – weight coefficients of appropriate indicators, $v_i^{(j)}$ – components of j-th eigenvector V_j, that correspond to initial indicators X_i from the G_j , K_j – set of indices for those indicators X_i, that belong to subset G_j.

Equation (18) meets the condition, that the sum of weight coefficients should be equal to 1. This condition with the normalization procedure provides the location of partial composite indicators values in the range [0; 1].

We propose to calculate the final economic development comprehensive index I_{COM} using partial composite indicators I_j based on one of the convolution's forms like (12), (13), (14). For example, for linear weighted convolution appropriate expression has a form:

$$I_{COM} = \sum_{j=1}^{s} \beta_j I_j. \tag{19}$$

Weight coefficients β_j are calculated in proportion to the eigenvalues λ_j that correspond with G_j, $j = 1, 2, \ldots, s$:

$$\beta_j = \frac{\lambda_j}{\sum_{k=1}^{s} \lambda_k}. \tag{20}$$

Under such conditions, the values of the I_{COM} will also be in the range from 0 to 1. This approach to calculations simplifies the interpretation of the result.

To assess the studied objects' structure, the range of values of the comprehensive index should be divided into ranges. Dividing the range [0; 1] of values of the comprehensive index into intervals of the same length to achieve this goal is impractical.

First, ranges can be formed that don't cover any of the objects under study.

Second, the latent characteristic under study is usually nonlinear, and the use of intervals of the same length can disrupt the true structure of objects.

Third, such a division can be led to a situation where one group includes objects that have significant differences in the values of the integrated indicator, while two neighboring objects belonging to different groups may have a slight deviation of the values of the comprehensive index.

To solve the problem of grouping, you can also use the approach presented in [17], in which the definition of the boundaries of the ranges is carried out by calculating the ratios of two adjacent values of the integrated indicator:

$$\delta_j = \frac{I_j}{I_{j-1}}, \tag{21}$$

$j = 2, 3, \ldots, m$.

The basis for the transition to a new range of values of the comprehensive index is a significant rise in the change of values of δ_j. The grouping objects is executed according to the level of the corresponding values of the comprehensive index. This approach also has drawbacks. Given the slight difference in the values of the integrated indicator, which are in the middle of the range of all its possible values, one of the groups can have a very large number of objects, which will be significantly different from the content of other groups. Besides, in the case of a slight discrepancy in the values of the comprehensive index for neighboring objects, a significant rise in the values of δ_j may not be observed. Thus, all objects can belong to one group. It is also necessary to take into account the fact that the value of δ_j is also affected by the level of values of the comprehensive index for which this value is calculated. And the closer these values are to 0, the smaller should be the hike in the change of values of δ_j, which decides on the formation of a new range.

The iterative procedure presented in [31] can be used to determine the limit values of the comprehensive index' ranges. Its advantage is the "adjustment" of grouping ranges to the value of a specific sample, which makes its application more practical. However, the disadvantage of this approach is the use of a training sample.

Another approach that allows you to solve this problem is the use of desirability scales, which allow you to match the quantitative and qualitative levels and group objects according to the level of studied quality. One such scale is the Harrington scale. The use of this scale involves the transformation based on Harrington's function [39]:

$$H(Z_i) = exp(-exp(-Z_i)), \quad (22)$$

where Z_i is the value of the indicator on the scale of partial indicators Z. The values $d = H(Z)$ of the Harrington's function form the desirability scale.

The correspondence between the values of Z_j and the values of the initial indicators I_j is determined by the formula:

$$Z_j = (Z^* - Z_*)\frac{I_{COM_j} - I_{min}}{I_{max} - I_{min}} + Z_*, \quad (23)$$

where Z_j – current value of the Z-scale, corresponds with the value of I_{COM_j}; I_{COM_j} – current value of comprehensive index I_{COM}; Z_* and Z^* – low and high bounds of Z-scale, which define the workspace of Z_j; I_{min}, I_{max} – minimum and maximum of I_{COM}; $j = 1, 2, \ldots, m$.

Transformation (23) is required to match the value of the comprehensive index I_{COM} and Z-scale with the correspondence of the minimum and maximum values of both indicators.

Next step, we identify the value of $d_j = H(Z_j)$, $j = 1, 2, \ldots, m$, and distribute objects under study into five groups by qualitative development level of the group (table 1).

This approach allows taking into account the nonlinear nature of the studied characteristic, in this case, the economic development level, as well as to investigate changes in the structure of the objects under study by the values of the comprehensive index calculated for different periods.

Table 1. The relationships between the quantitative values of the desirability scale and qualitative development levels of group

Qualitative levels of development	The range of quantitative values on the desirability scale
relatively high	0.80..1.00
above average	0.63..0.80
average	0.37..0.63
below average	0.20..0.37
relatively low	0.00..0.20

Table 2. The relationships between the quantitative values of the desirability scale and qualitative development levels of group

Code	Region	Code	Region
r-01	Vinnytsia	r-13	Mykolaiv
r-02	Volyn	r-14	Odesa
r-03	Dnipro	r-15	Poltava
r-04	Donetsk	r-16	Rivne
r-05	Zhytomyr	r-17	Sumy
r-06	Zakarpattia	r-18	Ternopil
r-07	Zaporizhzhia	r-19	Kharkiv
r-08	Ivano-Frankivsk	r-20	Kherson
r-09	Kyiv	r-21	Khmelnytskyi
r-10	Kyrovohrad	r-22	Cherkasy
r-11	Luhansk	r-23	Chernivtsi
r-12	Lviv	r-24	Chernihiv

4 Findings

Let us consider the practical testing of the proposed approaches to the calculation of t economic development comprehensive index for Ukraine's regions, grouping regions based on their values, and the study of structural changes in the resulting grouping caused by the COVID-19 pandemic. We choose the data of the State Statistics Service of Ukraine [7] and the Ministry of Development of Communities and Territories of Ukraine [40] for the period of the first 9 months of 2019 and the first 9 months of 2020 as the information base for the calculations. We choose the following initial indicators:

X_1 – Volume of sold industrial products per capita, UAH;

X_2 – Volume of agricultural production per capita of the rural population, UAH;

X_3 – Volume of construction works performed per capita, UAH;

X_4 – Volume of capital investments per capita cumulatively since the beginning of the year, UAH;

X_5 – Exports of goods per capita, USD;

X_6 – An unemployment rate of the population aged 15-70 years (according to the ILO's Methodology), %;

X_7 – Employment rate of the population aged 15-70 years (according to the ILO's Methodology), %;

X_8 – Index of real wages, %;

X_9 – The volume of housing commissioned per 10 thousand people, sq. meters of the total area;

X_{10} – The volume of freight turnover of road and rail transport, thousand ton-kilometers per 1000 population, thousand ton-km.

Table 3. Indicator's values for data for first three quarters of 2019

Code	Values									
	X_1	X_2	X_3	X_4	X_5	X_6	X_7	X_8	X_9	X_{10}
r-01	39228.9	20678.0	3774.4	5730.1	697.8	9.8	57.9	112.1	1248.9	2721.7
r-02	21736.3	11004.0	1322.5	8039.0	503.6	11.7	50.2	109.1	2978.8	2261.9
r-03	110916.2	25975.0	4163.1	13679.4	1971.8	7.7	59.5	112.9	858.1	3143.8
r-04	53475.4	17444.0	911.7	4294.1	804.4	13.7	50.9	108.9	105.4	2583.4
r-05	28004.0	15169.0	1301.6	3781.7	438.3	9.8	57.3	107.1	1121.4	2631.3
r-06	14401.8	3895.0	1008.4	3135.1	906.9	9.1	55.3	106.3	2852.6	3646.2
r-07	86201.6	22827.0	1456.1	5135.8	1369.6	9.5	57.8	111.7	351.4	2866.1
r-08	37051.0	5895.0	1522.7	4051.9	491.2	7.6	55.8	108.1	4151.6	1987.5
r-09	50780.4	17359.0	3627.6	17018.5	815.3	5.9	59.3	111.7	6897.8	2146.4
r-10	24742.5	28345.0	1361.0	4134.5	508.9	11.2	55.6	107.9	598.2	7529.0
r-11	7917.8	14675.0	175.7	898.1	57.2	14.2	58.4	108.8	59.5	5611.3
r-12	30915.6	8160.0	2744.3	6599.9	631.4	6.8	57.5	107.4	3842.8	1373.6
r-13	40158.9	24129.0	2075.6	7038.5	1451.0	9.7	58.9	112.4	598.3	4675.4
r-14	18915.3	11939.0	4653.9	5677.4	468.8	6.1	57.9	106.3	2844.4	5052.0
r-15	91151.2	22940.0	3673.1	10075.9	1121.6	11.1	56.1	109.7	1118.2	3515.0
r-16	26703.4	8958.0	1545.4	3649.4	277.9	8.6	57.9	111.0	2019.0	3480.3
r-17	33304.1	21871.0	878.7	4306.1	547.1	8.0	58.7	108.0	722.6	2697.3
r-18	14411.8	12093.0	1428.1	5437.1	308.9	10.4	53.4	109.9	2571.6	1119.1
r-19	51750.2	23955.0	3793.0	5187.6	360.9	5.0	62.2	108.3	984.1	1620.8
r-20	22025.6	24547.0	1092.2	5644.6	220.1	10.3	58.3	107.1	885.8	1692.7
r-21	25488.3	17527.0	1860.6	4677.0	356.6	8.7	56.6	109.7	1544.0	1870.5
r-22	45162.6	20718.0	1054.1	5544.1	483.4	8.5	58.7	109.6	944.5	3133.7
r-23	11193.4	6784.0	1330.5	2501.1	166.1	7.2	58.6	109.4	3952.4	1954.4
r-24	24933.9	20188.0	1212.6	5185.2	547.7	10.5	58.3	109.3	1150.5	1644.5

Table 4. Indicator's values for data for first three quarters of 2020

Code	Values									
	X_1	X_2	X_3	X_4	X_5	X_6	X_7	X_8	X_9	X_{10}
r-01	36446.7	44134.0	4023.4	4345.4	679.2	10.4	56.9	109.0	720.5	3528.4
r-02	21419.8	25428.0	1376.7	7016.7	449.0	12.3	49.2	101.8	1806.3	2728.5
r-03	95582.3	58710.0	3737.0	10444.6	1755.7	8.2	58.4	106.5	235.0	5894.9
r-04	43994.1	43567.0	1300.5	3345.8	700.3	14.5	49.8	103.1	45.5	1188.8
r-05	26948.8	33675.0	1028.8	3118.3	397.7	10.5	55.8	108.7	438.6	5683.4
r-06	13637.5	7090.0	861.8	1956.5	769.9	10.2	54.3	104.3	1430.0	3680.1
r-07	78074.0	56801.0	1011.5	3907.7	1267.3	10.4	56.5	108.4	150.0	4121.6
r-08	30883.1	12247.0	1581.1	2212.1	400.2	8.1	54.4	107.4	1833.7	1031.1
r-09	49227.9	32672.0	4524.2	8784.3	762.2	6.6	58.2	105.1	3932.9	5024.2
r-10	26339.1	56737.0	958.0	3910.2	706.2	12.3	54.2	111.7	231.0	17025.4
r-11	6154.2	37587.0	169.3	620.4	45.8	15.2	57.3	112.5	51.0	299.9
r-12	30522.8	17182.0	3226.0	3772.5	646.3	7.4	56.2	106.4	1734.9	3386.5
r-13	40592.0	49388.0	1732.1	3632.5	1311.1	10.3	57.9	110.9	205.1	8912.5
r-14	20213.1	18536.0	6105.9	4248.0	400.4	6.9	57.0	109.4	1391.7	5460.3
r-15	80188.6	48838.0	3956.8	8287.4	1163.0	11.7	55.2	106.6	457.1	4618.8
r-16	28737.4	19118.0	1496.5	2010.7	297.3	9.1	56.7	109.2	1080.6	5255.5
r-17	30338.2	48048.0	1008.9	2983.4	596.6	9.1	57.2	110.1	282.9	2086.5
r-18	13649.8	26998.0	1472.1	3532.0	300.9	11.3	52.1	108.4	1543.0	1236.3
r-19	45981.1	61006.0	3608.1	3819.2	381.7	5.8	60.8	105.5	870.3	3537.0
r-20	22620.1	56237.0	763.4	2250.0	198.3	11.1	57.2	110.5	355.6	841.6
r-21	26941.7	39572.0	2763.3	4151.3	355.2	9.5	55.3	109.0	1069.1	3948.0
r-22	44996.7	43113.0	1112.1	3232.8	523.3	9.3	57.6	107.0	398.9	4440.9
r-23	10588.4	14647.0	1553.1	1448.2	126.7	8.6	57.2	107.7	1670.9	729.0
r-24	24360.0	43737.0	1687.1	3600.9	554.4	11.5	56.8	109.1	577.4	2961.3

We assigned to each of Ukraine regions' names the corresponding code which we used for the designation of each of them to further use (table 2).

The values of initial indicators to provide calculations are shown in tables 3 and 4.

Let's group the initial indicators by the method of an extreme grouping of parameters. To determine the correlations between the initial indicators and the latent characteristics of each group in the context of maximizing the expression (3), we use the method of principal components. Taking into account expression (10), it is necessary to calculate the factor loadings for the selected principal components and choose the largest from them in absolute value. The number of groups is defined as the number of principal components that explain a given level of variance of the initial indicators following expression (11).

We choose the level of explanation of the variance of the initial indicators as $\gamma = 0.80$. Under such conditions, it is necessary to choose the first four principal components. The values of the eigenvectors and the eigenvalues of the corresponding correlation matrices of the initial indicators are given in Tables 5 and 6, and the values of the factor loadings – in tables 7 and 8.

Table 5. Most significant eigenvalues of correlation matrix and values of appropriate eigenvectors for data 2019

Eigenvalues			
λ_1	λ_2	λ_3	λ_4
3.64	2.42	1.36	0.87
Eigenvectors			
V_1	V_2	V_3	V_4
0.45	-0.16	0.13	-0.11
0.29	-0.39	-0.28	-0.14
0.38	0.23	-0.14	0.32
0.42	0.16	0.27	0.15
0.42	-0.17	0.23	0.14
-0.20	-0.49	0.35	0.02
0.22	0.12	-0.69	-0.20
0.12	-0.12	0.21	-0.26
0.01	0.57	0.25	0.23
-0.02	-0.34	-0.25	0.81

Table 6. Most significant eigenvalues of correlation matrix and values of appropriate eigenvectors for data 2020

Eigenvalues			
λ_1	λ_2	λ_3	λ_4
3.35	2.53	1.76	0.92
Eigenvectors			
V_1	V_2	V_3	V_4
0.49	-0.11	0.14	-0.26
0.28	-0.44	-0.05	-0.20
0.34	0.33	-0.19	0.06
0.46	0.14	0.24	0.09
0.45	-0.15	0.19	0.05
-0.22	-0.38	0.43	0.06
0.20	-0.04	-0.64	-0.27
-0.11	-0.36	-0.49	0.20
0.00	0.57	-0.02	0.21
0.22	-0.21	-0.10	0.85

Analysis of Tables 7, 8 allows us to formulate a conclusion, that we have the following distribution of initial indicators between subsets G_j:

Table 7. Values of factor loadings for data 2019

Initial indicators	Values of factor loadings			
	w_1	w_2	w_3	w_4
X_1	0.86	0.24	0.16	0.10
X_2	0.56	0.61	0.32	0.13
X_3	0.73	0.36	0.16	0.30
X_4	0.81	0.25	0.31	0.14
X_5	0.79	0.27	0.27	0.13
X_6	0.38	0.77	0.41	0.02
X_7	0.42	0.19	0.80	0.19
X_8	0.24	0.19	0.25	0.24
X_9	0.02	0.88	0.29	0.22
X_{10}	0.04	0.53	0.29	0.76

Table 8. Values of factor loadings for data 2020

Initial indicators	Values of factor loadings			
	w_1	w_2	w_3	w_4
X_1	0.89	0.17	0.19	0.25
X_2	0.52	0.70	0.06	0.20
X_3	0.62	0.52	0.25	0.06
X_4	0.84	0.22	0.31	0.09
X_5	0.83	0.23	0.25	0.05
X_6	0.40	0.61	0.57	0.05
X_7	0.36	0.07	0.85	0.26
X_8	0.21	0.58	0.66	0.19
X_9	0.01	0.90	0.03	0.20
X_{10}	0.40	0.33	0.14	0.81

$G_1=\{X_1, X_3, X_4, X_5\}$;
$G_2=\{X_2, X_6, X_8\}$;
$G_3=\{X_7, X_9\}$;
$G_4=\{X_{10}\}$.

Note, that subset G_4 consists of one initial indicator X_{10}, so, partial composite index I_4 coincides with this indicator.

To calculate patrial composite indices I_j, $j = 1, 2, 3, 4$, we conduct a normalization procedure for initial data. In this case, we execute this step using formula (16), because this way allows keeping the proportions between the values of the indicator, which is important in the calculation of composite index's values.

We also take into account, that indicator X_6 is a disincentive, and other indicators are incentives. Weight coefficients we calculate, using formula (18). Values of composite indices I_j, $j = 1, 2, 3, 4$, have been calculated using linear convolution by expression (12). To calculate the comprehensive index, we also use weighted linear convolution like (19) with weight coefficients, obtained by the formula (20).

To correctly compare the results of calculations and identify changes in the levels of the comprehensive index for the relevant periods and the regions' structuring, the values for 2019 and 2020 will be combined into one sample. The normalization procedure is performed for the combined data.

Further calculations of both partial composite indices and comprehensive indices are executed for each period separately. The values of the selected eigenvector elements for calculating the weights by formula (18) are different, as well as the corresponding eigenvalues that will be used to calculate the weights of the generalized indicator by formula (20). So, for the data of 2019 and the data of 2020, we obtain different values of weight coefficients, which means that both composite and comprehensive indicators will be calculated according to different rules.

Therefore, for a more accurate comparison of the results, we propose to calculate corresponding weights as the average values of the appropriate components obtained for 2019 and 2020.

Weight coefficients for calculation of composite indices I_1-I_4 in accordance with distribution initial indicators to G_j have such values: $w_{11} = 0.30; w_{13} = 0.18; w_{14} = 0.27; w_{15} = 0.25; w_{22} = 0.25; w_{26} = 0.28; w_{29} = 0.47; w_{37} = 0.77; w_{38} = 0.23; w_{4,10} = 1.00$. Values of weight coefficients for calculation of comprehensive index are: $w_1^{COM} = 0.42, w_2^{COM} = 0.28, w_3^{COM} = 0.19, w_4^{COM} = 0.11$. The results of the calculations of comprehensive index values are presented in table 9.

Table 9. Comprehensive index values

Code	2019	2020	Code	2019	2020
r-01	0.45	0.46	r-13	0.48	0.49
r-02	0.40	0.38	r-14	0.48	0.46
r-03	0.67	0.65	r-15	0.55	0.55
r-04	0.37	0.37	r-16	0.38	0.37
r-05	0.37	0.38	r-17	0.39	0.39
r-06	0.39	0.34	r-18	0.36	0.34
r-07	0.49	0.50	r-19	0.48	0.50
r-08	0.44	0.36	r-20	0.36	0.35
r-09	0.68	0.59	r-21	0.38	0.41
r-10	0.40	0.48	r-22	0.41	0.42
r-11	0.28	0.26	r-23	0.39	0.32
r-12	0.47	0.43	r-24	0.38	0.39

A comparison of the calculation results of the comprehensive index shows that for most of Ukraine's regions there is a decrease in its values. In our opinion, this fact indicates a negative impact of the pandemic COVID-19 on economic development. At the same time, for some regions, in particular, Vinnytsia, Zhytomyr, Zaporizhia, Kirovohrad, Mykolaiv, Kharkiv, Khmelnytskyi, and Chernihiv regions, there is an increase in the values of the indicator in 2020. This increase is especially noticeable for the Kirovohrad region. This can be explained by the fact that for a long time in this area was the best epidemiological situation in Ukraine. Also, the growth of industrial production, in particular pharmaceuticals one in Vinnytsia, Kirovohrad, Zaporizhia, and Kharkiv regions.

Let us consider the changes in the structure of Ukraine's regions in 2020 compared to 2019 in terms of the economic development comprehensive index. Given the relatively high density of values of the comprehensive index for different regions, the use of the approach to the grouping of regions, based on the analysis of the values of the delta, calculated by the formula (21), doesn't allow to determine their structure. Therefore, to solve this problem, we apply an approach based on the use of the Harrington desirability scale. For this purpose, we transform the values of the integrated indicator according to formulas (22) and (22). The distribution of regions by groups is executed according to the rules given in table 1. The results of the calculations are listed in table 10.

Table 10. Identifying the structure of regions for data 2019 and 2020

Code	$d = H(Z)$		Level of development	
	2019	2020	2019	2020
r-01	0.72	0.75	above average	above average
r-02	0.47	0.35	average	below average
r-03	0.99	0.99	relatively high	relatively high
r-04	0.29	0.28	below average	below average
r-05	0.27	0.34	below average	below average
r-06	0.44	0.15	average	relatively low
r-07	0.86	0.88	relatively high	relatively high
r-08	0.68	0.25	above average	below average
r-09	0.99	0.97	relatively high	relatively high
r-10	0.46	0.84	average	relatively high
r-11	0.00	0.00	relatively low	relatively low
r-12	0.81	0.66	relatively high	above average
r-13	0.84	0.86	relatively high	relatively high
r-14	0.84	0.77	relatively high	above average
r-15	0.95	0.95	relatively high	relatively high
r-16	0.37	0.31	average	below average
r-17	0.42	0.43	average	average
r-18	0.24	0.13	below average	relatively low
r-19	0.83	0.88	relatively high	relatively high
r-20	0.22	0.18	below average	relatively low
r-21	0.37	0.56	average	average
r-22	0.56	0.58	average	average
r-23	0.39	0.06	average	relatively low
r-24	0.34	0.42	below average	average

The analysis of results obtained shows, that the first group with a relatively high level of economic development is quite large. Traditionally, this group includes Kyiv, Kharkiv, and Dnipropetrovsk regions, which in the "pre-pandemic" period had a fairly high level of economic development. These regions have a fairly high production potential, they account for a significant share of foreign investment in 2020 and therefore the pandemic has not had such a destructive impact on the economic development of these regions. Zaporizhia and Poltava regions also have significant potentials and were distributed to this group. The lowest level of economic development is in the Luhansk region, and in 2020 the situation has not changed.

It should be noted that for many regions there have been no changes in the level of economic development, although there has been a decrease in the value of the corresponding comprehensive index. For those regions where changes are taking place, they are usually associated with a decline in economic development. The only exception is the Kirovohrad region.

The most significant decrease in the level took place in Zakarpattia, Ternopil, and Chernivtsi regions. These are the regions that were the first to suffer from the pandemic and were in the "red" zone for a long time, which negatively affected all indicators of economic development and led to a significant reduction in the corresponding comprehensive index values.

Thus, the results of the research demonstrate the fundamental possibility of applying the proposed approach to the study of economic development of regions by constructing an integrated indicator. The analysis of the structure of the regions showed the real impact of the pandemic on the development of almost all regions, which led to the corresponding structural changes.

5 Conclusions

The study of economic development trends both in the economic system of the country as a whole and at the level of individual regions remains the focus of the most significant research. The results of such studies are especially important in periods of global challenges, one of which at this stage of the world community development was the COVID-19 pandemic. Solving the problems of assessing the level of development of regions, their structuring, identifying gaps and breaks in the development of individual territorial units is complicated by the significant multidimensionality of their description. The use of analytical methods of information processing based on economic and mathematical models allows us to present it in a concentrated form without significant losses, which contributes to the adoption of sound management decisions and the development of strategic plans for regional development. Therefore, models that allow for a significant reduction in baseline and identify latent characteristics of the studied phenomena are important for studies. In particular, such approaches include models based on the comprehensive index assessment technology.

The approaches offered in the article allow estimating the level of economic development of regions by block convolution of the set of initial indicators into a single complex measure – an economic development comprehensive index. Thus, the toolkit which allows to carry out a grouping of initial indicators to take into account the weights of components at the construction of such indicators, and also the weights of partial composite indices at their convolution into the economic development comprehensive index is offered. The article proposes some approaches to grouping regions by the level of economic development. An approach based on the transformation of the comprehensive index values with the projection of the result on the desirability scale is chosen for practical implementation. This way allows to rank regions, determine their structure by this characteristic and assess structural changes over time.

According to the research outcomes, it can be concluded that the COVID-19 pandemic has a destructive impact on the economic development of the vast majority of Ukraine's regions, which was reflected both in changes in the values of the comprehensive index and in the regions' structure.

The direction of further research is the development of analytical tools to take into account indicators of non-metric origins in the assessment procedures for the identifying level of economic development.

References

[1] S. Semerikov, S. Chukharev, S. Sakhno, A. Striuk, V. Osadchyi, V. Solovieva, T. Vakaliuk, P. Nechypurenko, O. Bondarenko, H. Danylchuk, E3S Web of Conferences **166** (2020)

[2] S. Semerikov, H. Kucherova, V. Los, D. Ocheretin, CEUR Workshop Proceedings **2845**, 22 (2021)

[3] International Labour Orgaization, *ILO monitor: COVID-19 and the world of work. Sixth edition: Updated estimates and analysis* (2020), `https://cutt.ly/GjAdvEa`

[4] The World Bank, *The Global economic outlook during the COVID-19 pandemic: a changed world* (2020), `https://cutt.ly/ff75GUY`

[5] The World Bank, *Global economy to expand by 4% in 2021; vaccine deployment and investment key to sustaining the recovery* (2021), `https://cutt.ly/0jAdKpT`

[6] H. Ellyatt, *'don't be under the illusion' that an economic recovery will come quickly with a vaccine, UBS' Weber says* (2020), `https://cutt.ly/ijAdCIO`

[7] *Derzhavna sluzhba statystyky Ukrainy (State statistics service of Ukraine)* (2020), `http://www.ukrstat.gov.ua/`

[8] *Derzhavna stratehiia rehionalnoho rozvytku na period 2021-2027 (State strategy for regional development for 2021-2027)* (2020), `https://cutt.ly/bjAfuSS`

[9] A. Williams, *EBRD revises down economic forecasts amid continuing coronavirus uncertainty* (2020), `https://cutt.ly/qjAd1F9`

[10] M.V. Moroshkina, Finance and Credit **10** (2020)

[11] E. Dhimitri, A. Gabeta, M. Bello, The USV Annals of Economics and Public Administration **15**, 51 (2015)

[12] A. Ramadhan, A. Sutandi, A.E. Pravitasari, IOP Conference Series: Earth and Environmental Science **556**, 012008 (2020)

[13] A. Soejoto, D. Fitrayati, L.R.N. Sholikhah, International Journal of Applied Business and Economic Research **14**, 9493 (2016)

[14] J. Li, Y. Chen, Z. Xi, X. Zou, Journal of Physics: Conference Series **1574**, 012170 (2020)

[15] T.A. Munandar, Azhari, A. Musdholifah, L. Arsyad, Journal of Computer Sciences **14**, 81 (2018)

[16] A.F. Doni, Y.D.P. Negera, O.A.H. Maria, Journal of Physics: Conference Series **1569**, 022078 (2020)

[17] P. Hryhoruk, N. Khrushch, S. Grygoruk, Periodicals of Engineering and Natural Sciences **7**, 712 (2019)

[18] L. Chagovets, S. Prokopovych, V. Kholod, Development Management **18**, 43 (2019)

[19] S. Eshugova, International Journal of Engineering & Technology **7**, 340 (2018)

[20] D. Cziraky, J. Sambt, J. Rovan, J. Puljiz, European Journal of Operational Research **174**, 427 (2006)

[21] P. Hryhoruk, N. Khrushch, S. Grygoruk, International Journal of Industrial Engineering & Production Research **31** (2020)

[22] B. Bal-Domanska, in *Advanced Studies in Classification and Data Science. Studies in Classification, Data Analysis, and Knowledge Organization*, edited by T. Imaizumi, A. Okada, S. Miyamoto, F. Sakaori, Y. Yamamoto, M. Vichi (Springer, Singapore, 2020), pp. 19–29

[23] V. Babenko, A. Panchyshyn, L. Zomchak, M. Nehrey, Z. Artym-Drohomyretska, T. Lahotskyi, WSEAS Transactions on Business and Economic **8**, 209 (2021)

[24] T.A. Munandar, R. Wardoyo, Intrnational Journal of Computer Applications **123**, 17 (2015)

[25] K. Gorbatiuk, O. Mantalyuk, O. Proskurovych, O. Valkov, SHS Web of Conferences **65**, 04008 (2019)

[26] K. Gorbatiuk, O. Mantalyuk, O. Proskurovych, O. Valkov, *Application of Fuzzy Clustering to Shaping Regional Development Strategies in Ukraine*, in *Proceedings of the 6th International Conference on Strategies, Models and Technologies of Economic Systems Management*, edited by P. Hryhoruk, N. Khrushch (Atlantis Press, Paris, 2019), Vol. 95 of *Advances in economics, business and management research*, pp. 271–276

[27] X. Wang, in *Big Data Analytics for Cyber-Physical System in Smart City*, edited by M. Atiquzzaman, N. Yen, Z. Xu (Springer, Singapore, 2020), Vol. 1303

[28] J. Gurna, K. Gurna, Statistics in Transition New Series **16**, 265 (2015)

[29] D.F. Meyer, J. De Jongh, N. Meyer, International Journal of Business and Management Studies **8**, 100 (2016)

[30] J. De Jongh, D. Meyer, Administratio Publica **27**, 122 (2019)

[31] P. Hryhoruk, N. Khrushch, S. Grygoruk, *An approach to design a composite index of economic development and identifying the bounds of its levels*, in *Proceedings of 9th International Conference on Advanced Computer Information Technologies* (Ceske Budejovice, 2019), pp. 48–51

[32] M. Stanichkova, L. Melecky, Regional Studies, Regional Science **5**, 231 (2018)

[33] V. Royuela, Social Indicators Research **148**, 831 (2020)

[34] A. Al-Baidhani, *COVID-19 economic impact on developed and developing economies* (2020), https://ssrn.com/abstract=3700352

[35] G. Gondwe, *Assessing the impact of COVID-19 on Africa's economic development* (2020), https://cutt.ly/ijAkGWt

[36] M. Gallagher, *COVID-19 economic impact assessment* (2020), https://cutt.ly/1jAkKmX

[37] M.A.R. Estrada, Estudios Economicos **37**, 139 (2020)

[38] P. Hryhoruk, N. Khrushch, S. Grygoruk, E3S Web of Conferences **166**, 13023 (2020)

[39] E. Harrington, Industrial Quality Control **21**, 494 (1965)

[40] *Ministerstvo rozvytku hromad ta terytorii Ukrainy. Reitynhova otsinka rehioniv (Ministry for communities and territories development of Ukraine. Rating assessment of regions)* (2020), https://cutt.ly/gjJ5Hp0

Multicriteria optimization of oil and gas enterprises financial stability using the genetic algorithm method

Marta Shkvaryliuk[1,*], *Liliana* Horal[1,**], *Inesa* Khvostina[1,***], *Natalia* Yashcheritsyna[1,****], and *Vira* Shiyko[1,†]

[1]Ivano-Frankivsk National Technical University of Oil and Gas, 15 Karpatska Str., Ivano-Frankivsk, 76019, Ukraine

Abstract. The article considers the problem of optimizing the financial condition of oil and gas companies. The offered methods of optimization of a financial condition by scientists from different countries are investigated. It is determined that the financial condition of the enterprise depends on the effectiveness of the risk management system of enterprises. It is proved that the enterprises of the oil and gas complex need to develop a system for risk management to ensure the appropriate financial condition. The financial condition is estimated according to the system of certain financial indicators, the integrated indicator of financial condition assessment is constructed using the method of taxonomy. According to the results of the calculation of the integrated indicator, it is concluded that this indicator does not have a stable trend. On the basis of the conducted researches it is offered to carry out optimization of an integral indicator of a financial condition with use of genetic algorithm in the Matlab environment. Based on the obtained results, recommendations of the management of the researched enterprises on increase of management efficiency are given.

1 Introduction

The loss of financial stability of any enterprise in a turbulent environment, which is exacerbated by negative external factors is a reality today. This fact leads to many negative consequences, one of which is bankruptcy and liquidation of the enterprise. Due to the fact that the reduction of financial stability of enterprises that provide Ukraine's economic and energy security has become a reality, there is an urgent need to optimize their financial stability. Examining various scientific sources related to the solution of this problem, we can conclude that there are many ways to optimize financial stability. However, we propose to use the genetic algorithm method in the process of the studied enterprises financial stability optimizing , which by optimizing the financial stability will generate optimal values of the enterprise internal factors. Which in the future will allow the company's management to make optimal management decisions and reduce the risk of its loss..

2 Background

Important contribution to the study of financial sustainability was made authors of [1–3].

The authors of [1] propose an economic and mathematical model of the assessing financial stability process by calculating the integral indicator of the service sector financial stability. To study the stability and controllability of the assessing financial stability process, the types of control maps for each of the coefficients were determined. Proposed apparatus of neural networks makes it possible not only to determine the most profitable activity of an enterprise but also to assess the financial condition of each of its research objects.

The main results of the G. Azarenkova, O. Golovko and K. Abrosimova study [2] are following: the theoretical and essential characteristics of enterprise financial sustainability has been determined; the financial status of PJSC "Turboatom" has been analyzed; the taxonomic index of financial sustainability has been calculated and the forecast of its significance has been made, the approaches to increase enterprise financial sustainability have been proposed.

The article [4] provides practical advice for enterprises to achieve the highest possible level of energy and financial security.

In work [3] an eigenstate method is proposed to analyse the basic indicators of the enterprise as it allows to construct an economical stability model of such enterprise, describe the methodology for analyzing the economic stability of an enterprise on the basis of eigenstate method, provides formulas for calculating the complex indicator of economic stability. The efficiency of the methodology is demonstrated with evidence from the economic stability analysis of the trading company.

W. Ma, M. Jin, Y. Liu and X. Xu in [5] analyze the relationship between enterprise management and financial performance, analyze the mean and heterogeneity of the enterprise management team characteristics, mathematically models its relationship, constructs fractional differential equations, and tests it through empirical research. The influence of the enterprise management age character-

*e-mail: marta.shkvaryliuk@gmail.com
**e-mail: liliana.goral@gmail.com
***e-mail: inesa.hvostina@gmail.com
****e-mail: yashcheritsyna@gmail.com
†e-mail: vnkShiyko@gmail.com

istics, international experience, education level, team size and government background on the financial performance of the company.

A positive aspect of modern research on solving the problem of optimizing financial stability is also the fact that in many works the problem is proposed to be solved by building an effective system of risk management in enterprises. In particular, this issue is considered in the [1, 6–9].

Research results of D. M. Sprčić, A. Kožul and E. Pecina [6] have revealed low levels of ERM development in listed Croatian companies. Managers are focused on financial and operative risk management, while strategic and other risks have been neglected. Regression analysis has indicated somewhat unexpected but important conclusion – the explored risk management rationales have weak predictive power in explaining corporate risk management decisions in Croatian companies. The level or risk management system development is dependent only on the size of the company and value of the growth options [6].

J. Cohen, G. Krishnamoorthy and A. Wright in [7] distinguish three major findings from our study. First, importantly, all three types of participants see a strong link between ERM and the financial reporting process. Second, despite recognition of the broad nature of ERM, the predominant experiences of the actual roles played by triad members center on agency theory, while resource dependence may be relatively underemphasized by all triad members. Finally, CFOs and AC members indicate that auditors may be especially underutilizing ERM in the audit process, suggesting an "expectations gap" [7].

The paper [10] proposes an approach to modeling the business climate of the country, which is based on the financial and economic indicators, and makes it possible to assess the development trends of the studied indicator. The proposed approach is based on the taxonomy method.

The analysis of sustainability and security of enterprises was carried out using a wide range of classical and advanced modeling methods, in particular, in article [11].

In our previous studies [12] it was possible to achieve an increase in the efficiency of modeling financial risk through the formation of an ensemble of models.

In article [13] is demonstrated the possibility of studying complex socio-economic systems as part of a network paradigm of complexity.

After conducting research on scientific papers that solve the problem of the economic entities financial stability loss risk reduction, it can be concluded that the optimization of financial condition through the use of a genetic algorithm has not been carried out. Therefore, this issue is relevant and needs research.

The analysis of methodological approaches to assessing the financial stability of economic entities shows the lack of a single methodological basis for assessment. Moreover, differences are manifested both in the components of financial stability, the system of indicators, and in the method of their consideration in the analysis of financial stability. All this necessitates the development of fundamentally new approaches and tools for assessing the financial stability of enterprises.

3 Methodology

Consider the dynamics of average indicators of liquidity, solvency, profitability and business activity of Ukraine's oil and gas industry, table 1.

Table 1. Dynamics of financial indicators' average values of Ukrainian oil and gas industry

Indicators	Years				
	2015	2016	2017	2018	2019
Current Ratio	89,6	93,3	98,7	10 3	84,7
Equity-to-assets	18,1	16,9	18,4	3	1
ROA	-0,1	0,1	0,8	3,4	1,6
WCP	-0,1	0,1	1,5	6,9	3,8
CashRatio	1	1	1,9	1,3	1
NPM	0	0,3	1,8	4,5	4,4
ROTA	0	0,5	1,4	4,7	3,1
Tot. Ass Turn.	0,8	0,7	0,8	1	0,7
Rec. Turn.	3,4	2,9	4,1	4,5	3,5

Table 2. Growth rates (basic) of financial indicators' average values of Ukrainian oil and gas industry, %

Indicators	Years				
	2015	2016	2017	2018	2019
Current Ratio	1,00	1,04	1,10	1,15	0,94
Equity-to-assets	1,00	0,93	0,10	1,86	1,67
ROA	1,00	1,00	8,00	34,00	16,00
WCP	1,00	1,00	15,00	69,00	38,00
CashRatio	1,00	1,00	1,90	1,30	1,00
NPM	1,00	1,00	6,00	15,00	14,67
ROTA	1,00	1,00	2,80	9,40	6,20
Tot. As. Turn.	1,00	0,88	1,00	1,25	0,88
Rec. Turn	1,00	0,85	1,21	1,32	1,03

According to table 1, it was established that the current ratio does not exceed the regulatory value of 100% in contrast to 2018, where the value of the indicator is 103.0%. Current ratio characterizes the ability of enterprises to repay their current liabilities for up to 1 year through current assets. The liquidity indicator shows that the analyzed enterprises do not have enough resources that can be used to repay short-term creditors' claims, the change in the studied indicator ranges from 84.7% to 98.7%. In 2018, the overall increase over four years increased by 14.96% and in 2019 there is a decrease of 5.47% compared to the base value.

The value of the Cash ratio did not reach the normative value of 20%. During 2015-2019 the value of the indicator ranges from 1.0% to 1.9%, i.e., money and their equivalents are not enough to meet the current liabilities of economic entities. The Equity-to-Assets – the solvency indicator characterizes the share of equity of enterprises in the total amount of funds invested in its activities. The value of the Equity-to-Assets does not exceed the normative by 50% and fluctuates during the study period in the

https://doi.org/10.1051/shsconf/202110706005

range of 16.9-33.8%, so there is every reason to believe that enterprises are not solvent.

According to table 2, return on assets (ROA) determines the return of 1 hryvnia assets of economic entities, its value increases rapidly from 0.1% in 2016 to 3.4% in 2018, then decreases to 1.6%. The dynamics of working capital profitability is similar, the value of the indicator increases from 0.1% in 2016 to 6.9% in 2018, decreases to 3.8% in 2019. The profitability indicator is the net margin (NPM), which reflects the ratio of net profit to the total revenue of the enterprise whose value increased from 0.3% in 2016 to 4.5% in 2018 and decreased to 4.4% in 2019. The ratio of operating profit to assets of oil and gas companies (ROTA) characterizes the return on total assets, the value which also increased from 0.5% in 2016 to 4.7% in 2018 and decreased to 3.1% in 2019.

The state of business activity in the context allows you to determine the productivity of assets of enterprises. The value of the indicator in the industry ranges from 0.7-1 for the study period. Working capital turnover – an indicator of business activity, which shows the efficiency of enterprises working capital usage in terms of revenue generated by them. According to the values of the indicator, the efficiency of its generation for the period under study is 0. The turnover of receivables shows how many times during the year receivables are repaid. The ability of entities to repay receivables during the period under review changes abruptly. The highest value is observed in 2018 and is 4.5 and the lowest value is 2.9 in 2016, due to the crisis of payments in the country.

Given the results of the calculated average values of Ukraine's oil and gas industry financial indicators and their dynamics, it is possible to draw conclusions about the instability of their trends, which necessitates the calculation and modeling of an integrated indicator of financial condition. The taxonomy method was used for its construction, the results of its calculation and forecasting are shown in figure 1.

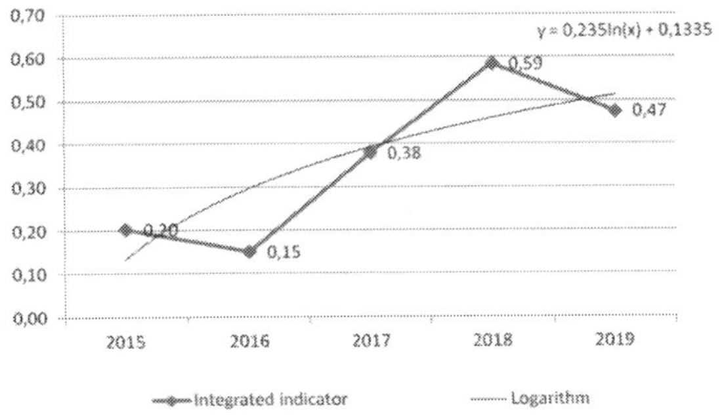

Figure 1. Dynamics of the integrated indicator of Ukrainian oil and gas enterprises' financial condition

According to figure 1, it is determined that the value of the integrated indicator of Ukraine's oil and gas companies financial condition increases from 2016 to 2018, then decreases to 0.47 in 2019. The decrease in the integrated indicator shows a certain signaling ability to reduce financial stability of oil and gas companies increase in the general level of financial risks.

Variation of indicators acquires a difference due to a significant difference between the maximum and minimum values of the sample. It can be noted that the taxonomy index in the industry was unstable during the analyzed periods: the closer this indicator is to 1, the lower the level of risk. As you can see, the level of risk of the financial condition is quite high for the analyzed period. According to the results of the forecast, the negative trend of the taxonomic indicator can be stated, so if we do not change the conditions of operation and development of oil and gas companies in the future, negative trends will further worsen the financial condition of companies in Ukraine's oil and gas industry. There is a need to develop effective management solutions to provide enterprises with a vector of positive development, this problem is of national importance and needs proper attention.

To begin with, it is necessary to find the optimal values of input parameters that form a stable financial condition of the enterprise, provide a sufficient level of the studied industry financial condition integrated indicator at a level above average, 0.6-1.0. Thus, we have the first restrictions on the function of the financial condition integrated indicator.

The problem of optimization problems has been in the field of view of domestic economics representatives for a long time. To determine the extreme values classical methods of higher mathematics are widely used, consider one of the most interesting and modern options – genetic algorithms. The most popular from a scientific point of view is, the use of an algorithm for finding optimization solutions using mathematical modulation of genetic processes. In his works, he shows the possibilities and patterns of heredity and variability in genetics in the transfer to the problem of determining the optimization values. Ideas and methods of genetics play an important role in genetic engineering and are applied to economic problems. The mechanism of heredity means the role of genes as elementary carriers of hereditary information. Scientists showed the work of the so-called "genetic" operators of ascent, mutation, mathematical implementation of single-point and multi-point crossover, the search for the most adapted individual.

In mathematics, the problem of stability arises when a physical object is perturbed in phase space, that is, when external forces take it out of equilibrium. As a result, the object can: move away from equilibrium; be in a slight deviation from it; return to equilibrium, withstanding adverse fluctuations.

In fact, the behavior of an object in an disturbed state determines the stability or instability of its undisturbed equilibrium state. Thus, the equilibrium state of an object can be considered stable when, after perturbation, it enters a state close to equilibrium or when it returns to it.

To study the phenomenon of stability in more detail, it is necessary to use the concept of "area of stability". It is often necessary to determine the effect of changing certain parameters on stability. To do this, build the stability area of the object in the space of changing parameters. The area of stability is determined by the set of values of the parameters of the object for which it is stable. Going beyond the parameter limit limits the object from steady to unstable. When the limit of stability is exceeded, the level of risk increases significantly. It is clear that the transition from the zone of stability to the unstable position is determined not by the boundary line, but by some area that can be called transitional.

Drawing analogies between economic and mechanical equilibrium, we should pay attention to the differences between static and dynamic equilibria. At static equilibrium the motion of an object ceases, whereas at dynamic equilibrium the physical body continues to move, but at the same time certain total characteristics of the system remain unchanged. An example is the flow of water in the riverbed: the height of the water and the speed of the flow can be constant, and its parameters, such as inflow and outflow, can change. In other words, static equilibrium implies the ability of the system in it, after minor deviations to return to the previous state, and dynamic equilibrium can be interpreted as the ability of a mechanical system in motion under the influence of certain forces, not to deviate from a given trajectory at insignificant accidental stresses or deviations.

Speaking of the enterprise financial stability, it is advisable to draw analogies with the dynamic stability, because the functioning of the oil and gas sector, its functions, the implementation of the whole complex of active and passive operations is nothing but a dynamic process. Thus, the financial stability of enterprises – one of the key dynamic characteristics of its activities, which largely reveals its viability. In the future, we will consider that the financial stability of the enterprise is a dynamic category, which is the ability to return to equilibrium after leaving it as a result of a certain impact. Sustainability in economic systems, despite some similarities with technical ones, is a much more complex concept. In view of this, analogies can be made for economic systems only conditionally. Due to the fact that a universal approach to assessing financial stability as a single scalar indicator has not yet been developed, we propose to use the tools of multi-criteria optimization, which can be used to implement the concept of economic equilibrium.

We believe that a financially stable enterprise must achieve a certain equilibrium – the optimal ratio between risk, return, liquidity and other key financial performance indicators, on which depends its financial stability. As target functions we will take the key financial indicators of the oil and gas company: current liquidity, autonomy ratio, net margin and receivables turnover, which according to the correlation-regression analysis have the greatest impact on the integrated indicator of financial stability. To achieve a certain equilibrium of the bank should optimize all these criteria, taking into account its real financial condition. In addition, the task of constructing Pareto-effective financial indicators that optimize the integrated indicator of financial stability is proposed.

To find solutions to the multicriteria optimization problem, we use the method of genetic algorithm, which has proven itself well for solving this class of problems. A genetic algorithm is a heuristic search algorithm that is applied to optimization and modeling problems by random selection based on the use of mechanisms resembling evolutionary processes in nature. They are a kind of evolutionary methods of calculation. Genetic algorithm – a method of optimization based on the concepts of natural selection and genetics. In this approach, the variables that characterize the solution are represented as genes on the chromosome. The genetic algorithm, operating on a finite number of solutions (population), generates new solutions in the form of various combinations of parts of the solutions of this population. Operators such as selection, recombination and mutation are used for this purpose.

In a genetic algorithm, a chromosome is a numerical vector that corresponds to a variable. Each of the chromosome vector positions is called a gene.

The genetic algorithm is actually a kind of random search and is based on approaches that resemble the mechanism of natural selection. In a genetic algorithm, some random set of initial data, called a population, is first formed. Each element of the population is called a chromosome and represents some solution of the problem in the first approximation, i.e. satisfies the system of constraints of the problem. Chromosomes evolve during iterations called generations (or generations). During each iteration, the chromosome is evaluated using some degree of compliance (fitness function), which is also called compliance function. A mutation is an operation that implements random changes in different chromosomes.

The simplest mutation is to randomly alter one or more genes. In a genetic algorithm, a mutation plays an important role in restoring genes dropped from a population during a selection operation so that they can be used in new populations. In addition, it allows the formation of genes that were not present in the original population. The intensity of mutations is determined by the mutation rate, which is the proportion of genes that are mutated in this iteration. Too small a value of this factor means that many genes that could be useful will never be considered. At the same time, too large a value of the coefficient will lead to large random perturbations. Descendants will no longer be like their parents and the algorithm will lose the ability to learn while maintaining hereditary traits.

We used the Matlab Optimization Toolbox to find Pareto-effective sets of unit coefficients. The standard adaptive feasible mutation function was chosen as the mutation operator, which is used for constrained tasks and allows you to randomly generate directions based on the most recent successful or unsuccessful generations. To perform the crossover operation, the Scattered method was used, which involves creating a random binary vector and selecting genes from the first parent chromosome for which the corresponding value is 1, or from the second parent chromosome when the value is 0 when combining these genes to form a new offspring.

A multicriteria problem is often understood not as a verbal description of the problem, but as its model, namely: a multicriteria problem is a mathematical model of making an optimal decision based on several criteria. These criteria may reflect assessments of the different qualities of the object or process about which the decision is made. Formally, the multicriteria problem as a model is given in the form:

$$F(x) \to \max \text{ for all } x \in \mathbb{D}, \quad (1)$$

where $\mathbb{D}$ is the set of valid solutions; $F(x)$ is a vector function of the argument x (integral indicator of financial condition), which can be represented as follows:

$$F(x) = [f_1(x), f_1(x) \ldots, f_k], \quad (2)$$

where $f_1(x), f_2(x) \ldots f_k(x)$ – scalar functions of the vector argument x each of which is a mathematical expression of one optimaliti criterion.

Since this model uses a vector objective function, it is often called the problem of vector optimization. Obviously, problem (1) does not belong to the class of mathematical programming problems, because the models of this class of problems always contain only one objective function of the vector argument.

Here we consider a complex vector criterion, which can be used to achieve the maximum effect, without necessarily reaching the extreme in all functions. The existence of a solution that literally maximizes all target functions is a rare exception. The problem of vector optimization in the general case does not have a clear mathematical solution. To obtain a solution, it is necessary to use additional subjective information of a specialist in this subject area, which is commonly referred to as a decision maker. This means that when solving the problem by different specialists with the involvement of different sources of information, most likely different answers will be received. Problems of vector optimization are currently considered in the framework of decision theory [10], the main feature of which is the presence of uncertainty. This uncertainty cannot be ruled out by various modeling techniques and objective calculations. In multicriteria problems, the uncertainty is that it is not known which criterion to prefer and to what extent. To eliminate this uncertainty, it is necessary, firstly: to formulate a special principle of optimality, and secondly: to involve additional subjective information of the decision-maker based on his experience and intuition.

Therefore, in accordance with the above information, we formulate the objective functions and conditions of optimization. The function $G(x)$ is defined as an integrated indicator of financial condition calculated by the taxonomy method and takes into account the levels of unit indicators that were previously selected to be included in the integrated indicator of financial condition $Y(x)$:

$$Y(x) = -0,17432 - 0,00076x_1 + 0,00095x_2 + \\ + 0,055039x_3 + 0,125108x_4 \quad (3)$$

and Finscore ($G(x)$):

$$G(x) = 3,639721 + 0,009527x_1 - 0,03471x_2 + \\ + 0,13336x_3 - 0,16618x_4. \quad (4)$$

Therefore, taking into account the constructed relationship between the studied enterprise financial condition $Y(x)$ and Finscore $G(x)$ integrated indicator and its unit indicators, it is necessary to optimize the complex indicator of financial condition, maximize it by optimizing independent variables.

First, let's set the problem by describing the stages of optimization. Figure 2 shows the stages of the optimization process.

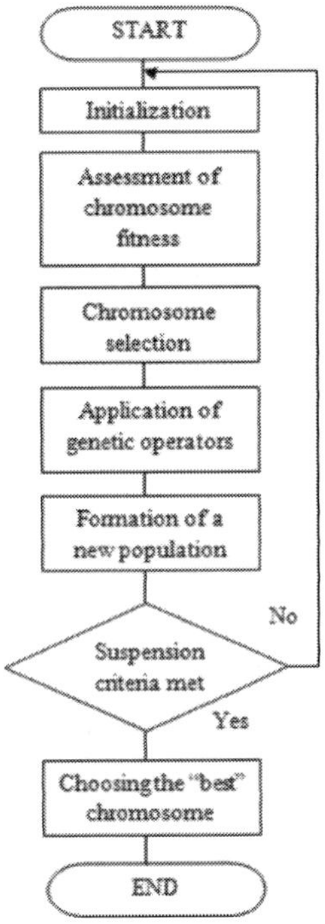

Figure 2. Conditional block diagram of the genetic algorithm

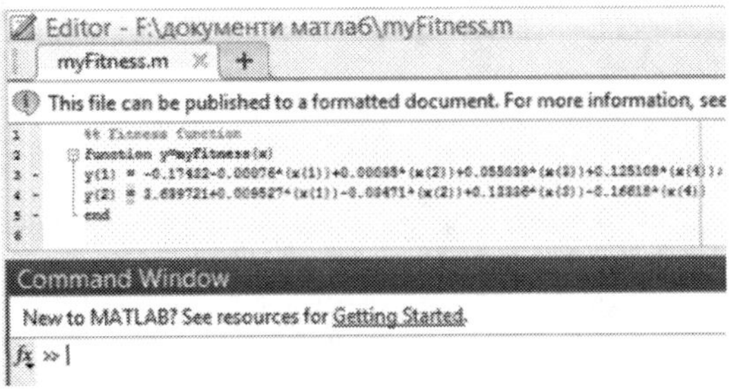

Figure 3. M-file in Matlab

Therefore, we need to optimize, namely to maximize the financial condition, it should be as close as possible to 1, but there are some limitations, which will be described below. Next, we will implement the financial condition integrated indicator function optimization in MatLab with pre-imposed restrictions on independent variables. First, we form an m-file in which we introduce the optimized function, in the economic-mathematical model of the integrated indicator of the financial condition of the object

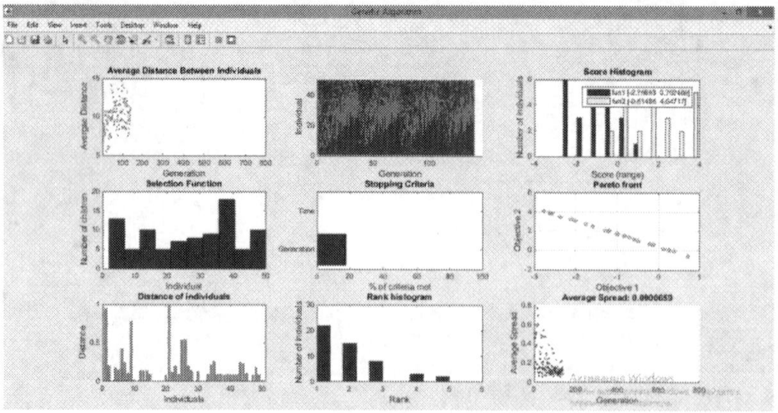

Figure 4. Program results

under study. The M-file with the optimized function with the given restrictions is shown in figure 3.

To implement a multi-criteria task, use the built-in Optimization ToolBox. The obtained optimization results are shown in figure 4.

4 Conclusions

With regard to the optimization of financial stability, this means that we have obtained many key indicators that shape the financial condition of the enterprise and it provides an opportunity to form an effective management strategy aimed primarily at achieving optimal values of indicators that together and determine the economic essence of the financial stability of the enterprise.

The analysis of existing approaches to assessing the financial condition of enterprises shows the lack of a unified methodological basis for this issue. The key problem is the lack of a single indicator that would accumulate all aspects of financial condition. The article proposes a method of assessing and optimizing the financial condition based on the concept of maximizing the financial stability of enterprises Finscore. The formulated problem of multicriteria optimization of indicators CurrentRatio, Equity-to-Assets, NPM and ReceivablesTurnover allows to obtain Pareto-optimal combinations, which achieve financial stability as a maximized value (optimal ratio between key indicators of the financial condition of the enterprise).

References

[1] S. Drobyazko, A. Barwinska-Małajowicz, B. Slusarczyk, O. Chubukova, T. Bielialov, Journal of Risk and Financial Management **13**, 300 (2020)

[2] G. Azarenkova, O. Golovko, K. Abrosimova, Accounting and Financial Control **2**, 1 (2018)

[3] V. Mokeev, E. Bunova, A. Perevedentceva, Procedia Engineering **129**, 681 (2015), international Conference on Industrial Engineering (ICIE-2015)

[4] O. Mykoliuk, V. Bobrovnyk, V. Fostolovych, H. Kucherova, P. Nataliia, *Modelling the Level of Energy Security at Enterprises in the Context of Environmentalization of Their Innovative Development*, in *2020 10th International Conference on Advanced Computer Information Technologies (ACIT)* (2020), pp. 621–625

[5] W. Ma, M. Jin, Y. Liu, X. Xu, Chaos, Solitons & Fractals **125**, 17 (2019)

[6] D.M. Sprčić, A. Kožul, E. Pecina, Zagreb International Review of Economics and Business **20**, 25 (2017)

[7] J. Cohen, G. Krishnamoorthy, A. Wright, Contemporary Accounting Research **34**, 1178 (2017)

[8] N. Shmygol, O. Galtsova, O. Solovyov, V. Koval, I.W.E. Arsawan, E3S Web Conf. **153**, 03001 (2020)

[9] K. Kostetska, N. Khumarova, Y. Umanska, N. Shmygol, V. Koval, Management Systems in Production Engineering **28**, 15 (2020)

[10] D. Ocheretin, V. Los, K. Hanna, O. Bilska, E3S Web of Conferences **166**, 13024 (2020)

[11] A. Matviychuk, O. Novoseletskyy, S. Vashchaiev, H. Velykoivanenko, I. Zubenko, SHS Web of Conferences **65**, 06005 (2019)

[12] I. Khvostina, A. Matviychuk, S. Savina, O. Yatsiuk, *Analysis of the Dependence of the Performance of the Ensemble on the Accuracy of its Individual Models*, in *Proceedings of the 2019 7th International Conference on Modeling, Development and Strategic Management of Economic System (MDSMES 2019)* (Atlantis Press, 2019), pp. 371–375, `https://doi.org/10.2991/mdsmes-19.2019.70`

[13] V. Soloviev, V. Solovieva, A. Tuliakova, M. Ivanova, *Construction of crisis precursors in multiplex networks*, in *Proceedings of the 2019 7th International Conference on Modeling, Development and Strategic Management of Economic System (MDSMES 2019)* (Atlantis Press, 2019), pp. 361–366, `https://doi.org/10.2991/mdsmes-19.2019.68`

Improvement of the methods of assessing the influence of external factors on the strengths and weaknesses of the enterprise

Yriij Lozovik[1,*], *Bogdan* Tishkov[1,**], *Oksana* Pomazun[1,***], *Vasily* Derbentsev[1,****], and *Alexey* Hostryk[2,†]

[1]Kyiv National Economic University named after Vadym Hetman, 54/1 Peremogy Ave, Kyiv, 03057, Ukraine
[2]Odessa National Economic University, 8 Preobrazhenskaya Str., Odessa, 65082, Ukraine

Abstract. In the article considered the issues of constructing a classic SWOT analysis matrix, forming a list of key factors of the external and internal environment to assess the degree of their impact on the activities of real enterprises of Ukraine, improving the method of conducting SWOT analysis through the use of scientifically based scales and indicators. The thesis justified that SWOT analysis and other classic strategic management tools are conceptually described, and their use causes many approaches to their mathematical formalization and use. We propose our own approach to quantifying the strengths and weaknesses of the enterprise, external threats and opportunities at the main ten stages. The focus is on step-by-step calculation of integrated indicators for the main groups of SWOT analysis and their qualitative interpretation. Examples of implementation of SWOT analysis for a real enterprise are given. The company's strengths and weaknesses were identified. External capabilities and threats were analyzed. To assess the importance of indicators, a pairwise comparison technique was used, which provides for the estimation of indicators based on the average geometric value according to T. Saati. This made it possible to prioritize factors within the SWOT analysis methodology. Four quadrants are filled at the intersection of strengths/weaknesses, opportunities/threats and four types of potential strategies for the enterprise are described in the field defining: guidelines for the strategic development of the company; strategic advantages of the company; guidelines for the internal transformation of the enterprise; Limiting strategic development. Using a quantitative approach made it possible to build a SPACE matrix of strategy selection.

1 Introduction

An important tool of enterprise management at the stage of strategic planning is to identify and study the available internal and external resources, study the main opportunities and threats that may have a positive and negative impact on the studied enterprise. The lack of constant control over the factors of the environment of the enterprise leads to subjectivity and ambiguity of decisions, which creates significant uncertainty and risks in business. There is a need to improve the current methodology of strategic management.

The classical method of SWOT analysis greatly simplifies the reality and is aimed at one-time diagnosis of the enterprise environment, its capabilities and threats. It is most often used using brainstorming techniques, which leads to considerable subjectivity. In addition, the list of factors may differ significantly in different sectors of the economy. There is a need for a scientifically sound approach to setting up a model that can generate source information in the form of quantitative and qualitative data, can normalize them, process different types of data and work with different amounts of information. The use of incomplete and untimely information, sometimes even distorted information, leads to a significant violation of the quality of the model. At the same time, existing practice proves that the effectiveness of these methods of strategic management depends on the experience and intellectual potential of analysts, who actually play a key role in the quality of the study. Given that the head also acts as an expert, there is another problem regarding the adequacy of the study of available information and the correctness of the management decisions made by him based on the results of the analysis.

We have proposed our own approach to solving the problem of improving the classical SWOT analysis through the formation of a database of indicators, which is determined at the beginning of the study and can always be supplemented and improved. We also proposed a new stage and technology for determining the weight values and the formation of a qualitatively new model of SWOT analysis in order to process and analyze the factors of neutral, positive and negative effects. The developed measures are aimed at solving the problem of constant monitoring of the external environment of the enterprise, which can be integrated over time into the information system of enterprise management. It is expected that this will allow

*e-mail: yuralm@ukr.net
**e-mail: tishcov_b@ukr.net
***e-mail: oksp@ukr.net
****e-mail: derbv@kneu.edu.ua
†e-mail: alexeyGostrik@gmail.com

managers to develop operational and timely strategies for businesses.

Identification of key strengths and weaknesses of enterprises, external and internal factors of influence should be carried out using the classic tools of strategic management, which have gained wide popularity among marketers and strategists. Despite their more than half a century history, some of them have become classics, however these tools are described conceptually, their use leads to many approaches to their mathematical formalization and usage. The analysis of scientific works of domestic and foreign researchers dealing with SWOT analysis showed that the method has advantages and disadvantages. Since enterprises in different industries have different lists of factors and their weight, it is not possible to form a universal list of factors for evaluation.

1.1 Brief literature review

SWOT analysis is one of the most common methodologies for analysing the activities of the enterprise, which is the subject of many publications [1–8]. The classical theory of development of SWOT analysis methodology is constantly evolving. In the works [9, 10] a list of general factors without reference to a particular field is given. In the work of P. Mikhnenko [9] an attempt was made to build a dynamic modification of the SWOT analysis.

Currently, SWOT analysis remains one of the most popular tools of strategic management, which is often used at the stage of strategic planning for enterprises in various sectors of the economy. In particular, the works of well-known scientists include work on resource assessment, study of the main opportunities and threats, both for enterprises of traditional sectors of the economy, and for the evaluation of educational institutions, hospitals and other social facilities. The works are aimed to study the current situation and prospects for the development of various enterprises. In [11] a new SWOT model for digital ecosystem methodologies is proposed, in which the business and organizational issues, challenges and priorities are addressed. In [12] presented report on SWOT and PEST analyzes for implementation of reuse practices according to SuWaNu Program. In [13] the authors show the use of SWOT for a systematic analysis of the COVID-19 epidemic prevention.

The main purpose of the paper [14] is to present the rationale, the methodological details and a practical example of the application of the enhanced SWOT analysis in the context of technological foresight. The authors take into account and propose an innovative extension of SWOT analysis by an additional dimension: the assessment of factor significance in two time perspectives: the current state and the hypothetical horizon. As a result, a thirty-two-field SWOT diagram is obtained. The paper alsopresents the practical implications of the proposed methodology by offering a case study.

Scientists have paid attention to the improvement of methods of SWOT analysis, detailing the influence of factors and their quantitative assessment in the recent publications [15, 16]] but the issues of qualitative SWOT analysis remain relevant today. In particular, the lack of a unique approach to the description of factors, evaluation of weights and selection of scales for quantitative measurement and conclusion of the study leads to different modifications of these conceptual models and improvement of methods of strategic business analysis, identification of strengths and weaknesses [14, 15], a description of the situation in the industry [16] and determine the impact of major counter-parties on the diagnostic company and the industry as a whole.

Analysis of the development of the methodology shows that modern approaches to management provide opportunities for the application of advanced methods of SWOT analysis, in particular with the use of modern mathematical methods of modelling [17].

1.2 The purpose and objectives of the study

The purpose of the study is to improve the well-known and popular model SWOT analysis for assessing the strengths and weaknesses of the enterprise, its external capabilities and threats through the use of scientifically sound scales and indicators, which, in turn, will allow future users to reduce the time for analytical research and get quality conclusions and proposals (strategies) for the development of the enterprise.

To achieve this goal, the following tasks are formulated:

1. To improve the main stages of SWOT analysis for continuous monitoring of the impact of internal and external factors on the enterprise.
2. To identify a standard set of factors for assessing the internal and external environment.
3. To make a choice of scientifically substantiated scales and weight values of an estimation of the basic indicators for construction of mathematical model and its use for a substantiation of strategy of the enterprise.
4. To provide for the assessment of factors taking into account neutral, strong and weak action.
5. To test the proposed method of SWOT analysis on the example of a Ukrainian brewery.

2 Methodology

2.1 SWOT analysis

SWOT analysis today can be attributed to the classic tools of strategic management. In the first concept, strategic analysis was called SOFT analysis, in which existed and expected positive external factors were collected in Satisfactory and Opportunity clusters, and internal – in Fault and Threat clusters. The transition from SOFT to SWOT is not semantic, but conceptual, because its essence is the transition from the interface "good – bad" to the interface "internal-external". This methodological concept of strategic marketing analysis was first tested by Albert Humphrey [18]. The main elements of SWOT analysis are: Strengths, Weaknesses, Opportunities, Threats.

More than half of a century of history of this very popular tool and the accumulated experience of marketing research vast experience, include a number of significant changes in SWOT analysis, allow you to more effectively perform strategic business analysis.

The technology (algorithm) of SWOT analysis involves: first, the construction of a matrix, and secondly, the development of business strategies. At the intersection of exogenous and endogenous factors, the strategy and risks for the studied enterprise are determined.

In the scientific works which were analysed the sequence of SWOT analysis was as follows [1, 2, 4, 5]:

Stage 1. Formation of a hierarchy of indicators by groups of SWOT analysis.

Stage 2. Selection of scientifically sound scales and weight values are to assess the main indicators by groups.

Stage 3. Calculations and construction of a consolidated matrix of SWOT analysis. Defining guidelines for strategic development.

Stage 4. Development of strategies aimed at mitigating the effects of negative factors and strengthening the strengths of the enterprise.

Stage 5. Evaluation and control of the achieved results of the enterprise.

The above phasing of the SWOT analysis does not involve the formation of a database of key factors that may have different degrees of influence in the dynamics and has a number of problems with the assessment of neutral factors. A whole group of indicators that need constant monitoring is ignored. Therefore, we propose to supplement the methodology with new stages and tasks. The main stages of the SWOT analysis include the following ten stages:

Stage 1. Formation of the general base of factors for an estimation of groups of internal and external environment. Estimation of weights.

Stage 2. Definition of criteria (norms) of conformity on the given factors.

Stage 3. Assessment of internal factors in relation to deviations from the set norms (division of factors into neutral, strong and weak).

Stage 4. Assessment of external factors in relation to deviations from the set norm (division of factors into neutral, opportunities and threats).

Stage 5. Formation of a hierarchy of indicators by groups of SWOT analysis.

Stage 6. Selection of scientifically sound scales and weight values for evaluation of key indicators by groups.

Stage 7. Definition of strategic landmarks in development.

Stage 8. Carrying out calculations and construction of the consolidated matrix of SWOT analysis.

Stage 9. Development of strategies aimed at mitigating the effects of negative factors and strengthening the strengths of the enterprise.

Stage 10. Evaluation and control of the achieved results of the enterprise.

This approach provides an opportunity to conduct a SWOT analysis not once, but in the dynamics – tracking changes in key factors. The methodology must begin with the identification of all possible factors of influence from the internal and external environment and evaluation.

2.2 Mathematical description

For a mathematical description of the proposed approach to quantify the indicators of SWOT analysis, we introduce the following notation:

$IF_{i(j)}$ – (j) is the factor of the i group of the internal environment IF.

$EF_{i(j)}$ – (j) is the factor of the i group of the external environment EF.

The weight values of estimates of all factors of the internal / external environment, respectively 100%:

$$\sum_{(j)} W_{i(j)} = 1. \tag{1}$$

Depending on the actions of all the factors of the internal environment *(IF)* on the studied enterprise, we propose to divide into three groups:

- neutral $IF^{(0)}$;
- strengths $IF^{(+)}$;
- weaknesses of $IF^{(-)}$.

Depending on the actions of environmental factors *(EF)* on the studied enterprise, they can be divided into three groups:

1) neutral $EF^{(0)}$;
2) positive influence (opportunities) $EF^{(+)}$,
3) negative impact (threat) $EF^{(-)}$.

To assess only the strengths / weaknesses of the company, it is advisable to list the weight values for these groups, so that their share was 100%.

$$IF^{(+)*}_{i(j)} = \frac{w^{(+)}_{i(j)}}{\sum_{(j)} w^{(+)}_{i(j)} IF^{(+)}_{i(j)}}; \tag{2}$$

$$IF^{(-)*}_{i(j)} = \frac{w^{(-)}_{i(j)}}{\sum_{(j)} w^{(-)}_{i(j)} IF^{(-)}_{i(j)}}. \tag{3}$$

Where to get:

$$IF^{(+)} = \sum_{i} \sum_{(j)} IF^{(+)*}_{i(j)}; \tag{4}$$

$$IF^{(-)} = \sum_{i} \sum_{(j)} IF^{(-)*}_{i(j)}. \tag{5}$$

This recalculation will allow the interpretation of integrated indicators of certain groups within certain research scales.

To assess the opportunities / threats of the environment, we calculate the weight values for these groups, so that their share was 100%.

$$EF_{i(j)}^{(+)*} = \frac{w_{i(j)}^{(+)}}{\sum_{(j)} w_{i(j)}^{(+)} EF_{i(j)}^{(+)}}; \quad (6)$$

$$EF_{i(j)}^{(-)*} = \frac{w_{i(j)}^{(-)}}{\sum_{(j)} w_{i(j)}^{(-)} EF_{i(j)}^{(-)}}. \quad (7)$$

And:

$$EF^{(+)} = \sum_{i} \sum_{(j)} EF_{i(j)}^{(+)*}; \quad (8)$$

$$EF^{(-)} = \sum_{i} \sum_{(j)} EF_{i(j)}^{(-)*}. \quad (9)$$

Calculation of the balance between the strengths ($IF^{(+)}$) and weaknesses ($IF^{(-)}$) of the internal environment which is carried out without taking into account the action of neutral factors:

$$IF = IF^{(+)} - IF^{(-)} \quad (10)$$

Under this approach, the share of factors of strengths and weaknesses should be 100%, so the values of the weights of the relevant group of indicators (strong or weak action) should be adjusted to the total value of weights for these two groups.

Similarly, the balance between the external capabilities of ($EF^{(+)}$) and the threats to ($EF^{(-)}$) of the environment is calculated based on the following equation:

$$EF = EF^{(+)} - EF^{(-)} \quad (11)$$

3 Empirical results

3.1 Factors for assessing the internal and external environment of the enterprise

According to the results of research of basic factors for the assessment of the internal environment of the enterprise, also a list of the most used was formed, presented in table 1.

The following list of key factors for assessing the degree of environmental impact on the enterprise is shown in table 2.

Since the list of factors is quite large (more than 30 factors for the assessment of the internal environment and more than 20 – for the external), the application of the method of pairwise comparisons, Fishburn's method or other methods of setting weights becomes problematic. The solution of such a problem is possible through the decomposition of factors in new subsystems. To distribute the factors influencing the internal environment of the enterprise, we provide the allocation of such subsystems as: innovation and development, logistics, production, sales and promotion, enterprise finance, personnel management and subsystems of strategic and tactical management. And to assess environmental factors, it is proposed to use two subsystems: intermediate environmental factors and macro environment.

3.2 Determination of weights by the method of pairwise comparisons

Using the method of analysis of hierarchies by T. L. Saati[17] allows you to distribute the factors in these groups, to determine the weight of each subsystem, and the factors themselves within these groups. The transition from a two-level hierarchy to a one-level one using linear convolution is achieved by multiplying the weight value of a certain factor by the weight value of its subsystem. This will determine the weight of all factors within the internal / external environment. Note that the number of levels of indicators is determined by the objectives of the study. Different indicators of F_{ij} have their significance in the system of SWOT analysis. Determination of weights (w_i) should be carried out by the method of pairwise comparisons by T. L. Saati. Its essence is to establish weights based on a system of pairwise comparisons of indicators with each other. We calculate the weights for certain groups of indicators, which involves comparing the indicators with each other on the following 5-level scale: 1 – EI (equally important), 2 – SA (slight advantage), 3 – BA (big advantage), 4 – HA (huge advantage), 5 – MA (maximum advantage).

Inversely proportional estimates between indicators are as follows: 1/2 = 0.5 (SA^{-1}); 1/3 = 0.33(3) (BA^{-1}); 1/4=0.25 (HA^{-1}); 1/5=0.2 (MA^{-1}).

Table 3 shows the calculation of weights by internal factors of the subsystem "Innovation and Development" (F_1).

Similarly, weights were obtained for other groups of indicators of internal and external environment. The results of calculations of weights for groups of factors of the external environment are shown in table 4 and the internal environment in table 5.

3.3 Weights of indicators of the internal environment

The use of linear convolution allows the weight values of indicators within the studied subsystems to be converted into weight values of indicators within a single system of internal indicators (table 6).

3.4 Weights of environmental indicators

Similarly, make the transition from a two-level hierarchy to a one-level hierarchy of weights of the environment (table 7).

In the second stage, researchers will be able to assess these factors in terms of their deviation from the norm. Deviations in the study process can be neutral (in case of compliance with the norm), or have negative and positive deviations. This assessment allows us to divide the factors into three groups:

- neutral;
- strengths;
- weak sides.

Table 1. List of factors for assessing the internal environment of the enterprise

The name of the factor $IF_{i(j)}$	Strengths, $IF_{i(j)}^{(+)}$	Weak sides, $IF_{i(j)}^{(-)}$
	Subsystem "Innovation and Development" (IF_1)	
Experience in developing new products and brands, $IF_{1(1)}$	Significant experience in developing new products and brands	Lack of experience in developing new products and brands
Reputation and image of the company, $IF_{1(2)}$	Good reputation and image of the company	Insufficient reputation and image of the company
Accreditation and certification, $IF_{1(3)}$	Availability of accreditation and certification	Lack of accreditations and certifications
Financing of development programs, $IF_{1(4)}$	Available funds for financing development programs	Lack of funds to finance development programs
	Subsystem "Logistics" (IF_2)	
Location of the enterprise, $IF_{2(1)}$	Good placement	Bad location
Logistics, $IF_{2(2)}$	High quality logistics	Bad (poor quality) logistics
Assets and resources, $IF_{2(3)}$	Adequacy of assets and resources	Lack of assets and resources
	Subsystem "Production" (IF_3)	
Technologies and equipment, $IF_{3(1)}$	Modern technologies and equipment	Outdated technologies and equipment
Reserve of production capacities, $IF_{3(2)}$	Available reserves of production capacity	There is no reserve for production capacity
The level of production costs, $IF_{3(3)}$	Below the industry average	Above the average industry level
Ensuring business processes, $IF_{3(4)}$	High quality organizational support	Bad(poor) organizational support
	Subsystem "Sales and Promotion" (IF_4)	
Brand awareness, $IF_{4(1)}$	Growing popularity and level of brand recognition	Declining brand popularity
Product differentiation, $IF_{4(2)}$	Wide product differentiation	Narrow product differentiation
Product quality, $IF_{4(3)}$	High quality	Low quality
Product sales channels, $IF_{4(4)}$	Developed sales channels	Undeveloped product sales channels
Pricing policy, $IF_{4(5)}$	Existence of a reasonable pricing policy	Lack of sound pricing policy
	Subsystem "Enterprise Finance" (IF_5)	
Equity, $IF_{5(1)}$	Sufficient share of equity	Lack of equity
Financial results, $IF_{5(2)}$	High financial results	Unsatisfactory financial results
Profitability of brands, $IF_{5(3)}$	High profitability of existing brands	Availability of non-profit brands
Financial liquidity and solvency, $IF_{5(4)}$	Ability to repay accounts payable in a timely manner	Lack of funds to repay accounts payable
Financial stability, $IF_{5(5)}$	Ability to finance activities from internal sources	Unsecured financing of activities from internal sources
	Subsystem "Personnel Management" (IF_6)	
Provision of management staff, $IF_{6(1)}$	Competence of management staff	Lack of management staff
Provision of industrial and production personnel, $IF_{6(2)}$	Provision of industrial and production personnel	Insecurity of industrial and production personnel
Competence of management staff, $IF_{6(3)}$	High qualification of management staff	Low qualification of management staff
Qualification of engineering and technical workers, $IF_{6(4)}$	High qualification of engineering and technical workers	Low qualification of engineering and technical workers
The level of wages, $IF_{6(5)}$	Above the average industry level	Below the average industry level
Remuneration mechanisms, $IF_{6(6)}$	Increasing the incentive for high quality work	Lack of incentives for quality of work
Opportunities for professional and career growth $IF_{6(7)}$	Opportunities for professional and career growth	Lack of opportunities for professional and career growth
	Subsystem "Strategic and tactical management" (IF_7)	
Innovation and creativity of managers, $IF_{7(1)}$	High level of innovation and creativity of managers	Low level of innovation and creativity of managers
Strategic management mechanism, $IF_{7(2)}$	Developed mechanism of strategic management	Undeveloped mechanism of strategic management
Tactical control mechanism, $IF_{7(3)}$	Developed mechanism of tactical control	Undeveloped mechanism of tactical control
Corporate culture, $IF_{7(4)}$	Positive corporate culture	Ineffective corporate culture

Table 2. A list of key factors for assessing the degree of environmental impact on the enterprise

The name of the factor $EF_{i(j)}$	Opportunities, $EF_{i(j)}^{(+)}$	Threats $EF_{i(j)}^{(-)}$
Subsystem "Intermediate environment (external environment of direct action)" (EF_1)		
Competition in the industry, $EF_{1(1)}$	Stabilization of competition in the industry	Exacerbation of competition in the industry
Production capacity in the industry, $EF_{1(2)}$	Increasing production capacity in the industry	Reducing production capacity in the industry
Economic contractors and contracts, $EF_{1(3)}$	Emergence of new contractors and the possibility of signing important contracts	Loss of new important contracts and contractors
Competition with manufacturers of substitute products, $EF_{1(4)}$	Stabilization of competition with producers of substitute goods	Exacerbation of competition with producers of substitute products
New product sales channels, $EF_{1(5)}$	The emergence of new available marketing tools to promote products	Reduction of existing sales channels
New marketing tools for product promotion, $EF_{1(6)}$	The emergence of new available marketing tools to promote products	The emergence of new unavailable marketing tools to promote products
Market demand, $EF_{1(7)}$	Rising average industry price	Reduction of demand in existing markets
Price, $EF_{1(8)}$	Rising average industry price	Decrease in the average industry price
New markets for products, $EF_{1(9)}$	Emergence of new markets	Loss of existing markets
Staffing, $EF_{1(10)}$	Training of a significant number of qualified personnel	Lack of qualified specialists in the labor market
Prices for raw materials and electricity, $EF_{1(11)}$	Reduction of prices for raw materials and electricity	Rising prices for raw materials and electricity
Terms of external financing, $IF_{1(12)}$	Attractive conditions of external (loan) financing	Unavailability or unattractive conditions of external (loan) financing
Product advertising, $EF_{1(13)}$	Expanding opportunities for product advertising	Narrowing opportunities for product advertising
Transformation of tastes and preferences of consumers, $EF_{1(14)}$	Correct assessment of new tastes and preferences of consumers	Incorrect assessment of new tastes and preferences of consumers
Subsystem "Macro environment (external environment of indirect action)" (IF_2)		
Political situation, $EF_{2(1)}$	Improving the military-political situation	Exacerbation of the military-political situation
Legislation, $EF_{2(2)}$	Stability, completeness and internal consistency of the legal framework	Instability, incompleteness and internal contradictions of the legal framework
New production / technologies, $EF_{2(3)}$	Emergence of new available types of equipment / production technology	Emergence of new inaccessible types of equipment / technologies of production
New information technologies, $EF_{2(4)}$	Emergence of new available information technologies	The emergence of new inaccessible information technology
Training and retraining, $EF_{2(5)}$	Extensive opportunities for training and retraining	Reduction of bases for training and retraining
Raw material base, $EF_{2(6)}$	Development of the national raw material base	Dependence on imported suppliers of raw materials
Intellectual Property, $EF_{2(7)}$	Development of services for the protection of intellectual property	Lack of intellectual property protection services
Weather, $EF_{2(8)}$	Improving climatic conditions	Deterioration of climatic conditions
Welfare of the population, $EF_{2(9)}$	Improving the welfare of the population	Deterioration of the welfare of the population
Export quotas, $EF_{2(10)}$	Increase in the export quota	Reduction of export quota
Excise duty, $EF_{2(11)}$	Increase in excise duty on beer	Reduction of excise duty on beer
Pandemic, $EF_{2(12)}$	Improving the pandemic situation	Deterioration of the pandemic situation

Table 3. Matrix of pairwise comparisons for estimation of weighting factors on internal factors of the subsystem "Innovation and development" (F_1)

Subsystem	$IF_{1(1)}$	$IF_{1(2)}$	$IF_{1(3)}$	$IF_{1(4)}$
Experience in developing new products and brands, $IF_{1(1)}$	EI	SA^{-1}	BA	EI
Reputation and image of the company, $IF_{1(2)}$	SA	EI	SA	SA
Accreditation and certification, $IF_{1(3)}$	BA^{-1}	SA^{-1}	EI	SA^{-1}
Financing of development programs $IF_{1(4)}$	EI	SA^{-1}	SA	EI
Weight values	0.256	0.389	0.124	0.231

Table 4. Calculation of weights by groups of environmental factors (EF)

Subsystem		$EF_{(1)}$	$EF_{(2)}$
Intermediate environment	$EF_{(1)}$	EI	SA
Macroenvironment	$EF_{(2)}$	SA^{-1}	EI
Weight values	1.00	0.666	0.334

Neutral factors are important in terms of their changes in dynamics and require constant monitoring of their deviations.

The use of 3, 5 or 7 level scales, which are always the most convenient for the expert, will determine the degree of deviation from the norm in a positive or negative direction. Experts will be able to use different types of scales – current, interval and fuzzy research scales. An important condition is compliance with a single scale in terms of expert evaluation. Weight values within a given hierarchy of indicators can be determined on the basis of Fishburne's rule or the method of pairwise comparisons by Saati. It should be noted that for a qualitative interpretation of strengths and weaknesses, for example, on a 5-level scale, it will be necessary to list in a given proportion the weight values of a particular group up to 100%. If this is not done, the proportion may be different and the results will not be subject to a 5-level study scale. And to assess the balance between strengths and weaknesses, it will be necessary to list in a given proportion of weight values by factors of strengths and weaknesses (excluding the action of neutral factors), so that their share was 100%.

4 Discussion of research results

4.1 Key influencing factors for public partnership "Carlsberg Ukraine"

The implementation of the proposed approach to SWOT analysis will be carried out for public partnership "Carlsberg Ukraine" – one of the largest players in brewing in Ukraine in 2020. The study was conducted on the basis of reporting information of the enterprise and information from external sources, in particular [19].

Systematization and generalization of information on the internal and external environment allowed to identify a list of key factors of influence. The strengths of public partnership "Carlsberg Ukraine" include the following factors ($IF^{(+)}$):

- $IF^{(+)}_{1(1)}$ – extensive experience in developing new products and brands;
- $IF^{(+)}_{1(2)}$ – good reputation and image;
- $IF^{(+)}_{2(2)}$ – logistics;
- $IF^{(+)}_{3(2)}$ – significant scale of production and available reserves of production capacity;
- $IF^{(+)}_{4(1)}$ – increasing the popularity and recognition of brands;
- $IF^{(+)}_{4(2)}$ – wide product differentiation;
- $IF^{(+)}_{4(3)}$ – high quality products;
- $IF^{(+)}_{4(4)}$ – developed product sales channels (retail chains, individual outlets, restaurants, cafes, pubs);
- $IF^{(+)}_{4(5)}$ – the existence of a reasonable pricing policy (law mainstream, mainstream, premium, super premium);
- $IF^{(+)}_{5(2)}$ – financial results;
- $IF^{(+)}_{6(2)}$ – provision of industrial and production personnel;
- $IF^{(+)}_{6(5)}$ – the level of wages corresponds to the average industry level;
- $IF^{(+)}_{6(7)}$ – opportunity for professional and career growth;
- $IF^{(+)}_{7(4)}$ – positive corporate culture.

And the weaknesses of public partnership "Carlsberg Ukraine" are the following factors ($IF^{(-)}$):

- $IF^{(-)}_{1(4)}$ – insufficient funding for development programs;
- $IF^{(-)}_{2(3)}$ – lack of assets and resources;
- $IF^{(-)}_{3(1)}$ – availability of outdated technologies and equipment;
- $IF^{(-)}_{3(3)}$ – the level of production costs is higher than the industry average;
- $IF^{(-)}_{5(1)}$ – lack of equity;
- $IF^{(-)}_{5(3)}$ – the presence of non-profit brands;
- $IF^{(-)}_{5(4)}$ – lack of funds to repay accounts payable;
- $IF^{(-)}_{5(5)}$ – unsecured financing of activities from internal sources;
- $IF^{(-)}_{6(1)}$ – staff turnover and insecurity of management staff;
- $IF^{(-)}_{6(3)}$ – competence of management staff;
- $IF^{(-)}_{6(7)}$ – lack of proper incentives for quality of work;
- $IF^{(-)}_{7(1)}$ – insufficient level of innovation and creativity;
- $IF^{(-)}_{7(2)}$ – an underdeveloped strategic management mechanism.

Table 5. Calculation of weights by groups of factors of the internal environment (IF)

Subsystem		IF_1	IF_2	IF_3	IF_4	IF_5	IF_6	IF_7
Innovation and development	IF_1	*EI*	*SA*	*SA*	*SA*	SA^{-1}	SA^{-1}	*EI*
Logistics	IF_2	SA^{-1}	*EI*	*BA*	SA^{-1}	SA^{-1}	BA^{-1}	SA^{-1}
Production	IF_3	SA^{-1}	BA^{-1}	*EI*	*SA*	*SA*	SA^{-1}	SA^{-1}
Sales and product promotion	IF_4	SA^{-1}	*SA*	SA^{-1}	*EI*	SA^{-1}	BA^{-1}	*EI*
Enterprise finances	IF_5	*SA*	*SA*	SA^{-1}	*SA*	*EI*	*SA*	SA^{-1}
HR	IF_6	*SA*	*BA*	*SA*	*BA*	SA^{-1}	EI	SA^{-1}
Strategic and tactical management	IF_7	*EI*	*SA*	*SA*	*EI*	*SA*	*SA*	*EI*
Weighting factor	1.00	0.151	0.092	0.106	0.096	0.166	0.187	0.203

Table 6. The results of the evaluation of the weights of the internal environment

Factor / subsystem	Weight within subsystems	Normalized weight values within the system
Subsystem "Innovation and Development" (IF_1)		
Experience in developing new products, $IF_{1(1)}$	0.256	0.039
Reputation and image of the company, $IF_{1(2)}$	0.389	0.059
Accreditation and certification, $IF_{1(3)}$	0.124	0.019
Financing of development programs, $IF_{1(4)}$	0.231	0.035
Subsystem "Logistics" (IF_2)		
Location of the enterprise, $IF_{2(1)}$	0.260	0.024
Logistics, $IF_{2(2)}$	0.327	0.030
Assets and resources, $IF_{2(3)}$	0.413	0.038
Subsystem "Production" (IF_3)		
Technologies and equipment, $IF_{3(1)}$	0.238	0.025
Reserve of production capacities, $IF_{3(2)}$	0.142	0.015
The level of production costs, $IF_{3(3)}$	0.337	0.036
Ensuring business processes, $IF_{3(4)}$	0.283	0.030
Subsystem "Sales and Promotion" (IF_4)		
Brand awareness, $IF_{4(1)}$	0.134	0.013
Product differentiation, $IF_{4(2)}$	0.167	0.016
Product quality, $IF_{4(3)}$	0.192	0.018
Product sales channels, $IF_{4(4)}$	0.192	0.018
Pricing policy, $IF_{4(5)}$	0.315	0.030
Subsystem "Enterprise Finance" (IF_5)		
Equity, $IF_{5(1)}$	0.254	0.042
Financial results, $IF_{5(2)}$	0.124	0.021
Profitability of brands, $IF_{5(3)}$	0.164	0.027
Financial liquidity and solvency, $IF_{5(4)}$	0.094	0.016
Financial stability, $IF_{5(5)}$	0.364	0.061
Subsystem "Personnel Management" (IF_6)		
Provision of management staff, $IF_{6(1)}$	0.140	0.026
Provision of industrial and production personnel, $IF_{6(2)}$	0.100	0.019
Competence of management staff, $IF_{6(3)}$	0.192	0.036
Qualification of engineering and technical workers, $IF_{6(4)}$	0.077	0.014
The level of wages, $IF_{6(5)}$	0.104	0.019
Remuneration mechanisms, $IF_{6(6)}$	0.212	0.040
Opportunities for professional and career growth, $IF_{6(7)}$	0.174	0.032
Subsystem "Strategic and tactical management" (IF_7)		
Innovation and creativity of managers, $IF_{7(1)}$	0.161	0.033
Strategic management mechanism, $IF_{7(2)}$	0.270	0.055
Tactical control mechanism, $IF_{7(3)}$	0.299	0.061
Corporate culture, $IF_{7(4)}$	0.270	0.055
Total by groups		1

Table 7. The results of the assessment of the weights of the environment

Factor / subsystem	Weight within subsystems	Normalized weight values within the system
Subsystem "Intermediate environment (external environment of direct action)" (EF_1)		
Competition in the industry, $EF_{1(1)}$	0.090	0.060
Production capacity in the industry, $EF_{1(2)}$	0.090	0.060
Economic contractors and contracts, $EF_{1(3)}$	0.095	0.063
Competition with manufacturers of substitute products, $EF_{1(4)}$	0.079	0.053
New product sales channels, $EF_{1(5)}$	0.067	0.045
New marketing tools for product promotion, $EF_{1(6)}$	0.074	0.049
Market demand, $EF_{1(7)}$	0.057	0.038
Price, $EF_{1(8)}$	0.066	0.044
New markets for products, $EF_{1(9)}$	0.102	0.068
Staffing, $EF_{1(10)}$	0.050	0.033
Prices for raw materials and electricity, $EF_{1(11)}$	0.086	0.058
Terms of external financing, $EF_{1(12)}$	0.032	0.022
Product advertising, $EF_{1(13)}$	0.046	0.030
Transformation of tastes and preferences of consumers, $EF_{1(14)}$	0.065	0.043
Subsystem "Macro environment (external environment of indirect action)" (EF_2)		
Political situation, $EF_{2(1)}$	0.087	0.029
Legislation, $EF_{2(2)}$	0.100	0.033
New production / technologies, $EF_{2(3)}$	0.087	0.029
New information technologies, $EF_{2(4)}$	0.060	0.020
Training and retraining, $EF_{2(5)}$	0.077	0.026
Raw material base, $EF_{2(6)}$	0.061	0.020
Intellectual Property, $EF_{2(7)}$	0.058	0.019
Weather, $EF_{2(8)}$	0.066	0.022
Welfare of the population, $EF_{2(9)}$	0.095	0.032
Export quotas, $EF_{2(10)}$	0.110	0.037
Excise duty, $EF_{2(11)}$	0.095	0.032
Pandemic, $EF_{2(12)}$	0.104	0.035
Total by groups	-	1

Neutral factors of the internal environment included such factors of influence as: location of the enterprise ($IF_{2(1)}$); accreditation and certification ($IF_{1(3)}$); business process support ($IF_{3(4)}$); qualification of engineering and technical staff ($IF_{6(4)}$) and tactical management mechanism ($IF_{7(3)}$).

Let's identify possible threats for public partnership "Carlsberg Ukraine" ($EF^{(-)}$):

- $EF^{(-)}_{1(1)}$ – intensification of competition in the industry;
- $EF^{(-)}_{1(2)}$ – increase production capacity in the industry;
- $EF^{(-)}_{1(4)}$ – intensification of competition with producers of substitute products (spirits);
- $EF^{(-)}_{1(7)}$ – reduction of demand in available markets;
- $EF^{(-)}_{1(11)}$ – reduction of demand in available markets;
- $EF^{(-)}_{1(13)}$ – narrowing of opportunities for product advertising (ban on outdoor advertising of soft drinks);
- $EF^{(-)}_{1(14)}$ – incorrect assessment of new tastes and preferences of consumers;
- $EF^{(-)}_{2(1)}$ – aggravation of the military-political situation (war in the East of the country and annexation of Crimea);
- $EF^{(-)}_{2(2)}$ – instability, incompleteness and internal contradiction of the legal framework;
- $EF^{(-)}_{2(3)}$ – emergence of new inaccessible types of machinery / technologies of production;
- $EF^{(-)}_{2(6)}$ – underdeveloped national raw material base and dependence on imported suppliers of raw materials (barley, malt, cork, etc);
- $EF^{(-)}_{2(8)}$ – deterioration of climatic conditions;
- $EF^{(-)}_{2(9)}$ – deterioration of the welfare of the population;
- $EF^{(-)}_{2(11)}$ – increase in excise duty on beer;
- $EF^{(-)}_{2(12)}$ – deterioration of the pandemic situation.

Let's define the main opportunities for public partnership "Carlsberg Ukraine" ($EF^{(+)}$):

- $EF^{(+)}_{1(3)}$ – the emergence of new contractors and the possibility of signing important contracts;
- $EF^{(+)}_{1(5)}$ – the emergence of new sales channels;
- $EF^{(+)}_{1(6)}$ – the emergence of new available marketing tools to promote products;
- $EF^{(+)}_{1(8)}$ – growth of the average industry price;

- $EF^{(+)}_{1(9)}$ – the emergence of new markets;
- $EF^{(+)}_{1(10)}$ – training of a significant number of qualified personnel;
- $EF^{(+)}_{1(12)}$ – attractive external financing conditions;
- $EF^{(+)}_{2(5)}$ – wide opportunities for training and advanced training;
- $EF^{(+)}_{2(7)}$ – development of intellectual property protection services.

Neutral environmental factors included information factors such as: information technology ($EF_{2(4)}$) and export quotas($EF_{2(10)}$).

The level of influence of factors of positive / negative action on the functioning of the enterprise will be assessed using a five-level scale. We have identified such levels of influence as: "1" – very insignificant, "2" – insignificant, "3" – moderate, "4" – significant and "5" – very significant.

In order to assess the performance of strengths / weaknesses on a 5-level scale, we recalculated the weights of factors up to 100% and conducted a qualitative assessment.

4.2 Analysis of strengths and weaknesses of public partnership "Carlsberg Ukraine"

Analysis of strengths and weaknesses of public partnership "Carlsberg Ukraine" are shown in tables 8 and 9.

The results of the calculation allow us to state that the influence of strengths ($IF^{(+)}$=3.043) on the internal subsystem of public partnership "Carlsberg Ukraine" is medium (moderate), and the influence of weaknesses($IF^{(-)}$=1.814) – insignificant.

As noted above, to assess the balance between strengths and weaknesses, recalculation was performed in a given proportion of weight values by factors of strengths and weaknesses (excluding the action of neutral factors), (equations 3, 4): $IF^{(+)} = 1.371$; $IF^{(-)} = 0.997$.

Since a single scale is used to analyse the influence of internal environmental factors, the balance between the action of strengths and weaknesses will be determined by equation (10), the result is: $IF = 1.371 - 0.997 = 0.374$.

The obtained indicator shows that the strengths of public partnership "Carlsberg Ukraine" outweigh its weaknesses by +0.374. The company has its own capacity, financial resources, experience and capabilities in developing new brands, developed product promotion channels, flexible pricing policy and wide product differentiation. Weaknesses of the enterprise that weaken the balance include outdated technologies and non-profit brands, insufficient funding for development projects, indicators of financial stability.

4.3 Analysis of the impact of external opportunities and threats for public partnership "Carlsberg Ukraine"

Next, an analysis of the impact of external opportunities and threats to public partnership "Carlsberg Ukraine". The results of the analysis are shown in tables 10 and 11.

Table 8. The results of the assessment of the strengths of public partnership "Carlsberg Ukraine" on a 5-level scale for 2020

Factor	The weight of the factor in a single scale	The weight of the factor in assessing strengths	Factor impact assessment	Weighted assessment
Experience in developing new products and brands, $IF_{1(1)}$	0.039	0.100	4.000	0.402
Reputation and image of the company, $IF_{1(2)}$	0.059	0.153	3.000	0.458
Logistics, $IF_{2(2)}$	0.030	0.078	2.000	0.157
Reserve of production capacities, $IF_{3(2)}$	0.015	0.039	3.000	0.117
Brand recognition, $IF_{4(1)}$	0.013	0.033	4.000	0.134
Product differentiation, $IF_{4(2)}$	0.016	0.042	4.000	0.166
Product quality, $IF_{4(3)}$	0.018	0.048	3.000	0.143
Product sales channels, $IF_{4(4)}$	0.018	0.048	3.000	0.143
Pricing policy, $IF_{4(5)}$	0.030	0.079	4.000	0.314
Financial results, $IF_{5(2)}$	0.021	0.054	2.000	0.108
Provision of industrial and production personnel, $IF_{6(2)}$	0.019	0.049	4.000	0.195
The level of wages, $IF_{6(5)}$	0.019	0.051	1.000	0.051
Career opportunities, $IF_{6(7)}$	0.032	0.085	1.000	0.085
Corporate culture, $IF_{7(4)}$	0.055	0.143	4.000	0.571
Total	0.384	1.000	-	3.043

These calculations allow us to state that the impact of external threats ($EF^{(-)} = 2.338$) on the activities of public partnership "Carlsberg Ukraine" is moderate. and external opportunities ($EF^{(+)} = 2.941$) can be interpreted as moderate (average).

Similar to the calculation of the balance between strengths and weaknesses. Let's calculate the balance between external opportunities and threats to the environment (equations 8, 9): $EF^{(+)} = 1.150$; $EF^{(-)} = 1.423$.

To analyze the balance between the action of external opportunities and threats. We use the equation (11): $EF = 1.150 - 1.423 = -0.273$.

The obtained negative value indicates that the balance is shifted towards the threats of the enterprise. and proves

Table 9. The results of the assessment of the weaknesses of public partnership "Carlsberg Ukraine" on a 5-level scale for 2020

Factor	The weight of the factor in a single scale	The weight of the factor in assessing strengths	Factor impact assessment	Weighted assessment
Financing of development programs, $IF_{1(4)}$	0.035	0.074	2.000	0.149
Assets and resources, $IF_{2(3)}$	0.038	0.081	1.000	0.081
Technologies and equipment, $IF_{3(1)}$	0.025	0.054	2.000	0.107
The level of production costs, $IF_{3(3)}$	0.036	0.076	1.000	0.076
Equity, $IF_{5(1)}$	0.042	0.090	1.000	0.090
Profitability of brands, $IF_{5(3)}$	0.027	0.058	2.000	0.115
Financial liquidity and solvency, $IF_{5(4)}$	0.016	0.034	2.000	0.068
Financial stability, $IF_{5(5)}$	0.061	0.130	3.000	0.391
Provision of management staff, $IF_{6(1)}$	0.026	0.055	2.000	0.111
Competence of management staff, $IF_{6(3)}$	0.036	0.077	2.000	0.153
Remuneration mechanisms, $IF_{6(6)}$	0.040	0.085	2.000	0.169
Innovation and creativity of managers, $IF_{7(1)}$	0.033	0.070	1.000	0.070
Strategic management mechanism, $IF_{7(2)}$	0.055	0.117	2.000	0.234
Total	0.468	1.000	-	1.814

the need for constant monitoring of changes in the action of environmental factors.

4.4 SPACE matrix for public partnership "Carlsberg Ukraine"

The SPACE matrix was built based on the values of the balance between strengths and weaknesses, and the balance between opportunities and threats. Thus, based on two values of strategic dimensions EF and IF, we divided the space into four types of strategies: aggressive; conservative; protective or competitive.

For proposed research scale the following combinations are possible:

Table 10. The results of the assessment of the main external opportunities for public partnership "Carlsberg Ukraine"

Factor	The weight of the factor in a single scale	The weight of the factor in assessing strengths	Factor impact assessment	Weighted assessment
Economic contractors and contracts, $EF_{1(3)}$	0.063	0.171	3.000	0.513
New product sales channels, $EF_{1(5)}$	0.045	0.121	3.000	0.363
New marketing tools for product promotion, $EF_{1(6)}$	0.049	0.134	2.000	0.267
Price, $EF_{1(8)}$	0.044	0.120	2.000	0.240
New markets for products, $EF_{1(9)}$	0.068	0.184	4.000	0.737
HR, $EF_{1(10)}$	0.033	0.090	3.000	0.271
Terms of external financing, $EF_{1(12)}$	0.022	0.059	2.000	0.117
Training and retraining, $EF_{2(5)}$	0.026	0.069	4.000	0.278
Intellectual Property, $EF_{2(7)}$	0.019	0.052	3.000	0.155
Total	0.369	1.000	-	2.941

if $EF > 0$ and $IF < 0$ we choose a conservative strategy;
if $EF \geq 0$ and $IF \geq 0$ we choose the offensive strategy;
if $EF < 0$ and $IF < 0$ we choose the defence strategy;
if $EF < 0$ and $IF > 0$ then choose a competitive strategy.

Thus, according to the SPACE matrix public partnership "Carlsberg Ukraine" is characterised by a competitive strategy (figure 1).

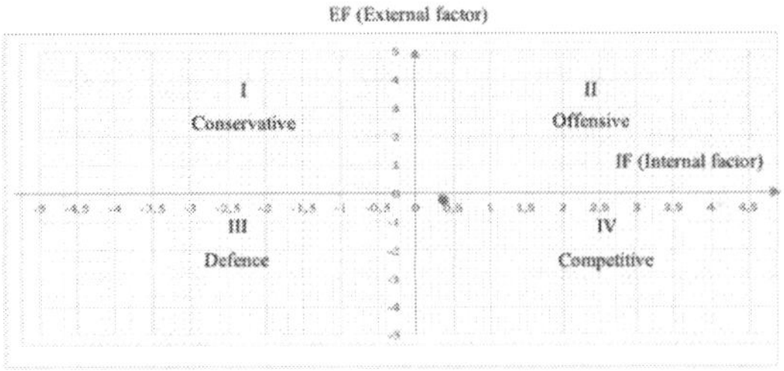

Figure 1. SPACE strategy selection matrix for public partnership "Carlsberg Ukraine"

Filling the four quadrants at the intersection of strengths / weaknesses, opportunities / threats will identify and describe four types of potential strategies for public partnership "Carlsberg Ukraine". In particular, we define:

- potential strategic advantages of the company (ST);

Table 11. The results of the assessment of the impact of external threats on the activities of public partnership "Carlsberg Ukraine"

Factor	The weight of the factor in a single scale	The weight of the factor in assessing strengths	Factor impact assessment	Weighted assessment
Competition in the industry, $EF_{1(1)}$	0.060	0.105	3.000	0.314
Production capacity in the industry, $EF_{1(2)}$	0.060	0.105	1.000	0.105
Competition with manufacturers of substitute products, $EF_{1(4)}$	0.053	0.092	3.000	0.276
Market demand, $EF_{1(7)}$	0.038	0.066	2.000	0.133
Prices for raw materials and electricity, $EF_{1(11)}$	0.058	0.100	4.000	0.401
Product advertising, $EF_{1(13)}$	0.030	0.053	2.000	0.106
Transformation of tastes and preferences of consumers, $EF_{1(14)}$	0.043	0.076	2.000	0.151
Political situation, $EF_{2(1)}$	0.029	0.051	2.000	0.101
Legislation, $EF_{2(2)}$	0.033	0.058	1.000	0.058
New production / technologies, $EF_{2(3)}$	0.029	0.050	2.000	0.101
Raw material base, $EF_{2(6)}$	0.020	0.035	2.000	0.071
Weather, $EF_{2(8)}$	0.022	0.038	3.000	0.115
Welfare of the population, $EF_{2(9)}$	0.032	0.055	2.000	0.110
Excise duty, $EF_{2(11)}$	0.032	0.055	1.000	0.055
Pandemic, $EF_{2(12)}$	0.035	0.060	4.000	0.241
Total	0.574	1.000	-	2.338

- guidelines for strategic development of the company (*SO*);
- limitation of strategic development (*WT*);
- landmarks of internal transformations of the enterprise (*WO*).

At the intersection of strengths and opportunities we substantiate the guidelines for strategic development for public partnership "Carlsberg Ukraine" (table 12).

The analysis of behavioral strategies of Public partnership "Carlsberg Ukraine" is aimed at improving strengths and eliminating weaknesses through external opportunities. In particular: the formation of new sales channels, expanding opportunities for external financing of enterprise development projects, signing new important contracts in the field of logistics, creating effective marketing communications and management training. Much attention should be paid to monitoring changes in consumer tastes and preferences, the behavior of direct competitors and competitors producing substitute products, the price situation in the market. The company must pay attention to the negative factors caused by the pandemic, military-political problems and instability of legislation.

5 Conclusion and future work

According to the results of the study we obtained the following generalizations.

The method of conducting SWOT analysis based on the formation of a basic set of indicators, selection of scientifically sound scales and the possibility of assessment taking into account the neutral, strong and weak action, which allowed to address the application of this model for continuous monitoring of the enterprise environment.

The selection of key indicators for assessing the impact of internal and external factor, which reduced the time for their formation. It should be noted that the proposed list of factors can always be supplemented by a market analyst and adapted for the analysis of enterprises taking into account industry characteristics.

It is determined that the estimation of weighting factors more than 20 indicators can be done only through the decomposition of factors in new subsystems. For this purpose, the following subsystems of internal environmental factors are identified: innovation and development, logistics, production, sales and promotion, enterprise finance, personnel management and subsystems of strategic and tactical management. And for the external environment – two subsystems of factors: the intermediate environment and the macroenvironment. This allowed us to apply the method of pairwise comparisons by T. L. Saati to estimate the weights.

The proposed implementation of continuous monitoring of factors of neutral action cannot be ignored, because over time they can move to the category of both negative and positive factors.

Approbation of the proposed SWOT analysis methodology, carried out on the example of assessment of the environment of "Carlsberg", its capabilities and threats, allowed to identify measures to improve its activities, to

Table 12. SWOT matrix of strategic decisions for public partnership "Carlsberg Ukraine"

		External factors	
		Threats	Opportunities
Internal factors	Strength	ST_1 – vertical back integration strategy (with suppliers); ST_2 – production of own raw materials; ST_3 – strategy of diversification of activity (focus on new types of low-alcohol and soft drinks); ST_4 – flexible pricing policy for dealers and agents; ST_5 – opening of own sales centers in regional markets of the world; ST_6 – organization of cooperation with trade representatives abroad; ST_7 – creation of more advanced types of packaging and labels that would meet the new requirements of consumers.	SO_1 – formation of a new project team to address issues of further diversification of the company; SO_2 – development of new. more advanced types of product packaging. which would correspond to the best world analogues; SO_3 – access to new competitive markets; SO_4 – development of long-term partnerships with suppliers of raw materials and equipment; SO_5 – formation of new sales channels abroad; SO_6 – interaction with leading research institutes and equipment manufacturers to update beer production technologies; SO_7 – interaction with universities and institutes of Ukraine and the world to find new creative workers; SO_8 – involvement of partners in financing their own projects.
	Weakness	WT_1 – monitoring the situation on the beer market in Ukraine and the world; WT_2 – monitoring the desires and attitudes of beer consumers; WT_3 – insurance and reinsurance of risks; WT_4 – entering the world market of new types of premium products; WT_5 – creating more comfortable working conditions in the company; WT_6 – intangible methods of staff retention.	WO_1 – training in advanced marketing technologies to promote products in regional markets (trainings and seminars); WO_2 – using of external sources to finance projects to restructure the main business processes of the enterprise; WO_3 – development of partnerships with educational institutions to attract new employees; WO_4 – involvement of SEO optimizers. content managers. specialists in contextual and media advertising to improve search engine results abroad; WO_5 – involvement of leading brewing laboratories in the implementation of own projects; WO_6 – sale of obsolete equipment; WO_7 – strengthening the company's corporate culture.

carry out quantitative and qualitative assessment of external and internal environment.

The use of the method of pairwise indicators and scientifically sound research scales of the SWOT analysis method allowed to significantly improve the quantitative assessment of indicators, to carry out a qualitative assessment of the results. But this technique will continue to be subjective, the success of which will always be derived from the analytical skills of those who build matrices and interpret indicators.

References

[1] L. Breiman, J. Friedman, C.J. Stone, R.A. Olshen, *Classification and Regression Trees* (Chapman and Hall/CRC, 1984)

[2] S.E. Jackson, A. Joshi, N.L. Erhardt, Journal of Management **29**, 801 (2003)

[3] A.R. Ommani, African Journal of Business Management **5**, 9448 (2011)

[4] B.S. Sigmon, Community and Economic Development Publications **16** (2014)

[5] S. Brad, E. Brad, Procedia Engineering **131**, 616 (2015)

[6] E. Gürel, M. Tat, The Journal of International Social Research **10**, 994 (2017)

[7] S. Ivanov, CEUR Workshop Proceedings **2422**, 385 (2019)

[8] O. Kravchenko, I. Borisyuk, Z. Vakolia, O. Tretyak, O. Mishchenia, International Journal of Higher Education **9**, 94 (2020)

[9] P. Mykhnenko, Ekonomycheskyi analyz: teoryia y praktyka **18**, 60 (2015)

[10] V. Podgornov, Vestnik Adygeiskogo gosudarstvennogo universiteta. Seriia 5: Ekonomika (2012)

[11] C. Namugenyi, S.L. Nimmagadda, T. Reiners, Procedia Computer Science **159**, 1145 (2019)

[12] SuWaNu Europe, *Report on SWOT and PEST analyses for implementation of reuse practices* (2019), https://suwanu-europe.eu/wp-content/uploads/2020/01/SWOT-PEST_-Antwerp-and-Limburg.pdf

[13] J. Wang, Z. Wang, International Journal of Environmental Research and Public Health **17** (2020)

[14] J. Nazarko, J. Ejdys, K. Halicka, A. Magruk, Łukasz Nazarko, A. Skorek, Procedia Engineering **182**, 482 (2017), 7th International Conference on Engineering, Project, and Production Management

[15] D. Pyzhlakov, Rossyiskoe predprynymatelstvo **6**, 133 (2008)

[16] Y. Lozovik, T. Lozovik, Vcheni zapysky Universytetu KROK **20**, 105 (2009)

[17] T.L. Saati, K.P. Kearns, *Analytical Planning: The Organization of Systems* (Pergamon, 1985), ISBN 0-08-032599-8

[18] A. Humphrey, *SWOT Analysis for Management Consulting* (SRI Consulting, 2005)

[19] *PAT Karlsberg Ukraina* (2021), https://carlsbergukraine.com

Environmental investments as the basis of "green" economy: empirical evidence from Ukrainian companies

Iryna Vasylchuk[1,*], *Anatolij* Suprun[1,**], *Kateryna* Slyusarenko[1,***], and *Maryna* Sadovenko[1,****]

[1]State University of Economics and Technology, 16 Medychna Str., 50005, Ukraine

Abstract. Investments in environmental protection are crucial for achieving the goals of building green economy. The purpose of the study is, firstly, to assess the state and trends of Ukrainian companies investments for environmental purposes; secondly, to identify the determinants of the impact on the volume of environmental protection investments; finally, to determine presence / absence of a link between environmental investments and financial performance of companies. The initial data for the study were obtained from published statistical reports for the period from 2010 till 2019. Methods of analysis for testing hypotheses are descriptive and correlation-regression ones. It was found out that the selected independent variables (belonging to environmentally damaging activities and revenue of the companies) are positively correlated with the volume of costs and investments in environmental protection of Ukrainian companies. However, for the independent variable "financial results" a negative correlation was obtained. The hypothesis of environmental investments positive impact on the financial profitability of companies has not been confirmed. Environmental policy in Ukraine concerning implementation of "green" growth strategy is defined as a reflective one due to a number of economic and political factors.

1 Introduction

The "green" economy is a key factor for the implementation of sustainable development concept on the background of more efficient resource and energy consumption, cutting of CO_2 emissions, reduction of harmful effects on the environment and progress of a socially integrated society [1]. The OECD estimates that investment in modern, smart and clean infrastructure in the next decade is a critical factor for sustainable economic growth. Achieving the set environmental goals of sustainable development requires 6.3 trillion USD of investment in infrastructure annually until 2030 in the global dimension [2]. Destructive climate change on the planet can be prevented under the condition of joint efforts only.

The analysis of professional publications revealed that scientists use several concepts to describe the phenomenon of investment in environmental measures, which can be combined under the "umbrella" brand "green investment": ecological investment [3–5], environmental protection investment [6], "green" investment [1, 7–10], environmental protection expenditure [11], etc. Until recently, Ukrainian authors' research on environmental issues was based on terminology from statistical reports, where environmental expenditures are considered to be the sum of current expenditures and capital investments directed to environmental protection. Capital investments in environmental protection are treated as investments in the manufacture for own use or purchase of tangible and intangible assets, the cost of capital repairs and modernization, carried out to protect the environment [12].

In recent years, with the increase of environmental awareness and business activity, the expansion of environmental protection investment areas range and the emergence of new "green" financial instruments, the concept of "green" investment was introduced in the academic discourse. Consequently, they are currently viewed from the standpoint of an entrepreneurial approach, which provides long-term benefits from environmental activities, both financial and non-financial ones. There is an increasing tendency to interpret such investments as entrepreneurial capital (material, intellectual, raw materials, fixed assets and technology), which is directed to the construction of a facility or production of goods / services / technologies with ultimate goal to have a positive impact on the environment (e.g. level of pollution) with the simultaneous restoration or maintenance of the natural capital level [1].

At the same time, the "green investment" brand has a broader understanding and includes both investments in "green" companies through financial intermediaries through various financial instruments and public investments in environmental projects and direct business investments in environmental protection projects. For the purposes of this study, the generalized term "environmental protection investment" is used, meaning the companies expenses for their own environmental protection projects in.

*e-mail: vasylchuk_ip@duet.edu.ua
**e-mail: suprun_aa@duet.edu.ua
***e-mail: slusarenko_kv@duet.edu.ua
****e-mail: sadovenko_mm@duet.edu.ua

The analysis of the professional literature dealing with the problem revealed that domestic scientists focus their attention on the analysis of statistical data on trends in environmental investment in Ukraine [4], sources of their funding [3], [13], state of environmental taxation [14], models for optimizing environmental investment strategies of industrial enterprises [15], etc. The researches of foreign scholars generally concern not only the analysis of environmental protection expenditure trends in Europe and the world [11], but also the relationship between "green" investments and firm performance in the context of Corporate Social Responsibility (CSR) on the example of different jurisdictions and activities [9, 10, 16, 17], the impact of financial markets on development of "green" economy [18]. Ukrainian studies turned out to have no determinants of environmental investment in business and their impact on the financial performance of companies. Besides filling this gap it is important to identify the latest trends in Ukrainian companies environmental protection investment and their activities aimed at meeting environmental requirements of sustainable development, which is manifested through increased investment.

Therefore, the overarching research issues to be studied are as follows:

1. What is the state and trends in the Ukrainian companies investment for environmental purposes?
2. What factors affect the amount of costs (investments) in environmental protection?
3. Do the volumes of investments in environmental protection affect the profitability of Ukrainian companies?

Our study examined relationships among ecological investment and financial indicators for Ukrainian companies on the basis of proposing appropriate hypotheses and testing them using econometric analysis methods, which is new for the Ukrainian discourse, but not new for the world practice.To obtain answers to above mentioned questions, two methods of analysis are used, such as descriptive analysis of aggregated statistics for the period 2010–2019 and correlation-regression analysis of statistics in terms of economic activities of companies from 2015 till 2019.

The theoretical underpinnings for analysis come in the form of legitimacy theory, institutional theory, stakeholder theory, resource-based view, slack resources theory and the good management theory.

2 Theoretical framework and literature review

With the spread of the of sustainable development concept, the management of many companies are convinced that the integration of environmental goals in strategic decision-making and environmental measures on an ongoing basis encourage the creation of economic benefits for the company. At the same time, some managers still consider investing in the environment to be a "necessary evil" and wonder why they are to carry out environmental activities and incur unnecessary costs. The answer to this question can be obtained through taking into consideration the provisions of legitimacy theory, institutional theory, stakeholder theory and resource based view.

The theory of legitimacy substantiates that the implementation of environmental activities and coverage of its results in non-financial reporting depends on the intensity of social and political pressure that companies face in the process of their activities. This is especially true for companies that place a significant burden on the environment and are eco-sensitive activities. Companies strive for a balance between their own and society values. In a situation when such a balance is achieved, a "social contract" between the company and society virtually exists. If society has evidence that a company does not work according to the terms of the "social contract", then such a company will have a negative assessment by society [19]. With such behavior, society may terminate the contract with the company. The canceled social contract is called the "legitimacy gap" [20–22]. Therefore, the company must strive to ensure that its activities meet social norms and expectations. Responsible company behavior can reduce possible fines and ensure the company right for existence [23]. It has been established that while attempting to prevent the loss of legitimacy, companies with poor environmental performance try to publish as much positive information about environmental protection in non-financial statements as far more environmentally friendly companies [20].

While the theory of legitimacy focuses on communication with society, the theory of stakeholders focuses on communication with different groups of stakeholders. Stakeholder theory is the most important approach to explain how measures in the company environmental activities provide its higher financial performance [24]. This theory postulates that it is not enough for managers to focus solely on the needs of shareholders, it is of paramount importance to take into account the interests of internal and external stakeholders. They affect the company reputation, confirm its legitimacy and affect its financial performance directly due to mechanisms to reducing / increasing demand for products and providing / not providing funds to financial activities. It is maintained [25, 26] that, the company is able to achieve an acceptable level of financial performance only through active management of relations with key stakeholders.

Institutional theory provides further explanation at the intersection of stakeholder theory and legitimacy theory and argues that institutions are an important components of business environment. It describes the mechanisms by which a company seeks to reconcile society's perceptions of its activities with societal values and norms, that is, which way the company seeks to institutionalize them. In this context, the argument of M. Porter is important, who affirms that profitability and pollution reduction are not mutually exclusive goals. In his view, pollution is a waste of resources (e.g. energy, materials), and, therefore, efforts to reduce pollution (e.g. through improved products or processes) can not only reduce the negative impact on the company environment, but also strengthen its competitiveness. He suggests that well-designed environmental legislation can stimulate innovation and increase competitiveness [27]. In other words, the more we institutionalize "Be Green!" narrative as a society norm and the more environmentally friendly behavior is spread among companies, the more severe the stakeholders punishment is for a company with environmentally harmful behavior. This issue is especially relevant for large companies, involved in mining, agriculture, transport, processing which are, undoubtedly, significant "polluters" of the environment.

The resource based approach postulates the positive impact of the company environmental performance on its financial performance. According to this approach, companies consider stakeholder activities as a strategic investment [28, 29]. By investing in such a strategy, companies create assets that are valuable, rare, and have no substitutes, such as leadership and positive reputation. These assets, in their turn, provide the company with a competitive advantage and potentially higher profitability [30]. The company efforts aimed at reducing environmental impact and improving relationships with environmental stakeholders can also create a competitive advantage. A company which is successful in managing natural resources through innovation in environmental products can reduce costs, cut risks, improve reputation, increase employee motivation and, thus, rocket revenues.

3 Hypotheses development

Several hypotheses are to be tested in order to answer the second research question on environmental determinants of environmental protection investment (EPI) determinants and the presence / absence of relationship between environmental protection investment and financial performance.

According to the reviewed literature, the relationship between environmental protection investment and companies sales and profitability could be positive, negative or neutral. Thus, reducing energy use and pollutant emissions gives opportunity for more efficient use of resources [27, 31], and increasing environmental protection investment has a positive effect on company product sales and favors increasing its efficiency [17, 32]. One of the most important aspects of the relationship between environmental investment and company financial performance is its causality, i.e. the direction of causation. There are two management theories explaining the causality of this relationship such as the theory of good management (Good Management Theory) and the theory of surplus resources (Slack Resource Theory) [33]. Both theories suggest that there is a positive relationship between environmental investment and a company financial performance. The Good Management Theory is based on the assumption of high relationship existence between environmental performance and effective stakeholder management. Effective environmental measures improve relationships with key stakeholder groups, such as the local community, customers, employees, investors or local government. Unlike the theory of good management, the theory of surplus resources has the opposite argument. According to it, the company high revenues and its financial results contribute to increased investment. The availability of surplus financial resources provides opportunity for the company to use these resources for investing in environmental projects [33–36].

Taking into consideration the above mentioned theoretical provisions and the results of previous empirical research [9, 28, 32, 37], we postulate that: 1) Ukrainian companies can enjoy positive relationship between financial performance and environmental protection investment (based on the slack resource theory and on good management theory); 2) companies which carry out environmentally non-friendly activities, increase the volume of environmental protection investment.

It should be pointed out that two types of dependent variable are to be tested such as total environmental protection expenditures and capital investments for environmental protection. The argument in favor of such a division is that the vast majority of companies, one way or another, incur current environmental costs under the pressure of external stakeholders and regulations. It can be interpreted as reflexive behavior. At the same time, long-term capital investments in environmental protection and increasing their volume may be the evidence of serious companies intentions to upgrade business processes for future resource saving. Therefore, such activities can be perceived as the integration of environmental goals of sustainable development into the companies long-term strategy characterizing their proactive position.

3.1 Industry profile and environmental protection investment

Companies, carrying out activities significantly damaging the environment, are under considerable pressure from stakeholders, such as government environmental inspectorates, local communities, politicians, and workers. As a result, corporate management must bring its corporate standards and values into line with public expectations in order to obtain a "license to operate" according to the theory of legitimacy. The increase of civic consciousness due to the spread of the paradigm of sustainable development contributes to the transformation into the norm of environmentally friendly behavior of corporations, which corresponds to the principles of institutional theory.

Ukraine is an industrialized country with a powerful mining and processing complex, represented by companies of relevant activities. Such companies make significant emissions of harmful substances into the air, pollute the land and water, require significant amounts of water consumption and energy resources. Besides, Ukraine has developed agriculture, which requires a large area of arable land, water abstraction from rivers for irrigation, application of herbicides and mineral fertilizers, violating the environment severely. Therefore, agricultural companies are expected to increase investment in environmental protection by all means. The transport industry also belongs to environmental pollutants. It is no wonder that society expects such companies to do much more to protect the environment. Therefore, the classification of the company as involved in environmentally harmful activities is an important factor influencing the volume of environmental investment [9]. The study suggests the following hypotheses:

H1.a: there is a positive relationship between the company's involvement in environmentally damaging sectors of the economy and total environmental protection expenditure (determinant – environmentally damaging economic activities).

H1.b: there is a positive relationship between the company's involvement in environmentally damaging sectors of the economy and long-term environmental protection investment (determinant – environmentally damaging economic activities).

3.2 Sales and environmental protection investment

In Ukrainian reality, companies with significant revenues have more financial resources which can be used for environmental protection expenditure. First of all, revenue is a source of covering the current costs of environmental protection. Previous studies showed that, in the periods when revenue is falling, even large companies reduce capital investment, especially in environmental protection [3, 38]. At the same time, according to the theory of legitimacy and resource based view, companies strive to gain a reputation as a "good citizen", so they pursue environment-related policies [29] and invest significant resources in protecting the environment of areas where their production is located.

With the adoption of the Sustainable Development Goals (Sustainable Development Goals – 2030, adopted by the UN in 2015), more and more large Ukrainian companies are reporting an increase in both current expenditures and long-term capital investments in environmental protection [38]. From the standpoint of the slack resource theory, companies should direct more financial resources to achieve certain strategic goals of sustainable development, including environmental ones. Taking into consideration all the above mentioned, we can derive the hypotheses:

H2.a: there is a positive relationship between the volume of revenue from companies sales and the volume of total environmental protection expenditure (determinant – company sales).

H2.b: there is a positive relationship between the volume of revenue from companies sales and the volume of long-term environmental protection investment (determinant – company sales).

3.3 Financial performance and environmental protection investment

As well as in case of significant revenues, a company that consistently generates high profits similarly owns available financial resources. These surplus resources can be used to increase capital investment in environmental protection, as the most common source of financing for long-term investment is long-term borrowings and equity. Therefore, in the "replete" years, companies begin projects on resource efficiency, modernization of equipment and improvement of technological processes. Thus, the volume of profits has a positive effect on the volume of environmental protection investment. From the standpoint of the slack recourse theory, this study proposes the following hypotheses:

H3.a: there is a positive relationship between volumes of net profits and volumes of total environmental protection expenditure (determinant – company net profit).

H3.b: there is a positive relationship between volumes of net profit and volumes of long-term environmental protection investment (determinant – company net profit).

3.4 Environmental protection investment and financial performance

From the viewpoint of good management theory, increasing investment in environmental protection will help save natural resources through the use of more efficient resource-saving technologies. This will have a positive effect on reducing costs and, consequently, will increase the financial efficiency of the company. In addition, the reduction of pollution and waste leads to a reduction in environmental fines and, consequently, increases its financial results. Moreover, the constant attention of company management to environmental issues contributes to the formation of a positive image of the company and enhances its reputation. In turn, positioning the company as one that sells environmentally friendly products increases the commitment of a wider range of customers, making a positive impact on the company volume of production. These arguments give us opportunity to put forward the following hypothesis:

H4.a: there is a positive relationship between long-term environmental protection investment and financial performance (determinant - environmental protection investment).

4 Research Method

This study employed six variables: industry profile, company sales, company net profit, return on assets, long-time environmental protection investment, total environmental protection expenditure. The sample is formed on the basis of data from statistical reports on the activities of economic entities in Ukraine for the period 2015–2019 in terms of economic activities [39].

In this study, the dependent variable, which is environmental protection investment (EPI) is considered in two contexts: both as total environmental protection expenditure and as long-term component environmental protection expenditure (long-term environmental protection investment). Accounting of these costs as capital expenditures is based on the effect they produce over more than one year. On the contrary, current environmental expenditure is accounted for as intermediate consumption (they have effect only within a current year). Therefore, long-term environmental protection investment indicators were used in the analysis as indicators with lag $(n + 1)$.

In order to test hypothesis *H4* financial performance (FP) is measured by ROA (after tax earnings divided by total assets). It is postulated that it is positively affected by already made long-term capital investments, so ROA was presented as an indicator with a lag $(n + 1)$.

The independent variables selected for regression analysis have already been tested for other jurisdictions and are based on the results of previous empirical studies published [9, 10, 16, 17, 33, 40, 41]. Since the analysis is performed on indicators from statistical reports, the analogue for the dependent variable company sales used company turnover indicator. Industry profiles (environmentally damaging industry) are considered as a dummy variable which refers to low environmentally damaging industry or highly environmentally damaging industry (highly environmentally damaging industry is scored by one, otherwise zero). According to the NACE classification highly environmentally damaging industries (EDI) include agriculture (A), industry (B, C, D, E), transport (H), construction (F). The data were analyzed using linear regression based on the following models:

Model 1:

$$EPI = \alpha_i + \beta_{1i} * EDI + \beta_{2i} * S + \beta_{3i} * NP + \varepsilon_i \quad (1)$$

Model 2:

$$FP = \alpha_i + \beta_{1i} * EPI + \varepsilon_i \quad (2)$$

where EPI represents environmental protection investment; EDI shows environmentally damaging industry; S is a company' sales; NP is a company' net profit; FP is financial performance.

Correlation-regression analysis was used to test hypotheses about the relationship between environmental protection investment and their determinants.

5 Results and discussion

Descriptive data analysis. In order to answer the first research question we used the method of indicators descriptive analysis characterizing state and trends of environmental protection expenditure in Ukraine. Indicators are derived from statistical reports for a period from 2010 till 2019 [12, 39]. The analysis shows that there is a positive trend in environmental protection expenditure (in nominal terms), both in aggregate ones and in terms of components over the past 10 years (table 1). It should be emphasized that that table 1 shows the data on the total costs incurred both within government programs and by businesses. In the total environmental protection expenditure, the largest share is made up of current expenditures, which ranges from a minimum value of 62.8% in 2019 to a maximum value of 78.96% in 2010. Correspondingly, there is a tendency to increase the share of long-term capital investments, the share of which made up a record of 37.17% in 2019. It serves as evidence that priorities of environmental policy in the country have changed towards increasing projects creating long-term effects.

We can observe very small share of total environmental protection expenditure in the total GDP of the country – from 1.5% to 1% with a downward trend (table 2). At the same time, for Europe (EU-27) such expenditures are steadily 2% of GDP [42]. In 2018 national expenditure on environmental protection as a percentage of GDP decreased to 1%, which is less than 2 times compared to the European Union. This situation does not allow us to talk about the seriousness of the government efforts aimed at transition Ukraine to a "green" agenda for the period up to 2030. By no means we can observe the improvement of the situation in terms of "total environmental costs per capita", as there is a steady trend towards depopulation in Ukraine. Over ten years, nominal environmental spending per capita has increased almost fivefold, but real growth, taking into account depopulation and inflation, is only 3 times as high. In fact, environmental spending per capita in Ukraine makes as little as 35 US dollars for the whole period of 2019. Most scholars and practitioners point out that the reasons for this situation are the presence of "chronic" social problems in the country that require priority funding, permanent crisis in Ukrainian economy, production falling in recent years, which all results in lack of financial resources to solve environmental problems.

It is a common knowledge, the main "agent of change" in the economy of each country is the business that implements, develops and finances innovative projects and activities. As part of total capital investment in environmental protection, business investment accounted for the vast majority until 2018, although since 2016, there was a period of significant decrease in business investments in long-term environmental projects (Table 3). In 2019 EU-27 invested EUR 51 billion into assets essential for providing environmental protection services and about EUR 31 billion (60% of total environmental protection investments) was spent by corporations [42].

If we compare the share of environmental capital investments of economic entities with total investments for the same period, we can see that their share is insignificant and decreases from 4.5% in 2016 to 1.7% in 2019. The share of such investments in the GDP of Ukraine (0.23% in 2019) and in the total environmental protection expenditure of economic entities (0.8% in 2019) is too small and tends to decrease. In Europe, the share of environmental protection investments in total investments of corporations was also relatively low in this period and amounted to 1.7% only in 2019, which is, nevertheless, much more than in Ukraine.

Thus, we can make a conclusion that there is chronic underfunding of environmental spending in the country and decrease of business interest in implementing long-term environmental projects. This does not give us opportunity to claim about active implementation of environmental goals of sustainable development by Ukrainian companies.

Hypotheses testing is based on regression analysis. To answer the questions about the determinants defining the amount of costs and investments in environmental protection, a regression analysis was performed according to the identified *Hypotheses 1-3*. Descriptive statistics are given in table 4. According to statistics on economic activities of business entities in Ukraine, we were able classified as few as 4 types of ecologically damaging industries out of

Table 1. Dynamics of environmental protection expenditures in Ukraine from 2010 till 2019

Year	Total environmental protection expenditure (TEPE)	Capital investments for environmental protection		Operating cost for environmental protection	
		UAH million	% from total costs	UAH million	% from total costs
2010	13128.1	2761.5	21.04	10366.6	78.96
2011	18490.5	6451.0	34.89	12039.4	65.11
2012	20514.0	6589.3	32.12	13924.7	67.88
2013	20377.9	6038.8	29.63	14339.1	70.37
2014	21925.6	7959.9	36.30	13965.7	63.70
2015	24591.1	7675.6	31.21	16915.5	68.79
2016	32488.7	13390.5	41.22	19098.2	58.78
2017	31492.0	11025.6	35.01	20466.4	64.99
2018	34392.3	10074.3	29.29	24317.9	70.71
2019	43735.9	16255.7	37.17	27480.2	62.83

Table 2. Indicators of environmental productivity in Ukraine for 2010–2019

Year	Total environmental protection expenditure (TEPE), UAH million	Nominal GDP, UAH million	TEPE as percentage of GDP, %	Average annual population, thousands people	Total environmental expenditure per capita, UAH
2010	13128.1	1082569	1.2	45865	286.2
2011	18490.5	1316600	1.4	45693	404.7
2012	20514	1408889	1.5	45577	450.1
2013	20377.9	1454931	1.4	45483	448.0
2014	21925.6	1566728	1.4	43722	501.5
2015	24591.1	1979458	1.2	42836	574.1
2016	32488.7	2383182	1.4	42668	761.4
2017	31492	2982920	1.1	42477	741.4
2018	34392.3	3558706	1.0	42269	813.7
2019	43735.9	3974564	1.1	42019	1040.9

Table 3. Indicators of capital investments for environmental protection by business entities in Ukraine, 2015-2019

Year	Environmental protection investment (EPI), UAH million	Nominal GDP, UAH million	EPI as percentage of GDP, %	Capital investments, UAH million	EPI as percentage of capital investments, %	Total environmental protection expenditure (TEPE), UAH million	TEPE as percentage of GDP, %
2015	7268.7	1979458	0.37	213478.1	3.40	23780.5	1.20
2016	12810.9	2383182	0.54	281667.9	4.55	31082.6	1.30
2017	9866.6	2982920	0.33	359159.8	2.75	29276.1	0.98
2018	8985.3	3558706	0.25	471115.5	1.91	32368.6	0.91
2019	8990.4	3974564	0.23	524474.1	1.71	32359.6	0.81

19 according to the NACE (31% of all observations), but in 2019 their expenditures for environmental protection amounted to more than 90% of all the annual expenditures. The standard deviation is quite significant for all indicators, due to the above mentioned problems and significant variation in data.

The next step is the results of multiple regression analysis for the most common environmental investment factors in the scientific literature. It is worth emphasizing that revenues and net income, which are the main source of financing capital investment and current expenditures, were identified as determinants of the impact on environmental expenditures. Correlation analysis did not reveal a close relationship between them (multiple $R = 0.139$, $R^2 = 0.019$), which allows these indicators to be included as independent variables.

While characterizing the results of regression analysis with three independent variables for all costs (table 5), it should be pointed out that significant factors influencing the volume of total environmental protection expenditure were sales revenue (Coef = 0.0045833***, Standard error = 0.0008195), net profit (Coef = -0.034506**, Standard error = 0.0168399) and companies belonging to environmentally dirty industries (Coef = 5372.1979***, Standard error = 1694.7341). The following factors are also significant for long-term environmental protection investment: sales revenue (Coef = 0.001506***, Standard error = 0.00028678), net profit (Coef = -0.01929**, Standard error = 0.00589308) and belonging of companies to envi-

Table 4. Descriptive statistics

Variable	Min	Max	Mean	SD
Sales	0	3764364.90	540815.25	962630.03
Net Profit	-188267.9	109288.81	2323.16	45818.90
ROA	-8.1	34.7	8.94	9.34
Environmentally damaging industry (Dummy)	0	1	0.308	0.466
Total environmental protection expenditure	0	30174.50	2405.15	7639.96
Long-term environmental protection investment	0	12563.20	650.90	2455.81

ronmentally dirty industries (Coef = 1945.76***, Standard error = 593.067134). However, the adjusted coefficients of determination (Adj. R^2) for the model at 49% level indicate a sufficient explanatory power of the estimated model. The acceptability of this model for this study is based on the view [43, 44] that a high R^2 value is not necessary for the social sciences and humanities. According to [44] R^2 value 0.12 or below indicate low, between 0.13 to 0.25 values indicate medium, 0.26 or above and above values indicate high effect size.

It should be borne in mind that that although all three ratios are significant, the relationship between environmental investment and revenue and between environmental investment and net profit is low. The reason is previously mentioned trend to finance environmental costs on a residual basis. There is a negative relationship for the net profit determinant, which refutes *hypothesis H3*. This means that an increase in profit per unit of measurement leads to a slight decrease in the level of environmental investment.This result can be explained by the low volume of capital investment in environmental protection (60-70% of total expenditures), the chronic unprofitability of companies, and the financing of current expenditures mainly by revenues.The most influential determinant is belonging to environmentally damaging industries, which can be explained from the standpoint of legitimacy theory, stakeholder theory and slack resource theory. Based on the calculations, we can derive a model of multiple regression with weight coefficients:

$$\text{Total environmental protection expenditure} = \\ = -1646.389 + 5372.1979EDI + \\ +0.0045833S(t-1) + (-0.034506)NP(t-1) \quad (3)$$

$$\text{Longterm environmental protection investment} = \\ = -591.893 + 1945.76EDI + \\ +0.001506S(t-1) + (-0.01929)NP(t-1) \quad (4)$$

Thus, we can conclude that *hypotheses 1-2* are confirmed and *hypothesis 3* is disproved for Ukrainian companies. The main type of analysis used to test the determinants was based on the methods of linear regression by the method of least squares. Least squares regression is based on several assumptions, in particular, such as linear dependence between variables, independence of regression residues, normality of distribution of variables and regression residues, homoskedasticity and absence of multicollinearity. These assumptions were verified by appropriate tests. The main problem that causes the lack of independence between the regression residues is the presence of autocorrelation, which is inherent in regression based on time series. It was considered in this analysis, the autocorrelation did not come out. White's test was used to check for heteroskedasticity, and the variance inflation factor was used to check for multicollinearity, whose value should not exceed 10 in accordance with a rule. The calculated values of the variation factor showed no problems with multicollinearity.

Testing of *Hypothesis 4* on the impact of environmental investment (capital investment) on the financial performance of companies by the method of pairwise correlation gave negative results, indicating a lack of relationship (multiple R = 0.086, R^2 = 0.007). Thus, the hypothesis is rejected, which means that company managers do not take into account the provisions of the theory of good management in their decisions to implement investment projects aimed at environmental protection, and still do not consider such investments to be a factor improving the reputation and profitability of companies.

It should be pointed out that the determinants selected for analysis were limited by the type of collected data, as statistical reports in Ukraine do not provide more detailed and diverse information. At the same time, there are individual companies, for which other important factors may be taken into account, in particular, such as form of ownership, belonging to the signatories of the UN Global Compact, certification according to environmental standards, etc. The fact that companies are classified as environmentally damaging industries is not controversial, although previous research [27, 31] shows that such companies may resort to a "greenwashing" strategy instead of real activity.

6 Conclusion

The research is devoted to finding answers to questions about the state of environmental costs and investments in Ukraine and the impact of factors on environmental investments on the example of Ukrainian companies. The analysis revealed that despite the nominal increase in environmental spending, their share in GDP is insignificant. The share of capital investment by companies has dropped dramatically in recent years. It does not give opportunity to draw optimistic conclusions about the creation of a basis for the implementation of the policy of "green" growth in Ukraine.

The main factors influencing the volume of environmental investment were the following: the attitude of companies to environmentally damaging industries, income

Table 5. Results of multiple regression (Dependent = environmental investment)

Variables	Coef	SD	t-stat	p-value	Result
Dependent variable (total environmental protection expenditure)					
Constant	-1646.389	957.36622	-1.7197067	0.0919285*	No hypothesis
Environmentally damaging industry	5372.1979	1694.7341	3.16993559	0.0026548***	Confirmed
Sales (t-1)	0.0045833	0.0008195	5.59291762	$1.043{\cdot}10^{-6}$ ***	Confirmed
Net Profit (t-1)	-0.034506	0.0168399	-2.0490573	0.0459441***	Not confirmed
Adj. R^2			0.497		
F-stat			17.78		
F-stat (p-value)			$6.7{\cdot}10^{-8}$		
Dependent variable (Long-term environmental protection investment)					
Constant	-591.893	335.027448	-1.7667	0.083636469*	No hypothesis
Environmentally damaging industry	1945.76	593.067134	3.280843	0.001932279***	Confirmed
Sales (t-1)	0.001506	0.00028678	5.251055	$3.41097{\cdot}10^{-6}$ ***	Confirmed
Net Profit (t-1)	-0.01929	0.00589308	-3.27306	0.001976216***	Not confirmed
Adj. R^2			0.497918		
F-stat			17.85903502		
F-stat (p-value)			$6.38{\cdot}10^{-8}$		

Note: *Significant at 10%, **Significant at 5% , ***Significant at 1%

(revenue) of companies and net financial results. The hypothesis on the impact of capital investment in environmental protection on the financial performance of companies was also tested. Regression analysis methods were used to test the hypotheses. According to the results of the analysis, all three identified factors have an impact on the volume of environmental investment of companies, but the most important factor is the attitude of companies to environmentally damaging industries. The hypotheses about the relationship between the financial results of companies and environmental investments and the impact of environmental investments on the financial performance of the companies were not confirmed.

Summing up the study, we can state that the results of environmental activities of economic entities in Ukraine can be explained only from the standpoint of slack resource theory, since companies invest in long-term environmental projects only in the presence of slack financial resources. None of the provisions of the theories of legitimacy theory, stakeholder theory and institutional theory are used as fundamental ones in the development of corporate strategies. Only a small proportion of companies in Ukraine develop sustainable development strategies and compile relevant reports. In Ukraine, when determining the remuneration for managers, appropriate indicators are not set for the implementation of the Sustainable Development Goals, and therefore, the theory of good management does not work. The environmental activity of the vast majority of Ukrainian companies is mostly declarative, and the behavior of companies from these positions is reflexive. In total, we can conclude that Ukraine belongs to the poor countries, whose main task is to survive. Under such conditions, a country lacks the funds and political will to pursue a genuine environmental policy, and the vast majority of companies are guided by the interests of profit maximization.

To draw more substantiated conclusions as a direction of future research, it is advisable to conduct an analysis based on non-financial reporting of a large sample of companies, which is to provide a more accurate picture.

References

[1] K. Markevich, ed., *Zeleni investytsi u stalomy rozvytku: svitovij dosvid ta ukrayinskyj kontekst (Green Investments in Sustainable Development: World Experience and Ukrainian Context* (Razumkov centre, Kyiv, 2019), ISBN 978-966-2050-13-4, `https://razumkov.org.ua/uploads/article/2019_ZELEN_INVEST.pdf`

[2] OECD, Tech. rep., OECD Publishing (2017), `http://dx.doi.org/10.1787/9789264273528-en`

[3] I.P. Petrenko, Naukovyj visnyk Mizhnarodnogo gumanitarnogo universytetu **2**, 62 (2018)

[4] O.V. Latysheva, Ekonomichnyj visnyk Donbasu **51**, 59–65 (2018)

[5] O.I. Vyshnytska, Visnyk Sumskoho derzhavnogo universytetu, Seriya "Ekonomika" **2**, 51–58 (2013)

[6] C. Zhao, T. Xiang, H. Peng, Theoretical Economics Letters **8**, 3395–3410 (2018)

[7] G. Inderst, C. Kaminker, F. Stewart, Tech. Rep. 24 (2012), `https://www.oecd.org/environment/WP_24_Defining_and_Measuring_Green_Investments.pdf`

[8] N.V. Zakharchenko, Ekonomichnyj visnyk **1**, 166–180 (2019)

[9] A. Chariri, G. Ratna Sari Br Bukit, O. Bethary Eklesia, B.U. Christi, M. Tarigan D, *Does Green Investment Increase Financial Performance? Empirical Evidence from Indonesian Companies* (E3S Web of Conferences, 2018), Vol. 31, p. 09001

[10] X. Su, Emerging Markets Finance and Trade **0**, 1 (2020)

[11] E. Broniewicz, ed., *Environmental Management in Practice* (IntechOpen, 2011), ISBN 978-953-307-

358-3, http://www.intechopen.com/books/environmental-management-in-practice/

[12] Tech. rep., State Statistics Service of Ukraine (2019), http://www.ukrstat.gov.ua/druk/publicat/

[13] S.Y. Yadukha, Visnyk Xmelnyczkogo nacionalnogo universitetu **3**, 277–281 (2013)

[14] U.Z. Kostyuk, Innovacijna ekonomika **1**, 181–186 (2015)

[15] N.V. Kameneva, Aktualni problemy ekonomiky **1**, 348–360 (2016)

[16] S.I. Kim, H. Shin, H. Shin, S. Park, Sustainability **11**, 4445 (2019)

[17] C. Bühringer, U. Moslener, U. Oberndorfer, A. Ziegler, Research Policy **41**, 442–451 (2012)

[18] R. Adamo, D. Federico, A. Notte, Investment Management and Financial Innovations **11**, 134–145 (2014)

[19] M. Milne, D.M. Patten, Accounting Auditing & Accountability Journal **15**, 372–405 (2002)

[20] J.C. Chen, D.M. Patten, R. Roberts, Journal of Business Ethics **82**, 131–144 (2008)

[21] C.H. Cho, D.M. Patten, Accounting, Organizations and Society **32**, 639–647 (2007)

[22] C.H. Cho, D.M. Patten, R.W. Roberts, The Journal of Business Ethics **67**, 139–154 (2006)

[23] Y. An, H. Davey, I.R. Eggleton, Journal of Intellectual Capital **12**, 571–585 (2011)

[24] R.E. Freeman, *Strategic management – A stakeholder approach* (Boston, 1984)

[25] S. Miles, Business and Society **1**, 21 (2017)

[26] J.S. Harrison, R.E. Freeman, M.C.S. de Abreu, Review of Business Management **17**, 858 (2015)

[27] M.E. Porter, C. van der Linde, Harvard Business Rewiew pp. 120–134 (1995), reprint 95507

[28] B.M. Ruf, K. Muralidhar, R.M. Brown, J. Janney, K. JandPaul, Journal of Business Ethics **32**, 143–156 (2001)

[29] M.V. Russo, P.A. Fouts, Academy of Management Journal **40**, 534–559 (1997)

[30] X. Luo, C.B. Bhattacharya, Journal of Marketing **70**, 1 (2006)

[31] S. Ambec, P. Lanoie, Academy of Management Perspectives **22**, 45–62 (2008)

[32] S. Brammer, A. Millington, Strategic Management Journal **29**, 1325–1343 (2008)

[33] H. Fauzi, K. Idris, Issues in Social and Environmental Accounting **3** (2009)

[34] J.E. Mattingly, L. Olsen, Journal of Leadership & Organizational Studies **25**, 481–498 (2018)

[35] S.A. Waddock, S.B. Graves, Strategic Management Journal **18**, 303–319 (1997)

[36] S.A. Waddock, S.B. Graves, Business & Society **36**, 250–279 (1997)

[37] G.A. Mill, Journal of Business Ethics **63**, 131–148 (2006)

[38] I. Vasylchuk, A. Sokolov, Investyciyi: praktyka ta dosvid **1**, 46–52 (2021)

[39] Tech. rep., State Statistics Service of Ukraine (2020), http://www.ukrstat.gov.ua/druk/publicat/kat_u/2020/zb/11/DSG_2019.pdf

[40] T. Artiach, D. Lee, D. Nelson, J. Walker, Accounting & Finance **50**, 31 (2010)

[41] I. Vasylchuk, Biznes-Inform **1**, 153–162 (2016)

[42] Tech. rep., Eurostat (2020), https://ec.europa.eu/eurostat/statisticsexplained/

[43] J. Cohen, Current Directions in Psychological Science **1**, 98 (1992)

[44] F. Moksony, Szociolygiai Szemle pp. 130–138 (1990)

Scientific approach to quantitative measurement and economic processes research in corporate management

Oliver Orlov[1,*], *Yevheniya* Ryasnykh[1,**], *Kathrin* Dumanska[1,***], and *Olena* Savchenko[1,****]

[1]Khmelnytskyi National University, 11 Instytutska Str., Khmelnytskyi, 29000, Ukraine

Abstract. Theoretical and methodological studies of quantitative measurement problems of the outcome economic indicators in corporate management are examined in the article. The author's vision of scientific exploration as one of the corporate management optimization directions is offered and the basis of such direction as profit margin calculation methodology is presented. It is also suggested to solve problems of quantitative measurement of basic economic processes by using allocation of fixed costs in proportion to profit margin. The methodological basis for evaluating cost-effectiveness with the help of profit margin ratio is developed by the authors. It was revealed that fixed costs are the conditions for doing business and the functional purpose of profit margin is the fixed costs coverage. It is proved that in case of multi-assortment production one of the most effective corporate management approaches to cost estimation and pricing is the profit margin concept provided in this article. One of the advantages of proposed profit margin concept is the availability of special tools revealing the resources that allow covering fixed costs and generating profit with a given level of cost-effectiveness. The obtained results of further adoption can be applied as element of corporate management of manufacturing companies and enterprises in conditions of multi-assortment production and changeable economic environment

1 Introduction

The quantitative measurement problems in corporate management have been studied by different economists during decades. Great interest among scientists was aroused by book "Economics: Principles and Policy" [1]. According to W. J. Baumol and A. S. Blinder, the degree of dissatisfaction with the theory is growing among practitioners, which does not equip us with tools for analysis and does not lend itself to empirical testing. A similar position is taken by H. A. Simon, a Nobel prizewinner (1978), who underlined, that: "It would be a gross delusion to believe that scientific research cannot be fundamental if its results are primarily of an applied nature or if it is carried out in response to problems raised by everyday life. The real world is probably the most abundant of all sources of serious problems requiring fundamental scientific research" [2]. The same idea is carried out in the Nobel lecture of M. Allais: "In my approach, I never went from theory to facts, but, on the contrary, tried to deduce an explanatory thread from the facts, without which they seem incomprehensible and useless" [3]. And R. Coase directly demands: "What we need is an increase in empirical research" [4].

In its turn, the problem of quantitative measurement of economic calculations was put forward in his Nobel lecture by F. A. von Hayek and accordingly his lecture was called "Pretence of Knowledge". In his opinion, unlike the natural sciences, in economics all the circumstances that will determine the outcome of the process will never be fully known or measured. Moreover, he is negative about the requirement of quantitative measurements in economics. That's why in his opinion, the problems are not connected with theory, but with attempts to apply theories to a specific situation in the real world. "If we want to protect the reputation of science and prevent groundless claims of knowledge based on superficial similarities with the methods of physical sciences, it is necessary to direct great efforts to expose such claims, some of which have now turned into the group interests of authoritative university departments" [5]. F. A. von Hayek's views on the role of economic and mathematical modeling as an instrument of economic science are of interest. Further, F. A. von Hayek expresses his gratitude to K. Popper for the proposed test, by means of which it is possible to distinguish what deserves a scientific status from what is not. According to F. A. von Hayek, many of the doctrines that were recognized as scientific would not have passed this test. F. A. von Hayek is a supporter of extreme liberalism, an opponent of any state regulation. In his opinion, the results of research on such a complex phenomenon as the market will never be fully known and measured. Indeed, at the macroeconomics level, it is very difficult to achieve quantitative measurements of economic processes, as these studies are based on the use of the method of oversimplification and these results are very difficult to use in real practice.

It is necessary to emphasize that the problem of cost allocation in profit margin formation process, as one of

*e-mail: pro_univer@ukr.net
**e-mail: pro_univer@ukr.net
***e-mail: KathrynD@bigmir.net
****e-mail: savchenkoov@i.ua

the basic problems of quantitative measurement in corporate management, is supposed to be one of the most important for manufacturing companies. Methods of costs allocations are noticed as the most problematic in corporate management of multi-assortment production, but at the same tame, it can be considered as the basis of profit margin formation methodology. There is no generally accepted overall method of the cost allocation with some degree of grounded justification for its effectiveness. Many Ukrainian companies (as well as Czech Republic ones, for example) keep in use the traditional techniques of fixed cost allocation using wages, production volumes etc. as the basis for allocating these costs.

According to scientific view of P. Novák and B. Popesko: "companies are now very often lacking sophisticated tool for proper identification and allocation of costs and hence their proper factored into the price of products" and as they summarize: "most companies, up to 70%, allocate their overhead costs on the basis of historical overhead costs evolution. This situation is most striking especially in small and medium-sized companies. And a fluctuation in capacity utilization often leads to inadequate coverage of fixed (overhead) costs" [6]. One of the objectives of their research was to find out the costing methods used by companies in Czech Republic. In fact, many companies keep in use the traditional techniques known from 1970s, as well as companies in Ukraine.

At the same time, R. Lin and Z. Chen revealed two kinds of fixed input allocation problems: the problem of fixed cost allocation across a set of comparable decision making units (DMUs) and that of fixed resource allocation with sharing a common output target among DMUs. As the result, the authors adopt the data envelopment analysis technique to build models for solving two problems of allocating the above two kinds of fixed inputs among all DMUs and common output target sharing problem that is accompanied with the fixed resource allocation, respectively. This method is quite specific and based on special algorithms [7].

Considering another's scientific view points it is necessary to admit that K. Janovská, S. Vilamová, M. Piecha, J. Kutác, R. Kozel and J. Citbajová proposed model for allocation of overhead activity costs which focuses on the allocation of overhead costs and calculation of assignment rates of overhead activities, which mutually cooperate. The model enables the objective stipulation of assignment rates of mutually cooperating overhead activities using exact economic-mathematical methods – the Leontief structural model [8]. Some studies do not pay attention to cost allocation methods, giving general review of CVP analyses and break-even calculations [9]. The practical application of breakeven analysis in production planning and control in different countries is presented by other authors [10], but the problem of overheated costs as usual is out of consideration. M. Potkanya and L. Krajcirova [11] supposed to calculate break-even point of each product line as product line revenue in total company revenue, divided by average selling price of the product, multiplied to BEP of Company. These calculations will carry over revenue of certain product in the whole structure to weight of the product line in break-even point of company. These authors also suggest distributing fixed costs on the base of variant costs, which we believe in totally wrong.

Such authors as P. T. M. Ingenbleek, R. T. Frambach and T. M. M. Verhallen [12] believe that they provide insight into the relationship between pricing practices and new product performance by difference between new product market performance and price level as two major outcomes of new product pricing. The authors consider that value-informed pricing has an unjustified positive impact on relative price level and market performance. The effects of cost-informed and competition-informed pricing may differ depending upon the objective (market performance or higher prices), product conditions (product advantage and relative product costs), and market condition (competitive intensity). As the result, it is suggested to consider different factors in new product pricing: the explicit pricing objective; product advantages, relative product costs, and competitive intensity. The regression analyses was used to determine interact of variables. The research was based on collected data from managers who recently had been engaged in such a decision-making process.

Authors group of D. De Toni, G. S. Milan, E. B. Saciloto and F. Larentis [13] consider suggesting and testing a theoretical model showing the impacts of pricing policy on corporate profitability. For this purpose 150 companies of Northeast of Rio Grande do Sul State, Brazil were under the consideration. The authors used 15 questions related to the aspects considered or not in the price defining process. The factor analysis was made. The data analysis became the basis of conclusions made by researches. As we consider, this can not be understood as methodology of pricing policy.

In resent article A. Farm offers a theory of pricing in consumer markets that views cost-plus pricing and value-based pricing in conjunction with price competition and price leadership, including, in particular, competitive price leadership. Various theories of pricing were given with cost-plus pricing, value-based pricing etc. It is necessary to notice, that "cost-plus pricing also implies, – what authors agree with, – that indirect costs and normal profits are allocated to a firm's products in proportion to their direct costs" and it also stated in the article, that: "the markup is obtained by summing two ratios, namely the ratio of indirect costs to direct costs for estimated sales, and the ratio of normal profits to direct costs for estimated sales" [14]. The next calculations we believe were mistaken because of using direct costs as a base for allocation of indirect costs. On the other hand it is underlined that there are many variants of cost-plus pricing depending on the systems of cost accounting, different definitions of normal profits, and different markup techniques.

Taking into account the view point of B. Cohen and M. Neubert, who emphasize that product-oriented pricing, company-oriented, competition-oriented pricing should be considered in multi-assortment production. The authors refer to R. Luostarinen and M. Gabrielsson [15] who studied 89 companies in Finland, "these companies often bypassed cost-informed pricing (setting a product's price floor) by choosing below cost pricing for their first international customer, and thereafter applying value-added pricing based on the benefits brought to its customers" [16]. That means the company-oriented pricing (cost informed pricing to be more precise) with its pros and contras is used very often in lot of industries and our developments can be useful there as well.

"In this case, it should be underlined three main systems necessary to solve company's or enterprise's sustainability tasks connected with accumulation of potential of success in managerial decision making. Such systems include: strategic planning, implementation and control; these systems are respectively responsible for the development, implementation and revision of current strategy" [17] and manufacturing activity of legal entity as well. Thus, complex application of these systems in corporate managerial decisions is one of the effective company's management directions in machine-building production.

Practically, view point of these authors on the economic processes research in corporate management is capable of future gradual implementation of theory into real life. In this article we try, through empirical research, to consider specific problems that are encountered in real practice and to propose methods for solving these problems. Quantitative measurements are especially difficult in a multi-assortment manufacturing environment. Obviously, therefore, scientists try to carry out their calculations in the single-item production conditions. According to the research, for a market economy, especially at the present stage, characteristic feature is the rapidity of structural changes that relate to all areas of the enterprise, and this is especially true for the problems of calculating the prime cost, profit, cost-effectiveness and prices. These indicators are mutually influenced by changes in the scale of production, the structure of the assortment, prices, the value of fixed and variable costs. In this situation, it is very important to clearly define the components of this relationship and which of these elements play a decisive role in corporate decision making process.

The importance of break-even calculations distribution of fixed costs proportionally to profit margin and the necessity of pricing on the break-even ratio basis in the frame of corporative management is observed and grounded in comparative calculation examples of the article. Taking into consideration that large enterprises with comparative management systems are interested in high volumes of multi-assortment production and therefore in large sales revenue we have chosen two profitably known enterprises in our region for break-even calculations, because such indicators are more evident for companies. In conditions of multi-assortment production it is the corporations, large enterprises that should take into account target profits, pricing, fixed costs and cost-effectiveness since these indicators cannot be calculated in isolation from output structure and scale production changes.

Before starting the specific study it is necessary to define clearly the list of problems that interfere with making effective management decisions and, after forming the correct problem view point, the necessary decision-making tools can be developed.

2 Methodology of researching

The goal of the research is the elaboration of methodological decision making apparatus for corporate management on the basis of economic-mathematical modeling that can accumulate information on assortment structure, pricing, fixed and variable costs and cost-effectiveness.

At most enterprises in Ukraine, fixed costs are distributed in proportion to direct wages and our break-even calculations at a number of machine-building enterprises showed that using the distribution of wages as a base, in order to achieve break-even, it is necessary to increase the production volume of unprofitable products and reduce profitable ones, which clearly contradict common sense. That's why we can confirm that break-even calculations are not carried out at Ukrainian enterprises at all.

Thus, the profit margin calculation methodology bases on the concept of fixed costs distribution in proportion to the profit margin is proposed to Ukrainian corporate management and uses in calculations of this research.

3 Results

3.1 Proofs of the concept of fixed costs distribution in proportion to the profit margin

Long-term research on the examples of machine-building enterprises led us to the conclusion that in the multi-assortment production conditions on the way of real calculations of profit, prime cost, cost-effectiveness and the formation and adjustment of prices, there is a controversial problem of fixed costs distribution between types of products. In general, the choice of the base distribution is subjective and, therefore, choosing the base, we set the prime cost value in advance. Depending on the distribution base, we can get several prime cost values. Does this mean that there can be several managerial decisions?

In Economist magazine we published an article "Do the ends always justify the means? How to distribute conditionally fixed costs" [18], where for the first time it was proposed to distribute fixed costs in proportion to the profit margin. What considerations did we proceed when justifying this method from? Margin theory underlines that at the break-even point the profit margin is equal to fixed costs. Therefore, having determined the break-even points and the profit marginal amount for these volumes, we can actually determine the amount of fixed costs for each group of products. Since a priori it cannot be that for the calculation of break-even (as the balance of fixed costs to relative profit margin) there can be one value of fixed costs, and another for calculating the cost.

As a matter of fact, a number of proofs that for the distribution of fixed costs only the profit marginal should be used as the distribution base is given in addition to this provision, and some of these proofs are presented below.

The first proof. Foreign economists in their works suggest break-even calculations using the average weighted profit marginal, which are based on the relative weight of each product in physical or value terms [19] and similar calculations are presented in many works of Ukrainian and Russian scientists. They are extremely complicated and it is almost impossible to produce them if the number of products exceeds ten. But the most important thing is that the results of these calculations completely coincide with the break-even calculations using the break-even ratio (formula 1):

$$BER = FC/PM, \quad (1)$$

where BER – break-even ratio,

FC – fixed costs,

PM – enterprise profit margin.

At the same time, there is no need in the calculations to take into account the relative weight of each product group, as the requirement for the invariability of the assortment structure is observed in calculations, then to achieve break-even it is necessary to increase the production of each group of products by 10%.

The second proof. A. Upchurch, and after him many other authors, in their calculations give break-even indicators for types of products, but do not bring these calculations to indicators of profit and cost-effectiveness, as they practically do not calculate the value of fixed costs by product types. These calculations are not difficult at all, since they have data on the values of direct costs, prices and break-even points by product types.

Taking into account calculations, it is known that the break-even point for product types is determined by the formula 2:

$$BEP_i = FC_i/(P_i - VC_i), \quad (2)$$

where BEP_i – break-even point for product types,

FC_i, VCi, P_i – fixed costs, variable costs and price for product types.

Since in these calculations the values of BER, P and VC are known and calculated, it is easy to calculate the value of fixed costs (formula 3):

$$FC_i = BEP_i \cdot (P_i - VC_i) \quad (3)$$

Therefore, it isn't also difficult to calculate the prime cost and profit by product types as information on price and variable costs is available in the original version.

In is to understand, why do foreign and domestic scientists do not use the break-even calculations to the indicators of prime cost, profit and cost-effectiveness by product types? Perhaps, because in this case it would be necessary to recognize that the most optimal method of allocating fixed costs is a marginal approach.

The third proof. The formula of BER indicator can be derived from the well-known formula of "break-even turnover", widely used in economic literature:

$$BET = FC/PMR, \quad (4)$$

where BET – break-even turnover;

FC – fixed costs;

PMR – profit margin ratio for the whole enterprise, which is calculated as the relation of profit margin (PM) to sales revenue (SR) that is (formula 5):

$$PMR = PM/SR \quad (5)$$

Therefore, BET indicator formula can be transformed as follows (formula 6):

$$BET = \frac{FC}{PM} \times SR \quad (6)$$

According to proposal the break-even ratio formula will look like formula 7:

$$BER = FC/PM \quad (7)$$

Thus, the break-even calculations method can be reflected as formula 8:

$$BET = BER \times SR \quad (8)$$

The results of the break-even calculations according to formulas 4 and 8 are identical, but if formula 4 allows you to calculate the break-even turnover only for the whole enterprise, then using profit margin you can calculate the break-even turnover in physical and value terms both for the whole enterprise and for individual products types, and also to distribute fixed costs by product types. Thus, on the one hand, with the help of the margin of safety (MOS), which is directly related to BER, we can calculate the enterprise profit (loss) as a whole by formula 9:

$$MOS = 1 - BER \quad (9)$$

And on the other hand, we can calculate profit (loss) for individual product types by formula 10:

$$P_i = MOS \times PM_i, \quad (10)$$

where PM_i – is the profit margin for the i-product type.

And **the fourth proof**. Only with the distribution of fixed costs in proportion to the profit margin using BER the observance of the CVP system limitation is ensured, i.e. the invariance of the assortment structure when calculating the break-even by product. All other distribution bases distort the assortment structure.

3.2 Calculation system of profit margin and price indicators according to the profit margin concept

The concept of fixed costs distribution in proportion to the profit margin has its own assumptions. It is assumed that all fixed costs (including production costs) are irrelevant for making management decisions. Since fixed costs can be partially semi-variable or semi-fixed, this approach can affect the accuracy of product prime cost calculations. It should be noted that all methods of allocating fixed costs always allow some element of inaccuracy. "The best that can be counted on," notes A. Upchurch, "is to get an acceptable result that minimizes such inaccuracy" [20].

But the essence of the problem lies in the fact that the accuracy of calculating the prime cost, though it may sound paradoxically, does not really matter in managing costs, profits, cost-effectiveness and prices by product types. Actually, the purpose of this article is to disclose the relationship between all elements of the calculation in this system and identify those of them that play the main role. But to find out these relationships, and at the same time to check the realistic assumptions in the distribution of fixed costs, it is possible only by conducting empirical studies on examples close to reality, which will ensure the reliability of the recommendations proposed on their basis. The results of calculations are presented in table 1.

According to the table 1, for this example $BER = 1104/1380 = 0.8$; $MOS = 0.2$; $P = 1380 - 1104 = 276$.

With the help of MOS we can easily calculate the profit on products without preliminary calculation of the prime cost. So, for product "A" the profit is: $0.2 \times 8 = 1.6$ and for the volume: $0.2 \times 480 = 96$. But for our research it is very important to find out how the cost will behave under certain changes. In table 1 it is necessary to note that for products "A" and "C" the indicators of PMR and cost-effectiveness are 0.4 and 8.69, respectively.

Due to market conditions changes, the following modifications have occurred in the product range:

1. Product "D" was discontinued.
2. Variable costs for product "C" have decreased by 1.2 units. Thus, there are two possible options. The first is to keep the same price, which will lead to an increase of cost-effectiveness, but taking into account the current market situation, it will lead to sales decrease. The second is to reduce the price, keep the same costeffectiveness with $PMR = 0.4$ and thereby sales will increase by 30 units and the enterprise has chosen the second option.
3. The enterprise has accepted an order for the product "E" with variable costs of 14, production of 20 units and decided to include cost-effectiveness into price at the level of product "B" with $PMR = 0.5$.

According to these calculations we have such results:

For product C. The price is calculated using the formula:

$$PR_{ini} = \frac{VC_i}{1 - PMR}. \qquad (11)$$

Thus, the results of calculations according to this formula, in which fixed cost are not taken into consideration and hence the prime cost too, are: $PR_C = 4.8/(1\text{˘}0.4) = 8$; $PM_{iC} = 8\text{˘}4.8 = 3.2$; $PM_C = 3.2 \times 50 = 160$.

So, for the newly ordered product "E" we have the following results: $PR_E = 14/(1\text{˘}0.5) = 28$; $PM_{iE} = 28\text{˘}14 = 14$; $PM_E = 14{\times}20 = 280$. As a result, the profit margin for the enterprise will amount: $480 + 720 + 160 + 280 = 1640$; fixed costs increased by 44 units and amounted about 1148. And thus we can observe new values of BER and MOS: $BER = 1148/1640 = 0.7$; $MOS = 0.3$; $P = 1640\text{˘}1148 = 492$.

According to calculations we can analyze the results. For products "A" and "B" practically no changes occurred either in prices, sales volumes, or in the value of variable costs, but there was a redistribution of fixed costs. The decrease in fixed costs for these products by 120 units was taken over by products "C" and "E", which led to a decrease in the prime cost and increase in profit for products "A" and "B" by 120 units and a significant increase in their cost-effectiveness. For product "C" the price was reduced by 20%, which ensured the volume of sales up to 50 units, the previous value of $PMR = 0.4$ remained and cost-effectiveness increased from 8,69% to 13,6%. New product "E", in price of which we have included $PMR = 0.5$, i.e. at the product "B" level, is sold with cost-effectiveness of 17,6%, as in the case with product "B". Thus the question can be put, what conclusions can be drawn from these examples?

First of all, in the conditions of multi-assortment production, the cost of goods by type of product is not a subject, but an object of calculations, as its value cannot be determined in isolation from changes in the structure of product assortment. Therefore, in these conditions, it is impossible to set the task of determining the "target cost" and "target profit" and, naturally, the cost price by type of product cannot be the basis for the prices formation and adjustment.

Secondly, these examples clearly show that the decisive role in the calculations is played by the "margin profit ratio" by product type. With the help of PMR certain level of comparative cost-effectiveness, can be included in calculations, price changes can be made and what is of great importance, there can be formed prices for new orders. The proposed pricing formula fully meets the requirements of anti-cost (value) pricing, the supporters of which completely reject participation of full cost in pricing. "If, when setting prices, one proceeds from full costs," R. J. Dolan note, "as it occurs in practice, the price is determined by fixed costs, and this is logically incorrect" [21]. Fixed costs do not take part in the proposed pricing formula (11), but with the help of PMR, such a level of marginal profit is included in the calculation, which allows you to cover fixed costs and generate a profit with a given level of cost-effectiveness.

Thirdly, it should be noted that, in fact, using PMR, we form a comparative cost-effectiveness by product types, regardless of changes in the assortment structure. Scientific approach towards economic situation demands thorough analysis of all indicators and components but in practice employees of industrial enterprises are not aware of the existence of such indicator as "profit margin ratio" (PMR), and in the economic literature and textbooks its role is not given much attention.

Quantitative measurements suggested by our author's group as a margin calculation methodology are developed specially for the most profitable machine-building enterprises of Khmelnytskyi region with the aim of their corporative management improvement.

Analyzing proposed materials of Public Joint-Stock Company "Temp" it was realized that it has purchased a laser machine at about UAH 5 million and usage of this equipment has led to a reduction of materials cost, wages and technological energy for products manufacture. Fixed costs distribution proportionally to wages led to an increase of distribution ratio, because fixed costs have increased due to depreciation, and enterprise wages decreased as a whole. In this case, fixed costs for the products processed on this machine decreased, and accordingly the entire "load" of fixed costs increase due to depreciation

Table 1. Calculation of prime cost, profit and cost-effectiveness in the distribution of fixed costs proportionally to the profit margin

Products	Quantity	Price and revenues		Variable costs		Profit margin		Fixed costs	Prime cost	Profit		Cost-effectiveness
		PR	R	VC_i	VC	PM_i	PM	FC	PC	P	PMR	$CE, \%$
A	60	20	1200	12	720	8	480	384	1104	96	0.4	8.69
B	40	36	1440	18	720	18	720	576	1296	144	0.5	11.1
C	30	10	300	6	180	4	120	96	276	24	0.4	8.69
D	20	12	240	9	180	3	60	48	228	12	0.25	5.26
$\sum P$	150		3180		1800		1380	1104	2904	276	0.436	9.5

Table 2. Calculation results with account changes

Products	Quantity	Price and revenues		Variable costs		Profit margin		Fixed costs	Prime cost	Profit		Cost-effectiveness
		PR	R	VC_i	VC	PM_i	PM	FC	PC	P	PMR	$CE, \%$
A	60	20	1200	12	720	8	480	336	1056	144	0.4	13.6
B	40	36	1440	18	720	18	720	504	1224	216	0.5	17.6
C	50	8	400	4.8	240	3.2	160	112	352	48	0.4	13.6
D	20	28	560	14	280	14	280	196	476	84	0.5	17.6
$\sum P$	170		3660		1960		1640	1148	3108	492	0.455	15.8

has moved to the of products prime cost that were not processed on this equipment. With the distribution of fixed costs proportionally to the profit marginal, the situation is exactly opposite. Since the increase of profit margin (and profit) was greater than the increase of fixed costs, then the profit margin ratio and cost-effectiveness increased for the products processed on this machine, the profit margin despite the fact that they have taken on the "load" of fixed costs. However, for other products such like situations have led to a positive profit increase.

Thorough research of Public Joint-Stock Company Krasilovagromash economic situation, we carried out similar calculations in connection with the introduction of new equipment that was used for some products and obtained similar results. With the distribution of fixed costs proportionally to wages some of low-profit products, in production of which no new equipment was used, turned into unprofitable ones.

These two examples (and there could be given much more of them) prove certain threats for enterprise financial condition in case wrong decisions would be taken and, at the same time, such examples can be of significant importance in making business decisions. And vivid proof of such like conclusions can be the advice of E. A. Helfert, who wrote: "... common sense suggests that most of the analytical efforts should be directed to those areas where the probability of insufficient analysis is the greatest ..." [22].

Generally speaking, it is very difficult to break the stereotypes that have evolved over the centuries, and that the proposal, when setting prices and profits, does not take into account the production prime cost as the main component is rather provocative, but at the same time, we are sure that in the sphere of manufacture of production technically purposed products, this technique has the right to life and that in the near future it can be widely used.

4 Conclusions

The corporate management mechanism provides such conditions of its functioning that ensure stability of the companies and enterprise development. Corporate management contains a wide complex of tasks, especially in conditions of multi-assortment production but absence of analytic instruments doesn't support the precise evaluation of decision making results of well grounded managerial solution. That's why the variability directions of corporate management should be considered at the new and, at the same time, effective basis as profit margin concept with fixed costs distribution in proportion to the profit margin methodology proposed in this article for Ukrainian enterprises.

Summarizing research results the following conclusions can be given:

1. The proposed base for fixed costs distribution in proportion to the profit margin transforms the distribution system into a system for managing costs, prices, profits and cost-effectiveness by product types.

2. Fixed costs and marginal profit are a kind of "black boxes" that collect and store all information on volume changes and assortment structure, prices and prime costs, which can be used in any moment to carry out the necessary calculations with the help of BER and MOS.

3. In the conditions of multi-assortment production, the prime cost by product type cannot be formation and adjustment basis of prices, since it strongly depends on changes of production volume and assortment structure of the whole enterprise.

4. Products with the same PMR will be of the same cost-effectiveness regardless their production volume.

5. The proposed pricing formula on anti-cost basis can be used for "value" pricing. The degree of "value" should be correlated with the value of *PMR* in price calculation.

6. Prices changes are advisable by product types in case of variable costs increase or decrease using the profit margin ratio in order to ensure given cost-effectiveness.

Thus, it is necessary to admit that the most important tool in proposed profit margin concept is the "margin ratio" by product type, because it provides price formation on the anti-cost basis and with its help the required level of cost-effectiveness is set. The role of this indicator in the economic literature is underestimated and in practice it isn't simply used. The proposed calculation formulas have a direct effect, do not require additional time and resources are easy in use and have internal logic. Consequently, different companies can profitably use suchlike methodology for expenditure decrease and therefore in profit and production volume increase. Suggested methodology is advisable for any corporate management system as it can assist in improvement of such indicators as profit margin and cost-effectiveness.

References

[1] W. Baumol, A. Blinder, *Economics: Principles and Policy* (New York, 1979)

[2] H.A. Simon, The Quarterly Journal of Economics **69**, 99 (1955)

[3] M.H. Birnbaum, Journal of Mathematical Psychology **48**, 87 (2004)

[4] R. Coase, Journal of Economics & Management Strategy **15**, 255 (2006)

[5] F.v. Hayek, American Economic Review **79**, 3 (1989)

[6] P. Novák, B. Popesko, Economics & Sociology **7** (2014)

[7] R. Lin, Z. Chen, Applied Mathematical Modelling **40**, 5377 (2016)

[8] K. Janovská, S. Vilamová, M. Piecha, J. Kutác, R. Kozel, J. Citbajová, E3S Web Conf. **134**, 03015 (2019)

[9] A. Birca, Problemy sovremennoi nauki i obrazovaniia **5**, 55 (2017)

[10] A. Ogbo, C. Orga, T. Adibe, Industrial Engineering Letters **2**, 47 (2012)

[11] M. Potkany, L. Krajcirova, Procedia Economics and Finance **26**, 194 (2015)

[12] P.T.M. Ingenbleek, R.T. Frambach, T.M.M. Verhallen, Journal of Product Innovation Management **30**, 560 (2013)

[13] D. De Toni, G.S. Milan, E.B. Saciloto, F. Larentis, Revista de Administração (São Paulo) **52**, 120 (2017)

[14] A. Farm, Journal of Post Keynesian Economics **43**, 61 (2020), `https://doi.org/10.1080/01603477.2019.1616562`

[15] R. Luostarinen, M. Gabrielsson, Thunderbird International Business Review **48**, 773 (2006)

[16] B. Cohen, M. Neubert, ISM Journal of International Business **II**, 27 (2017)

[17] O. Orlov, K. Dumanska, N. Ponomaryova, D. Kobets, E3S Web of Conferences **166**, 12002 (2020)

[18] O. Orlov, Y. Ryasnykh, Economist pp. 45–49 (2001)

[19] A. Atkinson, R. Kaplan, E. Matsumura, S. Young, *Management Accounting: Information for Decision-Making and Strategy Execution* (Pearson, London, 2012), ISBN 9780137024971

[20] A. Upchurch, *Management Accounting: Principles and Practice* (Ft Pr, 1998)

[21] R. Dolan, *Evernote (B): Announcing a Pricing Plan Change* (2019), harvard Business School Supplement 519-036

[22] E.A. Helfert, *Financial Analysis Tools and Techniques: A Guide for Managers* (McGraw-Hill Education, New York, 2001)

The impact of inflation on the unemployment rate in Egypt: a VAR approach

Emad Attia Mohamed Omran[1,*] and *Yuriy* Bilan[1,**]

[1]Faculty of Management and Economics, Tomas Bata University in Zlin, nam. T.G. Masaryka 5555, 76-001 Zlin, Czech Republic

Abstract. Unemployment and inflation are among the most critical phenomena facing both developed and developing countries due to their harmful social, economic, and political effects. The Egyptian monetary policy's main objective is to maintain a low inflation rate in the medium run to keep the confidence and a high rate of investment and economic growth. At the same time, economists argue that targeting a low-rate of inflation may increase unemployment. Although the classical Philips curve indicates a trade-off between inflation and unemployment, several empirical studies have argued that the relationship between inflation and unemployment depends on the shocks' source and lagged responses. The main objective of this paper is to examine the relationship between inflation and Egypt's unemployment rate. We used time-series data from 1980 to 2019, where a vector autoregressive (VAR) model and the Impulse response function tool (IRF) were employed. The results show that inflation has a positive relationship with GDP while negatively affecting the unemployment rate.

1 Introduction

One of the most critical phenomena facing both developed and developing countries is unemployment due to its harmful social, economic, and political effects. The catastrophic consequences of the COVID-19 crisis have taken the global economy to a recession which caused a high unemployment rate [1]. On the other hand, inflation volatility can cause harmful effects on economic growth [2]. Therefore, keeping a low inflation rate is one of the monetary policy's primary goals in most countries [3]. Nevertheless, according to PhilipsâĂŹs curve [4], there is a trade-off between inflation and unemployment. The main implication of this relationship is that if a country wants to keep a low inflation rate, it must accept a high unemployment rate [5]. This paper's main objective is to examine the relationship between inflation and the unemployment rate in Egypt. The research hypotheses can be formalized as follows:

H1: There is a negative relationship between inflation and the unemployment rate in Egypt.

H2: There is a negative relationship between inflation and gross domestic product (GDP) in Egypt.

2 Literature review

Philips [4] reported a negative relationship between wage inflation and the unemployment rate in the UK. Similarly, Samuelson and Solow [6] confirmed this relationship using data from the US. Nevertheless, some empirical studies challenged these hypotheses [7–12]. According to Berentsen et al. [13], the is a positive association between inflation in the long run. Ho and Iyke [14] found that the relationship is harmful only when the unemployment is less than 5%. These investigations are in line with the general idea of links between the unemployment rate and economic growth, proved in empirical research in different countries [15–18].

The empirical studies showed different results regarding the relationship between inflation and unemployment in both the short and long- run [19]. According to Hussain and Saaed [20], this relationship varies from one economy to another. Ahiadorme [21] found that the impact of inflation on output and unemployment rate is consistent with the Philips curve predictions and Okun's law in the short run. It should be considered in territorial development planning due to the essential impact of unemployment on their development [22, 23], and appropriate influence on the business environment [24, 25], level of life of the population [26] and widening the gap in employersâĂŹ value proposition [27]. Nevertheless, the new Keynesian Philips curve exists, and Okun's relationship is positive in the long run. Sasongko and Huruta [28] showed that there is a one-way causality between inflation and unemployment. Moreover, unemployment causes inflation, but not vice-versa. These patterns may be hidden in terms of the severe informal economy share; however, their impact on inflation and level of life remain in general [29].

Some studies reported an inverse relationship between inflation and the unemployment rate [30–33]. Other studies showed a positive relationship between inflation and the unemployment rate in the long run [13, 19, 34, 35]. Korkmaz and Abdullazade [33] found a trade-off between inflation and the unemployment rate. Mukherjee [32] found an inverse relationship between inflation and the unemployment rate in India. On the contrary, Touny [19] showed a positive relationship between inflation and un-

*e-mail: omran@utb.cz
**e-mail: bilan@utb.cz

employment in Egypt, while Vermeulen [36] found no evidence of the trade-off between inflation and the unemployment rate in the short run.

3 Methodology

To investigate the relationship between inflation and employment rate, we applied a three-variables vector autoregressive model (VAR) with the impulse response function (IRF) and time-series data for the period 1980-2019. It was collected from the central bank of Egypt and the world bank table 1 shows the variables of our VAR model.

Table 1. Definition of variables

Variable	Definition
GDP	Log (Real gross domestic product)
UNEMP	Log (Unemployment rate %)
INF	Log (Inflation rate%)

According to Sims [37], we can present the VAR model as follows:

$$Y_t = \alpha_0 + \alpha_1\, y_(t-1) + \ldots + \alpha_m\, y_(t-m) + u_t \quad (1)$$

where:

α_i denotes the coefficient matrices;

y_t denotes a set the endogenous variables;

u_i denotes the stochastic error term for the VAR model.

Therefore, the three variables VAR models can be specified as follows:

$$GDP_t = \sigma + \sum_{i=1}^{k} \beta_i GDP_{t-i} + \sum_{j=1}^{k} \emptyset_j UNEMP_{t-j} + \sum_{m=1}^{k} \delta_m INF_{t-m} + u_{1t} \quad (2)$$

$$UNEMP_t = a + \sum_{i=1}^{k} \beta_i GDP_{t-i} + \sum_{j=1}^{k} \emptyset_j UNEMP_{t-j} + \sum_{m=1}^{k} \delta_m INF_{t-m} + u_{2t} \quad (3)$$

$$INF_t = d + \sum_{i=1}^{k} \beta_i GDP_{t-i} + \sum_{j=1}^{k} \emptyset_j UNEMP_{t-j} + \sum_{m=1}^{k} \delta_m INF_{t-m} + u_{3t} \quad (4)$$

To get the impulse response functions, we set the variables according to Cholesky ordering: we start with INF, then GDP, and finally the $UNEMP$.

Table 2 presents the descriptive statistics of our three variables. To measure the Kurtosis and skewness of each series from the normal distribution, we used the Jarque-Bera test. The results show that at 1%, 5%, and 10% levels of significance, all the three variables follow the normal distribution.

Table 2. Descriptive statistics

	GDP	INF	UNEMP
Mean	11.11337	1.000868	0.955840
Median	11.13466	1.051948	0.958406
Maximum	11.45659	1.469919	1.119051
Minimum	10.67888	0.355979	0.696356
Std. Dev.	0.228200	0.280783	0.119838
Skewness	-0.192637	-0.744492	-0.738313
Kurtosis	1.880566	2.756127	2.755662
Jarque-Bera	2.277549	3.699394	3.640206
Probability	0.320211	0.157285	0.162009
Sum	433.4214	39.03385	37.27776
Sum Sq. Dev.	1.978867	2.995876	0.545723
Observations	39	39	39

4 Results

Before testing our vector autoregressive model (VAR), we want o to make sure that there is no unit root in our variables. To do so, we employ the Augmented Dickey-Fuller test (ADF) at 1%, 5%, and 10% levels of significance. The null hypothesis is that there is a unit root in the variables, meaning that they are nonstationary. Contrarily, the alternative hypothesis is that there is no unit root in the variables, meaning they are stationary. Table 3 shows the augmented dickey fuller test with intercept. AS we can see at levels, all the variables were nonstationary. When we took the first difference, UNEMP and INF became stationary. Furthermore, when we took the second difference, all the variables became stationary.

Table 3. Augmented Dickey-Fuller test (Intercept)

Variable	P-value	Unit root	Stationary
	Levels		
GDP	0.0321	Yes	No
UNEMP	0.3095	Yes	No
INF	0.1806	Yes	No
	First difference		
GDP	0.1002	Yes	No
UNEMP	0.0019	No	Yes
INF	0.0000	No	Yes
	Second difference		
GDP	0.0000	No	Yes
UNEMP	0.0000	No	Yes
INF	0.0000	No	Yes

Table 4 shows an augmented dickey fuller test with the trend and intercept. As we can notice, all the variables were nonstationary at levels. Nevertheless, they became stationary ant the first and second difference. Table 5 shows the augmented dickey fuller test with no trend and intercept, as we can see. All the variables were nonstationary at levels, but they became stationary except GDP when we took the first difference. Nevertheless, taking the second difference converted all the variables to stationary.

Table 6 shows the lag length criterion tests. As we can see, LR, FPE, AIC, SC, and HQ test suggest using one lag in our vector autoregressive model (VAR).

Table 4. Augmented Dickey-Fuller test with trend and intercept

Variable	P-value	Unit root	Stationary
	Levels		
GDP	0.9092	Yes	No
UNEMP	0.2265	Yes	No
INF	0.4999	Yes	No
	First difference		
GGDP	0.0635	No	Yes
UNEMP	0.0131	No	Yes
INF	0.0000	No	Yes
	Second difference		
GDP	0.0000	No	Yes
UNEMP	0.0000	No	Yes
INF	0.0000	No	Yes

Table 5. Augmented Dickey-Fuller test (no trend and intercept)

Variable	P-value	Unit root	Stationary
	Levels		
GDP	1.0000	Yes	No
UNEMP	0.8150	Yes	No
INF	0.3931	Yes	No
	First difference		
GDP	0.0834	No	Yes
UNEMP	0.0000	No	Yes
INF	0.0000	No	Yes
	Second difference		
GDP	0.0000	No	Yes
UNEMP	0.0000	No	Yes
INF	0.0000	No	Yes

Table 7 presents the diagnostics tests for our VAR model. As we can see, there is no serial correlation and Heteroskedasticity in the residuals. Moreover, the model satisfies the normality and stability conditions.

Figure 1 shows the response of each variable to a one standard deviation shock to the inflation rate. We calculated the impulse response function for ten years ahead. The first panel shows the response of GDP to a one standard deviation shock to the inflation rate. As we can see, inflation has a positive impact on GDP. This result is consistent with Semuel and Nurina [38], but it is inconsistent with Faria and Carneiro [39]. When we look at the second panel, we can see that unemployment responds positively to a one standard deviation shock to inflation; there is a trade-off between inflation and the unemployment rate. This result is consistent with Muhammad [40]. Finally, we look at the third panel; we see that inflation responds negatively to a one standard deviation shock.

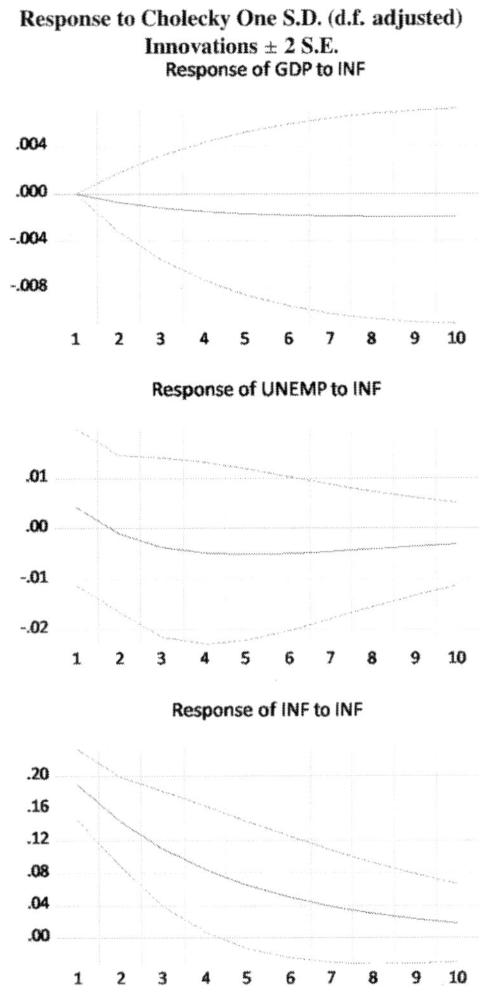

Figure 1. The impulse response functions

5 Conclusion

The Egyptian monetary policy's main objective is to maintain a low inflation rate in the medium run to keep the confidence and a high rate of investment and economic growth. At the same time, economists argue that targeting a low-rate of inflation may pursue or stifle employment. This article contributes to this debate by investigating the relationship between inflation and Egypt's unemployment rate. The authors employed annual time series data from 1980 to 2019, where a vector autoregressive model (VAR) and the Impulse response function tool (IRF) were used. The results show a positive relationship between inflation and GDP, while there is a negative relationship between inflation and the unemployment rate. These results are consistent with the results of Philips [4]. The main recommendations of these results to Egypt's policymakers are that they need to implement their monetary policy considering the unemployment pressures in short-run.

This article was supported by the Internal Grant Agency of Tomas Bata University in Zlin, Project No. IGA/FaME/2021/005, "Significant factors in the sustainability of economic growth with a focus on the SME segment".

Table 6. Lag length criterion test

Lag	LogL	LR	FPE	AIC	SC	HQ
0	44.40942	NA	$1.63 \cdot 10^{-5}$	-2.509662	-2.373615	-2.463886
1	169.8185	220.4159*	$1.41 \cdot 10^{-8*}$	-9.564755*	-9.020570*	-9.381653*
2	176.9738	11.27512	$1.61 \cdot 10^{-8}$	-9.452959	-8.500636	-9.132531
3	185.0917	11.31585	$1.76 \cdot 10^{-8}$	-9.399498	-8.039036	-8.941744

Table 7. Diagnostic tests for the VAR model

Diagnostic test	Test statistic	P-value
Residual Serial Correlation LM Test	2.328545	0.1137
Residual Heteroskedasticity Test	0.167819	0.8462
Stability Test	Satisfied	No roots outside the unit circle.
Normality Test (Jargue-Bera)	0.24032	0.886779

References

[1] A. Suomi, T.P. Schofield, P. Butterworth, Frontiers in Psychology **11**, 2745 (2020)

[2] P. Rother, *Fiscal policy and inflation volatility* (2004), `https://www.ecb.europa.eu/pub/pdf/scpwps/ecbwp317.pdf`

[3] F.S. Mishkin, M.A. Savastano, Journal of Development Economics **66**, 415 (2001)

[4] A.W. Phillips, Economica **25**, 283 (1958)

[5] F. Furuoka, Economics Bulletin **5**, 1 (2007)

[6] P.A. Samuelson, R.M. Solow, The American Economic Review **50**, 177 (1960)

[7] T.I. Palley, The Manchester School **71**, 35 (2003)

[8] A.Ö. Önder, Applied Economics **41**, 2597 (2009)

[9] A. da Silva Correa, A. Minella, Revista Brasileira de Economia **64**, 231 (2010)

[10] V. Nevena, Industrija **48**, 73 (2020)

[11] D. Byrne, Z. Zekaite, Economics Letters **186**, 108521 (2020)

[12] K.J. Forbes, J. Gagnon, C.G. Collins, *Low inflation bends the phillips curve around the world* (2020), `https://tinyurl.com/dx2zrhvv`

[13] A. Berentsen, G. Menzio, R. Wright, American Economic Review **101**, 371 (2011)

[14] S.Y. Ho, B.N. Iyke, The Journal of Developing Areas **53** (2019)

[15] C. Blázquez Fernández, D. Cantarero Prieto, M. Pascual Sáez et al., Economics and Sociology **11**, 263 (2018)

[16] H.A. Bobenič, M. Bruothová, Z. Kubíková, R. Ručinskỳ, Journal of International Studies **11**, 222 (2018)

[17] O.B. Soylu, I. Çakmak, F. Okur, Journal of International Studies **11**, 93 (2018)

[18] A. Widarjono, Economics & Sociology **13**, 203 (2020)

[19] M.A. Touny, International Journal of Economics and Finance **5**, 115 (2013)

[20] M.A. Hussain, A.A.J. Saaed, Archives of Business Research **6** (2018)

[21] J.W. Ahiadorme, Macroeconomics and Finance in Emerging Market Economies pp. 1–20 (2021), `https://doi.org/10.1080/17520843.2021.1901347`

[22] L.Đ. Akimova, N.L. Khomiuk, I.M. Bezena, I.L. Lytvynchuk, O. Petroye, International Journal of Management **11**, 567 (2020)

[23] R. Kostiukevych, H. Mishchuk, A. Zhidebekkyzy, J. Nakonieczny, O. Akimov, Economics & Sociology **13**, 46 (2020)

[24] Y. Bilan, H. Mishchuk, I. Roshchyk, O. Joshi, Business: Theory and Practice **21**, 780 (2020)

[25] K. Karamanis, C. Beneki, M. Ioakimidis, Journal of International Studies **11**, 93 (2018)

[26] H. Mishchuk, N. Samoliuk, Y. Bilan, Administratie si Management Public pp. 63–76 (2019)

[27] Y. Bilan, H. Mishchuk, N. Samoliuk, V. Mishchuk, Entrepreneurial Business and Economics Review **8**, 189 (2020)

[28] G. Sasongko, A.D. Huruta, Business: Theory and Practice **20**, 1 (2019)

[29] H. Mishchuk, S. Bilan, H. Yurchyk, L. Akimova, M. Navickas, Economics & Sociology **13**, 289 (2020)

[30] A. Wajid, R. Kalim, *The impact of inflation and economic growth on unemployment*, in *Proceedings of 3rd International Conference on Business Management* (2013), `http://escholar.umt.edu.pk:8080/jspui/handle/123456789/928`

[31] R. Singh, Impact of GDP and Inflation on Unemployment Rate: A Study of Indian Economy in 2011-2018 **8**, 329 (2018)

[32] A.K. Mukherjee, International Journal of Educational Research and Development **1**, 56 (2019)

[33] S. Korkmaz, M. Abdullazade, Advances in Economics and Business **8**, 303 (2020)

[34] A. Orji, I. Orji-Anthony, J. Okafor, Asian Economic and Financial Review **5**, 766 (2015)

[35] A. Afonso, H. Şen, A. Kaya, Review of Public Economics **235**, 133 (2021)

[36] J.C. Vermeulen, Southern African Business Review **21**, 20 (2017)

[37] C.A. Sims, Econometrica **48**, 1 (1980)

[38] H. Semuel, S. Nurina, *Analysis of the effect of inflation, interest rates, and exchange rates on Gross Domestic Product (GDP) in Indonesia*, in *Conference on Global Business, Economics, Finance and Social Sciences, 20-11-2014 - 21-01-2015, Bangkok - Thailand* (2014), `http://repository.petra.ac.id/17008/`

[39] J.R. Faria, F.G. Carneiro, Journal of applied economics **4**, 89 (2001)

[40] M. Shahid, Journal of Economics and Sustainable Development **5**, 1 (2014)

Modeling of strategies for the development of the resort-recreation sphere of Ukraine

Pavel Zakharchenko[1,*], *Ganna* Kostenko[1,**], *Tatyana* Kungurtseva-Mashchenko[1,***], *Svitlana* Zhvanenko[1,****], and *Viktor* Mukhin[1,†]

[1]Department of Economics, Entrepreneurship and Finance, Berdyansk State Pedagogical University, 4 Schmidta Str., Berdyansk, 71100, Ukraine

Abstract. In modern world economy resort recreations – one of the most high-profitable industries of managing. Ukraine owns the powerful resort and recreational potential, effective development of which can bring a real economic benefit. For this purpose, it is necessary to form a system concept for the development of such industry, which are integral part of the economic transformations. The purpose of the article consists in development of approach to modeling of transformations strategy development of resort-recreation systems in which transformation acts as their internal and necessary part. As a result of research, the concept of transformation strategy development was grounded, as a certain period of cyclic dynamics, and the scenarios of origin of catastrophe, development and introduction of innovations is got. The offered approach assumes opportunity to consider development of economy of resort-recreation systems as process of transformation change of strategies. On this basis the model based on theory of catastrophes, which allows carrying out the scenario description of transformation strategies of resort-recreation systems is constructed.

1 Introduction

Recreation and tourism is one of the most profitable area of the modern world economy. For many countries, it has become not only a constantly growing source of financial income, but also due to the attraction of millions of tourists, the infrastructure of these areas is beginning to develop more actively and new additional jobs are creating. Ukraine has a strong resort and recreational potential, the effective development of which can bring real economic benefits. Therefore, the recreation industry in the process of market transformation of the economy should take one of the leading places in the structure of the economic complex of the country [1, 2].

However, despite the rich recreational and tourism resources base and a wide network of tourism entities [3], there is still lack of a clear modern strategy for the development of resorts that meets global and European standards in Ukraine. As a rule, the functional strategies and development plans of resort and recreational complexes are of local character and are focused mostly on the industry of rest and, and only then on the rehabilitation and curative forms of travel. As a result, the level of development of the recreation industry in Ukraine is one of the last in Europe, and the degree of compliance of its resort sector with environmental requirements for the environment and cultural and historical heritage is quite low.

Decreased real incomes, weakened coordination of the industry, as well as lack of control over the use of natural healing resources have led to a number of negative results including a reduction in the number of visitors by more than 50%, a significant reduction in bed capacity, high prices for resort and recreational products. The result was a difficult socio-economic situation of national resorts, which for many years chosen the strategy of survival as the main direction of its development. Therefore, one of the current areas of research in this field is the development of models of strategies for socio-economic development of resorts based on modern market requirements [4, 5]. Thus, the identification of determinants and modern mechanisms of sustainable strategic development based on economic transformations is becoming increasingly apparent and important.

The concept of transformation of economic systems has deep grounds in the general economic theory. It is directly related to the problems of formation, development, change of strategies of economic systems, which have been studied in the works of many economists over a long period of time. In the process of studying of the economic transformation of resort and recreational systems [6], the authors relied on the theoretical developments contained in the works of J. Stiglitz [7], J. Shumpeter [8] and other scientists. The works [9–11] are devoted to the general problems of transition processes in the economies of different countries, as well as to the problem of transformational economy.

At the same time, many aspects of the transformation of economic systems are still insufficiently studied. In par-

*e-mail: pvzz1957@gmail.com
**e-mail: annkosten@rambler.ru
***e-mail: tanyakungurtseva2704@gmail.com
****e-mail: zhvanenko_sveta@ukr.net
†e-mail: mukhinviktors@gmail.com

ticular, significant acceleration and complication of world economic processes, strengthening of intersystem integrations intensify transformational changes in resort and recreational systems, which requires their careful study. To date, the process of strategy transformation, taking into account its complexity and uncertainty of future stages of development, has not been fully disclosed. And a holistic concept of transformation of resort and recreational systems, which would be based on a clear understanding of its parameters and their organic relationship in a single holistic system, has not yet been developed.

2 Methodological aspect of a research of strategies for the development of the resort-recreation sphere

Modern trends in the development of resort and recreational systems require a systematic approach to understanding their essence, which consists in the processes of creation, operation, interaction and transformation of various systems, subjected to a single purpose of operation – resort and recreational activities. This approach involves a fundamental change in development strategy. A change in strategy is an inevitable reaction to adaptation to dynamically changing environmental influences. It entails structural changes aimed at modifying resort and recreational products, optimizing the business model and transforming the development mechanism [12, 13].

Trends in systemic transformations, economic crises, dissipative phenomena, increased risks are accompanied by changes in strategies for the operation of resort and recreational systems. The basis for such changes is the transformations of economic processes inherent in the resort and recreational economy. Thus, there is a need to develop tools for analysis and practical application of transformational strategies for the development of resort and recreational systems.

The purpose of the study is to develop and analyse a model of transformation of strategies for the development of resort and recreational system based on the theory of disasters. Taking into account the results of modelling in practice allows to ensure the viability of resort and recreational systems in a dynamically changing environment and market transformations, and also shifts the emphasis from the priority of analysis and adaptation to the priority of "constructing" their own development strategies.

A vast variety of approaches to determining an effective strategy for the development of resort and recreational systems is associated with the specifics of the formation of the national economic model, which is characterized by historical traditions and professional experience in using strategic tools to achieve goals. Accordingly, the transformation strategy of resort and recreational systems will be understood as consistent management action, which is produced and is subjected to constant adjustment as a result of environmental uncertainty, and which leads to sustainable qualitative changes in the resort and recreational system in the long run by creating and strengthening competitive advantages. This definition of strategy practically means the priority of treatment and rehabilitation activities in broad integration with tourism industry. This allows us to move to a methodology focused on meeting the demand of both individual consumers and the general population by creating popular competitive offers and increasing sales, which, accordingly, leads to a significant increase in profits of resort and recreational systems [14].

It should be noted that the national economy is currently in the process of transformation, i.e. in the process of transition from one socio-economic state to a qualitatively different state. Such economic transformation has a number of specific characteristics and effects that distinguish it from economies that are relatively stationary and develop on their own basis, through self-improvement and partial changes in its inherent states, relationships and relations. Such changes are based on the modern concept of smart and sustainable growth, which was formed after the crisis of 2008 and was reflected in the report of the UN Conference on Trade and Development on world investment and investment policy of the new generation [15].

At the core of the problems of transformation of the national economy and, accordingly, the resort and recreational sphere, are issues related to the change of existing technologies, standards of functioning and the established economic order, in order to ensure economic development. Thus, when choosing a strategy, the resort and recreational system actually sets its trajectory of economic transformations to achieve economic development. But it should also be understood that the strategies that now operate successfully in foreign resort and recreation companies will not necessarily also work effectively in the national environment. Therefore, each resort and recreational system must develop and implement its own economic transformation strategy, which combines elements of different approaches to economic development [16].

Today, many economists explicitly or implicitly use the synergetic paradigm in their researches. According to its principles, the resort and recreational system is considered as an open financial and economic structure. Its development is based on the nonlinear laws and is subject to the general evolutionary laws of complex nonequilibrium systems. External disturbances and risk phenomena that accompany the development and implementation of development strategies can create situations of dynamic uncertainty and can lead to instability of the market position of such systems due to unpredictability of consumer behaviour and the complex impact of other external and internal factors. The combination of instability and risks of transformational development with the instability of the external environment, due to the dynamic nature of economic evolution and transformational cyclicality, necessitate the development of models for forecasting possible scenarios for the development of resort and recreational systems.

Intensive research on the dynamics of socio-economic systems is currently associated with the theory of economic dynamics – an actively developing area of modern economic analysis. Methods of qualitative study of dynamical systems, which constitute the content of the theory of catastrophes and chaos, are an effective tool for ob-

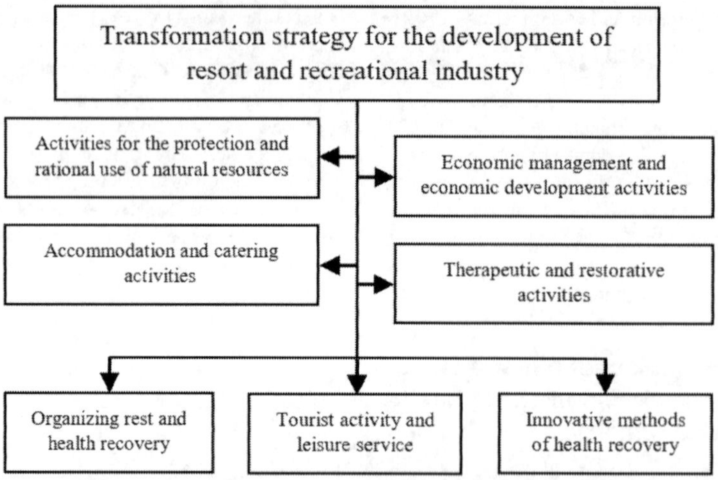

Figure 1. The structure of the transformation strategy of the resort and recreational system

taining information about the processes occurring in the simulated real systems. The use of this methodology allows us to conduct a fairly complete analysis of the dynamics of the market position of economic systems as a result of the transition to a new development strategy [17].

3 Modeling of strategies for the development of the resort and recreational industry

We will use the methods of catastrophe theory to model the transformation of strategies for the development of resort and recreational systems. Catastrophes are abrupt changes that occur in the form of the system's response to a gradual change in external conditions. Catastrophe theory studies sudden abrupt qualitative changes in the states of dynamical systems as a result of slow, smooth and small changes in parameters. Its approaches are essentially a methodology for modelling the transformations of evolving processes, allowing to describe the diversity of gaps from a mathematical point of view and to obtain generalizations of these phenomena. Catastrophe theory methods can be used to qualitatively characterize different types of economic dynamics by identifying structures that reflect both slow changes and the emergence of features in dynamical systems described by smooth functions with a limited number of parameters for any number of variables [18].

Consider the situation when the resort and recreational industry, which operates within the old strategy and, accordingly, produces the old resort and recreational product, is transformed according to a new development strategy and begins to produce a new resort and recreational product. Investments to expand the production of a new product will be received at the expense of the costs of modernizing the old resort and recreational product. The mathematical model of the resort and recreational system, working under the old strategy and producing only the old product, can be represented as:

$$K\frac{dy}{dt} = -y^3 + \alpha_1 y + \alpha_2 \tag{1}$$

where K – coefficient of proportionality; y – a volume of output of the old resort and recreational product during period t; $\frac{dy}{dt}$ – the rate of change of a volume of output of the old resort and recreational product with the respect to time t; α_1 – parameter that specifies the cost (investment) to expand the output of the old product; $\alpha_2 = p - q$, where p – the sale price of the old resort and recreational product, q – costs for the development of a new resort and recreational product.

Due to the implementation of economic transformations, the resort and recreational industry is moving to a new development strategy and a new resort and recreation product is beginning to produce. The mathematical model of the resort and recreational industry, working under a new strategy and producing a new type of product, can be represented as:

$$K\frac{dx}{dt} = -x^3 + \alpha_3 x + \alpha_4 \tag{2}$$

where x – a volume of output of the new resort and recreational product during period t; $\frac{dx}{dt}$ – the rate of change of a volume of output of the new resort and recreational product with the respect to time t; α_3 – parameter that specifies the cost (investment) to expand the output of the new product; $\alpha_4 = p_n$, where p_n – the sale price of the new resort and recreational product.

Since the costs q go to expanding the production volume of a new product, we will assume that $\alpha_3 = q$. Then the system of equations describing the functioning of the resort and recreational industry when producing the old and new product, will look like:

$$\begin{aligned} K\frac{dy}{dt} &= -y^3 + \alpha_1 y + (p - q) \\ K\frac{dx}{dt} &= -x^3 + qx + p_n \end{aligned} \tag{3}$$

https://doi.org/10.1051/shsconf/202110707001

Assume that the costs q begin to be incurred at some point of time t_i in the operation of the resort and recreational industry and continue over time Δt according to the law $q = ut$, where u – the rate of change of investment in the release of a new resort and recreational product. Then, varying the price of the old resort and recreational product p, the price of a new product p_n, investment to expand the production of the old product α_1, time when investing in the release of a new product starts t_i, their duration Δt and speed u, you can get different scenarios of working of the resort and recreation industry.

4 Simulation results

Consider the main scenarios obtained as a result of modeling the transformation of the strategy of development of the resort and recreational system at different values of the parameters.

Scenario 1. The evolution of the economy of the resort and recreational industry, operating within the old development strategy.

One of the scenarios obtained in modeling the strategy of development of the resort and recreational industry was the scenario corresponding to the situation when the system continues to operate under the old development strategy (figure 2).

As can be seen from figure 2 the economy of the resort and recreational industry, which operates under the old strategy is developing rapidly in the early stages, and this corresponds to the life cycle of resort destinations. The growth of profits and production of the old resort and recreational product in the leisure industry is facilitated by stable demand and low prices ($p = 3$). However, since time ($t = 4.8$) growth has slowed significantly, investment in the old product is significantly reduced ($\alpha_1 = 3$).

Since time ($t = 6$), there is a sharp drop in profits of the resort and recreational industry, which leads to the transformation of development strategy and the beginning of the release of a new product. Due to the low rate of investment ($u = 2.25$), late transformation ($t_i = 6$), low prices for the new product and a sharp drop in profits to zero by the time ($t = 8$) resort and recreation system is in disaster.

Scenario 2. Transformational growth of the economy of the resort and recreational industry, operating under the new development strategy.

Another scenario obtained in modeling the strategy of development of the resort and recreational industry was the scenario corresponding to the situation of completion of economic transformations and transition to a new strategy (figure 3).

The scenario presented in figure 3, confirms the concept of the need to move to a transformational development strategy. In the context of such a strategy, there is a post-transformational increase in profits ($t_i = 8$), which corresponds to the course of the transformation cycle and the growth of market confidence. The resort and recreational system goes from the point of catastroph. The growth of demand and the release of a new resort and recreational product of a therapeutic and restorative nature is facilitated by the growth of demand and price increase ($p_n = 4$). This reduces investment in the old product ($\alpha_1 = 2$) and increases the rate of investment in the new product ($u = 4.3$). It should be noted that the economic strategy for managing the development of the resort and recreational industry should include the growth of the prices of the providing services through the implementation of the additional benefits for consumers, as well as the improving service quality.

Scenario 3. Shifting the transformation of the strategy of development of the resort and recreational system in time.

Under the conditions of the transformation cycle, earlier transformations are considered as a way to mitigate the effects of the crisis [19]. To simulate such a scenario, we reduce the time parameters of the model. In figure 4 presents a scenario of early transformation of the strategy of development of the resort and recreational system ($t_i = 3$).

The scenario presented in figure 4, demonstrates a situation when the early start of the transformation of the strategy of development of the resort and recreational industry contributes to the growth of market confidence, which, accordingly, leads to earlier and much greater investment in innovation. The resort and recreation system starts from the point of catastrophe, but the average level of profit decreases. This scenario can be called a "reversal strategy", i.e. if it is known that a catastrophe will occur, the transformation should be carried out at an early stage. In this case, the early start of investment will be made by reducing the average level of profit.

Scenario 4. Reducing the speed of investment in the transformation of the strategy of development of the resort and recreational industry.

Another scenario obtained in modelling the transformation of the strategy of development of the resort and recreational industry was the scenario corresponding to the situation of "overheating of the resort and recreational economy" (figure 5).

This scenario reflects the decline of the rate of investments from to. The resort and recreation industry starts from the point of catastrophe, but the level of profit decreases sharply. Scenario analysis shows that there is some critical rate of investment, above which the resort and recreational industry suffers a catastrophe, i.e. overheats. A similar situation is observed with the investment interval: the smaller the investment interval at a constant speed, the higher the probability of a catastrophe, i.e. it is necessary to increase investment, but for a longer time interval.

5 Conclusions

Analysis of the current state and trends of the world and domestic economy led to the conclusion that the resort and recreational sphere is one of the most profitable and most dynamically developing, which is one of the main sources of revenue for a government budgets of many developed countries. However, the available resource potential of domestic resort and recreational systems is insufficiently

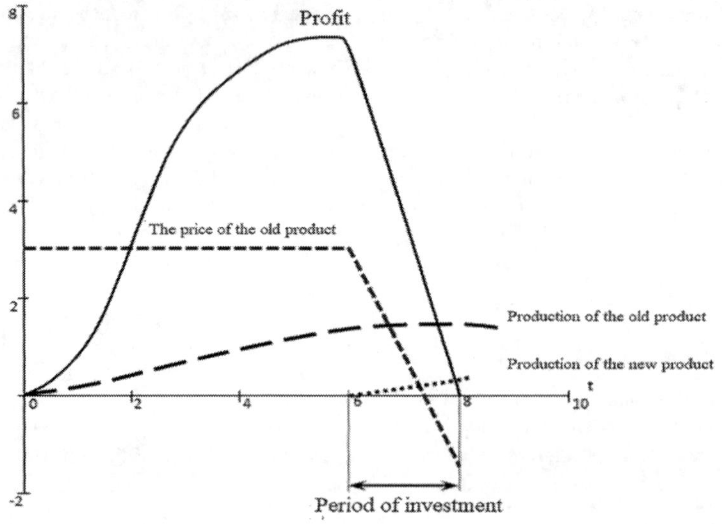

Figure 2. The scenario of a catastrophe in the economy of the resort and recreational system at $\alpha_1 = 3$, $p = 3$, $p_n = 2$, $t_i = 6$, $\Delta t = 2$, $u = 2.25$

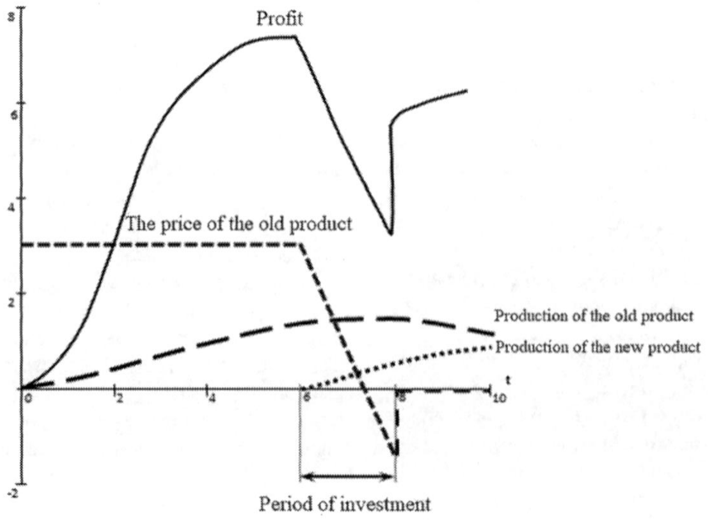

Figure 3. The scenario of economic growth of the resort and recreational system at $\alpha_1 = 2$, $p = 3$, $p_n = 4$, $t_i = 6$, $\Delta t = 2$, $u = 4.3$

used, and the state and effectiveness of recreational development not only lags far behind the leading countries of the world, but also does not meet the needs of modern consumers.

The conducted study on the development of economic mechanisms and models for the effective development of the resort and recreational sphere has identified a set of basic modern approaches and methodologies for managing the development of resort and recreational systems. Their systematization proves that there is some unique predominant characteristic that acts as a key parameter, a factor that determines the level of development of the system. At the same time, there is no systematic approach to this issue, which makes it impossible to reproduce sustainable competitive advantages, form a set of modern competitive strategies and create tools for choosing economic action in multi-alternative situations caused by high and uncertain external and internal environment. In order to eliminate the existing discrepancies, we have used a model approach to the creation of a mechanism for the transformation of strategies for the development of organizations of the resort and recreational complex based on the theory of disasters.

Analysis of the results of modelling the transformation of strategies for the development of resort and recreational system at different values of the parameters allows

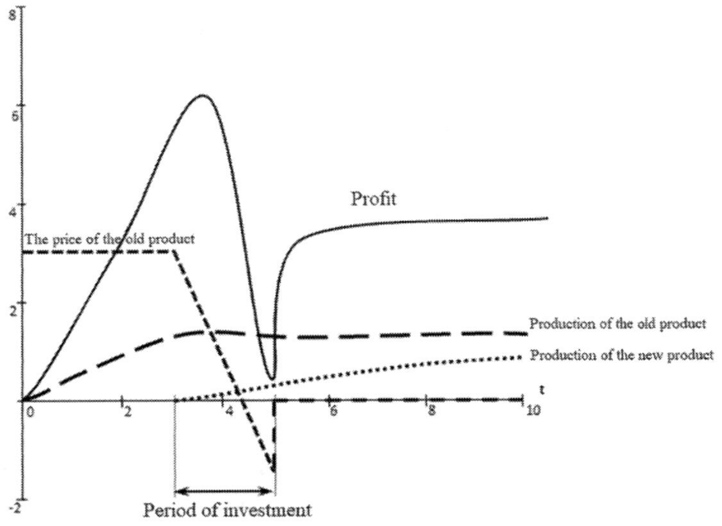

Figure 4. The scenario of shift in time of transformation of strategy of development of resort and recreational system at $\alpha_1 = 3$, $p = 3$, $p_n = 4$, $t_i = 3$, $\Delta t = 2$, $u = 2.25$

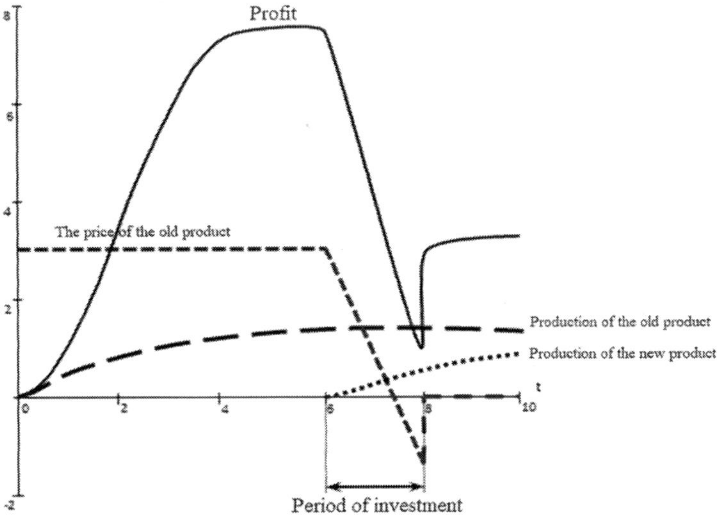

Figure 5. The scenario of reducing the rate of investment at $\alpha_1 = 3$, $p = 3$, $p_n = 4$, $t_i = 6$, $\Delta t = 2$, $u = 2$

us to conclude that transformational changes are one of the most important prerequisites to reaching sustainable economic development and, consequently, effective functioning of resort and recreational economy as a whole. The generalization of the results of model experiments in the presence of several strategies for the development of the resort and recreational system allowed to obtain quantitative and qualitative values of factors, to explore the trajectories of movement depending on the impact of economic transformation.

The transformation of strategies should correspond to the second scenario of the model, i.e. the transformational growth of the economy of the resort and recreational system, which operates within the new development strategy. In this case, there is a significant increase in profits of the resort and recreational system and an increase in demand for resort and recreational products.

References

[1] G.M. Romanova, *Strategicheskoe planirovanie razvitija kurortov i turizma v regione* (SPbGUEF, Saint Petersburg, 2018)

[2] M. Bokov, *Strategic management of recreational enterprises in conditions of transitional economy* (National Institute for Strategic Studies, Kyiv, 2011)

https://doi.org/10.1051/shsconf/202110707001

[3] O. Hanchuk, O. Bondarenko, I. Varfolomyeyeva, O. Pakhomova, T. Lohvynenko, E3S Web of Conferences **166**, 09005 (2020)

[4] E.A. Makarycheva, Ekologija i zhizn **4**, 19 (2017)

[5] M. Amirkhanov, A. Tatarinov, *Economic problems of development of recreation regions* (Helios, Kyiv, 2017)

[6] P. Zakharchenko, A. Kostenko, T. Kungurtseva-Mashchenko, I. Gorbachova, CEUR Workshop Proceedings **2422**, 5 (2019)

[7] J.E. Stiglitz, *People, Power, and Profits: Progressive Capitalism for an Age of Discontent* (W. W. Norton and Company, New York, NY, 2019)

[8] J.A. Shumpeter, *The Theory of Economic Development: An Inquiry Into Profits, Capital, Credit, Interest, and the Business Cycle. Transl. by Redvers Opie* (Oxford University Press, 1961)

[9] V. Heiets, *Transformatsiini peretvorennia. Ekonomika Ukrainy: stratehiia i polityka dovhostrokovoho rozvytku* (Feniks, Kyiv, 2013)

[10] W. Strang, *Recreation and the Local Economy; an Input-output Model of a Recreation-oriented Economy*, Technical report (University of Wisconsin Sea Grant Program, 1970)

[11] O. Kovtun, A. Opalenko, O. Ivanylova, CEUR Workshop Proceedings **2422**, 27 (2019)

[12] G.B. Klejner, Vestnik RAN **9**, 41 (2018)

[13] A.A. Thompson, A.J. Strickland, *Strategic management: Concepts and cases* (McGraw-Hill/Irwin, 1992)

[14] A.T. Bykov, *Kurortno-turistskij kompleks kak obekt upravlenija* (SPbGUEF, Saint Petersburg, 2017)

[15] R. Nuree, Voprosy jekonomiki **9**, 136 (2018)

[16] E. Inskeep, *National and regional tourism planning: methodologies and case studies* (Routledge, London, 1994)

[17] R.E. Lucas, T.J. Sargent, *Rational expectations and econometric practice*, Vol. 2 (U of Minnesota Press, 1981)

[18] W. Sanns, *Catastrophe theory with Mathematica: A geometric approach* (Der Andere Verlag, 2000)

[19] G.G. Malinetsky, *Mathematical Foundations of Synergetics: Chaos, structures, computational experiment* (LIBROKOM, Moscow, 2012)

Modeling the development of the tourism industry in the smart age of globalization through transnational cooperation and capacity building

Nadiia Shmygol[1,*], *Olena* Cherniavska[2], *Dariusz* Pawliszczy[3], *Yeugeniia* Shmygol[4], and *Alexandra* Cherniavska[5]

[1]Department of Management, Zaporizhzhia Polytechnic National University, 64 Zhukovskoho Str., Zaporizhzhia, 69600, Ukraine
[2]National Aviation University, 1 Liubomyra Huzara Ave., Kyiv, 03058, Ukraine
[3]Gromadka Commune Office, 9 General Wł. Sikorskiy Str., 59-706 Gromadka, Poland
[4]Warsaw University of Technology, 1 Politechniki Sq., 00-661 Warsaw, Poland
[5]Kyiv National University of Building and Architecture, 31 Povitroflotskyi Ave., Kyiv, 03680, Ukraine

Abstract. The relevance of the chosen topic is related to the growing interest of researchers in technologies and mechanisms of growth of new touristic opportunities in the smart era of society, features of its impact on national policies in order to achieve strategic socio-economic goals of the countries both for domestic territories and along the path of development of geoeconomic strategies in the global space. In addition the relevance of the research direction is enhanced by the fact that tourism is one of the sectors of the world economy that has suffered the most from restrictions on movement during the pandemic crisis. The European Commission has recently presented a series of proposals setting out the foundations of the EU economy and society suitable for the digital age, outlining the development of a genuine European data space and offering a European approach to new technologies. The new industrial strategy and the SME strategy (adopted on March 10, 2020) identify the path to digital transformation to empower industry and small and medium-sized enterprises. The latest strategic communication of the European Commission "Tourism and Transport for 2020 and beyond" emphasizes the role of digital transformation and sustainability. This involves investing in digital skills and promoting digital innovation, as well as linking the tourism business and businesses to existing data spaces, technology providers, and community actors at the local and regional levels. To determine the possibilities of further evelopment of the industry, ithe article proposes a factor model of the total revenue from tourism activities, taking into account regional changes in tourist flow, provide recommendations for the effective development of the industry with account of future trands of tourism industry and ecosystem development. These issues are equally pertinent both for tourism, given the advantages innovative, new technologies can bring to tourism destinations and for businesses, local communities and travellers. They can facilitate tourism management both at destination and business level, contributing to balanced and sustainable growth of tourism in post Pandemic time. New smart technologies can help businesses provide more personalised services and therefore enhance the tourism experience.

1 Introduction

The European Commission has recently presented a series of proposals [1–8] setting out the foundations of the EU economy and society suitable for the digital age, outlining the development of a genuine European data space and offering a European approach to new technologies.

The new industrial strategy [9] and the SME strategy [10] (adopted on March 10, 2020) identify the path to digital transformation to empower industry and small and medium-sized enterprises. The latest strategic communication of the European Commission "Tourism and Transport for 2020 and beyond" [11] emphasizes the role of digital transformation and sustainability. This involves investing in digital skills and promoting digital innovation, as well as linking the tourism business and businesses to existing data spaces, technology providers, and community actors at the local and regional levels. However, the real volume of domestic tourism is much higher than official statistics show.

The majority of the population, engaged in domestic tourism, prefer individual search for accommodation, food, vehicles, sightseeing facilities, etc. It is not possible to take into account the relevant tourist flows in the calculations.

However, to determine the possibilities of further development of the tourism industry, the article proposes a factor model of the total revenue from tourism activities, taking into account regional changes in tourist flow, provide recommendations for the effective development of the industry with account of future trends of tourism industry and ecosystem development.

Tourism is one of the major components of economic growth for communities worldwide. A key requirement of tourism has been to attract more and more tourists from different parts of the world. Important component of smart city is smart tourism [12]. Smart tourism refers to the ap-

*e-mail: nadezdash@ua.fm

 https://doi.org/10.1051/shsconf/202110707002

plication of information and communication technology, such similar to the smart cities, for developing innovative tools and approaches to improve tourism [13, 14].

Tourism ecosystem coversa range of activities such as travel, transport, accommodation, food, recreation on land and water, culture or nature. Directly and indirectly, it contributes close to 10% to EU GDP and has made the EU the world's leading tourism destination, with 563 million international arrivals and 30% of global receipts in 2018.

The Treaty on the Functioning of the European Union ("TFEU") outlines the Union's competences in this field (Title XXII Tourism, Article195), specifying that EU measures to support, coordinate or supplement the actions of the Member States shall be especially aimed at: a) encouraging the creation of a favourable environment for the development of undertakings in this sector; b) promoting cooperation between the Member States, particularly by the exchange of good practices [15].

Digital technology is changing people's lives. The EU's digital strategy aims to make this transformation work for people and businesses, while helping to achieve its target of a climate-neutral Europe by 2050. The European Commission is determined to make this Europe's "Digital Decade" [1].

Smart tourism is understood as any public or private actor of the tourism ecosystem facilitating access to tourism and hospitality products and services through making better use of data, innovative digital solutions and new technologies, such as artificial intelligence, blockchain, Internet of Things (IoT), communications systems, cyber security, etc. The overall objective of smart tourism is to offer a more efficient, enhanced and more customised tourism product or service, for the benefit of all actors involved in its development and implementation as well as of travelers and local communities [16].

According with five-year plan on smart tourism released The China National Tourism Administration, all 4A and 5A-rated scenic spots in China are able to provide online booking services, electronic tour guides and free Wi-Fi to all tourists in 2020, tour buses, cruise ships and scenic spots with large numbers of tourists will have real-time monitoring systems set up to address security concerns, digital technologies will be adopted to improve tourists' experiences, from smart hotels offering check-in via mobile devices to real-time collection of tourist data for tailoring personalized hotel or sightseeing services [17, 18].

A significant contribution to the development of scientific approaches to the development of the tourism industry and the search for effective ways to manage the industry through the development of new marketing methods made in [19–34] and others. However, modern approaches to the management of the tourism industry, with the emergence and development of new organizational forms of management, are constantly changing.

Nowadays, multimedia methods and information tools allow to create more and more ideas about specific travel services. Some travel agencies see a chance to survive in the market by transferring activities to the virtual world. Undoubtedly, this is a way to increase the cluster of potential customers, but it will remain a dubious way to survive for the travel services market, as it concerns only the type of activity and not sales mechanisms [19–21].

However, customers often complain about the lack of flexibility in their offers by virtual travel agencies. More than 70% of respondents emphasize that there are restrictions imposed on them by sites that sell travel services, and do not recommend using their offers. Thus, thanks to the existence of the Internet, information about travel destinations, for example, can be found more easily, in part, you can choose providers of other services, including globally. Currently, about one third of international tourist trips are organized through the use of online resources, and half of them also involve reservations through the online network [32].

From the point of view of partial service providers, the spread of Internet access opens up many opportunities: new perspectives for action, better access to information, more effective promotion and, finally, direct (cheaper) contact with potential recipients. This allows you to get feedback and respond more quickly to possible changes in the market – that is, the tasks that were previously the responsibility of travel agencies [33].

2 The main problems of the tourism industry and areas for their solution

According to official statistics, tourism in Ukraine is about 3-4% of GDP, which is much lower than the international average of 10%. However, the latest calculations of the UN World Tourism Organization, UNWTO have determined this share at about 9% of Ukraine's GDP.

Prior to the COVID-19 crisis, tourism in Ukraine was improving after a significant decline due to the events of 2014. Kyiv, Odesa and Lviv were the most developed tourist destinations in terms of the chain of value formation of tourist services, destination management and popularity for the vast majority of international tourists.

According to the State Statistics Service of Ukraine, employment in the field of tourism, excluding the strong transport sector and other components of the tourism economy, is 700-900 thousand jobs, which is the result of the interaction of the accommodation and catering sector with such industries as arts, entertainment and recreation. The structure of subsidiary tourism accounts corresponds to the structure of countries with similar tourism products, but with a much lower share of catering in the overall structure. The transport and culture sector provides a relatively higher level of contributions to Ukraine's economy compared to similar indicators in other countries. Tourism in Ukraine has a high potential for development in products such as weekend tours and business tourism, as well as in the field of natural and cultural resources (tourism, rural tourism, medical facilities/health centers).

With the help of smart technologies, you can really influence the form of the package, the choice of route, service provider, the ability to collect discounts on loyalty programs and, above all, independence in organizing the trip.

The choice of target segments to address the priority problems of the tourism industry of Ukraine in this study is proposed to perform using the methodology of factor analysis. The effective indicator will be the total cost of tours sold by tour operators and travel agents (TCTS), which must be maximized due to the identified reserves of its growth based on a comparative analysis of statistics for 2018 and 2019.

Thus, the total cost of vouchers sold according to 2019, in accordance with the areas of tourism, consisted of:

- the cost of vouchers sold to citizens of Ukraine for travel within the country – 1186749,4 thousand UAH or 6,34% of the volume of GRP, equal to 18726217,3 thousand UAH;
- the cost of sold vouchers to citizens of Ukraine for travel abroad – 17341997,2 thousand UAH or 92,61% of the volume of TCTS;
- the cost of sold vouchers to foreigners for travel within Ukraine – 197470,7 thousand UAH or 1,05% of the volume of TCTS.

As we can see, domestic tourism with the use of tourism entities in Ukraine needs purposeful development. The real volume of domestic tourism is much higher than official statistics show.

The majority of the population, engaged in domestic tourism, prefer to individually search for accommodation, food, vehicles, sightseeing, etc., in order to reduce the cost of recreation. It is not possible to take into account the relevant tourist flows in the calculations. However, we will try to calculate the impact of various factors on increasing sales in the tourism industry.

3 Construction of a factor model for assessing the prospects for the development of the tourism industry

The cost of realized vouchers for each direction of tourist activity depends on the corresponding volume of tourist flow, or the number of tourists and the average cost of the realized voucher. Thus, we will have:

$$TCTS = \sum_{i=1}^{3} (VTF_i \times ACT_i), \qquad (1)$$

where $TCTS$ – the total cost of tours sold by tour operators and travel agents in all areas of tourism, UAH;

VTF_i – the volume of tourist flow in the i-th direction of tourist activity, equal to the number of tourists, people;

ACT_i – the average cost of a ticket for the i-th direction of tourism.

The index $i = 1$ in formula 1 corresponds to the travels of Ukrainian citizens within the country; and $i = 2$ – abroad; and $i = 3$ – travel of foreigners in Ukraine.

According to the theory of economic analysis, when constructing factor models of multiplicative or additive-multiplicative type, the following rules must be observed:

1. The sequence of factors affects the final result of the analysis. Therefore, in the model it is necessary to place at first quantitative indicators, and then – qualitative. As can be seen, formula 1 corresponds to the specified restriction.
2. If the factor model contains several quantitative or qualitative indicators, then the sequence of their placement depends on the level of subordination to the performance indicator: first there must be factors of a higher level of subordination. As we can see, model 1 consists of two factors of the first level of subordination, so this rule is not relevant for it.

Further decomposition of the factor model 1 into components is due to the qualitative indicator of GRP, which requires consideration of the second rule:

$$ACT_i = T_i \times PTD_i = \\ = \sum_{j=1}^{m} \left(VTF_{Sj,ij} \times T_{ij}\right) \times \sum_{j=1}^{m} \left(TD_{Sh,ij} \times PTD_{ij}\right) \qquad (2)$$

where T_i, PTD_i – respectively, the average duration of the tour and the average price of one day in tour in the i-th direction of tourist activity;

T_{ij}, PTD_{ij}, – respectively, the duration of the tour and the price of one tour for the i-th direction of tourist activity in the j-th region;

$VTF_{Sh,ij}$ – the share of the volume of tourist flow of the j-th region in the i-th direction of tourist activity in VTF;

$TD_{Sh,ij}$ – the share of realized tourist days in the j-th region in the i-th direction of tourist activity;

m – the number of regions for which the assessment is performed.

Substituting formula 2 in 1, we obtain the final form of the factor model of the total cost of vouchers sold in all areas of tourism, taking into account regional changes in tourist flows, duration of vouchers and price levels of one tourist:

$$TCTS = \sum_{i=1}^{3} (VTF_i \times T_i \times PTD_i) = \\ = \sum_{i=1}^{m} (VTF_i \times \sum_{j=1}^{m} \left(VTF_{Sj,ij} \times T_{ij}\right) \times \\ \times \sum_{j=1}^{m} (TD_{Sh,ij} \times PTD_{ij})). \qquad (3)$$

To determine the influence of factors on the performance in model 3 in this study used one of the methods of deterministic factor analysis – the method of chain substitutions. The essence of this method is to consistently replace the basic values of the factors of the level of 2018 with the actual values of 2019 with the subsequent calculation of the changes.

The results of the factor analysis of changes in the total cost of sold vouchers in the areas of tourism are shown in table 1.

The calculations allow us to analyze the changes in the capacity of the market of tourist services in terms of individual factors and areas of tourism. As you can see from Table 1, during 2018-2019, the total cost of sold vouchers increased by $\Delta TSCTS = 472228,9$ thousands of UAH or 2,59%, all areas of tourism had an increase in market capacity.

Table 1. The results of factor analysis of changes in the total cost of sold vouchers in the areas of tourism during 2018-2019, thousand UAH

Change in performance $\Delta TCTS$ due to factors	Directions of tourist activity			Total changes $\Delta TCTS$
	Travels of citizens of Ukraine		Travels of foreigners within Ukraine	
	Within the country	Abroad		
$\Delta TCTS$	148109,6	251407,3	72712,0	47228,9
$\Delta TCTS_{VTF}$	50687,2	1214068,0	17648,1	1282403,2
$\Delta TCTS_{ACT_i}$	97422,4	-962660,7	55063,9	-810174,3
$\Delta TCTS_{T_i}$	-155564,1	-1278001,1	-61141,8	-1494707,0
$\Delta TCTS_{VTF_{Sh,ij}}$	-116207,2	175084,4	-23561,5	35315,7
$\Delta TCTS_{T_{ij}}$	-39356,9	-1453085,6	-37580,2	-1530022,7
$\Delta TCTS_{PTD_i}$	252986,5	315340,5	116205,7	684532,7
$\Delta TCTS_{TD_{Sh,ij}}$	192696,9	-58958,7	10549,8	144288,0

These changes were due to two factors:

1. The volume of tourist flow, which in 2019 increased by 149,101 people. The corresponding increase in market capacity was equal to $\Delta TCTS_{VTF} = 1282403,2$ thousands of UAH. Moreover, only due to the growing number of trips of Ukrainian citizens abroad, this figure increased by $\Delta TCTS_{VTF2} = 1214068,0$ thousands of UAH.

2. The average cost of sold vouchers, which during 2018-2019 decreased by UAH 319.5. from UAH 8,236.1. up to UAH 7,916.6 per person. These changes had a negative impact on the capacity of the market of tourist services, which decreased by $TCTS_{ACT} = -810174,3$ thousands of UAH. And, on directions of tourist activity the specified situation is ambiguous:

- due to the travels of Ukrainian citizens within the country, the market capacity increased by $TCTS_{ACT_1}$=97422, 4 thousands of UAH;
- due to the travel of Ukrainian citizens abroad, the cost of vouchers sold tended to decrease significantly by $TCTS_{ACT_2} = -962660,7$ thousands of UAH, which is negative;
- due to the travel of foreigners within Ukraine, the target also increased by $TCTS_{ACT_3} = -55063,9$ thousands of UAH.

Thus, the generalized factor analysis using model 1 showed an increase in tourist activity of the population of Ukraine with a simultaneous desire to minimize the cost of recreation, primarily by reducing the cost of travel abroad.

The study of the factors that led to the reduction of revenues of the tourism industry in previous periods is an important step in finding reserves for its development in the future. Therefore, the factor of the average cost of sold vouchers is considered in more detail in model 2. The corresponding analytical calculations are given in table 1, indicate that:

1. The average cost of sold vouchers, which during 2018-2019 decreased by UAH 319,5 from UAH 8236,1. up to UAH 7916,6 per person. These changes had a negative impact on the capacity of the market of tourist services, which decreased by $TCTS_T = -1494707,0$ thousands of UAH. And, on directions of tourist activity the specified situation is ambiguous:

- due to the travels of Ukrainian citizens within the country, the market capacity increased by $TCTS_{T_2}$=−1278001, 1 thousands of UAH;
- due to the travel of foreigners within Ukraine, the target also increased by $TCTS_{PTD_1} = 252986,5$ thousands of UAH.

2. The average duration of vouchers during the study period decreased in Ukraine from 4,70 to 4,29 days. This led to a reduction in performance by $TCTS_{PTD_2} = 315340,5$ thousands of UAH. All, without exception, areas of tourism have reduced their income due to this factor. The largest losses occurred due to travel of Ukrainian citizens abroad, as $TCTS_{PTD_3} = 116205,7$ thousands of UAH.

3. The average price per day in 2019 tended to increase:

- travel of citizens of Ukraine within the country – from UAH 586,8 up to 745,8 UAH. As a result, the total cost of sold vouchers increased by $TCTS_{PTD} = 684532,7$ thousands of UAH;
- travel of citizens of Ukraine abroad - from 2016,8 UAH up to 2054,1 UAH;
- travel of foreigners within Ukraine – from UAH 720,6 up to 1751,1 UAH. This led to an increase in revenues by $TCTS_{VTFnT} = 35315,7$ thousands of UAH.

Accordingly, the average price of one tour in Ukraine as a whole increased from UAH 1752,3 up to UAH 1845,6.

Thus, the increase in ticket prices in 2019 did not lead to a reduction in tourist traffic. However, the limited holiday budget has led to a reduction in the average duration of vouchers.

Regional structural changes in the volume of tourist flow allowed the tourism industry of Ukraine to receive additional income in the amount of thousands of UAH. by redistributing the number of tourists from those areas where the reduction in the average duration of trips was the most significant, in favor of other areas where a similar reduction was less noticeable.

4 Consequences of the pandemic and measures to mitigate them

Tourism is one of the sectors of the world economy that has suffered the most due to restrictions on movement (figure 1). Governments responded immediately to the need to minimize the economic impact of the COVID-19 pandemic, based on two common approaches to managing the situation: the first is to provide affordable credit lines for business, and the second is to defer payment of debts and taxes.

The issue of opening borders for the tourism sector is still unclear. A number of tourism-oriented countries are urging governments to open borders and provide technical solutions that will allow for regular tourist visits. Because decision-making balances economic factors with potential

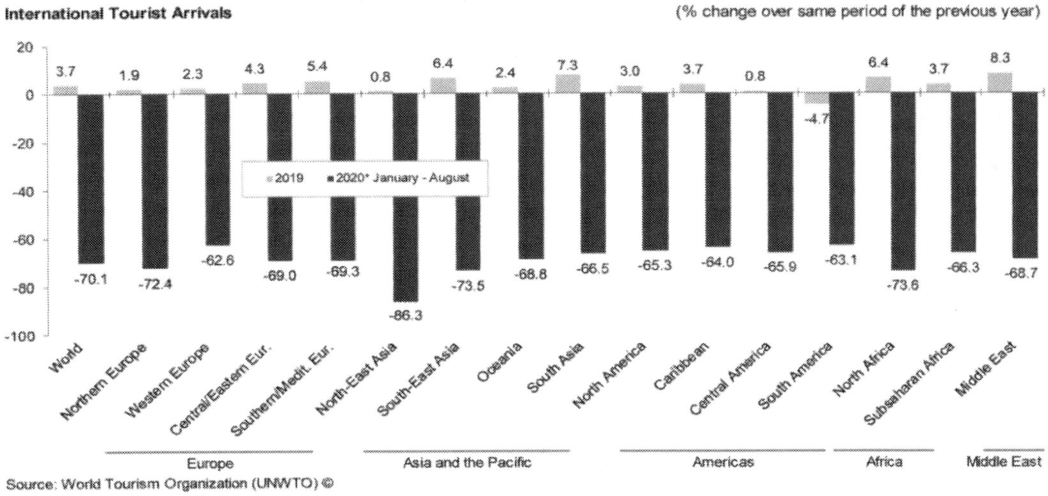

Figure 1. International tourist arrivals [35]

risks to human health, the measures taken by governments can be divided into three main categories: ensuring a fair balance between the protection of tourists and the interests of tourism workers; providing conditions for business survival throughout the supply chain with a special focus on SMEs; focusing on coordination mechanisms to better address and support the recovery of the tourism sector.

Countries that have already overcome the peak of COVID-19 are gradually lifting restrictions on business and services, but with regard to tourism, it is recovering locally. Under favorable conditions, it is assumed that the next step in the recovery of the tourism industry will be the opening of borders between neighboring countries, where the situation is relatively similar (for example, Australia and New Zealand, Thailand and China, etc.).

The European Commission has issued a number of recommendations for the opening of borders in two stages, and the second stage is planned to lift all coronavirus-related restrictions between member states. Regarding the opening of the EU's external borders to third-country nationals (including Ukraine), the situation remains uncertain.

In general, the losses of the tourism industry in Ukraine are estimated at more than 1.5 billion dollars USA [36]. Along with outbound tourism, domestic tourism also suffered significant losses. The introduction of restrictive measures has negatively affected the leisure and travel industry, as well as related industries – hotel and restaurant business, transport (passenger traffic), retail, entertainment and cultural institutions.

The Government of Ukraine continues to work to minimize the effects of the pandemic on the national economy and has proposed a number of measures, both financial and non-financial, aimed at mitigating the impact of these effects on public life, business and the economy. It was be allocated UAH 4 billion to compensate for interest on existing loans for micro and small enterprises, as well as an expanded program of available loans (5-7-9%); UAH 24 billion to provide loans with state investment guarantees; UAH 1.6 billion in support of creative industries.

The deadline for payment of taxes on land, land rent and real estate tax, which was due in April 2020, was extended to June 30, 2020, and similar taxes for March 2020 were abolished. Temporary unemployment benefits were provided for those who lost their jobs as a result of the COVID-19 pandemic, in the amount of two thirds of the amount of wages for each reduced working hour, but not more than the established minimum monthly wage (currently UAH 4,723 or € 150).

The government reorganized the Department of Tourism and Resorts of the Ministry of Economy and Trade of Ukraine and established the State Agency for Tourism Development. Currently, efforts are aimed at strengthening the position of the State Agency through human resources and financial support from the state budget.

A draft law was submitted to the Verkhovna Rada of Ukraine, which provided for the exemption of tourism service providers from VAT, payment of income tax and the Single Social Contribution, payment of land tax and tax on nonresidential real estate; cancellation of payment of tourist tax; exemption from taxation when paying rent and land use, tenants of state and public property.

On May 27, 2020, the Government of Ukraine approved a program to stimulate the economy to overcome the effects of the COVID-19 epidemic. The program includes initiatives in the following areas: access to finance, access to markets, deregulation, modernization and development, access to infrastructure.

However Ukraine is still significantly behind in implementing measures to support the tourism sector which seriously jeopardizes the industry's competitiveness in the global market during the projected recovery in 2021 [36].

5 Conclusions

Thus, based on data from factor analysis and trends in Ukraine's tourism industry, traditionally the most effective reserves for growth in tourism revenues was to increase the average duration of trips, expanding the geography of tourist destinations, including foreign ones. According to the scenarios for this sector, the number of international tourists could decrease by 58- 78%, leading to a decrease in tourism revenues from $ 1.5 trillion. In such conditions, domestic tourism, including informal tourism, is acquiring new significance. Vodafone Ukraine noted a significant increase in the number of subscribers on Ukrainian southern resorts in summer 2020 compared to summer 2019: Odessa, Mykolaiv and Kherson regions (by 30%); Kirilivka resorts (by 55%); Berdyansk resorts (by 37%); Azov resorts of the under-control part of Doneck region resorts (by 100-170%). The calculation of real income from domestic tourism is difficult due to the existing practice of statistical accounting, as well as the fact that not all touristic places pay a tourist tax, the bulk of domestic travel is organized independently without contacting travel agencies.

Smart technologies make people more flexibile in touristic activities. New meaning and distribution acquire also such alternative directions as smart tourism, staycation (day trip distance without overnight accommodation) and workation (working vacation) that blends leisure and productive time.

Workations is emerging as a popular engagement strategy as collaboration tools and advancements in workplace productivity technologies (laptops, project management software, and teleconferencing) have made it easier combine work and tourism to stay in the loop on work while enjoying time off in remote locales. Organisation and strategic communication of the European Commission "Tourism and Transport for 2020 and beyond" involves future investing in digital skills and promoting digital innovation, as well as linking the tourism business and businesses to existing data spaces, technology providers, and community actors at the local and regional levels. So modeling the development of the tourism industry became in the smart age of globalization through transnational cooperation and capacity building in the touristic and non touristic industries.

Digital technologies and smart tourism techniques can help reduce the environmental impact of the travel and tourism value chain and green the industry. Because of the horizontal nature of tourism activities, effective support can actually generate sustainable jobs and growth for the entire ecosystem, with positive impacts on countries and local communities

The latest strategic communication from the European Commission, Tourism and Transport 2020 and Beyond, highlights the role of digital transformation and sustainability. Both were cited as key elements of immediate recovery and long-term balanced development and growth in tourism.

This includes investing in digital skills and stimulating digital innovation, as well as connecting tourism businesses and participants to existing data spaces, technology providers, and government actors at the local and regional level. As the Communiqué highlights, in this transition period, small and medium-sized enterprises will need special attention and assistance in understanding the benefits of digital transformation and innovation in order to become more resilient and competitive.

References

[1] European Commission, *A Europe fit for the digital age: Empowering people with a new generation of technologies* (2021), https://ec.europa.eu/info/strategy/priorities-2019-2024/europe-fit-digital-age_en

[2] European Commission, *Data governance and data policies at the European Commission*, https://ec.europa.eu/info/sites/default/files/summary-data-governance-data-policies_en.pdf

[3] European Commission, *Communication: Shaping Europe's digital future* (Publications Office of the European Union, Luxembourg, 2020), https://tinyurl.com/3xjaj5tt

[4] European Commission, *On Artificial Intelligence - A European approach to excellence and trust* (2020), https://tinyurl.com/4tpka957

[5] European Commission, *COMMUNICATION FROM THE COMMISSION TO THE EUROPEAN PARLIAMENT, THE COUNCIL, THE EUROPEAN ECONOMIC AND SOCIAL COMMITTEE AND THE COMMITTEE OF THE REGIONS: A European strategy for data* (2020), https://eur-lex.europa.eu/legal-content/EN/TXT/?qid=1582551099377&uri=CELEX:52020DC0066

[6] European Commission, *REPORT FROM THE COMMISSION TO THE EUROPEAN PARLIAMENT, THE COUNCIL AND THE EUROPEAN ECONOMIC AND SOCIAL COMMITTEE: Report on the safety and liability implications of Artificial Intelligence, the Internet of Things and robotics* (2020), https://eur-lex.europa.eu/legal-content/en/TXT/?qid=1593079180383&uri=CELEX%3A52020DC0064

[7] European Commission, *Factsheets: a europe fit for the digital age* (2020), https://ec.europa.eu/info/publications/factsheets-europe-fit-digital-age_en

[8] European Commission, *COMMUNICATION TO THE COMMISSION: Data, information and knowledge management at the European Commission* (2016), https://ec.europa.eu/info/sites/default/files/communication-data-knoweldge-management_en.pdf

[9] European Commission, *European industrial strategy* (2020), https://ec.europa.eu/info/strategy/priorities-2019-2024/

europe-fit-digital-age/european-industrial-strategy_en

[10] European Commission, *COMMUNICATION FROM THE COMMISSION TO THE EUROPEAN PARLIAMENT, THE COUNCIL, THE EUROPEAN ECONOMIC AND SOCIAL COMMITTEE AND THE COMMITTEE OF THE REGIONS: An SME Strategy for a sustainable and digital Europe* (2020), https://ec.europa.eu/info/sites/info/files/communication-sme-strategy-march-2020_en.pdf

[11] European Commission, *COMMUNICATION FROM THE COMMISSION TO THE EUROPEAN PARLIAMENT, THE COUNCIL, THE EUROPEAN ECONOMIC AND SOCIAL COMMITTEE AND THE COMMITTEE OF THE REGIONS: Tourism and transport in 2020 and beyond* (2020), https://eur-lex.europa.eu/legal-content/EN/TXT/PDF/?uri=CELEX:52020DC0550&from=EN

[12] Z. Lijing, P. Yanrong, H. Jianhua, *The Development Strategy for the Tourism in Hebei under the Background of Smart City Based on Data Mining*, in *2014 7th International Conference on Intelligent Computation Technology and Automation* (2014), pp. 991–994

[13] A.K. Tripathy, P.K. Tripathy, N.K. Ray, S.P. Mohanty, IEEE Consumer Electronics Magazine **7**, 32 (2018)

[14] R. Smith, *SMART Tourism Tools: Linking Technology with the Touristic Resources of City Destinations* (2015), https://documents.nhtv.nl/load.aspx?doc=thesis%5C2015%5CAT%5CITMC%5CSmithR.pdf

[15] World Tourism Organization, *European Union Tourism Trends* (UNWTO, 2018), ISBN 978-92-844-1946-3, https://www.e-unwto.org/doi/pdf/10.18111/9789284419470

[16] Europian Commission, *Boosting the uptake of digitalisation, innovation and new technologies in tourism through transnational cooperation and capacity building* (2020), https://ec.europa.eu/easme/en/

[17] S. Zhou, *Smart tourism plan to focus on digital technologies*, https://www.chinadaily.com.cn/china/2017-03/08/content_28472460.htm

[18] O. Cherniavska, Y.R. Liu, O. Cherniavska, I. Khan, O. Zham, International Journal of Management **11**, 185 (2020)

[19] K. Zaychuk, *Directions of state regulation of tourist markets*, in *Intern. Research Practice Conf.* (2017), pp. 13–15

[20] O. Strishenetz, *Theoretical aspects of formation and development of the cross-border market of tourist and recreational services. Structural Changes in the Economics of Environmental Management: Theoretical Foundations and Applied Aspects* (2016)

[21] V. Sarichev, Statistics of Ukraine **2014**, 80 (2014)

[22] I. Matyushenko, S. Berenda, V. Reznikov, *Ukraine's European Integration in the System of International Economic Integration* (V.N. Karazin KhNU, Kharkiv, 2015)

[23] *On tourism: The law of ukraine of 15.09.1995. № 324/95*, https://zakon.rada.gov.ua/laws/show/324/95-%D0%B2%D1%80#Text

[24] N. Alpatova, All about accounting **31**, 7 (2008)

[25] S. Shevchenko, Scientific Bulletin of NLTU of Ukraine **23**, 124 (2013)

[26] O. Fedak, Debit-Credit **13**, 51 (2008)

[27] V. Vitlinsky, A. Skrypnyk, Finances of Ukraine **12**, 19 (2005)

[28] O. Davidova, An efficient economy **10**, 21 (2012)

[29] V. Goncharov, M. Zos-Kior, B. Rakhmetulina Zhibek, Current economic issues **10**, 118 (2013)

[30] M. Zos-Kior, I. Kuksa, V. Ilyin, A. Chaikina, Economic Annals-XXI **161**, 43 (2016)

[31] Y. Bilan, M. Zos-Kior, V. Nitsenko, U. Sinelnikau, V. Ilin, Journal of Security and Sustainability **7**, 287 (2017)

[32] V. Gryshko, M. Zos-Kior, O. Zerniuk, International Journal of Engineering & Technology **7**, 131 (2018)

[33] M. Zos-Kior, I. Kuksa, I. Samoilyk, M. Storoška, Economic Annals-XXI **168**, 4 (2018)

[34] I. Kuksa, L. Vasyurenko, A. Litvinov, O. Lytvynova, I. Shtuler, M. Zos-Kior, O. Naholiuk, Accounting **6**, 913 (2020)

[35] L. Pasternak, *UNWTO: International tourism plunges 70% as travel restrictions impact all regions* (2020), https://tinyurl.com/vmwdp9ux

[36] Hotel & Destination Consulting, *The COVID-19 pandemic and its consequences in the field of tourism in Ukraine. Update to the document "Road map of competitive development of tourism in Ukraine* (2020), http://www.ntoukraine.org/assets/files/EBRD-COVID19-Report-UKR.pdf

Human potential and the system of its commercialization management

Tetiana Shkoda[1,*], *Iuliia* Gernego[1,**], *Mariia* Tepliuk[1,***], and *Anastasia* Stefanita[2,****]

[1]Kyiv National Economic University named after Vadym Hetman, 54/1 Peremohy Ave., Kyiv, 03057, Ukraine
[2]Academy of Public Administration, 100 Ialoveni Str., Chisinau, MD 2070, Moldova

Abstract. The important factor of business sustainability is human capital. In this regard, there is need to estimate human potential, concerning the effectiveness of its management and commercialization probability. The current research problem lays upon the human potential management and the possibility of its commercialization. The paper aims to provide scientific support on human potential and human capital theories, concerning their connection within the system of a sustainable business development. The research methodology is based on theoretical and methodological approaches of Ukrainian and foreign experts from the last ten years, reflecting human capital and human potential concepts, taking into account business needs. The main part of research highlights the special status of human potential, based on its specific characteristics, including network effects, the ability to convert human potential benefits into significant social and economic effects, hybrid and multiplicative. The paper proposes a new methodological approach on technology for human potential commercialization and management. Human potential commercialization measurement is offered as a new model to assess the significance of human potential and its managerial effectiveness estimation. The study results can be used within programs for sustainable and social business development. More extensive study is required to prove the proposed methodological approach effectiveness within different sectors and branches.

1 Introduction

1.1 Research question

Nowadays, our society is under the influence of innovation processes, which require its transformation into an information society. Thereby, the process of added value formation is based on specific information and knowledge resources, created by human beings. The human capital formation, development and management, along with innovative transformations and human potential support, are playing the strategically decisive role in ensuring economic stabilization and social welfare.

The human development has received increased attention across a number of researches in recent years, because traditional economic concepts based on obtaining the maximum possible benefits from limited resources, including natural resources, labor and capital, are losing relevance.

One of the most significant current discussions in social and economic sciences concerns social progress as one of the main objectives and driver of post-industrial society growth. In particular, the enhancement of human potential is deeply connected with competitive business development, causing a high level of social and economic development. On the one hand, the rich human potential creates the basis for dynamic growth and rapid business development [1], providing a basis for high competitiveness and further growth. On the other hand, the successful and sustainable business creates the framework for the systematic reproduction, promoting the positive dynamics of human development. This shows the availability of a system of direct and inverse relationships between human development and business profitability.

Thereby, currently, successful business faces an urgent need to increase its human resources along with more traditional material, financial, informational resources etc. The above-mentioned combination has a pivotal role in accomplishing the mission and strategic objectives around which its entire activity revolves.

Thereby, to appreciate a business in an appropriate way and its potential for further development, there is a need to estimate its human, material, financial, and other potential. Evidence suggests that the growing production capacity of business is possible in case of using efficiently all means of production that it possesses. Although technical progress has substantially reduced human presence in a measure of production processes, no economic process has been noticed at any stage in the evolution of humanity that can replace by human contribution. Thereby, human potential involvement becomes more substantial, providing an important requirement to meet diverse economic and social needs.

Thereby, the key factor of sustainable development of a modern enterprise is not only the modernization of its own production and updating of the scientific and technical base, but also the increase of competitiveness. The search for ways to effectively develop the human capital

*e-mail: shkoda-tetiana@kneu.edu.ua
**e-mail: iuliiagern@ukr.net
***e-mail: mteplyuk@gmail.com
****e-mail: iuliia88iuliia@gmail.com

of the organization is directly related to the management quality. This is becoming especially important in the context of the systemic social and economic crisis that is observed in globally and in our country nowadays. This requires fast decision-making, flexibility, efficiency in the enterprise. Consolidation of resources and clear become fundamentally important performing management functions of the organization, which directly depend on the quality policy of capitalization of all available organizational resources.

In light of recent trends of sustainable human development, it is rather difficult to find studies that provide a comparative analysis of indicators of both human potential and the system of its commercialization management. Thus, the current study tends to highlight the tendencies of human potential development within innovative society, considering also its systematic management at business level. The genesis of this thesis can be traced back to the time the authors became interested in human capital and development, including methodological approach to their estimation.

1.2 Previous researches

Several previous studies and international organizations reports have found that human potential plays a significant role in building business capacity and competitiveness both at the national level and globally.

A number of cross-sectional studies proved the existence of the concept of human development [2, 3].

There are numerous and varied approaches to provide human potential definition. The conceptual studies of human potential are represented both at the macro and micro level. In accordance with the topic of current study, the human potential concept is considered at the microeconomic level. This contains the dynamic factor, representing working capacity, knowledge and potential for further development. The study of the human resources at microeconomic level, conducted by Isaic-Maniu et al., reflects labour as an active production factor. Thereby, to characterize labour, the researchers define necessity to highlight the following aspects, namely: volume, structure, dynamics, movement and use of labour [4].

However, there are also another points of view on human potential essence and its managerial capacity. For instance, the authors of the paper on entrepreneur strategies consider the human potential as presentation of the total number of employees that the company needs in order to carry out its activities. The employees are considered as a common factor, which has considerable impact on enterprise development. At the same time, human potential also includes other members outside the unit, which through the board of directors, consultants or other forms use a part of the time budget for works and actions in the interest of the respective organizations [5].

There is also a growing body of literature that recognizes the importance of the following aspects analysis, including: the size and structure of human resources as the basis for an appropriate potential formation; the analysis of human resource behavior as the static factor of human potential development; analyzing the efficiency of human resource growth and utilization as the dynamic factor of human potential development. The above-mentioned aspects are considered at the microeconomic level [6].

Understanding the complexity of human potential causes the widespread of researches on its main components, including the use of knowledge and working capacity. For instance, the fundamental concept of knowledge use was considered by Hayek [7]. Machlup also studied human potential to show the particular importance of information and knowledge [8]. Shumpeter defined human potential in the context of business development and entrepreneur capacity growth [9]. Bell has examined the role of human potential within post-industrial society [10]. Several studies of Porter have documented the influence of human potential on social responsible business strategies and its competitive advantages [11]. Existing research by Castells recognizes the critical role played by human development and potential in our information age [12].

A considerable amount of literature has been published on fundamental aspects of attracting investment for human development and potential [13]. In particular, such aspects are considered within particular human development and potential concepts. Factors found to be influencing human capital and human potential development have been explored in several studies by Becker [14]. Data from several studies of Mincer suggest that investment in human capital plays a significant role in income formation [15]. Results from studies of Schultz demonstrate a strong and consistent association between investment in human potential and social sector development [16]. Several attempts have been made by Sharp to consider human potential as one of important object for investment activity [17].

Experts from international and regional organizations have made a significant contribution to the research and development of a new paradigm of sustainable human development and human potential. In particular, some parts of above-identified problems have been reflected in: UN programs [18] and reports (Transformation of our world: Sustainable Development Agenda 2030 [19]; World Investment Report 2017 [20]); NATO (The Shared Perspective of the World in 2030 and Beyond: Themes and Drivers [21]; Multiple futures project – Navigating towards 2030 [22]); World Business Council for Sustainable Development (WBCSD) (Vision-2050. The new agenda for business.

The above-mentioned researches mostly consider the nature of human development and human potential and their influence on social and economic development. However, there is still a rather little published researches on possibility to combine the peculiarities of human potential and studies on system of human potential management. Thus, the chosen problem is of a significant scientific interest nowadays.

1.3 The scientific problem

The scientific problem of this article lays upon the human potential as a driver of sustainable economy development, concerning methodology on human potential commercialization management. The hypothesis of the paper has both theoretical and applied background, concerning the systematic approach on human potential management, appearing both in theoretical concepts and sustainable business. This aims to activate sustainable human development both at business and national levels.

The aim of the current study is to identify the social (human development) dimension of building a systematic approach to sustainable business process management. The aim is specified in the form of tasks, including the theoretical foundations of human potential concept, providing theoretical and applied framework in human potential commercialization management, assessment of existing experience and opportunities for further implementation of human potential commercialization management ideas in the context of social responsible business and sustainable national economy.

The *object* of the research is the human potential boost for business activity that provide basis for its sustainable development.

The *subject* is human potential commercialization management, contributing the social and economic growth of business, national economy and their global competitiveness.

1.4 The research methodology

The *research methodology* is based on foreign and Ukrainian scientific literature and statistical data analysis (within the last ten years) to show a link between human potential and effectiveness of its commercialization management. The primary data of international and Ukrainian social responsible business development reports within the last five years are reproduced to confirm appropriate human potential trends. The methods of economic and mathematic modelling and dynamics are used to consider human potential effectiveness and its commercial nature. A combination of quantitative and qualitative approaches was used in the innovative methodology of building the system of human potential commercialization analysis. In particular, the theoretical framework of human potential commercialization system is based on management technologies. The possibility of the system application is considered upon its beneficiaries' requirements. This provides robust evidence for consideration of human potential commercialization management as a driver of sustainable business development.

The scientific novelty of the methodology is as follows: despite a large number of foreign and domestic approaches to the measurement of human capital as a criterion of its management, the issues of its measurement are incomplete, the indicators are contradictory, and therefore requires further elaboration in order to create a universal and integrated model. This research proposes a new methodology to provide the system of human potential commercialization management.

Both the human potential development indicators and indicators of its commercialization management are presented. On the one hand, this provides opportunity to define the role of human potential for business development. On the other hand, this allows to estimate the possibility to increase human potential by sustainable business entities.

1.5 The current research structure

The current research consists of the following parts:

- primarily, the theoretical and methodological framework based on human potential concept and its managerial background is considered;
- then, the distribution of institutions by levels of human potential formation and commercialization is proposed;
- finally, the research conclusions on the value of commercialized human potential are provided.

The research practical implication is that results can be used within business strategies and sustainable development models to accept social and economic challenges both on business and national levels.

2 Main theoretical and methodological assumptions of the research

2.1 The conceptual framework of human potential sustainability

The widespread of human potential ideas within the works of economists, as well as the relevant processes practical implementation caused the necessity to build an appropriate conceptual framework, considering the possible sustainability and commercialization.

Sustainability is a manifestation of the state of a phenomenon or process, which presupposes the existence of equilibrium over a long period of time. At the same time, sustainability is a manifestation of the result of the interaction of elements or certain components of the object. Therefore, it is logical to assume that the basis of balancing a particular system are internal (endogenous) factors. Accordingly, changes in parameters of sustainability under the influence of destructive exogenous factors can be counteracted by strengthening the system from the inside. Sustainability is a systemic manifestation of a phenomenon or process that reflects the ability of the research object to function in conditions of external and internal influences [2, 23].

At the same time, the probability of human potential commercialization concerns meaning and manifestation of "human development", "human growth" and "human progress". The term "growth" is associated with the indicator of national income per capita, the transition from lower to higher quality. At the same time, "development" is broader, because it involves covering various aspects of the change process. In the context of human potential, attention should be paid to the level of health and education, quality of work and social stability, changes in the environment. Development is defined as the process during which changes occur, the transition from one state to another, more perfect; transition from the old qualitative state to the new, from simple to complex, from lower to higher.

 https://doi.org/10.1051/shsconf/202110708001

The above-mentioned definition of development includes processes that improve the objects of research that may be associated with it [24]. "Progress" involves progressive development, and "modernization" characterizes the dynamic aspect of development, providing a kind of modifications (figure 1).

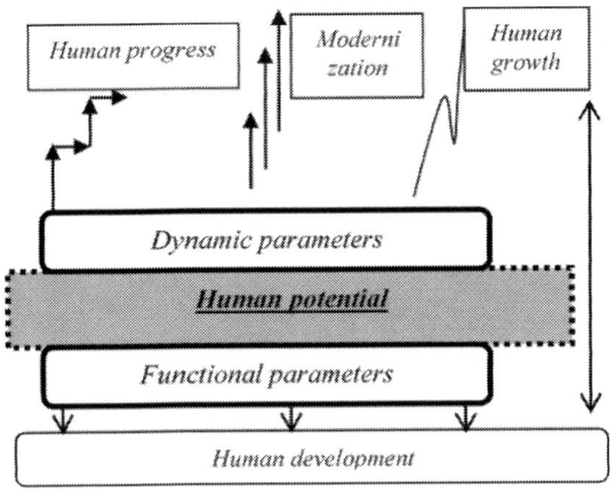

Figure 1. Schematic correlation of human potential parameters

The range of research on above-mentioned questions is reflected in the work of UNDP experts. The relevant analytical findings were primarily summarized in the Global Report on Human Development in 1990 [25].

According to UNDP experts, human potential considers a process that allows people to make more choices. In principle, this choice is likely to be endless and will change over time. However, at all levels of development, key aspects of human potential create opportunities to live healthy lives, acquire knowledge and access the resources needed to ensure a decent standard of living. If such opportunities are not available, the person will not have access to other opportunities, concerning the possibilities of commercialization.

2.2 The system of human potential management commercialization

In the context of business management, the available practical experience indicates the need to provide mandatory procedures for the commercialization of human potential and its transformation into human capital as a newly created added value. This may also be confirmed by the results of a sociological study on the prospects for the commercialization of human potential.

Therefore, we propose to manage the commercialization of human potential in accordance with the technology shown within figure 2.

One of the important system-forming components in the organizational and economic mechanism of strategic management of human capital is the functional component of the management of commercialization of human potential (figure 3).

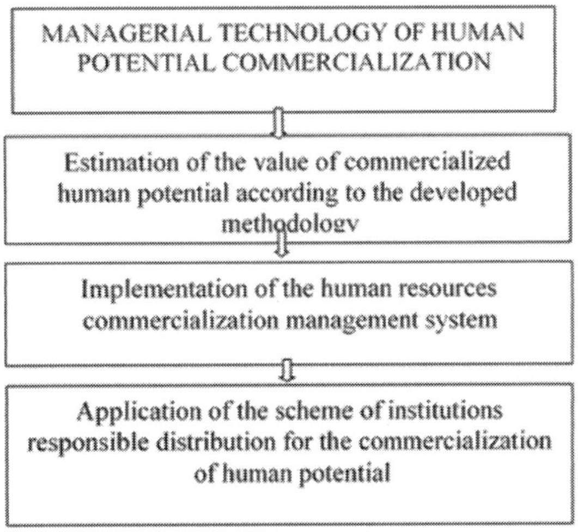

Figure 2. Technology for managing the commercialization of human potential

In our opinion, it is expedient to apply a system-synergetic scientific approach to its construction. Its systemic component involves the use of a set of organizational structures, forms and methods of management, legal norms that allow the implementation of economic laws and the process of commercialization of human potential.

The implementation of the system-synergetic approach requires the establishment of links between the components in the organizational and economic mechanism of human capital strategic management, which provides the coordinated interaction of the main subsystems, namely:

- the control subsystem provides expertise on a resonant managerial influence according to the goals and principles of operation, as it is the subject of management, which is the most mobile and flexible element of the system. The supporting subsystem is also included in the framework of the control subsystem, within which the normative-legal, information-methodical and methodological support is carried out;
- managed subsystem is a managerial object that implements organizational and managerial, economic and socio-psychological relations. A regulatory-analytical subsystem is also included in the managed subsystem, which establishes a correspondence between the goals and results for the implementation of feedback between the subsystems.

Supply and regulatory and analytical subsystems are tools for the influence of the control subsystem on the managed subsystem. This provides a distinction between internal mechanisms of integrity and mechanisms of interaction with the external environment. At the same time, the inverse effect of the controlled subsystem on the control subsystem in the form of transmission of certain information.

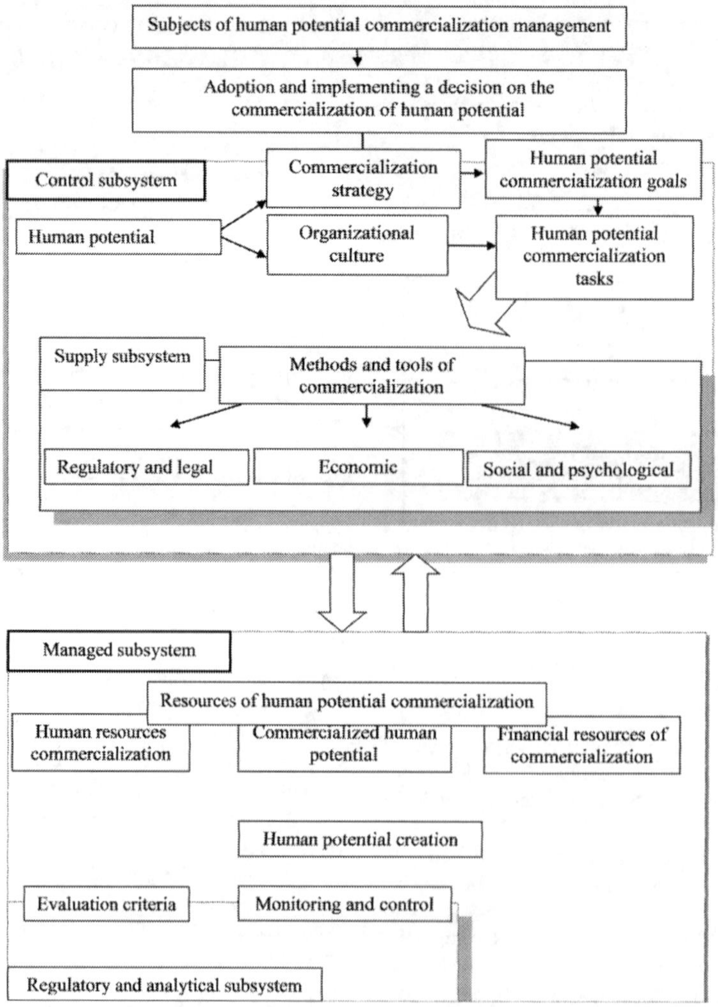

Figure 3. System for commercialization of human potential management

3 Research results

3.1 The conceptual framework of human potential sustainability

Determining the connection between human capital and human potential is an important practical aspect for business development, considering its competitiveness and sustainable development at regional and national levels (figure 4).

The element-by-element division of human potential and human capital created the possibility to be carried out on the basis of institutional economic theory. The theoretical background of this part of research lays in works of such Ukrainian scientists as Grishnova [26], who associates human capital at the microeconomic level with the production and commercial capital of the enterprise, as the source of profit is all types of capital; and Nikolaichuk [27], who considers the problem of human potential commercialization, its conversion into the human capital.

The problem of the relationship between the owner and the exploiter of human capital is both important for researchers and practical application within business entities.

Thereby, the human potential commercialization provides the boost for personal development, business effectiveness increase and national economy sustainability. Thus, there is a practical need of human potential commercialization measurement.

3.2 Human potential commercialization measurement and its managerial effectiveness

The strategic concepts of human potential development and human capital increase provide evidence that the result of the commercialization of human potential is human capital in the form of added value. In this case, a realistic relationship emerges, namely: the value added increase causes the higher intensity of human capital growth (figure 5).

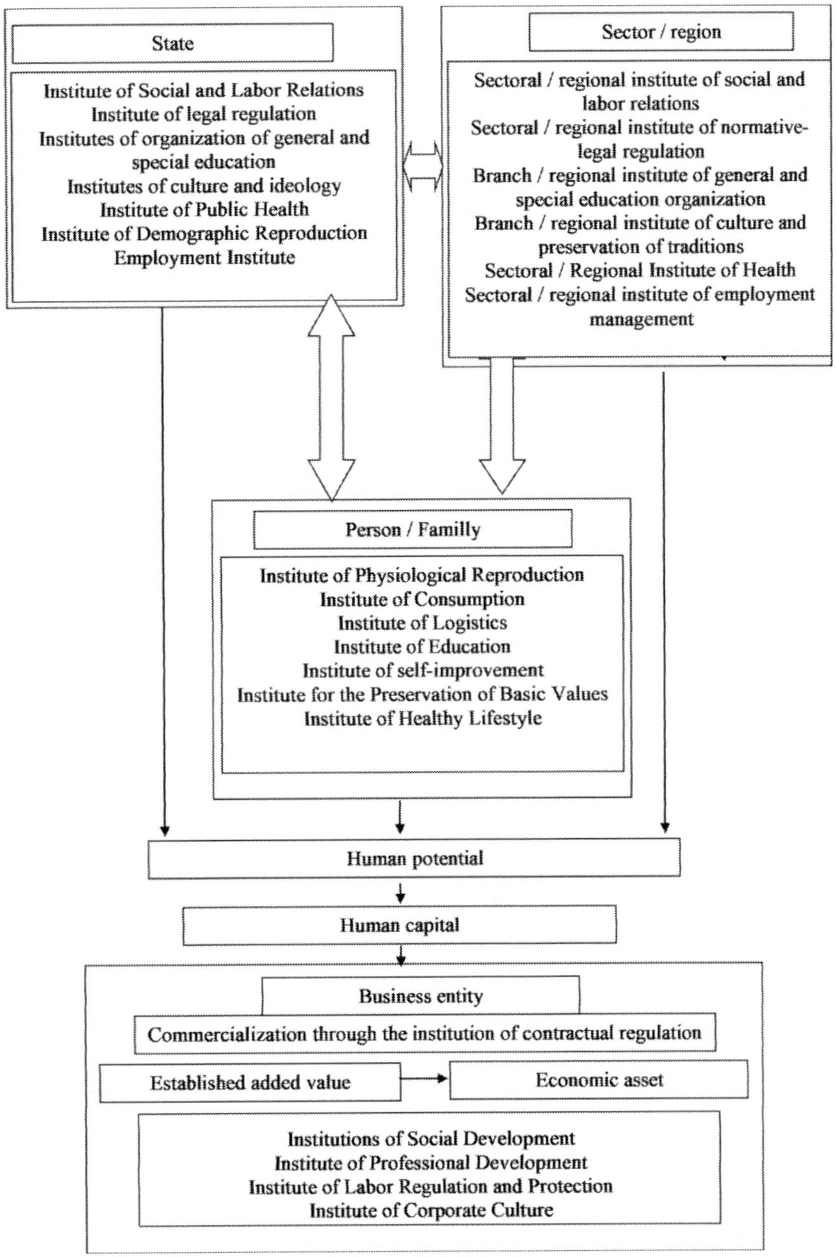

Figure 4. Distribution of institutions by levels of formation and commercialization of human potential

To calculate the value of commercialized human potential, it is necessary to use the technique of determining future value, based on the logic of complex interest (figure 5). This represents a geometric relationship between the initial cost of human potential suitable for commercial use, interest rate and accumulation period:

$$FV_{chp} = V_{chp} * (1 + r)^n. \qquad (1)$$

FV_{chp} – the future value of commercialized human potential;

V_{chp} – the initial cost of human potential suitable for commercial use;

r – interest rate or discount rate;

n – the number of interest accrual periods.

To make the appropriate calculations according to the above-mentioned formula, it is necessary to determine the amount of human potential suitable for commercial use and the interest rate. It is proposed to calculate the initial cost of human resources suitable for commercial use,

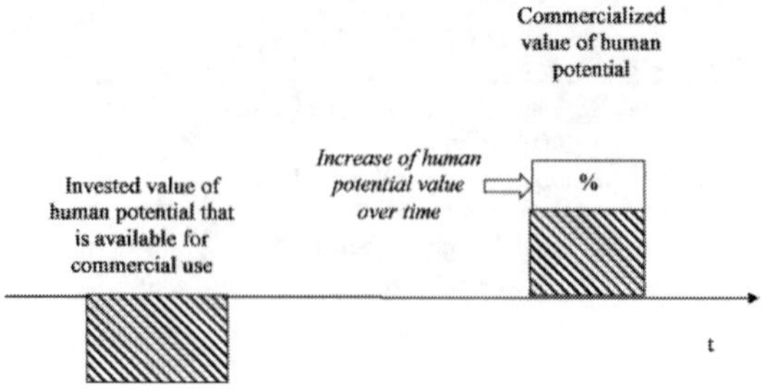

Figure 5. Increase in the principal amount of commercialized human potential as a share of human capital of employees hired during the last year

based on the use of the cost method according to the following algorithm:

1. First stage

The cost of developing human potential suitable for commercial use is determined:

$$C_{chp} = (C_{hP} + C_{Person}) * (1 + (P/100)) * K_{cf}. \quad (2)$$

where C_{hP} is the cost of developing human potential suitable for commercial use;

C_{Person} – staff costs;

P – labor profitability;

K_{cf} – conversion factor, by which the time costs are reduced to a single point in time (adjusted costs for the value of the labor price index (wages) on the date of assessment. The coefficient should be determined by formula (3).

$$K_{cf} = W_{mb}/W_{msfe}. \quad (3)$$

where W_{mb} – an average monthly salary of full-time employees by enterprise / business segment;

W_{ms} – the average monthly salary in the equivalent of full employment.

2. Second stage

The value of the coefficient is taken into account, considering the degree of frequency of human potential use (K_{HpF}) that is suitable for commercial use during the first year of operation:

$$K_{HpF} = 1 - (T_A/T_N). \quad (4)$$

where T_A – actual working time fund;

T_N – nominal fund of working time.

3. Third stage

The final cost of human potential suitable for commercial use is calculated in the following way:

$$V_{chp} = K_{cf} * K_{HpF}. \quad (5)$$

The results of the authors' calculations according to the proposed algorithm at the example of air transport enterprises are represented in table 1 [28].

Based on the data of table 1, the authors should admit that the lower is labor profitability, then the higher is the amount of the cost of developing human potential suitable for commercial use. To determine the discount rate, it is proposed to adapt the formula of the weighted average cost of capital to the needs of our study [29]. The adaptation is possible on the basis that we determine the share of human capital of employees hired during the last year.

Thus, for the cash flow of all human capital, a discount rate is applied equal to the sum of weighted rates of return on equity and borrowed human capital, where the weight is the share of equity and borrowed human capital in the structure of human capital. This discount rate is called the weighted average cost of human capital ($WACHC$).

The weighted average cost of human capital is calculated by the following formula:

$$WACHC = khc_{ft} \cdot (1 - t_{w1}) \cdot whc_{ft} + \\ +(khc_{pt} \cdot (1 - t_{w2}) \cdot whc_{pt} + khc_c \cdot (1 - t_{w3}) \cdot whc_c + \\ +khc_l \cdot (1 - t_{w4}) \cdot whc_l) \quad (6)$$

where khc_{ft} – the cost of human capital;

t_{w1} – the amount of tax rates withheld from the salaries of key employees;

whc_{ft} – share of own human capital in the structure of human capital;

khc_{pt} – the cost of attracting human capital from external partners;

t_{w2} – the amount of tax rates withheld from the salaries of external part-timers;

whc_{pt} – specific share of attracted human capital of external part-timers;

khc_c – the cost of attracting borrowed human capital under civil law contracts;

t_{w2} – the amount of tax rates withheld from wages under civil law contracts;

whc_c – specific share of attracted human capital under civil law contracts in the structure of human capital;

khc_l – the cost of attracting borrowed human capital under leasing agreements;

t_{w3} – the average interest paid by the lessee to the lessor under leasing agreements;

whc_l – specific share of attracted human capital under leasing agreements.

In contrast to the classical formula of the weighted average cost of capital [29], in the proposed version of the formula for the weighted average cost of human capital, it is proposed to apply multipliers of tax rates withheld from the salaries of key employees, part-time employees and those working under civil law contracts. This multiplier is of great importance for the part of the attracted human capital, which works under leasing contracts, because there the lessee pays interest for the provision of leasing services to the lessor.

We believe that the scientific basis for the division of human capital into own and involved ones is the so-called "shamrock theory" of Charles Handy [30]. The scientist presented the organization in the form of a shamrock, the first leaf of which consists of key employees, the second

Table 1. Calculations according to the proposed algorithm at the example of air transport enterprises

Indicator	Year					
	2013	2014	2015	2016	2017	2018
Net Income	115056,8	61492,8	102152,9	73093,46	54303,81	35514,15
Total average number of employees, persons	14167	12731	13540	15789	16100	16410
Total amount of labor costs, thousand UAH	268267,1	729970,4	389582,6	588721,58	646796,65	704871,73
Total salary fund (full-time and part-time employees)	1217366,7	1065527,0	1561911,3	1699528,38	1827792,78	1956057,18
Nominal fund of working time (T_A)	17491943	14133377	13461641	15437586	14085205	12732824
Actual working time fund (T_N)	15277142	11874427	11530893	13095421	11864246	10633070
Average monthly salary in the equivalent of full employment, UAH (W_{ms})	5205,23	7023,6	7655,69	8087,15	8762,29	9437,42
Average monthly salary of full-time employees by enterprise / business segment, UAH (W_{mb})	4538,88	6393,9	6983,62	7319,68	7909,69	8499,71
K_{cf}	0,8720	0,9103	0,9122	0,9051	0,9027	0,9006
Labor profitability (P)	0,43	0,08	0,26	0,12	0,08	0,05
K_{HpF}	0,8734	0,8402	0,8566	0,8483	0,8423	0,8351
C_{chp}	1301005,72	1635899,22	1784845,95	2073665,95	2235680,30	2397742,68

– external contractors, and the third – temporary and part-time employees. Kravchenko refers to the attracted human capital that part of which is involved for performance of one-time or special works, and also - human capital of external part-timers [31].

Therefore, there is a need to apply the formulas of such indicators as equity and borrowed human capital. Thereby, the formula for calculating equity [32] will be adopted in the following way:

$$B_{own_HC} = NI_e/p_m \tag{7}$$

where NI_e – the amount of net profit per employee;

p_m – market price of one employee[1].

It is proposed to calculate the amount of attracted human capital taking into account the number of employees working under civil law contracts and on lease terms.

Moreover, the proposed formula makes it possible to calculate the invested human capital, as well as to offer another option for calculating the economic value added of human capital. Today, the classical formula given by Fitzenz [33] makes it possible to determine how much economic added value of the organization accounts for the average amount of labor spent to obtain the result.

The return on assets of human capital can be calculated by the formula of the ratio of profit of the enterprise and invested human capital. This formula is very close to the formula of return on investment in human capital [34], which Zakharova defines through the ratio of net profit of the enterprise and the amount of funds that were invested in human capital for a certain period. However, the values for the calculation in both variants are different.

Thus, the economic added value of human capital is proposed to be calculated in the following way:

$$EVA_{HC} = (PA_{HC} - CB_{HC}) * K_{HC} \tag{8}$$

where EVA_{HC} – economic added value of human capital;

PA_{HC} – return on assets, human capital;

CB_{HC} – invested human capital (asset value human capital).

Since we consider human capital as an asset of the enterprise, it is important to note that by European standards it belongs to intangible assets and includes: personal goodwill; skilled and assembled employees; lease agreements; favorable employment contracts; know-how, inseparable from the person [35].

In general, the established method of assessing human capital as a result of commercialization of human potential of the enterprise is considered one of the innovative approaches to the implementation of the socio-commercialization component of human capital of enterprises. It is proposed by the authors for practical implementation on the example of enterprises.

3.3 Human potential of business development in Ukraine

Social responsible business is considered as the business with the highest probability of human potential commercialization in Ukraine. In accordance with applied researches of Ukrainian analytics, the main driver for the corporate social development policy implementation is

[1]under the market price of one employee, we mean the salary of one employee, the cost of hiring, training, etc.

moral consideration and – according to representatives of the companies – the main barrier for non-implementation of the CSR policy is lack of financial resources, volatile political situation in the country, inadequate legal framework which would encourage this activity, and tax pressure [36].

However, 76% of business representatives are considering their social responsibility as the policy for development and improvement of conditions for staff. A rather significant part of companies concerns the reason of social responsibility as assistance for loyalty of staff increase (figure 6) [28, 36–39].

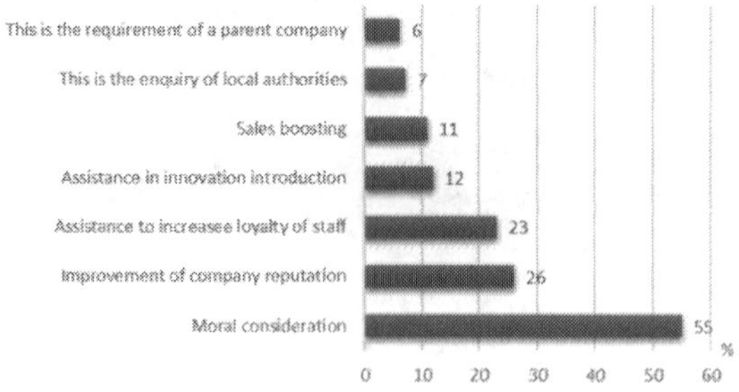

Figure 6. The reasons to develop the social responsible business

Almost half of the Ukrainian companies provide their employees with an opportunity to work under flexible working schedules and implement professional development programs. Salary rise, payment of declared salary and bonuses are also considered as social responsible practices, concerning labor relations (figure 7) [28, 36–39].

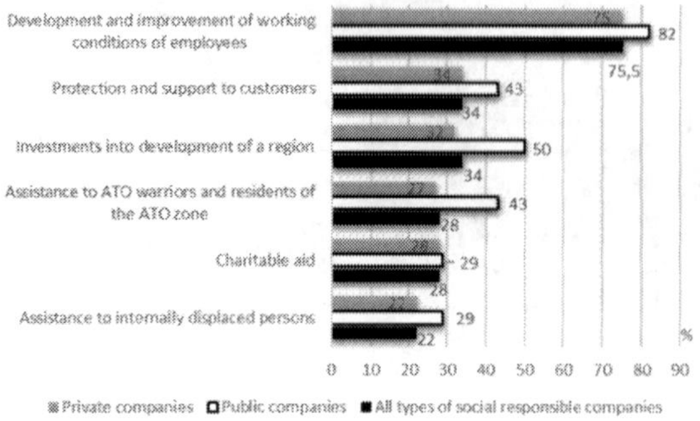

Figure 7. The meaning of social responsibility

This provides the background for human potential increase in our country, increasing also the probability of its commercialization.

4 Conclusions

Given all that has been mentioned so far, one may state that human potential is getting growing importance within business sustainable development. The primary analytical research of social responsible statistics shows the increasing trends of business staff support, increasing the probability of human potential commercialization.

At the same time, the study results show the lack of methodological basis on human potential and human capital research, considering a reflection of interrelations between them. Thereby, the commercialization criteria are chosen as the most appropriate for possible comparison. For measuring the concentration of efforts for increasing effectiveness of commercialized human potential, the theoretical basis should be appropriate for its practical implementation to increase business activity.

It has been conclusively shown that the effective management of the human potential cannot be achieved without establishing the effective use of the human resources and its appropriate measurement. A new approach of human potential measurement clarifies what indicators can be considered to precisely measure more accurate human capital. On the basis of the results of quantitate and qualitative methods analysis, it can be concluded that the weighted average cost of human capital is calculated taking into account the multipliers of tax rates withheld from the salaries of key employees, part-time employees and those working under civil law contracts. Collecting various quantitative and qualitative data will allow for better control and use of human potential, because the effective manager should know the whole potential of employees.

There is an urgent need to concentrate more efforts on the human potential commercialization by Ukrainian business. Thereby, the study results can be used by the business structures to provide managerial and financial support for human potential commercialization in part of the company staff support, stimulating its capacity growth.

In particular, the Ukrainian business stakeholders tend to pay attention on the following activities:

- stimulating the development of human potential and human capital, considering also their management and commercialization, in the works of Ukrainian scientists;
- creation of a centralized system of informing the national business entities and investors regarding the human potential development advantages;
- intensification of cooperation between public and private companies, who are ready to support human potential development both at business and national levels.

The essence of human potential is both applied and theoretical question. Despite there is already a measure of researches on human potential, experts still have a lack of information on some applied aspects, including:

- a measure of long-term results from human potential commercialization;
- lack of information on successful human potential and human capital development practises;
- lack of human capital and human potential statistic both at business and national levels.

Thereby, there is a need for further research, considering human potential commercialization and its practical implication within modern society. Thus, this is the subject for future researches.

References

[1] T. Shkoda, M. Tepliuk, I. Riepina, I. Semenets-Orlova, O. Domina, Financial and Credit Activity: Problems of Theory and Practice **4**, 522 (2020)

[2] M. Stavridou, S. Vangchuay, African Technology Development Forum Journal **9** (2017)

[3] T. Shkoda, Rynok pratsi ta zajniatist naselennia **1**, 37 (2016)

[4] B. Ileanu, A. Isaic-Maniu, C. Herteliu, T. Andrei, Journal of Applied Quantitative Methods **10**, 84 (2015)

[5] O. Nicolescu, C. Nicolescu, P. Truică, D. Urâtu, S. Corcodel, *White Charter of SMEs* (Sigma Publishing House, Bucharest, 2016)

[6] A. Skibiński, IBIMA Business Review **2019** (2019)

[7] W. Butos, Advances in Austrian Economics **13** (2010)

[8] F. Machlup, *Knowledge: Its Creation, Distribution and Economic Significance, Volume I: Knowledge and Knowledge Production* (Princeton University Press, Princeton, 2014)

[9] J. Schumpeter, in *Joseph Alois Schumpeter: Entrepreneurship, Style and Vision*, edited by J. Backhaus (Springer US, Boston, MA, 2003), pp. 5–59, ISBN 978-0-306-48082-9, https://doi.org/10.1007/0-306-48082-4_2

[10] D. Bell, D. Morrison-Love, D. Wooff, M. McLain, International Journal of Technology and Design Education **3** (2017)

[11] M. Porter, J. Heppelmann, Harvard Business Review (2017)

[12] K. Kirtiklis, Lingua Posnaniensis **59**, 65 (2017)

[13] I. Gernego, O. Dyba, L. Petrenko, Financial and credit activity: problems of theory and practice **3** (2019)

[14] G.S. Becker, *Human capital: A theoretical and empirical analysis, with special reference to education* (University of Chicago press, 2009)

[15] B. Chiswick, J. Mincer, Review of Economics of the Household **1**, 343 (2003)

[16] T. Schultz, Tech. rep., Center Discussion Paper, Yale University, Economic Growth Center, New Haven (1987)

[17] W. Sharpe, G. Alexander, J. Bailey, *Investments* (Prentice Hall, New Jersey, 1999)

[18] *Human Development Report* (2020), http://hdr.undp.org/en/content/human-development-index-hdi

[19] *Transforming our World: The 2030 Agenda for Sustainable Development* (2015), https://sustainabledevelopment.un.org/post2015/transformingourworld/publication

[20] *World Investment Report 2017* (2017), http://unctad.org/en/PublicationsLibrary/wir2017_en.pdf

[21] *The shared perspective of the world in 2030 and beyond: Themes and drivers* (2013), http://www.act.nato.int/images/stories/events/2012/fc_ipr/sfa_ws2_rap.pdf

[22] *Multiple futures project – Navigating towards 2030* (2010), http://www.act.nato.int/images/stories/events/2009/mfp/20090503_MFP_finalrep.pdf

[23] M. Dyba, I. Gernego, O. Dyba, A. Oliynyk, Management Theory and Studies for Rural Business and Infrastructure Development **42** (2020)

[24] P. Maskell, A. Malmberg, Cambridge Journal of Economics **23** (1999)

[25] *Human Development Report 1990. Concept and Measurement of human development* (1990), http://hdr.undp.org/en/reports/global/hdr1990/chapters/

[26] O. Grishnova, Social and Labour Relations: Theory and Practice **1**, 34 (2014)

[27] M. Nikolaychuk, *Human capital management in the conditions of knowledge economy development: problems of theory and practice* (KhNU, Khmelnytskyj, 2012)

[28] *State Statistics Service of Ukraine. Official site. Statistical information. Science and innovation. Scientific and innovative activity (1990 – 2019)* (2020), http://www.ukrstat.gov.ua/

[29] T. Davydiuk, *Analiz pidkhodiv do otsinky liudskoho kapitalu yak obiekta bukhhalterskoho sposterezhennia (Analysis of approaches to the assessment of human capital as an object of accounting observation)* (DonNUET, Kryvyi Rih, 2017)

[30] C. Handy, *Gods of Management: The Changing Work of Organizations* (Macmillan, New York, 1979)

[31] T. Kravchenko, State and regions. Series: Economics and Business **6**, 142 (2013)

[32] A. Zlenko, O. Isaikina, University Economic Bulletin **46**, 115 (2020)

[33] J. Fitzenz, *The ROI of Human Capital. Measuring the Economic Value of Employee Performance* (AMACOM, New York, 2009)

[34] O. Zakharova, N. Podluzhna, Socio-labor relations: theory and practice **2**, 90 (2016)

[35] G. Zhavoronkova, V. Zhavoronkov, D. Sokovnina, L. Melnyk, M. Gomeniuk, *Intellectual capital of enterprises of APC in the regional innovative system* (Vydavets Sochinskyj, Uman, 2012)

[36] A. Zinchenko, M. Saprykina, *CSR development in Ukraine: 2010–2018* (Yuston, Kyiv, 2017)

[37] E. Hontz, O. Okuniev, M. Saprykina, *Company transparency index 2019 – Ukraine* (2020), https://drive.google.com/file/d/1CjQ-gDivH30WDx-QSZysDsUFCtJHeeoe/view

[38] *Biznes, shcho zminiuie krainu (Business that changes the country)* (2019), https://project.liga.net/projects/business_changing/#17

[39] *Best practices in corporate social responsibility have been identified in ukraine* (2017), https://cutt.ly/hlx1OmX

Global migration processes analysis and modelling of migration attractiveness of countries through fuzzy logic

Hanna Danylchuk[1,*], *Liubov* Kibalnyk[1,**], *Oksana* Kovtun[2,***], *Oleg* Pursky[3,****], and *Zenon* Stachowiak[4,†]

[1]The Bohdan Khmelnytsky National University of Cherkasy, 81 Shevchenko Blvd., Cherkasy, 18031, Ukraine
[2]State Institution of Higher Education "University of Educational Management", 52A Sichovykh Striltsiv Str., Kyiv, 04053, Ukraine
[3]Kyiv National University of Trade and Economics, 19 Kyoto Str., Kyiv, 02156, Ukraine
[4]Cardinal Stefan Wyszyński University, 5 Dewajtis Str., 01-815 Warsaw, Poland

Abstract. The article deals with the analysis the current state of migration in the context of globalization and identifies the most important corridors for the labour movement. The main donor countries of migrants are developing countries, with low socio-economic indicators, difficult environmental conditions and high levels of poverty. According to forecasts, the most migratory flows will take place in the countries of North America and in Europe, which is due to rising trends in unemployment in the countries of the "third world" and the demand for cheap labour, changes in the structure of the economies of developed countries, changes in labour market demand. The main world regional corridors in 1990–2019 have been identified through statistical analysis. And their growing and declining trends. The need to use economic and mathematical modelling techniques to analyse and determine the migration attractiveness of recipient countries in an uncertain environment has been substantiated. It has been shown that fuzzy logic tools are the most effective in this case. Based on the results of the simulation using the Mamdani method, the world's attractiveness rating for migration is calculated, which with a "high" thermo leads such countries as Italy, France, United Arab Emirates. The findings suggest that migrants are attracted by countries with the lowest inflation rates, high and average GDP per capita and average or low taxation levels.

1 Introduction

Since the 19th century, international labour migration has been the subject of scientific research in various fields. Since the late 1980s and early 1990s, population migration as a social and economic phenomenon has become particularly relevant and global in the context of the development of integration processes which have led to a significant labour movement. Today, migration is characterized by the permanence of territorial flows and the diversity of expression forms. Changes in migration flows are mainly reflected in population growth, territorial movements and the emergence of new types of migration movements. Migration processes have a decisive impact on the socio-economic development of countries and regions as a whole.

By D. Massey [1] the period of post-industrial migration (since the mid-1960s) has been associated with its emergence as a global phenomenon cause of the number and diversity of donor and recipient countries dramatically increasing. Developing countries, third world countries, have intercepted the baton of dominant donors of migrants. The main flows of migrants were from industrialized countries to post-industrial ones.

The second wave of globalization (from about 1950-s to 1980-s) led to the migration movement from less developed countries (Afghanistan, Pakistan, India, Vietnam, Morocco, Egypt, Turkey, etc.) to more developed West and East countries (Europe, USA, Canada, Japan, etc.). Most migrants were low-skilled workers, employed under temporary employment programmes. In the early 1980s such Asian countries as Korea, Taiwan, Hong Kong, Singapore, Malaysia and Thailand experienced migration.

During the second and the beginning of the third wave of globalization, along with the traditional countries of immigration like the United States, Canada, and Australia, some European countries (Germany, France, etc.) began to accept migrants actively. During that period, the number of migrants in the world rose sharply. From 1970-s to 1980-s there were 82 million people migrants and, in 1980, there were 99 million people. Over the next 10 years, the number of international migrants increased by 56 million people [2].

Recently, many scholars believe that the end of the twentieth and the beginning of the twenty-first century is a phase of unprecedented international migration. However, other researchers, notably M. J. Miller and S. Castles refers to this assumption as a "myth", as the number of migrants in the world is growing, but their share in the world

*e-mail: abdanilchuk@gmail.com
**e-mail: liubovkibalnyk@gmail.com
***e-mail: kovtun.oa71@gmail.com
****e-mail: Pursky_O@ukr.net
†e-mail: zenstach@wp.pl

https://doi.org/10.1051/shsconf/202110708002

population does not exceed 2-3 % in the history of migration [3]. More than a century ago, during the period of mass migration to the New World, the share of migrants in the world did not exceed 3 % of the world's population, so authors [4] refer to the opinion of M. J. Miller and S. Castles about false migration. It should be noted, however, that the scale of international migration in recent decades has never been greater in the history of human development.

In our view, international migration should be seen as a widespread socio-economic process with many causes and consequences, playing a significant role in the economic, inter-ethnic and demographic changes in the development of communities and society as a whole. It have a projection in the social, political and cultural life.

Considering the relationship between socio-economic and migration processes, it is necessary to take into account the theory of factors that are the causes of "repulsion" or "attraction". The theory of factors of "attraction – repulsion" was developed by E. Ravenstein [5], who formulated the "Laws of migration" on the basis of census data in England and Wales at 19th century. E. Ravenstein concluded that migration can be explained by factors of "attraction – repulsion" and unfavourable conditions in one region (strict legislation, excessive taxes, etc.), "displace" people from their residence place as well as favorable conditions created in other regions, which "attract" them.

E. Ravenstein divided all the factors of migration into internal (push factors) and external (pull factors), which can be divided into five groups, namely: economic, social, cultural, political and environmental. Among the economic factors, it is worth mentioning, first of all, the level of wages, the quality of life, the level of unemployment, the stability of economic development and the type of tax system.

The migration laws formulated by E. Ravenstein were as follows: the main reason for migration is better conditions in another locality than in the one where the person lives; the extent of migration decreases with increasing distance; migration often occurs in several stages; population migration is two-way; people's mobility is determined by their personal characteristics (gender, age, social class, etc.).

Also, the density and volume of migration flows are influenced by the structural changes that occur in social and economic systems related to the cyclicality of the economy, the technological transition and the mismatch of requests (needs) of the National Labour Market Structure for Highly Qualified Professionals [6].

Consequently, the impact of migration processes on the social and economic development of certain regions, countries and their associations is beyond doubt, and scientists and public administrators need to have tools, which would make it possible to develop effective employment policies, to identify possible losses, imbalances and profits for the State from migration flows, to form social packages of support for forced migrants, to improve migration policies etc.

In view of the above, the issues of modelling and forecasting migration flows in the face of globalization challenges are of particular relevance. Considerable attention is paid to the outlined scientific problem in the research of foreign and domestic scientists. Thus, in the work of [7] using the simulation method, namely system dynamics, a system of causal relationships has been constructed, and the main factors influencing the decision of individuals to travel abroad have been identified. In the proposed model two feedback loops are presented, one of them reinforcing (reinforcing loop) and the other – counteracting (counteracting loop). It is the counteracting loop that allows the system after numerous iterations, occurring continuously dynamically in each period, to reach a balance.

Econometric methods and models, namely regression analysis methods, are the most commonly used in studies on migration processes at the regional and national levels. This method is the basis for forecasting and for studying the impact of various objective factors on the dynamics and volumes of migration flows. In scientific researches [8, 9] in constructing regression models, it has been proved that the main factors influencing migration processes are the index of average wages, GDP per capita, labour demand, employment and unemployment, Provision of social benefits, employers' need for workers to replace vacant jobs.

In [10] an approach is proposed for modeling the business climate of countries using the taxonomy method. The resulting model can be used to increase the choice of influencing factors when modeling the attractiveness of countries for migrants.

The authors [11] investigate the possibilities of applying machine learning methods to modeling macroeconomic indicators which have impact on migration processes.

Also in the research [12] the authors applied the method of statistical spatial modelling, taking into account heterogeneous influences on migration flows, visualized the spatial distribution of net migration, indexes of the internal and external flows among the municipalities of the Brazilian Amazon. The authors have shown that the selected variables demonstrate spatial relationships, and spatial regression models provide more accurate estimates of the indexes by including autoregression with spatial lag.

In some researches [13, 14] authors use a cluster analysis to study migration flows, the results of which confirm that the development of individual regions (countries) is uneven and asymmetrical in terms of the main indicators of labour migration, as well as being the basis for the search for cluster convergence directions.

In the research [15] the author applies web-based analysis techniques to research international migration flows when the network is seen as a set of nodes (countries) and arcs with some directions. The migration process is viewed as a socio-spatial network that has a set of nodes located in a geographical space and interconnected ribs of arcs with a certain length. This method allowed the author to identify certain patterns of international migration (asymmetric and reciprocal) and migration clusters.

Through the dynamic multi-factor model based on the assumptions of the theory of positional games, the authors of the study [16] carried out a prediction of the migration behaviour of the individual depending on the economic situation. The impact of migration on the socio-economic development of the recipient country through the Cobb-Douglas production function is also assessed.

Using the agent modelling methodology, the authors of other research [17] proved that it is a reliable method of modelling autonomous human behaviour in migration decisions, taking into account not only socio-economic influences but also environmental ones.

It is also suggested that agent models should be applied in situations where human behaviour needs to be investigated in conflict situations affecting migration decisions – internally displaced persons, refugees, undocumented migrants [18]. This concept makes it possible to construct adequate models of the movement of forced migrants and to forecast their likely places of residence, which will enable the public authorities of the State to deal effectively and in a timely manner with the problems of housing and social guarantees, employment and so on.

Using the structural vector autoregression and the estimated model of the dynamic stochastic general equilibrium (DSGE) of a small open economy, the authors of the research [19] conclude that migration shocks have a significant impact on the volatility of GDP per capita, to replace investment income per capita as well as investment housing and housing.

In summary, contemporary globalization processes have led to the emergence of a new pattern of historical migratory movements and flows, which has led to the emergence of new mega trends. Therefore, traditional methods of analysis and modelling do not always allow for an adequate assessment and forecasting of these processes, hence the need to apply fuzzy logic methodology, that would allow ranking countries according to their migration attractiveness in an uncertain environment.

2 Research methods

As noted above, it is useful to use a fuzzy inference engine in the form of a fuzzy set, which corresponds to the current values of the input variables, using fuzzy knowledge base and fuzzy operations, to assess migration processes. Fuzzy sets theory is used specifically to solve problems in which the input data is unreliable and poorly formalised. Currently, fuzzy logic is used in the construction of neural networks, genetic algorithms, and the design of fuzzy systems. Fuzzy logic provides effective means of representation of uncertainty and inaccuracies of the real world, and the presence of mathematical means of representation of uncertainty of input information makes it possible to construct models corresponding to realities [20].

The foundations of the fuzzy sets theory of and fuzzy logic were laid in the 1960s by Lutfi A. Zadeh [21]. Thanks to this research a new scientific branch has appeared, which received the name "fuzzy logic". His work laid the foundations for the approximate human reasoning modeling and gave impetus to the development of a new mathematical theory. L. Zadeh introduced the term "fuzzy set", suggesting that the ownership function can accept any value within [0; 1] not just the values of 0 or 1. Also, a series of operations on fuzzy sets has been defined and a generalization of known methods of logical inference has been proposed by introducing the notion of a linguistic variable.

A fuzzy inference engine used in expert and knowledge-based management systems, established subject matter experts or learning neural networks. In turn, the training set of networks is based on experimental data as a set of fuzzy predicate rules of the form:

Rule 1: if $x \in A_1$, than $y \in B_1$,

Rule 2: if $x \in A_2$, than $y \in B_2$,

...

Rule N: if $x \in A_n$, than $y \in B_n$,

where x – input variable, y – output variable, A and B – membership functions defined accordingly x and y.

Expert $A \to B$ knowledge, reflecting the unclear causal relationship between input and output, is called fuzzy connections R:

$$R = A \to B, \qquad (1)$$

where "$\to$" is called fuzzy implication.

The relation R can be seen as a fuzzy subset of a direct product $X \times Y$ of a complete set of assumptions X and inferences Y. Thus, the process of obtaining a fuzzy output B' with observation A' and knowledge A B represented as follows:

$$B' = A' \cdot R = A' \cdot (A \to B), \qquad (2)$$

where "$\cdot$" – convolution operation [22].

Fuzzy inference algorithms differ in the type of rules, logical operations, and dephasation methods. The most common modifications to the fuzzy inference algorithm are the Mamdani and Sugeno algorithms. The main difference between the two is the way the values of the output variable in the rules are specified, and the knowledge base. In Mamdani-type systems, the values of the input variables are given by fuzzy terms, in Sugeno-type systems it is as a linear combination of the input variables. For tasks where identification is more important, it is useful to use Sugeno algorithm, and for tasks where explanation and justification are more important, Mamdani algorithm will have the advantage.

Mamdani algorithm was one of the first to be used in fuzzy output systems [23]. Formally, Mamdani algorithm can be defined as follows.

Let the knowledge base contain only two fuzzy rules of the kind:

Rule 1: if $x \in A_1$ and $y \in B_1$, than $z \in C_1$,

Rule 2: if $x \in A_2$ and $y \in B_2$, than $z \in C_2$,

where x, y – names of the input variables, z – name of the output variable, A_1, A_2, B_1, B_2, C_1, C_2 – some fuzzy sets, assigned by membership functions $\mu_{A_1}(x)$, $\mu_{A_2}(x)$, $\mu_{B_1}(y)$, $\mu_{B_2}(y)$, $\mu_{C_1}(z)$, $\mu_{C_2}(z)$, the precise value of z_0 should be determined on the basis of the information given and the clear values x_0, y_0.

The operation of implication of fuzzy sets consists of the following four steps:

1. *Fuzziness*: The degrees $\mu_{A_1}(x_0)$, $\mu_{A_2}(x_0)$, $\mu_{B_1}(y_0)$, $\mu_{B_2}(y_0)$ of each premise of each rule. The ownership functions defined on the input variables apply to their actual values to determine the degree of truth of each premise of each rule.

2. *Fuzzy inference*: there are "cut" level for the assumptions of each of the rules (using min operation). The calculated meaning of truth for the assumptions of each rule applies to the conclusions of each rule. This results force one fuzzy subset that will be assigned to each output variable for each rule.

$$\alpha_1 = \mu_{A_1}(x_0) \cap \mu_{B_1}(y_0), \alpha_2 = \mu_{A_2}(x_0) \cap \mu_{B_2}(y_0), \quad (3)$$

where "$\cap$" is the operation of the logical minimum (*min*).

Then there are "cut" membership functions:

$$\mu'_{C_1}(z) = (\alpha_1 \cap \mu_{C_1}(z)), \mu'_{C_2}(z) = (\alpha_2 \cap \mu_{C_2}(z)). \quad (4)$$

3. *Composition*: using the maximum transaction (*max*, designated as "$\cup$") to find the found cut functions. Result is the resulting fuzzy subset for the output variable with the membership function:

$$\mu_{\epsilon}(x) = \mu_C(z) = \mu'_{C_1}(z) \cup \mu'_{C_2}(z) \quad (5)$$

4. *Clarity* (to find z_0) by centroid method:

$$y = z_0 = \frac{\alpha_1 z_1^* + \alpha_2 z_2^*}{\alpha_1 + \alpha_2} \quad (6)$$

where "ω" – function domain of $\mu_{\Sigma}(x)$ [22].

3 Results and discussions

The United Nations estimates the number of international migrants in 2019. It has reached 272 million people. Consider the structure and trends of global labour migration processes from 1990 to 2019 (table 1), using the geographical topic [24]. In 2019, more than half of all international migrants lived in North America (82.3 million people) and Europe (59 million people). North Africa and West Asia ranks third with the largest number of international migrants (49 million people), Sub-Saharan Africa (24 million people), Central and South Asia (20 million people), and East and South-East Asia (18 million people). Latin America and the Caribbean (12 million people) and Oceania (9 million people) recorded low numbers of international migrants. Through the indexes of the average absolute increase we generated the forecast for 2025 and 2030 (table 1).

There is an upward trend in international labour migration in North America, Europe, Latin America, Oceania, Central and South Asia and South and North Africa, and a downward trend only in East and South-East Asia.

The 10 largest regional migration corridors in 2019. Presented in table 2, five of which represent nearly half of the world's migration flows (124 million people).

The "Europe to Europe" direction, which has 4.19 million people international migrants, is the largest regional migration corridor in the world. A large proportion of these migrants have moved between European Union countries. In 2010–2019, it increased by more than 5 million people international migrants, compared to 2000–2010, with an average annual increase of more than 0.5 million people.

The direction "Latin America and the Caribbean to North America" is the second largest regional migration corridor in 2019 (26.6 million people people). During the period 1990–2000, in this direction the number of migrants increased by 0.9 million people per year, but the growth slowed between 2000–2010 and 2010–2019 (0.5 and 0.3 million people per year, respectively).

The next three largest regional migration directions were almost the same in 2019 (18-19 million people international migrants). The number of international migrants inside the corridor "North Africa and West Asia" increased by 7.3 million people in 2010–2019, while the corridor "Central and South Asia to North Africa and West Asia" increased by 5.4 million people.

While international migration is a global phenomenon, only 20 countries received two thirds of all international migrants in 2019. Almost half of all international migrants live in 10 countries only. The largest number of migrants is in the United States of America, with 51 million people migrants, or about 19% of the world's total number of migrants admitted in 2019. The most attractive countries for migration in 2019 were Germany (13.1 million people) and Saudi Arabia (13.1 million people), the Russian Federation (12 million people) and the United Kingdom (10 million people). Of the 20 most attractive countries for migration, seven are in Europe, four – in North Africa and Western Asia, three – in Central and South Asia, two – in East and South-East Asia and North America, and one each in Oceania and Sub-Saharan Africa.

Between 1990 and 2019, the United States recorded the largest absolute increase in international migrants (27.4 million people). The countries with the largest increases were Saudi Arabia (8.1 million people per year), the United Arab Emirates (7.3 million people per year), Germany (7.2 million people per year) and the United Kingdom (5.9 million people per year).

In 2019 one third of all international migrants come from only 10 countries. In 2019 India has become the leading country of the international migrants origin (17.5 million people). The second largest migrants contributor was Mexico (11.8 million people), followed by China (10.7 million people), the Russian Federation (10.5 million people) and the Syrian Arab Republic (8.2 million people).

Thus, based on data on the volume and intensity of migration processes, it can be concluded that different regions and countries have different attractions for migrants.

The level of the migration attractiveness of countries was determined through the Mamdani fuzzy inference.

Table 1. Labour migration by geography, 1990-2019, mil. people and forecast for 2025, 2030 [24]

Years	Countries							
	North Africa and West Asia	South Africa	Central and South Asia	East and Southeast Asia	Oceania	Latin America and the Caribbean	Europe	North America
1990	4.73	7.16	6.84	26.17	13.29	17.61	27.61	49.61
1995	5.02	6.69	8.34	21.25	14.28	18.91	33.34	53.49
2000	5.36	6.57	10.51	20.47	13.15	20.32	40.35	56.86
2005	6.02	7.22	12.95	18.96	14.22	23.28	45.36	63.59
2010	7.13	8.26	15.75	19.58	15.86	32.56	50.97	70.68
2015	8.07	9.44	17.87	19.44	21.34	42.05	55.63	75.01
2020	8.93	11.67	18.30	19.63	23.57	48.59	58.65	82.30
2025 (Forecast)	9.63	12.42	20.21	18.54	25.28	53.75	63.82	87.75
2030 (Forecast)	10.33	13.17	22.12	17.45	26.99	58.91	68.99	93.20

Table 2. Largest regional labour migration corridors in the world in 2019, mil. people [24]

№	Contributing countries groups	Recipient countries groups	Number of migrants
1	Europe	Europe	41.86
2	Latin America and the Caribbean	South America	26.58
3	North Africa and West Asia	North Africa and West Asia	18.93
4	Central and South Asia	North Africa and West Asia	18.52
5	South Africa	South Africa	18.31
6	East and Southeast Asia	East and Southeast Asia	14.32
7	North Africa and West Asia	Europe	13.05
8	Central and South Asia	Europe	11.20
9	East and Southeast Asia	North America	10.24
10	Latin America and the Caribbean	Latin America and the Caribbean	8.24

The fuzzy inference simulation process was conducted in the Matlab environment using the Fuzzy Logic Toolbox. To construct the Mamdani fuzzy inference system, four input and one output (linguistic) variables in the fuzzy inference system were specified: GDP (gross domestic product per capita), IR (inflation rate), UR (unemployment rate), PIT (personal income tax) and EMA (Evaluation of migration attractiveness) (figure 1). These variables were selected from correlation and regression analysis as relevant to the migration attractiveness of countries.

The phasing of the introduced linguistic variables and the definition of their terms have been carried out. The parameters of the membership functions for these term sets are shown in table 3 and the structure of the functions are shown in figure 2.

Based on the rules, a decision-making mechanism is created, predicts the value of the performance variable. The peculiarity of this model is its flexibility, it can be filled with other rules, its content and quantity adjusted.

On the basis of the input variables statistics an evaluating of migration attractiveness was calculated. Figure 3 illustrates Mamdani fuzzy inference with the example of France. The results of the evaluation of migration attractiveness of selected countries are presented in table4.

The analysis of the recipient countries rating allows to draw conclusions that none of the studied countries has received the score «ultrahigh». Italy, France and the United Arab Emirates are among the top three most attractive countries with a «high» term. The middle-level attraction cluster is Saudi Arabia, the UK, Canada, the US, and Australia. Russia has the lowest rating. The results are largely consistent with the statistical analysis, but allow for a clearer definition of a country's ranking. It should be noted that the proposed fuzzy model for the evaluating the migration attractiveness of recipient countries can be fur-

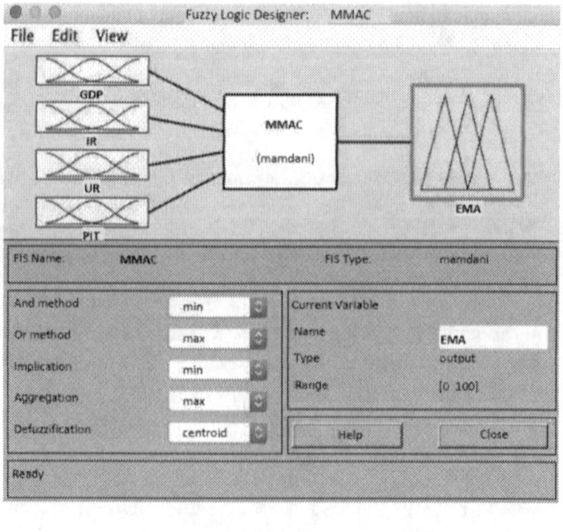

Figure 1. Structure of the model of the evaluation the labour migration attractiveness of countries in 2019 through the Fuzzy Logic Toolbox

 https://doi.org/10.1051/shsconf/202110708002

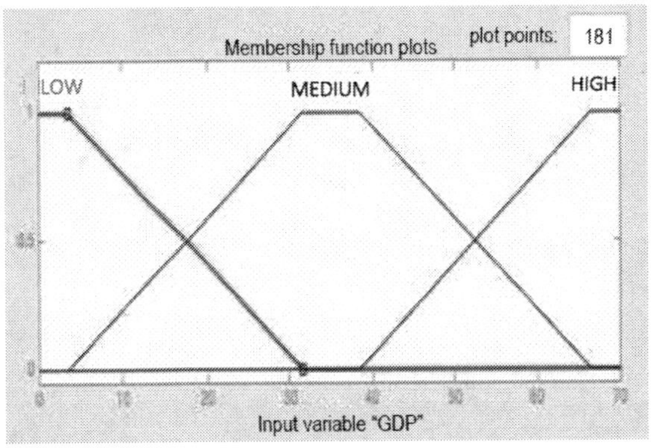

a)

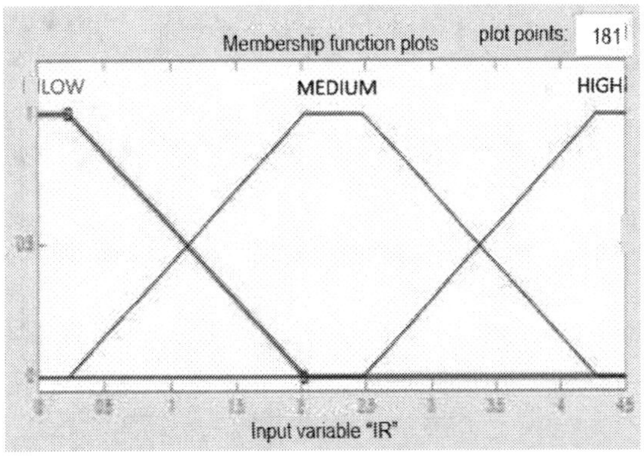

b)

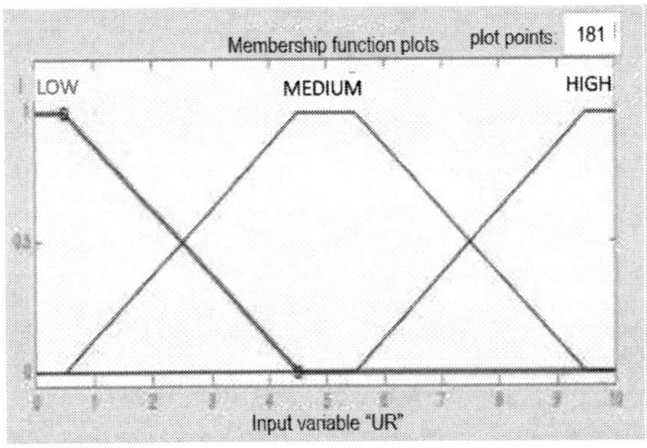

c)

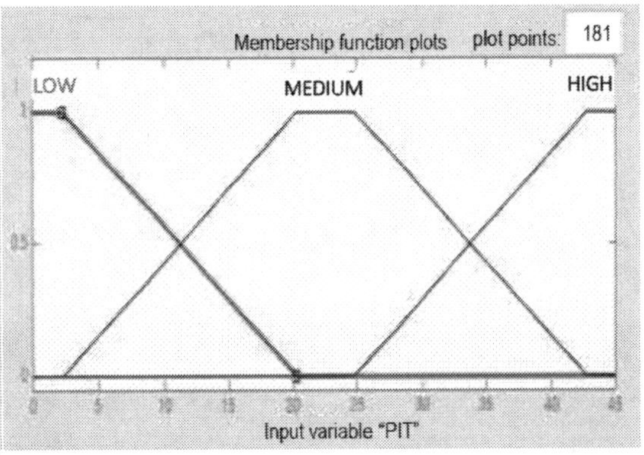

d)

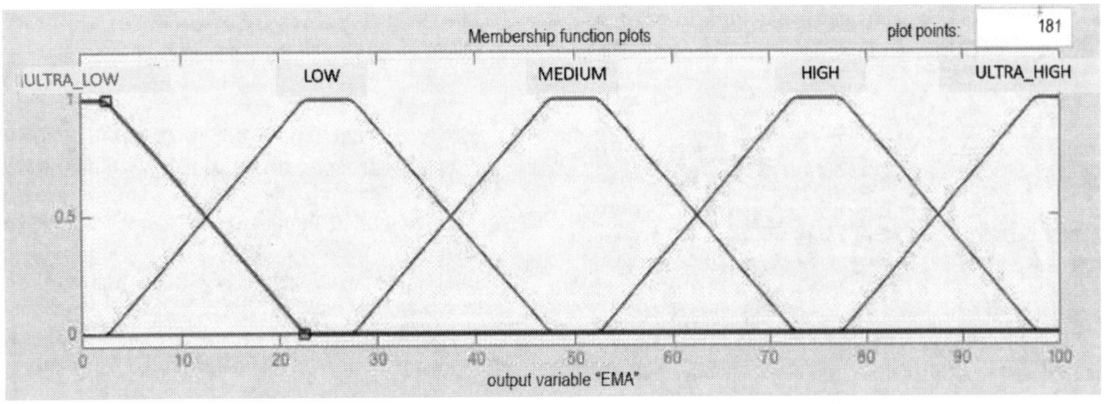

e)

Figure 2. Membership function graphs for input (a, b, c, d) and output (e) linguistic variables

ther developed and refined by introducing additional input variables.

4 Conclusion

Thus, using the Mamdani fuzzy inference method, a system has been developed that allows to make a decision on

Table 3. Attributes of the attribution model of the evaluation the labour migration attractiveness of countries in 2019

Linguistic variable	Evaluating interval	Term	Evaluating rate
		low	0 – 23.3
GDP (thousand $)	[0; 70]	medium	23.3 – 46.6
		high	46.6 – 70
		low	0 – 1.5
IR (%)	[0; 4.5]	medium	1.5 – 3
		high	4 – 4.5
		low	0 – 3.3
UR (%)	[0; 10]	medium	3.3 – 6.6
		high	6.6 – 10
		low	0 – 15
PIT (%)	[0; 45]	medium	15 – 30
		high	30 – 45
		ultralow	0 – 20
		low	20 – 40
EMA	[0; 100]	medium	40 – 60
		high	60 – 80
		ultrahigh	80 – 100

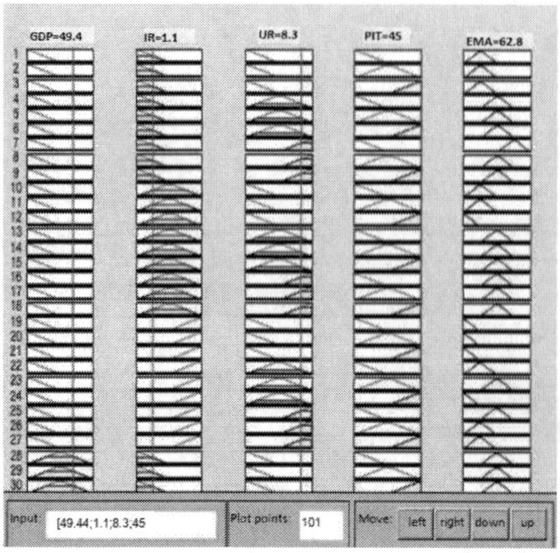

Figure 3. Implementation of Mamdani fuzzy inference using Fuzzy Logic Toolbox of Matlab package

the selection of the optimal country for labour migration based on the analysis of the most popular recipient countries in 2019. The parameters of the Mamdani model are interpreted quite easily, and the use of fuzzy logic makes it possible to model economic problems effectively in order to analyse the economic indicators of the migration attractiveness of countries. As a result of the implementation of this model, recipient countries are ranked as being the most attractive for migrant employees. Such countries as Italy, France and the United Arab Emirates, which have medium and high levels of per capita GDP, low inflation, and a "loyal" level of tax, were in the first three positions.

Table 4. Rating of recipient countries' attractiveness to international labour migration, 2019 [24–28]

Country	Input variables				Output variable	Rank
	GDP, thousand $	IR, %	UR, %	PIT, %	EMA	
Italy	44.20	0.6	9.8	43	68.9	1
France	49.44	1.1	8.3	45	62.8	2
United Arab Emirates	69.90	1.9	2.4	0	60.3	3
Saudi Arabia	48.91	2.1	5.9	0	58.3	4
Great Britain	48.71	1.7	4.1	45	56.3	5
Canada	51.34	1.9	5.4	33	52.9	6
USA	65.12	1.8	3.9	37	52.5	7
Germany	56.05	1.4	3	45	51.1	8
Australia	53.32	1.6	5.3	45	50	9
Russia	29.18	4.5	4.4	13	46.6	10

The model could be improved to reflect global trends in 2020 and the impact of the pandemic on global migration flows.

References

[1] D.S. Massey, *Patterns and processes of international migration in the 21st century*, in *Conference on African Migration in Comparative Perspective, Johannesburg, South Africa, 4-7 June, 2003* (2003), `http://citeseerx.ist.psu.edu/viewdoc/download?doi=10.1.1.473.925&rep=rep1&type=pdf`
[2] E. Scherbakova, Demoskop Weekly (2006)
[3] M.J. Miller, S. Castles, *The age of migration: International population movements in the modern world* (Palgrave Macmillan Basingstoke, Hampshire, 2009)
[4] T.J. Hatton, J.G. Williamson, NBER Working Paper (2002)
[5] E.G. Ravenstein, Journal of the Royal Statistical Society **52**, 241 (1889)
[6] L. Serhieieva, O. Kovtun, A. Opalenko, O. Ivanylova, International Journal of Industrial Engineering & Production Research **31**, 625 (2020)
[7] A. Novik, Ekonomika ta derzhava **4**, 119 (2019)
[8] M. Averkyna, A.Kudrei, Pryazovskyi ekonomichnyi visnyk **1** (2020)
[9] O. Ovchinnikova, Intelekt XXI **4**, 27 (2017)
[10] D. Ocheretin, V. Los, H. Kucherova, O. Bilska, E3S Web of Conferences **166**, 13024 (2020)
[11] V. Babenko, A. Panchyshyn, L. Zomchak, M. Nehrey, Z. Artym-Drohomyretska, T. Lahotskyi, WSEAS Transactions on Business and Economics **18**, 209 (2021)

[12] D. Sathler, S. Adamo, E.E. Lima, D. Macedo, A. de Sherbinin, P. Kim-Blanco, Applied Geography **109** (2019)

[13] I. Aksonova, V. Derykhovska, in *State management and administration, service sector, economy and international relations as the driving force of economic growth states of the XXI century*, edited by M. Goldenblatt (European science platform, Vinnytsia, 2020), 1st edn.

[14] I. Porat, L. Benguigui, Physica A: Statistical Mechanics and its Applications **561**, 125210 (2021)

[15] H. Akbari, Social Networks **64**, 181 (2021)

[16] A.A. Tarasyev, J.B. Jabbar, IFAC-PapersOnLine **51**, 407 (2018)

[17] D. Kniveton, C. Smith, S. Wood, Global Environmental Change **21**, 534 (2011)

[18] C. Searle, J. van Vuuren, Computers, Environment and Urban Systems **85**, 101568 (2021)

[19] C. Smith, C. Thoenissen, Journal of Economic Dynamics and Control **109**, 103781 (2019)

[20] A.G. Korchenko, *Building information security systems on fuzzy sets. Theory and practice* (EMK-Pres, 2006)

[21] L.A. Zadeh, Information and Control **8**, 338 (1965)

[22] A.V. Leonenkov, *Fuzzy modeling in MATLAB and FUZZY TECH*, 736 (BHV Peterburg, SPb, 2005)

[23] E.H. Mamdani, Proc. IEEE **121**, 185 (1974)

[24] United Nations, *International migrant stock: the 2019 revision* (2020), https://www.un.org/en/development/desa/population/migration/data/estimates2/estimates19.asp

[25] World Bank, *GDP per capita from world development indicators database* (2020), https://data.worldbank.org/indicator/NY.GDP.PCAP.PP.CD?end=2019&start=2019

[26] World Bank, *Inflation of consumer prices from International Monetary Fund* (2020), https://data.worldbank.org/indicator/FP.CPI.TOTL.ZG

[27] World Bank, *Unemployment rate from International Labour Organization, ILOSTAT database* (2020), https://data.worldbank.org/indicator/SL.UEM.TOTL.ZS

[28] TRADING ECONOMICS, *List of countries by personal income tax rate* (2020), https://tradingeconomics.com/country-list/personal-income-tax-rate

https://doi.org/10.1051/shsconf/202110708003

Foreign direct investment trends: an analysis of the structure and dynamics in the context of globalization

Liubov Kibalnyk[1,*], *Maryna* Leshchenko[2,**], *Olena* Feshchenko[3,***], and *Haiyu* Wang[2,****]

[1]Department of Modeling of economy and business, The Bohdan Khmelnytsky National University of Cherkasy , 81 Shevchenko Ave., Cherkasy, 18031, Ukraine
[2]Department of International Economics and Business, Cherkasy State Technological University, 460 Shevchenko Ave., Cherkasy, 18030, Ukraine
[3]Department of Economics, Entrepreneurship and Management, University of Educational Management, 52-A Sichovykh Striltsiv Str., Kyiv, 04053, Ukraine
[4]Department of Marketing and Information Science, School of Economics and Management, Qingdao University of Science and Technology, 99 Songling Rd., Laoshan District, Qingdao, Shandong, China

Abstract. The phenomenon of the international investment activity in the context of globalization is investigated in the article. Foreign direct investment is the most important basis for further analysis of the world economy. The article discusses the current trends of the foreign direct investment flows with an emphasis on their geographical location. In particular, the inflows of FDI by region and national economies are considered. The specific features of modern factors of the foreign investments' distribution by regions are determined. The study shows that some methodological approaches are useful in determining the level of national economies' interconnection of the linked processes of international capital flows. The clustering method was used for the analysis of foreign investments and the minimum spanning trees for the selected groups of countries were constructed. From the defined list of countries, one group with similar trends in the FDI movement has been distinguished. The article stipulates that countries should consider the need for their active involvement in globalization processes and contribute to the formation of a favourable investment environment within the country.

1 Introduction

In recent decades, an important and defining feature of forming the world economic system is economic globalization, against which there is an active search for competitive models of national economies that would meet modern challenges of combining global trends and development patterns with national interests and goals. Modern processes of globalization are directly related to the internationalization of markets for services and capital, accompanied by increasing international mobility of national resources and the interdependence of national economies. International investment is an independent and especially important area of international economic relations being characterized by not only rapid growth and significant completion of the world trade but also its contribution to the economic growth of national economies and the increase of their competitiveness. As the processes of the world economy globalization have contributed to a dramatic increase in capital mobility and the intensification of international investment, studying and modelling of the globalization effect on the flows and structure of foreign investment are extremely relevant.

*e-mail: liubovkibalnyk@gmail.com
**e-mail: mari.leshchenko@gmail.com
***e-mail: elf-imef@ukr.net
****e-mail: haiyuwang@hsln.cn

Most researchers, interpreting the concept of "the world economy globalization", are based on the processes of interdependence of national economies in the world. For example, Director-General of the International Labour Office, in his report to the 85th session of the International Labour Conference (Geneva, June 1997) notes that globalization is a complex phenomenon of economy interdependence arising from the exchange of goods and services as well as capital flows [1]. At the same time, globalization as a world-class process determines the intensification of interrelations, interaction and interdependence of states [2]. The formation of the metaspace in the conditions of forming and functioning of transnational financial and other networks formed on the basis of international capital flows, should be distinguished among the main manifestations of globalization. As a result, the manifestations of globalization for specific economic entities depend on the extent to which the carriers of capital flows – international companies – have diversified their revenues and placed their own assets in different countries in order to increase exports of goods and services and to use local advantages [3]. That is to say, the flow of foreign investment becomes the foundation of forming a unified system of relations of the new configuration of the global economy.

Foreign investment processes have always been the focus of economists' researches. Indeed, FDI is key flows in

https://doi.org/10.1051/shsconf/202110708003

the international economic relations and reflects the level of economic development of an individual country. Moreover, it is important whether the country acts as a host economy or the home country of the investor. J. H. Dunning's concept of foreign investment determines that investments are made outside the host country but inside the investor's company; and investor has the control over resources (assets, capital, technology, management skills and knowledge, market access and entrepreneurship) [4] The eclectic theory of J. H. Dunning [5] determines that the beginning of the process of investing abroad is usually preceded by the analysis of three main factors of entering the international capital market: the availability of valuable assets (property benefits), location benefits and internationalization priorities [6]. Other factors that also need to be carefully analyzed include the need to control subsidiaries, the availability of resources, and company's overall strategy [7]. All these factors must be considered when choosing priorities for the development of international investment activity of the company.

M. Porter complemented the research of J. H. Dunning and made a significant contribution to the development of the theory of foreign direct investment. His theory of countries' competitive advantages assumes the ability of national firms to use internal resources in such a way as to have a competitive advantage in the world market while focusing on innovation and strategic behaviour of companies [8]. In continuation, S. Hymer defines the process of foreign investment as an operation to obtain physical assets abroad, and the current control remains with the multinational company in the home country [9]. Capital proved to be the most mobile factor of production, so it was international investment that became the dominant form of economic globalization.

Investment attraction involves justification and selection of investment priorities using transparent and clear tool set for investors. The concept of investment attractiveness can be interpreted from the point of view of both investors and recipients [10]. In modern conditions, attracting investment contributes to the development of national economies, the introduction of new technologies, the renewal of worn-out fixed assets, the creation of new jobs, and therefore, is a necessity for the proper functioning and development of the state. The activities of modern companies that are carriers of foreign investment are based on the desire to combine the benefits of efficiency through the globalization of operations and the benefits of delegating authority to subsidiaries [11].

The level of national investment risk is also an important factor in attracting foreign direct investment. Based on measurement of investment risk for different countries and the extended gravity model, the influence of national investment risk on foreign direct investment is discussed. Investment risks in various countries are totally different and instable because of the complexity of international situation. The market size of the host country, bilateral investment agreements, resource endowments and tax burden levels can also significantly affect foreign direct investment.

In today's development of global economy, its main characteristic is a distinctive trend, resulting in convergence of economic development of different countries, enhancement of industrial relations, rapid growth of international trade, migration of capital, expansion of international relations in banking and insurance spheres, intertwining of financial and stock markets, etc. [12].

To analyze the dynamics and structure of international investment, as well as other macroeconomic indicators, researchers use a number of methods and techniques. One of them is cluster analysis, which is a set of models and methods of aggregating (combining) rows of data matrix. The term of "cluster analysis" was coined by R. C. Tryon in 1939. Cluster implies the accumulation (condensation) of points-objects (rows of the data matrix) in the space of variables (columns of the data matrix) [13].

The use of cluster analysis was considered by P. E. Green, R. E. Frank and P. J. Robinson, discussed problems with determining the appropriate measure of similarity and the appropriate number of clusters [14]. J. Inglis and D. Johnson [15], D. G. Morrison [16], L. A. Neidell [17], and A. Shuchman [18] also expressed concern about the use of cluster analysis. Moreover, W. D. Wells [19] expressed reservations about the use of cluster analysis unless very different, homogeneous groups could be identified [20].

Cluster analysis is a purely empirical method of classification and as such is primarily an inductive technique [21]. Though some theorists have not been favorably disposed toward the use of cluster analysis, and criticism of the ad hoc nature of clustering solutions is common, classification is an important and frequently overlooked tool of science [20].

A. Wolf [22] has suggested that classification is both the first and last method employed by science. The essence of classification is that certain things are thought of as related in a certain way. Indeed, the final outcome of other methods of study may well be a new classification. Finally, cluster analysis has been used as a general data reduction technique to develop aggregates of data which are more general and more easily managed than individual observations [20].

In [23], the author defines cluster analysis as a multidimensional statistical procedure that collects data containing information about a sample of objects and then organizes objects into relatively homogenous groups – clusters (Q-clustering, or Q-technique, actually cluster analysis). The main purpose of cluster analysis is the distribution of many objects and features under study into homogeneous, in the appropriate sense, groups or clusters.

2 Research methods

In the context of our study, it should be noted that cluster analysis allows processing a large amount of information, reducing and compressing large data sets making them compact and clear. Cluster analysis is important for the set of time series that characterize economic development (e.g., investment and commodity conditions). The use of this method makes it possible to identify the periods when the values of the relevant indicators were quite close, as well as to determine the groups of time series, the dynamics of which is the most similar.

A cluster is a group, a class of homogeneous units of the population. The main task of cluster analysis is the formation of such groups in multidimensional space. The homogeneity of the data set is given by the rule of calculating a certain metric, which characterizes the degree of similarity of the *j*-means and *k*-means units of the data set.

Such a metric can be the distance C_{jk} between them or the similarity coefficient r_{jk}. Similar metrics for the selected units deemed to be uniform. Choice metric is the central point of the cluster analysis, which determines the final version of the division into classes together [24].

During the use of cluster analysis, it should be emphasized that this technique is based on two assumptions. The first assumption is that the considered features of the object, in principle, allow the desired division of the pool (set) of objects into clusters. The second assumption is the correct choice of scale or units of measuring features. In general, the method of cluster analysis involves: selecting a sample of objects for clustering, determining a set of variables, by which objects will be evaluated in the sample, if necessary, normalizing the values of variables, calculating values of similarities between objects, creating groups of similar objects (clusters), interpreting the results of the analysis.

After obtaining and analyzing the results, it is possible to adjust the selected metric and clustering method to obtain the optimal result.

To determine the "similarity" of objects, it is necessary to make a vector of characteristics for each object first; as a rule, it is a set of numerical values. However, there are also algorithms that work with qualitative (so-called, categorical) characteristics. In the process of normalization, all values are reduced to some range, e.g., $[1, -1]$ or $[0, 1]$. Finally, for each pair of objects, the "distance", the degree of similarity, between them is measured. There are many metrics; here are just the main ones.

Characteristically, that the choice of cluster analysis tools entirely depends on the purpose of the study and the individual preferences of a researcher, since the results of clustering can significantly differ when using different methods.

In our study, we use dendrograms and a minimum spanning tree (MST) to show results. These methods are characterized by sequential combination of initial elements and corresponding decrease in the number of clusters. They look at the matrix of similar dimensions $N \times N$ (where N – the number of objects) and gradually combine the most similar objects [25].

The dendrogram shows the distances or similarities corresponding to the construction of new clusters for each step of the agglomerative hierarchical clustering algorithm on the vertical axis on the left, and the objects combined in accordance with the analysis – on the horizontal axis. Important steps are to determine the parameters of the dendrogram, its construction and interpretation of the choice and number of clusters. Based on the proximity table and the chosen aggregation strategy, the objects in the table are gradually, step by step, merged into clusters. Hierarchical agglomerative cluster analysis uses the following strategies to work with the proximity matrix: the strategies of the nearest neighbour, far neighbour, group average, centroid strategy based on the increase of the sum of squares, flexible strategy [26].

The initial actions in the procedures of hierarchical agglomerative cluster analysis are the same. First, a pair of objects with the smallest degree of distance is looked for among all the objects of the proximity matrix and combined into one group. The columns and rows of these objects are removed, and a new column and row with the listed attribute values are inserted in their place, so as not to violate the diagonal of zeros. As a result, the size of the proximity matrix is reduced by one, and the smallest value found becomes a parameter of the dendrogram, as it determines the distance between these objects, and the group itself is denoted by the number of $n + 1$. At each subsequent step, there is a merger of two objects or an object and a group or two groups, for which the degree of closeness is minimal; a similar recalculation is performed; and the combined groups are denoted as $n+2, n+3, \dots, n+(n-1)$. The procedure is completed when the dimension of the proximity matrix is 2×2 [27]

Minimum spanning tree shows the location of system elements and their optimal combinations. Graphically, the minimum scanning tree is represented as a connected graph consisting of n vertices (nodes) and $n-1$ edges. The minimum scanning tree has the shortest length among all the trees, based on the sum of distances between two elements. The minimum scanning tree reflects the hidden information contained in economic time series. Pairwise correlation coefficient is used to quantify the degree of similarity of system elements:

$$C_{ij} = \frac{(Y_i Y_j) - (Y_i)(Y_j)}{\sqrt{((Y_i^2) - (Y_i)^2)((Y_j^2) - (Y_j)^2)}} \qquad (1)$$

where i, j – variable index, $Y_i = lnP_i(t) - lnP_i(t-1)$ and $P_i(t)$ – values of i variable at time t. The matrix of $n \times n$ size is composed of C_{ij} correlation coefficients.

It is known that the correlation coefficient can range from -1 (completely uncorrelated pair) to 1 (completely correlated pair). The matrix of correlation is a symmetric matrix with units on the main diagonal. To understand and interpret the topological structure of the studied system, a generalized metric is used; it is determined by the formula:

$$d(ij) = \sqrt{2(1 - C_i j)} \qquad (2)$$

By this definition, $d(i, j)$ numerically satisfies the following axioms:

(i) $d(i, j) = 0$, if and only if $i = j$, i.e., the axiom is fulfilled under the condition of full correlation;

(ii) $d(i, j) = d(j, i)$ – the second axiom is fulfilled because we have a matrix of cross-correlation coefficients and, accordingly, a matrix of distance D symmetric by definition;

(iii) $d(i, j) \leqslant d(i, k) + d(k, j)$ – which is numerically verified – the third axiom is fulfilled [28].

MST and the associated hierarchic tree show the existence of clusters of any market assets, information about which is important from economic point of view. The obtained taxonomy allows grouping economic objects being homogeneous in terms of economic activity [29].

When interpreting the results of cluster analysis, it is worth considering that cluster analysis has some shortcomings and limitations. In particular, the composition and number of clusters depend on the selected criteria of distribution and grouping. When reducing the original data array to more compact one, certain distortions may occur, as well as individual features of individual objects may be levelled by replacing their characteristics with generalized values of cluster parameters. Besides, when classifying objects, the possibility of the absence of any cluster values in the considered set is often ignored.

3 Results and discussions

The globalization of the world economy has contributed to the intensification of international investment processes and influenced the formation of the structure and dynamics of foreign direct investment. The influence of the external factors on the development of national economies has been intensified; the factors include the world markets for goods, services and production factors, global competition, economic policy of the subjects of the world economic processes involving states, regional integration groups, international organizations, transnational corporations, etc. As a result, it is worth noting the similar dynamics of gross domestic product (GDP) and foreign direct investment (FDI), as the most important indicators of economic development in general, and international investment activity, in particular (figure 1).

It should be noted since the amounts of attracting FDI provide control over business and are often associated with the ownership of tangible assets, such as equipment, buildings or other real estate, they are the main flows in the international investment processes. FDI is a determining factor that affects the rate of economic growth, and, therefore, is characterized by a high level of sensitivity to changes and transformations occurring in the global economy.

The development of the international investment under the influence of global transformations should be divided into three conditinal waves, in fact, with a 7-year period of changes, except for the last continuing till nowadays.

The first wave of investment inflows (figure 1, *b*) is a period of stable growth from 1994 to 2001, characterized by a sixfold increase in FDI (from 278.76 billion US dollars to 1569 billion US dollars in 2000) and the beginning of short-term decline to 895.37 billion US dollars in 2001.

The second wave of investment inflows is a period of rapid growth from 2002–2003 to 2008, characterized by more than a fourthfold increase in FDI (from 737.03 billion US dollars to 3,136.1 billion) and a two-year decline to 1,447.4 billion US dollars in 2009.

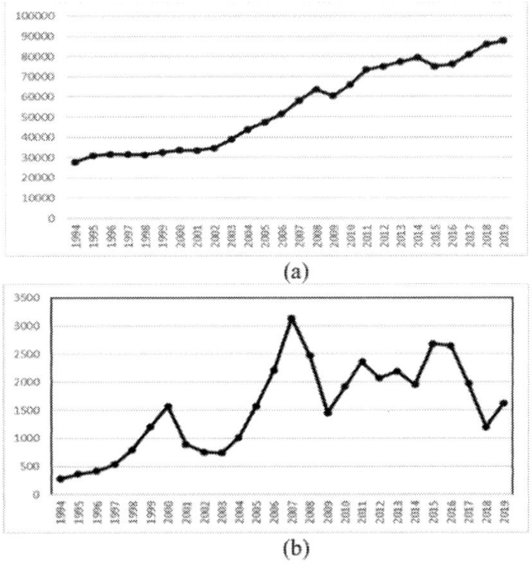

Figure 1. GDP (*a*) and FDI (*b*) dynamics in the world, 1994–2019 [30]

The third wave of investment inflows can be divided into 2 periods. The first period took place from 2009 to 2014, with a gradual recovery from the previous wave and 2184.5 billion US dollars FDI in 2013. The second period is the growth of FDI during 2014–2016 and a sharp decline during 2017–2018. In 2018, FDI was much lower than in the beginning of this period and even lower than in the post-crisis period in 2009.

The analysis of the FDI dynamics showed a growing trend of FDI with wave-like movement and minimum volume in 2018. A recovery begins in 2019 and is projected to have values of 2015 in the near future.

Under the influence of global transformational shifts, the redistribution of the international investment markets occurs. The formation of regional and sectoral structure of FDI and their dynamics are influenced by a number of factors related to structural changes in the economy in the transition to a new technological level of development that, in its turn, leads to a new stage of global competition, stimulates the aggravation of geopolitical and geoeconomic contradictions (table 1) [31].

The developed countries are among the leaders in attracting FDI; however, developing countries have much higher growth rate of incoming FDI. The attracted investments in less developed and transition economies remain volatile due to their low attractiveness to investors and volatile demand for the products and resources they can offer. The FDI flows to the developed countries were over 800 billion US dollars in 2019; they exceeded the indicators of the previous year being 557 billion US dollars in 2018. The United States of America remained the most attractive investment country attracting 251 billion US dollars in FDI flows, followed by China – 140 billion US dollars and Singapore – 110 billion US dollars [32].

Table 1. FDI flows by region, 2017-2019 (USD billion and per cent)[31]

Region	1994	2000	2003	2007	2010	2013	2016	2019
	US dollars at current prices in billions							
Developed countries	150.6	1119.1	337.9	1282.1	710.4	716.5	1265.2	800.2
Developed countries: *America*	53.3	380.9	60.6	333.4	226.7	270.8	507.9	296.7
Developed countries: *Asia*	1.4	15.3	9.6	31.3	5.7	14.1	31.3	32.8
Developed countries: *Europe*	34.7	52.1	48.0	46.1	31.6	25.6	34.0	27.9
Developing countries	102.4	231.6	194.9	522.4	622.0	656.0	652.0	684.7
Developing countries:*Africa*	6.1	9.7	18.2	51.1	46.6	52.1	46.0	45.4
Developing countries: *America*	27.7	79.8	45.6	116.9	160.7	185.4	136.6	164.2
Developing countries: *Asia*	68.4	142.0	130.7	353.2	412.8	415.7	468.4	473.9
Transition economies	1.9	5.9	17.8	87.2	63.8	83.9	66.3	54.9
	percent							
Developed countries	59.1	82.5	61.4	67.8	50.9	49.2	63.8	52.0
Developed countries: *America*	20.9	28.1	11.0	17.6	16.2	18.6	25.6	19.3
Developed countries: *Asia*	0.5	1.1	1.8	1.7	0.4	1.0	1.6	2.1
Developed countries: *Europe*	34.7	52.1	48.0	46.1	31.6	25.6	34.0	27.9
Developing countries	40.2	17.1	35.4	27.6	44.6	45.0	32.9	44.5
Developing countries: *Africa*	2.4	0.7	3.3	2.7	3.3	3.6	2.3	2.9
Developing countries: *America*	10.9	5.9	8.3	6.2	11.5	12.7	6.9	10.7
Developing countries: *Asia*	26.9	10.5	23.7	18.7	29.6	28.5	23.6	30.8
Transition economies	0.8	0.4	3.2	4.6	4.6	5.8	3.3	3.6

However, it is worth noting that even the recorded FDI values for these countries remained low compared to historical levels.

In particular, the FDI inflows to North America remained at 298 billion US dollars level, and total investment flows to the developed countries fell by 6% (about 643 billion US dollars being only half of the peak recorded in 2007).

Trends for the developed countries are mostly determined by the FDI dynamics in the European Union countries where the foreign investment flow fell by 15% to the indicator of about 305 billion US dollars. Although the trends for the largest players-countries were extremely diverse, FDI in the UK fell by 6% during the development of Brexit; the sale of Hong Kong and China assets caused the FDI decrease by 48% in the turbulence conditions, and the foreign investment inflows to Germany nearly tripled as multinational corporations lent to foreign affiliates during a period of slow growth.

Among developing countries, the growth engine of the group of countries remains the region of Southeast Asia where the growth of the attracted FDI was over 30% in 2019 compared to the previous year. This increase was due to the positive results of economy growth in some countries including Indonesia, Malaysia, Singapore and Vietnam especially due to the significant investment inflow to the branches of manufacturing sector. Among the countries of South Asia, the FDI distribution in the region is significantly influenced by the indicators of India where the growth of the attracted FDI was in the fields related to information technologies. In particular, the increase in the FDI inflows to Bangladesh by 50% in 2019 was due to the attraction of capital to infrastructure facilities, mainly from China. For the countries of West Asia, the average decrease of FDI was 10% except for Saudi Arabia where there was a steady increase of FDI mainly due to capital investment to chemical industry [33]. It is worth noting that most countries of West Asia continued to attract investment to the oil and gas industries.

In the region of Latin America and Caribbean countries, unstable dynamics of the FDI volumes should be noted due to the slowdown in economic development of some countries. Thus, investment flows in South America have not fluctuated dramatically as compensation for declining FDI to Brazil and Argentina and growing volumes in Chile, Peru and Colombia. Moreover, a significant increase in the FDI flows to Brazil did not ensure the introduction of new rules to facilitate sales of state-owned subsidiaries; therefore, this factor is expected to be an additional one in attracting new FDI in the coming years. In the countries with positive dynamics, growth was mainly due to additional public investment in infrastructure (Peru, Chile, Colombia), additional investment flows to mining companies (Peru, Chile), due to the efforts to improve the investment climate in national economies (Chile, Colombia). It is also worth noting a certain decrease in the volume of FDI attracted to Mexico in recent years against the background of investor's uncertainty regarding domestic policy towards the ratification of the new trade USMCA (United States-Mexico-Canada Agreement) [33]. Concerning other Central America countries, the FDI flows grew but not significantly.

Following the results of 2019, African countries slightly reduced the indicators of FDI attracted by about 2%. Unstable economic development, constant tension in trade policy, political instability in several countries of the continent act as a deterrent to the FDI growth, even against the background of ratification of a new AfCFTA (African Continental Free Trade Area) [32]. As a positive result, a number of positive reforms to the rules for oil and gas

https://doi.org/10.1051/shsconf/202110708003

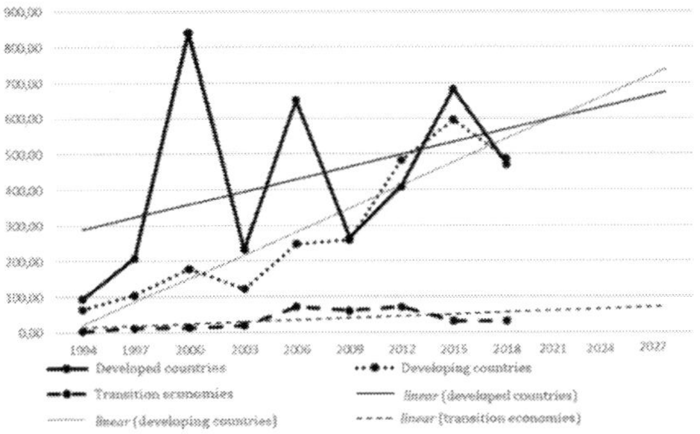

Figure 2. Wave changes in FDI according to groups of countries, 1994–2018 [30]

companies to reduce mandatory requirements to the state ownership, particularly in Nigeria, should be noted.

In countries with transition economies, the volume of FDI attracted in 2019 increased compared to 2017–2018 but were lower by 18% than the indicators of 2016. The agreements of cross-border mergers and acquisitions involving companies from the Russian Federation, the region's largest economy mainly in the mineral resources and Internet services sectors, played a significant role in the slight increase in the indicators.

While maintaining the leadership of the developed countries, the FDI geography has significantly changed in favour of developing countries (figure 2). This can be explained by their active involvement in various international activities and a number of geopolitical and trade risks.

Based on the data of figure 2, it can be argued that since 2009, developing countries have been increasing their potential to attract FDI and may acquire the status of a leader in terms of investments attracted in the next 5–10 years.

A number of restrictions to international capital movement applied by national governments have played a significant role in changing the intensity of FDI flows. Almost all countries in the world today are members of one or more transactions on regional integration or trade agreements. These agreements focus on cross-border measures concerning services, investment and competition policy, capital movements, intellectual property rights, government procurement, standards, labour and the environment [34].

Thus, only 10% of protectionist measures adopted by countries restricted foreign investment 15 years ago, and 34% – according to the results of 2018. Moreover, most restricted measures (21% from 31%) in 2018 were at the expense of the developed countries. National governments banned international mergers and acquisitions being worth of about 153 billion US dollars with these instruments under the pretext of a threat to national security or for antitrust reasons.

In particular, the USA and Germany have become more cautious about foreign and especially Chinese investment due to fears that after the arrival of foreign investors to national economy, foreign investors will be able to access important technologies and assets making economic security vulnerable. This trend has become characteristic to the countries of the European Union. For example, in the spring of 2018, the European Commission adopted regulations to verify any foreign investment in the countries of the union. China, in its turn, tried to limit the outflow of capital abroad that led to a reduction of Chinese foreign investment in recent years. In addition, as a result of Donald Trump's reform, American corporation repatriated profits, as a result of which the US national economy lost its leading positions in terms of investment abroad.

This structure and dynamics of the FDI movement shows a special role of public policy of countries in shaping the investment attractiveness of their own economy. We should note that the distribution of the global FDI flows is based on the factor of investment attractiveness of countries. This is especially important in the context of globalization of the world economy and profound structural changes occurring under the influence of technological innovations, and requires all countries to mobilize their own resources and capabilities. Researches show that different methods of assessing investment attractiveness that are mainly developed for certain levels of economic system, are based on processing the indicators of attracting FDI. Ultimately, national economies with favourable investment climate determined by a set of legal, financial, political, social, and cultural factors stipulate the feasibility of investing in a particular economic system; and a change in the investment climate leads to redistribution of investment flows in the world economy and is a determining factor in the activity of investors.

In order to identify the relationships between different countries with different rates and indicators of economic development and to determine their future prospects, we use the method of cluster analysis of indicators of attracting FDI and the generalized indicator – the annual index of actual foreign direct investment (according to UNCTAD). This index estimates and classifies 140 countries by comparing FDI and GDP of each country and is a ratio of a country's share of world FDI flows to its share of the world GDP [35]. A country with higher ratios of GDP, employment and exports in the corresponding aggregate (global) indicators will receive higher share of FDI inflows. If the inflow index of an investment country is higher than 1, it means that the country receives more foreign direct investment than could be predicted by comparing the above ratios.

To implement clustering, we chose the index of the actual inflow of foreign direct investment for 15 countries belonging to different groups of countries in terms of economic development (the USA, Brazil, UK, China, Hong Kong, Singapore, Japan, Canada, Germany, Mexico, Russia, Poland, Belarus, Ukraine, and Azerbaijan) and the volume of FDI received by these countries for the same period of time, namely for the period of 1994–2019. The investigation period was chosen considering the following features: first, all countries had to be independent; second, the correct construction of trees required databases to be large enough to analyze the obtained results accurately; third, since 1994, all selected countries began the

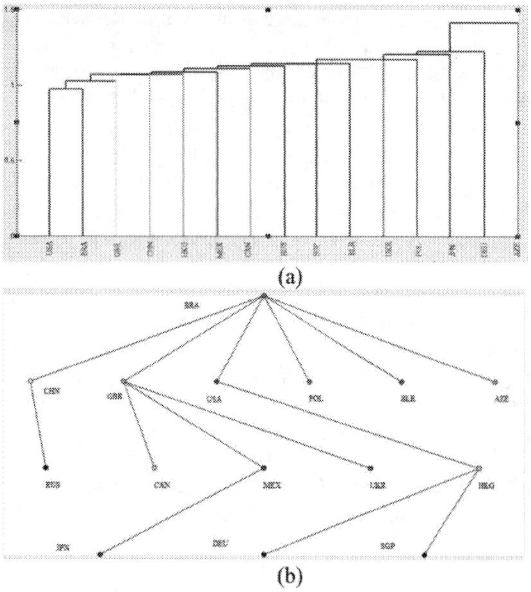

Figure 3. Dendrogram (*a*) and minimum scanning tree (*b*) for countries in terms of attracting FDI, 1994–2019

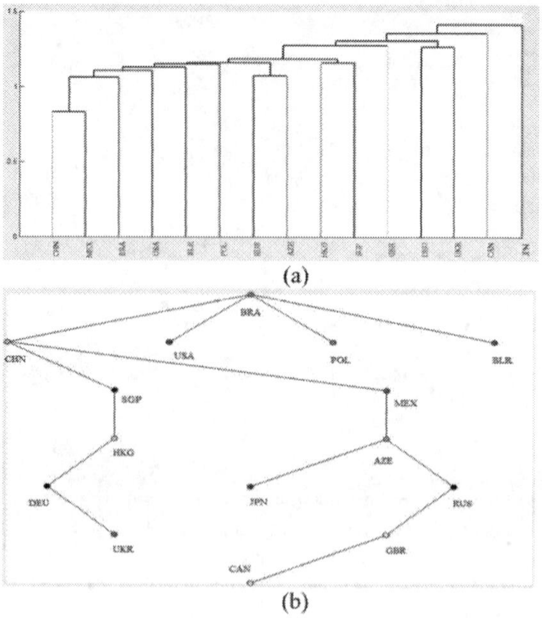

Figure 4. Dendrogram (*a*) and minimum scanning tree (*b*) for countries on the index of actual FDI inflow, 1994–2019

stable development of foreign economic relations with the rest of the world, resulting in the formation of a modern economic space.

In the process of analyzing the graph of hierarchical tree, one should understand that initially, each country is a separate cluster. When building a tree, data with similar trends are combined in separate clusters depending on the strength of the relationship between them. This analysis continues until all clusters are combined into one. Ideally, the dendrogram clusters are determined (separated) by jumps at a distance from each other. If such a distance is insignificant, the clusters are close; if the jumps in distances are significant, the objects at a great distance from each other, i.e., dissimilar clusters, begin to unite into clusters. As a result of calculations, we constructed a dendrogram (*a*) and minimum spanning tree (*b*) for the selected countries according to both defined indicators (figures 3, 4).

Data analysis of figure 3 (*a*) makes it possible to separate one cluster, which is clearly demarcated. This cluster involves the USA and Brazil. The values of this cluster are close to 1; therefore, it can be concluded that their relationship is also quite insignificant. All other clusters are formed indistinctly; a consistent overlap is observed. The analysis of the relationships between other countries shows that the separated cluster is a core, which other clusters gradually join (a core and periphery that joints this core alone). The overlap of each subsequent country with the previous one means that there is no strong connection between the rest of the countries that could unite them. In addition, the analysis of distances between objects in the feature space allow us to conclude that there is a very slight difference in features for countries. The number of dendrogram levels is quite large that indicates that many steps need to be taken to form clusters.

The location of these countries in the same cluster can be explained by the fact that the USA and Brazil are quite close to each other compared to the rest of the countries; and the United Kingdom has strong relations with both countries through the so-called "traditions of the past", when it had direct influence on these countries, and later had stable economic relations with them. The constructed minimum scanning tree gives a visualization of the obtained results of cluster analysis on the plane. The main concentrators are Brazil, United Kingdom and Hong Kong (figure 3, *b*).

Dendrogram and minimum scanning tree were constructed according to the index of actual inflow of foreign direct investment in order to identify changes in relations between the studied countries (figures 4, (*a*, *b*)).

Since dendrograms allow us to show the relationship between the studied objects, we can assume according to our calculations that the pattern in terms of the level of investment attractiveness is observed between the Russian Federation and Azerbaijan due to the significant influence of the Russian Federation on the economic development of Azerbaijan since its independence. In the cluster Hong Kong – Singapore, Hong Kong plays the main part; its policy influences the investment activity and development of Singapore, therefore, the changes in the indicators of investment attractiveness of the first country immediately influence the other country. It is also worth noting that most of the formed clusters are interconnected by geographical location; so, it is very important factor in the movement of FDI and the formation of the influence of more powerful countries on their neighbours. Among other countries, clustering with the intensity it should be due to the usual generalization of economic indicators, is not observed. The data of figure 4 (*b*) shows the formation of three clusters. The first one involves Brazil, China, Poland, and Belarus; the second – Azerbaijan, Mexico, Japan, United Kingdom, Canada, and the Russian Federation. The changes in the indicators of investment attractiveness within these clusters will be similar.

4 Conclusion

Thus, liberalization of international capital movements, the intensification of FDI flows are important factors of the globalization of the world economy, since in the conditions of uneven distribution of capital in the world economy, open access to additional financial resources for developing countries, and gives additional opportunities for productive use of surplus capital using the advantages of local markets in the host countries, for the developed countries. At the same time, it has been proven that FDI contributes to the development of stable and long-term economic relations between countries. Indicators of the relative attractiveness of the national economy for foreign investors include absolute and relative values of the FDI flows and characterize current changes of the FDI flows in the globalization levels for a definite period of time. Quantitative indicators allow monitoring of manifestations of globalization processes, in particular, international investment activity that is important not only for research but also for business and public administration in order to make decisions in response to changes in the dynamic economic environment. According to the obtained analysis results, it can be stated that the international exchange of foreign investments is an unstable cyclical process. In particular, so far, most players in the international capital market have failed to restore pre-crisis level of investment activity. Asian and Latin American developing countries remain the most attractive countries for foreign investment in recent years. We can conclude that such trends show the long-term prospects of economic development of these regions, and geopolitical uncertainty in the world significantly affects the dynamics of the volume and structure of international investment.

The analysis of the results shows that on the basis of the index of actual inflows of foreign direct investment, it is possible to track the trends of dynamics of international investment in the world. The results of the analysis show that this country is Brazil; it is the country that has the greatest number of connections with the rest analyzed countries. This makes it possible to distinguish one group of countries from a set analyzed, which have similar trends of the FDI movement. The group involves the USA, China, Brazil, Poland, Belarus. That is to say, if there is a change in the indicator in a positive or negative direction, similar changes should occur for all countries within the cluster. However, for more accurate results, it is necessary to expand the number of countries studied. It has been determined which of the countries selected for analysis with different level of development in the context of globalization has the greatest impact on the rest.

According to the FDI dynamics, the main concentrators are Brazil, United Kingdom and Hong Kong. But results of the FDI index of the actual inflow show that the main concentrators are Brazil, China and Mexico. These results confirm one of the main modern trends, a trend of economic growth of emerging markets and developing economies. The presence of Brazil and Mexico, Singapore and Hong Kong in the same cluster, according to the dendrogram, is a confirmation of this fact. The global share of developing economies in FDI inflows reached 54% in 2018, where half of the top 10 host economies are developing ones. At the same time, developed countries, such as the United States, Britain, Germany, Canada and Japan retain economies that significantly affect the dynamics and structure of FDI.

The results of cluster analysis confirm the similarity of FDI dynamics between the countries. However, they are ambiguous in defining the interactions between home and host economies. Prospects for further research are the use of other methods of economic and mathematical modeling for comprehensive analysis and identification of patterns in the structure and dynamics of FDI.

References

[1] *World Labour Report 1997-1998: Industrial Relations, Democracy and Social Stability* (International Labour Organization, Geneva, 1998), ISBN 9221103315, `http://web.archive.org/web/20020414004243/http://www.ilo.org:80/public/english/dialogue/govlab/publ/wlr/97/index.htm`

[2] A. Kolot, *Globalization as a factor in the transformation of social and labor relations* (2001), pp. 55–67

[3] A. Filipenko, *Global forms of economic development: history and modernity* (Znannya, Kyiv, 2007)

[4] J.H. Dunning, *Multinational Enterprises and the Global Economy* (Addison Wesley, New York, 1993)

[5] J.H. Dunning, International business review **9**, 163 (2000)

[6] O. Rogach, *International investments: theory and practice of business of transnational corporations* (Lybid, Kyiv, 2005)

[7] R.W. Griffin, M.W. Pustay, *International business* (Pearson, 1995)

[8] M. Porter, *The Competitive Advantage of Nations* (The MacMillan Press Ltd., London, 1990)

[9] S. Hymer, The American Economic Review **60**, 441 (1970)

[10] H. Danylchuk, N. Chebanova, N. Reznik, Y. Vitkovskyi, Global Journal of Environmental Science and Management **5**, 227 (2019)

[11] I. Sazonets, M. Varich, *The essence of the basic concept of the activity of TNCs in the minds of globalization* (Yugo-Vostok, Donetsk, 2011)

[12] M. Leshchenko, Y. Vdovychenko, Scientific Journal of Polonia University **28(3)**, 23 (2018)

[13] R.C. Tryon, *Cluster Analysis; Correlation Profile and Orthometric (Factor) Analysis for the Isolation of Unities in Mind and Personality* (Edwards Brother, Inc., lithoprinters and Publishers, Ann Arbor, Michigan, 1939)

[14] P.E. Green, R.E. Frank, P.J. Robinson, Management Science **13**, B387 (1967)

[15] J. Inglis, D. Johnson, Market Research Society. Journal. **38**, 1 (1996), https://doi.org/10.1177/147078539603800410

[16] D.G. Morrison, Management Science **13**, B775 (1967)

[17] L.A. Neidell, *Procedures and Pitfalls in Cluster Analysis*, in *Proceedings, Fall Conference, American Marketing Association* (1970), p. 107

[18] A. Shuchman, Management Science **13**, B688 (1967)

[19] W.D. Wells, Journal of Marketing Research **12**, 196 (1975)

[20] G. Punj, D.W. Stewart, Journal of Marketing Research **20**, 134 (1983)

[21] R.W. Gerard, Science **125**, 429 (1957)

[22] A. Wolf, *Essentials of Scientific Method* (Macmillan Company, New York, 1926)

[23] M. Pistunov, *Cluster analysis in economics* (National Mining University, Dnepropetrovsk, 2008)

[24] O.V. Balueva, N.M. Chynkulyak, Efektyvna ekonomika (2013)

[25] Y. Williams, *Methods of hierarchical classification. Statistical methods for computers* (Nauka, Moscow, 2006)

[26] M. Malyutova, *Statisticheskie metody dlya E'VM* (Nauka, Moscow, 1986)

[27] N. Kunanets, Bulletin of the National University "Lviv Polytechnic" **783: Information systems and networks**, 435 (2014)

[28] V. Eleyko, *Fundamentals of econometrics* (MARKA Ltd., Lviv, 1995)

[29] V. Derbentsev, O. Serdyuk, V. Solovyov, O. Sharapov, *Synergetic and econophysical methods of research of dynamic and structural characteristics of economic systems* (Brama-Ukraine, Cherkasy, 2010)

[30] The World Bank, *World bank open data* (2021), https://data.worldbank.org/

[31] UNCTADstat, *Beyond 20/20 WDS - Table view - Foreign direct investment: Inward and outward flows and stock, annual* (2021), https://unctadstat.unctad.org/wds/TableViewer/tableView.aspx?ReportId=96740

[32] UNCTAD, *Investment trends monitor* (2019), https://unctad.org/system/files/official-document/diaeiainf2019d2_en.pdf

[33] UNCTAD, *Investment trends monitor* (2020), https://unctad.org/system/files/official-document/diaeiainf2020d1_en.pdf

[34] M. Leshchenko, D. Palamarchyk, International Journal of Economics and Society pp. 280–283 (2015)

[35] M. Harrison, International Business & Economics Research Journal **2**, 93 (2003)

https://doi.org/10.1051/shsconf/202110709001

Analysis of monetary regulation of the investment sphere in the Republic of Belarus

Marina Zelenkevich[1,*] and *Natallia* Bandarenka[1,**]

[1]School of Business of Belarusian State University, 7 Oboinaya Str., Minsk, 220004, Belarus

Abstract. The purpose of the article is to substantiate the possibility and necessity of the central bank's monetary policy to stimulate investment and economic growth for developing economies on the example of the investment sphere and monetary policy in Belarus. It was determined that the impact of monetary regulation on investment and economic growth is achieved in the course of the central bank's activities to maintain indicators of price and financial stability which reflect favourable conditions for investment. Price stability is achieved through the implementation of various central bank strategies such as targeting the exchange rate, money supply and inflation. These strategies are defined as the objectives of monetary policy. The article discusses the advantages of monetary regulation in comparison with fiscal regulation, and also contains an analysis of its practical implementation in the Republic of Belarus in the period 2000–2019. As a result of the study the economic and financial results of the strategies applied at different stages were determined, their consequences for the economy were substantiated, and the strategies that best affect the financial and economic indicators in the country were identified. For countries with a small open economy which includes Belarus maintaining price and financial stability is complemented by a set of measures to reduce the devaluation expectations of market entities and create a favorable foreign economic environment.

1 Introduction

The main driver of innovative economic growth is investments that help maintain the country's competitiveness in the context of unstable development. To the global threats of climate change, environmental and demographic problems, poverty, terrorism and others, new challenges are added which also significantly affect the investment decisions of market actors and, accordingly, economic development. We are talking, in particular, about the new challenge of the XXI century – the COVID-19 pandemic which since the beginning of 2020 has acquired a global scale and devastating consequences for the world economy. According to experts from international organizations the fall in world GDP in 2020 amounted to 3–5% which is stronger than in the crisis year of 2009, when it reached 1.3%. Today traditional investment promotion instruments are no longer sufficient or effective. One of the urgent problems of the modern economy characterized, on the one hand, by globalization and integration, and, on the other hand, by high risks is the search for new ways to achieve its stable growth. In this regard scientific research is being updated to substantiate the transformation of approaches to determining ways to effectively implement the investment process, substantiating the methods and tools for its regulation taking into account new realities.

Monetary regulation of economic processes is actively used in world practice mainly for solving short-term problems of countercyclical policy. Target orientation of monetary regulation to stimulate investment and economic growth, i.e. the solution of structural, long-term problems from monetary instruments is still the subject of scientific controversy among economists and financiers. The research presented in the article corresponds to the following scientific directions of the conference: monitoring, modeling and forecasting in the banking sector and the dynamics of emerging markets in the crisis and post-crisis period. The subject of the analysis is to determine the features and problems of the investment sphere development in Belarus in crisis and post-crisis periods, as well as to substantiate the place and significance of monetary methods for stimulating investment in the system of state management of the economy.

Scientific works of scientists such as J.M. Keynes [1], J.R. Hicks [2], M. Friedman [3], F.S. Mishkin [4], P.A. Samuelson [5], J. Tobin [6] and others are devoted to identifying characteristics investments, forms and methods of their regulation, including financial and monetary regulation. The most famous works of Belarusian and Russian scientists in which the investment process and methods of its regulation in developing economies are analyzed belong to E.B. Dorina [7], V.N. Shimov [8], S.Yu. Glazyev [9], L.N. Krasavina [10], A.L. Kudrin [11], S.R. Moiseev [12], V.N. Shenaev [13] and others. These scientific works substantiate certain areas of financial and monetary regulation of investments, as well as their impact on economic development. This is mainly research on the strategy and tactics of implementing economic, in-

*e-mail: marina.zelenkevich@gmail.com
**e-mail: bondnata@mail.ru

novation and social policies using the potential of national financial and credit systems. However, there are no substantiated scientific approaches to determining the forms and directions of the banking system's impact on the investment process in a developing economy which creates obstacles to optimizing the methods and sources of investment growth taking into account its specifics. This does not allow adapting the existing theoretical and practical studies of monetary regulation of the investment process to the modern conditions of its implementation taking into account the increased external and internal risks.

The scientific literature presents studies of some directions of the influence of monetary policy on investment. A feature of emerging economies is the proof of the limited investment bank lending to the economy. The reasons for this situation are substantiated in some scientific works [14]. These include the following:

- Second-tier banks cannot concentrate and provide loans in an amount greater than the current state of the economy allows. This is determined by the low degree of monetization of the economy the indicator of which is the ratio of the monetary aggregate to GDP. The highest values of the monetization indicator are observed in the developed countries of Europe (from 86 to 116%). In the countries of Central and Eastern Europe the level of monetization ranges from 20.8% in Serbia to 84.6% in Bulgaria. In the CIS countries this indicator has a lower value and varies from 15.4% in Belarus to 44.6% in Russia [15];
- Due to the specifics of the emerging economy not all free funds go to banks (for example, economic agents keep part of their money in the form of cash);
- The long-term nature of the resources attracted by banks does not always depend on the banks themselves: this parameter, first of all, is set by the inflationary and devaluation expectations of economic agents, the level of their income and other economic and psychological factors;
- The money supply in circulation should be provided with competitive goods and services demanded by national financial assets as a store of value.

Thus, the noted features of the monetary sphere in developing economies limit the possibilities of monetary expansion as a basis for growth and often have the opposite effect expressed in the of inflation growth.

2 Analysis of the investment sphere in the Republic of Belarus

The relevance of the search for new forms and methods of investments regulation in the Republic of Belarus is confirmed by the presence of problems that exist in the investment sphere of the country. Figures 1 and 2 provide information characterizing the investment process in Belarus from 2010 to 2019 [16, 17].

The data in figures 1 and 2 shows that the share of fixed capital investments in GDP decreased from 32.5% in 2010 to 21% in 2019. At the same time, the growth rates of fixed capital investments over 9 years increased slightly by 5%; and in certain periods (2012, 2015–2016) had negative values.

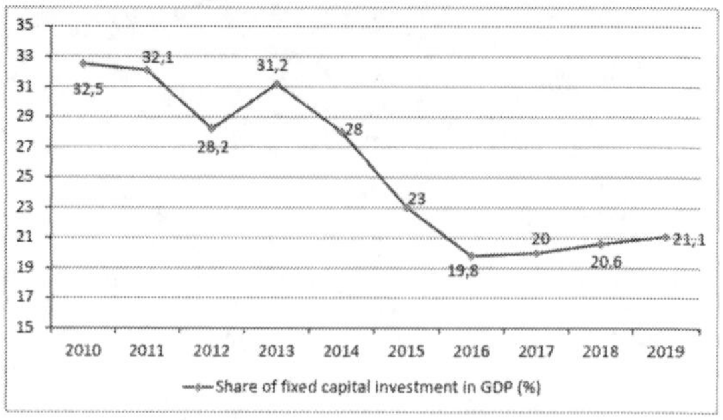

Figure 1. Share of fixed capital investment in GDP in Belarus

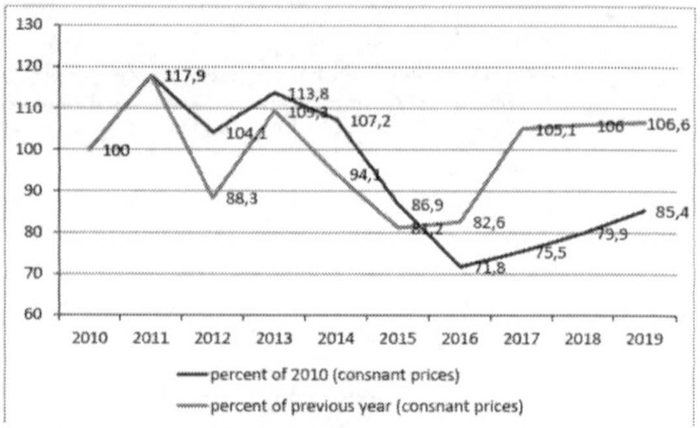

Figure 2. Growth rate of fixed capital investment in Belarus

Along with the negative trends characterizing the investment sphere in Belarus, it should be noted that there was a positive trend in attracting foreign investment to the country in the period from 2002 to 2019 (figure 3). In some time intervals a decrease in foreign investment was observed, for example, from 2012 to 2016 which was associated with the crisis in the Belarusian foreign exchange market in 2011 and 2014. By 2017 the volume of foreign investment began to grow gradually reaching the level of 2015 [18].

In the structure of foreign investments the main share is occupied by direct investments in the real sector and their volume during the analyzed period increased almost 2 times. Portfolio investments, on the other hand, occupy an insignificant share in the total volume of investments: their annual volume is less than 1.0% during the analyzed period (figure 4) [18].

The volume of other investments during the analyzed period increased until 2006 and in the subsequent period had an almost constant downward trend. Other investments include: trade credits (prepayment for exports and provision of loans or exports); loans received from international financial organizations; bank deposits (these include own accounts of foreign legal entities in Belarusian banks (investments in Belarus from abroad) and Belarusian legal entities in foreign banks; receivables and payables related

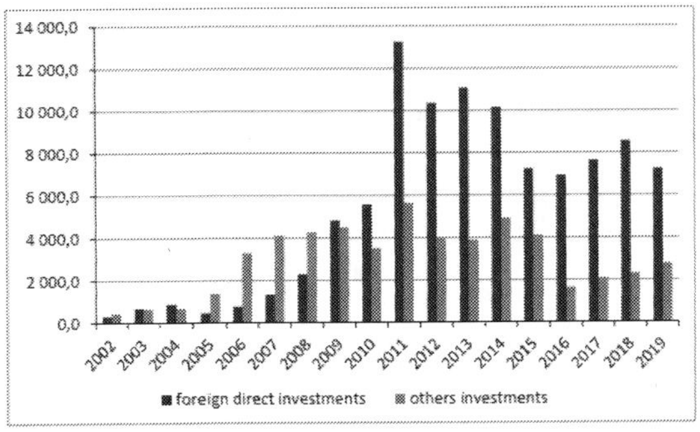

Figure 3. Foreign investment in real sector of the economy in the Republic of Belarus (million US dollars)

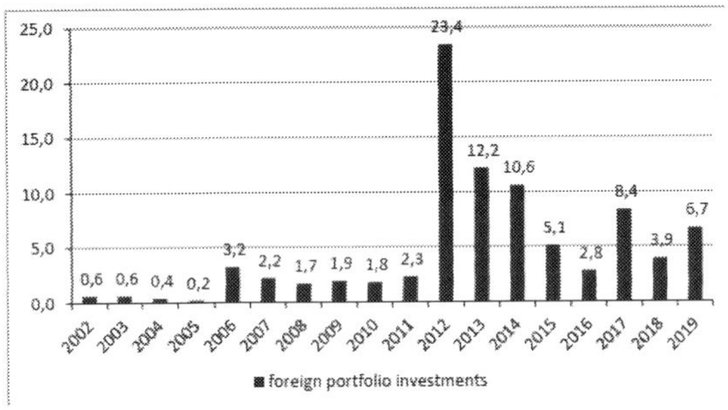

Figure 4. Foreign portfolio investment in Belarus (million US dollars)

to overdue interest payments, etc. A feature of the investment sphere of Belarus is the formation of investment resources mainly due to internal gross savings.

In the structure of investment resources organizations' own funds are in first place, and their share has practically not changed over the analyzed period: in 2005 it was 44.0%, and in 2019 – 40.6%. In second place by sources of financing are bank loans (13.6% in 2019), in third and fourth places are funds from the republican budget and savings of the population (11.4% and 10.4% in 2019. At the same time the share of budgetary funds remained practically unchanged while the share of population funds increased from 7.9% in 2005 up to 10.4% in 2019 [19].

Analysis of fixed capital investment by type of ownership showed the following trends. Despite the significant share of the state budget in the structure of investment resources their share as consumers of investment decreased from 50% in 2010 to 39.0% in 2019 [20]. At the same time the shares of private and foreign organizations in fixed capital investment are steadily growing (table 1).

From the data presented in figure 5 it can be seen that the organizations' own funds significantly exceed other sources of investment. In the structure of external sources a significant share is taken by bank loans followed by household savings and budget financing [20].

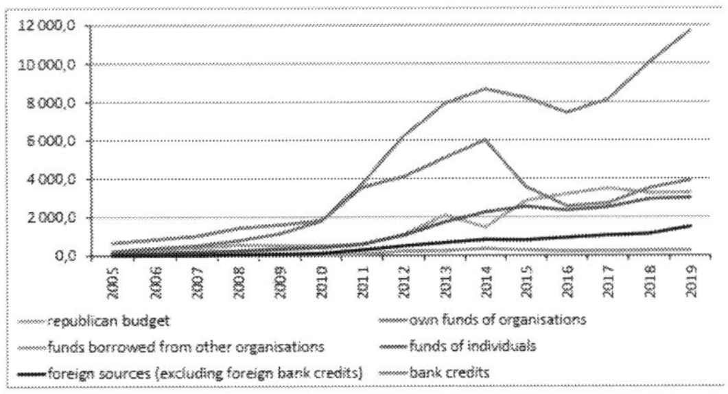

Figure 5. Fixed capital investment in Belarus by source of financing (million belarusian rubles)

Complex trends in the investment sphere of Belarus are due to the peculiarities of the country's socio-economic development. These include:

- low level of development of the financial market, in connection with which enterprises are forced to use their own resources;
- predominance of the state form of ownership in the capital of the banking system;
- low level of interconnection between the banking sector and the real sector of the economy;
- dependence of the volume of foreign investments on the situation in the world commodity and foreign exchange market;
- high susceptibility of the economy to socio-economic and political processes in Russia which remains the main economic partner of Belarus.

In addition, studies of Belarusian scientists confirm that unfavorable tendencies in the Belarusian economy are a consequence of the tendencies of decreasing the efficiency of production factors that have developed in the country over a long period [21].

The revealed features of the investment sphere of Belarus necessitate the identification of effective ways to stimulate the investment process.

3 Methodology of monetary regulation of investments

The most general classification of financial regulation methods presented in scientific sources is their division into two groups – fiscal methods and monetary methods. This generalized approach is based on the fundamental findings of Keynesian and monetarist theoretical directions.

Fiscal regulation which includes two directions – tax and budget, has an indirect impact on the development of investment processes at all levels of management. At the national level the tax and budget systems are formed through which the state determines and implements socio-economic development goals. Rational distribution of the tax burden between legal entities and individuals taking into account the strategic goals of economic development

Table 1. The structure of fixed capital investment by type of ownership

	2010	2012	2014	2016	2017	2018	2019
Total fixed capital investment	100	100	100	100	100	100	100
on which by							
– state ownership	50.0	40.0	37.0	41.0	40.7	39.5	39.0
– private ownership	47.6	53.9	54.1	51.2	51.8	52.7	51.7
– foreign ownership	2.4	6.1	8.8	7.8	7.5	7.8	9.3

gives the state the opportunity to accumulate financial flows and direct them to finance priority areas and industries.

Monetary regulation reflects the impact of the central bank on the activities of banks and other credit institutions through monetary instruments and methods. It is associated with changes in the volume and structure of the money supply in the country, interest rates and exchange rates in the credit and foreign exchange markets. The central bank influences money supply and demand and, thus, the degree of investment activity of organizations. The growth or decline of investments in the economy will depend on the cost of resources in the money market. The central bank having legal, organizational and economic powers regulates the value of credit money.

In the context of the globalization of the economy *foreign economic regulation* is of great importance. It is a tool for influencing participants in foreign economic relations and includes such methods as customs procedures, payments, control and the legally established procedure for interaction between them. Foreign economic regulation is considered as a kind of fiscal regulation which is implemented in the context of the country's participation in international economic relations. A feature of its implementation is the need to coordinate national goals with the functioning of international institutions and the system of organizing world economic relations formed on the basis of international law and interstate agreements. Foreign economic regulation affects the import of goods and services, the export of manufactured goods and services, foreign exchange earnings, as well as the attraction and use of various forms of foreign investment.

Monetary regulation plays an important role in the development of the investment process. The first effect of monetary regulation on investment is short-term. It consists in ensuring optimal operating expenses of organizations and the population which can provide a potential growth in savings. The monetary regulation in this case is aimed at rational management and stimulating investment activity of market entities at the micro level. The second effect of the monetary regulation has a long-term character which consists in creating conditions for attracting domestic and foreign investments. In fact, it is aimed at creating and maintaining the investment climate in the country. This long-term effect is realized at the macro level through the state investment policy.

The advantages of monetary regulation over fiscal regulation are still the subject of scientific controversy. Scientific discussions are built on the basis that along with the positive aspects, monetary policy also has disadvantages. In particular, one of the problems of monetary policy implementation is cyclical asymmetry in investment and banking operations. We believe that it is necessary to determine the merits of monetary policy as the basis for the implementation of the state's investment policy.

First, the advantage of monetary policy is its *flexibility and efficiency*. This means that when external economic, political and other conditions change, monetary policy can be transformed rather quickly taking into account these changes.

The parameters of monetary policy are set by the Central Bank endowed with the necessary powers to make quick decisions. The Central Bank responds promptly to the market situation based on information about the current situation and forecasts of the development of financial markets and the economy as a whole. The Central Bank makes daily decisions on the use of open market operations, changes in reserve requirements, interest rates and other instruments. Thus, it influences the volume and structure of money supply on a daily basis.

The property of efficiency which fully relates to monetary policy cannot be attributed to fiscal one. As is known from practice the implementation of fiscal policy is a rather complicated and cumbersome process. This is due to the fact that making different kinds of decisions such as directions on budget expenditures, changes in tax rates, tax incentives, etc. may be postponed for a long time due to public discussions and approval of these decisions in the legislature.

In addition, the methods and instruments of monetary policy are softer and politically more conservative than those of fiscal policy. For example, an increase in government spending has a direct impact on the volume and structure of the distribution of budgetary resources which are approved by the law on the state budget. Consequently, these changes must be made to the law in accordance with the approved procedure. Tax changes are usually adopted for a long time and require public and parliamentary discussion. However, they can be unpopular and have political consequences. Monetary policy appears to be more politically acceptable, as it indirectly affects the social sphere and vulnerable sectors of society. Influencing the conjuncture of the money market the Central Bank provides the population with an opportunity to choose assets for savings and investment methods.

The *second advantage* of monetary policy is that Central Banks have so-called political independence from the executive authorities which is seen as a guarantee of the effectiveness of their activities. In world practice it is believed that the most independent is the US Federal Reserve

https://doi.org/10.1051/shsconf/202110709001

System. In most countries of the world central banks report directly to parliament or to a special parliamentary commission. However, it should be understood that the independence of the central bank is relative – it is independence "within the government structures". The political independence of the central bank consists in formal independence from government instructions, legislative prohibition of direct government loans from the central bank, protection of the central bank management from dismissal, and others. This relative freedom makes it easier for the central bank to make the unpopular decisions needed to restore the economy in the long time period than it is for the government.

The independence of the central bank is seen as an important advantage of monetary regulation over fiscal, since the subject of fiscal regulation, the Ministry of Finance, being a government agency by its status expresses the interests of the government which can have a political overtone and contribute mainly to the achievement of short-term goals.

The *third advantage* of monetary regulation over fiscal regulation which is considered in the scientific literature is the effectiveness of its impact on economic stability. For example, Russian scientists Zubov and Inozemtsev note that since the monetary regulation mechanism is implemented through banks which are private structures their decisions are more effective (speed of adoption and implementation of decisions) in comparison with government organizations that implement fiscal policy. In private banks and other private credit institutions, responsibility for spending their own funds is personified, and not blurred as when using state (budget) funds [22].

Foreign sources also substantiate the advantages of monetary regulation. So in the work of I. Stiglitz and B. Greenwald [23] it is proved that monetary policy based on the credit paradigm is more effective in terms of impact on the stability of the economy. There are studies that prove that the modern structure of banking systems with large, state-owned banks is highly capable of absorbing the consequences of crisis impacts [24]. These theoretical conclusions are confirmed by empirical data that characterize the tendency to increase the degree of government regulation of banking systems in both developed and developing countries. State regulation in the banking system is ensured by the mandatory presence of the first link in its structure – the Central Bank, as well as the presence of a system of state banks. Moreover, in times of crisis the number and importance of state-owned banks increases.

The fourth advantage of monetary regulation is that the volume of bank investment resources significantly exceeds budgetary financing of the economy. This is due to the fact that the volume of budget financing is limited by the volume of the expenditure side of the budget and has a strictly targeted direction. This conclusion is confirmed by the indicators for the Republic of Belarus, presented in figure 6 [25].

As can be seen from the data presented, the volume of bank lending in the Republic of Belarus consistently exceeds the volume of budget financing by an average of 8-10 times annually.

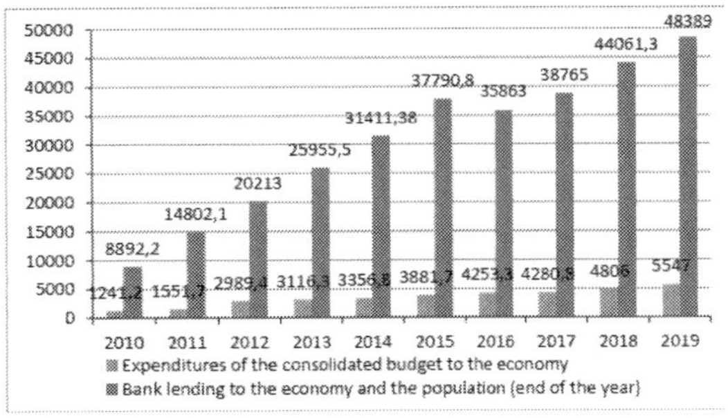

Figure 6. Banking and budgetary investment resources for the economy of the Republic of Belarus (million belarusian rubles)

Modern financial systems of developing countries are largely based, first of all, on the banking (credit) system, which is more developed, concentrated and stable.

4 Monetary policy strategies and its economic results in the Republic of Belarus

In world practice the evolution of goals, objectives and strategies of central banks went through three stages starting from the Great Depression of the 1930s. and ending with the Great Inflation period of the 1970s:

- at the first stage the tasks of the central banks were determined by the requirements of the gold standard. The main functions of central banks were to ensure the convertibility of the national currency into gold, ensure the stability of the payment system, and provide loans to government bodies;
- the consequences of the Great Depression focused the attention of central banks on the need to maintain employment and price stability, ensure the stability of the balance of payments and the exchange rate of national currencies;
- lafter the Great Inflation of the 1970s. the main goal of central banks is to achieve price stability.

At present it is reasonable to single out the fourth stage in the evolution of monetary policy the beginning of which was the global financial and economic crisis of 2008–2009.

The consequences of the crisis determined the need for central banks to resolve two key issues: improving the mechanisms of banking regulation and financial supervision and achieving not only price, but also financial stability.

Central banks pursue a policy aimed at creating conditions for long-term sustainable economic growth by ensuring low and predictable inflation. Another important indicator of the conditions for long-term growth is the provision and maintenance of financial stability in the country. From the standpoint of strengthening the investment orientation of monetary policy price stability is necessary for the following reasons: first, it stimulates business to long-term planning and implementation of the investment process, since it reduces uncertainty in economic relations. Secondly, it increases confidence in the national currency. If the population does not have confidence in the national currency problems arise with the flow of savings into the banking system. In turn, the decline in the stability of the resource base of banks limits their ability to lend to the economy for a long time. Third, price stability affects the stability of the financial sector, which ultimately leads to the formation of an affordable price for credit resources.

Achieving price stability is based on the central bank's policy to control inflation. This is achieved by regulating the volume of money supply and, accordingly, the credit activity of the banking system. In the scientific literature there is evidence that inflation control is an important, but not the only goal of monetary regulation [26]. As is known from scientific sources and economic practice hyperinflation fetters economic activity, moderate inflation often promotes or accompanies growth, and zero inflation or deflation are companions of depression and subsequent stagnation. A decrease in inflation may be associated not only with an increase in economic activity which is a positive result, but also with a drop in consumer demand for goods and services, as well as a low level of lending to the real sector of the economy. Therefore, it is inappropriate to assert that the main task of monetary policy is exclusively to control inflation since with this approach it can turn into a brake on economic growth.

In order to substantiate the type of monetary policy strategy that best affects the growth of investments in the Republic of Belarus it is necessary to analyze the target strategies of the National Bank. So, in the period 2000–2011 in the Republic of Belarus the exchange rate targeting regime was applied and from 2012 to 2014 this strategy was applied in combination with inflation control. During this period was carried out a policy aimed at stimulating domestic demand through credit expansion. At the same time it was supplemented by active budget support from the public sector and a soft wage policy which led to the formation of significant macroeconomic imbalances. As a result, the possibilities of further application of the exchange rate targeting regime were exhausted after the currency crisis of 2011 and the external shock of 2014.

A special stage in the economic development of Belarus is occupied by the favorable period of 2005–2010 when the growth of investments was 120% in 2005, 132% in 2006, 117% in 2007, 123.5% in 2008 and 115.8% in 2010 compared to the previous year. The share of fixed capital investments in GDP increased from 23.2% in 2005 up to 33.7% in 2010. According to international organizations high growth in this period in Belarus was achieved mainly due to the rapid accumulation of fixed capital. This became possible due to the high ratio of investment to GDP (the growth of fixed assets based on a high share of investment in GDP accounted for approximately 70%) [27]. However, during this period there was a bias in favor of building and assembly works as a result of the implementation of government programs for housing construction while the growth of investments in machinery and equipment in 2009 turned negative. Based on the

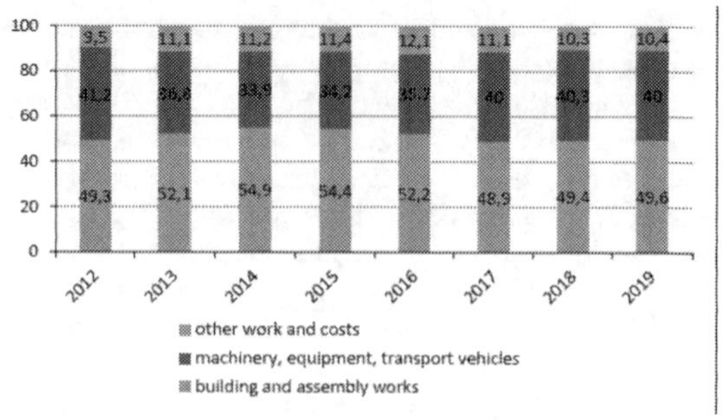

Figure 7. The structure of fixed capital investment in Belarus (%)

theoretical concept of endogenous growth developed in the second half of the 1980s capital accumulation (especially investments in technological innovation) is considered the most important determinants of long-term economic growth [28]. As a result of the research scientists have proven that the ratio of investment in machinery and equipment to GDP has a positive effect on economic growth. At the same time some scientific sources proved that the share of investment in housing (building and assembly works) in GDP does not have a significant effect on economic growth [29]. Figure 7 shows the structure of fixed capital investments in the Republic of Belarus from which it can be seen that almost half of the investments are just building and assembly works. The share of investments in machinery and equipment varied slightly from 33.9% (2014) to 41.2% (2012) and 40% in 2019 which indicates the absence of progressive changes in this indicator [19].

In the early 2000s there was an artificial stimulation of investment. This was reflected in the infusion of large sums of money into the economy in the form of directed loans, as well as in the form of a large volume of housing construction. Such artificial stimulation has become the main source of high rates of economic growth. High investment activity distinguished this period from subsequent ones and continued until the global financial and economic crisis of 2008-2009. Then the main destabilizing factor was the drop in demand for Belarusian products in foreign markets. The financial results of the organizations' work have worsened. The rate of GDP growth began to decline [30]. The monetary policy pursued at that time by the National Bank of the Republic of Belarus was based on the exchange rate strategy which contributed to economic growth until 2011.

Over the next 2011-2015 the Belarusian economy functioned in unfavourable external conditions which, first of all, were predetermined by the slowdown in economic growth in Russia which is the main economic partner of the Republic of Belarus.

Since 2010 the policy of stimulating economic growth with monetary and fiscal policy instruments has been resumed which limited the effectiveness of stabilization measures and exacerbated the existing imbalances. This led to an increase in the current account deficit of the balance of payments to 14.5% of GDP and led to a significant adjustment in the nominal exchange rate of the Belarusian ruble to the US dollar in 2011 [31].

In 2012 the economy returned to a state of temporary stabilization. This was due to an exchange rate adjustment and a significant increase in the export of petrochemical goods to the EU countries and the machine-building goods to Russia. Despite the presence of obvious imbalances in the economy, further economic policy focused on increasing domestic demand and stimulating production, which required external borrowing.

In 2013 the volume of GDP in Belarus increased by 1% in comparable prices and was provided by expanding domestic demand, which contributed to the accelerated formation of added value in retail trade and construction. The increase in activity in these segments was due to a fairly high household incomes growth and a fixed capital investment increase. At the same time there was a noticeable decrease in production volumes in industry (primarily in its export-oriented industries) and agriculture of the country.

In 2014 real GDP growth in the Republic of Belarus amounted to 1.6%. The national economy growth was due to the recovery in industrial production which was achieved through an increase in the production of the petrochemical, chemical and mining industries. At the same time observed a slowdown in the household income growth and a decline in fixed capital investment.

A decrease in the capacity of traditional sales markets for domestic products and an unfavourable pricing environment for exported goods led to a sharp decline in exports (compared to 2012 by almost 20%). This led to a negative trade balance of $ 10.2 billion (in total for 2013–2014). This, in turn, created an additional burden on the domestic foreign exchange market and limited the ability to maintain the exchange rate stability of the national currency.

The deterioration of the macroeconomic environment had a negative impact on the financial results of the functioning of the real sector of the economy. For 2014–2015 the number of unprofitable organizations in the country increased 2.7 times, and the total value of their net loss increased 3.9 times. Agriculture was among the sectors that most acutely faced the growing financial problems: in 2015 (compared to 2012) the number of unprofitable agricultural enterprises increased 4.9 times, and the total amount of their loss increased 10.7 times.

In 2014 two loans were received from the Russian Government for a total of $ 2.0 billion. In 2015 two more loans were received in the amount of $ 870 million which helped to support GDP growth.

Monetary policy based on the exchange rate strategy in the Republic of Belarus in the period 2000–2014 in the context of accelerated economic growth and declining efficiency of accumulation had negative consequences.

Such growth prompting the resort to excessive emission lending to business entities and external borrowing negatively affects the state of the country's financial system, does not stimulate investment activity. In the future this trend continued. However, according to the Belarusian scientist and the practice of the monetary sphere D. Kalechits, the reason for the low rates of equilibrium GDP is explained not so much by the expansionary monetary policy as by the fall in the return on factors of production. This is confirmed by the fact that the average contribution of total factors of production to GDP decreased from 4.1% in 2001-2009 down to minus 0.5% in 2010–2018 [32].

Structural weaknesses constrain economic growth and thereby increase the vulnerability of a country's economy to the crisis. After the initially successful stabilization of the post-crisis 2011 in 2012–2014 a policy of alternating stimulating and restraining economic measures was implemented This was mainly expressed by an increase in wages and an increase in directed lending, did not lead to the restoration of stability, and inflation remained at a high level [16] (figure 8).

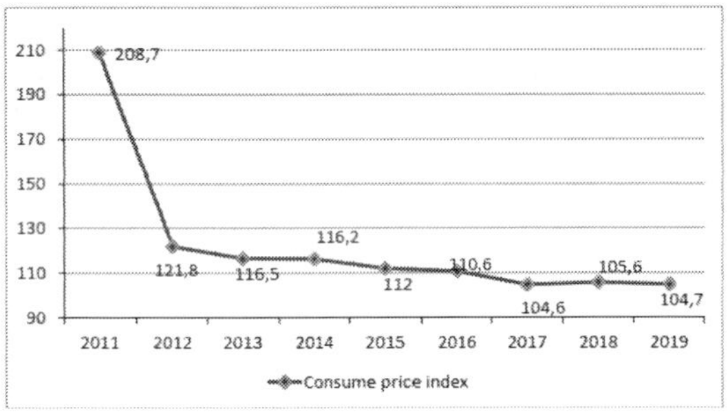

Figure 8. The trend of consumer price index in Belarus (December to December of previous year) (%)

One of the rational steps of the central bank in the following years was the monetary policy aimed at the development of the financial sector, as well as at its orientation towards the long-term development of the Belarusian economy. Since 2015 the National Bank has been implementing a monetary targeting strategy which is based on the thesis of the quantitative theory of money: in the long term the growth of commodity prices is determined solely by the supply of money.

A significant deterioration in external economic conditions in 2015 caused by the fall in world prices for raw materials, as well as geopolitical tensions led to a decrease in economic activity in Russia and Ukraine, a sharp drop in investment demand, and devaluation of national currencies in the EAEU member states. This led to a drop in Belarus' GDP in real terms by 3.8%. Subsequently, the situation improved slightly until 2018 but in 2019 there was a decrease in GDP growth rates from 3 to 1.2% [16] (figure 9).

In 2016 Belarus was launched a three-year program of the Eurasian Fund for Strategic Development providing for a financial loan of $ 2 billion. In 2017 the Russian Government provided a loan to the Republic of Belarus for refinancing the external public debt in the amount of $ 700 million. In addition, in June 2017 Belarus placed a double tranche of Eurobonds in international capital markets for a total of $ 1.4 billion.

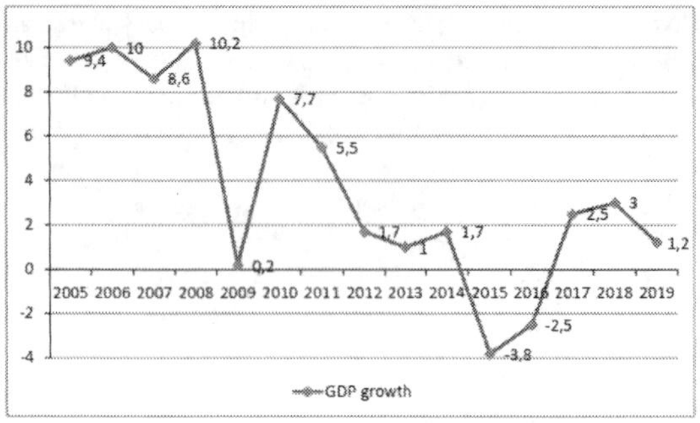

Figure 9. Gross Domestic Product in Belarus (at constant prices, percent to the previous year)

During 2017 sustained economic growth (102.5%) and a reduction in the current account deficit were achieved.

Leverage was actively used to attract external investment resources in the form of foreign investment. This policy, however, entailed a significant growth of external debt. The external public debt of the Republic of Belarus as of 01.01.2019 increased to 39288.8 million US dollars which amounted to 65.9% of GDP against 18.5% in 2007 [33].

In 2015–2018 the National Bank pursued a tough emission policy. The credit issue was carried out exclusively within the framework of using standard bank refinancing operations at market interest rates and for short periods.

The monetary targeting strategy in the Republic of Belarus had positive results such as stabilization of the economy, partial improvement of the main macroeconomic indicators. Nevertheless, there were no fundamental changes in the development of the investment process and economic growth during this period. Table 2 shows the results of the National Bank's credit issue in order to stimulate economic growth at various stages of monetary policy.

As can be seen from the data presented in the table 2 the expansionary monetary policy pursued in the Republic of Belarus had different results for inflation and the exchange rate. The best period can be considered 2004–2007 when the growth of money supply contributed to economic growth through improved conditions for economic activity (relative stability of prices and exchange rates) (figure 9). The next two periods when the National Bank used the exchange rate strategy in combination with maintaining low inflation had negative results, such as high inflation and devaluation of the Belarusian ruble. In the period from 2008 to 2015 external economic factors had a significant impact on the domestic economic situation: the global financial crisis and the decline in demand for Belarusian products in Russia after 2014. The monetary targeting policy implemented in the country since 2016 demonstrated the achievement of price and currency stability which positively influenced the investment activity of organizations and the population and, accordingly, was reflected in the indicators of economic growth.

The National Bank of the Republic of Belarus plans to apply a monetary targeting strategy was carried out until 2020 which created the basis for the transition to 2021. The inflation targeting regime in international practice is recognized as the most effective tool for managing inflation expectations of economic entities, thereby acquiring a long-term nature. A distinctive feature of this regime is a clear and transparent definition of the goals and priorities of monetary policy, its openness for all economic agents. This contributes to the formation of favourable expectations of the population and business, increases their confidence in the central bank and its monetary policy.

The transition to an inflation targeting strategy in the Republic of Belarus is provided for by the Program of the country's socio-economic development for the new five-year period 2021–2025. In this regard, since 2015 the National Bank has already implemented a number of measures to form the initial economic conditions for the implementation of this strategy. Among them is the transition to flexible exchange rate formation. A floating rate is especially necessary in case of deterioration of external conditions, since it is the exchange rate that assumes the main negative consequences, thereby protecting the domestic financial market and the economy.

Along with achieving price stability another monetary goal for long-term growth is to ensure financial stability which is a prerequisite for low inflation and unemployment. In the economic literature "financial stability refers to the state of the financial system in which banks and non-bank financial institutions, other financial intermediaries, the payment system and the financial market properly perform their inherent functions and retain the ability to perform them in the event of the destabilizing impact of internal and external factors" [34].

The stable development of the financial sector includes measures to strengthen the degree of state regulation in the financial market which was implemented in the activities of central banks in the following areas:

A. *Reorienting the objectives of central banks.* Since the post-crisis period a transition in the activities of central banks from solving purely monetary problems to general economic ones has been planned. General economic objectives are designed to solve the problems of improving the financial and investment climate, thereby contributing to financial stability, sustainability and economic growth.

B. *Reducing the degree of freedom and autonomy of central banks.* The provisions on the need for stricter regulation of the activities of central banks and strengthening their interaction with the government were developed. At present the central bank is becoming not so much an economic as a political and partly ideological institution. At the national level in developed countries a new function of the central bank is to provide financial assistance and support to banks and other financial companies.

C. *Concentration of supervisory functions in one place and creation of a single mega-regulator of the financial*

Table 2. The results of monetary policy strategies of the National Bank in the Republic of Belarus

Period	The aim of monetary policy	Emission growth (%)	Consumer price index growth (%)	Exchange rate trend (%)
2000-2003	Exchange rate	+819,0%	+194,97%	-
2004-2007	Exchange rate	+336,45%	+27,93%	-0,66%
2008-2011	Exchange rate	+132,73%	+86,6%	+116,43%
2012-2015	Exchange rate and inflation	+110,09%	+58,6%	+90,32%
2016-2018	Money supply	+101,62%	+11,1%	+2,42%

market. In the process of development of the financial market the deepening of specialization of individual financial transactions and tools has led to the emergence of multiple organizational forms of autonomous state supervisory and regulatory institutions. Among them such as the Central Bank, the Securities and Stock Exchange Commission, the Commodity Exchange Commission, the Deposit Insurance Corporation, the Mortgage Lending Commission, etc. Each country has developed its own system of state organizations associated with the regulation of individual segments of the financial market. A situation arose when several government institutions controlled the financial market which accordingly led to inconsistency in their actions and results. In the process of reforming the financial system after the crisis of 2008 the functions of the mega-regulator of the national financial markets were gradually legislatively assigned mainly to the Central Bank of the country.

Macroprudential policy of the National Bank plays a key role in maintaining financial stability. This policy solves the following tasks:

- *encouraging financial intermediaries* to build up additional capital and liquidity reserves in advance for those times when they may be required;
- *adjustment and alignment of credit flows in the economy.* An imbalance can cause not only a shortage of credit resources but also their excess. Unequal distribution and wasteful use of bank loans (when they are cheap and available) can disrupt the financial market and cause instability;
- *stimulating the efficient distribution of financial resources in the economy.* There is always the possibility that resources will be directed to those economic entities that will not be able to pay off their obligations. Detecting and correcting such distortions can reduce financial risks;
- *increasing the stability of the financial infrastructure,* limiting the level of concentration of risks in certain segments, and preventing possible situations of liquidity shortage in the financial market.

Macroprudential policy has an obvious advantage: it helps to reduce risks precisely in problem segments, as it acts point-wise maintaining macro-financial stability and minimally affecting the economic growth.

5 Conclusions

The results of the analysis of the investment sphere, the stages of evolution of monetary policy, its types and strategies in the Republic of Belarus in the period 2000-2019 allow us to conclude that:

- in developing economies which are characterized by a deficit of domestic investment, insufficient foreign investment, and low rates of development of the investment sphere in general, monetary methods and instruments should be actively used to solve long-term tasks of investment development and create favourable conditions for stimulating investment. The monetary policy of the Central Bank is aimed at creating and maintaining price and financial stability which ultimately form the conditions conducive to attracting investment in the economy;
- the choice of strategies (targets) and types (monetary expansion or restriction) of monetary policy which are established and supported by the Central Bank demonstrates the different nature of their impact on price and financial stability, as well as on the results of economic development as a whole. For Belarus the monetary targeting policy implemented since 2016 demonstrated the achievement of the best indicators of price and currency stability which positively influenced the investment activity of organizations and the population and, accordingly, was reflected in the indicators of economic growth;
- for countries with small open economies such as Belarus the long-term investment effect of the national monetary policy is largely determined by external economic factors and the situation in the foreign exchange market which have now become unfavourable due to the increased epidemiological and political risks in the world and in the country.

References

[1] J.M. Keynes, The quarterly journal of economics **51**, 209 (1937)
[2] J.R. Hicks, The Economic Journal **49**, 696 (1939)
[3] M. Friedman, in *Corporate ethics and corporate governance* (Springer, 2007), pp. 173–178
[4] B.S. Bernanke, T. Laubach, F.S. Mishkin, A.S. Posen, *Inflation targeting: lessons from the international experience* (Princeton University Press, 2018)
[5] P.A. Samuelson, in *The world scientific handbook of futures markets* (World Scientific, 2016), pp. 25–38

[6] J. Tobin, Journal of money, credit and banking **1**, 15 (1969)
[7] E. Dorina, E. Kadovba, University Economic Bulletin pp. 47–57 (2019)
[8] V.N. Shimov, Problems of Economic Transition **48**, 10 (2005)
[9] S.Y. Glazyev, Russia and new world order. Strategy of economic growth on a threshold of the XXI century pp. 164–165 (1997)
[10] L. Krasavina, Studies on Russian Economic Development **21**, 426 (2010)
[11] G.I. Idrisov, V.N. Knyaginin, A.L. Kudrin, E.S. Rozhkova et al., Voprosy Economiki **4** (2018)
[12] S.R. Moiseev, Voprosy Ekonomiki pp. 112–122 (2019)
[13] L. Krasavina, A. Galchinskiy, V. Shenaev, *Monetary circulation and credit under capitalism* (1989)
[14] O.I. Rumyantseva, Belarusian Economic Journal **2**, 61 (2014)
[15] Exchange portal TAKE-PRIFIT.BY, *Rating of countries by money supply M2* (2018), https://take-profit.org/statistics/money-supply-m2/
[16] I. Medvedeva, ed., *Statistical Yearbook of the Republic of Belarus* (National Statistical Committee of the Republic of Belarus, Minsk, 2019), p. 472
[17] *Macroeconomics and finances of the CIS and other countries of the world 2014–2018: statistical review* (Interstate stat. com. Commonwealth of Independent States, Moscow, 2019), p. 59
[18] National Bank of the Republic of Belarus, *Direct foreign investments* (2018), http://www.nbrb.by/statistics/foreigndirectinvestments
[19] I. Medvedeva, ed., *Investment and construction in the Republic of Belarus: statistical review* (National Statistical Committee of the Republic of Belarus, Minsk, 2019), p. 134
[20] National Statistical Committee of the Republic of Belarus, *Investments in fixed capital* (2018), http://www.belstat.gov.by/ofitsialnaya-statistika/realny-sector-ekonomiki/investitsii-i-stroitelstvo/investitsii-v-osnovnoy-kapital/godovye-dannye/
[21] V.N. Komkov, Bankauski Vesnik **29** (2011)
[22] V. Inozemtsev, V.M. Zubov, Problems of Economics **3**, 76 (2015)
[23] J.E. Stiglitz, B. Greenwald, *Towards a new paradigm in monetary economics* (Cambridge Univ. Press, Cambridge, 2003)
[24] A. Yudanov, *Experience of competition in Russia: reasons for success and failure* (KnoRus, Moscow, 2008)
[25] I. Medvedeva, ed., *Statistical Yearbook of the Republic of Belarus* (National Statistical Committee of the Republic of Belarus, Minsk, 2020)
[26] M. Ershov, Economic Issues **12**, 4 (2010)
[27] E. Trofimova, *Banking systems of the cis countries: different perspectives but common risks* (2004), http://data.cbonds.info/comments/8055/CIS_banks_021204_R.pdf
[28] P. Romer, Journal of Political Economy **XCIV**, 1002 (1986)
[29] H. Jalilian, M. Odedokun, Applied Economics **32**, 289 (2020)
[30] E. Petrovich, Economic Bulletin of the NIEI of the Ministry of Economy of the Republic of Belarus **6**, 4 (2016)
[31] M.N. Vlasenko, Belarusian Economic Journal **3**, 43 (2018)
[32] D. Kalechits, *Transformation of the goals of the National Bank at the present stage*, in *Intern. Scientific and practical conf. Economic growth of the Republic of Belarus: globalization, innovation, sustainability* (BSEU, Minsk, 2019), http://www.nbrb.by/top/pdf/transformaciya-celej-deyatelnosti-kalechits.pdf
[33] National Bank of the Republic of Belarus, *Gross external debt of the Republic of Belarus* (2020), http://www.nbrb.by/statistics/externaldebt
[34] S. Kalechits, Bankauski Vesnik **3**, 8 (2019)

Econometric modelling of bank activities: value-based approach to the problem loans terms' rescheduling

Serhii Onikiienko[1,*], *Mykhailo* Dyba[1,**], and *Iuliia* Gernego[1,***]

[1]Kyiv National Economic University named after Vadym Hetman, 54/1 Peremohy Ave., Kyiv, 03057, Ukraine

Abstract. The permanent state of the financial crisis has predictably brought to the forefront such traditional problem of banking as problem loans. This research aims to work out an econometric approach to the solution of the problem of loans terms' rescheduling. For this purpose, we, firstly, treated credit as a bank's investment project with cashflows' chart including initial outflow (principal) and following inflows represented by loan payments. Secondly, we combined the schematic representation of loan's cashflows with NPV formula accustomed to loan's cashflows and it allowed to create the econometric models for three types of loan: classic, annuity, serial. Thirdly, for the case when borrower breaks a loan's payment schedule and it leads to the reduction of loan's NPV and loss of the wealth of bank's shareholders, respectively, we outlined special compensative models of cashflows where default in payment is interpreted by the lender as an additional forced loan. We suggested modifying the loan terms (interest rate or effective period of the loan agreement) for the rest of payment periods. Fourthly, we laid the special compensative models of forced loans' cashflows a top corresponded initial cashflows of loans and this has made it possible to get formulas calculating the modified interest rate and the additional number of loans' payment periods with the aid of backward calculation. As a result, we developed the econometric models of the loan terms' modifications based on the prolongation of the initial credit period and the increasing of the initial interest rate.

1 Introduction

1.1 Research question

Investigating the problem loan management is a continuing concern within modern banking. There is a growing body of literature that recognizes the importance of this problem related to such aspects as the definition and identification of problem loans, detection of reasons and circumstances that lead to the occurrence of problem loans, problem loans' management including loans' portfolio management, mitigation of related risks etcetera.

Most of these studies have been conducted and represented by governmental or supranational bodies charged with financial supervision and regulation (World Bank, IMF, EBRD, and Basel Committee). Leading financial corporations and banks also joined with this process. As a result, it was developed standards, practices and other recommendatory documents related to the management of problem loans at the level of the banking system or bank. Much of these documents are concerned with the necessity to maintain the recommended level of capital adequacy.

In the case of any given loan banks are free to act at its option. The bottom line is that bank has only two actual options to manage the existing problem loan: either to use the right of pledge via forfeiture of the assignable assets or to reschedule the loan. The first approach is well-known and widely used in practices. The second one can be treated as less harmful to both bank and borrower. In this case, the initial loan terms should be changed in such a way as to save the expected economic effect projected by the bank at the moment of the loan issue. However, the value saving approach that can be used by the bank for handling the problem of problem loans' rescheduling has not yet been characterized in detail and provided with econometric tools. The successful fulfilment of the value saving requirement will enable the bank to prevent the undesired reduction of the credit portfolio's market value. At the same time for the borrower, it could give a handle to save the business and gain time to adapt it to the new loan terms.

1.2 Previous researches and scientific problem

The conceptual framework of this research was based on the wealth maximisation theory [1]. Within the context of banking, this theory got reinterpretation where the wealth of the bank's shareholders directly depends on the credit portfolio's economic performance [2, 3]. A value-based approach was used for the twofold purpose: first to develop the econometric models for the main types of credits to estimate the economic effectiveness of credit for the lender [4], , and second to design compensative models assuming the problem loans terms' rescheduling to grant the lender the planed economic effect [5]. A wide range of

*e-mail: onikiien@gmail.com
**e-mail: dyba_m@ukr.net
***e-mail: iuliiagern@ukr.net

aspects on problem loans terms' rescheduling is considered in [6, 7]. The methodological basis on econometric models is provided in [8].

For the achievement of the intended outcomes, we proceeded four successive steps.

Firstly, credit was treated as a bank's investment project with its own cashflows' chart including initial outflow (principal) and following inflows in the form of loan payments.

Secondly, the usage of cash-flow diagrams for a schematic representation of loan's cashflows combined with NPV formula accustomed to loan's cashflows allowed to create econometric models for three main types of loan: classic loan, annuity loan and serial loan.

Thirdly, for the case when borrower breaks a loan's payment schedule and it leads to the reduction of loan's NPV it was outlined special compensative models of cashflows where default in payment is considered by the lender as an additional forced loan. We suggested modifying the loan terms (interest rate or effective period of the loan agreement) for the rest of payment periods.

Fourthly, special compensative models of forced loans' cashflows were laid atop corresponded initial cashflows of loans and this combination made it possible to get formulas calculating the modified interest rate and the additional number of loans' payment periods with the aid of backward calculation.

Thus, the chosen problem is of a significant scientific interest nowadays.

2 Main methodological assumptions of the research

2.1 Classic loan

A classic loan is characterized by the regular interest payments and a bullet repayment of the principal's amount at the end of the loan term [9]. Figure 1 illustrates this case.

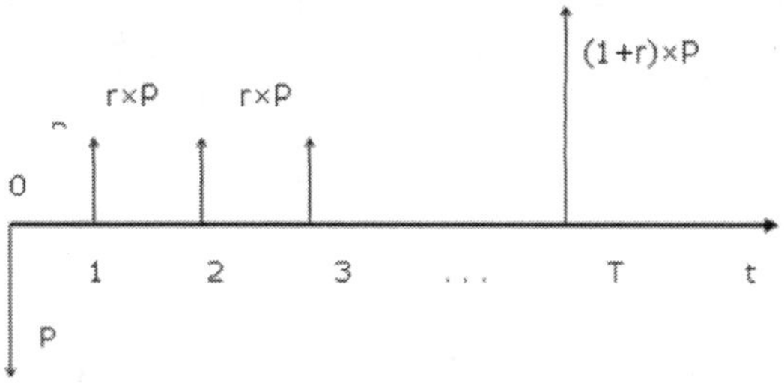

Figure 1. Cash flows' chart for classic loan

The net present value (NPV) of the classic loan is:

$$NPV = -P + r \times P \times \sum_{t=1}^{t} \frac{1}{(1 + WACC)^t} + \frac{P}{(1 + WACC)^T} \quad (1)$$

where P is the principal amount, r is the interest per period, T is the term of the loan and $WACC$ is the weighted average cost of capital for the lender.

Table 1 gives the initial loan's terms for the classic loan.

Table 1. Initial terms of the classic loan

Terms of the loan	Value
Principal amount (P)	UAH 10 000,00
Interest rate (r)	13%
Term of the loan (T)	8 years
The weighted average cost of capital for the bank ($WACC$)	8%

We can calculate the expected loan's NPV using formula (1) and it is UAH 2,873.32.

Let's assume that borrower has made payments for the first four years and is failed to pay interest instalment in the fifth period. This situation leads to the reduction of the expected NPV. Of course, we can recalculate the new value for NPV using formula (1) but for descriptive reasons we do it in a tabular form (table 2).

Table 2. Numerical illustration: cash flows and NPV calculation for the classic loan with the missed payment, thousand UAH

Period (t)	Interest rate (r), %	Cash flow (CF_t)	Present value factor $\left[\frac{1}{(1+WACC)^t}\right]$	The present value of cash flow $\left[\frac{CF_t}{(1+WACC)^t}\right]$
0	–	-10	1	-10
1	13	1,3	0.925	1,2
2	13	1,3	0.857	1,11
3	13	1,3	0.793	1,03
4	13	1,3	0.735	0,9
5	13	-	0.68	-
6	13	1,3	0.63	0,82
7	13	1,3	0.58	0,76
8	13	10,3	0.54	6,1
				NPV= 1,98

As we can see the single missed payment decreases loan's NPV by UAH 884.76 (UAH 2,873.32 – UAH 1,988.56).

Taking into account that in this case, we deal with the unsecured loan without opportunity for the lender to get the principal of the loan back, we have two *economically viable* alternatives to offset a loss of the loan's value (NPV): either to increase the interest rate for the rest of periods or to prolong the effective period of the loan agreement.

2.2 Increasing of interest rate

In this case, the statement of the problem is simple.

It is necessary to offset a loss of the loan's value (caused by missed interest payment (figure 2)) using the modified (increased) interest rate for the rest of the loan period.

To solve this problem, it is proposed to formalize the problem by replacement of actual data with disposal variables.

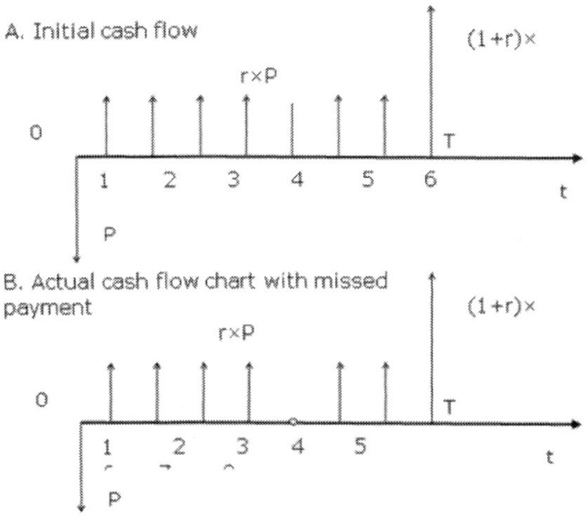

Figure 2. Cash flows charts

The comparison of this cash flow charts shows the absence of one payment (figure 2).

The missed payment ($P \times r$) can be treated by bank both as cash outflow and as a forced extraordinary loan at this period marked as T'.

In our case the principal amount of this forced credit can be compensated by annuity payment (marked as A) for the rest of periods ($]T'; T]$) (figure 3).

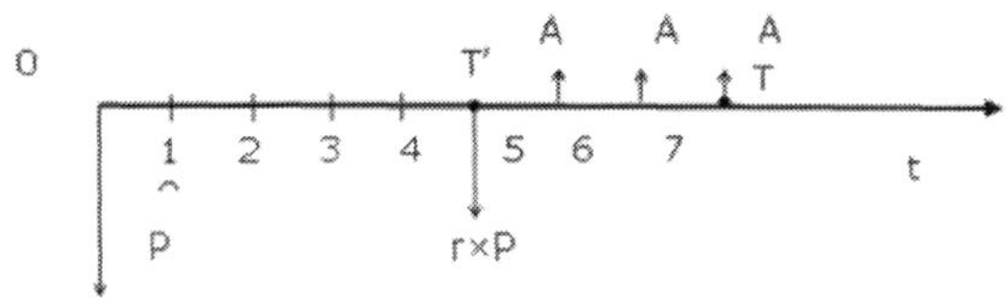

Figure 3. Cash flow for the rest of periods

The amount of this additional annuity payment can be derivable from the equation:

$$r \times P = A \times \left[\frac{1}{WACC} - \frac{1}{WACC \times (1 + WACC)^{T-T'}}\right] \quad (2)$$

The additional annuity payment is:

$$A = \frac{r \times P}{\frac{1}{WACC} - \frac{1}{WACC \times (1+WACC)^{T-T'}}} \quad (3)$$

The amount of this additional annuity payment can be added to the original loan's interest payment.

This is shown within the figure 4.

In this particular case, the amount of an additional annuity payment that is to be added to the initial interest payment is UAH 504.44.

The modified interest rate (r') for the rest of periods ($]T'; T]$) after the missed interest payment at the period T'

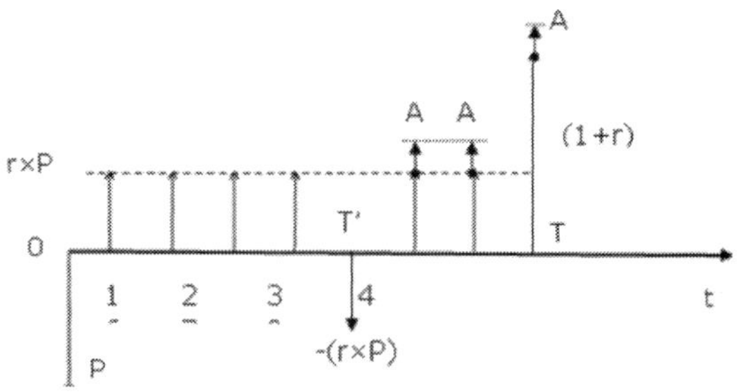

Figure 4. Cash flow for the modified payment

is:

$$r'_{[T';T]} = \frac{A + rtimesP}{P} = \frac{A}{P} + r \quad (4)$$

On substituting (3) into (4) we get the primary case

$$r'_{[T';T]} = \frac{\frac{r \times P}{\frac{1}{WACC} - \frac{1}{WACC \times (1+WACC)^{T-T'}}}}{P} + r \quad (5)$$

and these after simplification:

$$r'_{[T';T]} = r \times \left[\frac{WACC}{1 - \frac{1}{1+WACC}^{T-T'}} + 1\right] \quad (6)$$

In the case under consideration, the modified interest rate (r') for the rest of periods ($]5; 8]$) after the missed interest payment at the period 5 is 18.04% and checking calculation is represented in table 3.

Table 3. Numerical illustration: cash flows and NPV calculation for the classic loan with the missed payment and modified interest rate, thousand UAH

Period (t)	Interest rate (r), %	Cash flow (CF_t)	Present value factor $\left[\frac{1}{(1+WACC)^t}\right]$	The present value of cash flow $\left[\frac{CF_t}{(1+WACC)^t}\right]$
0	–	-10	1	-10
1	13	1,3	0.925	1,2
2	13	1,3	0.857	1,11
3	13	1,3	0.793	1,03
4	13	1,3	0.735	0,9
5	13	-	0.68	-
6	18,04	1,8	0.63	1,13
7	18,04	1,8	0.58	1,05
8	18,04	11,8	0.54	6,377
				NPV= 2,873

2.3 Prolongation of the effective period of the loan agreement

An alternative solution to the problem is to prolong the effective period of the loan agreement.

Consequently, it is necessary to compensate for the loss loan's NPV (UAH 884.76) caused by the missed interest payment at the 5th period.

The simplest solution is to use the step-by-step prolongation procedure by adding one after one payment periods until the loan's NPV reach the target value (UAH 2,873.32).

This approach represented in table 4.

As table 4 shows that it is possible to compensate the loss loan's NPV caused by the missed interest payment on the basis of adding four new payment periods and the new loan term is meant to be almost 12 years.

The above method can be treated as a one based on empirical judgments but not on value-based computational approach.

It needs a great number of iterations to get the result and does not provide the problem's solution in a formalized manner.

In the furtherance of this goal, the first thing we have to do is to chart additional cash flows that can arise out of the loan's prolongation (figure 5).

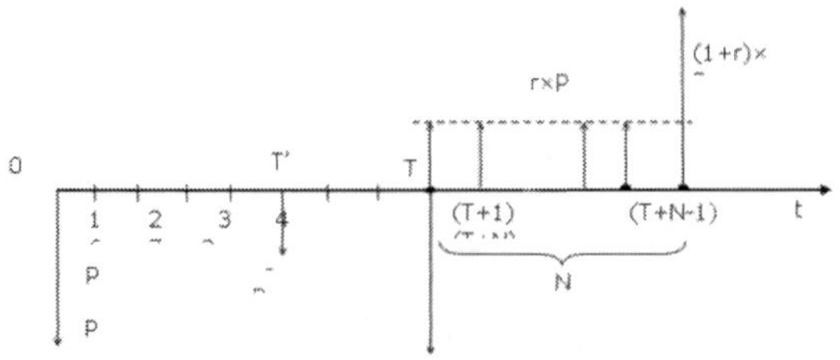

Figure 5. Cash flow for the forced loan in the case of loan prolongation

This chart depicts only those cash flows which first, occurred as a result of failure to pay on the due date (outflows): an interest payment on period T' (in this case $T' = 5$) and absence of repayment of principal P on a loan at period T. Secondly, it shows additional interest payments ($r{\times}P$) over the time of N additional payment periods added to the initial effective period of the loan agreement and repayment of principal (P) at the end ($T + N$).

The loss of the loan's original NPV due to the missed interest payment is:

$$NPV_{loss} = \frac{r \times P}{(1 + WACC)^{T'}} + \frac{P}{(1 + WACC)^{T}} \quad (7)$$

NPV of additional cash flows over the time of N additional payment periods is:

$$NPV_{CFadd} = r \times P \times \left[\sum_{t=T+1}^{T+N} \frac{1}{(1 + WACC)^{T}} \right] + \frac{P}{(1 + WACC)^{T+N}} \quad (8)$$

Taking into account that interest payments are annuities, formula (7) may be expressed as follows:

$$NPV_{CFadd} = \frac{1}{(1 + WACC)^{T}} \times \left[\frac{1}{(1 + WACC)^{T}} - \frac{P}{(1 + WACC)^{T+N}} \right] + \frac{P}{(1 + WACC)^{T+N}} \quad (9)$$

Now it is possible to equate the loss of the loan's original NPV due to the missed interest payment (6) with NPV of additional cash flows over the time of N additional payment periods:

$$NPV_{loss} = NPV_{CFadd} \quad (10)$$

or

$$\frac{r \times P}{(1 + WACC)^{T}} + \frac{P}{(1 + WACC)^{T}} = \frac{r \times P}{WACC * (1 + WACC)^{T}} \times \left[1 - \frac{1}{(1 + WACC)^{N}} \right] + \frac{P}{(1 + WACC)^{T}} \times \frac{1}{(1 + WACC)^{N}} \quad (11)$$

The deduced equation (10) contains the only one unknown variable we need – the number of additional payment periods (N):

$$N = \frac{\ln\left(\frac{WACC - r}{WACC \times (r \times (1 + WACC)^{T-T'} + 1) - r}\right)}{\ln(1 + WACC)} \quad (12)$$

In the case under consideration the number of additional payment periods $N = 3.95$.

Thus, the lender must prolong the effective period of the loan agreement up to 11.95 years to compensate the loan's loss of NPV (UAH 884.76) caused by the missed interest payment at the 5th period on the assumption that interest rate on this loan stays fixed.

2.4 Annuity loan case

An annuity loan is characterized by the instalment in all periods being of an equal amount CF_{ann}. Period by period the loan is settled and because the loan outstanding thereby is being reduced, the interest will period by period become less and the repayments equally higher. Figure 6 illustrates this case.

The expected net present value of the annuity loan is:

$$NPV = -P + CF_{ann} \times \sum_{t=1}^{T} \frac{1}{1 + WACC)^{t}}, \quad (13)$$

where CF_{ann} is equal in all periodic instalments.

Table 5 gives the initial loan's terms for the annuity loan.

We can calculate the expected loan's NPV using formula (12) and it is 2,873.32.

Let's assume that borrower has made payments for the first four years and is failed to pay an instalment at the

Table 4. Numerical illustration: cash flows and NPV calculation for the classic loan with the missed payment and prolongation of the loan's effective period of the agreement (step-by-step prolongation procedure), thousands UAH

Period (t)	Additional period ($t+1$)	Cash flow (CF_t)	Present value factor $\left[\frac{1}{(1+WACC)^t}\right]$	The present value of cash flow $\left[\frac{CF_t}{(1+WACC)^t}\right]$	Cumulated NPV $\left[-K+\sum_{t=1}^{T+N}\frac{CF_t}{(1+WACC)^t}\right]$
0		-10	1	-10	-10
1		1,3	0,92	1,2	-8,79
2		1,3	0,85	1,1	-7,68
3		1,3	0,79	1,03	-6,64
4		1,3	0,73	0 ,9	-5,69
5		0	0,68	0	-5,69
6		1,3	0,63	0,8	-4,87
7		1,3	0,58	0,75	-4,11
8		1,3	0,54	0,7	-3 41
9	1	1,3	0,5	0,65	-2,76
10	2	1,3	0,46	0,6	-2,16
11	3	1,3	0,42	0,55	-1,6
12	4	11,3	0,39	4,487	2,883
					NPV = 2,883

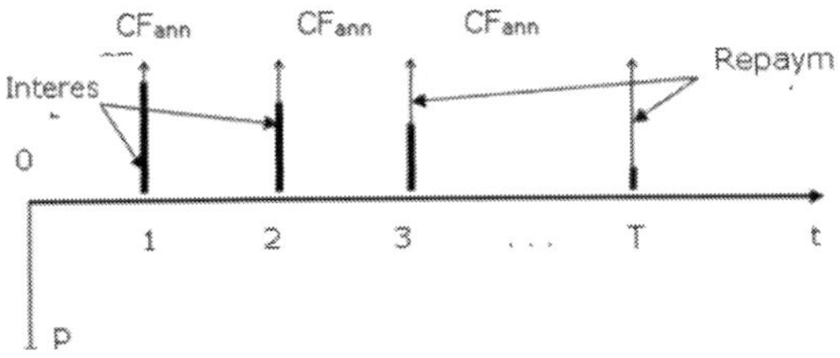

Figure 6. Cash flow for the annuity loan

Table 5. Initial terms of the annuity loan

Terms of the loan	Value
Principal amount (P)	UAH 10
Installment per period (CF_{ann})	UAH 2,24
Term of the loan (T)	8
The weighted average cost of capital for the bank ($WACC$)	8%

5th period. This situation leads to the reduction of the expected NPV. We could recalculate the new value for NPV using formula (12) but for descriptive reasons we did it in a tabular form (table 6).

As we can see the single missed payment decreases the loan's NPV by UAH 1524.61 (UAH 2873.32 – UAH 1348.71). Considering that in this case, we deal with the unsecured loan without opportunity for the lender to get the principal of the loan back, we have two economically viable options to compensate the loss of the loan's value: either to increase the instalment for the rest of periods or to prolong the effective period of the loan agreement.

Table 6. Numerical illustration: cash flows and NPV calculation for the annuity loan with the missed payment, thousands UAH

Period (t)	Cash flow (CF_t)	Present value factor $\left[\frac{1}{(1+WACC)^t}\right]$	The present value of cash flow $\left[\frac{CF_t}{(1+WACC)^t}\right]$
0	-10	1	-10
1	2,24	0,925	2,07
2	2,24	0,857	1,92
3	2,24	0,793	1,78
4	2,24	0,735	1,65
5	0	0,680	0
6	2,24	0,630	1,41
7	2,24	0,583	1,3
8	2,24	0,540	1,21
			NPV = 1348,71

2.5 Increasing of instalments

In this case, the statement of task is simple. There is a need to compensate a loss of the loan's NPV (caused by missed instalment) using the increased value of instalment.

To solve this problem, it is proposed to formalize the problem by replacement of actual data with disposal variables.

The comparison of cash flow charts (figure 7) shows the absence of one payment (instalment). The missed payment (CF_{ann}) can be treated by bank both as a cash outflow and (in our case) as a *forced* extraordinary loan at this period marked as T' in the case when the lender is ready to increase the instalment.

Similarly, with the classic loan's approach it is possible to derive an equation of the additional annuity payment (A) for the rest of periods ($]T';T]$):

$$CF_{ann} = A\left[\frac{1}{WACC} - \frac{1}{WACC\times(1+WACC)^{T-T'}}\right] \quad (14)$$

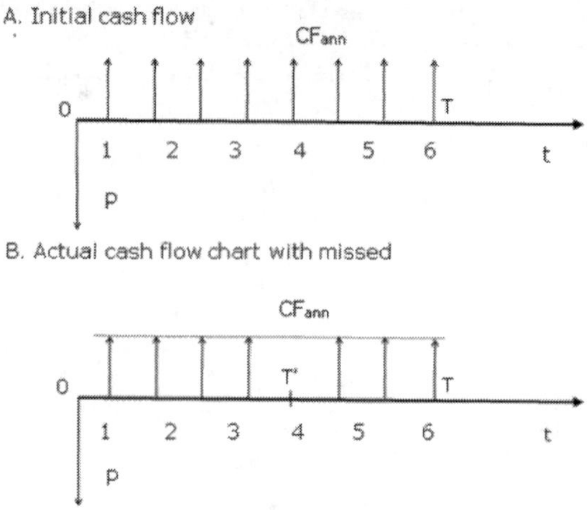

Figure 7. Cash flows

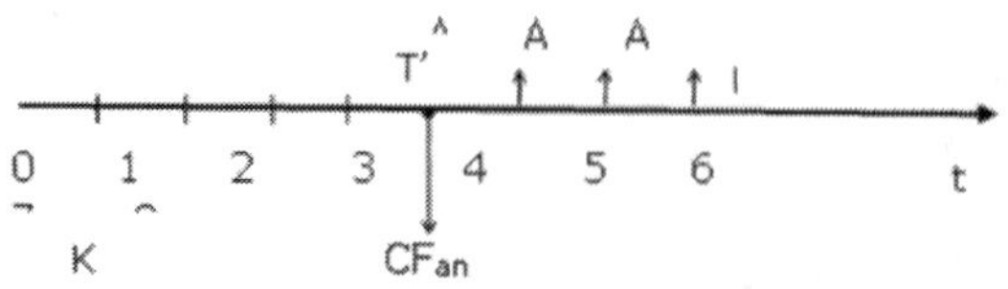

Figure 8. Cash flow for the forced loan to compensate the loss of the loans' NPV caused by missed instalments

The additional annuity payment is:

$$A = CF_{ann} \times \left[\frac{WACC}{1 - \frac{1}{(1+WACC)^{T-T'}}} \right] \tag{15}$$

The amount of this additional annuity payment (A) can be added to the original loan's instalment (figure 9).

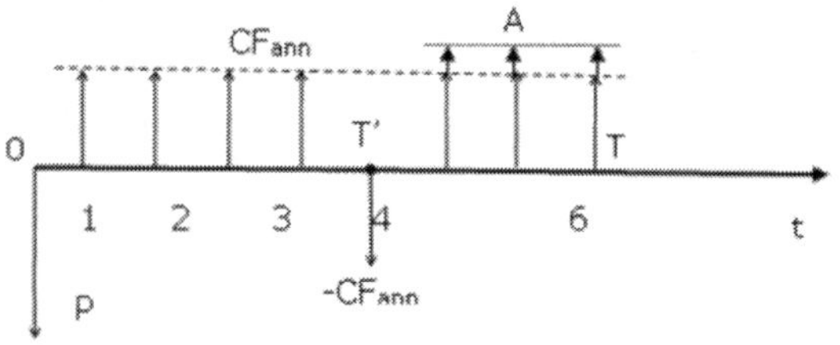

Figure 9. Cash flow for the modified annuity

In the case under consideration (table 5), the additional annuity payment for the rest of periods ($]5'; 8]$) after the missed instalment (2,240.15) at the period 5 is 869.25.

Checking calculation is represented in table 7.

Table 7. Numerical illustration: cash flows and NPV's calculation for the annuity loan with the missed payment and modified instalment, thousand UAH

Period (t)	Cash flow (CF_t)	Compen-satory cash flow (A)	Summary cash flow ($CF'_t = CF_t + A$)	Present value factor $\left[\frac{1}{(1+WACC)^t}\right]$	The present value of cash flow $\left[\frac{CF_t}{(1+WACC)^t}\right]$
0	-10	-	-10	1	-10
1	2,24	-	2,24	0,92	2,07
2	2,24	-	2,24	0,85	1,92
3	2,24	-	2,24	0,79	1,78
4	2,24	-	2,24	0,73	1,64
5	-	-	-	0,68	0,00
6	2,24	0,86	3,1	0,63	1,95
7	2,24	0,86	3,1	0,58	1,81
8	2,24	0,86	3,1	0,54	1,67
					NPV= 2,873.32

2.6 Prolongation of the effective period of the loan agreement

Let's assume that borrower has made payments for the first four years and is failed to pay an instalment at the 5th period. This situation leads to the reduction of the loan's expected NPV (table 6). The cash flow model of the compensatory *forced* loan is displayed graphically in figure 10.

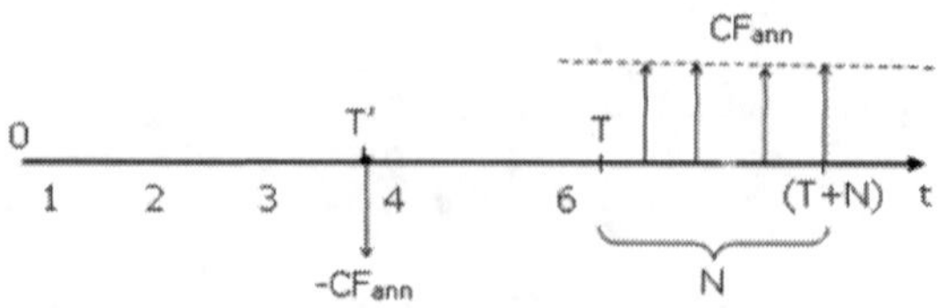

Figure 10. Cash flow for the forced loan to compensate the loss of the loans' CF caused by missed instalments via prolongation

This chart depicts only those cash flows which first, occurred as a result of failure to pay on the due date (outflow): an instalment payment ($-CF_{ann}$) on period T' (in this case $T' = 5$). Secondly, it shows additional instalment payments (CF_{ann}) over the time of N additional payment periods added to the initial effective period of the loan agreement $T + N$.

The loss of the loan's original NPV due to the missed instalment is:

$$NPV_{loss} = \frac{CF_{ann}}{(1 + WACC)^T} \tag{16}$$

NPV of additional cash flows (instalments) over the time of N additional payment periods ($[T + 1; T + N]$) is:

$$NPV_{CFadd} = \frac{CF_{ann}}{WACC \times (1 + WACC)^T} \times \left[1 - \frac{1}{(1 + WACC)^N}\right] \tag{17}$$

Now it is possible to equate the loss of the loan's original NPV due to the missed payment (15) with NPV of additional cash flows (instalments) over the time of N additional payment periods (16):

$$NPV_{loss} = NPV_{CFadd}, \tag{18}$$

or

$$\frac{CF_{ann}}{(1+WACC)^{T'}} = \frac{CF_{ann}}{WACC\times(1+WACC)^{T}}\times \times\left[1-\frac{1}{(1+WACC)^{N}}\right] \tag{19}$$

After simplification

$$\frac{1}{(1+WACC)^{N}} = 1 - WACC\times(1+WACC)^{T-T'} \tag{20}$$

The deduced equation (19) contains the only one unknown variable we need – the number of the additional payment periods (N):

$$N = \frac{\ln\left(\frac{1}{1-WACC\times(1+WACC)^{T-T'}}\right)}{\ln(1+WACC)}. \tag{21}$$

This formula shows that the number of additional payment periods (N) depends only on the original term of the loan (T), lender's cost of capital ($WACC$) and the missed payment's numerical order (T').

In the case under consideration (missed instalment at the period $T' = 5$) the number of additional payment periods (prolongation) (N) is 1.38.

Thus, the lender must prolong the effective period of the loan agreement up to 9.38 years to compensate the loan's loss of NPV caused by the missed interest payment at the 5th period on the assumption that interest rate on this loan stays fixed.

2.7 Increasing of interest rate

It is to be noted that we have the missed payment at the period T' that is composed of the fixed part of the principal amount $\frac{P}{T}$ and accrued interest.

Hence, this sum of the missed payment at the period T' can be treated as a cash outflow $CF_{T'}$:

$$CF_{T'} = \frac{P}{T} + r\times\left(P-(T'-1)\times\frac{P}{T}\right) \tag{22}$$

or

$$CF_{T'} = P\times\left(\frac{1}{T} + r\times\left(1-\frac{T'-1}{T}\right)\right) \tag{23}$$

Compensation for these losses, considering the time value of money, is to be made by increasing the loan's interest rate accruing on the rest of the main debts' payments over a period $]T';T]$ (figure 11).

This statement can be written as:

$$PV^{r'}_{]T';T]} - PV^{r}_{]T';T]} = CF_{T} \tag{24}$$

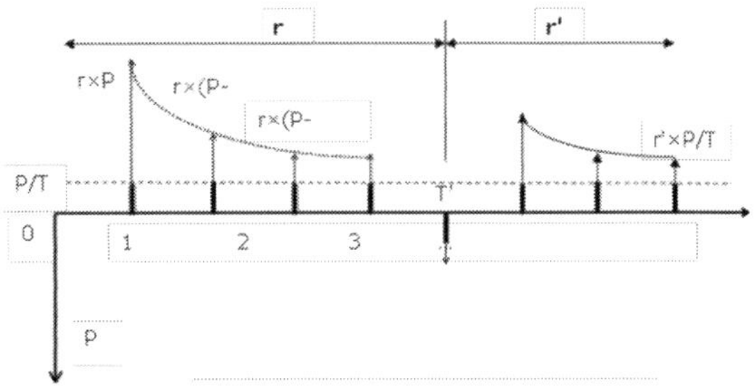

Figure 11. Cash flow chart

In equation (24) $PV^{r'}_{]T';T]}$ is the present value of the loans' cash flows that will be generated using modified interest rate r' for payment periods that are followed by the missed payment ($]T';T]$):

$$PV^{r'}_{]T';T]} = \sum_{t=1}^{T-T'}\left[\frac{\frac{K}{T} + K\times r'\times\left(1-\frac{T'-1+t}{T}\right)}{(1+WACC)^{t}}\right] \tag{25}$$

$PV^{r}_{]T';T]}$ is the present value of the loans' cash flows that could be generated using the original interest rate r for payment periods that are followed by the missed payment ($]T';T]$):

$$PV^{r}_{]T';T]} = \sum_{t=1}^{T-T'}\left[\frac{\frac{K}{T} + K\times r\times\left(1-\frac{T'-1+t}{T}\right)}{(1+WACC)^{t}}\right] \tag{26}$$

Consequently, the left side of an equation (25) is

$$PV^{r'}_{]T';T]} - PV^{r}_{]T';T]} = P\times(r'-r)\times\sum_{t=1}^{T-T'}\frac{1-\frac{T'-1+t}{T}}{(1+WACC)^{t}} \tag{27}$$

Substitution of (26) and (22) for (25) yields equation (27) containing the only one unknown variable – the modified interest rate r' for payment periods that are followed by the missed payment ($]T';T]$):

$$r' = \frac{\frac{1}{T} + r\times\left(1-\frac{T'-1}{T}\right)}{\sum_{t=1}^{T-T'}\frac{1-\frac{T'-1+t}{T}}{(1+WACC)^{t}}} + r \tag{28}$$

where T is the loan's term, r is the original interest rate, $WACC$ is the bank's cost of capital, and T' is the payment period's consecutive number with the missed payment.

In the case under consideration, the modified interest rate (r') for the rest of periods ($]5';8]$) followed by the missed payment at the period 5 is 41,75% and checking calculation is represented in table 8.

3 Discussion

In this study the combination of econometric and value-based approaches was used to develop the cash-flow based econometric models for the main types of bank loans. The proposed models are based on the assumption that any loan can be presented as the bank's investment project with

Table 8. Numerical illustration: cash flows and NPV calculation for the serial loan with the missed payment and modified interest rate, thousands UAH

Period (t)	Repayment $\left[\frac{P}{T}\right]$	Remaining principal of the loan	Interest rate, %	Interest payment	Cash flow (CF_t)	Present value factor $\left[\frac{1}{(1+WACC)^t}\right]$	The present value of cash flow $\left[\frac{CF_t}{(1+WACC)^t}\right]$	Cumulated NPV
0	-	10	13	-	-10	1	-10	-10
1	1,25	10	13	1,3	2,5	0.92	2,36	-7,63
2	1,25	8,7	13	1,13	2,3	0.85	2,04	-5,59
3	1,25	7,5	13	0,97	2,2	0.79	1,76	-3,82
4	1,25	6,25	13	0,81	2,06	0.73	1,51	-2,3
5	-	6,25	13	-	0.00	0.68	0,00	-2,3
6	1,25	3,75	41,75	1,56	2,81	0.63	1,77	-0,5
7	1,25	2,5	41,75	1,04	2,29	0.58	1,33	0,8
8	1,25	1,25	41,75	0,52	1,77	0.54	0,95	1,76
								NPV = 1,760,44

such attributes as initial outflow (principal), following inflows in the form of loan payments and bank's cost of capital. We deduced the formulas for calculating the expected net present value for classic (1), annuity (12), and serial (19) loans that can be used as a targeted measure of the loan's economic effect for a bank at the moment of loan issue. For the cases of problem loans where borrower breaks the loan's payment schedule and it leads to the reduction of the loan's expected NPV we developed calculating formulas for the loan terms' rescheduling:

- recalculation of the interest rate (5) and prolongation of the effective period of the loan agreement for a classic loan (11);
- recalculation of the instalment (14) and prolongation of the effective period of the loan agreement for annuity loan (21);
- recalculation of the interest rate for a serial loan (27).

This study has several strengths. First, the proposed value-based approach to the problem loans terms' rescheduling by its nature is economically feasible and equitable upon both parties and excludes prejudiced attitude. Second, the developed calculating formulas are simple for use in practical work and can be easily automated and incorporated in the existing dashboard of credit monitoring. Third, there is no need for credit officer to prepare some special data for calculation – all necessary data are kept handy at all times.

The findings in this study are subject to several limitations. First, the assumption as to whether the cost of capital for the lender is non-fluctuating is quite disputable. Secondly, it is intentionally supposed that borrower is not able to meet the credit payment 1) in full, 2) only once, 3) at the definite period. For cases when default in payment happens more than once, the proposed in these study formulas can be used over and over again. Thirdly, we assumed that borrower would not be able to meet the credit payment only once and would be able to service a loan in the future. For confirming this favourable scenario bank needs to make a new long-time forecast of borrower's creditworthiness and liquidity and check it regularly.

It has previously been found that much of the current literature focuses on one particularly important aspect of the cause of bank insolvency: the relationship between non-performing loans and bank failure (e.g. [10, 11]). Some authors have considered the macroeconomic effects of the problem and nonperforming loans on banking systems [12–14]. The study [15] presents some bed loans' management methods that already exist in the literature and makes a clear distinction between the ex-post and ex-ante management of such loans. The different characteristics of informal restructurings and of enhanced and hybrid debt restructurings are covered by the study were represented in [16]. Kryklii et al. described the domain role of the bank's problem loans management system in the strategy of the problem debts' mitigation [17]. The restructuring techniques like debt-equity swap and write-offs to ease debtors' debts and to help viable businesses to successfully survive recession have been analyzed by Dedu [18]. The duration models to analyze the factors affecting the duration of private debt restructuring for distressed firms were represented by Jiang-Chuan et al [19].

In this research, we intentionally did not analyze the regulatory approaches related to the banking system entirely and focused attention only on the ex-post problem loans' management at the level of a given bank. Therefore, our study odds to the previous knowledge: banks can use the value-based econometric approaches in the process of problem loans' terms rescheduling to prevent the reducing their value.

These findings are likely to have practical consequences. First, the developed formulas for calculation of loan's NPV can help banks to assess the expected economic effect of the loan at the process of loan approval as well as to assess the value of credit portfolio for the analytical purposes. Second, the use of formulas related to the problem loan terms' restructuring can not only simplify the procedure and provide the economically feasible equitable decisions but also help viable businesses to successfully survive the recession.

The proposed econometric approach is based on the single-variable econometric models. For the future, more research is required to develop a deeper understanding of the relationships between two recommended approaches: increasing of interest rate and prolongation of the effective period of the loan agreement. These approaches are not mutually exclusive and can be combined. It allows setting basic parameters for the two-factor model where the fractional increase of one parameter leads to fractional reducing of another one. All one has to do is to solve the two-variable equation with a modified interest rate (r) and additional loan term (N):

$$E(NPV) = f(N, r'), \tag{29}$$

where $E(NPV)$ is the expected initial NPV of a loan for the lender.

The solution of this equation will be a feasible solution set where all optimal combination presented on the trending graph (figure 12).

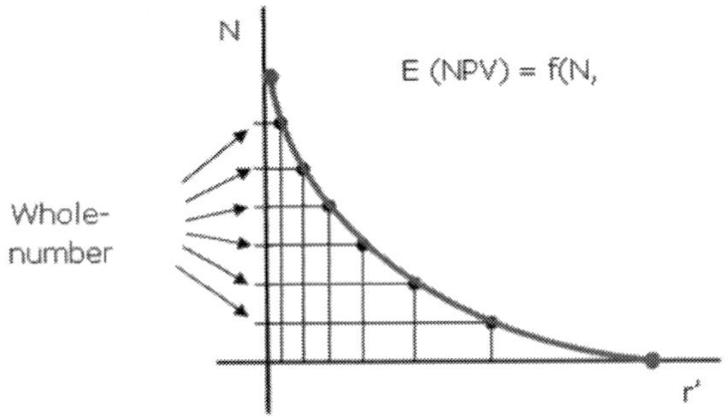

Figure 12. Two-factor model for loan terms' reformating

To the right of trending graph there is an area of suboptimal solutions where loan's virtual NPV for lender outnumbers the expected initial $E(NPV)$ of a loan and to the left of trending graph vice versa. For practical application, the feasible solution set can be reduced through the use of whole-number values for additional payment periods (N) according to the repayment terms of a loan: month, quarter or year.

4 Conclusions

This study set out to develop the econometric models and applied approaches to an issue of the problem loan terms' rescheduling. This study has shown that that the NPV criteria can be used for credit's performance evaluation and provided a way not only calculate projected increment of credit's present value for the lender but recalculate loan terms if needed as well.

The key advantage of the suggested method of the loan's terms renegotiation consists in its focus on the use of the value-based approach in cases where loan terms can be reformatted with the use of either increasing of interest rate or prolongation of the effective period of the loan agreement. This is the first study that has provided a lender with formulas for the loan terms' rescheduling.

At practice, restructuring or rescheduling exercise may result in variation to the existing terms of the loan/financing facilities. In particular, there is a need for further research, considering econometric modelling of bank activities in the era of Covid-19 [20]. Additionally, there is a need to consider value-based approach to the problem loans terms' rescheduling due to the national economies peculiarities. More extensive study is required. Thereby, this is the subject for future researches.

References

[1] M. Jensen, *Value Maximization and the Corporate Objective Function* (Business School Press, Harvard, 2000)
[2] J. Burchett, S. Brendan, *Valuation techniques for bank loan portfolios* (2019), `https://www.stout.com/en/insights/article/valuation-techniques-bank-loan-portfolios`
[3] P. Ahtiala, Scot. J. of Pol. Econ. **52** (2005)
[4] W. Thomas, *Presentation on theme: "Loan Valuation Using Present Value Analysis". U.S. Department of Treasury. Office of Technical Assistance (OTA)* (2015), `https://slideplayer.com/slide/4582574/`
[5] J. Trefzger, R. Cannaday, J. of Real Estate Pract. and Edu. **18** (2015)
[6] S. Chowdhury, M. Ullah, M. Amin, T. Rahman, M. Hossain, A. Talukder, *Implication of Loan Rescheduling and Write-Off on the Performance of Banks* (BIBM, Dhaka, 2016)
[7] S. Handayani, M. Nengsih, Managament Insight: J. Ilmiah Manajemen **14**, 210 (2019)
[8] L. Găban, I. Rus, A. Fetița, L. Bechiș, The Annals of the University of Oradea. Economic Sciences **26** (2017)
[9] S. Onikiienko, Bulletin of the National Bank of Ukraine **10**, 30 (2004)
[10] A. Campbell, J. of Bank. Reg. **9** (2007)
[11] H. Chaibi, J. Int. Bus. Res. **9** (2016)
[12] R. Espinoza, A. Prasad, IMF Work. Pap. **10** (2010)
[13] H. Fajar, Umanto, J. Banks and Bank Sys. **12** (2017)
[14] M. Nkusu, IMF Work. Paper **11** (2011)
[15] D. Anastasiou, MPRA Paper **77581** (2016)
[16] J. Garrido, *Out-of-court Debt Restructuring. World Bank Studies* (World Bank, Washington, D.C., 2012)
[17] O. Kryklii, O. Krukhmal, J. Act. Prob. of Econ. **5** (2014)
[18] V. Dedu, S. Lazarescu, D. Nitescu, Theoretical and Applied Economics, Asociatia Generala a Economistilor din Romania - AGER **11** (2009)
[19] J. Huang, C. Huang, J. Bank. and Bank Sys. **4** (2009)
[20] A. Bielinskyi, I. Khvostina, A. Mamanazarov, A. Matviychuk, S. Semerikov, O. Serdyuk, V. Solovieva, V. Soloviev, IOP Conference Series: Earth and Environmental Science **628** (2021)

Computational method for risk assessment of regional socio-economic development

Oleg Pursky[1,*], *Tatiana* Dubovyk[1,**], *Iryna* Buchatska[1,***], *Iryna* Lutsenko[2,****], and *Hanna* Danylchuk[3,†]

[1]Kyiv National University of Trade and Economics, 19 Kioto Str., Kyiv, 02156, Ukraine
[2]National Technical University of Ukraine "Igor Sikorsky Kyiv Polytechnic Institute", 37 Peremohy Ave., Kyiv, 03056, Ukraine
[3]The Bohdan Khmelnytsky National University of Cherkasy, 81 Shevchenko Blvd., Cherkasy, 18031, Ukraine

Abstract. In this study, we present the computational method for risk assessment of the socio-economic development of regions. An attempt has been made to develop a method for the determination of integral risk indicators of socio-economic development based on the joint use of the methods of factor analysis and expert evaluation. This approach has increased the reliability of the calculations and made it possible to analyze the influence of socio-economic indicators on the risk level of socio-economic development. The integral risk indicator shows the effect of the inconsistency in the level of factor provision on the socio-economic development of the *j*-th region (district) in comparison with the general situation in the country (regions). The closer the value of integral risk indicator is to 1, the higher the level of risk in this region. Using Kyiv region districts as an example, the process of risk assessment for regional socio-economic development has been considered. The results obtained in this investigation demonstrate that the presented computational method solves the problem of formalization of risk assessment for the socio-economic development of regions.

1 Introduction

The market today is functioning in a turbulent environment facing continuous change because of hyper-competition, changing demands of customers, regulatory changes, and technological advancement [1]. Modern world economic conditions, economic globalization, acceleration of market development processes, information technologies, socio-political factors require public administration new approaches to the formation of socio-economic strategies, development of adequate methodological solutions, and tools in the field of governance, especially it concerns socio-economic development management of regions [2, 3]. Using modern information technologies and new electronic communication channels significantly reduce costs related to organization and support social activity and business, and the new possibilities allow re-designing socio-economic development strategy at any moment [4]. In connection with the globalization and the processes of post-industrial economy development, the effect of unpredictability appears in the change of socio-economic systems state due to the increasing influence of economic crises, suddenly emerging threats and risks [5–9]. One of the urgent problems in the risk analysis of socio-economic systems is the construction of adequate methods. This is due to the multidimensionality of socio-economic systems, the stochasticity of their behavior, as well as the complex interaction between the elements of the systems [6, 8].

In socio-economic studies, to improve the reliability of the procedure for assess-ment of socio-economic development using mechanism for determining the integral indicators based on factor analysis, taking into account consider the knowledge and experience of experts [10]. Accordingly, the aim of this study is to develop a reliable computational method for risk assessment of regional socio-economic development on the basis of the joint use of the methods of factor analysis and expert evaluation. This article poses and solves the problem of formalization of risk assessment for regional socio-economic development with the Kyiv region districts as an example.

2 Computational method of risk assessment

This section presents a method for determining the integral risk indicator of regional socio-economic development. The method is based on a model for determination of the socio-economic development integral indicator which is described in detail in [10]. In this model, methods of factor analysis and expert evaluations are used to determine integral indicators. To reducing the dimension of the feature space (socio-economic indicators), one of the methods of factor analysis is used [11–13], the principal component analysis (PCA) [14, 15]. Based on the reduced set of independent factors, a single integrated indicator is obtained, which combine all these factors in the

*e-mail: pursky_o@ukr.net
**e-mail: tatiana_dubovik@i.ua
***e-mail: i.buchatska@knute.edu.ua
****e-mail: lutsenkois0802@gmail.com
†e-mail: abdanilchuk@gmail.com

best way [15, 16]. The main disadvantage of factor analysis methods is the reliability of the conclusions, in particular, in this model of determining the integral indicators [10], the weight of the factor is determined by the dispersion of initial indicators, which is not always reliable in socio-economic studies, since in this case the importance of indicators for the socio-economic system is not taken into account [10, 16, 17]. Therefore, within the framework of this model, in order to increase the reliability of the algorithm for determining the integral indicators based on factor analysis, expert evaluation procedures have been introduced in the mechanism of determining the weight of the factors [10, 18, 19]. In this case, the generalized weight of factors that takes into account both the weight of the factor, determined on the basis of expert evaluations, and the weight of the factor determined statistically, can be obtained as the weighted average of these two evaluations [10, 20]:

$$w_i = (\overline{q_i} + \overline{v_i}) \Big/ \sum_{i=1}^{n} (\overline{q_i} + \overline{v_i}), \tag{1}$$

where $\overline{q_i} = q_i \Big/ \sum_i^n q_i$, $\overline{v_i} = v_i \Big/ \sum_i^n v_i$ are expert and statistical (factor analysis) weighted coefficients of the factor, respectively. Thus, the complex indicator of socio-economic development for j-th region is calculated as the sum of factors with the corresponding weighted average weight coefficients w_i [10]:

$$I_j = \sum_{i=1}^{n} w_i F_{ij} \; (j = 1, 2, ..., n), \tag{2}$$

where n is a number of factors; F_{ij} is the value of the i-th factor for the j-th object (region). Taking into account the proposed complex indicators of socio-economic development (2), the value of the integral risk indicator R in the region can be calculated using the following formula [21, 22]:

$$R_j = 1 - \frac{\left(I_j / I_{avg}\right)}{m}, \tag{3}$$

where R_j is the integral risk assessment for socio-economic development of j-th region; I_{avg} is the numerical value of the complex indicator I in average in study regions; m is the number of factors. In this formula (3), the risk assessment is calculated taking into account the average value of the complex indicator I_{avg} for all regions and the number of factors. In this case, the risk R_j varies in the range from 0 to 1. The integral indicator R_j shows the effect of the inconsistency in the level of factor provision on the socio-economic development of the j-th region (district) in comparison with the general situation in the country (regions). The closer the value of R_j is to 1, the higher the level of risk in this region.

Figure 1 illustrates a general scheme of the developed computational method for determining the regional socio-economic development integral risk indicator. This method makes it possible to develop a procedure for automated data processing of socio-economic research and includes the following stages:

1 stage. The entering of values of indicators of socio-economic development and expert evaluations in the form of the matrix of indicators and the matrix of expert evaluations of indicators, followed by its normalization to a single scale of measurements.

2 stage. The calculation of the pairwise correlations matrix and determination of its eigenvalues and eigenvectors.

3 stage. Obtaining a matrix of factors by multiplying the normalized matrix of indicators and the matrix of eigenvectors, and normalization of factors and calculation of their variance.

4 stage. Determining the number of N factors included in the integral risk indicator (figure 1) on the basis of eigenvalues of the matrix of pairwise correlations of indicators and the given boundary value L dispersion of normalized indicators or in other words – a sampling of the minimum number of factors with maximal eigenvalues λ_i is made, the sum values of which are not less than nL [10].

5 stage. Determining the relative contribution $\%(F_i)$ of each of the N factors in the description of the total dispersion of all n indicators as the ratio of the eigenvalue λ_i of the factor F_i to the total dispersion of the features, which is also equal to n:

$$\sum_{i=1}^{N} \lambda_i \geq nL, \%(F_i) = \lambda_i \Big/ \sum \lambda_i = \lambda_i / n \tag{4}$$

6 stage. The determination of the experts competence and calculation of Kendall's coefficient of concordance for evaluation of the consistency of their conclusions.

7 stage. The determination of weighting coefficients of factors (1) included in the integrated indicators (2).

8 stage. The calculation of integral risk indicators of regional socio-economic development (3) and the visualization of the results of data processing.

3 Risk assessment for regional socio-economic development of the Kyiv region districts

In this section, we consider the process of risk assessment for regional socio-economic development in accordance with the developed computation method for integral risk indicators (figure 1) on the example of the Kyiv region districts. Listed in a single scale of measurements and normalized values of socio-economic indicators of districts are presented in table 1.

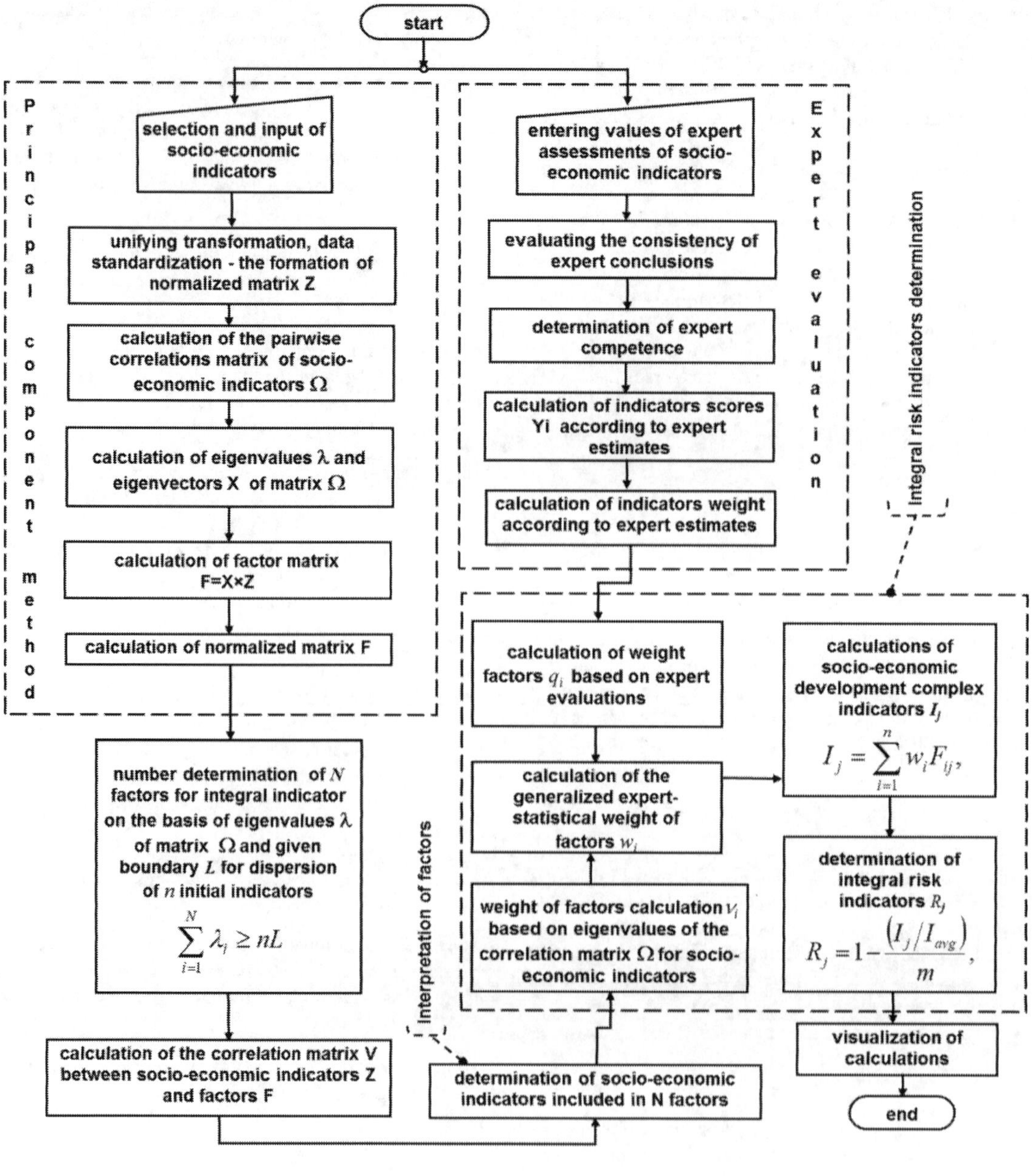

Figure 1. General scheme of the computational method for risk assessment of regional socio-economic development

According to the National State Statistics Service of Ukraine [23], one of the main indicators that characterize the level regional socio-economic development are (table 1):

(P1) Number of cars per 1000 people;

(P2) Services rendered per unit of population, UAH;

(P3) Natural increase (reduction) of the population;

(P4) Registered unemployment rate;

(P5) Average monthly salary, UAH;

(P6) Provision of housing by the population, m^2 per person;

(P7) The ratio of m^2 of built housing to the population;

(P8) Preschool establishments per unit of population;

(P9) General educational institutions per unit of population;

(P10) Number of crimes per 1000 people;

Table 1. Normalized values of socio-economic indicators

Districts	P1	P2	P3	P4	P5	P6	P7	P8	P9	P10	P11
D1	-0,209	-0,054	-0,094	-0,129	0,077	-0,270	-0,063	-0,157	-0,171	0,065	-0,310
D2	-0,069	-0,075	-0,461	0,186	-0,005	0,294	-0,049	-0,027	-0,062	-0,240	-0,301
D3	0,044	-0,050	-0,088	0,300	0,253	-0,108	-0,068	-0,252	-0,298	-0,465	-0,236
D4	-0,077	0,016	0,005	0,174	0,223	-0,005	0,005	-0,142	-0,221	-0,287	-0,328
D5	0,468	-0,026	0,027	0,199	0,004	0,046	-0,031	0,111	0,058	-0,220	0,129
D6	0,036	0,013	-0,130	-0,129	0,255	-0,228	0,011	0,018	-0,053	0,097	-0,063
D7	-0,109	-0,071	-0,312	0,174	-0,173	-0,330	-0,001	-0,084	-0,044	-0,167	-0,312
D8	0,048	0,219	0,127	0,161	0,026	0,251	0,038	-0,211	0,021	-0,172	0,004
D9	-0,035	-0,086	0,131	-0,192	0,360	-0,142	-0,066	-0,043	-0,072	0,110	-0,090
D10	-0,170	-0,064	0,110	0,123	-0,219	0,054	-0,069	-0,219	-0,139	0,076	-0,175
D11	-0,125	-0,061	-0,131	0,035	-0,259	0,037	-0,064	-0,243	-0,014	0,066	0,087
D12	0,475	-0,067	-0,030	0,110	-0,058	-0,288	-0,049	0,071	0,020	-0,071	0,209
D13	-0,193	0,928	0,466	-0,205	-0,129	0,089	0,948	0,594	0,671	0,200	0,250
D14	-0,128	-0,052	-0,102	0,186	0,331	0,037	-0,037	0,014	0,027	-0,147	-0,071
D15	0,085	-0,041	-0,006	-0,381	-0,219	0,183	-0,061	0,110	0,229	0,306	0,181
D16	-0,170	0,073	0,110	-0,230	0,067	-0,176	0,162	0,198	0,256	0,254	0,313
D17	-0,003	-0,066	-0,292	-0,230	-0,166	-0,091	-0,069	0,250	0,129	0,156	0,089
D18	-0,189	-0,086	0,338	0,060	-0,285	0,106	-0,069	-0,222	-0,288	0,159	0,206
D19	0,043	-0,077	0,226	0,161	0,151	-0,228	-0,068	-0,088	-0,190	-0,002	0,101
D20	-0,166	-0,066	-0,169	-0,167	0,270	-0,031	-0,066	0,108	0,076	0,096	-0,118
D21	0,274	-0,080	0,189	-0,066	-0,126	0,114	-0,069	0,034	-0,057	0,097	-0,081
D22	0,090	-0,067	0,043	0,174	0,084	-0,168	-0,069	-0,097	-0,166	-0,132	-0,045
D23	0,389	-0,057	0,165	0,211	-0,044	0,285	-0,069	0,056	-0,035	-0,268	-0,003
D24	-0,182	-0,082	-0,088	-0,419	-0,067	0,465	-0,063	-0,154	0,188	0,231	0,275
D25	-0,125	-0,020	-0,036	-0,104	-0,352	0,106	-0,062	0,375	0,136	0,259	0,291

(P11) Emissions of pollutants.

It should be noted that the list of indicators, depending on the goals and objectives of the risk assessing, may change, thereby changing its emphasis. Thus, for the Kyiv region we have a matrix of initial socio-economic indicators in the size of 25 × 11 (25 districts of the Kyiv region: Baryshivsky (D1), Bilotserkivsky (D2), Boguslavsky (D3), Boryspilsky (D4), Borodyansky (D5), Brovarsky (D6), Vasylkivsky (D7), Vyshgorod (D8), Volodarsky (D9), Zgurivsky (D10), Ivankivsky (D11), Kagarlytsky (D12), Kyiv-Sviatoshynsky (13), Makarivsky (D14), Myronivsky (D15), Obukhovsky (D16), Perejaslav-Khmelnytsky (D17), Polissya (D18), Rokytnyansky (D19), Skvyrsky (D20), Stavyshchensky (D21), Tarashchansky (D22), Tetiivsky (D23), Fastivsky (D24), Yahotynsky (D25)).

On the basis of the normalized matrix of socio-economic indicators (table 1), the pairwise correlations matrix of indicators is dimensioned 11 × 11. For the pairwise correlations matrix of indicators, we determine eigenvalues λ (table 2) and eigenvectors $\mathbf{X}$. The matrix of factors F is obtained by multiplying the normalized matrix of socio-economic indicators (table 1) into the matrix of the eigenvectors of the pairwise correlations matrix. The obtained factors are normalized. The normalized factor matrix is used to calculate the matrix of correlations between factors and indicators of socio-economic development that is required for the interpretation of factors.

On the basis of the calculated eigenvalues of the pairwise correlations matrix (table 2) and the given threshold L of the dispersion for normalized socioeconomic indicators (table 1), the formula (4) determines the number of N factors in the integral risk indicator. In this case, the number of main components (factors) must be used, which exhaust at least 60-70% of the variance of the initial random variables. For example, at a given threshold of 0,6 from table 2 it is necessary to select N factors with maximal eigenvalues, the sum of values of which is not less than $0,6 \times 11 = 6,6$. The sum of the first three eigenvalues λ is 7,49 that is the integral index consists of the first three factors ($N = 3$) that explain approximately 68% (see formula 4) of the variance of the initial data (table 2). The calculate matrix of correlations between the normalized socio-economic indicators and the factors shows, which indicators are included in the given three factors (with the value of the variance of the indicators should not be less than the given limit value of 0,6). Table 3 shows the structure of factors: the coefficient of correlation between the indicators and factors in which they are included, statistical and expert weights coefficients and weighted average weight coefficient of factors. The first factor included the first four indicators: 1) the number of cars per 1000 people; 2) services rendered per unit of population; 3) natural increase (reduction) of the population; 4) the level of registered unemployment. The second factor included the eleventh indicator – emissions of pollutants. The third factor entered the seventh indicator – the ratio m^2 of the built housing to the population.

By multiplying the obtained factors by the corresponding weighted average weight coefficients of factors, by for-

Table 2. Eigenvalues of the pairwise correlation matrix of indicators

λ	4,565	1,733	1,193	0,931	0,764	0,587	0,495	0,364	0,244	0,097	0,022
$\%(F_i)$	41,51%	15,76%	10,85%	8,46%	6,95%	5,34%	4,50%	3,31%	2,22%	0,89%	0,20%
$\sum\%$	41,51%	57,27%	68,11%	76,58%	83,53%	88,87%	93,37%	96,69%	98,91%	99,80%	100,00%

Table 3. Structure of factors and their weight coefficients

Factors	Socio-economic indicators	Correlation coefficient	Statistical weight coefficient	Experts weight coefficient	Weighted average weight coefficient
F1	P1	0,912	0,456	0,269	0,363
	P2	0,873			
	P3	0,741			
	P4	0,647			
F2	P11	0,794	0,293	0,086	0,191
F3	P7	0,663	0,251	0,644	0,446

mula (2) we obtain the values of integral risk indicators according to formula (3) that allow ranking the regions in terms of their risk assessment of socio-economic development (figure 2). In table 4 calculations results of integral risk indicators R_j of social and economic development of Kyiv region districts are presented. It is also worth noting that for a better understanding of the calculation procedures should be carefully study the model that presented in [10].

4 Conclusions

The obtained results demonstrate that the presented computational method solves the problem of formalization of risk assessment for the socio-economic development and can be used to analyze and predict the socio-economic situation in the region. In the framework of the presented method by changing the values of socio-economic indicators, it is possible to analyze and model the socio-economic situation in the region, which undoubtedly provides management with valuable information on possible risks and directions of effective strategies for socio-economic development. The management receives not only an adequate assessment of the risk level of socio-economic development of the region, but also the opportunity to determine the immediate causes and consequences that shape the current and future socio-economic situation in the region. The results of modeling the process of assessing the level of socio-economic development of Kyiv region showed that the main advantages of the method of determining the integrated risk indicators are the possibility of studying correlations between socio-economic indicators, between indicators and factors, interpretation of factors, determining negative and positive characteristics of socio-economic situations in context of research objects. The proposed method for risk assessment makes it possible to implement a unified approach to data analysis and to ensure the efficiency of constructing integral risk indicators.

This study was supported by the Ministry of Education and Science of Ukraine, Project No. 0112U000635, "Development and implementation of the modern information systems and technologies in the socio-economic activities".

References

[1] A. Raman, S. Bharadwaj, J. Mukherjee, International Journal Business Information Systems **27** (2018)

[2] S. Belew, J. Elad, *Starting an Online Business All-in-One for Dummies*, 3rd edn. (For Dummies, New Jersey, 2017)

[3] O. Pursky, T. Dubovyk, I. Moroz, I. Buchatska, A. Savchuk, CEUR Workshop Proceedings **2422**, 15 (2019)

[4] O.I. Pursky, B.V. Grynyuk, D.A. Shestopal, Aktual'ni Problemy Ekonomiky = Actual Problems in Economics p. 407 (2016)

[5] D. Chernov, D. Sornette, *Critical risks of different economic sectors* (Springer, 2020)

[6] K. Pitilakis, P. Franchin, B. Khazai, H. Wenzel, *SYNER-G: systemic seismic vulnerability and risk assessment of complex urban, utility, lifeline systems and critical facilities: methodology and applications*, Vol. 31 (Springer, 2014)

[7] M. Machina, W.K. Viscusi, *Handbook of the Economics of Risk and Uncertainty* (North Holland, Amsterdam, 2014)

[8] S. Ashmarina, M. Vochozka, *Sustainable Growth and Development of Economic Systems* (Springer, 2019)

[9] H. Danylchuk, L. Kibalnyk, O. Serdiuk, CEUR Workshop Proceedings **2224**, 211 (2019)

[10] O. Pursky, T. Dubovyk, I. Gamova, I. Buchatska, CEUR Workshop Proceedings **2393**, 919 (2019)

[11] K. Adachi, *Matrix-based introduction to multivariate data analysis* (Springer, 2020)

[12] K.A. Pituch, J.P. Stevens, *Applied multivariate statistics for the social sciences: Analyses with SAS and IBM's SPSS* (Routledge, 2015)

[13] K. Vehkalahti, B.S. Everitt, *Multivariate analysis for the behavioral sciences* (CRC Press, 2018)

Table 4. The value of integral risk indicators of socio-economic development calculated for Kyiv region districts

Districts	R_j	Districts	R_j	Districts	R_j	Districts	R_j	Districts	R_j
D9	0,805	D22	0,744	D15	0,684	D19	0,669	D8	0,567
D16	0,803	D20	0,722	D25	0,677	D3	0,661	D14	0,553
D23	0,758	D17	0,711	D2	0,675	D24	0,636	D6	0,529
D10	0,755	D18	0,709	D1	0,671	D7	0,620	D13	0,502
D21	0,752	D12	0,704	D5	0,670	D11	0,608	D4	0,483

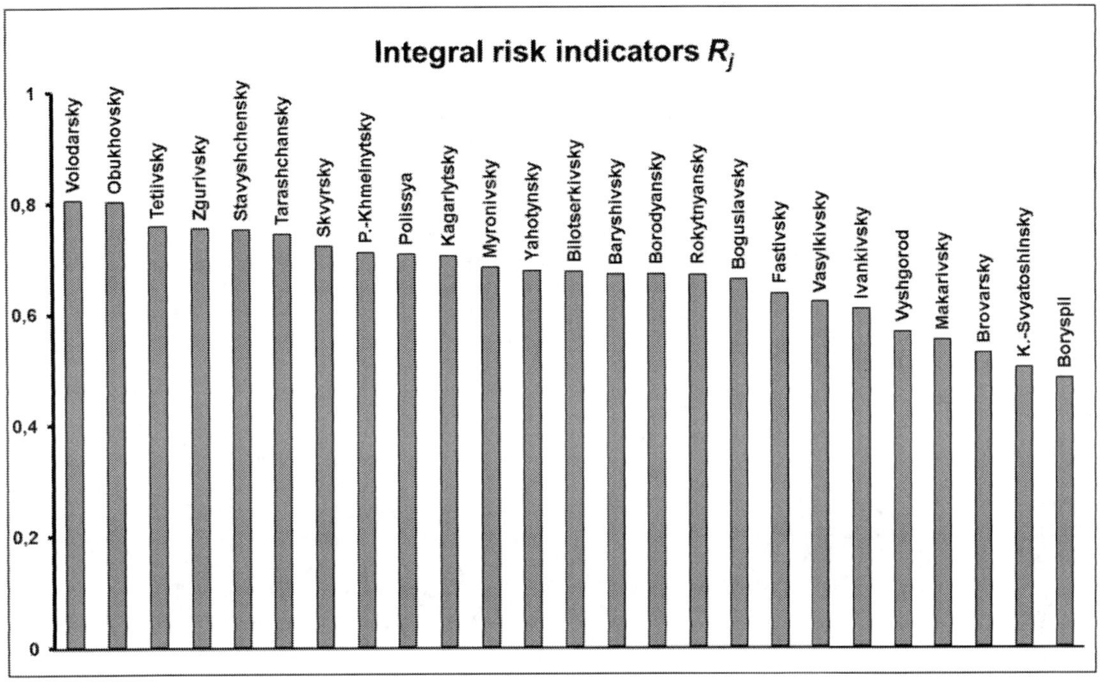

Figure 2. Ranking of regions in terms of their risk assessment of socio-economic development

[14] V. Gray, *Principal Component Analysis: Methods, Applications and Technology* (Nova Science Pub. Inc., 2017)
[15] S. Okafor, I. Okafor, *Factor Analysis in Economics: Theory and Applications* (LAP Lambert Academic Publishing, 2017)
[16] G. Di Franco, A. Marradi, *Factor analysis and principal component analysis* (FrancoAngeli, 2013)
[17] W.H. Finch, *Exploratory factor analysis* (SAGE Publications, 2019)
[18] S. Farhady, *Expert Evaluation Game Factors* (LAP Lambert Academic Publishing, 2013)
[19] J.J. Phillips, P.P. Phillips, *Handbook of training evaluation and measurement methods* (Routledge, 2016)
[20] L. Ustinovieius, Journal of Civil Engineering and Management **7**, 321 (2001)
[21] M. Rausand, S. Haugen, *Risk Assessment: Theory, Methods, and Applications (Statistics in Practice)* (Wiley, 2020)
[22] A. Kuklin, M. Pecherkina, A. Tyrsin, A. Surina (2017)
[23] *Statistical yearbook of Ukraine for 2017* (2017)

A forecasting the consumer price index using time series model

Volodymyr Shinkarenko[1,*], *Alexey* Hostryk[1,**], *Larysa* Shynkarenko[2,***], and *Leonid* Dolinskyi[3,****]

[1]Odessa National Economic University, 8 Preobrazhenskaya Str., 65082, Odessa, Ukraine
[2]International Humanitarian University, 33 Fountain Road Str., 65009, Odessa, Ukraine
[3]University of the State Fiscal Service of Ukraine, 31 Universytetska Str., Irpin city in Kiev Region, 08201, Ukraine

Abstract. This article examines the behavior of the consumer price index in Ukraine for the period from January 2010 to September 2020. The characteristics of the initial time series, the analysis of autocorrelation functions made it possible to reveal the tendency of their development and the presence of annual seasonality. To model the behavior of the consumer price index and forecast for the next months, two types of models were used: the additive ARIMA*ARIMAS model, better known as the model of Box-Jenkins and the exponential smoothing model with the seasonality estimate of Holt-Winters. As a result of using the STATISTICA package, the most adequate models were built, reflecting the monthly dynamics of the consumer price index in Ukraine. The inflation forecast was carried out on the basis of the Holt-Winters model, which has a minimum error.

1 Introduction

The pandemic caused by the spread of the COVID-19 virus around the planet has painfully affected the global and national economy [1–4]. According to the NBU [5], the gross domestic product of Ukraine decreased by 6% in 2020. In these conditions, one of the most pressing problems is the containment of price increases. In macroeconomics, the current situation is called stagflation [6], which is characterized by an increase in prices during a crisis in the economy. The consumer price index demonstrates the general level of inflation in the economy and is an indicator of the standard of living of the population and socio-economic development. To regulate price increases, it is necessary to model inflation based on a scientific analysis of the dynamics of the cost of goods and services.

M. Sarel [7] analyzed the possibility of a nonlinear impact of the CPI on economic growth, when this indicator is critical 108. Below this value, the CPI does not affect growth, or may even have a slightly positive effect.

M. Kuzheliev [8] examines the impact of inflation targeting and other key instruments of monetary policy on key economic indicators in Ukraine during periods of stability and crises. This study uses econometric methods (multidimensional regression and the model of the simultaneous equation), which are used for the overall and transmission impact of inflation on the assessment of economic growth. The results show that inflation does not affect (less than 0.46 linear correlation) the basic economic indicators during periods of real GDP growth and the quarterly CPI level is less than 2%. On the other hand, there are significant simultaneous regressions (over 0.8 coefficient of determination) between the unemployed, real final consumption expenditure, UAH exchange rate and monetary policy instruments (discount rate, international reserves, amount of government bonds) for periods when the quarterly CPI (consumer price index) is more than 2%.

In papers [9, 10] presented the influence of the consumer price index on the economic growth of Romania over a period of 28 years, based on the annual data, 1991-2018. The increase of the values of the gross domestic product ensures a sustainable economic growth in Romania. It is proposed to use the cubic analysis model. The results obtained from the analysis show the influence of the consumer price index (CPI) on the gross domestic product (GDP).

The mutual influence of the growth of consumer prices and inflationary expectations in Ukraine is studied in the work of D. Khokhych [11]. T. Gitis [12] raises the problem of the influence of the growth of consumer prices on the level of income of the population. The analysis of the use of the value unit index for curbing inflation in Latin American countries was carried out by Yu. Yereshko[13]. Forecasting the consumer price index in Ukraine for the period of the end of 2020, the beginning of 2021 was not previously carried out in the works.

2 Materials and methods

Inflationary processes are studied based on the monthly consumer price index in Ukraine for the period from January 2010 to September 2020 according to the website of the State Statistics Service of Ukraine [14]. The data array was 129 variants, the units of measurement are percentages up to the previous month.

The analysis of the dynamics of the series and the construction of models were carried out using the STATIS-

*e-mail: shinkar@te.net.ua
**e-mail: alexeyGostrik@gmail.com
***e-mail: larysoncka@gmail.com
****e-mail: dolinsky.com@gmail.com

TICA package, which made it possible to choose the optimal model with a minimum error of the forecast.

The use of the STATISTICA package to construct time series is carefully described in [15, 16].

2.1 The Holt-Winters model or exponential smoothing

The Holt-Winters model or exponential smoothing with a linear trend and seasonality assessment is the progenitor of adaptive forecasting methods [17, 18]. The complete procedure consists of four steps:

1. Calculation of exponentially smoothed series:

$$L_t = \alpha \cdot \frac{Y_t}{S_{t-s}} + (1-\alpha) * (L_{t-1} + T_{t-1}) \quad (1)$$

2. Determining the value of the trend:

$$T_t = \beta \cdot (L_t - L_{t-1}) + (1-\beta) * T_{t-1} \quad (2)$$

3. Seasonality assessment:

$$S_t = \gamma \cdot \frac{Y_t}{L_t} + (1-\gamma) * S_{t-s} \quad (3)$$

4. The forecast is carried out:

$$Y_{t+p} = (L_{t+p} \cdot T_t) + S_{t-s+p} \quad (4)$$

where are L_t – smoothed value at the time of calculation; α is the smoothing coefficient of the series; S_{t-s} – coefficient seasonality of the previous series; Y_t is the value of the period; L_{t-1} – smoothing value for the previous period; T_{t-1} – trend value for the previous period; T_t – the value of the trend at the time of calculation; β – trend smoothing coefficient; S_0 – seasonality factor at the time of calculation; γ is the coefficient seasonality smoothing; Y_{t-p} – Holt-Winters model forecast for *p*periods ahead; p is ordinal the number to which we make a forecast; S_{t-s+p} – seasonality factor for the same period in the last season.

Thus, on the one hand, it is necessary to increase the statistical range and more correctly choose the smoothing coefficients. Finding compromise values of α, β, γ is the task of model optimization. One of the important components is the choice of coefficients that affect the forecasting error and subsequently the accuracy of the forecast .

2.2 Box-Jenkins model

ARIMA * ARIMA or Box-Jenkins model is an integrated model of auto regression – variable mean and time series analysis. Is an extension of ARMA models for non- stationary time series that can be made stationary by calculating the difference of some order from the original time series. The ARIMA model (p, d, q) means that the time series differences of order d obey the ARIMA model (p, q). The methodology of the model is described in detail in [19–21].

We use the following Box-Jenkins additive models of time series:

$$Y_t = f(t) + S_t + e_t \quad (5)$$

where Y_t is a level of time series at time $t = 1, 2, ...$; $f(t)$ – trend is traced the long-term and evolution is deterministic in time; S_t – seasonal fluctuations; e_t – random fluctuations (white noise).

3 Results and discussion

3.1 Determining the dynamics of a series

In contrast to [22], the time series should not be reduced to a special form, but its main characteristics should be determined immediately.

To build a optimal model and to forecast inflationary processes, first, the main characteristics of the dynamic series of the consumer price index are found. They are shown in table 1.

Table 1. Dynamic series characteristics

Descriptive Statistics (CI.sta)					
Variable	Mean	Std.Dv.	Min	Max	N
CI	100.89	1.82	98.7	114.00	129

The characteristics of the series show that the indicator ranges from 98.7% to 114%, the standard deviation of 1.82% demonstrates that the spread of the consumer price index is quite small.

The presence of anomalous values of 112% and 114% in general does not affect the overall dynamics of the consumer price index, so the study was conducted without removing them.

To determine the general trend in the behavior of a series of data, a graph was built showing the consumer price index and the polynomial trend $n = 3$ (figure 1).

Figure 1. Graph the consumer price index and the polynomial

The coefficient of determination $R^2 = 0.0732$ shows that there is no obvious dependence of the consumer price index on the time of its measurement. The absence of a

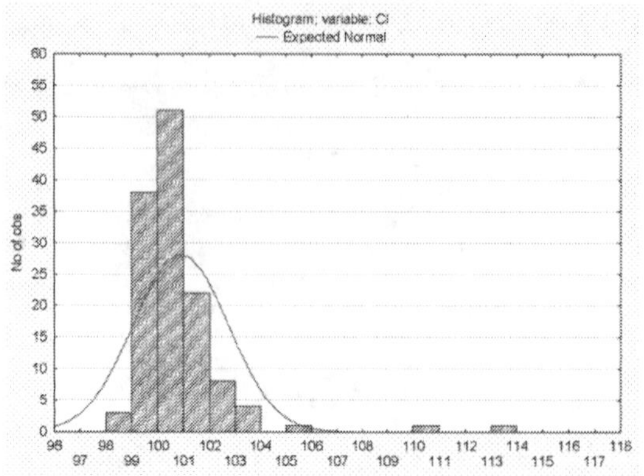

Figure 2. Histogram of the variable

trend relationship is also shown by the constructed histogram of the variable (figure 2).

Therefore, to model inflation and determine the indicator for subsequent periods, it is necessary to use the methods of Holt-Winters and Box-Jenkins.

3.2 The Holt-Winters model

To build the model, we use the "Time series and forecasting" module of the STATISTICA package. To find the optimal values of the model parameters, we use the "Search on the grid" command.

The method of random search is effective with the use of computers to solve optimization problems. The method is based on the generation of random numbers that have a uniform distribution in the interval [0, 1]. Random numbers obtained on a computer are called pseudo-random numbers. At the first stage, a step width of 0,1 is selected and all values of the coefficients α, β, γ are selected as $0, 1n$, $n = 0, 1, \dots, 10$. Determine the model values: among the set of values of the coefficients are solutions α, β, γ for which the objective function takes a high accuracy of value. If the forecast accuracy is insufficient, you should reduce the step width and find new optimal values of the coefficients.

As a result, the following optimal parameters were obtained: exponential smoothing $\alpha = 0,8$; trend component $\beta = 0,1$ and seasonality estimate $\gamma = 0,1$. A summary of the error is shown in the figure 3.

Mean absolute error

$$MAPE = \frac{1}{n} \cdot \sum_{i=1}^{n} \frac{|y_i - \overline{y}_i|}{y_i} = 0.81 \tag{6}$$

Mean squared error

$$MSE = \frac{\sum_{i=1}^{n} (y_i - \overline{y}_i)^2}{n} = 2.1 \tag{7}$$

is small, which indicates the adequacy of the model and high prediction accuracy. The forecast for 10 subsequent periods is shown in figure 4.

	Exp. smoothing: Multipl. season (12) S0=100,8 T0=-,004 Lin.trend,mult.season; Alpha= ,800 Delta=,100 Gamma=,100 CI
Summary of error	Error
Mean error	0,002744021551
Mean absolute error	0,807409687585
Sums of squares	271,007804741336
Mean square	2,100835695669
Mean percentage error	-0,006822170673
Mean abs. perc. error	0,792662442415

Figure 3. Summary of error Holt-Winters model

	Exp. smoothing: Multipl. season (Lin.trend,mult.season; Alpha= ,80 CI		
Case	CI	Smoothed Series	Resids
118	100,7000	100,2562	0,4438
119	100,1000	100,6236	-0,5236
120	99,8000	100,0304	-0,2304
121	99,8000	99,8512	-0,0512
122	99,9000	99,6222	0,2778
123	101,3000	100,4912	0,8088
124	100,4000	101,3446	-0,9446
125	100,1000	99,4287	0,6713
126	100,0000	99,6172	0,3828
127	99,9000	99,4727	0,4273
128	100,1000	100,1238	-0,0238
129	101,1000	101,5112	-0,4112
130		100,7576	
131		100,7824	
132		100,6869	
133		100,7884	
134		100,7019	
135		101,4397	
136		101,6210	
137		100,5662	
138		100,2405	
139		99,7890	

Figure 4. Prediction results using the Holt-Winters model

To assess the forecasting accuracy, let us compare the forecasted consumer price indices with the real ones for October and November 2020 and calculate the relative forecast error:

$$\delta_{10,2020} = \left| \frac{y_{10,2020} - \overline{y}_{10,2020}}{y_{10,2020}} \right| = \left| \frac{100,8 - 100,7576}{100,8} \right| = 0,0004 \tag{8}$$

$$\delta_{11,2020} = \left| \frac{y_{11,2020} - \overline{y}_{11,2020}}{y_{11,2020}} \right| = \left| \frac{100,7 - 100,7824}{100,7} \right| = 0,0008 \tag{9}$$

It should be noted that the forecasting accuracy of the constructed model decreases every month, since it is intended for short-term forecasts.

The relative forecast error is very small (less than 1%), therefore, forecasting the indicator using the Holt-Winters model shows good results. To illustrate the adequacy of the model, graphs of real and simulated consumer price indices are built. They are shown in figure 5.

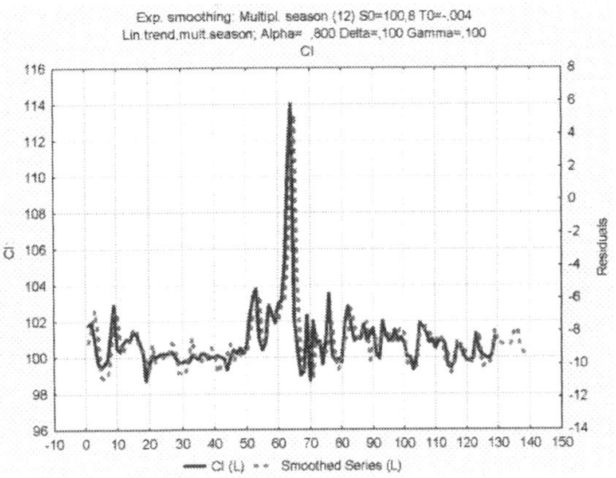

Figure 5. Graph of real indices and obtained by the Holt-Winters model

We are convinced of the adequacy of the obtained the Holt-Winters model of exponential smoothing by analyzing the histogram of residuals and Normal curve (figure 6).

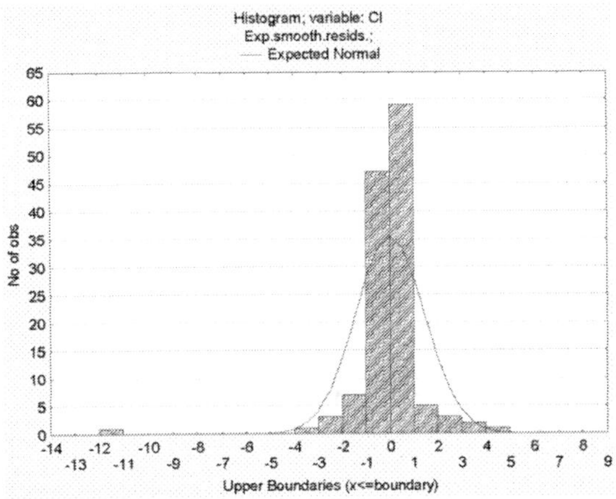

Figure 6. Histogram of residuals the Holt-Winters model

It should be noted that the forecasting accuracy of the constructed model decreases every month, since it is intended for short-term forecasts.

3.3 Box-Jenkins model

We will first define the time series. Analyzing the graph of autocorrelation (figure 7) and the partial autocorrelation function presented below, we conclude that there is a first-order autocorrelation with a lag of 1 and an annual seasonality with a delay of 12.

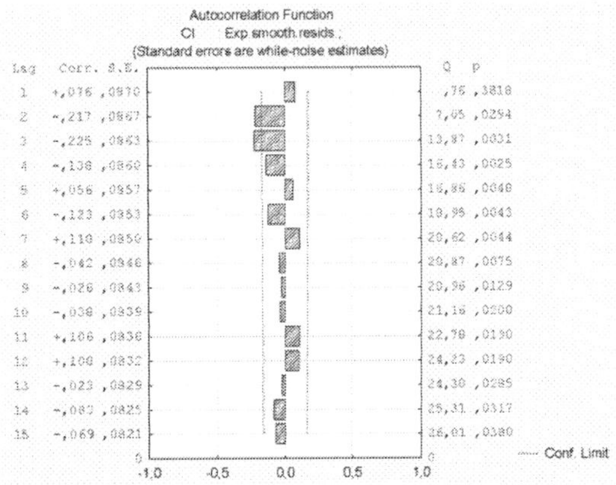

Figure 7. Graph of autocorrelation function

Partial autocorrelation function is presented below (figure 8).

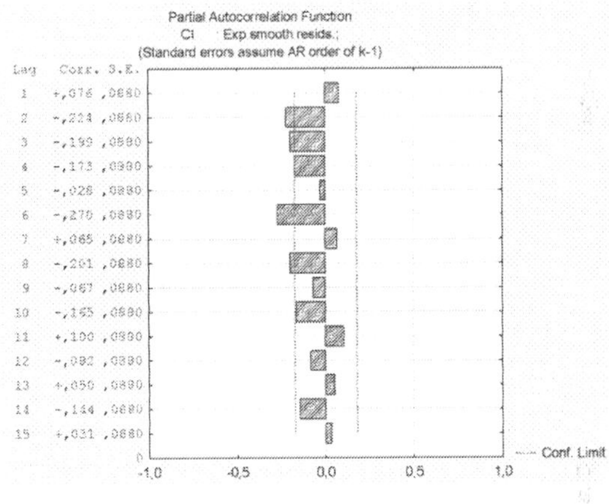

Figure 8. Graph of partial autocorrelation function

We will build an ARIMA*ARIMAS model using the "Automatic search for ARIMA model parameters". As a result, the model

$$p = 1, q = 1, P_s = 1 \qquad (10)$$

was proposed as the best program, seasonal lag = 12. The calculated parameters are shown in table 2.

Table 2. Estimates of parameters of the ARIMA model

Input: CI (CI2.sta) Transformations: D(1),D(12) Model:(1,1,1)(1,1,0) Seasonal lag: 12 MS Residual = 2.8764		
Parameter	Estimate	Standard deviation
p(1)	0.672278	0.075567
q(1)	0.986673	0.017673
Ps(1)	-0.342172	0.089743

For the built model $MSE = 2.8764$, which is slightly higher than the MSE for the previous model. Table 3 shows the forecasting results for the next 12 periods (for a year).

Table 3. Estimates of parameters of the ARIMA model

Forecasts; Model:(1,1,1)(1,1,0)
Seasonal lag: 12
Input: CI Start of origin: 1 End of origin: 129

Case No.	Forecast	Lower	Upper	Std.Err.
130	100.92	98.11	103.74	1.70
131	100.35	96.94	103.76	2.06
132	99.90	96.24	103.56	2.21
133	99.94	96.16	103.71	2.28
134	99.81	95.97	103.65	2.31
135	100.85	96.98	104.72	2.33
136	100.28	96.40	104.17	2.34
137	99.98	96.08	103.87	2.35
138	99.49	95.59	103.40	2.55
139	99.39	95.49	103.30	2.36
140	99.62	95.71	103.53	2.36
141	100.62	96.71	104.54	2.36

Calculate the relative forecast error:

$$\delta_{10,2020} = \left|\frac{y_{10,2020} - \overline{y}_{10,2020}}{y_{10,2020}}\right| = \left|\frac{100,8 - 100,92}{100,8}\right| = 0,001 \quad (11)$$

$$\delta_{11,2020} = \left|\frac{y_{11,2020} - \overline{y}_{11,2020}}{y_{11,2020}}\right| = \left|\frac{100,7 - 100,35}{100,7}\right| = 0,0035 \quad (12)$$

Comparing the predicted values with the real values, we obtain the relative forecast errors of 0.1% and 0.35%, which is worse than the Holt-Winters model, but it is quite acceptable for analyzing the dynamics of inflation. Figure 9 shows a plot of the ARIMA model with a predicted interval.

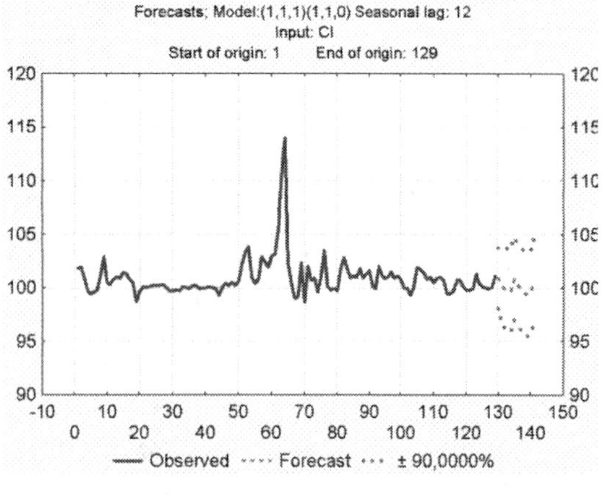

Figure 9. Plot of ARIMA model and 95% forecast interval

To clearly prove the adequacy of the selected model in figure 10, a graph of ARIMA model residues is constructed. Only in anomalous values the residues acquire rather large values, in others they do not exceed in absolute value 2%.

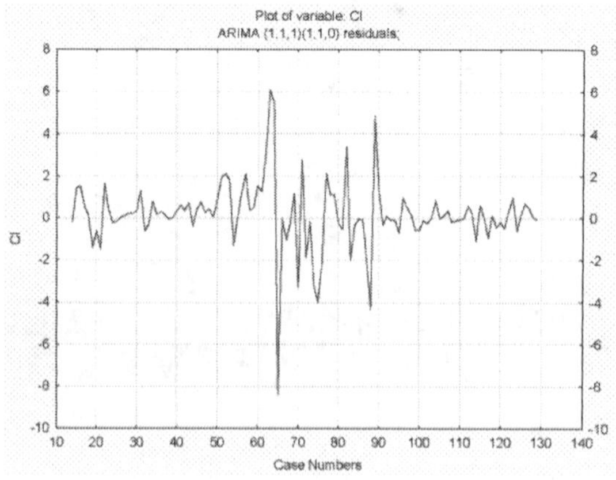

Figure 10. Graph of ARIMA model residues

We are convinced of the adequacy of the obtained the ARIMA*ARIMAS Box-Jenkins model by analyzing the histogram of residuals and Normal curve (figure 11).

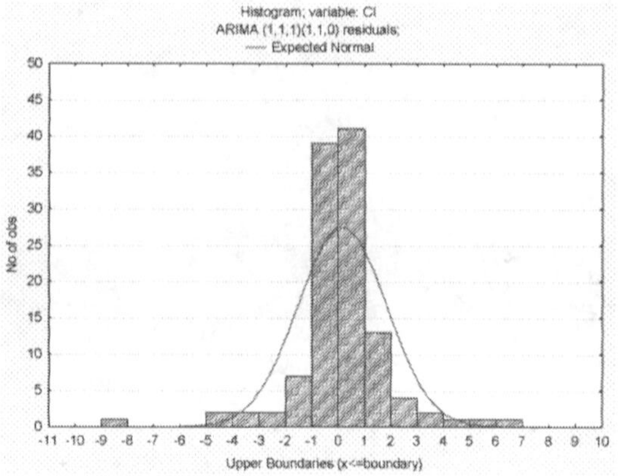

Figure 11. Histogram of residuals the ARIMA*ARIMAS model

4 Conclusion

Analyzing the dynamics of the consumer price index in the period from January 2010 to September 2020, it can be concluded that inflation at the level of 5-10% per year is typical for the Ukrainian economy. 2014 and 2015 were especially critical, after which the government managed not only to reduce the inflation rate, but also to maintain the growth of the consumer price index at 5% per annum.

The processing of the consumer price index array with the help of applied programs proved that the indicator is characterized by high levels of seasonality and trend component and cannot be approximated by elementary functions. At the same time, the use of Holt-Winters and Box-Jenicks methods made it possible to build models of inflationary processes with high accuracy.

Forecasting the indicator for the beginning of 2021 showed that, despite the crisis in the economy associated with the COVID-19 pandemic, there are no reasons for increase the inflation rate.

For a deeper study of the reasons for the increase in the consumer price index, one should consider the change in the value of its individual components, the relationship of the indicator with the exchange rate of the national currency and the gross domestic product.

References

[1] S. Semerikov, S. Chukharev, S. Sakhno, A. Striuk, V. Osadchyi, V. Solovieva, T. Vakaliuk, P. Nechypurenko, O. Bondarenko, H. Danylchuk, E3S Web of Conferences **166** (2020)

[2] A. Bielinskyi, I. Khvostina, A. Mamanazarov, A. Matviychuk, S. Semerikov, O. Serdyuk, V. Solovieva, V. Soloviev, IOP Conference Series: Earth and Environmental Science **628** (2021)

[3] V.N. Soloviev, A.O. Bielinskyi, N.A. Kharadzjan, CEUR Workshop Proceedings **2832**, 24 (2020)

[4] S. Semerikov, H. Kucherova, V. Los, D. Ocheretin, CEUR Workshop Proceedings **2845**, 22 (2021)

[5] NBU, *National Bank of Ukraine* (2020), `https://bank.gov.ua/ua/news/all/rishennya-oblikova-stavka`

[6] OED, *Online etymology dictionary* (2020), `https://www.dictionary.com/browse/stagflation#`

[7] M. Sarel, *Nonlinear Effects of Inflation on Economic Growth* (IMF Working Paper, IMF, 1995)

[8] M. Kuzheliev, D. Zherlitsyn, I. Rekunenko, A. Nechyporenko, G. Nemsadze, Banks and Bank Systems **15**, 94 (2020)

[9] A.G. Macovei, L. Scutaru, Academic Research International **7**, 95 (2016)

[10] A.G. Macovei, Ecoforum **9** (2020)

[11] D. Khokhych, Finance of Ukraine **4**, 64 (2020)

[12] T. Gitis, Y. Chemerys, V. Antonova, A. Nosanyova, Economic Bulletin of Donbass **1**, 64 (2020)

[13] J. Yereshko, E. Hafarov, Efektyvna Ekonomika **5** (2020)

[14] SSSU, *The state statistics service of ukraine* (2020), `https://www.ukrstat.gov.ua`

[15] C. Dzitsaki, American Journal of Applied Mathematics and Statistics **4**, 136 (2016)

[16] N. Dzitsakis, P. Klareglon, Internathional Journal of Economics and Financial Issues **8**, 9 (2018)

[17] S. Everette, J. Gardner, Journal of forecasting **4**, 1 (1985)

[18] G. Box, G. Jenkins, *Time Series Analysis, Forecasting and Control* (1976)

[19] Y. Kozak, V. Matskul, T. Shengelia, *Mathematical methods and models for master of economics. Practical applications* (2017)

[20] L. Dvořáková, SHS Web of Conferences **39**, 01005 (2017)

[21] V. Matskul, D. Okara, N. Podvalna, SHS Web of Conferences **73**, 01020 (2020)

[22] V. Shinkarenko, M. Matskul, D. Linok, CEUR Workshop Proceedings **2422**, 147 (2019)

Monitoring the quality of e-learning implementation in educational institutions

Valerii Habrusiev[1,*], *Hryhoriy* Tereshchuk[1,**], *Ivan* Tsidylo[1,***], *Serhii* Martyniuk[1,****], and *Olena* Kulyanda[2]

[1]Ternopil Volodymyr Hnatyuk National Pedagogical University, 2 M. Kryvonosa Str., Ternopil, 46027, Ukraine
[2]Ivan Horbachevsky Ternopil National Medical University, 1 Voli Sq., Ternopil, 46001, Ukraine

Abstract. The article analyzes some aspects of quality control and efficiency of electronic educational and methodological complexes of educational disciplines usage in the educational process. Analysis of the implementation of e-learning and of user work allows to build a model of student's work, identify the disadvantages of both organizational and technological plan and take timely steps to eliminate them. The solution to this problem should include a comprehensive approach to ensure an objective result. Google Analytics tools, Moodle Learning Management System (LMS) tools, the teacher's surveys to monitor the quality of e-learning in an institution proves its efficiency and allows to see the full advantages and disadvantages of ongoing e-learning in an institution, plan and implement further development and improving e-learning. The analysis of the effectiveness of e-courses usage should be carried out in a comprehensive manner, using the monitoring tools of the educational process adopted in the educational institution. Such as surveys for teachers and students, analysis of success, etc. The specifics of e-learning, the Moodle LMS tools and third-party services for collecting statistics and analyzing user activity in the LMS should be taken into account.

1 Introduction

At the present stage of information society development, the use of information and communication technologies (ICT) contributes to the globalization of education, the development of the international labour market, the growth of various types of personality mobility. An important result of globalization is the improvement of the mobility of students, university entrants and university graduates: a person with a high level of mobility can study, work, cooperate and be competitive in any country. Increase of the academic mobility, the introduction of international norms and standards by which academic qualifications from different countries can be compared and recognized, lead to increased competition among HEIs and improve the quality of higher education [1].

An indispensable condition for social and economic development of any country is investment in education of the population. In this context, the globalization of education contributes to the personal and professional development of professionals.

For Ukraine, informatization of education is extremely relevant in the context of its economic, social and cultural development [2]. The main focus of ICT use is to create an education system focused on the use of the latest ICT in the formation of a well-developed personality, which enables each person to independently acquire knowledge, skills and competences during education and training [3–5].

The purpose of the National Strategy for the Development of Education in Ukraine until 2021 is to: update the content, forms, methods and means of learning through the widespread introduction of modern ICT and e-learning in the educational process. The priority of the development of education is the introduction of modern ICT, which ensures the improvement of the educational process, accessibility and effectiveness of education, preparation of the young generation for life in the information society [6].

Introduction of e-learning involves not only the creation of e-courses and the orientation of students to use them during classroom or independent work [7–10], but also control over the effectiveness of the use of created courses in the educational process. Monitoring the implementation of e-learning, the work of users allows you to build a model of student work, identify the shortcomings of both organizational and technological plan and take timely steps to eliminate them. Such monitoring makes it possible to identify morally outdated or unsupported e-learning platforms, to avoid technical problems while users are operating, and to adapt to new software or technical conditions in a timely manner.

2 Literature review

The analysis of published research shows the presence of a number of publications on comparisons of learning management systems (LMS) in distance education, in particular with open source. In a study [11] by Cansu Cigdem

*e-mail: gabrusev@tnpu.edu.ua
**e-mail: g.tereschuk@tnpu.edu.ua
***e-mail: tsidylo@tnpu.edu.ua
****e-mail: sergmart65@ukr.net

Aydin and Guzin Tirkes, several analyzes and comparisons were conducted on open source LMS. The Moodle system is singled out as one that has many features among other LMS, which are aimed at improving the quality of education and include the tools that an e-learning system should have. In general, the use of open source software will ensure the development of teaching aids and improve the quality of education. Evaluation of usability in Moodle LMS was carried out in the works [9, 12–17]. The convenience of using standard modules in Moodle, one of the leading open source LMS, was evaluated. Thanks to this research, we get significant results and informational advice for administrators, teachers and students on how to improve the effective use of this system [12].

Using learning analytics to predict students' performance in Moodle was revealed in the work by Imani Mwalumbwe and Joel S. Mtebe. Therefore, this study designed and developed Learning Analytics tool and used the tool to determine the causation between LMS usage and students' performance. Data from LMS log of two courses delivered at Mbeya University of Science and Technology were extracted using developed Learning Analytics tool and subjected into linear regression analysis with students' final results. The study found that discussion posts, peer interaction, and exercises were determined to be significant factors for students' academic achievement in blended learning at MUST. Nonetheless, time spend in the LMS, number of downloads, and login frequency were found to have no significant impact on students' learning performance. The implications of these results on improving students' learning are discussed [18].

In the work [19] by I. N. Egorova and O. N. Kadushkevich modern Google Analytics tools have been investigated against effective attraction channels for users and bottlenecks detection. Conducted investigation allowed to suggest modern method for effective usage of Google Analytics tools. The method is based on main traffic indicators analysis, as well as deep analysis of goals and their consecutive tweaking. Method allows to increase website conversion and might be useful for SEO and Web analytics specialists.

T. S. Bondarenko proposes to organize quality monitoring on the basis of a process approach, in which monitoring of the quality of resources is carried out in three directions: the quality of the project, the quality of the process and the quality of the result of their use. In connection with the multidimensional nature of the quality of electronic educational resources, it is proposed to carry out quality assessment based on the use of a hierarchical quality model. At the top level of this model, constructed in accordance with the process approach to management of quality, there are directions of quality monitoring. There are evaluation criteria (didactic quality of the project, methodological quality of the project, etc.) at the next level of the hierarchy of the model. They are used to assess the quality of educational resources. The criteria determine by the indicators, which in turn measure the properties (attributes) of the educational resource. The principal possibilities for measuring of the quality of electronic educational resources in the course of monitoring are considered. To evaluate their values the following methods are suggested: measuring, registration, calculation and expert method [20].

Researchers do not pay attention to the possibility of using the distance learning platform and the Google Analytics service in the process of monitoring the quality of e-learning implementation in educational institutions. The purpose of the article is to highlight the capabilities of LMS tools and Google Analytics to monitor the quality of e-learning implementation in the educational process of educational institutions.

3 Problem solving

The analysis of the effectiveness of the use of e-courses should be carried out in a comprehensive way using the monitoring tools of the educational process adopted in the educational institution, such as surveys of teachers and students, analysis of progress, etc. The specifics of e-learning, the Moodle LMS tools and third-party services for collecting statistics and analyzing user activity in the LMS must be taken into account.

In this article, we will explore the use of the Moodle LMS [21], the Google Analytics Service [22], and a survey of teachers which allows to monitor e-learning.

3.1 Use of Moodle LMS tools to analyze the use of e-learning

Taking into account the importance of obtaining data on the operation and use of the learning resource management system, Moodle developers envisaged collecting certain data for further analysis. We shall consider the most commonly used metrics that can be obtained directly from Moodle [3]:

- *The most active courses for a given period.* A metric for determining the total number of completed actions (page loading, viewing of study material, submitting a task, completing a test) by students and teachers. This parameter is relative and does not allow for objective evaluation of course usage.
- *Total report on the most active courses for a given period.* This indicator is defined as the average activity of a student or teacher, calculated by:

$$K = \frac{D}{Nct}$$

  in which D is the total number of course users' actions, Nct is the number of registered users. This parameter allows for a more objective assessment of the activity of using e-courses (figure 1).
- *Courses with the highest number of students enrolled for a given period.* The student activity rate of a course is calculated as the ratio of the number of active course users to the total number of registered users.
- *Courses that lead by the number of views over a given period.* The activity rate of students in the use of training resources of the course is calculated as the ratio of

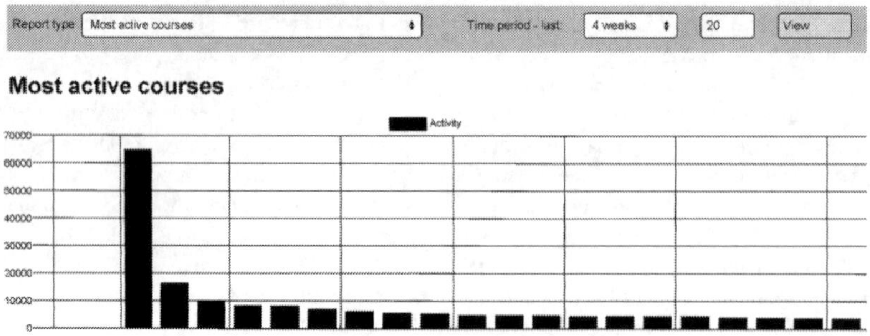

Figure 1. Total report on the most active courses

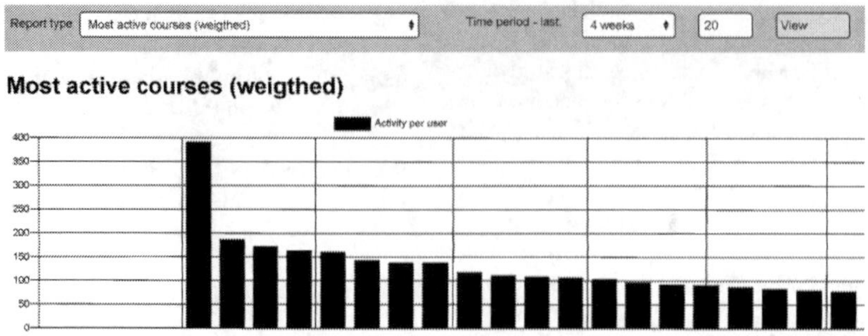

Figure 2. Most active courses, weighted

the total number of views of educational materials to the number of e-mail posts sent or posts in the forum.

- *Activity of teachers and students for a given course.* This indicator characterizes the activity of students and teachers over a certain period of time. This report allows you to estimate the number of user activities on the system over a given period, both for the system as a whole and for a particular course.
- *Event register (activity of students and teachers).* Detailed information about the activity of each user of the system allows to determine from which IP address the user worked, what actions they performed on the system, which pages they downloaded. Taking into consideration that the event register only contains information about individual actions of each user, including logging in, page loading, file uploading, etc., further systematization and analysis of this data is a difficult process.

To analyze the effectiveness of e-learning implementation, the most useful are the data on the dynamics of users' work with the e-courses server during a given period and a balanced report on the most active courses over a given period (figure 2).

3.2 Using Google Analytics tools to analyze the use of e-learning

Interpreting Google Analytics data from e-learning sites may be slightly different from the commonly accepted interpretation of data for business or social networking sites and Moodle LMS due to the specific features and organization of e-learning users. The main feature is that the number of potential users is known beforehand, this is the entire contingent of students studying at the institution or a separate group or stream. The purpose of the analysis is not to identify the problematic places why the client left and did not return for the second time or did not perform certain actions in the system [4, 20, 23], but to analyze how students and teachers use e-courses.

We ought to consider the main metrics that can be used to analyze the effectiveness of e-learning [21].

- *Referral sources.* The address of the page from which the user connected to the current page and data allow you to determine how the user got to the page. Additional information can be obtained from the address of the previous page: for example, a keyword that a user entered on a search engine and then saw a link to the e-courses site. This information is used to segment, analyze traffic sources, determine how students access the e-courses server, evaluate the demand of outsiders for distance learning services, etc.
- *The page address the user was visiting (page views).* Page visit information. This data allows you to determine which pages you visit most often, to determine in more detail what students do on the system, such as taking tests, viewing or downloading training materials, downloading system tasks, etc.
- *Exact request time.* The time users visit pages is saved. Query time information lets you determine what time

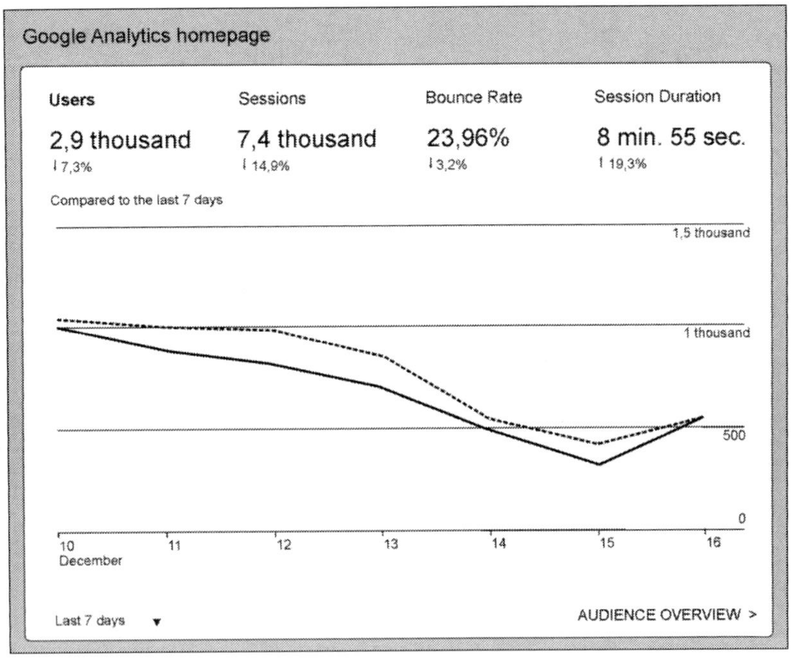

Figure 3. General information about using e-courses in Google Analytics

students are most active in e-courses, and schedule time for online events, such as webinars.

- *Average page view time.* Determining the average browsing time by page users. This parameter is relative, but it allows on average of how much time students use e-courses to prepare for classes, submit completed tasks, and perform in tests.
- *Bounce rate.* It determines the percentage of users who have started browsing the site from this page and have not gone to another site but immediately closed the site. This is the number of visitors (usually a percentage of total) who have visited only one page on the site (figure 3).

  The bounce rate for e-learning sites is no longer as critical a parameter as for other types of sites because the number of site users is fixed and known in advance. This parameter should be considered depending on the situation:

  1) bounce rate for the homepage;

  2) bounce rate for the courses page;

  3) bounce rate for the learning resources page (test, task, hyperlink, etc.).

  The actual setting allows you to estimate how many students have viewed this page and then left the system. In each case, the interpretation will be different.
- *Daily activity of students.* Information about the hourly activity of e-learning users during the day. The analysis of these data allows teachers to plan their time for conducting online activities, especially when conducting distance learning (figure 4).

  Therefore, it can be concluded, based on the dynamics of student activity, that e-courses are used for independent student work. Visitors (unique visitors, visitors, users, reach). The statistics system analyzes all the pageviews it records and tries to determine which ones were made from a single browser. Summing up the number of different browsers over a period of time, it counts the number of unique visitors, "meaning that one browser is used by one user. Google Analytics determines the "uniqueness" of a visitor by leaving a unique number of cookies on their browser during the first visit. All visits by this user will now be combined with this ID. With this metric, they can estimate the average number of students and teachers who use e-courses for the selected period.
- *Website presence time.* A metric that contains information about the amount of time a user has spent on the site. However, since the statistics system logs only the time the page is opened, it is not possible to determine the time the user spent on the last of the pages opened. Therefore, it is not possible to determine an average view time for users who have only one page. It is also impossible to estimate how long a user actually worked with the page and how much time it was opened in the background in the next window. Despite the relativity of this metric, some conclusions can be drawn about the work of students with e-courses.
- *User settings.* The statistics system determines which browser the client has, the type of operating system,

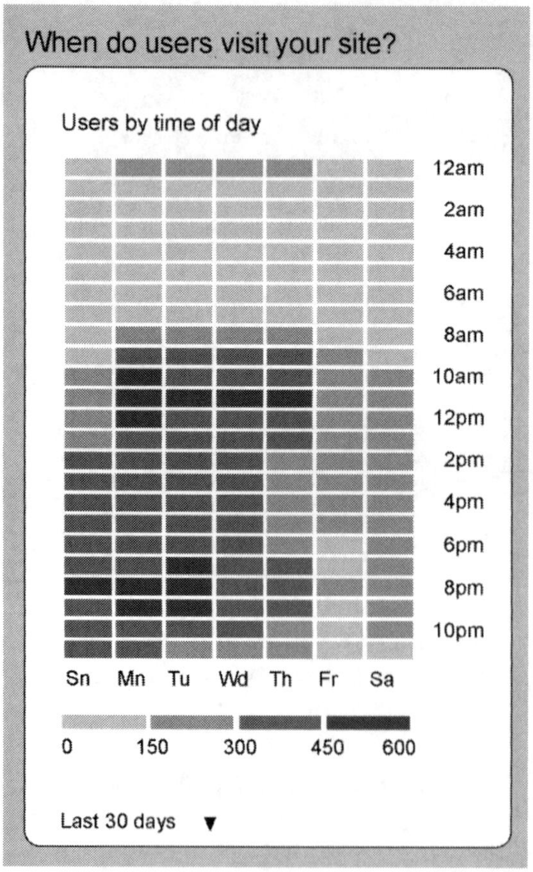

Figure 4. Daily activity of students

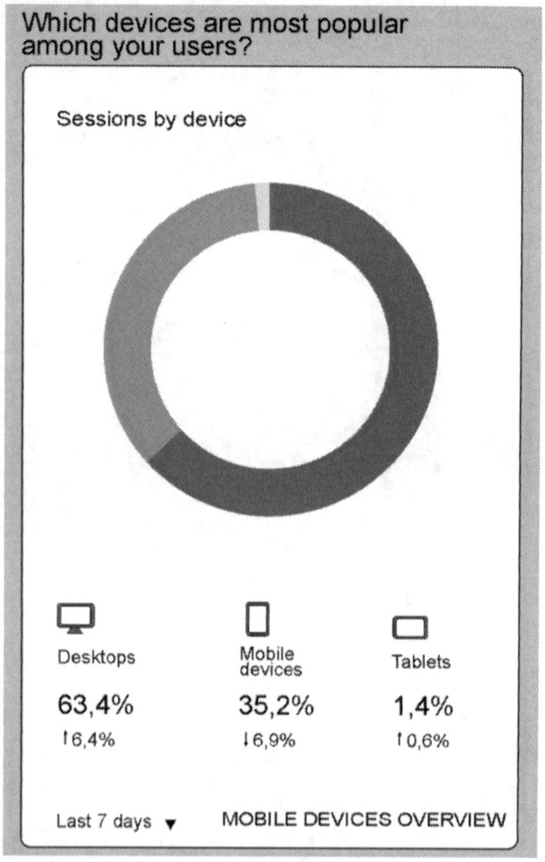

Figure 5. Devices used by users to work with e-courses

monitor resolution, etc. The results of the analysis of the equipment used by users for working with e-courses indicate that students are beginning to use mobile platforms (figure 5).

It should also be noted that with the trend of proliferation of tablets, smartphones, the number of users who use mobile technologies will only increase. This tendency must be taken into account during the development of e-courses and during the planned upgrades of the e-learning system software component.

3.3 Teachers survey to further analyze the use of e-learning

The survey is one of the most common methods of obtaining information about respondents. The survey was developed and conducted by the TNPU Distance Learning Center, author Valerii Habrusiev [24]. More than 200 teachers were involved in the survey, the survey is to ask people specific questions that answer the researcher to obtain the necessary information, depending on the tasks of the study. The characteristics of the survey can be attributed to its mass, which is caused by the specifics of the tasks that they solve. In pedagogical, sociological research, it is advisable to conduct a survey by means of a questionnaire.

Questionnaire is a method of obtaining information through the written answers of the respondents to the system of standardized questions of pre-prepared questionnaires. By generalizing the completed questionnaires objective information about the attitude of teachers to e-learning is obtained, problems of the implemented e-learning model are identified. The information obtained can be used in the development of new training programs, the choice of teaching methods, ways of presenting information during the implementation of e-learning.

During the evaluation of the e-learning conducted, a survey was conducted among university professors to find out information about the forms of the educational process during the use of e-courses, the number of students who use the e-courses etc.

1. How many students use your online courses?
   - less than 25 – 5.1%;
   - 25-100 – 12.8%;
   - 100-150 – 30.8%;
   - 150-200 – 25.6%;
   - more than 200 – 25.6%;

2. Choose the option for students to use the e-courses you have developed:
   - classroom work – 15.4%;
   - independent work – 53.8%;
   - individual work – 28.2%;
   - do not use – 7.7%;
3. Rate on a 3 point scale your level of e-learning use:
   - low level – 33.3%;
   - average level - 51.3%;
   - high level – 15.4%;
4. How do you assess the level of readiness of the university for the introduction of mixed (inverted) courses, mass online courses.
   - low – 30.8%;
   - average – 51.3%;
   - high – 17.9%;
5. What, in your opinion, is the most important thing to do in order to implement mixed (inverted) online mass courses?
   - to improve regulatory support – 20.5%;
   - to improve methodological support for e-learning – 30.8%;
   - to reduce the number of hours of classroom work – 23.1%;
   - to improve the material and technical base of the used equipment – 25.6%;
   - none of the above – 2.6%;
6. Choose the form of e-learning you use to work with students.
   - traditional (classroom work only) – 17.9%;
   - mass online courses – 2.6%;
   - mixed learning – 84.6%;
   - inverted education – 5.1%;
   - none of the above – 0%;
7. How do you evaluate your level of readiness for the implementation of mixed (inverted) training, mass online courses, distance learning.
   - low – 41%;
   - average – 46.2%;
   - high – 12.8%;

The results of the survey indicate that e-learning (e-courses) is used mainly to provide independent (53.8%) and individual work (28.2%) students. The identified problems mainly relate to the e-learning regulatory framework (56.4%) and the methodological training of teachers, which is also confirmed by the poor use of e-learning potential (51.3%).

4 Conclusions

Taking into account the increasing attention to e-learning technologies and their introduction into the educational process of educational institutions at all levels of accreditation, it can be stated that the quality control of e-learning, the development of methods and appropriate tools for such monitoring will be a constant focus of education methodologists and managers. The solution of this problem should include a comprehensive approach to ensure an objective result.

Using Google Analytics tools, Moodle LMS tools, surveying teachers to monitor e-learning quality in an educational setting proves effective and allows you to see all the benefits and disadvantages of ongoing e-learning in an educational setting, plan and implement further development and improvement of e-learning, allows to build in the future the model of work of the student, to reveal shortcomings of both organizational and technical planning and to take timely action to address them. Such monitoring makes it possible to identify morally outdated or unsupported e-learning platforms, to avoid technical problems while users are operating, and to adapt to new software or technical conditions in a timely manner.

Taking into consideration the increasing attention to e-learning technologies and their introduction into the educational process of educational institutions at all levels of accreditation, it can be stated that the issues of quality control of e-learning, the development of methods and appropriate tools for such control will constantly be the focus of methodologists and managers in the field of education. The solution to this problem should include a comprehensive approach to ensure an objective result.

Using Google Analytics tools, Moodle LMS tools, surveying university lecturers to monitor the quality of e-learning at educational establishments proves effective and allows them to see all the benefits and disadvantages of ongoing e-learning in the institution, plan and implement further development and improvement of e-learning.

The analysis of the effectiveness of the use of e-courses should be carried out in a comprehensive way using the monitoring tools of the educational process adopted in the educational institution, such as surveys of teachers and students, analysis of success, etc. The specifics of e-learning, the Moodle LMS tools and third-party services for collecting statistics and analyzing user activity in the LMS must be taken into account.

Analysis of the implementation of e-learning and of user work allows to build a model of student's work, identify the disadvantages of both organizational and technological plan and take timely steps to eliminate them.

References

[1] M.I. Striuk, S.O. Semerikov, A.M. Striuk, Information Technologies and Learning Tools **49**, 37 (2015)

[2] E. Fedorenko, V. Velychko, A. Stopkin, A. Chorna, V. Soloviev, CEUR Workshop Proceedings **2433**, 20 (2019)

[3] V. Habrusev, H. Tereshchuk, *MOODLE Learning Resources Management System* (TNPU after V. Hnatiuk, Ternopil, 2011)

[4] A. Kaushik, *Web Analytics 2.0: The Art of Online Accountability and Science of Customer Centricity* (Sybex, 2009)

[5] S.O. Semerikov, I.O. Teplytskyi, V.N. Soloviev, V.A. Hamaniuk, N.S. Ponomareva, O.H. Kolgatin, L.S. Kolgatina, T.V. Byelyavtseva, S.M. Amelina, R.O. Tarasenko, Journal of Physics: Conference Series **1840**, 012036 (2021)

[6] *National Strategy for the Development of Education in Ukraine until 2021* (2013), `https://zakon.rada.gov.ua/laws/show/344/2013#Text`

[7] V. Habrusev, H. Tereshchuk, *Regulations on the electronic educational-methodical complex of the discipline* (TNPU after V. Hnatiuk, Ternopil, 2014)

[8] O. Lavrentieva, L. Rybalko, O. Tsys, A. Uchitel, CEUR Workshop Proceedings **2433**, 102 (2019)

[9] V. Ustinova, S. Shokaliuk, I. Mintii, A. Pikilnyak, CEUR Workshop Proceedings **2433**, 308 (2019)

[10] V. Tkachuk, Y. Yechkalo, S. Semerikov, M. Kislova, V. Khotskina, CEUR Workshop Proceedings **2732**, 1217 (2020)

[11] C.C. Aydin, G. Tirkes, *Open source learning management systems in e-learning and Moodle*, in *IEEE EDUCON 2010 Conference* (2010), pp. 593–600

[12] G. Kakasevski, M. Mihajlov, S. Arsenovski, S. Chungurski, *Evaluating usability in learning management system moodle*, in *ITI 2008 - 30th International Conference on Information Technology Interfaces* (2008), pp. 613–618

[13] P. Nechypurenko, S. Semerikov, CEUR Workshop Proceedings **1844**, 319 (2017)

[14] I. Mintii, S. Shokaliuk, T. Vakaliuk, M. Mintii, V. Soloviev, CEUR Workshop Proceedings **2433**, 529 (2019)

[15] I. Mintii, CEUR Workshop Proceedings **2643**, 293 (2020)

[16] A. Abdula, H. Baluta, N. Kozachenko, D. Kassim, CEUR Workshop Proceedings **2643**, 306 (2020)

[17] K. Polhun, T. Kramarenko, M. Maloivan, A. Tomilina, Journal of Physics: Conference Series **1840**, 012053 (2021)

[18] I. Mwalumbwe, J.S. Mtebe, The Electronic Journal of Information Systems in Developing Countries **79**, 1 (2017)

[19] I.N. Egorova, O.N. Kadushkevich, ScienceRise **1**, 40–44 (2016)

[20] T.S. Bondarenko, Problems of engineering and pedagogical education pp. 32–44 (2016)

[21] *Moodle - Open-source learning platform* (2021), `http://www.moodle.org`

[22] *Google analytics* (2020), `https://www.google.com/analytics`

[23] *On the basic principles of development of information society in Ukraine for 2007-2015* (2007), `https://zakon.rada.gov.ua/laws/show/537-16#Text`

[24] *TNPU questionnaire*, accessed 2020, 11 Nov, `https://https://forms.gle/RJ2RVZYpu7BEuBBK9/`

Mathematical model for imitation of management of the enterprise's logistical system

Roman Ivanov[1,*], *Yuriy* Sherstennikov[1], *Vasyl* Porokhnya[2], and *Tetyana* Grynko[1]

[1]Oles Honchar Dnipro National University, 72 Haharina Ave., Dnipro, 49000, Ukraine
[2]Classical private university, 70B Zhukovskoho Str., Zaporizhzhia, 69002, Ukraine

Abstract. The study is devoted to solving the scientific problem of optimizing the retail trade in the production and sale of two types of products, taking into account the change in potential demand for products. The economic and mathematical model of the production activity of the enterprise was developed taking into account logistics and market demand. The logistics scheme takes into account all the main links of the logistics system, as well as the connections between them. The considered scheme makes it possible to take into account the diversification of products manufactured by the enterprise. The mathematical model is designed for discrete time. A numerical optimization method has been developed for this mathematical model. The optimal solutions for several cases are found and investigated. The dynamics of the main characteristics of drugs was calculated for all considered cases. A comparative analysis of economic efficiency for the studied cases has been performed. The economic efficiency of retail network optimization is proved.

1 Introduction

Today in the scientific literature, much attention is paid to the modeling of logistics processes and production. The main purpose of article was to present the possibilities and examples of the use of Tecnomatix (Siemens) plant simulation to simulate logistics and manufacturing processes. This tool allows you to simulate discrete events and create digital models of logistics systems (for example, manufacturing), optimize the operation of factories, production lines, as well as individual logistics processes. A review of the execution of a Tecnomatix plant simulation for simulating processes in manufacturing engineering and logistics was conducted and several selected simulations were presented.

Further research should elaborate on the few initial attempts to combine different modeling techniques with optimization [1]. Simulation is an appropriate technique to tackle unresolved issues where analytical computations fail [2].

Article [3] highlights the problems of mathematical modeling for a certain element of the logistics supply system, that is, the production system. The system of the production complex, consisting of a determining number of parallel subsystems, is modeled.

If a company is unable to adapt to changing market conditions, it cannot survive in today's market environment.

To survive in a highly competitive global economy, production systems must be able to adapt to new circumstances [4]. As it is clear from the above and other contexts, many different indicators must be taken into account for assessing the complexity of selected general process structures when designing a structure or optimizing production objectives [5].

The structure of production processes also depends on the production needs of certain products, which can show high variability. The organization of production processes is closely related to the process mappings and procedures responsible for production processes from individual components [6]. A key prerequisite for the effectiveness of the above and other production systems is the precise definition of the interaction between the links of the logistics system (LS) [6].

In addition, the requirements are growing in order to control possible destruction of production systems [7]. Rapid recovery of production in case of errors and other risks associated with the general I/O model, which is the production system, should be considered [8].

Production generally occurs as a series of individual actions that are performed manually, mechanically, or a combination of these. Optimization requires continuously processed orders in relation to individual projects. It can also be the sequential production of one type of product, in which we can easily identify a number of key performance indicators and manage and automate production [9, 10]. Today, the standard starting point for calculating and optimizing manufacturing systems is simulation. Computer simulations allow us to test various types of production quickly. Computer simulation makes it possible to check many consequences of changes in production, processes and selects the most efficient way to streamline logistics in the near future [11]. Simulation can be used both before calculating the design of the production system and in order to optimize the production system and in the design of

*e-mail: romanivanov1926@gmail.com

 https://doi.org/10.1051/shsconf/202110710004

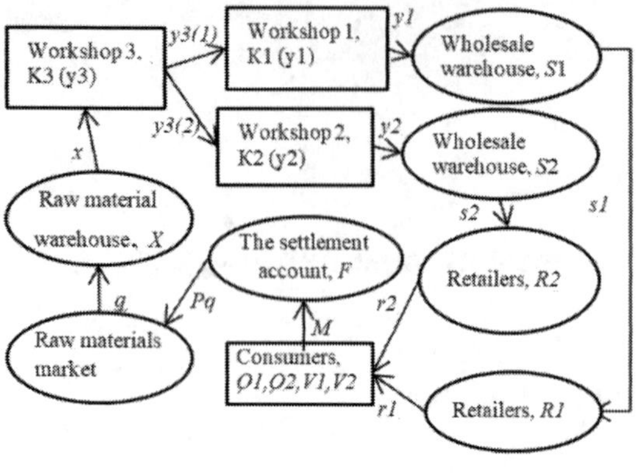

Figure 1. Scheme of the enterprise's logistics

production processes, respectively. In both cases, it is necessary to consider the simulation results as input and information for the design or redesign of a thoughtful system. In addition to defining the structure of production systems (in manufacturing and logistics systems, of a particularly cautious nature, see, for example, [12–14]), simulations are useful for planning production and its sustainability and continuity [15, 16]. Specifically, simulations can help coordinate the needs of different departments and open management bottlenecks and improve resource allocation, allocation of production between production lines or factories, testing strategies, performance dimension, and so on [17, 18]. The analysis of the articles makes it possible to draw the following conclusions. Simulation modeling is an urgent problem in planning and optimizing an enterprise's LS. Most of the research is devoted to a detailed study of certain links of LS. Studies of the entire logistics system from the purchase of raw materials to the delivery of finished products to the final consumer have been insufficiently completed. The article aims to fill in part the existing research gap.

The aim of the study is to develop an economic and mathematical model of the LS of an enterprise producing two types of products; using this model to develop a methodology for optimizing the work of such an important link in the LS as retail trade in the production and sale of two types of products.

To achieve the goal, the following tasks were set: draw up a system of equations for the proposed scheme of the enterprise's LS; on the basis of the obtained system of equations, develop a mathematical program for optimizing various modes of the LS operation; to compare the economic efficiency for different operating modes of the enterprise.

2 Results

We will consider the work of an enterprise that produces two types of products. Let the logistics system of an enterprise be represented by the diagram shown in figure 1.

Figure 1 shows that after the production link, each type of product is delivered to the end consumer via independent supply chains. In fact, wholesale warehouses for each type of product can be only certain areas, in some warehouse reserved for the first and second types of products. The same applies to retail chains. In part, these can be different outlets, to which either product 1 or product 2 is imported. It can also be supermarkets in which products 1 and 2 are sold in different (or even the same) departments. But in a mathematical description, it is convenient to represent these chains as independent. This presentation does not diminish the generality of the description.

In previous works [19–21] the production link was considered as an unstructured black box. Figure 1 represents the production process in two stages. Workshop 3 will carry out preliminary processing of raw materials and those operations that are common in the production of both types of products. Workshops 1 and 2 perform the final operations that are typical only for products of the first and second types, respectively. The presented scheme can also describe the joint work of a group of enterprises, some of them can be producers of products, while others can act as distributors of products.

Let us formulate a system of equations that describe the LS of the enterprise shown in figure 1. We use this notation. Each variable aj_i contains two indices. The j line index numbers the types of products ($j = 1, 2$); subscript i numbers time intervals (days) ($i = \overline{1, 730}$). We will consider a project with a planning horizon of two years.

1. A change in demand Qj_i for products on the market is an input effect for an enterprise whose task is to bring its output into line with demand.

$$rj_{i+1} = nj \cdot Rj_i \cdot (Qj_i - Vj_i) \qquad (1)$$

where rj_i is the rate of sales of the j product (pieces / unit of time) in the i-th; Rj_i – quantity j of goods in the retail trade network in the i-th period; Vj_i is the quantity j of the consumer's product (not yet consumed).

2. The quantity of goods in the retail network Rj_i is determined by the recurrent formula:

$$Rj_{i+1} = Rj_i + (sj_i - rj_i), \qquad (2)$$

where sj_i is the rate of deliveries (units per period) from a wholesale warehouse in a retail network.

3. The value Rj_i must be in the range $0 \leq Rj_i \leq Rmj$, where Rmj is the maximum possible quantity of the product in the retail system. The following formula for the rate of deliveries from the wholesale warehouse to the retail system corresponds to this requirement:

$$sj_{i+1} = \min\left[r_{ji} \cdot \left(1 + \frac{Rmj - Rj_i}{Rmj}\right), Rmj - Rj_i, Sj_i\right], \qquad (3)$$

where Sj_i is the stock of goods (quantity) in the j-th wholesale warehouse.

4. The rate of production yj_i is determined by the following formula:

$$yj_{i+1} = \frac{Yj_i}{tYj}, \quad (4)$$

where Yj_i – the value of work in progress in the i-th period; tYj is a production parameter.

5. The value of work in progress is determined by the formula:

$$Yj_{i+1} = Yj_i + a3j \cdot y3_i - yj_i \quad (5)$$

6. The value (quantity) of stock of goods in the wholesale warehouse Sj_i is calculated by the formula:

$$Sj_{i+1} = Sj_i + yj_i - sj_i, \quad (6)$$

where yj_i is the rate of flow that enters the warehouse from production.

7. The amount of raw materials purchased daily is determined by the following formula:

$$q_i = \begin{cases} q0, & \text{if } i < Tq, \\ kq \cdot q0, & \text{otherwise.} \end{cases} \quad (7)$$

Formula (7) allows you to describe the process of changing the volume of daily purchases of raw materials. Such a change may be necessary as a result of the project implementation during the initial period of time $[0; Tq]$.

8. The stock of raw materials in the raw material warehouse is calculated by the formula:

$$X_{i+1} = X_i + q_i - x_i. \quad (8)$$

9. The rate of supply of raw materials to the third workshop is calculated by the formula:

$$x_{i+1} = X_i / tX. \quad (9)$$

10. The amount of goods among consumers (not yet consumed):

$$Vj_{i+1} = Vj_i + rj_i - k1 \cdot Vj_i. \quad (10)$$

11. Such a formula has been adopted to determine the daily net profit of the enterprise, expressed in the corresponding monetary units (MU):

$$M_i = (1 - kp) \cdot [(1 - kad) \cdot (p1 \cdot r1_i + p2 \cdot r2_i) -$$
$$- p1 \cdot c1 \cdot y1_i - ks \cdot (S1_i + S2_i) -$$
$$- z \cdot (Rv1_i + Rv2_i) - Pq \cdot q_i], \quad (11)$$

where pj is the unit price (MU per unit of production); cj – the share of the cost in the cost of production; z, ks – payment for storage of a unit of goods for one period in a retail network and in a wholesale warehouse, respectively (MU per unit of production); kp is the income tax rate; kad is the rate of value added tax; Pq – raw material unit cost (MU per unit of raw materials).

The system of equations (1) – (11) is a mathematical model of the logistic system of an enterprise that manufactures two types of products and operates in accordance with the scheme shown in figure 1. The system of equations (1) – (11) contains quantities of two types: variables with a subscript and constants. All calculations by model (1) – (11) will be performed with the following values of constants, which units of measure are described above

$$z = 0,01; ks = 0,04; c1 = 0,4; c2 = 0,45;$$
$$Pq = 2; Q1 = 500; Q2 = 500; tX = 10;$$
$$tS1 = 3; tY1 = 3; tY2 = 4; tY3 = 5;$$
$$kp = 0,25; k1 = 0,33; k2 = 0,33; kq = 0,5; \quad (12)$$
$$a31 = 0,3; a32 = 0,7; p1 = 9; p2 = 11; pk = 10$$
$$Rm1 = 30; Rm2 = 58; tQ1 = 300; kQ1 = 0,8;$$
$$n1 = 6,05 \cdot 10^{-5}; n2 = 7,56 \cdot 10^{-5}; kad = 0,06$$

Initial values:

$$\begin{array}{llll} q_0 = 3,2 & x_0 = 0 & s1_0 = 0 & s2_0 = 0 \\ X_0 = 0 & Y1_0 = 0 & Y2_0 = 0 & M_0 = 0 \\ S1_0 = 0 & S2_0 = 0 & R1_0 = 0 & R2_0 = 0 \\ r1_0 = 0 & y1_0 = 0 & y2_0 = 0 & y3_0 = 0 \\ & V1_0 = 0 & V2_0 = 0 & \end{array} \quad (13)$$

The initial values (13) correspond to the case when a new project starts, so to speak, from scratch. Although the model allows you to describe an ongoing project.

Case 1.

We are considering an enterprise operating in accordance with the diagram in figure 1. We consider the demand for products unchanged throughout the entire life cycle T of the project. We are considering a project lasting two years (T = 730 days). The final products of the enterprise are produced only by the first and second shops. Parts of the production of the first and second shops make up 30% and 70% in the total amount of the final product, respectively. This means that in equation (5), the values of the parameters should be chosen as $a31$ = 0,3, $a32$ = 0,7. Figure 2 shows the ratio of work-in-process parameters for this case.

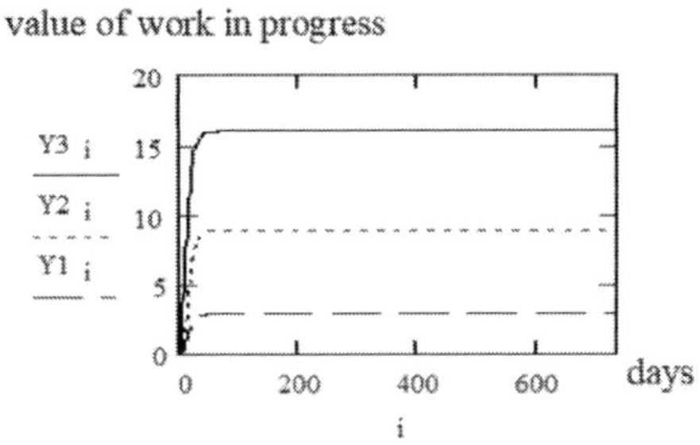

Figure 2. Dynamics of values of work in progress

yj_i- production rate

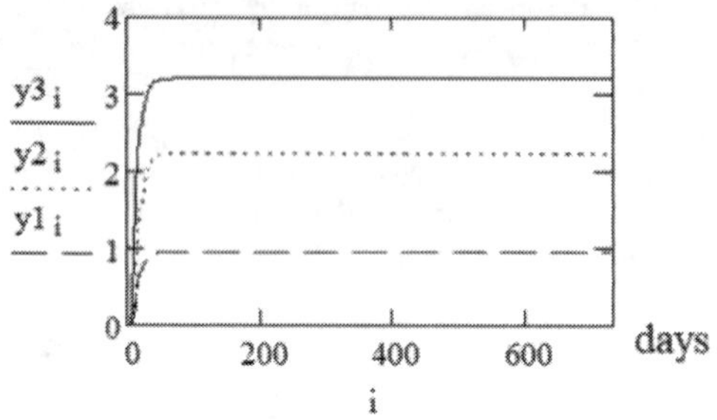

Figure 3. Dynamics of values of work in progress

$r1_i$ - first product sales rate
$s1_i$ - the rate of delivery of the first product at retail
$y1_i$ - production rate of the first product

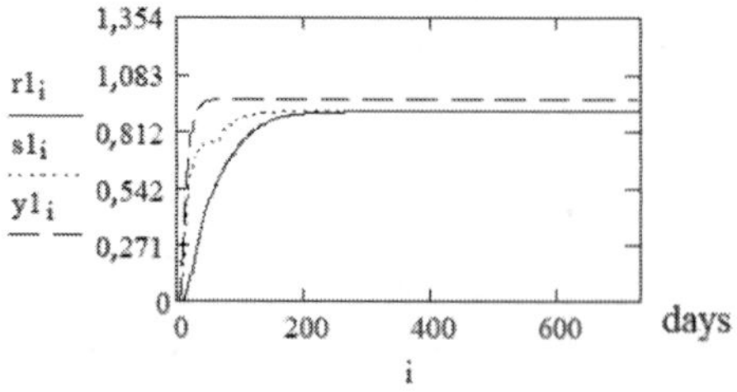

Figure 4. Dynamics of the main rates in LS for the first product

Figure 3 illustrates the dynamics of production capacities of each of the three shops, calculated in accordance with the system of equations (1)-(11).

Calculations have established that $y3_i = y1_i + y2_i$, as it should be.

Figure 4 shows that in the interval [0; 200] the rate of deliveries of the first product at retail $s1_i$ exceeds the rate of sales of the first product $r1_i$ and as a result the quantity of the first product in retail $R1_i$ in this time interval increases from 0 to $Rv1_i$, as can be seen from figure 5.

Then, by the 200th period, the rate of retail supply $s1_i$ becomes equal to the rate of sales of the first product $r1_i$ as a result of which the quantity of the first product $R1_i$ in retail stabilizes at the level $Rv1_i$ (figure 5).

Figure 4 shows that for all periods the rate of production $y1_i$ exceeds the rate of retail deliveries $s1_i$ and, as a consequence, the stock of goods in the wholesale warehouse $S1_i$ monotonically grows to a value of 51,3, which can be seen from figure 5.

Figure 6 shows the dynamics of the main indicators for the second product. The analogy between the behavior of the main indicators, which is visible from the comparison of figure 5 and figure 6 allows us to conclude that the dynamics of the main rates of both types of goods is similar.

This does not mean that you can arbitrarily set the rate of production of goods of each type. The rate of sales is the factor that determines the work of all parts of the LS.

$R1_i$ - quantity of the first product in retail;
$Rv1_i$ - level of stabilization of the quantity of the first product in retail;
$S1_i$- stock of the first product in the wholesale warehouse

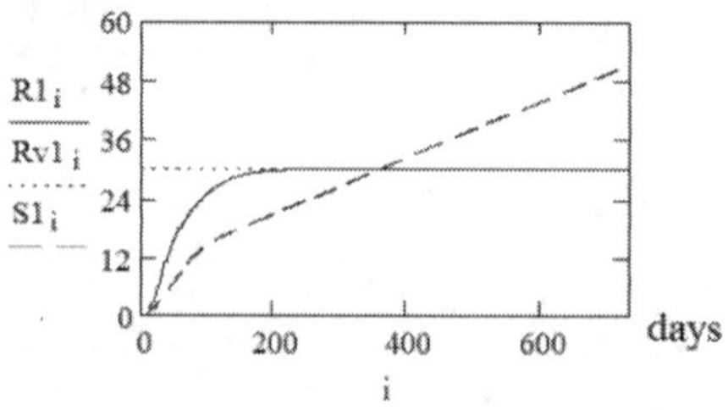

Figure 5. Dynamics of the main indicators in LS for the first product

$R2_i$ - quantity of the second product in retail;
$Rv2_i$ - level of stabilization of the quantity of the second product in retail;
$S2_i$- stock of the second product in the wholesale warehouse

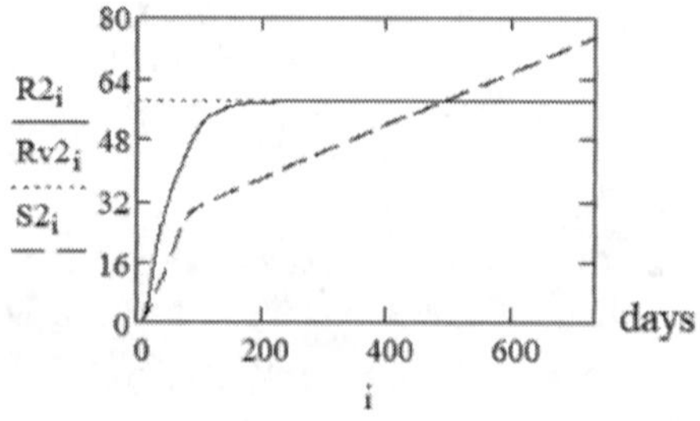

Figure 6. Dynamics of the main indicators in LS for the second product

Figure 7 illustrates the dynamics of daily profit over the entire life cycle of a project. Figure 7, it can be seen that after reaching the maximum value in the 200th period, the daily profit begins to decrease monotonically. This decline means that the project life cycle will be limited in time.

The total profit over the lifetime of the project will be $\sum_{i=1}^{T} M_i$=1712 (MU).

Case 2.

Above, we considered the case of constant demand for both types of products. Now let us consider the case when, in the 300th period, the demand for products of the first type decreases abruptly by 20%, and for products of the second type remains unchanged. Figure 8 shows the dynamics of the main rates for the first type of goods for this case. The sales rate $r1_i$ and the retail supply rate $s1_i$ decrease by 19,6% in the 300th period.

The main rates for the second product remain unchanged. Figure 9 shows the dynamics of the quantity of

M_i - daily profit

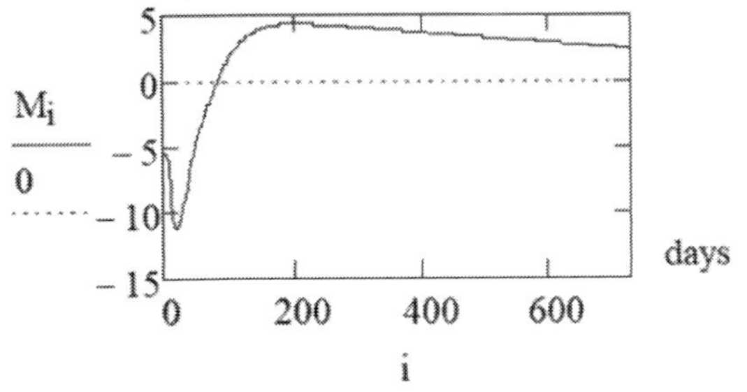

Figure 7. Dynamics of the daily profit of the enterprise

$r1_i$ - first product sales rate
$s1_i$ - the rate of delivery of the first product at retail
$y1_i$ - production rate of the first product

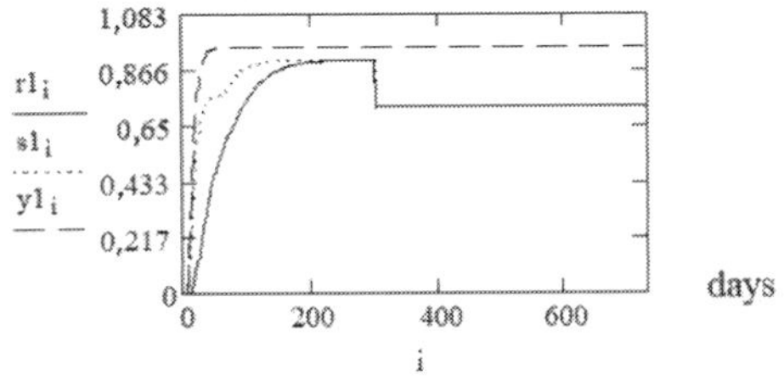

Figure 8. Dynamics of the main rates in LS for the first product in Case 2

goods in the wholesale warehouse $S1_i$ and in retail $R1_i$. The dynamics of the quantity of goods in retail $R1_i$ has not changed (figure 5). The dynamics of the quantity of goods in the wholesale warehouse $S1_i$ changes sharply after the 300th period. This abrupt change is explained in figure 8: there is a constant rate of deliveries y_i to the wholesale warehouse, while the rate of removal of goods $s1_i$ from the wholesale warehouse at the 300th period sharply decreases.

As seen from figure 10 the dynamics of the main indicators for the second product of goods remains unchanged (see figure 6).

Figure 11 illustrates the dynamics of daily profit throughout the entire life cycle of the project for the case under consideration. Figure 11 shows that with the 300th period, the daily profit decreases sharply. This decrease is due to a decrease in the rate of sales at a constant rate of production.

In this case, the total profit over the lifetime of the project will be $\sum_{i=1}^{T} M_i = 723,1$ (MU). This value is significantly less than case 1.

Case 3.

Now consider the following situation. The demand for products of the first type decreases sharply as in the previous case, and for products of the second type remains unchanged. However, the company, foreseeing significant loss of profit, decides to increase the retail network for

$R1_i$ - quantity of the first product in retail;
$Rv1_i$ - level of stabilization of the quantity of the first product in retail;
$S1_i$ - stock of the first product in the wholesale warehouse

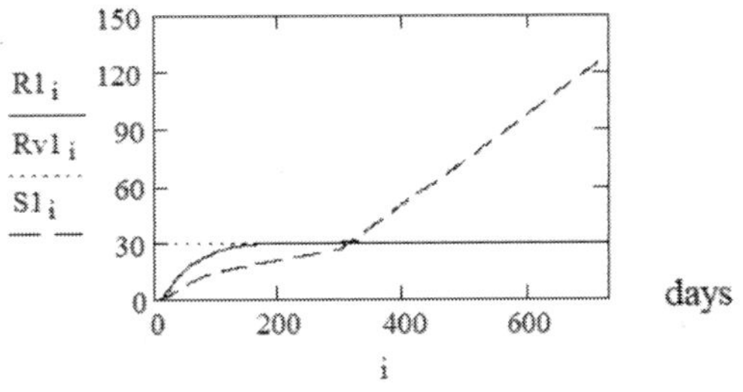

Figure 9. Dynamics of the main indicators in LS for the first product in Case 2

$R2_i$ - quantity of the second product in retail;
$Rv2_i$ - level of stabilization of the quantity of the second product in retail;
$S2_i$ - stock of the second product in the wholesale warehouse

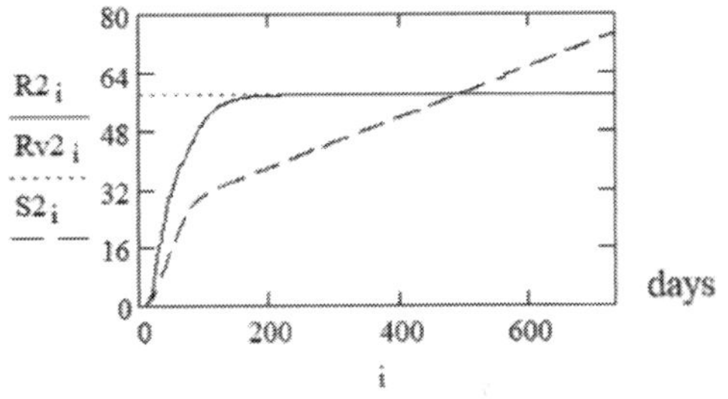

Figure 10. Dynamics of the main indicators in LS for the second product in Case 2

goods of the first type by 20% in the 300th period:

$$Rv1_i = \begin{cases} 30, & \text{if } i < 300 \\ 36, & \text{otherwise} \end{cases} \quad (14)$$

Figure 12 shows the result of calculating the main rates for the first product according to the model (1) – (11) taking into account (14).

Figure 12 it can be seen that at the 300th period the sales rate $r1_i$ of the first product decreases sharply, but, unlike the previous case (see figure 8), after the 300th period it begins to gradually increase and at the 500th period it reaches the rate of deliveries in retail $s1_i$.

Figure 13 shows the dynamics of the levels of stocks of goods in the sweat warehouse $S1_i$ and in the retail trade $R1_i$.

The balance of goods in the wholesale warehouse at the end of the project in this case is 67,6, which is significantly less than the balance in the previous case – 128,5 (see figure 9).

M_i - daily profit

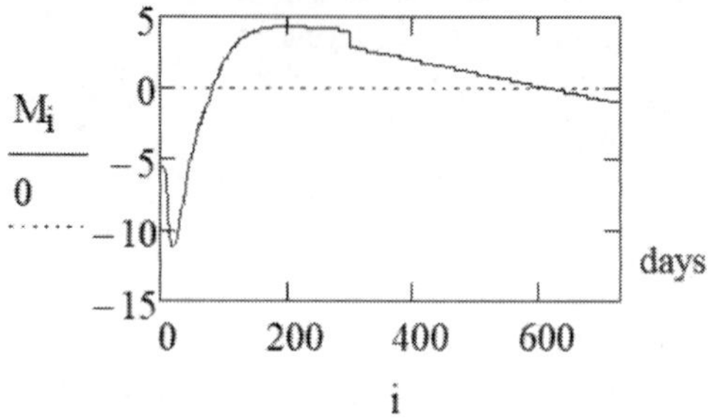

Figure 11. Dynamics of the daily profit of the enterprise in Case 2

$r1_i$ - first product sales rate
$s1_i$ - the rate of delivery of the first product at retail
$y1_i$ - production rate of the first product

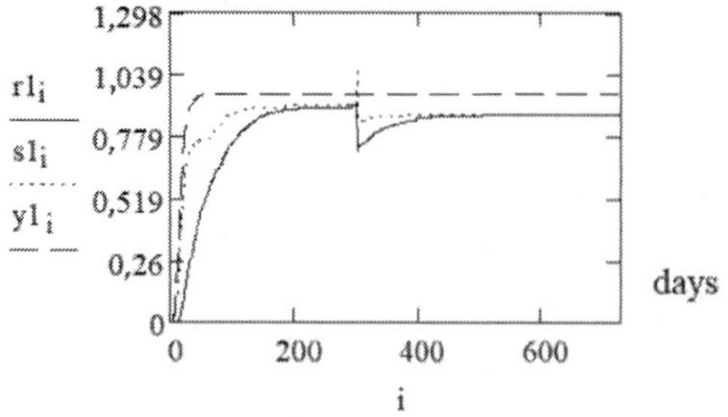

Figure 12. Dynamics of the main rates in LS for the first product in Case 3

This decrease is due to the increase in sales rates and is a favorable factor for increasing the total net profit, which in this case is $\sum_{i=1}^{T} M_i = 1442\,(MU)$, which is significantly more than case 2 (723,1 (MU)).

Comparison of economic results in the second and third cases shows the need to optimize drug parameters.

We will optimize the parameters of the retail trade, since it is the retail trade that directly affects the profit of the enterprise. We will assume that only the following retail transformation is available for both types of goods:

$$Rv1_i = \begin{cases} Rm1, & \text{if } i < tR1 \\ kR1 \cdot Rm1, & \text{otherwise} \end{cases}, \quad (15)$$

$$Rv2_i = \begin{cases} Rm2, & \text{if } i < tR2 \\ kR2 \cdot Rm2, & \text{otherwise} \end{cases}, \quad (16)$$

where $tR1 = tR2 = 300$.

Relations (15) and (16) mean that the company has the opportunity to choose the initial values of the retail capacity for each type of product and to perform the transformation of retail chains with a period of 300. Let us formulate an optimization problem for these conditions. Find the parameters $Rm1, kR1, Rm2, kR2$ at which the total net profit

$R1_i$ - quantity of the first product in retail;
$Rv1_i$ - level of stabilization of the quantity of the first product in retail;
$S1_i$ - stock of the first product in the wholesale warehouse

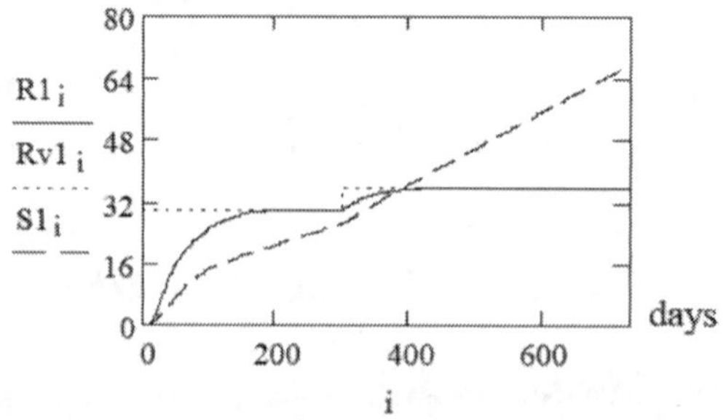

Figure 13. Dynamics of the main indicators in LS for the first product in Case 3

reaches its maximum:

$$F\,(Rm1, kR1, Rm2, kR2) = \sum_{i=1}^{T} M_i \rightarrow max. \quad (17)$$

The system of equations (1) - (11) serves as constraints for function (17). The solution to the optimization problem (17) system calculated Mathcad:

$$\begin{pmatrix} Rm1 \\ kR1 \\ Rm2 \\ kR2 \end{pmatrix} = \begin{pmatrix} 33,6 \\ 1,014 \\ 59,43 \\ 1,043 \end{pmatrix}, F = 2855\,(MU). \quad (18)$$

Figure 14 shows the dynamics of rates calculated for the optimal solution (18) for the goods of the first type. An analogy can be noted between figure 14 and figure 8.

$r1_i$ - first product sales rate
$s1_i$ - the rate of delivery of the first product at retail
$y1_i$ - production rate of the first product

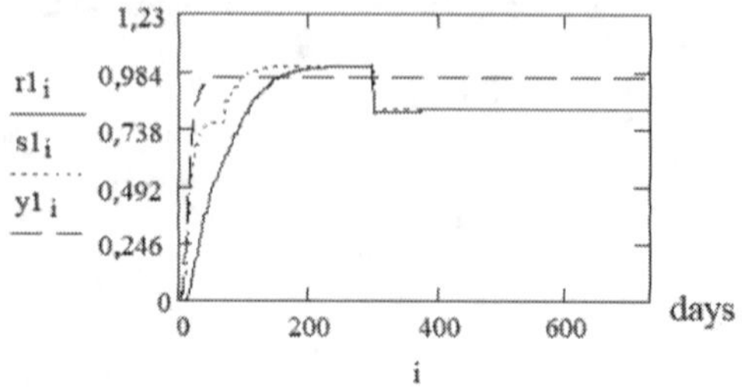

Figure 14. Dynamics of the main rates in LS for the optimal solution (18)

The total amount of products produced during the lifetime of the project in both cases is the same

$$\sum y1_i = 685,4.$$

But the number of products sold for the optimal solution is $\sum rl_i = 587,5$, which exceeds the number of products sold for the case corresponding to figure 8: $\sum rl_i = 536,4$. The fact that the amount of products sold is exactly in this ratio is directly visible from the comparison of figure 8 and figure 14.

Such a ratio of the quantities of produced and sold products leads to the fact that unsold products remain in the wholesale warehouse by the end of the project. Its number is $S1_{730} = 128$ for the case corresponding to figure 8 (see figure 9) and $S1_{730} = 64$ for the optimal solution (18), which is shown in figure 15.

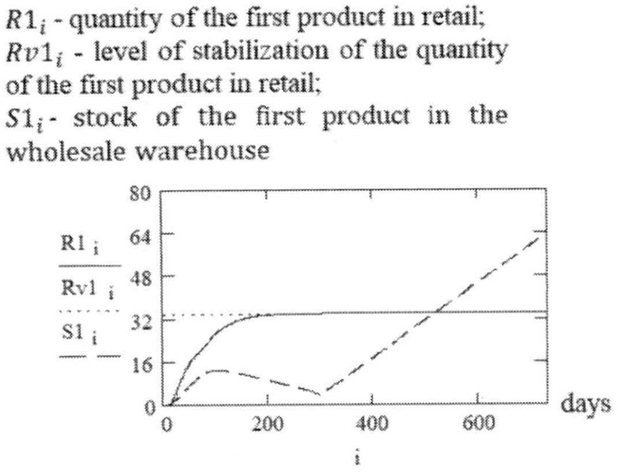

Figure 15. Dynamics of the main indicators in LS for the optimal solution (18)

Figure 16 displays the dynamics of the indicators of the second product for the optimal solution (18).

Comparing figure 10 and figure 16 it can be seen that in the first case, by the end of the project, unsold products of the second type remain in the wholesale warehouse in the amount of $S2_{730} = 74,7$, while for the optimal solution (figure 16) this quantity is practically zero.

Figure 17 shows the dynamics of daily net income. Calculation of net profit for the period under review gives the result $\sum_{i=1}^{T} M_i = 2852\ (MU)$.

This value is significantly higher than the previous ones, which justifies the need to develop mathematical models for the functioning of LS. Since the mathematical models containing the main parameters of the LS allow you to formulate and solve the optimization problem with a minimum cost of funds and time.

Comparison of figure 7 and figure 17 shows that for the optimal solution (figure 17), the decrease in daily profit over time is much slower, which means that it is possible to create longer-term projects using the optimal parameters of LS.

Case 4.

We considered above the case when the transformation of retail chains for both types of goods occurred simultaneously at a fixed point in time - at the 300th period (see formulas (15), (16)).

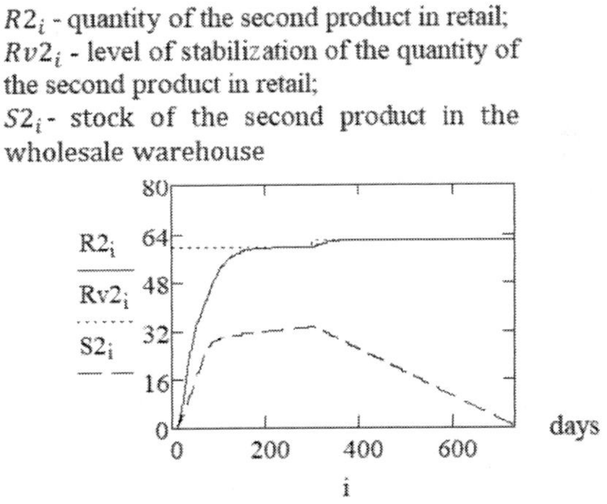

Figure 16. Dynamics of the main indicators for the optimal solution (18)

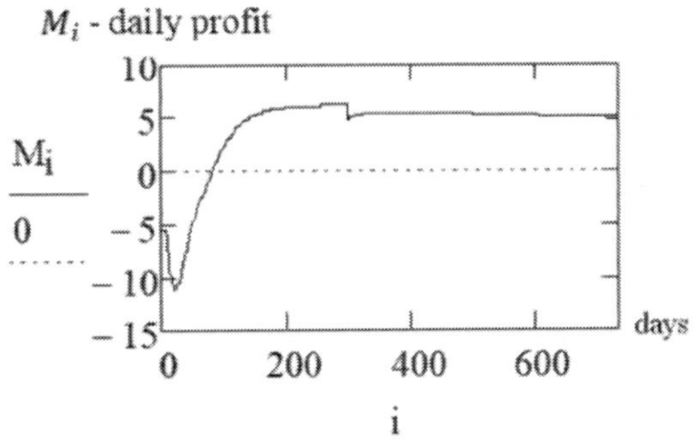

Figure 17. Dynamics of the daily profit of the enterprise for the optimal solution (18)

Now, suppose that the company can choose the moments of transformation $(tR1, tR2)$ of retail chains for each type of product independently.

The optimization problem in this case can be formulated as follows: find the parameters $Rm1, kR1, tR1, Rm2, kR2, tR2$ in formulas (15), (16) at which the total net profit reaches a maximum:

$$F(Rm1, kR1, tR1,\ Rm2, kR2,\ tR2) = \\ = \sum_{i=1}^{T} M_i \to max \tag{19}$$

The system of equations (1) – (11) serves as constraints for function (19). The solution to the optimization problem (19) system calculated Mathcad:

$$\begin{pmatrix} Rm1 \\ kR1 \\ tR1 \\ Rm2 \\ kR2 \\ tR2 \end{pmatrix} = \begin{pmatrix} 33,47 \\ 1,115 \\ 292,5 \\ 58,7 \\ 1,049 \\ 200 \end{pmatrix}, F = 3272\ (MU). \tag{20}$$

Comparison of solutions (20) with (18) shows that the possibility of independent changes in retail chains for each type of product leads to an increase in economic efficiency by 14,6%.

For a more complete comparison of solutions (20) and (18), the dynamics of the main levels of the drug is presented in figure 18 and figure 19 (compare with figure 15 and figure 16, respectively).

$R1_i$ - quantity of the first product in retail;
$Rv1_i$ - level of stabilization of the quantity of the first product in retail;
$S1_i$- stock of the first product in the wholesale warehouse

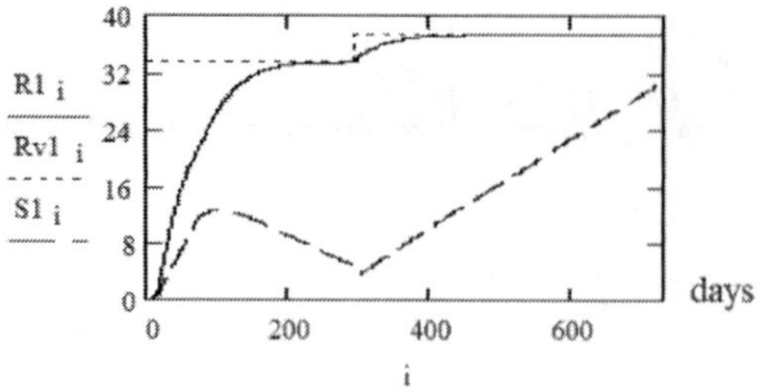

Figure 18. Dynamics of the main indicators in LS for the optimal solution (20)

$R2_i$ - quantity of the second product in retail;
$Rv2_i$ - level of stabilization of the quantity of the second product in retail;
$S2_i$- stock of the second product in the wholesale warehouse

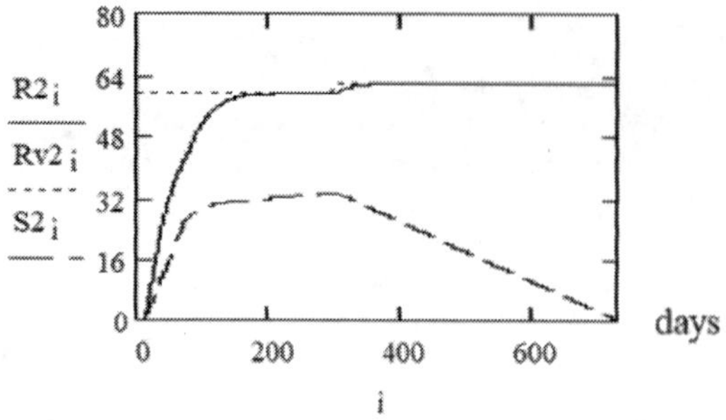

Figure 19. Dynamics of the main indicators for the optimal solution (20)

Figures 18 and 19 show that with the optimal solution (20), the capacities of wholesale storage facilities can be very small. They may not even exceed the retail capacity for both types of goods.

3 Conclusions

1. The work considers a universal scheme of the enterprise's LS, which contains all the main production stages, starting from the procurement of raw materials and components and uploading the supply of finished products to the retail network. A feature of the author's approach is a mutually consistent description of the work of all links of the enterprise's LS, taking into account the demand for products throughout the entire planning horizon. The proposed LS scheme allows describing the diversification of production.

2. The system of mathematical equations for the proposed scheme of the enterprise logistics system is compiled. The system of equations is written in the form of finite differences. Time is considered as a discrete variable. Such a description is more consistent with the real situation at the enterprise, since management decisions at the enterprise are made at discrete moments in time. The system of equations contains the main characteristics of LS as well as the number of potential buyers, which makes it possible to take into account the market demand for products.

3. The proposed model was used to analyze four situations affecting retail chain management. The formulated optimization problems were solved numerically in the Mathcad system. It has been proven that managing a retail network in accordance with the optimal solution can give a significant economic effect.

4. The fact that the proposed system of mathematical equations contains such market parameters as the number of potential buyers and the number of goods in the hands of buyers allows us to include in the description the influence of an advertising campaign on the efficiency of selling goods. It also allows you to assess the impact of advertising on production diversification. These questions can be topics for further research.

References

[1] C. Kogler, P. Rauch, Silva Fennica **52**, 29 (2018)
[2] A. Borshchev, A. Filippov, *From system dynamics and discrete event to practical agent based modeling: reasons, techniques, tools*, in *The 22nd International Conference of the System Dynamics Society, July 25 - 29, 2004, Oxford, England* (Oxford, England, 2004), `https://www.anylogic.com/upload/iblock/98a/98a7aebbe9e89099430f888f29a9715e.pdf`
[3] R. Bucki, P. Suchánek, Complexity p. 5286135 (2017)
[4] S. Žapčević, P. Butala, The International Journal of Advanced Manufacturing Technology **66**, 1725 (2013)
[5] V. Modrak, D. Marton, S. Bednar, Tehnički vjesnik **20**, 1057 (2013)
[6] V. Modrak, D. Marton, S. Bednar, Procedia CIRP **23**, 258 (2014)

[7] W. Chen, D. Huo, W. Xie, X. Teng, J. Zhang, Chinese Journal of Mechanical Engineering **29**, 1090 (2016)

[8] H.S. Park, N.H. Tran, Journal of Manufacturing Systems **31**, 337 (2012)

[9] R. Omega, V. Noel, J. Masbad, L. Ocampo, Advances in Production Engineering & Management **11**, 115 (2016)

[10] M.P. Brundage, Q. Chang, Y. Li, J. Arinez, G. Xiao, IEEE Transactions on Automation Science and Engineering **13**, 676 (2016)

[11] P. Chhaochhria, S. Graves, International Journal of Production Research **51**, 6860 (2013)

[12] G. Popovics, A. Pfeiffer, L. Monostori, International Journal of Computer Integrated Manufacturing **29**, 1272 (2016)

[13] A. Greasley, International Journal of Operations & Production Management **25**, 534 (2005)

[14] R. Malik, R. Leduc, Discrete Event Dynamic Systems **25**, 177 (2015)

[15] Y. Nonaka, Y. Suginishi, A. Lengyel, S. Nagahara, K. Kamoda, Y. Katsumura, *The S-Model: A digital manufacturing system combined with autonomous statistical analysis and autonomous discrete-event simulation for smart manufacturing*, in *2015 IEEE International Conference on Automation Science and Engineering (CASE)* (2015), pp. 1006–1011

[16] T. Aljuneidi, A.A. Bulgak, The International Journal of Advanced Manufacturing Technology **87**, 1585 (2016)

[17] B. Bako, P. Božek, Procedia Engineering **149**, 571 (2016), international Conference on Manufacturing Engineering and Materials, ICMEM 2016, 6-10 June 2016, Nový Smokovec, Slovakia

[18] C. Gershenson, D. Helbing, Complexity **21**, 9 (2015)

[19] U. Ghani, R. Monfared, R. Harrison, International Journal of Computer Integrated Manufacturing **28**, 844 (2015)

[20] Y.V. Sherstennikov, Problems of Economics **4**, 306 (2019)

[21] V. Porokhnya, Y. Sherstennikov, O. Ostapenko, R. Ivanov, E3S Web of Conferences **166**, 13001 (2020)

Analysis, forecasting, and management of indicators of the forest industry of the region

Olena Ovchynikova[1,*] and *Olha* Dupliak[1,**]

[1]Department of Automated Systems and Modeling in Economics, Khmelnytskyi National University, 11 Instytutska Str., Khmelnytskyi, 29016, Ukraine

Abstract. The article considers the theoretical and methodological foundations of forest industry research. The analysis of indicators of the forest industry of Khmelnytskyi region and Ukraine is carried out. To do this, a list of indicators and methods of their collection for the study of the forest industry. Since the state of ecology and forestry in Ukraine is not in the best condition, it is objectively necessary to use methods of economic and mathematical modeling of forest management, which would take into account environmental, economic and social factors, which will quickly implement the principles of sustainable development in forest management. The dynamics of forestry indicators is estimated. Models of dynamics of indicators of forest branch are developed; the received forecasts are analyzed and recommendations on forest management in the region are given. It is determined that with an increase in investment in the development of the forest industry by 1%, we can expect an increase in the volume of the logging market by 524632.5 points. The results of the study can serve as a basis for management decisions on the management of forestry indicators.

1 Introduction

Today, the domestic forest industry plays a significant role in the socio-economic life of society. At the same time, forest ecosystems perform a protective function of the environment, as well as designed to meet human ecological needs.

As the state of ecology and forestry in Ukraine is not in the best condition, it is objectively necessary to use methods of economic and mathematical modeling of forestry processes, which would take into account economic factors, which will quickly implement the principles of sustainable development in forestry [1].

The object of the study is the activities of forestry in Khmelnytskyi region and Ukraine.

The subject is theoretical, methodological and practical aspects of the effective functioning of forestry in the region, and economic and mathematical methods and models for forecasting forestry indicators.

The purpose of the study is to analyze the current state of forestry in Khmelnytskyi region and economic and mathematical forecasting to establish the dynamics and trends of the forest industry.

Prospects for the development of forestry and forestry activities largely depend on efficient and rationally organized forest management in accordance with modern requirements for economic activity. Forest resources should be considered as one of the main factors in ensuring the environmental, economic and social security of the region.

*e-mail: veleslavaovch@gmail.com
**e-mail: olgasvarychevska@gmail.com

2 Literature review

The problems of forest sector reform in recent years have gained unprecedented resonance due to the excessive interest of some political forces, businesses and NGOs in this segment of the national economy [2]. The forest theme became one of the main priorities of the economic program and a number of political parties, including parliamentary ones. Which began to create additional obstacles to the formation of modern forest policy aimed at ensuring continuous and inexhaustible forest use and increasing the impact of forestry and wood production on socio-economic the rise of the country and individual regions. Because of numerous political manipulations, promising areas of modernization of the forest management system do not receive adequate institutional support. Economic transformations in the country, which have been carried out for almost three decades in the forest sector, are accompanied by significant changes in the structure of domestic forestry, as well as affect the level of ecological and economic efficiency of forest use [3].

Theoretical and methodological bases of forest management, issues of efficiency of its functioning, assessment, complex use and restoration of forest resources are investigated in the works [4–14].

These works reflect research that contributes to the development of scientific understanding of the problems of organization of forestry production in general. Research [12] devoted to the study of management processes and environmental safety of forest resources of Ukraine. It is worth noting the work [9], in which the main attention is paid to the study of trends and the establishment of patterns of use of forest resources and the improvement of forest policy in Ukraine.

However, a number of issues related to the use and reproduction of forest resources and the formation of the information system of state support of forestry have not yet been properly reflected in scientific works.

Therefore, the effective functioning of the domestic forest industry implies the need for detailed economic and mathematical analysis in the context of sustainable development, further development requires forecasting of forest use taking into account the forest resource potential of individual regions.

The main tools for meeting information needs are forest inventory and monitoring [15]. Inventory provides a statistical assessment of resources, while monitoring assesses changes and identifies trends in the dynamics of the state of monitoring objects.

3 Problem description and methodology

In Ukraine, forest and wooded lands occupy about 17.6% (10.63 million hectares) of the total territory of the country 60.3 million hectares [16]. At the same time, the territory of Ukraine in comparison with many European countries is characterized by a low forest cover (16.7%), which is calculated as a percentage of the area of forested forest areas to land area (figure 1).

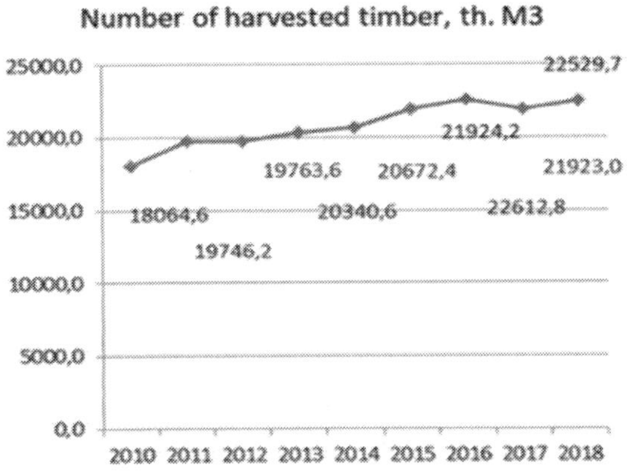

Figure 1. Dynamics of the amount of harvested wood

The problem of protecting forests from damage by pests and diseases remains difficult, as the general deterioration of the ecological situation in Ukraine has a negative impact on the condition of forests and reduces their natural resilience. In 2017, 2,371 fires were extinguished in the forests of subordinate enterprises on an area of 5,474 hectares, including 1,060 hectares on horseback.

This is 2.5 times more in number and 5 times in area compared to 2016. The average area of one fire doubled to 2.3 hectares. Losses from forest fires amounted to UAH 43.8 million. The main causes of fires (82%) are human factors. The problem of illegal deforestation remains unresolved [17].

Therefore, the process of reforestation is important, although the dynamics of indicators of forest pest control in the plantations of the State Forestry Agency of Ukraine indicates insufficient implementation of a set of measures aimed at protecting forest ecosystems from pests.

Harvesting of wood in protected areas is carried out exclusively in order to take measures to prevent changes in natural complexes due to anthropogenic impact, preservation and restoration of plant communities, fire and sanitation measures, disaster relief.

The vast majority of care felling (68% of the area) is carried out in operational forests [18]. The area on which pine and oak were harvested – in the Khmelnytskyi region in 2018 also has the largest share (figures 2, 3).

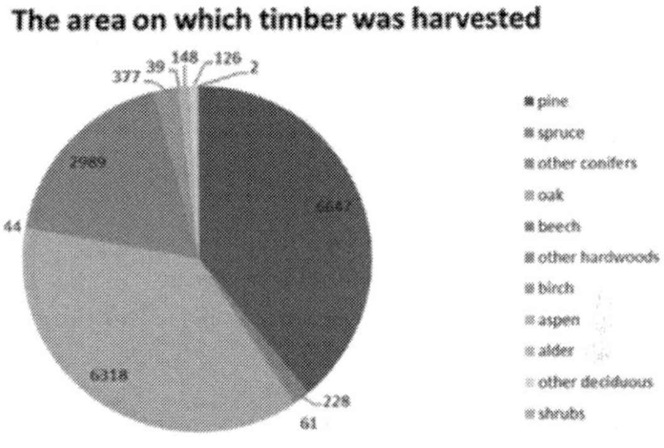

Figure 2. Harvesting of wood according to the species composition of stands in 2018

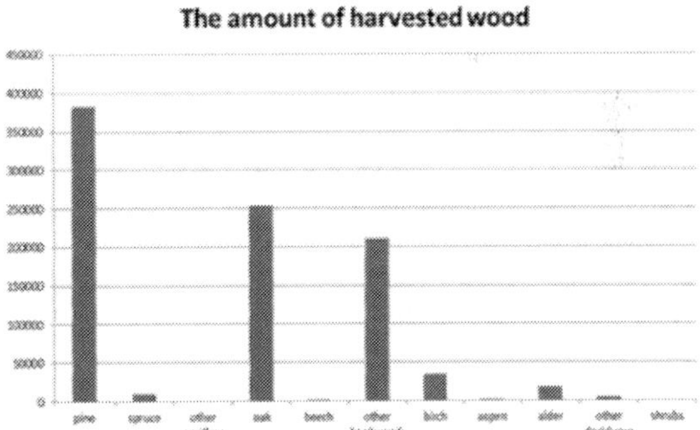

Figure 3. The amount of harvested wood by species composition in 2018

To bring the forests to the proper sanitary condition in the damaged plantations, measures are taken to improve the sanitary condition of forests.

Appropriate funding is also needed for the sustainable development [10, 11, 19, 20], and in particular, in the forest industry [21–23]. However, no funds have been provided for the financing of forestry and hunting activities for the last two years – analyzing the dynamics of financial support of the state, which had a growing trend in 2000, and since 2016 the funding of forestry and hunting tasks and measures from the state budget has been stopped (figure 4).

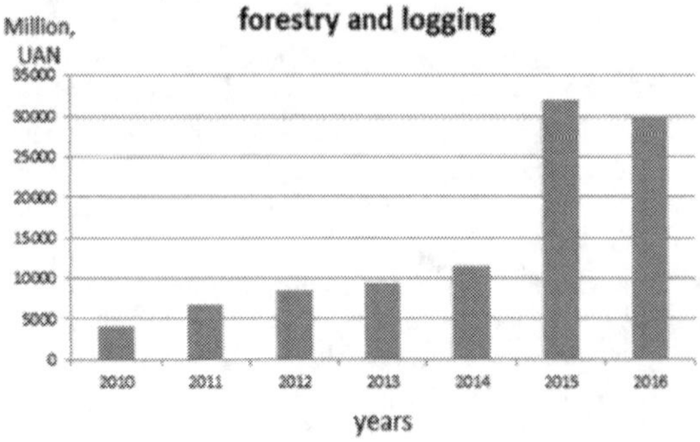

Figure 4. Capital investments in forestry and logging in Khmelnytskyi region

In modern conditions, the process of forestry research should reach a qualitatively new level – in particular, it is necessary to develop new approaches to the application of econometric models and methods for forecasting indicators at the regional level.

4 Findings

Statistical information describing the processes of the past years is used for forecasting, and the trends of changes in macroeconomic indicators are assessed by different methods. The following information was used to forecast the forestry indicators of Khmelnytskyi region (table 1) [18].

Table 1. Initial information for forecasting forest indicators

t	Years	Y_1	Y_3	Y_2	Y_4	Y_5
1	2013	1432	3095	596933	372308	9324
2	2014	1650	3156	615524	391478	11593
3	2015	1848	3690	698566	455775	31975
4	2016	2016	3961	750549	487522	29955
5	2017	1817	6624	781993	501765	33305
6	2018	1540	6577	836566	413171	37977

As there has been no capital investment in the forest industry for the last two years, we decided to forecast this indicator and use the forecast data in further forecasts, where

Y_1 – Forest reproduction area, hectares;

Y_2 – Logging by types of forestry products, business round wood, m^3;

Y_3 – Area of foci of forest pests and diseases eliminated by control measures and under the influence of natural factors, hectares;

Y_4 – The number of manufactured forestry products, Fuel wood, m^3;

Y_5 – Capital investments in forestry, thousand UAH.

This approach allows taking into account the nonlinear nature of the studied characteristic, in this case, the economic development level, as well as to investigate changes in the structure of the objects under study by the values of the comprehensive index calculated for different periods.

Table 2. The results of forecasting by the method of the average annual growth rate

	Y_1	Y_2	Y_3	Y_4	Y_5
$k-p$	1,06	1,10	1,20	1,08	1,38
			Forecast		
Y_{n+1}	1928,44	836608	8012	540629,7	45786,4
Y_{n+2}	2046,7	895038	9691	582504,6	62945,3
Y_{n+3}	2172,3	957549	11721	627623,0	86534,6

Table 3 presents the results of forecasting by the method of average absolute growth.

Table 3. The results of forecasting by the method of the average annual growth rate

	Y_1	Y_2	Y_3	Y_4	Y_5
$\Delta-y$	96,3	46265	5995,3	882,3	32364,3
			Forecast		
Y_{n+1}	1913,3	828258	7506	534129,3	39300,3
Y_{n+2}	2009,5	874523	8389	566493,5	45295,5
Y_{n+3}	2105,8	920788	9271	598857,8	51290,6

Table 4 presents the results of forecasting the indicators of the forest industry by extrapolation based on the moving average.

Table 4. The results of forecasting by the method of average absolute growth

	Y_1	Y_2	Y_3	Y_4	Y_5
Δx_t	-199	31444	2663	14243	3350
Δx_{t-1}	168	51983	271	31747	- 2020
Δx_{t-2}	198	83042	534	64297	20382
			Forecast		
Y_{n+1}	1812,8	800730,2	7227,7	513121,6	35062,7
Y_{n+2}	1795,4	813936,9	7795,9	520196,1	35463,0
Y_{n+3}	1755,7	820219,4	8328,0	523041,8	36132,3

Another method of forecasting is the FORECAST function. Table 5 shows the calculations of forecasts using this function.

However, for completeness of the analysis it is necessary not only to receive forecast values, but also to estimate, how much they are exact, that is, what error of the received results. To do this, use the ex post method.

For the indicator "Forest reproduction area" the best method of forecasting was the FORECAST function method, which gave only 9.6% error, and the worst – the method of average annual growth rate, which erred by 25.5%, as well as the method of average absolute growth, which gave an error of 24.2% (table 6). For the indicator "Logging by types of forestry products" the best was the method of average annual growth rate, which showed a deviation of only 0.005%, and the worst – the method of FORECAST function (6.1%). In general, for this indicator, all methods have an error of less than 10%, all methods can be used for further analysis (table 7).

Table 5. The results of forecasting by the method of average absolute growth

Years	Y_1	Y_2	Y_3	Y_4	Y_5
2010	2402	576092	3412	353913	4150
2011	1972	616736	4353	400612	6821
2012	1539	570052	5418	369453	8495
2013	1432	596933	3095	372308	9324
2014	1650	615524	3156	391478	11593
2015	1848	698566	3690	455775	31975
2016	2016	750549	3961	487522	29955
2017	1817	781993	6624	501765	33305
2018 actual meanings	1540	836566	6577	413171	37977
2018 forecast	1688,25	785511,3	4949,346	510227,4	37977,04
2019 forecast	1589,861	838136,8	5559,198	486121,9	43917,21
2019 forecast	1679,077	883086	5785,881	514158,9	50157,08

Table 6. Estimation of the error of forecasting methods for the indicator "Forest reproduction area"

Actual value	Forecast for 2018	Absolute deviation	Relative deviation, %	Forecasting method
1540	1913,2500	-373,25	-24,2	Average absolute increase
	1928,4479	-388,44788	-25,2	Average annual growth rate
	1812,8042	-272,8042	-17,7	Flowing average method
	1688,25	-148,25	-9,6	PREDICTION

Table 7. Error estimation of forecasting methods for the indicator "Logging by types of forestry products"

Actual value	Forecast	Absolute deviation	Relative deviation, per	Forecasting method
836566	828258	8308	1,0	Average absolute increase
	836608	-42,33	0,005	Average annual growth rate
	800730,24	35835,756	4,3	Flowing average method
	785511,3	51054,7	6,1	PREDICTION

For the indicator "Capital investment in forestry" we calculated the value for 2018 instead of the missing one, then the FORECAST function itself cannot be estimated. And among others, the best is the average absolute increase (3.5% error). The method of the average annual growth rate (20.6%) turned out to be the worst.

Table 8. Estimation of forecasting error for the indicator "Capital investment in forestry"

Actual value	Forecast	Absolute deviation	Relative deviation, per	Forecasting method
37977	39300,2500	-1323,25	-3,5	Average absolute increase
	45786,3719	-7809,3719	-20,6	Average annual growth rate
	35062,7072	2914,2928	7,7	Flowing average method
	37977,04	-0,04	0,0	PREDICTION

Table 9 shows the estimation of the forecast error of the indicator "Area of foci of forest pests and diseases eliminated by control measures and under the influence of natural factors". For this indicator, the best method was the current average (9.9% error), and the worst – the FORECAST function (24.7%) and the method of the average annual growth rate (21.8%).

Table 10 contains the estimation of forecasting of the indicator "Quantity of the made production of forestry". As can be seen for this indicator, all methods gave an error of more than 10%, so the methods of simple extrapolation can not accurately predict the behavior of it.

Consider another method of forecasting – a trend-based extrapolation. To determine the trend, it is enough to build a graph for the values of the studied indicator and on the basis of it to draw a conclusion about the nature of their change.

Figure 5 shows the trend models for the indicator "Forest reproduction area".

The quality of the selected model is evaluated by the value of the coefficient of determination: the closer it is to one, the better the model. As can be seen from the graph,

Table 9. Estimation of the error of forecasting methods for the indicator "Area of foci of forest pests and diseases eliminated by control measures and under the influence of natural factors"

Actual value	Forecast for 2018	Absolute deviation	Relative deviation, per	Forecasting method
	7506	-929,25	-14,1	Average absolute increase
6577	8012	-1434,8949	-21,8	Average annual growth rate
	7227,729	-650,729	-9,9	Flowing average method
	4949,346	1627,654	24,7	PREDICTION

Table 10. Estimation of forecasting error for the indicator "Quantity of forestry products"

Actual value	Forecast	Absolute deviation	Relative deviation, per	Forecasting method
	534129,25	-120958,25	-29,30	Average absolute increase
413171	540629,65	-127458,65	-30,80	Average annual growth rate
	513121,63	-99950,63	-24,20	Flowing average method
	510227,40	-97056,40	-23,50	PREDICTION

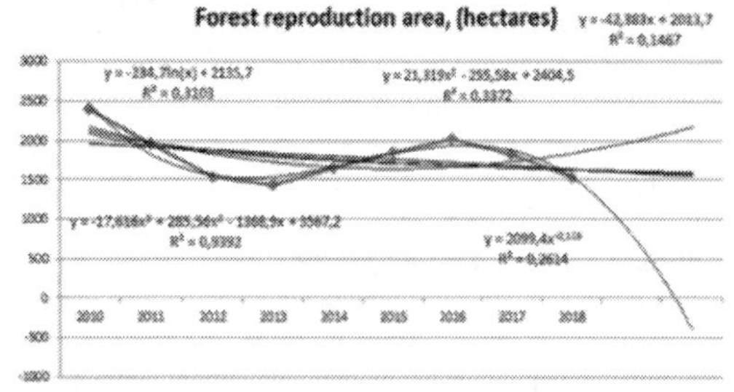

Figure 5. Trend models of the indicator "forest reproduction area" in Khmelnytskyi region

the dynamics of the indicator does not have a clear trend, so the trend models have a small coefficient of determination.

Table 11 shows that only a polynomial of the third degree was able to reflect trends with a high coefficient of determination (0,93), so only it can be used for forecasting.

Figure 6 shows the trend models for the indicator Area of pests and forest diseases, eliminated by control measures and under the influence of natural factors. This indicator also does not have a clear trend.

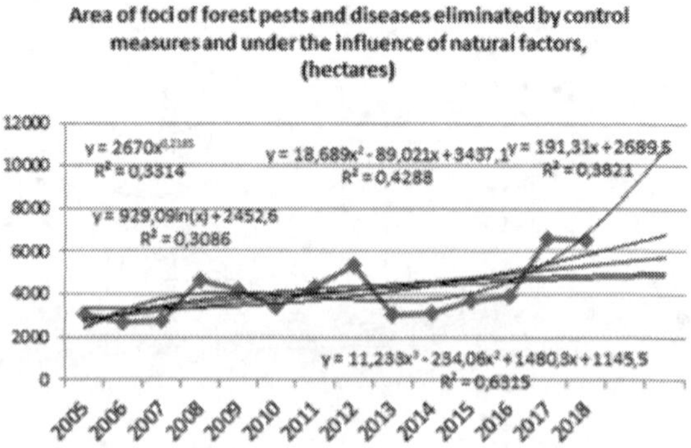

Figure 6. Trend models of the indicator "Area of foci of forest pests and diseases eliminated by control measures and under the influence of natural factors" in Khmelnytskyi region

Figure 7 shows the graph and trend equation for the indicator "Logging by type of forestry products, business roundwood". This indicator has a fairly pronounced dynamics, almost all trend lines have a coefficient of determination greater than 0,6. Therefore, all the obtained trend models can be used for further analysis.

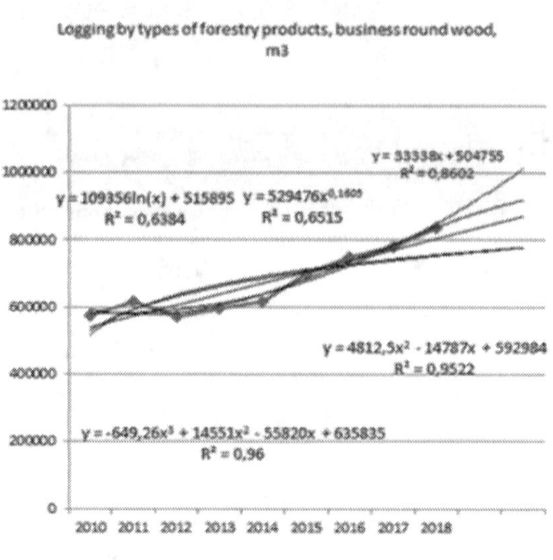

Figure 7. Trend models of the indicator "Logging by types of forestry products, business roundwood"

Figure 8 shows the dynamics and modeling of the indicator "The number of manufactured forestry products, fuel wood". The indicator of fuel wood production in the Khmelnytskyi region had a good trend line until 2017, and in 2018 decreased.

Table 11. Trend models and forecast for 2018 of the indicator "Forest reproduction area"

№	Type of dependence	Equation	Coefficient of determination	Forecast for 2018
1	Linear	$y = -42,383x + 2013,7$	$R^2 = 0,1467$	-
2	Polynomial of the third degree	$y = -17,616x^3 + 285,56x^2 - 1368,9x + 3567,2$	$R^2 = 0,9392$	1535,396
3	Polynomial of the second degree	$y = 21,319x^2 - 255,58x + 2404,5$	$R^2 = 0,3372$	-
4	Logarithmic	$y = -234,7\ln(x) + 2135,7$	$R^2 = 0,3103$	-
	Staid	$y = 2099,4x - 0,116$	$R^2 = 0,2614$	-

It is because of this decrease that we can see that not all models have clearly continued the trend, but all models predict an increase in this figure in the future. The coefficient of determination of the obtained trend models showed a high value, more than 0,6 – only in the case of a polynomial of the third degree. Therefore, for this indicator only this model can be used for analysis.

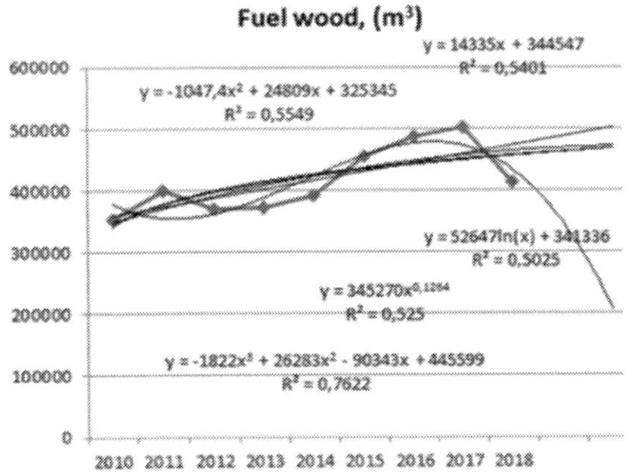

Figure 8. Trend models of the indicator "Logging by types of forestry products, business roundwood"

Table 12 calculates the forecasts of the indicator x_1 – Capital investment in forestry according to trend model, where

Y_1 – Forest reproduction area, hectares;

Y_2 – Logging by types of forestry products,business round wood, m^3;

Y_3 – Area of foci of forest pests and diseases eliminated by control measures and under the influence of natural factors, hectares;

Y_4 – The number of manufactured forestry products, Fuel wood, m^3;

Y_5 – Capital investments in forestry, thousand UAH.

Thus, we have developed forecasts of the main indicators of forestry in Khmelnytskyi region by simple extrapolation. Not all methods showed a qualitative result. The ex post method was used to estimate the forecast, which made it possible to estimate the forecasting error on the actual data for 2018. As can be seen from the forecasting analysis, not all time series of forest industry indicators have a clear trend, which can be estimated by forecasting extrapolation. Thus, the best trend is in the indicator "Logging by types of forestry products", its time series quite clearly reflects the constant increase in the indicator. So, it is for this indicator that our forecasts turned out to be the most accurate.

Correlation-regression method can also be used to predict the development of key indicators of forestry in Khmelnytskyi region . For modeling we will use the indicator Y – Logging by types of forestry products, business round wood, m^3, and x_1 – Capital investments in forestry, thousand UAH (table 13).

The main factor in the development of the forest products market during 2013 to 2018 was the volume of investment in forestry, which is confirmed by the study of correlations (table 14).

Based on the initial data, we build a regression model, starting with the assessment of the quality of modeling, which is shown in table 15. As can be seen from the table, the model of the impact of the amount of investment on logging showed a high correlation between the factor and the result.

The tabular value of the Fisher criterion is 5,98, so the obtained value of the Fisher criterion is greater than the tabular, which indicates the adequacy of the model.

The regression coefficients were obtained by the method of least squares (table 15). Based on the results of correlation-regression analysis, a model of the form (1) was constructed:

$$Y = 524632,5 * x_1 + 7,346678 \quad (1)$$

where Y – Logging by type of forestry products, business roundwood, m^3, x_1 – Capital investments in forestry, (thousand UAH).

Thus, with an increase in investment in the development of the forest industry by 1%, we can expect an increase in the volume of the logging market by 524632.5 points.

So, if we forecast logging according to the obtained model, we will get 864976.6 m^3 of roundwood logging in 2019 and 901939.1 m^3 in 2020.

Table 16 presents the empirical data obtained from the model, which for clarity. As can be seen from the figure, the values obtained for the model are almost indistinguishable from the actual ones, which once again confirms the quality of the simulation.

Table 17 presents all the forecasts obtained for this indicator.

Table 12. Results of forecasting

Years	Y_1	Y_2	Y_3	Y_4	Y_5
2010	2402	576092	3412	353913	4150
2011	1972	616736	4353	400612	6821
2012	1539	570052	5418	369453	8495
2013	1432	596933	3095	372308	9324
2014	1650	615524	3156	391478	11593
2015	1848	698566	3690	455775	31975
2016	2016	750549	3961	487522	29955
2017	1817	781993	6624	501765	33304,86
2018 actual meanings	1540	836566	6577	413171	37977,04
2018 forecast	1688,25	785511,3	4949,346	510227,4	37977,04
2019 forecast	1589,861	838136,8	5559,198	486121,9	43917,21
2020 forecast	1679,077	883086	5785,881	514158,9	50157,08

Table 13. Source information for modeling

Years	Y	x_1
2013	596933	9324
2014	615524	11593
2015	698566	31975
2016	750549	29955
2017	781993	33305
2018	836566	37977

Table 14. Matrix of the pairwise correlation coefficient

	Y	x_1
Y	1	0,941776
x_1	0,941776	1

Table 15. Regression statistics

Multiple R	0,941776
R-square	0,886942
Normalized R-square	0,858677
Standard error	35512,55
Observation	6

Table 16. Predicted values

Observation	Provided by Y	Leavings	Standard balances
1	593132,9	3800,10	0,1196
2	609802,5	5721,47	0,1801
3	739542,5	-60976,50	-1,9197
4	744702,2	5846,78	0,1841
5	789313,6	12679,41	0,3992
6	853637,3	32928,73	1,0367

Thus, the study showed that the development trend of forestry indicators is problematic and requires increased attention to management.

Studies have shown that the main factor influencing the development of the logging market is the financing of this industry, so it is necessary to pay attention to investment growth. First of all, it should be domestic investments and means of state regulation. One of the chosen ways is the one voted in the Verkhovna Rada and received the status of the Law of Ukraine "On Amendments to Certain Legislative Acts of Ukraine to Strengthen Forest Protection, Prevent Fires on Forest and Water Fund Lands, Peatlands and Other Categories".

Now it is necessary to act ahead, to carry out active explanatory work among the population. it is necessary to appeal to the representatives of the central executive bodies to do everything possible within their powers to prevent the possible consequences of inadequate use of forest resources. To instruct the heads of territorial bodies to apply preventive measures, including the elimination of natural dumps, sewage treatment, etc.

As stated in [23], the study of theoretical and practical features of the functioning of forestry requires additional research to determine the place of forestry in the system of the entire national economy, its impact on the development of the state economy. Further research in this area is to analyze specific species sources of forestry financing, determining the effectiveness of investments and the rationality of their use.

5 Conclusions

Forestry is one of the important areas of territorial development. It is very difficult to predict the natural phenomena with a high degree of probability, which is confirmed by the data of previous years [24]. In this regard, in the future it is planned to improve methods of preventive work, increase responsibility for violating safety rules, restrict unorganized recreation and access of vehicles to the forest fund during the growing season.

As stated in [12], "Forests also are central to maintaining the environmental commons. Nearly 90% of terrestrial biodiversity is found in the world's forests, with a disproportionate share in the forests of developing countries. Most of the carbon emissions of developing countries come from deforestation, which accounts for between 10% and 30% of global carbon emissions. Unfortunately, the lack of markets for the national and global environmental services offered by forests has contributed to high rates of deforestation in developing countries. Growing forests are a valuable resource not just for their timber and biodiversity values but also for their prospective value if a

Table 17. Forecast of roundwood logging, m^3

Year	Regression model	Trend model (best)	The method is average annual growth rate	The method is the average absolute increase	Extrapolation based on the flowing average	FORECAST
2019	864976	939295	895038	874523	813937	838136
2020	901939	1029961	957549	920788	820219	883086

global market emerges for the sequestering of carbon from forests".

Thus, the article develops forecasts of indicators of the forest industry of Khmelnytskyi region. These calculations can be used to carry out measures aimed at improving the efficiency of forestry.

The main directions of solving the problems of forest complex development (forestry and timber industries) are:

- improvement of forms of management of this complex;
- integrated use of available wood resources (own and imported);
- increasing financing and efficiency of investments in the forestry sector;
- intensification of forest reproduction and forest use.

In the current situation in our country, the decisive factor is the choice of forms of organization of production in integrated enterprises, including – ensuring the implementation of basic principles of forest reproduction, forest use, efficient processing of forest resources.

Continuation of scientific research on this issue will help further development of national forestry.

References

[1] J.P. Kimmins, *Forest ecology: A Foundation for Sustainable Forest Management and Environmental Ethics in Forestry* (Prentice Hall, Saddle River, New York, NY, 2003)

[2] A. Lysychko, Problems of theory and methodology of accounting, control and analysis pp. 74–82 (2018)

[3] L. Saharnatskaya, Bulletin of Sumy National Agrarian University. Series: Economics and Management **2**, 116 (2015)

[4] M. Shershun, Balanced nature using **1**, 5 (2013)

[5] O.I. Drebot, M. Shershun, O. Shkuratov, *Zbalansovanyi rozvytok lisovoho sektoru ekonomiky v konteksti yevropeiskoi intehratsii Ukrainy (Balanced development of the forest sector of the economy in the context of Ukraine's European integration)* (Ahrarna nauka, Kyiv, 2014)

[6] O.V. Sakal, *Efektyvne upravlinnia zemliamy lisohospodars'koho pryznachennia (Effective management of land for forestry purposes)* (State Institution 'Institute of Economics of Nature Management and Sustainable Development of the National Academy of Sciences of Ukraine', Kyiv, 2012)

[7] O.I. Furdychko, Ekonomika Ukrainy **5**, 56 (2013)

[8] E. Dhimitri, A. Gabeta, M. Bello, The USV Annals of Economics and Public Administration **15**, 51 (2015)

[9] Y. Mishenin, I. Kobylianska, T. Usik, I. Yarova, *Ekolohooriientovane lohistychne upravlinnia vyrobnytstvom (Eco-oriented logistics production management)* (Papirus, Sumy, 2013)

[10] P. Hryhoruk, N. Khrushch, S. Grygoruk, E3S Web of Conferences **166**, 13023 (2020)

[11] P. Hryhoruk, N. Khrushch, S. Grygoruk, *Assessing the Investment Capacity of the Agricultural Sector: Case of Ukraine*, in *Proceedings of the 2020 10th International Conference on Advanced computer information technologies ASIT'2020* (Deggendorf, 2020), pp. 183–187

[12] M. Dubyna, O. Zinkevych, Business Inform **2**, 187 (2020)

[13] M. Selinnyi, O. Korma, Modern Economics **17**, 211 (2019)

[14] B. Solberg, K. Rykowski, *Forest Policy Review and Strategy Development: Analytical Studies* (The World Bank, Washington, DC, 2000)

[15] V. Zhulkanych, *Osnovni pryntsypy i zavdannia staloho upravlinnia lisamy rehionu v systemi ratsionalnoho pryrodokorystuvannia (Basic principles and tasks of sustainable forest management of the region in the system of rational nature management)*, in *Proceedings of the International scientific-practical conference on The ecological, socio-economic and historical-cultural aspects of the Maramures border region development*, edited by F.D. Hamor (FOP Petryshyn, Khmelnytskyi, 2016), pp. 131–134, `http://cbr.nature.org.ua/conf2016/Pr_2016.pdf`

[16] *Derzhavne ahentsvo lisovykh resursiv Ukrainy (Sstate agency of forest resources of Ukraine)* (2020), `http://dklg.kmu.gov.ua/forest/control/uk/index`

[17] V. Doskich, *Ekolohichne lykho: Vynyshchennia lisiv v Ukraini (environmental disaster: Deforestation in Ukraine)* (2015), `https://cutt.ly/SbjbHM5`

[18] *Derzhavna sluzhba statystyky Ukrainy (State statistics service of Ukraine)* (2020), `http://www.ukrstat.gov.ua/`

[19] P. Hryhoruk, N. Khrushch, S. Grygoruk, IOP Conference Series: Earth and Environmental Science **628**, 012026 (2021)

[20] M. Nehrey, A. Kaminskyi, M. Komar, Periodicals of Engineering and Natural Sciences **7**, 702 (2019)

[21] I.I. Kychko, Y.O. Harus, Global and national economic problems **11**, 128 (2016)
[22] O. Iaremko, Efektyvna ekonomika **11** (2016)
[23] S. Makarenko, Efektyvna ekonomika **12**, 73 (2018)
[24] O. Ovchynnikova, O. Dupliak, O. Khan, E3S Web of Conferences **166**, 13004 (2020)

SHS Web of Conferences **107**, 11001 (2021) https://doi.org/10.1051/shsconf/202110711001
M3E2 2021

The concept of assessing the City Council management control level for effective decision making

Tetiana Melikhova[1,*] and *Olena* Mikhailutsa[1,**]

[1]Zaporizhzhia National University, 66 Zhukovkoho Str., Zaporizhzhia, 69600, Ukraine

Abstract.
In present paper, time series models of revenues and expenses of general and special local budget fund in Zaporizhzhia City Council for 2017-2020 are built. The possibility of forecasting time series has been investigated on the example of local budget revenues and expenses. The concept of assessing the level of management control which consists of information-analytical circuit, analysis and modeling circuit, and decision-making circuit has been proposed. The best options for absolute and comparative economic efficiency of accounting and economic expenses of management control have been substantiated. The forecast for calculating the absolute and comparative economic efficiency of management control on the data of Zaporizhzhia City Council is shown and the annual economic effect from its implementation is calculated. The methodological approaches to the analysis of income and expenses of the general and special fund have been improved. They, unlike existing methods, include time series modeling and studies describing complex changes in parameters over time. This made it possible to carry out high-quality management control and increase the efficiency of activities. The SPSS Statistics data analysis package is used for data processing, test assumption and building the future strategy.

1 Introduction

Successful management in the system of economic relations is directly dependent on specific conditions. The volatility in the financial sector in regions influences the growth in production expenses of enterprises, caused by increases in fuel, electricity and gas tariffs. The global financial crisis due to currency fluctuations negatively affects the confidence of individuals and legal entities in banks. Quarantining of establishments led to a halt of small and medium-sized businesses. Bankruptcy of banking institutions in previous years has increased the distrust of deposits and loans. Under current circumstances effective management of local authorities is becoming a priority. The dynamic operating conditions in the region with many different external and internal threats encourage the study of the conceptual frameworks of management control.

The modern concept of management accounting in the scope of control entered the theory and practice as a new concept of information and management – management control (controlling - in Germany, control de jestion - in France). Management control is an integral element of management process, it is control over the economic management, and not the management of activities for control. One of the main incentives for its implementation is the uncertainty of the results inherent in any type of activity that are envisaged, planned and designed. Striving for control unification corresponds to the desire for the integration of measures and weights in all countries, that is, it would be very important to have a generally recognized system of management control.

The implementation of local government activities is impossible without hazards, therefore, special attention of the management should be aimed at preventing such threats, as well as leveling their consequences. Optimization of resources and conducting of monitoring efforts aimed at the effective and stable operation of local council bodies, and risk management are approaches to the need for internal control. The systematic implementation of internal control is one of important management functions of the local council body. The control function involves the verification of planned and actually revenues and expenses incurred for general and special funds. Monitoring the optimal use of local budget funds, competent management of financial resources, rapid response to threats to minimize and prevent harm ensure effective city management.

2 Literature review

The main issues addressed in this paper are based on review and analysis of foreign and domestic publications. Information and analytical support of management control has been the subject of numerous studies during last decades in developed and developing countries. Definition of management control, as a process due to which managers ensure that resources come and are used efficiently and effectively to achieve the objectives set for the organization, is presented by S. Gschwantner and M. Hiebl [1].

*e-mail: tanyu_zp_zgia@ukr.net
**e-mail: elenamikhaylutsa7@gmail.com

Indeed, any research is based on the study of the conceptual and categorical apparatus. In foreign sources, management control is described as a function aimed at achieving certain goals within the established schedule. This process has three main components, such as taking corrective action, measuring of actual productivity and standard-setting. Management control is defined as a process that helps to achieve objectives of the organization [2]. But the author did not single out the functions of the management control itself. All functions of management control are important and closely interact with each other, passing into each other.

R. Savchenko and N. Savchenko believe that any economic entity operates under uncertainty, and the main task of managers and owners of an enterprise is to assess the uncertainty level and to make appropriate management decisions to achieve the strategic objectives. At the same time, the achievement of both strategic and operational objectives of enterprise under conditions of uncertainty and permanent risks is facilitated by the management control system. The presented classification division of management control [3] made it possible to prove the feasibility of identifying varieties of control depending on the levels of making management decisions.

However, a prerequisite for ensuring management control is the adoption of effective management decisions based on timely prevention and checking errors of accounting objects [4]. But the content of the author's definition of management control is very similar to the interpretation of internal accounting control. The threat of committing violations and mistakes can happen by the coincidence of subjective factors (conducting ineffective corporate management) and objective ones (changes in legislation, economic crisis). Based on a literature survey of the impact of management control, the authors [5] identified several factors that can affect the management control systems and activities of organizations.

Also, numerous factors enabling the correct focus of the management control system on management decision-making are provided as well. However, the author did not correctly identify the factors that positively or negatively affect the conduct of management control and management decision-making.

The study of management control features can be found in works of such scientists as S. Saliga [6], K. Saliga [7], Yu. Yarashykly [8], E. Antipenko [9] and others. The results of management control depend on its effectiveness. It is necessary to compare the benefits received and the expenses incurred to calculate the effectiveness of their implementations [8]. Especially noteworthy is the reasonable calculation of economic efficiency. The authors [6] developed a system of indicators to assess financial security and tax audit effectiveness, that is based on the principles for assessing the quality and effectiveness of tax audit as a financial security component and cost-benefit studies of economic activities of enterprises. The concept proposed by the author can be used to calculate the economic efficiency of any control object. Detailed consideration of the principle of economic efficiency assessment of the enterprise is given by K. Saliga. In [7] he analyzes the relationship of the indicator to the nature and effect of economic laws, correlation of economic efficiency indicators of production at the levels of the national economy and the economic entity, consistency, logical coherence, the link with partial indicators of economic efficiency, interdependence of changing indicators based on trends of living and materialized labor, making production self-sustainable as a constraint on changes in private performances, full accounting of production expenses and results, their correspondence to each other in time and content, embrace of all types of enterprise activities, availability of criteria for determination of private generalized productive indicators, efficiency of using all types of advanced, applied and used resources, the focus on achieving the main (general) production objective, logistic approach to assessing enterprise productivity, quantitative measurability of indicators and performance criteria and comparability of the options. So, the author is a well-grounded study based on the foundations of economic theory. These approaches will be used in the future.

In works [9, 10] the importance of evaluating the effectiveness of investment projects in market conditions is noted. The focus is on cash flow modeling, including all cash receipts and expenses related to the project for the pay period, taking into account the possibility of using different currencies. The authors highlighted the principle of positivity and maximum effect, the time factor, forthcoming expenses and revenues, comparisons "with the project" and "without the project", the presence of different project participants taking into account all the most real consequences of the project. They demonstrated multi-stage assessment accounting for the impact of inflation, the impact of uncertainties and risks (in quantitative form).

The project throughout its entire life cycle is under consideration and the conditions and the impact of working capital requirements on the efficiency of investment projects are compared.

E.G. Dolan and D.E. Lindsey point out that it is necessary to distinguish between external and internal production expenses. Full expenses (economic expenses), in addition to production expenses (accounting expenses), include normal profits as the minimum value, upon which it makes sense to deal with entrepreneurship [11]. The authors in their conceptual study substantiated the difference between accounting and economic costs.

However, despite significant scientific achievements, under modern conditions of management, state bodies of local authority generate new theoretical and practical problems that require solutions, including the improvement of work organization of their administrative apparatus and timely conducting the managerial control over income and expenses incurred for their maintenance.

Within the framework of this article, scientific works on the study of data mining methods, as well as decision-making based on these analyses, are considered. The authors of the article [12] applied a scientific econometric approach to determine the general presence and strength of the relationship between the economic indicators of large industrial enterprises in the region of Ukraine. The most significant coefficients of the analysis of the financial condition are revealed. Indeed, you need to have analyzed the object of research over a number of years to derive your own economic indicators. This allows you to justify the research results and make a prediction of the data.

A group of scientists [13] described the directions of promising research in the field of analysis and modeling of the dynamics of time series of processes in complex systems with the presence of a human factor. General information about time series and tasks of their analysis is given. The modern methods of analysis of time series of economic processes are considered. Scientists have proven that economic processes cannot be considered completely random, since they tend to self-organize and, moreover, are influenced by the memory of previous states.

The features of the main methods of analysis and forecasting of time series are presented in the works [14–23]. In addition, the authors highlight the problems and disadvantages of these methods that arise when they are applied to nonstationary time series. The article [14] provides a review of the literature on statistical methods for analyzing time series data. The author considers the following models: of exponential smoothing, of autoregressive integrated moving average, of unobservable components. The study [15] provides overview of statistical analysis of time series with an emphasis on methods of data interpretation, as opposed to methods of empirical forecasting. Scientists compare methods of processing non-stationary processes both from the point of view of non-stationarity on the average and non-stationarity in the structure.

Focusing on the fact that in the modern economy a much greater place is occupied by the estimation of dynamic models that reflect short-term and long-term relationships between economic variables, the author of the scientific study [24] analyzed the Box-Jenkins approach to identifying time series, estimated ARIMA models parameters and made forecast using these models.

Based on the analysis of literary sources, it can be concluded that there is no single view of the methodological approaches to management control in the city council for making management decisions. Thus, the implementation of theoretical achievements of scientists requires a deeper clarification of methodological developments, considering modern economic trends in the economic development and increased competition.

3 Materials and methods

In management control of the city council the presence of high-quality information and analytical support at all stages of its implementation is of great importance. A successfully built information and analytical support for management control will help to provide the quality of all its stages and timely process information, increasing the overall city council efficiency.

3.1 Data analysis

The use of various special processing techniques equipped with computer-assisted statistical information packages, such as spectral analysis, regression and correlation analyzes, guarantees a comprehensive and in-depth analysis of information. Universal mathematical approaches are successfully applied to analyze and perceive information in order to correctly forecast and develop a profitable investment strategy. Time series analysis provides insight into the causal mechanisms behind this series. The paper considers time series that describe revenues and expenses data for the local budget (general and special fund) with a complex nature of development. Simple smoothing method is used to find the evaluation of trend model parameters. The original time series is converted into a series, the values of which are the data averaged over three adjacent points in time series. The refined time series models of the special and general local budget fund revenue with the obtained trend lines is shown in figures 1, 2.

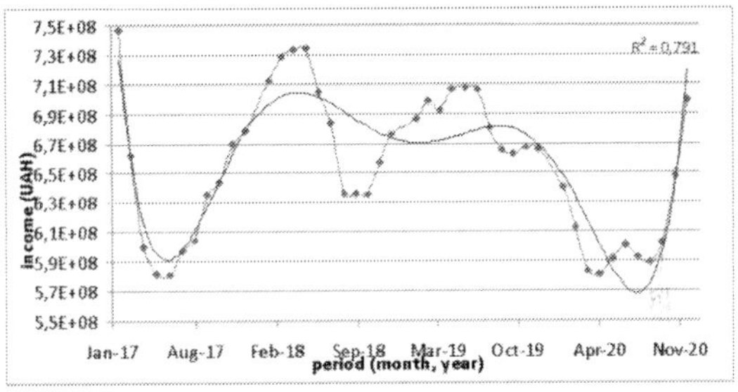

Figure 1. The refined time series model of the general local budget fund revenue

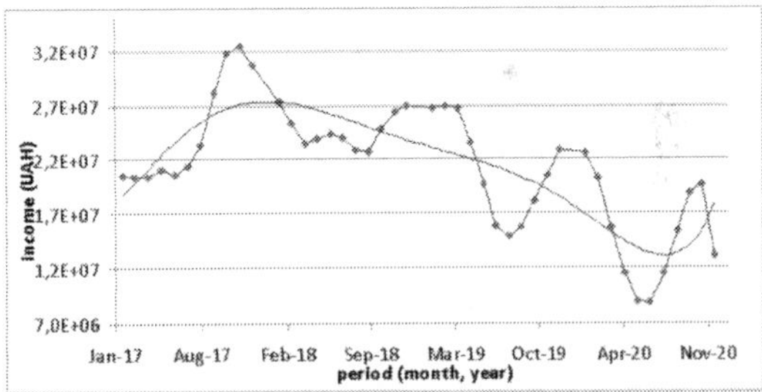

Figure 2. The refined time series model of the special local budget fund revenue

Based on the Fisher ratio (F-test), an analysis of time series stationarity describing the change in value of the completed general and special funds was carried out. Comparison of the obtained Fisher ratios for each of time series with table values for a specified value level of 0.05 made it possible to state that the value changing processes of completed general and special funds for 2017–2020 are stationary in the narrow sense.

The hypothesis of the stationarity of this process was tested on the basis of the Student's t-test [25]:

$$t = |E_1 - E_2| / \left(\sigma \sqrt{(1/n_1 + 1/n_2)}\right), \quad (1)$$

where n_1, n_2 – the number of elements included in both parts, E_1, E_2 – mathematical expectations, σ is a standard deviation of the difference in mathematical expectations. The Student's t-test of time series of general and specialized funds are, respectively, $t_g = 0.09$ and $t_s = 0.93$, while the table value of time series boundary is $t_{tabl} = 2.03$. Data analysis made it possible to conclude that this process is stationary in terms of mathematical expectation and there is no non-random component of considered time series.

Implementation of time series analyzing methods allows you to make a reasonable forecast of changes in the studied indicators under certain conditions and properties of time series. Using this forecast, it is possible to build a strategy of actions for the future, applying the obtained indicators to carry out management control of the targeted use of transfers, implementation of targets for the establishment of a special and general local budget fund (figure 3).

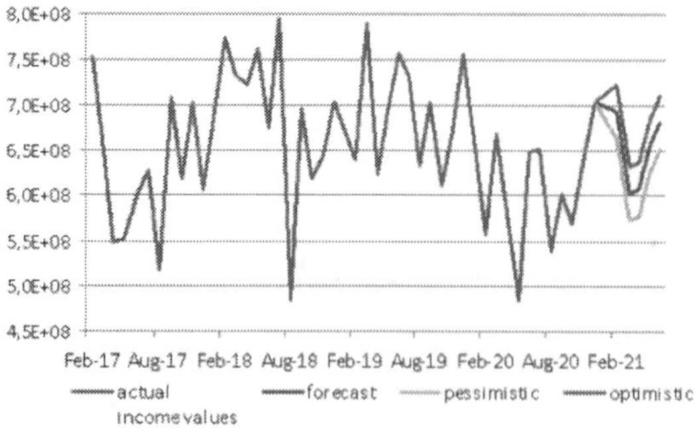

Figure 3. Actual and projected values of the general local budget fund (UAH)

Within the present study framework based on the data obtained on general and special funds expenses of the local budget in Zaporizhzhia City Council for 2017–2020, data forecast was carried out using the exponential smoothing method.

Figure 4 shows a plot of actual expenses (blue plot), an exponential forecast model with a smoothing factor of $k = 0.8$ (red plot), and a forecast value with $k = 0.1$ (green graph). The forecast model for $k = 0.1$ is smoother and reacts more slowly to bursts in recent periods than for k=0.8.

The calculated forecast accuracy for the two presented models is 97.39% and 97.99%, respectively. It is important to note that this model assumes regular recalculation of the forecast at the end of the last period and the appearance of new expenses data to forecast for the last period.

3.2 Economic efficiency for carrying out management control

Prerequisite for ensuring management control of the city council is making effective management decisions based on timely prevention and identifying errors in the formation of revenues and expenses of the general and special fund of local budget at all stages of management apparatus work.

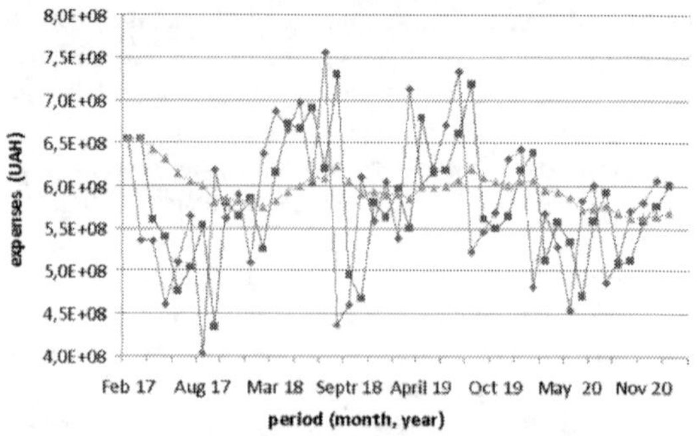

Figure 4. Actual and projected values of the general local budget fund (UAH)

The proposed concept of management control assessment, which includes information and analytical circuit, the analysis and modeling circuit, as well as a decision-making circuit, is shown in figure 5.

The absolute and comparative economic efficiency of accounting expenses and the economic expenses of management control are analyzed. Statistical data was taken from the general and special fund of the Zaporizhzhia City Council from the site zp.gov.ua, and the management information from the site smida.gov.ua, youcontrol.com.ua, and also used analytical internal and forecast information. For the optimal option, the specified indicators should exceed the corresponding standard indicator of the expenses of its implementation. The developed system of indicators for evaluating management control effectiveness is based on the formed quality and performance assessment principles of its implementation [26]. Let's apply the concept of assessing the level of management control in Zaporizhzhia City Council for January 2021 to calculate the absolute and comparative economic efficiency for its implementation (tables 1, 2) in order to determine the annual economic effect of its implementation (table 3). The data in tables 1–3 are derived from planned internal statistics. The information for January 2021 allowed us to expand the estimate of the annual effect for comparison with the actual data.

Economic effect of management control can be calculated as the difference between annual results and incremental expenses:

$$\Delta W_{mc} = (\Delta A_{mc2} - \Delta A_{mc1}) \cdot \left(\frac{Ex_{Emc1}}{\Delta A_{mc1}}\right) - (Ex_{Emc2} - Ex_{Emc1}) \qquad (2)$$

The assessment of the effectiveness of management control in the Zaporizhzhia City Council for January 2021 revealed the following:

1. The absolute economic efficiency in relation to the accounting costs of management control is: the planned 8.65, the actual 8.22 – a decrease of 0.43. The absolute economic efficiency in relation to the economic costs of management control is equal to: the planned 7.86, the actual 7.47 – a decrease of 0.39. Thus, the Zaporizhzhia

Table 1. Absolute economic efficiency for carrying out management control for January 2021, UAH

Indexes	Value	City Council plan	City Council fact	Deviation
Additional result obtained at the expense of management control, UAH / year	ΔA_{mc}	494 540	469 813	-24727
Cost savings		494 540	469 813	
Accounting expense for management control, UAH	Ex_{amc}	57 173	57 173	
Materials, UAH / year		2 703	2 703	
Salary, UAH / year		40 800	40 800	
Accrual per Individual Entrepreneur, UAH / year		8 976	8 976	
Amortization expense, UAH / year		2 214	2 214	
Others, UAH / year		2 480	2 480	
Indicator of absolute economic efficiency of accounting expense for management control, units	$E_{mc} = \Delta A_{mc}/Ex_{amc}$	8,65	8,22	-0,43
Regulatory profit from management control, UAH / year	P_r	5717	5 717	
Regulatory profitability of products, units	$PA_r = P_r/Ex_a$	0,10	0,10	
Economic expense of management control, UAH / year	Ex_{Emc}	62 890	62 890	
Indicator of absolute economic efficiency in relation to the economic expense for management control, units	$E'_{mc} = \Delta A_{mc}/Ex_{emc}$	7,86	7,47	-0,39
Criterion (normative indicator) of economic efficiency of expenses for management control	E'_{nmc}	7,5	7,5	

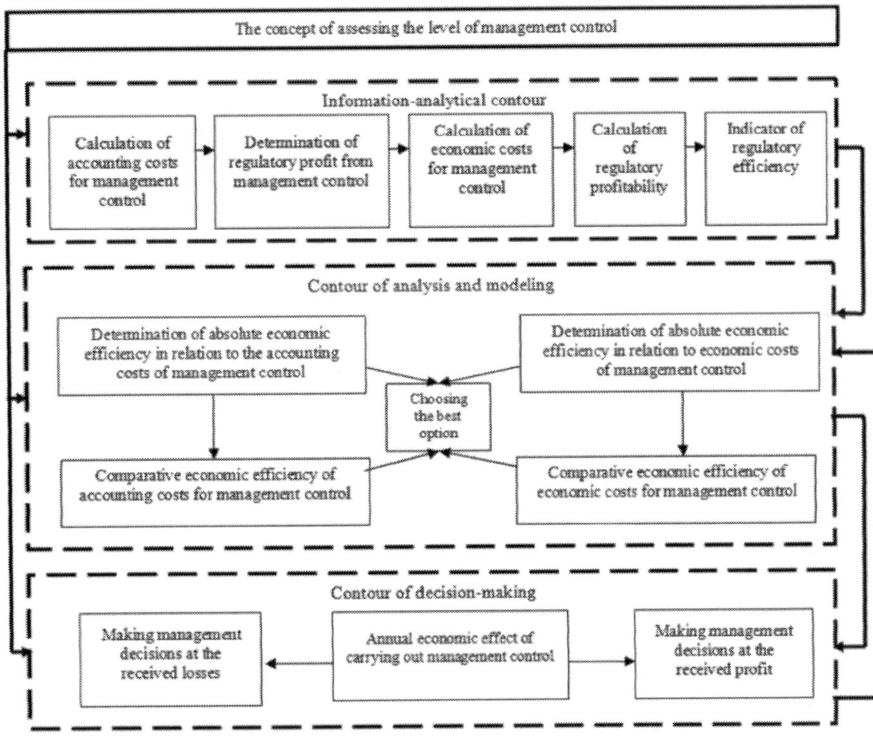

Figure 5. The concept of assessing the level of management control

City Council has a high absolute economic efficiency from the management control. The decrease in the actual figure compared to the planned was due to the lockdown, which occurred in January 2021 and reduced the amount of income received.

2. The comparative economic efficiency of accounting costs for management control is: planned 1.30, the actual 1.24 – a decrease of 0.6. The comparative economic efficiency of economic costs of management control is equal to: the planned 1.30, the actual 1.24 – a decrease of

Table 2. Comparative economic efficiency for management control in January 2021, UAH

Indexes	Value	City Council plan	City Council fact	Deviation
Additional result obtained at the expense carrying out of management control in the base period, UAH / year	ΔA_{mc1}	300 000	300 000	
Additional result obtained at the expense carrying out of management control in the new period, UAH / year	ΔA_{mc2}	494 540	469 813	-24727
Index of additional result obtained at the expense carrying out of management control, units	$I_{amc} = \Delta A_{mc2}/\Delta A_{mc1}$	1,65	1,57	-0,08
Accounting expense obtained at the expense carrying out of management control in the base period, UAH / year	Ex_{amc1}	45 000	45 000	
Accounting expense obtained at the expense carrying out of management control in the new period, UAH / year	Ex_{amc2}	57 173	57 173	
Index of accounting expense obtained at the expense carrying out of management control, units	$I_{Examc} = Ex_{amc2}/Ex_{amc1}$	1,27	1,27	
Indicator of comparative economic efficiency of accounting expense obtained carrying out of management control, units	$E_{Cmc} = I_{amc} I_{Examc}$	1,30	1,24	-0,06
Economic expense obtained at the expense carrying out of management control in the base period, UAH / year	Ex_{Emc1}	49 500	49 500	
Economic expense obtained at the expense carrying out of management control in the new period, UAH / year	Ex_{Emc2}	62 890	62 890	
Index of economic expense obtained at the expense carrying out of management control, units	$I_{ExEmc} = Ex_{emc2}/Ex_{emc1}$	1,27	1,27	
Indicator of comparative economic efficiency of economic expense obtained carrying out of management control, units	E'_{Cmc}	1,30	1,24	-0,06

Table 3. Annual economic effect of management control in January 2021, UAH

Indexes	Value	City Council plan	City Council fact	Deviation
Additional expense for obtaining additional economic result from carrying out of management control, UAH / year	$(Ex_{Emc2} - Ex_{Emc1})$	13 390	13 390	
Additional economic result from carrying out of management control, UAH / year	$\begin{bmatrix}(\Delta A_{mc2} - \Delta A_{mc1})\cdot \\ \cdot(Ex_{Emc1}/\Delta A_{mc1})\end{bmatrix}$	32 099	28 019	-4 080
Annual economic effect from carrying out of management control, UAH / year	ΔW_{mc}	18 709	14629	-4080

0.6. Thus, the Zaporizhzhia City Council has a high comparative economic efficiency from the management control. The decrease in the actual indicator compared to the planned one was due to a decrease in the amount of income, provided that the actual amount of planned expenses is observed.

3. The annual economic effect of management control is: according to the plan UAH 18,709, in fact UAH 14,629 – a decrease of UAH 4,080. Thus, the Zaporizhzhia City Council has a high annual economic effect from management control. In this study, the economic effect is achieved by obtaining a greater additional result (benefits) from managerial control in comparison with the additional costs that were spent on its implementation. As a result of the study, a relationship was revealed that shows that during management control errors are quickly identified and shortcomings in the work of administrative personnel are additionally reduced, which prevents financial sanctions, abuse and, in general, reduces the costs of the enterprise.

4 Conclusions

Having studied the subject area and detailed analysis of the presented issue, the following was established:

1. Time-series revenue and cost models of the general and special fund were built on the basis of the data obtained from the local budget in Zaporizhzhia City Council for 2017–2020. The time series modeling and research describing complex changes in parameters over time makes it possible to visually analyze and timely control the analytical material of revenues and expenses of general and special funds in order to make effective management decisions.

2. The concept of assessing management control is proposed, which, in contrast to the existing ones, consists of the following contours: information and analytical, analysis and modeling, decision making. It has been tested on data from the Zaporizhzhia City Council. It is universal and can be used to assess the level of management control in all regions of Ukraine. This will make it possible to conduct a timely assessment of management control, which will increase the efficiency of work.

3. The best options for absolute and comparative economic efficiency of accounting and economic costs of management control are proposed and substantiated. This will help to calculate the annual economic effect of management control, and the result (profit or loss) will help to make effective management decisions.

4. The concept of management control assessment was tested on Zaporizhzhia City Council data. The calculation results made it possible to conclude that Zaporizhzhia City Council has a high probability of absolute and comparative economic efficiency of management control which in turn gives a high economic effect.

In the future, the proposed methodology for assessing the level of management control can be used in the work of state bodies at all levels of government.

References

[1] S. Gschwantner, M. Hiebl, J Manag Control **27**, 371 (2016)
[2] R. Simons, *Levers of Control: How Managers Use Innovative Control Systems to Drive Strategic Renewal* (Harvard Business School Press, Boston, 1995)
[3] R. Savchenko, N. Savchenko, Ekonomika ta derzhava **9**, 70 (2015)
[4] A. Junevicius, R. Daugeliene, J. Jurkeviciene, P. Orzekauskas-Nussbaum, Engineering Economics **28**, 421 (2017)
[5] R. Chenhall, K. Langfield-Smith, European Management Journal **25**, 266 (2007)
[6] S. Saliga, *Formation of financial security of business entities on the basis of tax audit* (ZTSNTEI, Zaporizhzhia, 2011)
[7] K. Saliga, *The efficiency of economic activity of enterprises* (ZTSNTEI, Zaporizhzhia, 2005)
[8] T. Melikhova, Y. Yarashykly, Investytsiyi: praktyka ta dosvid **2**, 17 (2020)
[9] E. Antipenko, V. Donenko, V. Pokolenko, Y. Chupryna, D. Prykhodko, Upravlinnia rozvytkom skladnykh system **2**, 6 (2010)
[10] T. Baibakova, L. Suvorova, V. Epinina, Y. Azmina, Advances in Economics, Business and Management Research **38**, 34 (2017)
[11] E. Dolan, D. Lindsey, *Market: microeconomic model* (S-Petersburg, 1992)
[12] T. Melikhova, A. Makarenko, O. Mikhailytsa, A. Pozhuyev, CEUR Workshop Proceedings **2422**, 226 (2019)
[13] E. Andrianova, S. Golovin, S. Zykov, S. Lesko, E. Chukalina, Russian Technological Journal **8(4)**, 7 (2020)
[14] D. Rahardja, Journal of Statistics and Management Systems **12**, 1445 (2020)
[15] D.R. Cox, G. Gudmundsson, G. Lindgren, L. Bondesson, Scandinavian Journal of Statistics **8**, 93 (1981)
[16] A. Bielinskyi, V. Soloviev, CEUR Workshop Proceedings **2292**, 37 (2018)
[17] V. Soloviev, A. Belinskij, CEUR Workshop Proceedings **2104**, 116 (2018)
[18] V. Soloviev, N. Moiseienko, O. Tarasova, CEUR Workshop Proceedings **2393**, 905 (2019)
[19] V. Soloviev, A. Belinskiy, Communications in Computer and Information Science **1007**, 276 (2019)
[20] V. Soloviev, S. Yevtushenko, V. Batareyev, CEUR Workshop Proceedings **2546**, 87 (2019)
[21] V. Soloviev, A. Bielinskyi, O. Serdyuk, V. Solovieva, S. Semerikov, CEUR Workshop Proceedings **2732**, 455 (2020)
[22] V. Soloviev, O. Serdiuk, S. Semerikov, A. Kiv, CEUR Workshop Proceedings **2713**, 21 (2020)
[23] S. Semerikov, H. Kucherova, V. Los, D. Ocheretin, CEUR Workshop Proceedings **2845**, 22 (2021)
[24] G. Kantorovich, Higher School of Economics Economic Journal **6**, 85 (2002)
[25] V. Bukreev, S. Kolesnikova, A. Yankovskaya, *Identification of regularities in time series in problems of recognition of states of dynamic objects* (Tomsk Polytechnic University, 2010)
[26] T. Melikhova, in *Industrial management: theory and practice*, edited by V. Voronkova, N. Metelenko (ZNU, Zaporizhzhia, 2020)

The problem of estimating the sustainable development of the integrated technogenic industrial system

Sultan Ramazanov[1,*] and *Oleksandr* Honcharenko[1,**]

[1]Economics Information Systems Department, Kyiv National Economic University named after Vadym Hetman, 54/1 Peremohy Ave., Kyiv, 03057, Ukraine

Abstract. The purpose of this work is to research the problem of sustainability of the integrated technogenic system in an innovative economy and "Industry 4.0". The paper deals with the methods of assessing the level of sustainable development of the integrated technogenic system as a family of models for eco-economic and socio-humanitarian management of technogenic economic objects. The economic approach to modeling a sustainable development is considered, which consists of the optimal use of limited resources and applies natural, energy and resource-saving technologies to generate aggregate income. At the same time, the transition to the information society leads to a change in the structure of aggregate capital in favor of human, increasing intangible flows, flows of information and intellectual property. The problem of sustainable development through 7 main assets (capitals) that support the viability of socioenvironmental and economic systems is considered. The concept of sustainable development, the system of global dimensions of sustainable development and the level of sustainable development are defined. The presented methodology for the formation of humanitarian component of technogenic regional production uses available statistic data. Application of these methods will improve efficiency of managerial decisions in the technogenic regional production management, maximize benefits from the use of innovations and define strategic innovative directions for regional development.

1 Introduction: problem analysis and statement

The global goal of this research is to create a methodology for making managerial decisions for sustainable, safe, secure and viable development of environmental economy and integrated technogenic systems and objects.

The concept of sustainable and safe development of the state and individual regions allows to ensure stable and balanced development of four sectors: economic, ecological, social and spiritual and moral unity on the basis of innovative socio-humanitarian technologies, combining the principles of economic efficiency, social security and environmental safety.

Currently, the issues of ecological, economic and socio-humanitarian safe and sustainable development of civilization have come to the forefront of scientific research and public consciousness in general. Mankind has reached the point where modern civilization, often called technogenic-consumer, has found its hopelessness, when it is necessary to seriously reconsider its basis and consciously choose another, spiritual and ecological, development strategy, otherwise humanity may be destroyed from Earth due to global systemic crises that have erupted in recent years. To solve this problem, humanity needs to abandon a number of stereotypes and direct the vector of civilizational development to the formation of the sphere of mind ("noosphere" according to V.I. Vernadsky). The formation of the noosphere-ecological imperative is associated with the formation of a society that is able to ensure the coevolutionary and viable development of civilization for planetary integrity. Sustainable and safe development is impossible without the spiritual, cultural and educational improvement of man himself. The new model of civilizational development, which implements non-traditional ecological, economic and demographic imperatives, must have a deeply humanistic social orientation. Such an approach to understanding the noosphere requires the creation of a new model of science, which should be based not only on a rational intellectual approach to the ecosystem, but also based on its spiritual and cultural components. At the same time, the most important problem is the integration of science, education and innovative technology based on the noosphere paradigm of sustainable development.

The problem of studying the sustainable and safe development of the state and individual regions in socio-environmental and economic systems with humanitarian components, i.e. on the basis of integration of 4 spheres of activity and functioning of modern complex systems in innovative economy (presented both in phase space and in ESTI space – "education – science – technology – innovation"), as well as in modern science in general, is the main and relevant. Only the integration of methods of modeling socio-economic, ecological, cultural, spiritual and other

*e-mail: sramazanov@i.ua
**e-mail: alexgontcharenko@gmail.com

processes can ensure the sustainability and viability of the entire system [1–7].

Also, note that a system-thinking and active person usually forecasts and considers his/her performance results, compares his/her capabilities, considers environmental interests, those of future generations and the need for noospheric development.

The purpose of the paper is to study the problem of sustainability of integrated technogenic system in an innovative economy and to research the methods and methodologies for assessing the level of sustainable development of the state and individual regions in order to ensure stable and balanced development of four sectors: economic, ecological, social and spiritual and moral unity on the basis of innovative sociohumanitarian technologies, combining into a holistic system the principles of economic efficiency, social security and environmental security.

Description of key findings. Carrying out a comprehensive analysis of the sustainability of regional development involves the use of criteria and indicators that allow to obtain a reliable assessment of its level, direction and intensity of change. However, economics has not finally developed approaches to justify their choice, methods of calculation, determination of patterns of development, etc. The main disadvantage of the existing methodological provisions is that they do not allow to fully assess the sustainability of the region, the degree of influence of various factors on the level of its change and discrepancies with the limit values. Different views among scholars dealing with the problem of sustainable development of the region are manifested in the choice of criteria that should be the basis for assessing the sustainability of development.

We note that the general scheme of integration model of steady safe ecological, economic, social and humanitarian development of system is presented in [1–3, 6–8].

The paradigm of sustainable development, which provides a dynamic process of successive positive changes that ensure the balance of economic, social and environmental aspects, should be the basis for the formation of approaches to solving the problems of territorial entities. This is especially relevant today, when the focus of economic reforms is shifting to the level of regions and strengthening their role in the implementation of economic policy. The priority approach in the implementation of reforms at the regional level should be the belief that the development of the territory should not be equated with its economic development. It is impossible to consider the region developing steadily only on the basis of increase of economic indicators. Sustainable development should be aimed at achieving a high quality of life, with a positive dynamic of indicators. The priority of the new stage of economic change is the innovative modernization of the regional economy.

It is believed that the term "sustainable development" is inherently innovative, because for the continuous, stable operation of any system in changing environmental conditions, it must constantly increase the degree of organization, adapting to these changes, i.e. generate new forms and mechanisms of stability support.

The transition to sustainable development requires radical transformations, at the center of which is the socialization, greening and humanization of all major human activities, changing his consciousness and creating a new "sustainable society". Such changes must take place purposefully, consciously, taking into account socio-economic, political, technical and other conditions. In general, the level of sustainable development of the technogenic region is due to the interaction of factors' groups of socio-economic, environmental and innovative nature.

Socio-economic factors include: production and resource potential (labor, material, financial, raw materials, fixed capital, etc.), market (factors of demand, supply and distribution; interpenetration of regional, national and world markets), strengthening the social function of the state, income level (propensity to consume, save, invest), education and health care, social protection, level and quality of life, level of environmental awareness.

The environmental aspect of sustainable development implies the rational use of natural resources (subsoil, extraction of natural resources, forest, water, land fertility, wildlife) on the principles of their economy, efficiency of recreational capacity and investment in environmental activities. This approach to sustainability means the maximum reduction of their use, the search for substitutes, the widespread introduction of resource-saving technologies.

Innovative factors include intellectual potential (ideas and knowledge, which turn into new sources of raw materials and power; information), scientific and technical potential (substitution technologies, technological activity), investment in innovation.

Thus, sustainable development means that the socio-economic system ensures the dynamic stability of its properties, applying at the same time the whole set of factors that affect the level of competitiveness of the region's economy.

The system of indicators that characterize the sustainable development of the technogenic region can be identified as criteria for assessing the above factors. Leading international organizations are elaborating indicators of sustainable development: United Nations, World Bank, Organization of Economic Cooperation and Development (OECD), European Commission, Scientific Committee on Problems of the Environment (SCOPE) and others. Among individual countries, the experience, scale, duration and complexity of a system of sustainable development indicators in the United States and the United Kingdom should be particularly noted. These countries began this work at the governmental level in the mid-1990s.

2 Main results

The problem of sustainable development on the basis of the "hexagon" of fixed assets, which support the activity of SEESM [1]. The main assets that support the viability of integrated SEESM are the following 7 factors: S – Social capital; Φ – Financial capital; N – Natural capital (land, water, etc.); K – Physical capital (fixed assets); L – Labor resources; H – Human (intellectual) capital; I – Institutional factor (resource).

The generalized production-technological function (PTF) can be represented in general as a nonlinear function:

$$Y(t) = F[K(t, L(t), H(t), N(t), \Phi(t), S(t), I(t); \vec{c}]. \quad (1)$$

It can be used to analyze the sustainable development.

The level of sustainable development (SDL) will be assessed using the index I_{sdl}, which is calculated as the sum of indices for four areas: economic (I_{eco}), ecological (I_{ecl}), social (I_{soc}) and humanitarian (I_{hum}) with the appropriate weight coefficients, that is:

$$I_{sdl} = \alpha_1 I_{eco} + \alpha_2 I_{ecl} + \alpha_3 I_{soc} + \alpha_4 I_{hum},$$
$$\alpha_1 + \alpha_2 + \alpha_3 + \alpha_4 = 1, \alpha_i \geq 0,\ i = 1, ...4 \quad (2)$$

Each of the indices I_{eco}, I_{ecl}, I_{soc} and I_{hum} is calculated using well-known in international practice indices and indicators.

Using the arithmetic mean form of construction of the generalized indicator at decrease of social and economic indicators and at increase of ecological ones, the leveling of a threatening situation is possible. Therefore, the content of the considered indicator is more consistent with the geometric mean value, which reflects the proportionality between the indicators. In addition, the considered indicators will reflect the real situation in the region. The level of sustainable development of the region can be calculated by the following formula:

$$I_{sdl} = \sqrt[4]{[\alpha_1 I_{eco} \times \alpha_2 I_{ecl} \times \alpha_3 I_{soc} \times \alpha_4 I_{hum}]}. \quad (3)$$

In particular, the level of sustainability of economic development of the region can be determined by the following formula:

$$I_{eco} = \sqrt[5]{I_n \times I_u \times I_\Phi \times I_m \times I_{np}}, \quad (4)$$

where I_{eco} – the level of sustainability of economic development of the region; I_n – level of production component of regional development: innovation, labor, financial and natural resources; I_u – the level of innovation component of regional development; I_Φ – the level of the financial component of regional development; I_m – the level of the labor component of regional development; I_{np} – the level of natural resource component of regional development.

This construction of the indicator will reflect the importance of each of the considered components: eco-economic and socio-humanitarian subsystems (spheres) in the performance of the objective function. A change in any of the private indicators leads to a change in the value of the aggregate indicator and captures a change in the steady state of the region. In the general case, all indicators change over time, i.e. have a certain dynamics.

Simple conditions for sustainable development (SD) are defined as follows:

1) condition of weak stability:

$$\frac{dF[\cdot]}{dt} \geq 0 \quad or \quad F_{t+1}[\cdot] \geq F_t[\cdot], \quad (5)$$

where $F_t[\cdot] = F[K(t), L(t), H(t), N(t), \Phi(t), S(t), I(t), \vec{c}]$.

2) condition of strong stability:

$$\frac{dF[\cdot]}{dt} \geq 0,\ N = N^C + N^S \text{ and } \frac{dN^C}{dt} \geq 0, \text{or } N^C_{t+1} \geq N^C_t, \quad (6)$$

where N^C – critical part of natural capital, and N^S – natural capital, which can be replaced by artificial.

For example, given critical natural capital N^C, sustainable development can be supplemented by a time limit on depletion of this value. For a time-decreasing production function, the arguments of which are aggregated variables: labor – L, capital – K and natural resource N, we will have the ratio:

$$F_t(K, L, N) \leq F_{t+1}(K, L, N) \quad (7)$$

or, in the general case:

$$F(K(t), L(t), H(t), N(t), \Phi(t), S(t), I(t), \vec{c}) \leq$$
$$\leq F(K(t+1), L(t+1), H(t+1), N(t+1),$$
$$\Phi(t+1), S(t+1), I(t+1), \vec{c}) \quad (8)$$

And it also requires compliance with the condition of not decreasing in time the value of N^C, i.e. $N_t = N^C_t + N^S_t$, as well as the condition of partial replacement of natural capital N by artificial N^S (or non-renewable resource for renewable resource): $N_t = N^C_t + N^S_t$.

The integrated level of sustainable development for all capital (resources) can be defined, for example, in the case of linear dependence as:

$$Y_{sdl}(t) = c_1 K(t) + c_2 L(t) + c_3 H(t) +$$
$$+ c_4 N(t) + c_5 \Phi(t) + c_6 S(t) + c_7 I(t), \quad (9)$$

where $c_1, c_2, c_3, c_4, c_5, c_6, c_7$ weight (normalizing and scaling) coefficients.

In the general case, the integral level of sustainable development can be represented as a nonlinear function:

$$Y_{sdl}(t) = F_{sdl}[]K(t), L(t), H(t), N(t), \Phi(t), S(t), I(t), \vec{c}]. \quad (10)$$

Private versions of the PTF model:

a) Mankiw-Romer-Weil model. Option of accounting for human capital H in the production function (PF), along with physical capital (K), labor (L) and natural (N) resources:

$$Y(t) = K^\alpha(t) \cdot H^\beta(t) \cdot [A(t) \cdot L(t)]^{1-\alpha-\beta}, \quad (11)$$

where $\alpha, \beta > 0$, $\alpha + \beta < 1$; H; $A(t)$ – function of scientific and technological progress. Note that α is a part of capital provided by investment growth (capital costs); β is similar.

b) Model of accounting for all fixed assets:

$$Y(t) = A(t) K^\alpha(t) \cdot L^\beta(t) \cdot H^\gamma(t) \cdot S^\rho(t) \cdot \Phi^q(t) \cdot N^\tau(t) \cdot I^\nu(t), \quad (12)$$

where $\alpha, \beta, \gamma, \rho, q, \tau, \nu > 0$ and $\alpha + \beta + \gamma + \rho + q + \tau + \nu = 1$.

The following notations are also used here: K – physical capital, L – labor, H – human capital, S – social capital, Φ – financial capital, N – natural resources (land, water, etc.), $A(t)$ is a function of the level of scientific, technical and technological development, for example, $A(t) =$

$aT^S(t)$, where $T(t)$ – volume of innovative technologies (resources).

The system of global measurements of sustainable development. An important problem in implementing the concept of sustainable development is the formation of a system of measurements (indices and indicators) for quantitative and qualitative assessment of this complex process. The main requirements for this system of measurements are its completeness of information and the adequacy of the presentation of the interconnected triad of components of sustainable development. Well-known international organizations and numerous scientific teams are currently working in this direction, but unambiguous coordination of this system of measurements has not been achieved yet.

Here is a system of measurements of sustainable development, proposed by the Institute for Applied System Analysis of Igor Sikorsky Kyiv Polytechnic Institute [2, 9].

The level of sustainable development is assessed using the corresponding I_{sdl} index, which is calculated as the sum of indices for four areas: economic (I_{eco}), ecological (I_{ecl}), social (I_{soc}) and humanitarian (I_{hum}) with the appropriate weight coefficients.

Of course, all indicators that affect the components of these indices, as well as these indices themselves, are measured in different units and have different interpretations. Therefore, they are reduced to a normalized form so that their changes, as well as changes in the indices themselves, were in the range from 0 to 1. In this case, the worst values of these indicators will correspond to numerical values close to 0, and the best – will bring these values closer to 1. This rationing allows you to calculate each of the indices I_{eco}, I_{ecl}, I_{soc} and I_{hum} in the form of the average sum of its components with the appropriate weight coefficients.

1. The index of economic area (I_{eco}) is formed from two global indices:

a) the index of competitive development (the index of competitiveness – I_c), developed by the organizers of the World Economic Forum. This index is calculated annually for 117 economies of the world and is published in the form of the so-called "Global Competitiveness Report". The competitiveness index is formed from the following three indicators: the indicator of technological development of the country; indicator of civil institutions and indicator of macroeconomic environment. These three indicators are calculated based on the use of 47 data sets of the state of technology and innovation development, the level of information and communication technology, the level of R&D spending, the level of foreign investment, the level of business independence from government, the level of corruption in the country, etc.;

b) index of economic freedom (I_{ef}), which was developed by the intellectual center of the Heritage Foundation. It is published annually in the Wall Street Journal. The index of economic freedom is formed from the following ten indicators: the country's trade policy; fiscal burden from the government; government intervention in the economy; monetary policy; capital flows and foreign investment; banking and financial activities; pricing and remuneration policies; private property rights; regulatory policies; informal market activity. These ten indicators are calculated based on the use of 50 sets of various economic, financial, legislative and administrative data.

2. The index of ecological area (I_{ecl}) is assessed using the well-known ESI (Environmental Sustainability Index), calculated by the Center for Environmental Law and Policy at Yale University (USA) for 146 countries. The ESI index is formed of 21 environmental indicators, which were calculated based on the use of 76 sets of environmental data on the state of natural resources in the country, the level of environmental pollution in the past and today, the country's efforts to manage environmental conditions characteristics and more.

The ESI index quantifies a country's ability to protect its environment, both in the current period and in the long run, based on the following five criteria: the existence of a national environmental system; possibility to counteract environmental influences; reducing people's dependence on environmental influences; the country's social and institutional capacity to respond to environmental challenges; the possibility of global control over the environmental state of the country. In addition, this index can be used as a powerful tool for decision-making on an analytical basis, taking into account the social and economic areas of sustainable development.

3. The index of the social area (I_{soc}) is formed by averaging three global indices:

a) Index of Quality and Safety of Life (I_q), developed by the international organization Economist Intelligence Unit. This index is formed using the following nine indicators: GDP per capita at purchasing power parity; average life expectancy of the population; rating of political stability and security of the country; the number of divorced families per 1,000 population; level of public activity (activity of trade unions, public organizations, etc.); differences in latitude between warmer and colder regions of the country; unemployment rate in the country; the level of political and civil liberties in the country; the ratio between the average wages of men and women.

b) Human Development Index (I_{hd}), used by the United Nations Development Program. It is formed with the help of the following three indicators: the average life expectancy of the country's population; level of education and standard of living of the country's population, measured by GDP per capita at purchasing power parity (GDP per capita).

c) Index of a knowledge-based society, or K-society (I_{ks}), developed by the United Nations Department of Economic and Social Development. This index is determined by three main indicators: intellectual assets of society; development prospects of society and development quality of society, which are formed by 15 data sets on the level of youth education and information, investment climate in the country, corruption, inequality of material and social benefits (GINI-index), the level of child mortality, etc.

4. Methods of forming the humanitarian index (IDX^{HUM} or I_{hum}) [10] An urgent problem is the formation of a methodology for calculating the humanitarian index, which has an available statistical basis and, along with other indices, can be used in modeling and to assess the level of sustainable development of regions. Therefore, the task is to develop a methodology for forming a humanitarian index to assess the level of sustainable development of the region. The basis for the formation of the humanitarian index is directly the input data, which in some way is aggregated into parameters. The parameters are aggregated into indicators, and the indicators already make up the humanitarian index.

The information base for the formation of the humanitarian index was the data of the State Statistics Service of Ukraine for the period from 2000 to 2014. 87 items of input data were selected, which were aggregated into 22 parameters. The parameters were aggregated into 10 indicators, and the indicators were aggregated into one humanitarian index.

In order to be able to compare regions without being too subjective, the values of all input data were calculated per capita of the permanent population of a particular region for a particular year.

Input standardization was performed, after which the "worst" value of the input data corresponded to the value "0", and the "best" – the value "1". When the "best" value of the input data corresponds to their largest value, standardization is carried out according to the formula:

$$ID_i^{ST}(REG_jYR_k) = \frac{ID_i^{OR}(REG_jYR_k)}{\max\limits_{REG_j}(ID_i^{OR}(REG_jYR_k))}, \quad (13)$$

$$i = 1...87, j = 1...27, k = 2000...2014,$$

where ID^{ST} – standardized input data values, ID^{OR} – original input data values, REG – region, YR – year. In the case when the best values of input data correspond to their smallest values, standardization is carried out according to the formula:

$$ID_i^{ST}(REG_jYR_k) = 1 - \frac{ID_i^{OR}(REG_jYR_k)}{\max\limits_{REG_j}(ID_i^{OR}(REG_jYR_k))}, \quad (14)$$

$$i = 1...87, j = 1...27, k = 2000...2014,$$

where ID^{ST} – standardized input data values, ID^{OR} – original input data values, REG – region, YR – year.

"Preliminary" weight coefficients were calculated for the input data $WC^{PR}(ID_j^{ST}(REG_j))$ using the principal component method. The "preliminary" weight coefficients were reduced to a single value by the mean formula:

$$WC(ID_i^{ST}) = 1/27 \sum_j WC^{PR}(ID_j^{ST}(REG_j)), \quad (15)$$

$$i = 1...87, j = 1...27,$$

where WC – input data weight coefficients, WC^{PR} – "preliminary" weight coefficients of input data, ID^{ST} – standardized input data values, REG – region. The sum of the input data weight coefficients for each parameter is always equal to one.

Regarding the calculation of parameter values, the parameter value is equal to the sum of the values of all input data included in it, multiplied by the corresponding weight coefficients.

For example:

$$PRM_1(REG_j, YR_k) = \sum_{i=1}^{3} ID_i^{ST}(REG_jYR_k) \cdot WC(ID_i^{ST}),$$

$$PRM_4(REG_j, YR_k) = \sum_{i=9}^{18} ID_i^{ST}(REG_jYR_k) \cdot WC(ID_i^{ST}),$$

$$PRM_{12}(REG_j, YR_k) = \sum_{i=40}^{44} ID_i^{ST}(REG_jYR_k) \cdot WC(ID_i^{ST}),$$

$$j = 1...27, k = 2000...2014,$$

where PRM – parameter, ID^{ST} – the standardized values of the input data, WC – the weight coefficients of the input data, REG – region, YR – year.

A total of 22 parameters are used. As with the input data, "preliminary" weight coefficients $WC^{PR}(PRM_i(REG_j))$ were calculated for the parameters using the principal component method. The "preliminary" weight coefficients were also reduced to a single value by the mean formula:

$$WC(PRM_j) = 1/27 \sum_j WC^{PR}(PRM_i(REG_j)), \quad (16)$$

$$i = 1...22, j = 1...27,$$

where WC – weight coefficient of the parameter, WC^{PR} – the "preliminary" weight coefficient of the parameter, PRM – parameter, REG – region. The sum of the weight coefficients of the parameters for each indicator is always equal to one.

The values of the indicator are the values of the parameters included in it, multiplied by the corresponding weight coefficients. Here are some of the formulas:

$$IDT_1(REG_j, YR_k) = \sum_{i=1}^{3} PRM_i(REG_j, YR_k) \cdot WC(PRM_i),$$

$$IDT_4(REG_j, YR_k) = \sum_{i=10}^{14} PRM_i(REG_j, YR_k) \cdot WC(PRM_i),$$

$$IDT_{10}(REG_j, YR_k) = \sum_{i=21}^{22} PRM_i(REG_j, YR_k) \cdot WC(PRM_i),$$

$$j = 1...27, k = 2000...2014,$$

where WC – the weight coefficient of the indicator, WC^{PR} – the "preliminary" weight coefficient of the parameter, IDT – indicator, REG – region, YR – year.

For the indicators, "preliminary" $WC^{PR}(IDT_i(REG_j))$ weight coefficients were calculated using the principal component method. The "preliminary" weight coefficients were reduced to a single value using the average value formula:

$$WC(IDT_i) = 1/27 \sum_j WC^{PR}(IDT_j(REG_j)) \quad (17)$$

$$j = 1...10, j = 1...27,$$

where WC – the weight coefficient of the indicator, WC^{PR} – the "preliminary" weight coefficient of the indicator, IDT – indicator, REG – region. The sum of the weight coefficients of the indicators that make up the humanitarian index is always equal to one.

The value of the humanitarian index is equal to the sum of the values of all indicators included in it, multiplied by the corresponding weight coefficients, and is calculated by the formula:

$$IDX^{HUM}(REG_j, YR_k) = \sum_{i=1}^{10} IDT_i(REG_j YR_k) \cdot WC(IDT_i), \quad (18)$$

$$j = 1...27, k = 2000...2014,$$

where IDX^{HUM} – humanitarian index, IDT – indicator, WC – weight coefficient of the indicator, REG – region, YR – year. The values of humanitarian indices $IDX^{H}UM$ were calculated for the period from 2000 to 2014 for the regions of Ukraine.

3 Conclusion

The integration model of sustainable development is shown as a family of models for the creation of integrated information systems of ecological, economic and socio-humanitarian management of various socio-organizational systems and especially economic objects of technogenic nature to ensure sustainable and viable development. The economic approach to modeling sustainable development is considered, which consists of the optimal use of limited resources and applies natural, energy and resource-saving technologies to generate aggregate income. At the same time, the transition to the information society leads to a change in the structure of aggregate capital in favor of human, increasing intangible flows, flows of information and intellectual property.

The problem of sustainable development through 7 main assets that support the viability of socio-environmental and economic systems is considered. The concept of sustainable development, the system of global measurements of sustainable development and the level of sustainable development are defined.

The method of formation of the humanitarian component with the use of available statistics, which is described by the humanitarian index and is included in the models of quality condition indices (IQC), is presented. The use of these methods will increase the efficiency of solutions in the management of technogenic regional production, will increase the efficiency of the use of innovations and will identify areas of innovation strategies of the regions.

References

[1] S. Ramazanov, O. Honcharenko, O. Chernyak, B. Tishkov, R. Ahmedov, SHS Web of Conferences **65**, 06007 (2019)

[2] M.Z. Zghurovskyi, *Stalyy rozvytok u hlobal'nomu i rehional'nomu vymirakh: Analiz za danymy 2005 r. (Sustainable development in the global and regional dimensions: Analysis according to 2005)* (2006)

[3] A.V. Sergienko, Ph.D. thesis, Classic Private University (2017)

[4] K. Schwab, *The fourth industrial revolution* (Currency, 2017)

[5] G.B. Kleiner, Economics and Mathematical Methods **3**, 71 (2013)

[6] T.S. Klebanova, N.A. Kizim, *Models of estimation, analysis and forecasting of socioeconomic systems* (Inzhek, 2010)

[7] A. Bielinskyi, I. Khvostina, A. Mamanazarov, A. Matviychuk, S. Semerikov, O. Serdyuk, V. Solovieva, V. Soloviev, IOP Conference Series: Earth and Environmental Science **628** (2021)

[8] V.H. Pshihopov, M.Y. Medvedev, *Estimation and management in complex dynamical systems* (Fizmatlit, 2009)

[9] L.M. Savchuk, K.F. Kovalchuk, *Economic cybernetics: modeling of socio-economic systems* (Porogi, 2017)

[10] I.M. Lyashenko, *Fundamentals of mathematical modeling of economic, ecological and social processes* (Bogdan, 2006)

 https://doi.org/10.1051/shsconf/202110712001

Machine learning methods application for consumer banking

Andrii Kaminskyi[1,*], *Maryna* Nehrey[2,**], and *Larysa* Zomchak[3,***]

[1]Department of Economic Cybernetics, Taras Shevchenko National University of Kyiv, 64 Volodymyrska Str., Kyiv, 01026, Ukraine
[2]Department of Economic Cybernetics, National University of Life and Environment Science of Ukraine, 15 Heroyiv Oborony Str., Kyiv, 03041, Ukraine
[3]Department of Economic Cybernetics, Ivan Franko National University of Lviv, 1 Universytetska Str., Lviv, 79000, Ukraine

Abstract. Machine learning (ML) methods are effective tools for analysis of many actual problems in modern banking. Increasing growth of data and rapid digitalization underpin the acceleration of ML implementation. These processes are especially noticeable in consumer banking because banks have millions of the retail customers. The first goal of our research is to form an extended review ML application in consumer banking. From one side we have identified the most developed ML methods, which are applied in this segment (for example different types of regressions, fuzzy clustering, neural network, principal component analysis etc.). From the other side, we point out two multi-purpose tools used by banks in consumer segment intensively, namely scoring and clustering. Secondly, our goal is to present some innovative applications of ML methods to the analysis of each task. This includes several applications for scoring models and fuzzy clustering application. All applications are oriented to make banks business processes more effective. Considered applications were realised on real data from the Ukrainian banking industry.

1 Introduction

Machine Learning (ML) is a dynamically growing class of methods that has many successful applications [1, 2]. One of the areas of productive ML using is modern banking and more widely modern financial institutions. The basic reason for such use arises from digitalization and intensively implementing online technologies in the financial sphere. These processes generate Big Data that can be involved in ML processing. Data handling triggers for further development of existing methods and offers new ideas. We want to emphasize some directions in banking (especially in lending banking) in which ML methods were productively used and perspectives of development are fruitful.

The first direction is a concern to generalized scoring methodology. This methodology is highly implemented in different business processes in banking: estimation creditworthiness of borrowers (credit scoring), identifying potentially profitable customers (marketing scoring), anti-fraud systems (fraud scoring), and so on. Almost all ML methods were used for scoring (first of all credit scoring) construction. Today's package of scoring construction methods includes Multivariate Adaptive Regression Splines, Support Vector Machine (SVM), k-nearest-neighbors method, Random Forest (RF), Extreme Gradient Boosting (XGBoost), and, of course, Artificial Neural Network (ANN). In general, ANN maybe now one of the most popular methods in scoring building tools. We should highlight paper [3] where presented comprehensive results of the comparison of ML credit scorings with classical expert-based scorings. All types of scoring are generalized for involving new (online) data. It should be complemented that ordering banks them-selves also getting here.

The second direction where ML methods were multi-skilled implemented in banking is clustering. Modern banks operate with millions of customers. Each customer can be characterized by vectors with hundreds, maybe a thousand components (customer characteristics). Applying ML arises strong possibilities for clustering customers, identification of their behavior and, as consequence, improved productivity of banking services. When a bank wants to realize clustering, it needs to identify "hidden knowledge" that will help to divide customers effectively into a set of clusters. Here dominates the concept of K-Means algorithms. It involves k-mean, improved k-mean, k-medians, applied hierarchical clustering. We design fuzzy clustering in this paper. Of course, the task of clustering banks themselves is also treated by such an ML method.

The third direction is the cybersecurity of banks. This direction ever more actual through moving to online interaction with customers. The logic of applying ML methods for cybersecurity tasks may be based on Gartner's PPDR model [4]. This model points out five categories: prediction, prevention, detection, response, and monitoring. A good overview of cyber-risk and cyber-security problematics for financial institutions can be found in [5]. The

*e-mail: kaminskyi.andrey@gmail.com
**e-mail: marina.nehrey@gmail.com
***e-mail: lzomchak@gmail.com

spectrum of applying ML in this sphere is described here [6].

We tried to systematize methods of ML in the context of the above mentioned directions. The first two considered spheres have economic nature and the third is more technical (but of course concerns economic consequences!). Our focus was concentrated on the first two spheres which have economic nature and are closely connected with business processes.

Part 2 is devoted to the literature review and describes the application of the most effective methods. Part 3 contains some illustrations of applying ML, which was partially presented in our researches. The conclusion involves a point of view for development.

2 Materials and methods

Machine learning is the concept that a computer program can learn and adapt to new data without human interference [7]. Machine learning is a field of artificial intelligence that keeps a computer's built-in algorithms current regardless of changes in the worldwide economy. The application of machine learning methods in economics and banking discussed in researches [8–11]. Machine learning includes the next classes of methods: Supervised Learning, Unsupervised Learning, Reinforcement Learning, Ensemble methods, Neural Networks and Deep Learning (table 1).

Regression. Logistic regression is a classification method that constructs a binary variable result prediction (1/0, True / False, Yes / No, Good / Bad) for a given set of independent variables.

When constructing logistic regression use dummy variables. The maximum likelihood estimation method is used to estimate logistic regression parameters. Logistic regression calculates the probability of Y when realizing certain values of χ. Log-likelihood is the sum of the probabilities associated with the predicted and actual values of Y. Deviations in logistic regression have a distribution of χ^2. Performance indicators for the logistic regression model: Akaike Information Criteria; Null Deviance and Residual Deviance; the error matrix as a tabular representation of actual and predicted values; ROC curve (Receiver Operating Characteristic).

Fuzzy clustering. Fuzzy clustering is a class of algorithms in which the distribution of data points for clustering is not "clear" (but supposes "fuzzy). It is used in the Neural-fuzzy systems to determine fuzzy sets if they are unknown a priori. Fuzzy sets are like projections of clusters on each dimension. An a priori knowledge and cluster analysis combination lets to refine the parameters of the membership function. The disadvantage of this method of determining fuzzy sets is the complexity of their interpretation.

One of the most commonly used methods is the fuzzy c-means (FCM) method. It assigns a fuzzy membership value to each object, based on its distance to the cluster centers. The membership of the data point in that cluster will be higher than its membership in the other clusters if the data point is closer to the center of the cluster.

The FCM method is an iterative procedure of sequentially improving a certain fuzzy initial partition user-defined or automatically generated by a specific heuristic rule. According to the algorithm, on each iteration values of the membership functions of fuzzy clusters and their typical representatives are recursively listed. The FCM method will terminate when a specified a priori finite number of iterations is performed, or when the minimum absolute difference between the values of the membership functions on two consecutive iterations does not fall below some a priori setpoints.

Neural network technologies. Artificial Neural Networks (ANN) are widely used in finance and insurance problems. ANN using is credit scoring construction to good effect. Many different methods are applied in this direction. They are based on the representative types for supervised and unsupervised ANN. The main advantage of ANNs is that dependency between variables does not necessary to characterize. The quality of the ANN applying in credit scoring application can be explained by the Big Data of modern banks.

3 Results and discussion

3.1 Parametric scoring model based on the concept of survival

The classic scoring model is based on the linear function

$$Z = a_0 + a_1x_1 + a_2x_2 + \cdots + a_nx_n \quad (1)$$

where $x_1, x_2, ..., x_n$ – borrowers characteristics; $a_1, a_2, ..., a_n$ are the weights of characteristics that reflect their significance.

Such models have been implemented at all stages of relationships between borrowers and banks in the consumer lending segment. The illustration presents at figure 1.

Our practice indicates that all these scorings in differing degrees have been applied in Ukrainian consumer banking. This models actively enriches by new data from digitalization and online lending last five years. The data from customer behavior on websites, cells using types, and other data can be included in scorings.

One of the typical characteristics of above-mentioned scoring model is a static state. Below we want to present specific approach that includes dynamic consideration. The logic lies in inclusion time parameter t in scoring coefficients. This model will be "dynamic credit scoring model":

$$h(i, x, t) = ln\frac{P(i)}{1 - P(i)} = = a_0(t) + a_1(t)x_{i1} + a_2(t)x_{i2} + \cdots + a_{n(t)}x_{in} \quad (2)$$

$P(i)$ is the probability that the i-th customer (borrower) will be "Good" in the period from 0 to t; x_{ij} is the j-th characteristic of the i-th customer, where $i = 1, \ldots k, j = 1, \ldots n$; $a_j(t)$ is the coefficient of the model at time t, where $j = 1, \ldots n$; t is a time for estimation.

Table 1. Machine learning methods

Class	Subclass	Method	Researches
Supervised Learning	Regression	Linear regression Polynomial regression Ridge lasso/regression	[12–20]
	Classification	Logistic regression Decision trees Support vector machine Naïve Bayes k-nearest neighbor	
Unsupervised Learning	Clustering	k-means Agglomerative Mean-shift Fuzzy C-means DBSCAN	[21–29]
	Rule Engine	Association rules Eclat Apriori FP-growth	
	Dimensionality Reduction	Principal Component Analysis Partial Least Squares Regression Principal Component Regression T-distributed Stochastic Neighbor Embedding Singular Value Decomposition Mixture Discriminant Analysis Linear Discriminant Analysis	
Reinforcement Learning		Genetic algorithm Q-learning SARSA Deep Q-Network	[30, 31]
Ensemble		Boosting Gradient boosting machines Bagging Random forest AdaBoost Stacked Generalization	[32, 33]
Neural Networks and Deep Learning	Convolutional neural network	DCNN	[34–42]
	Recurrent neural network	Liquid State Machine (LSM) Long short-term memory (LSTM) Gated recurrent unit (GRU)	
	Generative Adversarial Networks	GAN	
	Autoencoders	Seq2seq	
	Perceptrons	MLP	

Model (2) differentiate customers much more accurately and qualitatively because it will show the likelihood of not just going out of time but getting out of time at some point in time. This will make it possible to build the financial model more accurately. In addition, if you apply this model to application scoring, it is possible to implement a specific strategy for working with customer (including monitoring, reminder system and other).

Let us illustrate this approach to collection scoring construction. This model was elaborated on Ukrainian data and applied to good effect for collection business processes [43]. Model was constructed on data pool of 50,000 debtors. The data include more than 30 characteristics, including the type and status of credit accounts, a partial history of payments to them, social and demographic parameters of debtors, information about their education and employment. Among the many factors, nine non-correlated indicators were identified, with a significant effect on the dependent variable (table 2).

Statistical significance was verified using χ^2 statistics, the Cramer coefficient, and the Information Value. The dependent variable assumed a value of 1 if the debtor made at least 1 payment during 8 quarters period. Value 0 was

Stages of credit granting processes	Identify profitable borrowers	Identify borrowers which repeat loan	Identify fraud-behavior for receiving loan	Differentiate borrowers on the base application data	Differentiate borrowers on the base credit history	Identify optimal strategy for collection
Scoring types	Marketing scoring	Lead-generation scoring	Fraud scoring	Application scoring	Behavioral scoring	Collection scoring

Figure 1. Stages of relationships between borrowers and banks

Table 2. Indicators of statistical significance of model parameters for the whole period

	Parameter name	Parameters description	χ^2	Kramer coefficient	Information value
x_1	Redemption model	The scheme of repayment 6 months before the purchase of debt portfolio. For example, 101010, where 1 is the payment and 0 is the non-payment. There are 64 such values in total, and they are ordered by probability of the first payment	7980	0.278	0.60
x_2	Non-payment days	The number of days since the last payment was made before buying debt portfolio	2842	0.166	0.50
x_3	Delay days	Number of delays days	6554	0.252	0.41
x_4	Principal and interest	The ratio of principal and interest to the total debt	5187	0.224	0.35
x_5	Term of the loan	Term of the loan at the context number of days after issuence	7373	0.268	0.32
x_6	Paying Principal	The Principal of the loan repaid	3968	0.196	0.30
x_7	Amount of penalties	Amount of all penalties and penalties for late payment	2608	0.159	0.20
x_8	Issue date	Date when loan was issued	2659	0.161	0.13
x_9	Product	Type of credit product (monetary/commodity)	1490	0.120	0.12

used otherwise. Correlation checking was performed using correlation analysis.

The effect of the characteristics influence (table 3) on the probability of debt payment in the dynamics is estimated. At each time moment t, "Information Value" metrics were calculated for each quarter. The probability of the event "Good" occurrence was taken as the dependent variable – the debtor made the first payment within a time period from 0 to t inclusive, (the alternative event "Bad" did not occur), where the parameter t in the model (2) denotes the quarter from 1 to 8, in which this event happened.

Debtors were then assigned scoring points on a scale of 0 to 100. Each group created scoring points: 0–20; 20–40; 40–60; 60–80; 80–100 corresponds to the probability $P(i)$ that the i-th debtor will make his payment in the period from 0 to t, which depends on both the score and the time.

In all models, the normalized R-square – correlation coefficient – close to 1, so the relationship in the models close. Each variable is significant and the model is adequate according to Student's and Fisher's criteria. The magnitude of all coefficients decreases with each quarter, except for the constant that compensates for this decrease.

The basic mathematical differences are time changing in weights. This can be used to collection strategies development while optimizing the costs of implementing it.

3.2 Neural network technologies in debt portfolio management

The ANNs usage provides an efficient classification of bank debtors into groups. Debtors are similar to each other in terms of risk characteristics in each group. The economic logic of this is the management of debt collecting strategies applying. Neural networks allow the classification of debtors from the training sample into groups of a more complex geometric shape than the classical linear discriminant function and visualize it. The most valuable property of ANNs is the ability to learn from multiple examples in cases where patterns are unknown and relationships between input and output are not obvious. In such cases, both traditional statistical and expert methods are ineffective.

Our research of the consumer debtors portfolios involves applying ANNs methodology to estimate debtors. The study founds effectiveness of problem structuring:

 https://doi.org/10.1051/shsconf/202110712001

Table 3. Dynamics of scoring factors over time

Parameter	Coefficient	1	2	3	4	5	6	7	8
Redemption model	$a_1(t)$	0,16	0,09	0,11	0,09	0,10	0,09	0,09	0,09
Non-payment days	$a_2(t)$	0,40	0,35	0,33	0,32	0,31	0,31	0,29	0,29
Delay days	$a_3(t)$	0,25	0,22	0,22	0,20	0,20	0,20	0,19	0,19
Principal and interest	$a_4(t)$	0,26	0,24	0,22	0,22	0,21	0,20	0,20	0,19
Term of the loan	$a_5(t)$	0,20	0,23	0,18	0,18	0,17	0,17	0,16	0,16
Paying Principal	$a_6(t)$	0,22	0,21	0,19	0,19	0,18	0,18	0,17	0,17
Amount of penalties	$a_7(t)$	0,19	0,20	0,17	0,17	0,16	0,16	0,16	0,15
Issue date	$a_8(t)$	0,16	0,20	0,15	0,16	0,14	0,15	0,14	0,13
Product	$a_9(t)$	0,21	0,19	0,18	0,17	0,17	0,16	0,16	0,16
Y-intersection	$a_0(t)$	-7,11	-6,13	-5,45	-5,16	-4,85	-4,71	-4,49	-4,30

– the contact problem with debtors;

– the debtors' insolvency.

The first problem is raised from the fact that a significant part of debtors is non-contact. This does not allow the application of soft-charging techniques and leads to high-value direct contact or using legal procedures, which are also costly. The practice has shown that the proportion of non-contact debtors approximately 70-80%.

The second problem is that some of the contact debtors refuse to pay for various reasons: lack of funds, unwillingness to pay due to high-interest penalties, etc. The implementation model in practice has shown that 40-50% are contact debtors. The ANN was applying twofold as pictured in the scheme (figure 2).

Based on the AANs using two different scorings were constructed: contact scoring and solvency scoring.

In doing so emphasis is put to such characteristics as socio-demographic (age, marital status, educational level, a region of residence, etc.), professional (employment status, vocational qualification, etc.), loan parameters (amount, interest rate, duration, etc.) and, of course, characteristics of overdue [44].

In both cases, scoring is the first step in applying Self Organizing Maps (SOM) or Kohonen cards to the training sample. Kohonen map is a special type of neural network that allows identifying hidden structures and patterns through learning neural networks. A special algorithm performs clustering based on two-dimensional visualization.

Neural networks technology creates a series of clusters which includes homogeneous debtors.

When using Kohonen maps, there is a problem of choosing between detail and visualization. The increasing details in one of these characteristics lead to the deterioration of the other. Really, more detailed consideration complicates economic analysis leads to the difficulty of visualization. On another side, reducing it can lead to the loss of important patterns. Our analysis showed that in considered cases it is suitable to divide from 6 to 9 clusters. The authors' experience shows that in most cases, splitting into 7 clusters is optimal.

The next question is concerned with correlation analysis of the debtor's characteristics. To avoid correlation problems and to include non-dependent characteristics, it is advisable to select only those with a correlation coefficient not exceeding 0.6. Another way concerns applying principal component analysis (PCA). However, the use of PCA, in this case, maybe difficult due to the complexity of their economic interpretation. After correlation analysis, the optimal levels of influence (significance) of characteristics are determined with the help of classification trees to improve model accuracy. Our analysis of different consumer loan debt portfolios leads to identify the most relevant contact characteristics (table 4).

The obtained scoring allows you to sort the debtors by the level of contact probability: the higher the scoring value, the more likely they are to contact him. Based on contact scoring, the following logic for managing arrears may be suggested. Before working remotely with a new portfolio of debtors, the probability of making contact is assessed by considered scoring. All debtors are allocated by three scoring classes: high-contact, medium-low and low-contact. After the introduction of remote work with the portfolio, the debtors are already factually divided into contact and non-contact. As a consequence, all contact information is being worked on further in debt collection. At the same time, debtors with a high value of contact scoring but no contact, in reality, should be continue elaborated for identifying contacts. It is logically not spending the time for those non-contracted debtors who have low contact scoring. Namely, for non-contact debtors with low contact scoring values, the following strategies may be employed:

– write-off strategy if the debt is not large;

– a strategy for obtaining additional information through a request to the credit bureaus. If there is a significant amount of debt, put a low priority on further work with his debtor. If there are no so much open loans, give high priority;

– transfer to legislative recovery, if the amount of debt is considerable.

Applying an approach based on these strategies creates frameworks for seeking optimal allocational resources. Summing up, we can conclude that ANNs are an effective technique for elaborating strategies of prioritizing collection efforts.

https://doi.org/10.1051/shsconf/202110712001

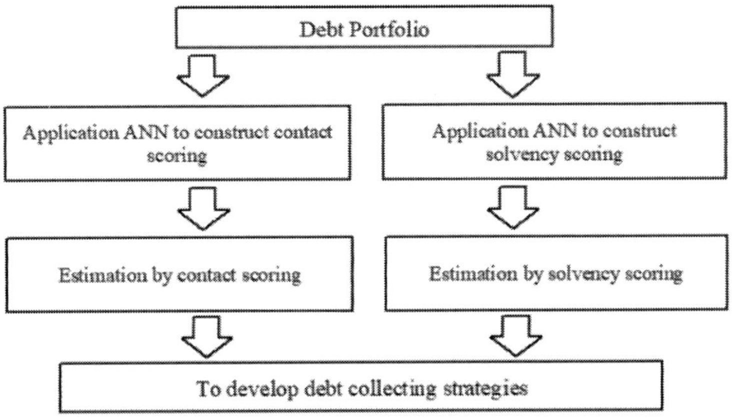

Figure 2. Applying ANN in debt portfolio

Table 4. Contact characteristics

Contact characteristics of the debtor	Medium level of significance debt portfolios
Principal debt and interest / monthly payment	36,55%
Principal debt and interest	22,25%
The time from the last payment to the beginning of the charge	13,04%
Debtor's age	6,54%
The amount paid by the debtor to late payment	6,33%
Loan amount	5,42%
Total / principal and interest	4,33%
The amount of the last payment made by the debtor	3,9%
Number of payments which were paid	1,11%
Become a debtor	0,41%

3.3 Fuzzy clustering of bank's consumer loan portfolio

The intensive development of consumer lending last decades leads to the fact that today banks possess hundreds of thousands or millions of borrowers in their credit portfolios. These are really Big Data. Credit portfolios involve different segments: mortgages, car loans, unsecured consumer loans, credit cards, and others. One of the crucial objectives consists of clustering borrowers. Typically, this corresponds to lead-generation. The objective is to find different clusters of borrowers who successfully paid previous loans. Such different clusters include borrowers with different characteristics, different behaviors, and preferences. So, it is logical to construct a corresponding marketing strategy. The clustering can be done by various approaches which involve choosing economic parameters for clustering basis and mathematical techniques.

Our research in clustering large credit portfolios of banks leads to forming an approach to applying ML for clustering. The first component of our approach corresponds to the identification of basic economic indicators for clustering. The second component is the fuzzy clustering application.

First component realization. We have chosen three indicators at the framework of the first component. They are some unification of risk and profitability indicators. By our logic, these indicators can be calculated for borrowers which have closed loans. The bank is planning to propose some lead-generating product. These indicators are:

– score of credit bureau(s);

– amount of loans which borrower was granted;

– level of overpayment for previous loans.

The economic logic of such indicators is the following.

The score of the credit bureau provides information about the risk level of the borrower. A lower score value (high risk) identifies a "bad" borrower which is not paid or overloaded. It may be logical to exclude such borrowers from further consideration or apply a special approach constructed for the corresponding cluster. A high score value (low risk) identifies a "good" borrower who paying off his/her loans. It will be a nice borrower for lead generation but they are not high profitably. Really, such borrower pays "accurately and timely". The average score value corresponds to the borrower which periodically hits in overdue but then pays all debt with a penalty, fees, and others. In reality, it is more profitable. Of course, it is average consideration. Different types may be in this category.

The second indicator is the loan amount. The basic economic logic here: low amount generates low profit for the bank but more often overpayment.

Third our indicator reflects the profitability of the borrower. We were following conceptually approach of Storbacka [45] for separate customers and provided extension it for borrowers. According to our approach borrowers can be divided into four classes:

A – borrowers with high overpayments,

B – borrowers which pay "accurately and timely",

C – there was some payments but the amount of the loan was not paid,

D – FPD (First payment default and no any payments).

Of course, A-type is more profitable. D, in any case, should be excluded.

Second component realization. The abovementioned estimations of borrowers arise tasks for clustering. This clustering should consider the risk of insolvency, profit from an overpayment, and the amount of loan. Clustering is economically significant because it is possible to construct some (marketing) strategy for each cluster. The consideration classical C-means approach leads to some "strict" separation. Very often it may be not optimal for customers. Because some borrowers can "fuzzy" belong to different clusters with different values of membership functions. So, we applied fuzzy C-means clustering. The conceptual difference between approaches is illustrated below figure 3.

The economic benefits are construction more effective lead-generating strategies for such customers. Really, the classical C-mean clustering will be 3 clusters (if we specify 3), but fuzzy C-mean clustering provides 7 clusters. This allows forming a more adequate set of strategies.

So, fuzzy clustering leads to a more advanced approach for cluster creation which adequate forming strategies of lead-generation.

4 Conclusion

Machine Learning has been successfully developing in many spheres. Sometimes, the applications are impressive. Modern banking really excellent "test site" for different ML methods and techniques. This is mainly due to Big Data and understanding the practical importance of applying ML. Really, the average retail bank typically has an intensive inflow of customers and very many customers in the portfolio. Each customer as from inflow as from portfolio can be characterized by hundreds of indicators. This is Big Data with hidden patterns of customer behavior and preferences.

Tools for business development in consumer banking are scorings and clustering. It is absolutely logical to apply ML and AI for the solution of objectives to construct effective scorings assessment and clustering procedures. The results are multifaced. Our paper illustrated a couple of solutions for consumer banking based on Machine learning applications. All these solutions were implemented and indicate effectiveness. The huge growth of data according to digitalization in banking and developing fintech produce new objectives and new spheres for applying ML and AI.

Machine learning methods are self-developing areas of researches with the synergetic interaction between them. This fact has been considered in the mentioned examples.

References

[1] A. Kiv, V. Soloviev, S. Semerikov, H. Danylchuk, L. Kibalnyk, A. Matviychuk, CEUR Workshop Proceedings **2422**, 1 (2019)

[2] A. Kiv, P. Hryhoruk, I. Khvostina, V. Solovieva, V. Soloviev, S. Semerikov, CEUR Workshop Proceedings **2713**, 1 (2020)

[3] L. Munkhdalai, T. Munkhdalai, O.E. Namsrai, J.Y. Lee, K.H. Ryu, Sustainability **11**, 699 (2019)

[4] P. Ganapathi, D. Shanmugapriya, *Handbook of Research on Machine and Deep Learning Applications for Cyber Security* (IGI Global, 2019)

[5] A. Bouveret, *Cyber risk for the financial sector: A framework for quantitative assessment* (International Monetary Fund, 2018)

[6] C. Chio, D. Freeman, *Machine learning and security: Protecting systems with data and algorithms* (O'Reilly Media, Inc., 2018)

[7] Investopedia.com, *Machine learning* (2021), `https://www.investopedia.com/terms/m/machine-learning.asp`

[8] M. Leo, S. Sharma, K. Maddulety, Risks **7**, 29 (2019)

[9] V. Babenko, A. Panchyshyn, L. Zomchak, M. Nehrey, Z. Artym-Drohomyretska, T. Lahotskyi, WSEAS Transactions on Business and Economics pp. 209–217 (2021)

[10] A. Kiv, V. Soloviev, S. Semerikov, H. Danylchuk, L. Kibalnyk, A. Matviychuk, CEUR Workshop Proceedings **2422**, 1 (2019)

[11] N. Volkova, N. Rizun, M. Nehrey, *Data science: Opportunities to transform education* (CEUR-WS, 2019), Vol. 2433, pp. 48–73, ISSN 16130073

[12] C. Brownlees, C. Hans, E. Nualart, Journal of Monetary Economics **117**, 585 (2021)

[13] S. Tungsong, F. Caccioli, T. Aste, *Relation between regional uncertainty spillovers in the global banking system* (2017)

[14] F. Shofiyah, A. Sofro, Journal of Physics: Conference Series **1108**, 012107 (2018)

[15] V. Granaturov, V. Kaptur, I. Politova, Economic Annals-XXI pp. 52–56 (2015)

[16] V. Granaturov, V. Kaptur, I. Politova, Economic Annals-XXI pp. 83–87 (2016)

[17] N. Davydenko, A. Buriak, Z. Titenko, Intellectual Economics **13** (2019)

[18] N. Klymenko, O. Nosovets, L. Sokolenko, O. Hryshchenko, T. Pisochenko, Academy of Accounting and Financial Studies Journal **23** (2019)

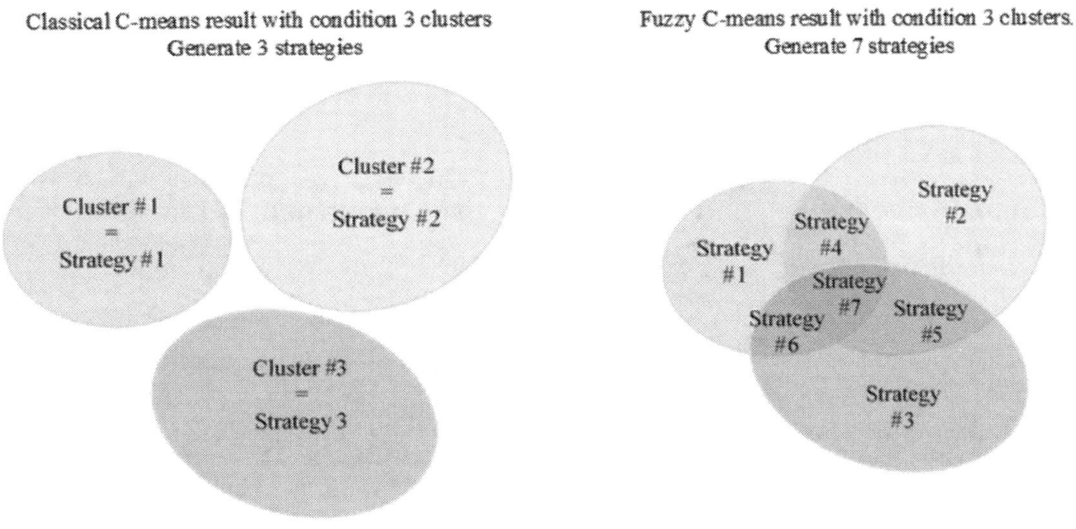

Figure 3. Classical vs Fuzzy C-means clustering

[19] M. Kuzheliev, D. Zherlitsyn, I.I. Rekunenko, A. Nechyporenko, G. Nemsadze, Banks and Bank Systems **15**, 94 (2020)

[20] O. Kuzmenko, P. Šuleř, S. Lyeonov, I. Judrupa, A. Boiko, Journal of International Studies **13**, 332 (2020)

[21] F. Butaru, Q. Chen, B. Clark, S. Das, A.W. Lo, A. Siddique, Journal of Banking & Finance **72**, 218 (2016)

[22] H. Ercan, S. Sayaseng, *The cluster analysis of the banking sector in Europe*, in *Economics and Management of Global Value Chains* (2016), pp. 111–127

[23] L. Fang, B. Xiao, H. Yu, Q. You, Physica A: Statistical Mechanics and its Applications **492**, 1997 (2018)

[24] P. Gogas, T. Papadimitriou, A. Agrapetidou, International Journal of Forecasting **34**, 440 (2018)

[25] H. Danylchuk, O. Kovtun, L. Kibalnyk, O. Sysoiev, E3S Web of Conferences **166** (2020)

[26] O. Kuzmenko, S. Kyrkach, Banks and Bank Systems **9** (2014)

[27] N. Khrushch, P. Hryhoruk, T. Hovorushchenko, S. Lysenko, L. Prystupa, L. Vahanova, CEUR Workshop Proceedings **2713**, 239 (2020)

[28] Y. Hrabovskyi, V. Babenko, O. Al'Boschiy, V. Gerasimenko, WSEAS Transactions on Business and Economics **17**, 231 (2020)

[29] K.B. Cyree, T.R. Davidson, J.D. Stowe, Journal of Economics and Finance **44**, 211 (2020)

[30] L. Guryanova, R. Yatsenko, N. Dubrovina, V. Babenko, CEUR Workshop Proceedings **2549**, 1 (2020)

[31] V. Derbentsev, A. Matviychuk, N. Datsenko, V. Bezkorovainyi, A. Azaryan, CEUR Workshop Proceedings **2713**, 435 (2020)

[32] K. Tanaka, T. Kinkyo, S. Hamori, Economics Letters **148**, 118 (2016)

[33] V. Ravi, H. Kurniawan, P.N.K. Thai, P.R. Kumar, Applied soft computing **8**, 305 (2008)

[34] N. Mohammadi, M. Zangeneh, IJ Information Technology and Computer Science **8**, 58 (2016)

[35] M. Alborzi, M. Khanbabaei, International Journal of Business Information Systems **23**, 1 (2016)

[36] S. Saha, S. Waheed, International Journal of Computer Applications **161**, 39 (2017)

[37] A. Matviychuk, I. Strelchenko, S. Vashchaiev, H. Velykoivanenko, CEUR Workshop Proceedings **2393**, 485 (2019)

[38] J. Asare-Frempong, M. Jayabalan, *Predicting customer response to bank direct telemarketing campaign*, in *2017 International Conference on Engineering Technology and Technopreneurship (ICE2T)* (2017), pp. 1–4

[39] W. Kanmani, B. Jayapradha, International Journal on Recent and Innovation Trends in Computing and Communication **5**, 293 (2017)

[40] E.A.E. Dawood, E. Elfakhrany, F.A. Maghraby, IEEE Access **7**, 109320 (2019)

[41] K. Lei, Y. Xie, S. Zhong, J. Dai, M. Yang, Y. Shen, Neural Computing and Applications **32**, 8451 (2020)

[42] A. Petropoulos, V. Siakoulis, E. Stavroulakis, N.E. Vlachogiannakis, International Journal of Forecasting **36**, 1092 (2020)

[43] A. Kaminsky, K. Pisanets, Formation of market economy in Ukraine pp. 136–142 (2012)

[44] A. Kaminsky, V. Sikach, Modeling and information systems in economics **84** (2011)

[45] K. Storbacka, Journal of Marketing Management **13**, 479 (1997)

Economic-mathematical model for complex risk assessment of the enterprise investment project using fuzzy logic

Inna Chaikovska[1,2,*], *Pavlo* Hryhoruk[2,**], and *Maksym* Chaikovskyi[2,***]

[1]Leonid Yuzkov Khmelnytskyi University of Management and Law, 8 Heroiv Maidanu str., Khmelnytskyi, 29000, Ukraine
[2]Khmelnytskyi National University, 11 Instytutska Str., Khmelnytskyi, 29016, Ukraine

Abstract. The article proposes an economic-mathematical model for determining a comprehensive risk assessment of the investment project of the enterprise which are based on the approaches of A. Nedosekin. The model is built using fuzzy logic and takes into account the probability of occurrence of each of the identified risks and the level of impact of each of them on the project. The probability of risk is set by experts in the form of points and converted into linguistic terms, and the level of influence of each of them on the project – the ratio of benefits and is determined using Fishburne scales. The proposed Project Risk Model consists of the following stages: formation of initial data using expert opinions; construction of a hierarchical project risk tree; determination of weight coefficients (Fishburne weights) of project risks; selection and description of membership function and linguistic variables; conversion of input data provided by experts from a score scale into linguistic terms; recognition of qualitative input data on a linguistic scale; determination of a complex indicator of investment project risks; interpretation of a complex indicator. The developed model allows managing the risks of the project to maximize the probability of its successful implementation, to compare alternative projects and choose less risky, to minimize the level of unforeseen costs of the project.

1 Introduction

1.1 Problem description

The following subsystems are distinguished in the project management system: time management [1], labor resources, cost, information and communications, quality, project risks, etc. Project risk management is one of the most important subsystems of project management because it allows at the planning stage of the project to identify problematic issues for its successful implementation. The comprehensive risk assessment of the project allows you to take into account the most significant risks for the project and quantify them to make an effectively-informed management decision. Quite often when assessing the risks of enterprise projects, the information is non-numerical, it is necessary to take into account both quantitative and qualitative information in one system and the formation of a single quantitative comprehensive indicator. To form a comprehensive risk assessment of the project, it is advisable to use fuzzy logic and obtain a comprehensive numerical risk assessment of the project.

If we consider the methods of risk identification, there will be many approaches only to the classification of risks. According to the common in foreign countries approach Construction Risk Management System (CRMS), proposed by American analysts, the process of risk identification consists of six stages: detection of uncertainties; compiling a preliminary checklist; consequence scenarios; a reflection of risks; systematic list of risks; total risk. Therefore, determining the total risk (complex value of project risks) is an urgent task. Besides, a comprehensive assessment should take into account both the probability of a risk event and the level of its impact on project implementation (weights). The obtained comprehensive assessment will determine whether it is appropriate to implement the project, what risks are most likely, and the level of impact on the successful implementation of the project, reduce the cost of the project, as well as make an effectively-informed management decision.

1.2 Literature review

The paper [2] identifies the need for effective methods of risk management in enterprises in modern conditions because the profitability of the enterprise largely depends on the level of risk. Also, risk allows you to assess the internal reserves of the enterprise and the level of risk depends on the feasibility of a particular financial and economic transaction. Risk management makes it possible to establish forecast quantitative estimates of economic performance of the enterprise.

The paper [3] states that in modern conditions it is necessary to use not only general methods of evaluation of investment projects. The use of various methods of evaluation of investment projects in one system will allow to make an informed management decision on the feasibility of their further implementation and ensure their flexibility.

*e-mail: inna.chaikovska@gmail.com
**e-mail: violete@ukr.net
***e-mail: larumlab@gmail.com

https://doi.org/10.1051/shsconf/202110712002

The paper [4] specifies that an investment project is a set of measures, works, and documents, the financial result of which is profit (income). The material result of the investment project is new or reconstructed fixed assets (objects). The project may also result in the acquisition and use of financial instruments, intangible assets followed by income or social impact. An investment project is an activity that involves the implementation of any action to achieve specific goals.

The paper [5] emphasizes that risk management requires a systematic assessment of the severity of risks affecting the project. The article notes the need to use different qualitative and quantitative methods of risk assessment of investment projects and their distinctive features, advantages, and disadvantages. The necessity of choosing the most expedient method of risk assessment for a specific project is determined.

The paper [6] assesses the risks to real estate with the construction of a fuzzy mathematical model. The authors emphasize the need to use several methods of quantitative risk assessment due to a certain subjectivity of the developed model.

The paper [7] considers the CAPM model for emerging capital markets and the DCF method, which allows assessing the attractiveness of both the business in general and the investment project of the enterprise in particular. The application of risk assessment methods to an investment project helps to assess its feasibility, the period when it will start to make a profit, and its level in the future.

In the paper [8] it is offered to use a method of expert estimations for estimation of risks of the investment project. This approach allows you to quantify the risks of the project, rank them and obtain a comprehensive indicator of risk assessment of the investment project.

The paper [9] proposes a methodology that modifies the capital pricing model (CAPM) using the discounted cash flow method. This approach allows you to assess the effectiveness of investment projects, risks, and management based on the information obtained.

The paper [10] analyzes the investment project for the construction of a greenhouse complex for growing vegetables. To achieve the planned key economic indicators of the project, it is necessary to reduce the negative impact of risk factors. Since the project did not provide for the analysis of project risks, namely there was no system of identified risks, qualitative and quantitative risk assessment, sensitivity analysis, break-even point, etc., the authors note the need for project risk management, namely mandatory funding conditions for project analysis risks.

The paper [11] proposes a mechanism for assessing and prioritizing risks. This will allow managers to make informed management decisions regarding risk prevention and speed up the recovery time from real risks.

In the paper [12] it is offered to consider risk in system stratification metamodeling system. The authors propose to increase the investment attractiveness, profitability, and competitiveness of the enterprise by reducing various types of costs by implementing a BPM-system based on the proposed stratification metamodeling system. A mandatory component of this system is a subsystem that takes into account the risk.

The paper [13] presents a model for assessing the risks of investment projects using Bayesian networks. The application of this approach is explained by the presence of different types of uncertainty and the need to formalize and process information taking into account uncertainty.

The paper [14] examines international risks and their impact on the implementation of the investment project. The authors compare the traditional NPV model and the NPV model taking into account the weight of entropy as a risk assessment of the investment project. The second model showed the best results and is offered as a basis for decision making.

The paper [15] uses a comprehensive assessment based on fuzzy logic to assess the security risk of an energy investment project. The model includes the definition of the factor characteristics of the object of evaluation, the establishment of a set of estimates, the establishment of a relationship matrix, the calculation of the weight of the index, fuzzy complex assessment. The practical implementation of the model demonstrated a risk value of 3.1154 (average risk).

In the paper [16], three models were developed to accurately predict the quality of project planning, based on both deterministic and fuzzy concepts, and the results show that the fuzzy model is more accurate and realistic than the deterministic one. Thus, the correct use of fuzzy theory will develop more accurate, realistic, and reliable models than deterministic ones.

Despite the significant achievements of scientists in the direction of risk assessment of investment projects of the enterprise, the further study requires a comprehensive assessment of investment risks using economic and mathematical modeling, namely fuzzy logic. After all, it is necessary to take into account both the probability of occurrence of each of the risks and their weights, the ability to take into account both qualitative and quantitative information in one system, comparing several alternative projects. This is necessary to manage the investment risks of the project, minimize unplanned costs and increase the level of competitiveness of the enterprise. In this regard, it is proposed to conduct a comprehensive risk assessment of the investment project using fuzzy set theory according to the methodology of determining a comprehensive assessment of the risk of bankruptcy of the enterprise A. Nedosekin [17], which is adapted to determine a comprehensive risk assessment of the investment project.

Fuzzy logic and fuzzy set theory have received a wide range of successful practical applications in various spheres of life. For example, in the paper [18] it is proposed to choose a solution in the presence of a large number of efficiency criteria based on fuzzy preference relations. This approach allows you to form a set of alternatives based on selected performance criteria, choose the best and make an effective management decision.

In the paper [19], economic and mathematical models for diagnosing the bankruptcy of the enterprise using the methods of fuzzy logic and developed a comprehensive analysis of the financial condition of the enterprise.

Also, the application of fuzzy logic in the financial sector is presented in [20].

Special attention in today's conditions deserves research, which reflects the use of fuzzy logic to stabilize the epidemiological situation with COVID-19 [21].

1.3 The aim and objectives of the research

The study aims to develop an economic-mathematical model for determining a comprehensive risk assessment of an investment project based on fuzzy logic, taking into account the probability of each of the identified risks and the level of impact of each of them on the project.

To achieve this goal the following tasks were solved:

1) set the task and develop a descriptive model for determining a comprehensive risk assessment of the project;
2) to build a mathematical model of the problem of determining a comprehensive risk assessment of the project and to develop an algorithm for its solution;
3) generate input data using the opinions of experts;
4) build a hierarchical project risk tree;
5) determine the weights (Fishburn weights) of project risks;
6) select and describe membership functions and linguistic variables;
7) convert the initial data provided by experts from a score scale into linguistic terms;
8) to recognize qualitative input data on a linguistic scale;
9) determine a comprehensive risk indicator of the investment project;
10) to analyze the obtained complex indicator.

2 Research methodology

The proposed model can be implemented with the following input data:

- a set of expert assessments of the probability of project risks;
- hierarchy of existing project risks (a hierarchical tree of logical conclusion);
- system of relations of advantages of some risks over others (for one level of hierarchy).

The initial data of the model are:

- comprehensive quantitative risk assessment of the project;
- interpretation of the obtained complex project risk indicator.

The model involves the use of elements of fuzzy logic (logical inference tree, membership function, linguistic terms). Expert knowledge and the Fishburn method of scales were also used to determine the project risk weights.

Problem statement and development of a descriptive model for determining a comprehensive risk assessment of the project.

The task of determining a comprehensive risk assessment of the project within this study is implemented by a construction company, which assesses the risks of a new investment project for the construction of a residential complex. The experts identified the following types of risk, which are typical for the project of residential complex construction, and their components that affect the results of the project and selected for the formation of a comprehensive assessment (R_0):

1. Technical R_1 (risks of reassessment of project sustainability $R_{1.1}$; risks associated with the reassessment of additional opportunities for project development $R_{1.2}$).
2. External R_2 (risks of incorrect assessment of demand for the project $R_{2.1}$; risks associated with the nature of competition in the market $R_{2.2}$; risks associated with the solvency of the customer $R_{2.3}$; risks of the uncertainty of the external environment of the project $R_{2.4}$).
3. Organizational R_3 (risks for estimating the costs of project commercialization $R_{3.1}$; risks of potential losses from project implementation $R_{3.2}$; risks of underestimation of project development costs $R_{3.3}$; risks of the uncertainty of the internal project environment $R_{3.4}$).

Each of the risks has its probability of occurrence and level of impact on the successful implementation of the project.

It is necessary to determine a comprehensive indicator of the risk level of the project to make a management decision on the feasibility of this investment project and the necessary actions to increase the probability of successful project implementation, taking into account existing risks.

Mathematical model of the problem of determining a comprehensive risk assessment of the project and the algorithm for its solution.

In this situation, for a comprehensive assessment of project risks, it is advisable to consider the PRM model (Project Risk Model):

$$PRM = < F, A, R >, \quad (1)$$

where F – the hierarchy of existing project risks (a hierarchical tree of logical conclusion);

A – a set of qualitative assessments of each factor in the hierarchy (linguistic terms);

R – a system of relations of advantages of some risks over others (for one level of hierarchy).

In this case:

$$A = ((L), (LM), (M), (HM), (H)), \quad (2)$$

where L - Low, LM – LowMedium, M – Medium, HM – HighMedium, H – High.

$$R = Ri(r)Rj|r \in (\succ, \approx), \quad (3)$$

where $\succ$ – the ratio of preference;

where $\approx$ – equilibrium ratio.

The proposed model consists of the following stages:

Stage 1. Formation of initial data with the use of expert opinions.

Experts estimate the probability of risks using a scale from 0 to 100 points. For each indicator, points are added and the average is determined C_i:

$$C_i = \frac{\sum_{i=1}^{n} C_{ij}}{N}, \quad (4)$$

where N – the number of interviewed experts;

C_{ij} – the sum of points for each indicator.

Next, the concordance coefficient is determined by the formula:

$$W = \frac{\sigma_f^2}{\sigma_{max}^2} = \frac{\sum_{i=1}^{m}\left(a_i - \frac{1}{2}n \cdot (m+1)\right)^2}{\frac{1}{12}n^2 \cdot m \cdot (m^2 - 1)}, \quad (5)$$

where σ_f^2 – the actual variance (SD) of the final (ordered, ranked) estimates provided by experts;

σ_{max}^2 – dispersion of final (ordered) assessments, provided that the opinions of experts completely coincide;

a_i – the total estimate obtained by the i-th object;

m – the number of studied objects;

n – the number of experts.

The materiality of the concordance coefficient is checked using χ^2 with $(m - 1)$ the number of degrees of freedom. Statistical characteristics are calculated by the formula:

$$\chi^2 = W \cdot n(m-1), \quad (6)$$

Stage 2. Construction of a hierarchical risk tree of the project using the system of relations of advantages.

Stage 3. Determination of weights (Fishburne weights). For a system of declining benefits N alternatives:

$$p_i = \frac{2(N-i+1)}{(N+1)N}, \qquad i = 1..N. \quad (7)$$

A system of equivalent N alternatives – a set of identical weights:

$$p_i = N^{-1}, \qquad i = 1..N. \quad (8)$$

Stage 4. Selection and description of membership function and linguistic variables. Selected triangular membership function:

$$\mu(x) = \begin{cases} 0, & x \le a \\ \frac{x-a}{b-a}, & a \le x \le b \\ \frac{c-x}{c-b}, & b \le x \le c \\ 0, & c \le x \end{cases}, \quad (9)$$

where $\mu(x)$ – the membership function of linguistic terms (0 – does not belong, 1 – belongs to all 100%);

a, b, c are some numerical parameters that take arbitrary actual values and are ordered by the relation $a \le b \le c$. Parameters a and c characterize the base of the triangle, and parameter b is its vertex.

Stage 5. Transformation of the initial data provided by experts on the probability of occurrence of each of the risks, from a score scale to linguistic terms. The obtained average scores of the probability of occurrence of each of the risks are translated into linguistic terms according to the selected membership function.

Stage 6. Recognition of qualitative input data on a linguistic scale:

$$Z^*(a) = (\mu_1^*(a), \mu_2^*(a), \mu_3^*(a), \mu_4^*(a), \mu_5^*(a)), \quad (10)$$

where a – the value of the factor to be recognized;

$\mu_i^*(a)$ is membership function with linguistic terms i;

$\mu_i^*(a)$ is determined by the formula (9).

Stage 7. Determination of a complex indicator.

First, you need to convert all vectors $Z^*(x^*)$ in the hierarchy F with weight P according to the formula:

$$\sum_{i=0}^{N} p_i \cdot (\mu_{i\cdot 1}, \mu_{i\cdot 2}, \mu_{i\cdot 3}, \mu_{i\cdot 4}, \mu_{i\cdot 5}) = \\ = (\sum_{i=1}^{N} p_i \cdot \mu_{i\cdot 1}, \sum_{i=1}^{N} p_i \cdot \mu_{i\cdot 2}, \sum_{i=1}^{N} p_i \cdot \mu_{i\cdot 3}, \\ \sum_{i=1}^{N} p_i \cdot \mu_{i\cdot 4}, \sum_{i=1}^{N} p_i \cdot \mu_{i\cdot 5}) \quad (11)$$

where i – possible options for determining the membership function for each linguistic term.

Next, the vector can be determined, which will characterize the complex risk assessment of the project (taking into account the selected triangular membership function):

$$A\_N = \sum_{i=1}^{5} (0.25i - 0.25) \cdot \mu_{0i}, \quad (12)$$

where $(0.25i–0,25) = (0; 0.25; 0.5; 0.75; 1)$ – nodal points of the triangular membership function, in which it is equal to 1 on a scale from 0 to 1 (nodal points at which the membership function refers to a certain linguistic term for 100 %) according to the formula (14 – 18) and figure 2.

Stage 8. Interpretation of the obtained complex indicator.

3 Results and discussion

Consider the implementation of the model according to the problem.

Stage 1. Formation of initial data with the use of expert opinions.

Table 1 provides information provided by experts regarding the assessment of the probability of occurrence of each of the selected risks of the project on a scale from 0 to 100 points (ascending).

The concordance coefficient according to formula (5) is 0.7048, which indicates that the opinions of experts are consistent.

https://doi.org/10.1051/shsconf/202110712002

Checking the materiality of the concordance coefficient according to the formula (6) $\chi^2 = 63.44$.

The data in table (χ^2) for (10-1) degrees of freedom and confidence probability (= 0.95, = 0.99, = 0.999) show that the calculated value of the Pearson criterion χ^2 is greater than the tabular (respectively 16.92; 21.67, and 27.88), which confirms the conclusion that experts agree.

Stage 2. Construction of a hierarchical tree of risks of the project using the system of relations of advantages.

The experts provided the following information on the risk-benefit ratios (figure 1).

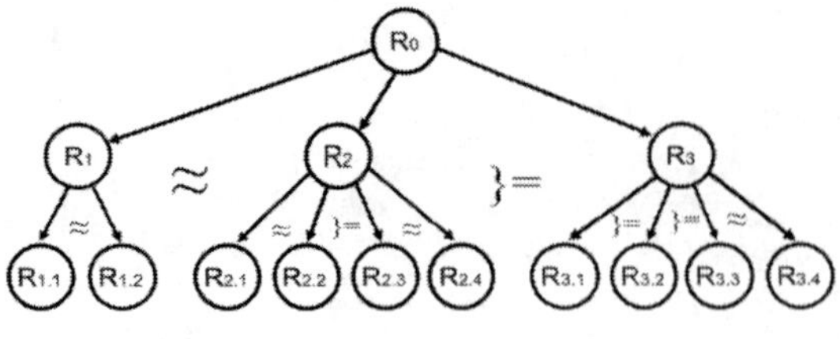

Figure 1. Hierarchical risk tree of the project with the indication of the system of relations of advantages

Figure 1 corresponds to the system of relations R (formula 13).

$$R = (R_1 \approx R_2 \succ R_3; R_{1.1} \approx R_{1.2}; R_{2.1} \approx \approx R_{2.2} \succ R_{2.3} \approx R_{2.4}; R_{3.1} \succ R_{3.2} \succ R_{3.3} \approx R_{3.4} \quad (13)$$

The obtained system of preference ratios can be used to determine the risk weights of Fishburne weights.

Stage 3. Determination of weights (Fishburne weights)

According to formulas (7) and (8), as well as the system of preference relations (figure 1) formed a system of Fisher scales (table 2).

After convolving the weights of different levels of the hierarchy, the following generalized weights (level of influence) were obtained: $R_{1.1}$ = 0.2000; $R_{1.2}$ = 0.2000; $R_{2.1}$ = 0.1333; $R_{2.2}$ = 0.1333; $R_{2.3}$ = 0.0667; $R_{2.4}$ = 0.0667; $R_{3.1}$ = 0.0857; $R_{3.2}$ = 0.0571; $R_{3.3}$ = 0.0286; $R_{3.4}$ = 0.0286.

The highest weights are in risks $R_{1.1}$, $R_{1.2}$, the lowest in $R_{3.3}$, $R_{3.4}$.

Stage 4. Selection and description of membership function and linguistic variables

The linguistic variable "Risk Level" was formed with a term set of values of A (formula 2). The triangular membership function (figure 2) with the following linguistic terms was chosen as membership functions: Low (L), Low Medium (LM), Medium (M), High Medium (HM), High (H), distributed on a scale from 0 to 100 points.

The following system of equations corresponds to this membership function with linguistic terms:

$$L : \mu_1(x) = \begin{cases} \dfrac{25-x}{25}, & 0 \le x \le 25 \\ 0, & 25 \le x \end{cases} . \quad (14)$$

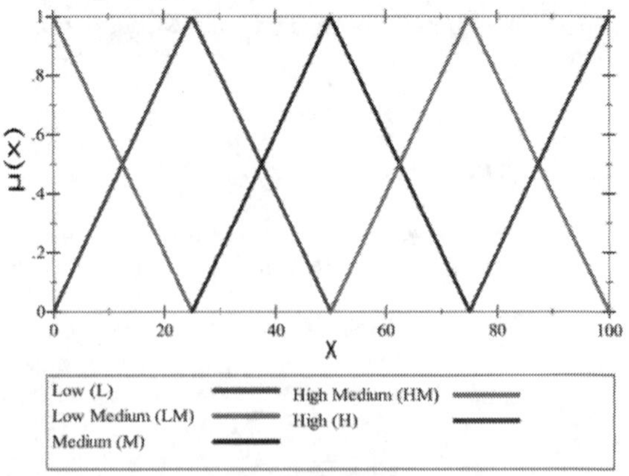

Figure 2. Triangular membership functions with linguistic terms Low (L), Low Medium (LM), Medium (M), High Medium (HM), High (H)

$$LM : \mu_2(x) = \begin{cases} \dfrac{x}{25}, & 0 \le x \le 25 \\ \dfrac{50-x}{25}, & 25 \le x \le 50 \\ 0, & 50 \le x \end{cases} . \quad (15)$$

$$M : \mu_3(x) = \begin{cases} 0, & x \le 25 \\ \dfrac{x-25}{25}, & 25 \le x \le 50 \\ \dfrac{75-x}{25}, & 50 \le x \le 75 \\ 0, & 75 \le x \end{cases} . \quad (16)$$

$$HM : \mu_4(x) = \begin{cases} 0, & x \le 50 \\ \dfrac{x-50}{25}, & 50 \le x \le 75 \\ \dfrac{100-x}{25}, & 75 \le x \le 100 \end{cases} . \quad (17)$$

$$H : \mu_5(x) = \begin{cases} 0, & x \le 75 \\ \dfrac{x-75}{25}, & 75 \le x \le 100 \end{cases} . \quad (18)$$

Formulas (14-18) are based on formula 9 and figure 2.

Stage 5. Transformation of the initial data provided by experts on the probability of occurrence of each of the risks, from the score scale to the linguistic terms.

The scores of the probability of occurrence of each of the risks from table 1 are translated into linguistic terms (according to figure 2): Low (L), Low Medium (LM), Medium (M), High Medium (HM), High (H) (table 3).

According to the table, it can be seen that among the risks are risks, both with a low probability of occurrence and with a very high probability. The indication of two linguistic terms for risk indicates that the obtained average scores of experts are on the border of two terms.

Stage 6. Recognition of qualitative input data on a linguistic scale

According to formulas (14-18) the input data was recognized according to the linguistic scale (table 4).

Table 1. Expert assessment of the probability of occurrence of each of the risks of the project in points

Risk	Ex1	Ex2	Ex3	Ex4	Ex5	Ex6	Ex7	Ex8	Ex9	Ex10	Values sum	Average value
$R_{1.1}$	40	45	55	60	50	60	40	55	45	60	510	51.0
$R_{1.2}$	60	65	75	70	80	70	75	65	55	60	675	67.5
$R_{2.1}$	15	20	20	30	30	35	40	45	45	35	315	31.5
$R_{2.2}$	60	80	75	65	85	100	85	90	95	100	835	83.5
$R_{2.3}$	10	15	15	25	40	35	30	45	25	15	255	25.5
$R_{2.4}$	60	45	55	50	60	40	40	55	60	70	535	53.5
$R_{3.1}$	70	80	90	60	55	85	70	65	60	80	715	71.5
$R_{3.2}$	55	60	60	45	80	65	45	75	80	85	650	65.0
$R_{3.3}$	25	15	15	10	20	30	35	15	40	40	245	24.5
$R_{3.4}$	95	90	100	70	80	65	55	85	90	100	830	83.0

Table 2. Fishburne system of scales

N	R	p_1	p_2	p_3	p_4
2	$R_{1.1} \approx R_{1.2}$	1/2	1/2	-	-
3	$R_1 \approx R_2 > R_3$	2/5	2/5	1/5	-
4	$R_{2.1} \approx R_{2.2} > R_{2.3} \approx R_{2.4}$	2/6	2/6	1/6	1/6
4	$R_{3.1} > R_{3.2} > R_{3.3} \approx R_{3.4}$	3/7	2/7	1/7	1/7

In the table, the cells corresponding to the values of linguistic variables obtained using risk analysis models are given the recognized value according to formulas (14-18). In other cells "0" is put.

Thus, the probability of risk R1.1 in accordance with expert opinions and linguistic terms refers to the level of Medium (96%), High Medium (4%); $R_{1.2}$ – Medium (30%), High Medium (70%); $R_{2.1}$ – Low Medium (74%), Medium (26%); $R_{2.2}$ – High Medium (66%), High (34%); $R_{2.3}$ – Low Medium (98%), Medium (2%); $R_{2.4}$ – Medium (86%), High Medium (14%); $R_{3.1}$ – Medium (14%), High Medium (86%); $R_{3.2}$ – Medium (40%), High Medium (60%); $R_{3.3}$ – Low (2%), Low Medium (98%); $R_{3.4}$ – High Medium (68%), High (32%).

Stage 7. Determination of a complex indicator

At the next stage, a comprehensive risk indicator of the investment project is determined, based on fuzzy sets. According to formula (11) found the vector:

$$\sum_{i=0}^{N} p_i \cdot (\mu_{i\cdot 1}, \mu_{i\cdot 2}, \mu_{i\cdot 3}, \mu_{i\cdot 4}, \mu_{i\cdot 5}) =$$
$$= (0.0006; 0.1920; 0.3802; 0.3727; 0.0545).$$

But according to the selected membership function, the nodal points in which the membership function is 1 are equal to (0; 25; 50; 75; 100). To find the integral exponent, use formula (12) and multiply the corresponding exponents of both vectors and find their sum.

$$A\_N = 51.8194 = R_0.$$

Thus, the complex risk indicator of the investment project is 51.8194 points.

Stage 8. Interpretation of a complex indicator.

According to formulas (14) – (18) and figure 2, the value of the complex risk indicator is average (93%) on the border of above average (7%). $R_{1.1}$ and $R_{1.2}$ risk have the greatest impact (according to weighting factors) (figure 3).

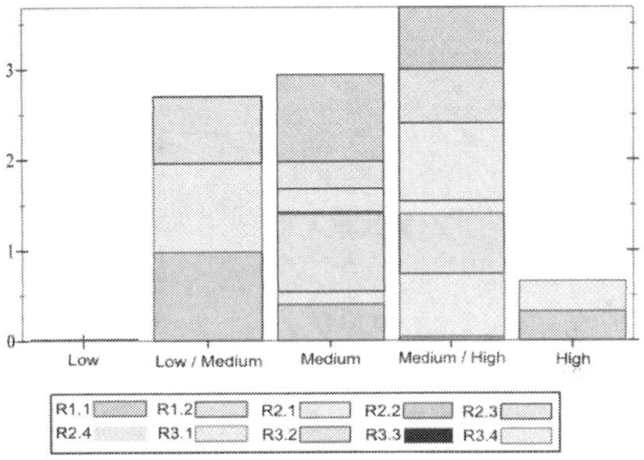

Figure 3. Components of a complex indicator of investment project risks and their compliance with linguistic terms

According to figure 3, the highest risks belong to the level of High Medium, in second place the level of Medium, and in third place – Low Medium.

Figure 4 shows a risk map of the project (according to table 4). The y-axis reflects the probability of risk (in ascending order), and the x-axis shows the risk weight (in ascending order). The bold line shows the critical limit of the level of risk. Those risks that are above this line are critical for the project and require a priority management decision to transfer them from the critical (red-purple) zone to the green zone.

Therefore, the most critical for the project are risks $R_{1.2}$ (risks associated with the reassessment of additional opportunities for project development (errors in the assessment of alternative technologies and choice of technology and equipment for the project, failure to develop project capacity), risk $R_{1.1}$ project sustainability: risk of confidence that the new project is guaranteed success due to its unique qualities, even when imitating it, risk of confidence in the company's potential for exclusive cooperation), $R_{2.2}$ (risks associated with the nature of competition in the market, namely risks strong competitive influence in the target markets of the enterprise), $R_{2.1}$ (risks of incorrect estimation of demand for the project), $R_{3.1}$ (risks for estimation

Table 3. Risks and their levels in linguistic terms

Risk number	Risk designation	Name of risk	Risk level (probability of occurrence)
1	$R_{1.1}$	Risks of reassessment of project sustainability	Medium, High Medium
2	$R_{1.2}$	Risks associated with the reassessment of additional project development opportunities	Medium, High Medium
3	$R_{2.1}$	Risks of incorrect assessment of demand for the project	Low Medium, Medium
4	$R_{2.2}$	Risks associated with the nature of competition in the market	High Medium, High
5	$R_{2.3}$	Risks associated with the solvency of the customer	Low Medium, Medium
6	$R_{2.4}$	Risks of the uncertainty of the external environment of the project	Medium, High Medium
7	$R_{3.1}$	Risks for estimating the costs of project commercialization	Medium, High Medium
8	$R_{3.2}$	Risks of potential losses from project implementation	Medium, High Medium
9	$R_{3.3}$	Risks of underestimation of project development costs	Low, Low Medium
10	$R_{3.4}$	Risks of the uncertainty of the internal environment of the project	High Medium, High

Table 4. Matrix of the actual distribution of values by fuzzy sets

Risk	Weights (level of influence)	x	Membership functions μ				
			Low ($\mu_1(x)$)	Low Medium ($\mu_2(x)$)	Medium ($\mu_3(x)$)	High Medium ($\mu_4(x)$)	High ($\mu_5(x)$)
$R_{1.1}$	0.2000	5.01	0	0	0.96	0.04	0
$R_{1.2}$	0.2000	67.5	0	0	0.30	0.70	0
$R_{2.1}$	0.1333	31.5	0	0.74	0.36	0	0
$R_{2.2}$	0.1333	83.5	0	0	0	0.66	0.34
$R_{2.3}$	0.0667	25.5	0	0.98	0.02	0	0
$R_{2.4}$	0.0667	53.5	0	0	0.86	0.14	0
$R_{3.1}$	0.0857	71.5	0	0	0.14	0.86	0
$R_{3.2}$	0.0571	65.0	0	0	0.40	0.60	0
$R_{3.3}$	0.0286	24.5	0.02	0.98	0	0	0
$R_{3.4}$	0.0286	83.0	0	0	0	0.68	0.32

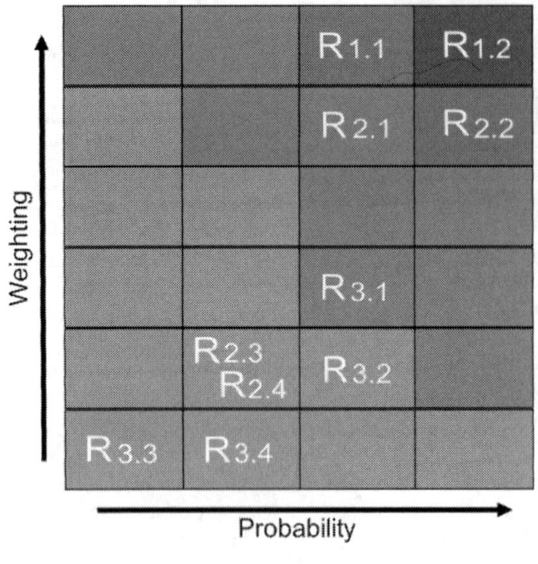

Figure 4. Risk map of the investment project

of expenses of commercialization of the project). To translate them into an acceptable green zone requires in-depth comprehensive marketing research, development, and implementation of additional measures to ensure the quality of raw materials; application of sanctions to suppliers, up to replacement of equipment; staff training during construction; search and implementation of reserves to reduce production costs; transition to alternative sources; reduction of the share of imported materials and spare parts due to the maximum use of domestic, primarily local, etc.

This approach to obtaining a comprehensive assessment should also be used to select one of several alternative projects because the best will be the one where the overall risk assessment of the project is less.

If we compare the obtained complex indicator with the complex indicator obtained by ordinary convolution (the sum of weights per indicator level), ie the average scores from table 1 and the corresponding weights, the complex indicator will be 57.2123. Thus, it is possible to notice the difference in obtaining a complex indicator by different methods, but the application of fuzzy logic allows to work with both quantitative and qualitative input indicators and to use easy-to-understand linguistic terms.

4 Conclusions

1. The task of determining a comprehensive risk assessment of the project within this study is implemented by a construction company, which assesses the risks of a new investment project for the construction of a residential complex. The experts identified the types of risk and their components that af-

fect the results of the project and selected for the formation of a comprehensive assessment. These include technical (risks of reassessment of project sustainability; risks associated with the reassessment of additional opportunities for project development), external (risks of incorrect assessment of project demand; risks associated with the nature of competition in the market; risks associated with the solvency of the customer; risks of the uncertainty of the external environment of the project), organizational (risks of estimating the costs of project commercialization; risks of potential losses from project implementation; risks of underestimation of project development costs; risks of the uncertainty of the internal environment of the project).

2. The mathematical model of the problem of determining a comprehensive assessment of project risks consists of the following components: hierarchy of existing project risks (a hierarchical tree of logical conclusion); a set of qualitative assessments of each factor in the hierarchy (linguistic terms); system of relations of advantages of some risks over others (for one level of hierarchy). The proposed Project Risk Model consists of the following stages: formation of initial data using expert opinions; construction of a hierarchical project risk tree; determination of weighting factors (Fishburne weights) of project risks; selection and description of membership function and linguistic variables; conversion of input data provided by experts from a score scale into linguistic terms; recognition of qualitative input data on a linguistic scale; determination of a complex indicator of investment project risks; interpretation of a complex indicator.

3. Initial data were formed using the opinions of experts: the experts provided an estimate of the probability of occurrence of each of the selected risks of the project on a scale from 0 to 100 points (ascending). The risks related to the nature of competition in the market and the risks of uncertainty in the internal environment of the project are most likely.

4. A hierarchical tree of risks of the project is constructed with the indication of the system of relations of advantages that gives the chance to form weighting factors using Fishburne weights.

5. According to the system of relations of preferences, Fisher's system of weights is formed. The highest weights are the risks associated with the revaluation of additional project development opportunities and the risks of revaluation of project sustainability.

6. The linguistic variable "Level of risk" with the term set of values A was formed. The triangular membership function with the following linguistic terms was chosen as membership functions: Low (L), Low Medium (LM), Medium (M), High Medium (HM), High (H), distributed on a scale from 0 to 100 points.

7. The initial data provided by experts on the probability of occurrence of each of the risks were converted from a score scale to the linguistic terms Low (L), Low Medium (LM), Medium (M), High Medium (HM), High (H).

8. Recognition of qualitative input data on a linguistic scale is carried out. The highest total value according to the triangular membership function in the linguistic term High Medium.

9. Determination of a complex indicator of the level of project risks. Its value is 51.8194 points.

10. The analysis of the obtained complex indicator shows that the total level of risk of the investment project is average (93%) on the verge of above average (7%). The constructed risk map shows that the most critical for the project implementation are the risks associated with the reassessment of additional opportunities for project development; risks of reassessment of project sustainability; risks associated with the nature of competition in the market; risks for estimating the costs of project commercialization. To reduce them requires in-depth comprehensive marketing research, development, and implementation of additional measures to ensure the quality of raw materials; application of sanctions to suppliers, up to replacement of equipment; staff training during construction; search and implementation of reserves to reduce production costs; transition to alternative sources; reduction of the share of imported materials and spare parts due to the maximum use of domestic, primarily local, etc.

The presented approach allows determining a comprehensive assessment of the risks of the investment project of the enterprise, which indicates its universality and creates conditions for its acceptability for different enterprises.

References

[1] I. Chaikovska, M. Chaikovskyi, Eastern-European journal of enterprise technologies **3**, 34 (2020)

[2] D. Peskova, J. Khodkovskaya, V. Charikov, *Risk assessment of investment projects in the digital economy*, in *European Proceedings of Social and Behavioural Sciences EpSBS*, edited by S.I. Ashmarina, V.V. Mantulenko (European Publisher, London, United Kingdom, 2019), Vol. 79 of *GCPMED 2019*, pp. 1170–1176

[3] M. Waszkiewicz, T. Grzeszczyk, Entrepreneurship and Sustainability Issues **7**, 2363 (2020)

[4] R. Shchur, S. Kropelnytska, I. Fufalko, Logos pp. 16–18 (2020)

[5] E. Bogomolova, MATEC Web of Conferences **239**, 08021 (2018)

[6] M. Zhang, W. Yang, Physics Procedia **24**, 1815 (2012)

[7] A. Voronov, M. Karmanov, I. Kiseleva, V. Kuznetsov, L. Leontieva, International Journal of Engineering And Technology **7**, 336 (2018)

[8] G.Z. Nizamova, M.M. Gayfullina, D.R. Musina, A.Y. Tumanova, A.A. Battalova, IOP Conference Series: Materials Science and Engineering **880**, 012111 (2020)

[9] S.G. Galevskiy, IOP Conference Series: Materials Science and Engineering **866**, 012040 (2020)

[10] L.Z. Bayguzina, G.A. Galimova, A.A. Sukiasyan, *Tools for Estimating the Risk Effect on the Investment Project Efficiency*, in *Proceedings of the International Scientific Conference "Far East Con" (ISCFEC 2020)* (Atlantis Press, 2020), pp. 529–536, ISBN 978-94-6252-929-8, ISSN 2352-5428, https://doi.org/10.2991/aebmr.k.200312.075

[11] S.E. Griffis, J.M. Whipple, Transportation Journal **51**, 428 (2012)

[12] V. Vitlinskyi, V. Glushchevsky, SHS Web Conf. **65**, 08002 (2019)

[13] O. Efimova, D. Koroleva, Humanitarian Balkan Research **4**, 57 (2019)

[14] C. Chong, Eurasia. Journal of Mathematics, Science & Technology Education **12**, 2451 (2016)

[15] Yang, Taihua, Li, Zhixiang, Qin, Jing, E3S Web Conf. **118**, 01039 (2019)

[16] H.H.E. Mousalami, MOJ Civil Engineering **5**, 5 (2019)

[17] A. Nedosekin, Ph.D. thesis, St. Petersburg University, St. Petersburg, Russia (2003)

[18] P.M. Hryhoruk, N.A. Khrushch, S.S. Grygoruk, Scientific Bulletin of Polissia **4**, 92 (2017)

[19] A. Matviychuk, Fuzzy Economic Review **XV**, 21 (2010)

[20] J. Diaz, E. Coba, P. Lopez, Contaduría y Administración **62**, 1687– (2017)

[21] Y. Fu, K. Liang, Computer Methods and Programs in Biomedicine **197** (2020)

AUTHOR INDEX

Attia Mohamed Omran Emad 241
Bakurova Anna 160
Bandarenka Natallia 287
Bayboltaeva Nesipkul 189
Berezina Olena 24
Bielinskyi Andrii 55, 114
Bilan Yuriy 241
Bilska Olha 15
Britchenko Igor 139
Buchatska Iryna 306
Chaikovska Inna 364
Chaikovskyi Maksym 364
Cherniavska Alexandra 253
Cherniavska Olena 253
Chernyak Oleksandr 1
Danylchuk Hanna 1, 270, 306
Derbentsev Vasily 211
Dolinskyi Leonid 312
Dubovyk Tatiana 306
Dumanska Kathrin 234
Dupliak Olha 334
Dyba Mykhailo 297
Feshchenko Olena 278
Gernego Iuliia 260, 297
Gorbatiuk Kateryna 152
Grygoruk Svitlana 195
Grynko Tetyana 325
Habrusiev Valerii 318
Honcharenko Iryna 24
Honcharenko Oleksandr 350
Horal Liliana 205
Hostryk Alexey 211, 312
Hryhoruk Pavlo 195, 364
Hushko Serhii 114
Imatayeva Aliya 189
Ivanov Mykola 146
Ivanov Roman 325
Ivanov Sergey 146
Ivashchenko Peter 139
Kainara Dmytro 133
Kaliuzhna Iuliia 133
Kaminskyi Andrii 356
Kaziyeva Assel 175
Khrushch Nila 195
Khvostina Inesa 205
Kibalnyk Liubov 1, 270, 278
Klyshbayeva Zaure 175
Kostenko Ganna 246
Koval Oleksii 47
Kovalenko Andrii 47
Kovtun Oksana 270
Kucherova Hanna 15
Kulyanda Olena 318
Kungurtseva-Mashchenko T. 246
Kutsenko Dmytro 47
Leshchenko Maryna 278
Liutyi Volodymyr 139
Los Vita 15
Lozovik Yriij 211
Lutsenko Iryna 306
Makazan Evgenia 15
Maksyshko Nataliia 31
Makulova Aiymzhan 175, 189
Mantalyuk Olha 152
Martyniuk Serhii 318
Matviychuk Andriy 114
Medvid Mykhailo 139
Melikhova Tetiana 343
Mikhailutsa Olena 343
Mukhin Viktor 246
Narivs'kyi Oleksii 182
Nehodenko Viktoria 40
Nehrey Maryna 356
Nezhyva Mariia 40
Ocheretin Dmytro 15
Onikiienko Serhii 297
Orlov Oliver 234
Osadchenko Ihor 24
Ovchynikova Olena 334
Pawliszczy Dariusz 253
Penev Vladyslav 167
Petkova Lesia 24
Pomazun Oksana 211
Porokhnya Vasyl 167, 325
Proskurovych Oksana 152
Pulina Tetiana 182
Pursky Oleg 270, 306
Radzivivska Svitlana 124
Ramazanov Sultan 350
Riabenko Anton 160
Ryasnykh Yevheniya 234
Sadovenko Maryna 225
Saparbayev Abdizhapar 175, 189
Saparbayeva Elmira 175
Sarsenbieva Nagima 189
Savchenko Olena 234
Semerikov Serhiy 1, 55, 114
Serdyuk Oleksandr 55, 114
Sherstennikov Yuriy 325
Shinkarenko Volodymyr 312
Shiyko Vira 205
Shkoda Tetiana 260
Shkvaryliuk Marta 205
Shmygol Nadiia 182, 253
Shmygol Yeugeniia 182, 253
Shynkarenko Larysa 312
Shyrokorad Dima 160
Slyusarenko Kateryna 225

AUTHOR INDEX

Soloviev Vladimir...1, 55, 114
Stachowiak Zenon..270
Stefanita Anastasia 260
Suprun Anatolij..225
Tepliuk Mariia 260
Tereschenko Elina..160
Tereshchuk Hryhoriy 318
Tishkov Bogdan..211
Trubavina Iryna..139
Tsidylo Ivan 318
Uczka W adys awa..182
Us Ivan 124
Vasylchuk Iryna...225
Vasylieva Oksana...31
Wang Haiyu..278
Yashcheritsyna Natalia..205
Yus'kiv Olesia 160
Zachosova Nataliia...47
Zakharchenko Pavel 246
Zaremba Olha 40
Zavhorodnia Tetiana 152
Zelenkevich Marina..287
Zhvanenko Svitlana 246
Zomchak Larysa 356